FORTSCHRITTE DER CHEMIE ORGANISCHER NATURSTOFFE

PROGRESS IN THE CHEMISTRY OF ORGANIC NATURAL PRODUCTS

BEGRÜNDET VON · FOUNDED BY

L. ZECHMEISTER

HERAUSGEGEBEN VON · EDITED BY

W. HERZ — TALLAHASSEE, FLA.
H. GRISEBACH — FREIBURG i. BR.
G. W. KIRBY — GLASGOW

VOL. 33

VERFASSER · AUTHORS

G. CIMINO · R. M. COATES · S. DE STEFANO · A. FONTANA
P. HEMMERICH · L. MINALE · K. L. RINEHART, JR. · L. S. SHIELD
G. SODANO · C. TONIOLO

1976

SPRINGER-VERLAG WIEN GMBH

Mit 48 Abbildungen · With 48 Figures

Originally published by Springer-Verlag Wien-New York in 1976
Softcover reprint of the hardcover 1st edition 1976

Library of Congress Catalog Card Number AC 39-1015

ISBN 978-3-7091-3264-7 ISBN 978-3-7091-3262-3 (eBook)
DOI 10.1007/978-3-7091-3262-3

Inhaltsverzeichnis. Contents

Mitarbeiterverzeichnis. List of Contributors

Cimino, Dr. G., Researcher, Laboratorio per la Chimica di Molecole di Interesse Biologico del C.N.R., I-80072 Arco Felice, Napoli, Italy.

Coates, Prof. Dr. R. M., Department of Chemistry, University of Illinois at Urbana-Champaign, Urbana, IL 61801, U.S.A.

De Stefano, Dr. S., Researcher, Laboratorio per la Chimica di Molecole di Interesse Biologico del C.N.R., I-80072 Arco Felice, Napoli, Italy.

Fontana, Prof. Dr. A., Istituto di Chimica Organica dell'Università, Via Marzolo, 1, I-35100 Padova, Italy.

Hemmerich, Prof. Dr. P., Fachbereich Biologie, Universität Konstanz, Postfach 733, D-7750 Konstanz, Bundesrepublik Deutschland.

Minale, Prof. L., Professor of Organic Chemistry, Director, Laboratorio per la Chimica di Molecole di Interesse Biologico del C.N.R., I-80072 Arco Felice, Napoli, Italy.

Rinehart Jr., K. L., B. S., Ph. D., Roger Adams Laboratory, University of Illinois, Urbana, IL 61801, U.S.A.

Shield, L. S., A. B., Sc. M., Roger Adams Laboratory, University of Illinois, Urbana, IL 61801, U.S.A.

Sodano, Dr. G., Researcher, Laboratorio per la Chimica di Molecole di Interesse Biologico del C.N.R., I-80072 Arco Felice, Napoli, Italy.

Toniolo, Prof. C., Istituto di Chimica Organica dell'Università, Via Marzolo, 1, I-35100 Padova, Italy.

Natural Products from Porifera

By L. Minale, G. Cimino, S. De Stefano and G. Sodano, Laboratorio per la Chimica di Molecole di Interesse Biologico del C.N.R., Arco Felice (Napoli), Italy

Contents

Acknowledgements

It is our privilege to thank Professors E. Lederer (Gif-sur-Yvette, France), president of the Scientific Committee of this Laboratory, for his encouragement and interest in this work, and R. H. Thomson (Aberdeen, Scotland) for advice and revision of the manuscript.

We are also obliged to the Zoological Station (Napoli) for collection of sponges and to Professor M. SARÁ (University, Genova) for identifying the sponges. Thanks are also due the technical staff of this Laboratory; special mention is due to Mrs. M. R. VACCARO (secretary) and Mr. R. TURCO (technical drawing-man). Finally we wish to thank Professor R. A. NICOLAUS (University, Napoli), director of this laboratory until August 1973.

I. Introduction

Since organic chemistry began, the chemistry of natural products from terrestrial organisms such as plants and fungi has been studied intensively; in contrast, marine species have received relatively little attention. However, in the last decade research in the field of marine products has increased substantially. The heightened interest in this area is attested by the appearance of the monumental treatise of HALSTEAD on Poisonous and Venomous Marine Animals in 1965 (*108*), BASLOW's review on "Marine Pharmacology" in 1969 (*14*) and SCHEUER's recent book "Chemistry of Marine Natural Products" (*160*) in 1973. In addition PREMUZIC's review devoted to the Chemistry of Natural Products Derived from Marine Sources, was published in volume *29* of this series in 1971 (*152*).

In the context of this more general and ecological interest in the marine environment, the sponges, very primitive multicellular animals, have received much attention. This has led to the discovery of many novel molecules. Since BERGMANN's review (*17*) in 1962 on the fatty acids and sterols in sponges about 100 unusual compounds have been isolated, mostly in the last five—six years, and this review is concerned with the study of these interesting molecules. Fatty acids and pigments are excluded as very little has been published on these topics (see *66* and *109*) since they were last reviewed in 1968 (*105* and *110*).

II. Structure Determination and Chemistry

1. Bromo-Compounds

All natural organobromocompounds are of marine origin and have been found especially in algae. The genus *Laurencia* has provided a series of brominated acetylenic cyclic ethers and bromosesquiterpenes, two of which, aplysin and aplysinol, were isolated earlier from the sea hare, *Aplysia kurodai*. Reviews by SCHEUER (*160*) and PREMUZIC (*152*) may be consulted and also the more recent papers (*87, 93, 102, 103, 107, 128, 129, 130*). A brominated bicyclic diterpenoid, aplysin-20, with unusual stereochemical features has been also isolated from *Aplysia kurodai* (*186*), and more recently unique brominated and chlorinated monoterpenes have been obtained from *A. californica* (*85*) (*86*). Two of them were also found as components of an extract of the alga *Plocamium coccineum*

(*85*). A brominated diterpene, concinndiol, an isomer of aplysin-20, has been recently described from the alga *Laurencia concinna* (*171*). Some simple bromophenols also occur in algae (*160*, p. 88) and a highly brominated phenolic pyrrole has been found in a marine bacterium (*111*, *132*).

During the past few years several reports have appeared which describe the occurrence of bromocompounds also in sponges and now the Porifera phylum appears as one of the richest sources of bromine-containing metabolites. The next subsection considers the tyrosine-derived bromo compounds; the bromopyrrole derivatives and miscellaneous metabolites are reviewed in subsection 1.2.

1.1 Tyrosine-Derived Bromo-Compounds

Sponges of the family Verongidae (Table III. Chapter III) provide a series of antibiotics and other closely related compounds which may be considered as metabolites of 3,5-dibromotyrosine. Chart 1 lists their structures.

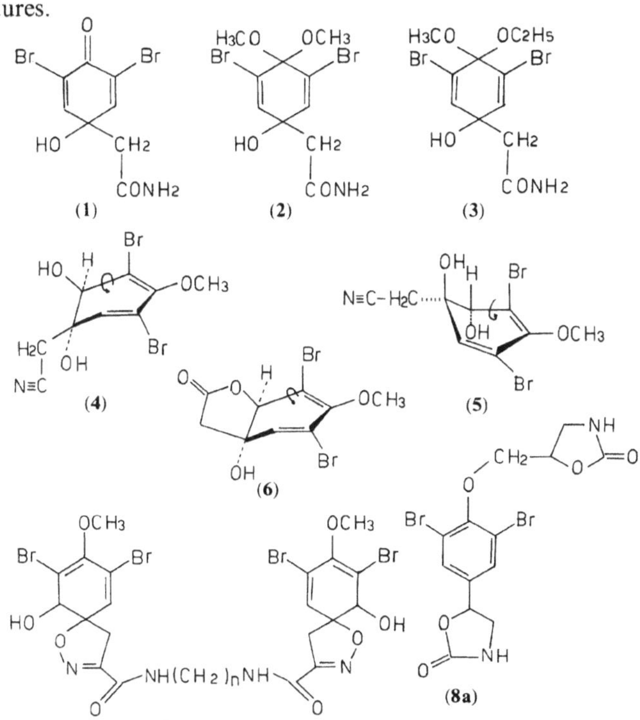

(**7**; n = 4)
(**8**; n = 5)

Chart 1. Tyrosine-derived bromo-compounds

In 1967 SHARMA and BURKHOLDER (*163*) in their continuing search for the antimicrobial activity of sponges succeeded in isolating the first two compounds of the series from the methanolic extracts of *Verongia fistularis* and *V. cauliformis*, these were assigned structures (**1**) and (**2**). Structure (**1**) was deduced (*164*) on the basis of spectral evidence, λ_{max} 257 nm, ν_{max} 3445, 1700, 1675 pnd 1660 cm^{-1}, δ 2.75 (2 H, s), 2.97 (3 H, s) and 7.59 (2 H, s), conversion of its acetate on lithium borohydride reduction to the 4-hydroxy-3,5-dibromophenylacetamide (**9**), and synthesis, which was accomplished by treating (**9**) in acetic acid with concentrated nitric acid.

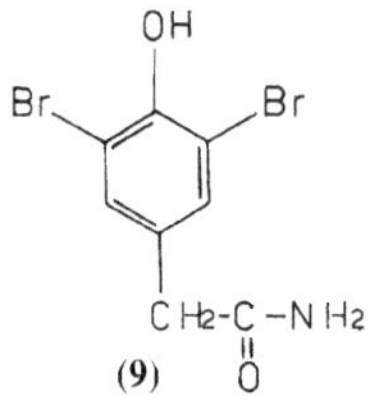

(**9**)

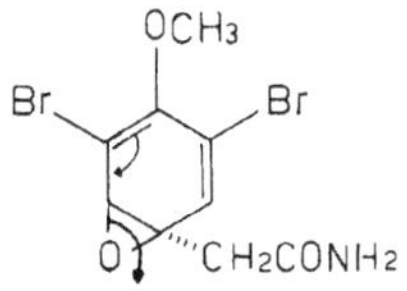

(**10**)

The relationship between the dienone (**1**) and the dimethoxyketal (**2**) has been established by converting the latter to the former, using sulfuric acid in ethanol (*168*). Failure to reconvert (**1**) into (**2**) by treatment with methanol under various conditions allowed the authors to assume that the ketal (**2**) was a genuine natural product and not an artifact generated during the extraction. Very recently ANDERSEN and FAULKNER (*11*) have reported the isolation from the ethanolic extracts of an unidentified species of *Verongia*, besides the dienone (**1**), of the mixed ketal (**3**), which was revealed to be a mixture of diastereoisomers (two methoxy signals in the 220 MHz n.m.r.). This suggested that the ketal (**3**) was not a natural product and led the authors to propose that the dienone (**1**), the dimethoxyketal (**2**) and the mixed ketal (**3**) may all be derived from a single intermediate, such as an arene oxide (**10**), by 1, 4 addition of water, methanol or ethanol during the extraction process. The recent work of KASPEREK *et al.* (*122*), showing that acid-catalyzed addition of methanol to 1,4-dimethylbenzene oxide give 4-methoxy-1,4-dimethyl-2,5-cyclohexadienol, was quoted by the authors in support of their argument.
Aeroplysinin-1 (**4**) was first isolated as the dextrorotatory isomer, $[\alpha]_D + 186°$, from *Verongia aerophoba* (*81*), which also contains the dienone (**1**), the lactone (**6**) and the more complex (**7**) and (**8**), which will be discussed below. Its gross structure was derived from spectral data, λ_{max} 284 nm (cisoid diene), ν_{max} 3365 (OH) and 2260 (CN) cm^{-1}, δ 2.28 (OH *tert.*) 2.74 (2 H, s, CH_2), 3.74 (3 H, s, OCH_3), 4.16 (2 H, m, CHOH), 6.34 (s, >C=CH) and conversion of its diacetate, on alkaline treatment

followed by strong acid hydrolysis, to the 3,5-dibromo-2-hydroxy-4-methoxyphenylacetic acid, identified as the methyl ester (**11**) by comparison with a synthetic sample. It should be noted that the conversion of (**4**) to an aromatic system can be most easily accomplished using the acetate; treatment of the parent compound with both alkali or acid results in the formation of a complex mixture.

Remarkably FULMOR *et al.* (*92*) isolated the laevorotatory antipode of aeroplysinin-1 from the sponge *Ianthella ardis*, for which they proposed the absolute configuration as shown in (**5**) on the basis of combined chemical, c. d. and n. m. r. data. The inability of (**5**) to dehydrate readily to the aromatic system suggested a *trans* relationship of the vicinal hydroxyl groups. The c. d. curve (strong negative Cotton effect) indicated left-handed helicity for the twisted diene, whereas the small spin-coupling (1.3 Hz) between olefinic and methine protons observed in the n. m. r. spectrum suggested a W relationship for these two protons. The absolute configurations of both antipodes shown in (**4**) and (**5**) have now been firmly established by two independent X-ray studies (*65, 134*). Because very few correlations between Cotton effects and X-ray data for skewed dienes have been published confirming the helicity rule (*144*), it is important to note that aeroplysinin-1 provides a clear-cut example. Aeroplysinin-1, which possesses, besides antibacterial properties, antitumor activity (see chapter IV), is the first example of a naturally occurring 1,2-dihydroarene-1,2-diol. Its biosynthesis could proceed *via* an arene oxide in agreement with the stereochemistry. Arene oxides have been proposed as intermediates in the biosynthetic oxidations of aromatic compounds; their formation and metabolism in biological systems have been reviewed recently by WITKOP and co-workers (*67*).

The two more complex brominated metabolites (**7** and **8**) obtained from *V. aerophoba* were also isolated by MOODY and THOMSON from *V. thiona* and accordingly named aerothionin and homoaerothionin. Structural elucidation of these two spirocyclohexadienylisoxazoles was the result of a collaborative effort between the Laboratories in Aberdeen (Scotland) and Naples (Italy) (*82*).

Aerothionin (**7**), the major component in both species (10% of dry *V. aerophoba*), $C_{24}H_{26}Br_4N_4O_8$, is optically active, $[\alpha]_D + 252°$, and shows λ_{max} 284 nm (cisoid diene, cf. aeroplysinin-1). The symmetrical structure is reflected in the simplicity of the n. m. r. spectrum, which shows singlets for olefinic, CHOH (singlet after D_2O exchange) and methoxy-protons at δ 6.50, 4.18 and 3.72. An isolated methylene group in an asymmetric environment is indicated by an AB quartet with line positions at δ 3.84 and 3.14 (J 18 Hz) p. p. m. A C_4 saturated chain is suggested by a methylene multiplet at δ 3.34 coupled to another methylene multiplet at δ 1.60. Mild basic treatment of aerothionin converts it quantitatively into the isomeric optically inactive dihydric phenol (**12**); it formed a tetramethyl derivative and unlike the other derivatives of aerothionin, its mass spectrum showed

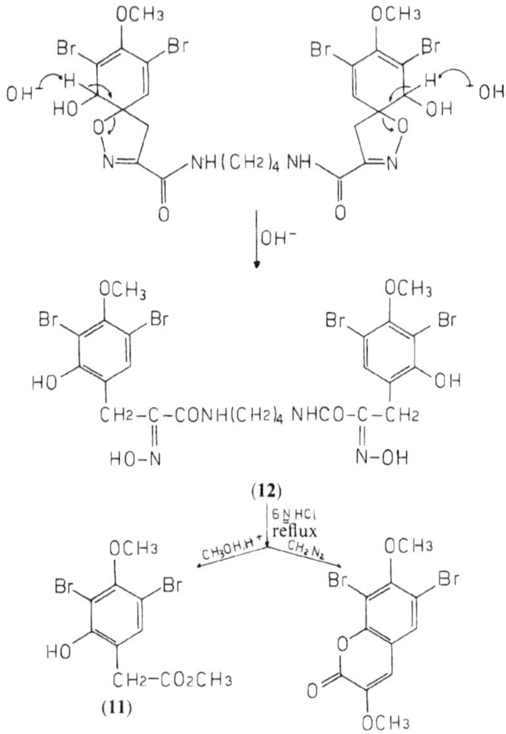

(**14**)

Chart 2. Degradation of aerothionin

a molecular ion, a quintet centred at m/e 874. Hydrolysis of the tetramethyl ether with 25% aqueous methanolic potassium hydroxide yielded the oximinopyruvic acid (**13**) identified by comparison with a synthetic sample, the acid chloride of which, when condensed with 1,4-diaminobutane, afforded back the tetramethyl derivative of (**12**). The relative positions of the aromatic substituents in (**12**) were established by degradative

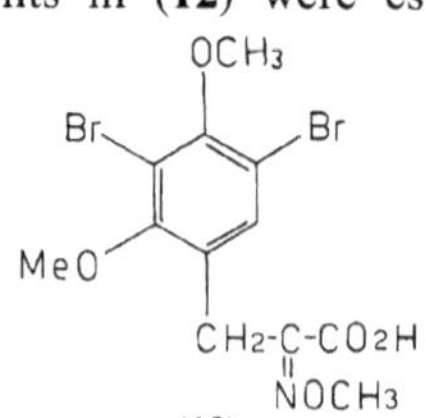

(**13**)

experiments outlined in Chart 2: hydrolysis with 6 N hydrochloric acid gave an arylacetic acid characterized as the methyl ester (**11**) and also a lactone converted by methylation with diazomethane into the coumarin (**14**).

Bearing in mind the structure and properties of aeroplysinin-1 (**4**), a co-metabolite in *V. aerophoba*, it was clear that aerothionin must have structure (**7**). Its rearrangement to the phenol (**12**) is straightforward (Chart 2, arrows).

The very closely-related homoaerothionin was found in smaller quantity in both *V. aerophoba* and *V. thiona*. Its structure (**8**) was established by analogy with that of aerothionin (*80*). A full account of the work on both aerothionin and homoaerothionin, which includes discussions of the mass spectra of the parent compounds and their derivatives, has been published (*143*). The structure of aerothionin has now been confirmed by X-ray analysis, which also revealed the relative stereochemistry (O – H and O – N *trans*) (*62a*). The spiro systems in (**7**) and (**8**) could arise in various ways including nucleophilic attack by an oxime function on an arene oxide (**15**). Following suggestions that nitriles may be derived *in vivo* from α-aminoacids by way of α-keto- and α-oximinoacids (*174*), we

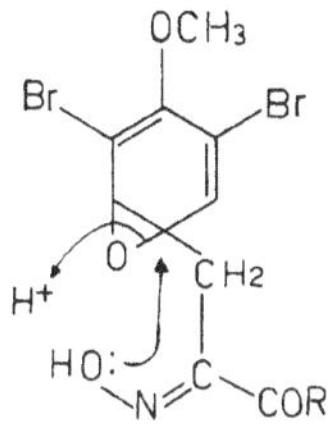

(**15**)

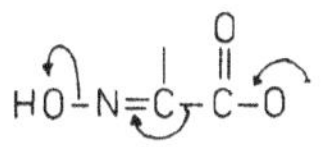

(**16**)

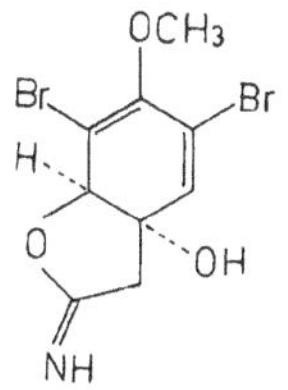

(**17**)

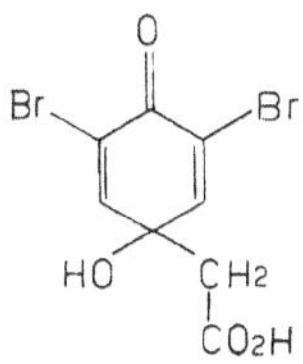

(**18**)

speculated that the oxime (**15**, R = OH) might be a likely precursor of the nitrile aeroplysinin-1 (**4** and **5**), as indicated in (**16**). There is no biosynthetic proof of this hypothesis but the co-occurrence of (**4**), (**7**) and (**8**) is a good indication of such interrelations.

Further examination of the extracts from *V. aerophoba* led to the isolation of an additional related dibromoconstituent, (+) aeroplysinin-2 (**6**), also obtained as racemate from a *Ianthella sp.* (*142*). It possesses a u. v. chromophore identical to that of aeroplysinin-1 and aerothionin and has an i. r. band at 1785 cm^{-1}, characteristic of a γ-lactone. This, along with the n. m. r. data and the conversion of its acetate into the ester (**11**), secured the gross structure. The absolute stereochemistry shown in (**6**) for the optically active isomer was based on the same criteria (n. m. r. and c. d. measurements) used by FULMOR *et al.* (*92*) to assign the absolute configuration of the (−) isomer of aeroplysinin-1. Aeroplysinin-2 was also found, together with the dienone (**1**) and the mixed ketal (**3**), in the ethanolic extracts of an unidentified species of *Verongia* (*11*). The coexistence of (**1**), (**3**) and (**6**) suggested that the lactone (**6**) might be an artifact and the iminoether (**17**) has been proposed as a possible intermediate leading to these compounds. Formation of the hydroxydienoic acid (**18**) from aeroplysinin-2 (**6**) on standing (*142*) seems in agreement with this latter suggestion.

All these *Verongia* metabolites seem fairly obviously biosynthetised from 3,5-dibromotyrosine or tyrosine *via* an epoxidation step and presumably the central C_4N_2 and C_5N_2 chains of aerothionin (**7**) and homoaerothionin (**8**) are derived from ornithine and lysine, respectively (*143*) [both dibromotyrosine and lysine have been found in sponge proteins (*9, 133*)].

Quite surprisingly *V. aerophoba* failed to incorporate radioactivity from [U-^{14}C]-L-tyrosine into aerothionin (**7**), aeroplysinin-1 (**4**) and the dienone (**1**); inactive aerothionin was also isolated when the animals were fed with [U-^{14}C]-L-ornithine (*71*). However the sponges utilized these aminoacids for the synthesis of fatty acids. A very slow rate of biosynthesis might account for these results. A dietary origin for these compounds can be also suspected; in this connection the recent report of the isolation of the brominated esters (**19**) and (**20**) (after methylation) from hydrolyzed extracts of the red alga *Halopytis incurvus* (*41*) seems relevant.

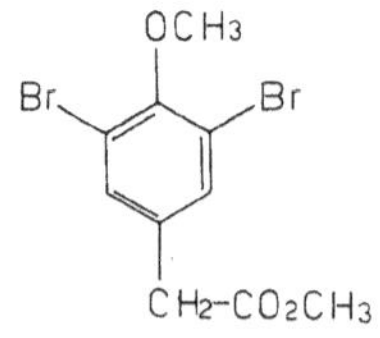

(**19**)

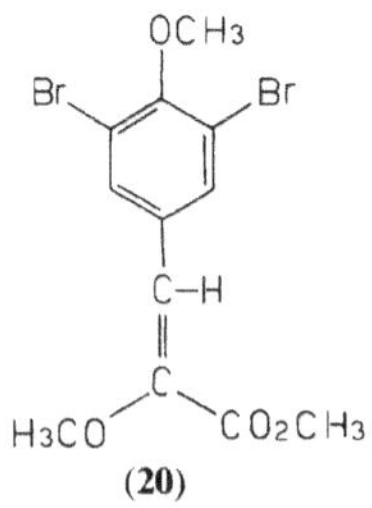

(**20**)

Very recently a further bromocompound, probably deriving biogenetically from dibromotyrosine, has been found in *Verongia lacunosa*. Structure (**8a**) has been assigned to this substance mainly on the basis of proton and ^{13}C n.m.r. data (*28a*). It is unique in that it appears to be the first bromocompound containing 2-oxazolidone rings isolated from a sponge.

1.2 Miscellaneous Bromo-Compounds

Simple dibromopyrrole derivatives have been isolated from *Agelas oroides* (*89*) and were shown to be the 4,5-dibromopyrrole-2-carboxylic acid (**21**), the corresponding amide (**22**) and nitrile (**23**), on the basis of their interconversion and synthesis of the methyl ester of (**21**). *Agelas oroides* also contains, in much larger amounts (2.3% of dry sponge), the

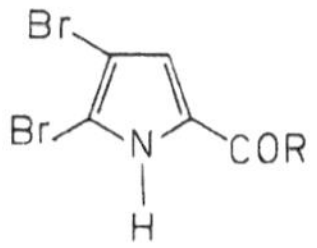

(**21**); R = OH
(**22**); R = NH_2

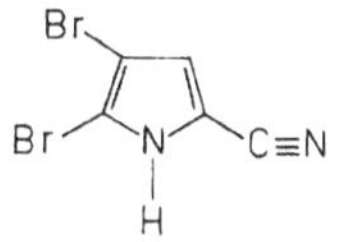

(**23**)

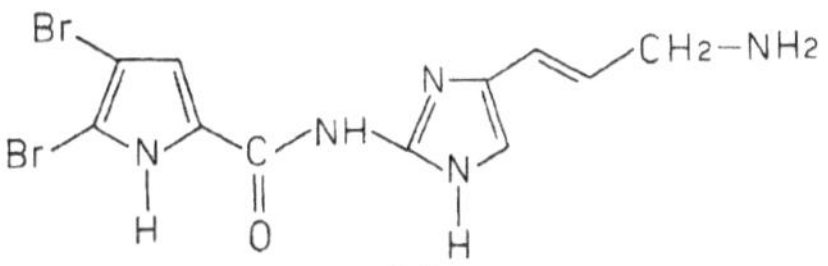

(**24**)

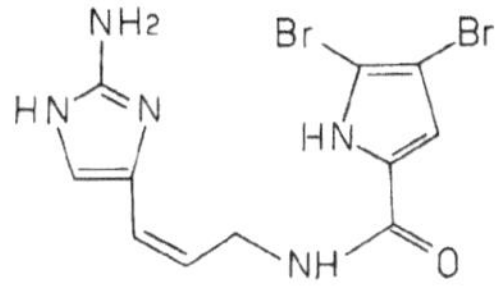

(**25**)

more complex oroidin, whose structure includes an amide derived from (**21**) and a substituted 2-aminoimidazole. Oroidin has been also recently found in *Axinella damicornis* and *A. verrucosa* (*61*). Structure (**24**) was first proposed (*89*) for this metabolite on the basis of spectral data and of degradative work carried out on the crystalline monoacetyl derivative; the degradative work is summarized in chart 3. The 2-amino-4 (or 5)-(3-aminopropyl)-imidazole (**26**) was identified as the diacetamide, m.p. 238 to 240°, λ_{max} 249 nm, by direct comparison with a synthetic sample. The formation of the 4,5-dibromopyrrole-2-carboxamide on alkaline treatment of both oroidin acetate and the dihydroderivatives, was the key argument which favoured formula (**24**) in preference to the alternative (**25**). However the synthetic work of Garcia *et al.* (*94*) indicated that formula (**24**) was incorrect; these authors suggested that alternative structure (**25**) be assigned to oroidin. This was confirmed by an 8 step synthesis of (**27**) which proved to be identical with acetyl dihydrooroidin.

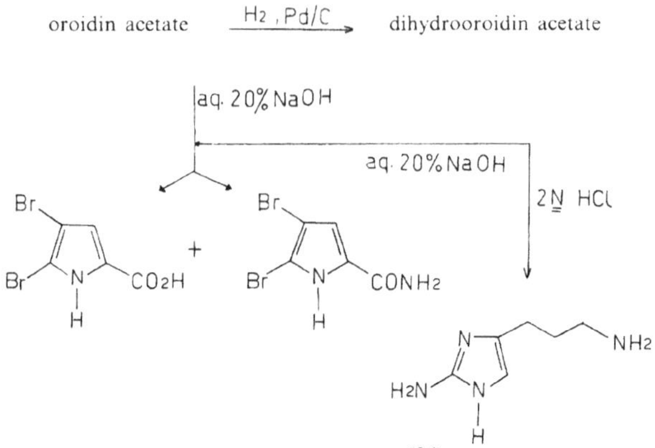

(26)

Chart 3. Degradations of oroidin

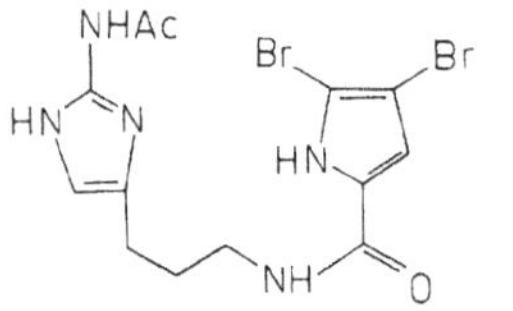

(27)

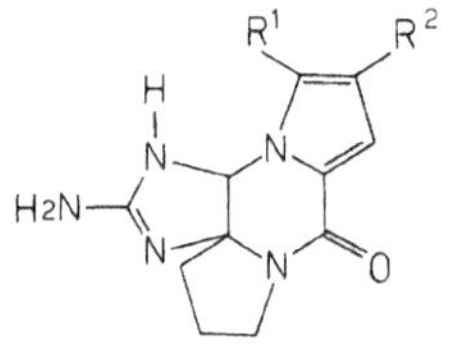

(28) $R^1 = R^2 = Br$
(29) $R^1 = R^2 = H$
(30) $R^1 = H$; $R^2 = Br$

Structure (**25**) is also related in a more logical way to the alkaloid dibromophakellin (**28**), isolated from *Phakellia flabellata* by SHARMA and BURKHOLDER (*165*). The i.r. spectrum of dibromophakellin revealed the presence of an amide function (ν_{max} 1675, 1587 cm^{-1}), whereas the presence of a pyrrole ring bearing a carbonyl function at the α-position was supported by u.v. absorption at 233 and 281 nm. A guanidin moiety was indicated by fragments corresponding to $M - NH_3$ and $M - NH_2CN$ in the mass spectrum. Dibromophakellin readily undergoes hydrogenolysis of the carbon-bromine linkages to give phakellin (**29**), whose skeleton was revealed by a detailed analysis of the 220 MHz n.m.r. spectrum. The complete structure of dibromophakellin, shown in (**28**) was definitively established by X-ray analysis of a single crystal of the monoacetyl derivative. *Phakellia flabellata* is also reported to produce the closely-related 4-bromophakellin (**30**). Both dibromophakellin and monobromophakellin showed broad spectrum antimicrobial activity (*167*).

More recently STEMPIEN *et al.* have announced (*173*) the occurrence in an unidentified species of *Agelas* of a further bromopyrrole derivative with antibiotic activity, to which they assigned the structure 4-bromo-pyrrole-2-carbonylguanidine (**31**).

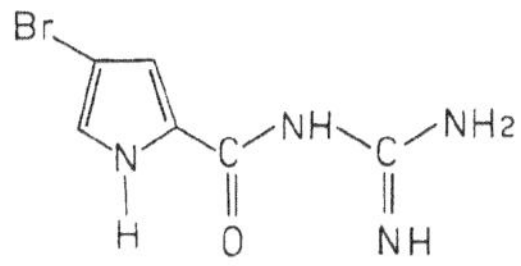

(**31**)

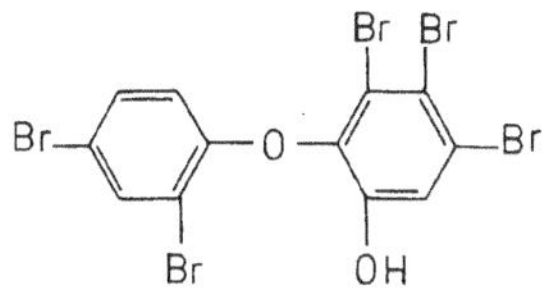

(**32**)

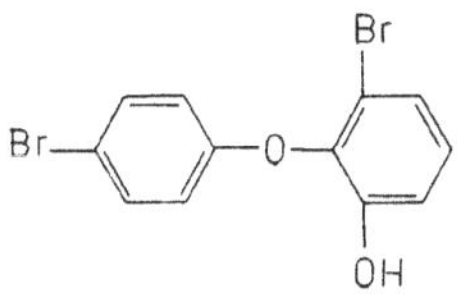

(**33**)

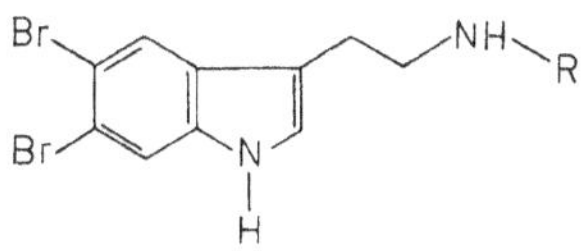

(**34**) $R=CH_3$
(**35**) $R=H$

Two brominated phenoxyphenols, active against both gram-neg. and gram-pos. organisms, have been found in *Disidea herbacea* (*166*) and structures (**32**) and (**33**) were assigned. Both structures were based on the conversion of these substances by catalytic hydrogenolysis into the known 2-hydroxy diphenyl ether, and on analyses of their n.m.r. spectra.

The structure of the pentabromo constituent was also confirmed by synthesis: 2,4-dinitrochlorobenzene was condensed with 3,4,5-tribromo-guaiol and the nitro groups of the resulting ether were reduced to give the diamino derivative. Sandmeyer reaction on the latter finally gave 1-(2′,4′-dibromophenoxy)-2-methoxy-4,5,6-tribromobenzene, which was identical with the methyl ether obtained from the natural substance.

Two new antibacterial bromoindole metabolites have been isolated from the Caribean sponge *Polyfibrospongia maynardii* (*179*). Structures (**34**) and (**35**) were assigned mainly by interpretation of their spectroscopic properties. In confirmation catalytic hydrogenation of (**34**) afforded the known 3-(2-methylaminoethyl)-indole.

2. Terpenes

Terpenes are among the most widespread groups of natural products. They are mainly of fungal and plant origin, but they have also been isolated from insects and marine animals. In marine organisms they have been reported to date from only six phyla, brown algae (Phaeophyta), red algae (Rodophyta), coelenterates (Cnidaria), molluscs (Mollusca), echinoderms (Echinodermata) and sponges (Porifera) (*160*).

Sponges have been found to contain terpenes in large amounts, some of them possessing unique structural features.

Furan rings occur frequently, although nearly all known furanoterpenes are plant products (*127*). In this group the linear furanoterpenes containing 21 carbon atoms are the most intriguing compounds from the biogenetic point of view.

Furthermore, sesterterpenes are relatively abundant in sponges in contrast with their very limited distribution elsewhere in insect protective waxes and fungi [the reviews of MCCRINDLE and OVERTON (*135*) and HANSON (*112*) may be consulted].

Even more interesting is the recent discovery of sesquiterpenes bearing an isonitrile function, a very rare feature in nature.

At the time of this writing neither monoterpenes nor triterpenes have been reported from sponge sources, apart from squalene which was found in *Ircinia spinosula* (*47*) and *I. muscarum* (*48*).

2.1 Sesquiterpenes

The chemical study of *Pleraplysilla spinifera* led to the isolation of six new furanosesquiterpenes. Two different samples identified as the same species (probably different varieties)*, both collected in the Bay of Naples, were analyzed and found to possess different sesquiterpene constituents. Chart 4 list their structures.

Dehydrodendrolasin (**36**), the bicyclic pleraplysillin-1 (**37**) and the ester pleraplysillin-2 (**38**) have been found as constituents of the first sample (*55, 62*).

Dehydrodendrolasin (**36**), the major component (5% of dry sponge), is closely related to dendrolasin, the odour-substance of the ant *Dendrolasius fuliginosus* (*153*). Its structure was established mainly by chemical interrelation with the latter (direct comparison of the perhydroderivatives). The all-*trans*-configuration of the conjugated triene system in (**36**) was suggested by a detailed analysis of the n.m.r. spectrum (*62*) using spin decoupling.

The second component, pleraplysillin-1 (**37**), is a new type of sesquiterpene with a carbon skeleton so far unique [the classical review of sesquiterpene biogenesis by PARKER *et al.* (*151*) and the comprehensive review of sesquiterpenes chemistry by BRYANT (*33*) may be consulted.

* In view of the well-known difficulties in the identification of sponges, it cannot be excluded that the two samples belong to different species. A comparative analysis revealed only slight morphological differences.

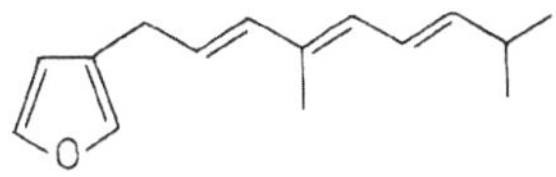

(36)

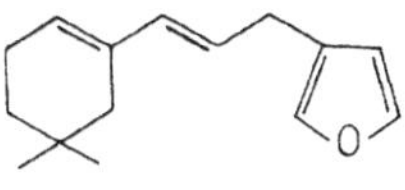

(37)

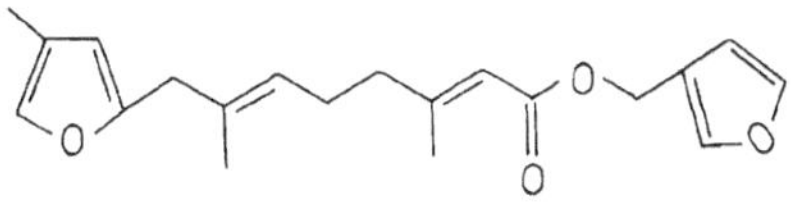

(38)

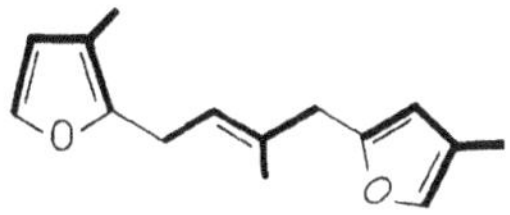

(39)

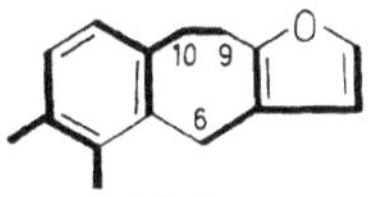

(40a)

OR

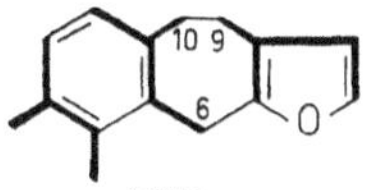

(40b)

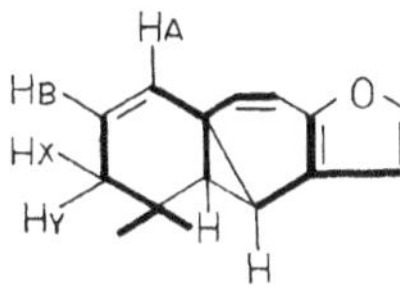

(41a)

OR

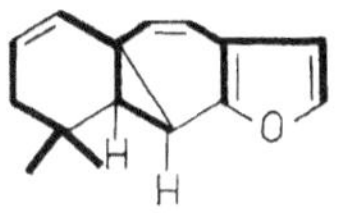

(41b)

Chart 4. Furanosesquiterpenes from *Pleraplysilla spinifera*

More recently a brief progress report by Rücker has appeared (*157*)], since it would seem to arise by a C–C cyclization involving a lateral methyl group of the polyisoprene chain.

Its structure (**37**) was deduced from spectral and degradative data (*62*). Ultraviolet [λ_{max}226, 234 and 241 nm] and n.m.r. [broad 1H singlets at δ 7.25, 7.13 and 6.17 p.p.m. and 2H doublet (J 6 Hz) at δ 3.15 p.p.m.] spectra disclosed the presence of a β-substituted furan ring isolated by a methylene from the 1,3-diene system; oxidative ozonolysis gave 3,3′-dimethyladipic acid. The position of the double bond in the cyclohexene ring was supported by a strong peak at m/e 160 in the mass spectrum corresponding to elimination of isobutene by the well-known *retro*-Diels-Alder process.

Structure (**38**) of the ester pleraplysillin-2 was essentially based on spectral studies (mainly n. m. r.) of both the parent compound and the corresponding acid (*55*). In accordance with the suggested structure, pleraplysillin-2, on oxidative ozonolysis, afforded levulinic acid.

The second sample of *Pleraplysilla spinifera* contains longifolin (**39**), previously found in a terrestrial plant, *Actinodaphne longifolia*, by HAYASHI *et al.* (*113*), and two more cyclic furanosesquiterpenes (*57*). On the basis of a careful n. m. r. examination, alternative structures (**40a**) and (**40b**) and (**41a**) and (**41b**), respectively, containing carbon skeletons of a new sesquiterpenoid have been proposed for the two polycyclic components which were designated spiniferin-2 and -1.

Spiniferin-2, $C_{15}H_{16}O$, the benzenoid component, is an optically inactive oil. In the n. m. r. spectrum the presence of a 2,3-disubstituted furan is indicated by doublets at δ 7.04 and 5.96 coupled to each other ($J = 2$ Hz); two aromatic protons are seen as a singlet at δ 6.82 and two aromatic methyl groups resonate at δ 2.23 and 2.27 p. p. m. An isolated methylene group between the aromatic and furan rings is indicated by a low field singlet at δ 4.04, while a C_2 saturated chain is suggested by an A_2B_2 system with line positions centred at δ 2.63 and 2.94. Decoupling experiments indicated that the isolated methylene is proximate to the methyl resonating at δ 2.27; furthermore both the furanoid-α and -β-proton signals were simplified on irradiation at δ 4.04 (H-6) and δ 2.63 (H-9), and the protons on C-10 (δ 2.94) are "long range" coupled with the benzenoid protons. Oxidation afforded a dicarboxylic acid whose spectral properties fully agreed with structure (**42**): the two aromatic protons now appears in the n. m. r. spectrum as an AB quartet with $J = 8$ Hz. Two structures (**40a**) and (**40b**) are compatible with the above spectroscopic evidence.

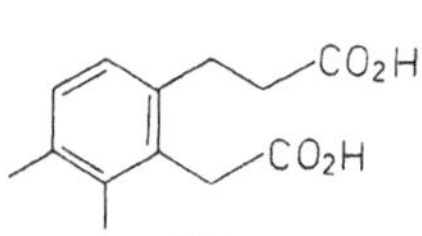

(**42**)

The isomeric spiniferin-1, $C_{15}H_{16}O$, the major furanoid constituent of the sponge, is optically active, $[\alpha]_D$-4.2°, and showed u. v. absorptions (λ_{max} 240 and 302 nm; ε 14,900 and 8,500) suggesting the presence a conjugated chromophore.

The two furan protons resonate at relatively low field, δ 7.25 and 6.50 p. p. m. (doublets, $J = 2$ Hz), in agreement with the presence of further unsaturation conjugated with the heterocyclic moiety. A signal centred at δ 6.26 is in accordance with the presence of conjugated CH=CH. The n. m. r. spectrum also displayed signals for two *tert*-

methyl groups (δ 0.8 and 1.4 p.p.m.), whose geminal nature was indicated by decoupling, an isolated $-CH=CH-CH_2-$ unit [ABXY system with line positions at δ 6.26 for Ha, 5.34 for Hb, 2.88 for Hx and 2.02 for Hy and coupling constants of 16 Hz (JXY), 10 Hz (JAB), 8 Hz (JBY), 4 Hz (JBX), 3 Hz (JAX)] and also significantly, two doublets at δ 3.62 and 0.75 with a coupling constant of 10 Hz attributable to the cyclopropane protons. All these assignements were confirmed by double irradiation experiments, which also showed that the C $(CH_3)_2$ grouping was attached to the methylene of the isolated $-CH=CH-CH_2-$ unit (irradiation of the methyl protons signal at δ 0.8 caused a sharpening of each line the Hx multiplet at δ 2.88). Bearing in mind the proposed alternative structures of the co-occurring spiniferin-2 (**40a**) and (**40b**), the spectral data can be reasonably interpreted in terms of the alternative structures (**41a**) and (**41b**). Catalytic hydrogenations of spiniferin-1 gave a dihydroderivative whose n.m.r. spectrum did not include signals corresponding to the ABXY system. Its u.v. spectrum (λ_{max} 266 nm) is in agreement with the presence of a furan conjugated with a double bond. By comparison the absorption of the parent compound (λ_{max} 302 nm) is consistent with the further conjugation of the double bond in the dimethylcyclohexene ring through the cyclopropane ring.

The proposed alternative structures (**41a**) and (**41b**) for spiniferin-1 were also consistent with the ^{13}C n.m.r. data which confirmed the presence in the molecule of eight sp^2 (two quaternary, six tertiary) and seven sp^3 (two quaternary, two tertiary, one secondary and two primary) carbon atoms.

Five more furanosesquiterpenes have recently been isolated from acetone extracts of the sponge *Disidea pallescens*, all of which, on a preliminary spectral examination, contain a 2,3-disubstituted furan ring (*61*). For the major components pallescensin-G and -F, the isomeric structures (**43**) and (**44**) have been proposed, respectively, on spectral grounds and biogenetic considerations (*46*). The spectral arguments are briefly summarized below.

The cisoid diene chromophore in pallescensin-G (**43**), $[\alpha]_D$-289°, was evident from the u.v. spectrum (λ_{max} 266 nm), which also suggested the presence of a furan ring (λ_{max} 220 nm) isolated from the diene. Additional evidence for the presence of a cisoid diene in an asymmetrical environment came from c.d. measurements (strong negative Cotton effect). The furan ring is 2,3-disubstituted (δ 7.10 and 6.02, 1 H each, J 2 Hz in C_6D_6). The n.m.r. spectrum also displayed signals for three olefinic protons appearing as an ABX system (see formula **43**; J AB 5 Hz, J AX 9 Hz, J BX ~ 0) and also two *tert*-methyl groups (δ 0.90 and 1.00 p.p.m.). The H_B proton is "long-range" coupled (J 1 Hz) to one of the two protons of the isolated methylene group between the furan ring and the diene system, which resonates as an AB quartet (J 17 Hz) at δ 3.49, while the Hx proton is "long-range" coupled to the *tert*-methyl protons resonating at δ 0.90. An analysis of the complex signals centered at δ 1.77 (3 H) and 2.30 (2 H) p.p.m. with double resonance experiments gave the sequence CH_2-CH_2-CH- of the remaining protons. Hydrogenation on palladium on charcoal yielded 2,3-dihydropallescensin-G, in the mass spectrum of which a strong peak at m/e 160 corresponding to elimination of isobutene from the dimethylcyclohexene ring by the well-known retro-Diels-Alder process supported the position of the double bond in the ring.

The isomeric pallescensin-F (**44**), optically inactive, has u.v. absorptions at 220 and 271 nm (furan and cisoid diene chromophores). The furan ring is 2,3-disubstituted (δ 7.03, 5.96, 1 H each, J 2 Hz). Furthermore in the n.m.r. spectrum a singlet at δ 1.03 (6 H, *tert*-Me's) and a "deceptively simple" ABX_2 system [see formula (**44**); JAB 10 Hz, JBX 4 Hz, JAX ~ OHz] together with the absence of any further olefinic signal were consistent with the presence in the molecule of a 1,2-disubstituted-6,6-dimethyl-cyclohexadiene moiety. An isolated methylene group between the furan ring and the diene system is indicated by a singlet at δ 3.39, while a saturated C_2 chain is suggested by a broad singlet at δ 2.47. The relationship between pallescensin-G and -F has been established by showing the identity (n.m.r., m.s. and g.l.c.) of 2,3-dihydropallescensin-G with the 1,4-hydrogenation product of pallescensin-F.

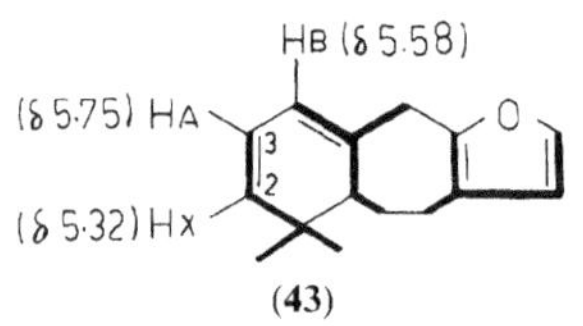

(**43**)

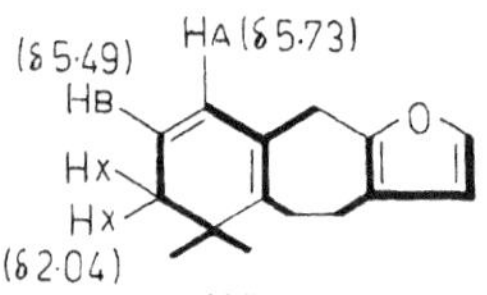

(**44**)

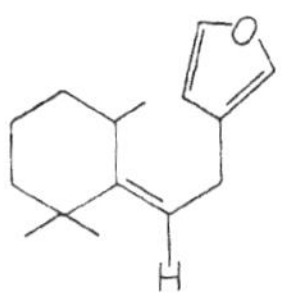

(**45**)

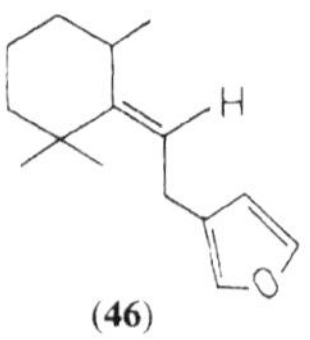

(**46**)

A series of monocyclic furanosesquiterpenes has now been obtained from the sponge *Microciona toxistyla* (*46*). Two of them, named microcionin-1, and microcionin-2, have been assigned the isomeric structures (**45**)—(**46**), respectively, with skeletons of the monocyclofarnesane type. The key argument in favour of the formulae (**45**) and (**46**) was the formation on ozonolysis of 2,2,6-trimethylcyclohexanone.

As pointed out before, the isonitrile function is a very rare feature in nature. Until 1973, only the fascinating metabolite xanthocillin, isolated from *Penicillium notatum*, has been described as a natural isonitrile (*1*). Very recently the isonitrile function has also been found in four sponge sesquiterpenes. Their structures are shown in Chart 5.

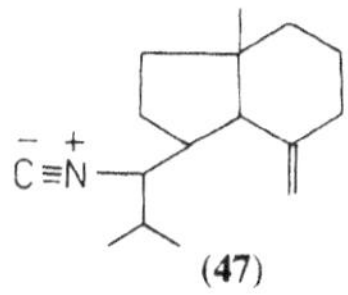

(47)

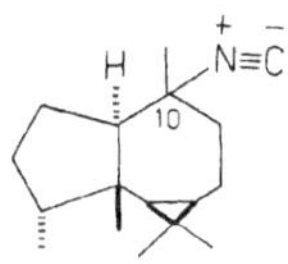

(48)

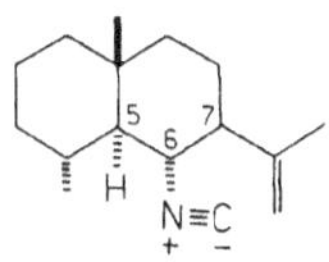

(49)

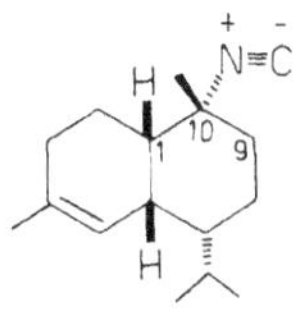

(49a)

Chart 5. Isonitrile sesquiterpenes from sponges

FATTORUSSO and his co-workers (*36*, *77*) obtained the first two from *Axinella cannabina* and proposed the names of axisonitrile-1 (**47**) and axisonitrile-2 (**48**), respectively.

Structure (**47**) suggested for axisonitrile-1 was deduced from spectroscopic data, along with extensive degradative work (*36*); the relevant chemical arguments which have contributed to structure elucidation are summarized in Chart 6. The presence of an exocyclic methylene group in axisonitrile-1 was supported by i.r. bands at 3050, 1640 and 895 cm^{-1}. The isonitrile $\nu_{\overset{+}{N}\equiv\overset{-}{C}}$ band was located at 2130 cm^{-1} and a doublet at 1385—1375 cm^{-1} is in accordance with the presence of an isopropyl group. The n.m.r. spectrum displayed signals for three methyl's, one tertiary and the other two secondary, an olefinic singlet (2 H) at δ 4.75 ($=CH_2$) and, significantly, a one-proton multiplet at δ 3.13 attributable to the CH – N proton. The presence of an isonitrile function was confirmed by conversion of axisonitrile-1 into the N-methylamine (**50**), which, on Hofmann exhaustive methylation followed by elimination, afforded the diene (**51**). The latter contains the $-CH=C(CH_3)_2$ grouping (n.m.r.). Treatment of (**47**) with Na/NH_3 led to N – C bond cleavage with formation of the olefin (**52**), which was converted, on ozonolysis, into the six-membered ring ketone (**53**) (ν_{max} 1707 cm^{-1}), which possesses three enolizable hydrogens. Baeyer-Villiger oxidation of the latter and hydro-

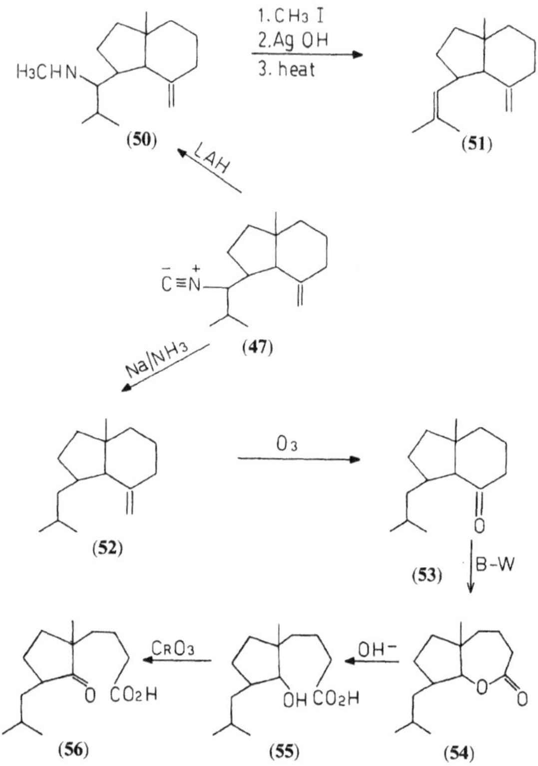

Chart 6. Degradative work on axisonitrile-1

lysis of the lactone (**54**) gave a hydroxyacid (**55**), which was oxidized to ketoacid (**56**). A band in the i.r. spectrum of (**56**) at 1738 cm^{-1} placed the ketonic group in a five-membered ring, while in the n.m.r. spectra of both the lactone (**54**) and the corresponding hydroxyacid (**55**), the $>CH-O$ protons appear as doublets, an observation which confirms the position of the isobutyl sidechain. The skeletal type of axisonitrile-1 appears to be new among the sesquiterpenoids although a closely related structure has been found in oppositol which occurs in the red alga *Laurencia subopposita* (*107*).

The second isonitrile component, axisonitrile-2, of *Axinella cannabina* is reported by FATTORUSSO and his co-workers (*77*) to be (**48**) based on the aromadendrane skeleton. They have shown that the hydrocarbon formed from axisonitrile-2 on Birch reduction is identical with the hydrogenation product of aromadendrene, whose stereochemistry shown in (**57**) has been defined by BÜCHI *et al.* (*34*). The stereochemistry at C-10 of axisonitrile-2 is not known. The authors have also reported (*36*, *77a*) the occurrence in the same sponge of axisothiocyanate-1 (**58**) and axisothiocyanate-2 (**59**). The relationship between the isothiocyanates and the corresponding isonitriles has been established by converting the latter to the former, using sulphur at 120°. Interestingly, *Axinella cannabina* also contains in low amounts the formamides, axamide-1 (**58a**) and axamide-2 (**59a**), which could be likely precursors of the corresponding isonitriles (*77a*).

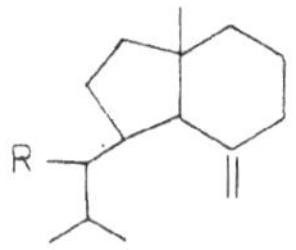

(**58**); R=N=C=S
(**58a**); R=NHCH=O

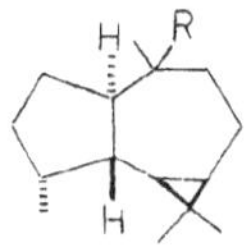

(**59**); R=N=C=S
(**59a**); R=NHCH=O

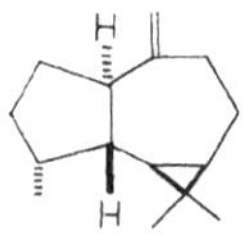

(**57**)

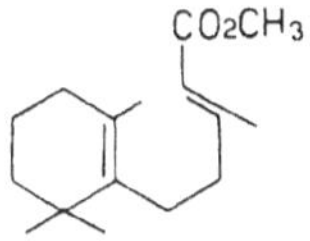

(**62**)

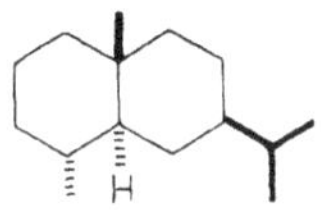

(**60**)

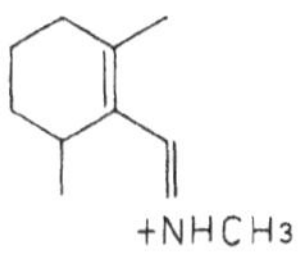

(**61**); m/e 152

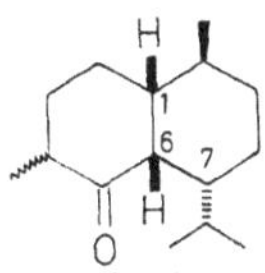

(**61a**)

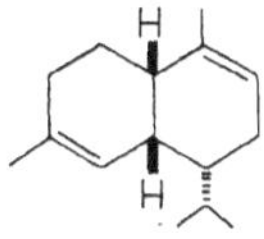

(**61b**)

The third sesquiterpene isonitrile in Chart 5 has been isolated from the extracts of *Acanthella acuta,* which also yielded a number of minor compounds (*138*). The major component, acanthellin-1, proved to have a 4-epi-eudesmane structure (**49**). The carbon skeleton was deduced by lithium/ethylamine reduction, which afforded a saturated bycyclic hydrocarbon (**60**) converted into eudalene by drastic dehydrogenation. The i.r.

spectrum suggested an isonitrile function (ν_{max} 2140 cm^{-1}), which was confirmed by reduction to an N-methylamine on litium aluminium hydride treatment. The n.m.r. spectrum indicated the presence of an isopropylidene group; spin-decoupling experiments suggested that the isonitrile group was in position 6 and proved its *trans* diequatorial relationship to the isopropylidene group. The methine proton at the nitrogen-bearing carbon appears as a triplet (J = 11 Hz) at δ 3.27 which is transformed into a doublet (J 11 Hz) by irradiation at H-7 (δ 2.15). Irradiation at δ 1.20 (H-5) also simplified the triplet at δ 3.27 into a doublet (J 11 Hz); the value of the coupling constants requires a *trans*-diaxial relationship for these three hydrogens and, accordingly, a *trans*-diequatorial relationship between the isopropylidene and isonitrile groups. Careful measurements of the n.m.r. line width from the angular methyl group at half height suggested a *trans*-ring junction. Concerning the stereochemistry of the *sec*-methyl group, its relatively low field reasonance (δ 1.22) implied that it is *cis* to the isonitrile group. Finally, characterization of the hydrocarbon (**60**) as 4-epieudesmane by direct comparison with an authentic sample definitively established the structure and stereochemistry (**49**) of acanthellin-1. The mass spectrum of the N-methylamine derivative which includes the major peak at m/e 152 corresponding to the structural fragment (**61**), confirmed the position of the isonitrile group.

The last sesquiterpene isonitrile (**49a**) characterized to date has been isolated from *Halichondria sp.*, where it occurs together with the corresponding isothiocyanate and formamide (*35a*). The relationship between these three compounds was proved by converting the isonitrile to the formamide (glacial HOAc) and to isothiocyanate (S, 120°). The cadalene skeleton for the *Halichondria* terpenoids was proved by lithium/ethylamine treatment of (**49a**), followed by dehydrogenation with palladium on charcoal. This resulted in the formation of 1,6-dimethyl-4-isopropylnaphthalene. Evidence for the location of the functional group rests on n.m.r. data (methyl singlet at δ 1.42; no signals between δ 2.5 and 5.5) and conversion of (**49a**) into the ketone (**61a**). The latter on reaction with methyllithium followed by treatment with formic acid and dehydrogenation with palladium on charcoal gave 1,5,6-trimethyl-4-isopropylnaphthalene. Ketone (**61a**) also provided evidence (n.m.r. and c.d.) for the assigned (+) amorphane stereochemistry at C-1, 6 and 7. The stereochemistry at C-10 in (**49a**) was inferred from lithium aluminium hydride reduction of the formamide to the corresponding N-methylamine, followed by methyl iodide quaternization and Hofmann degradation, which resulted in a mixture of Δ^1 and Δ^9 olefins. The latter can only have arisen from an axial amine by diaxial elimination. The Δ^9-olefin (**61b**; major olefin) was identical in g.l.c. with an authentic sample of (+)-α-amorphene.

References, pp. 64—72

A simple sesquiterpene, the methyl ester of the *trans*-monocyclofarnesic acid (**62**), has been isolated from *Halichondria panicea* (*51*) from which a group of triprenylphenols, the paniceins, have also been obtained (see page 35). Incidentally, this is the first time that monocyclofarnesic acid has been found in nature.

2.2 Diterpenes

Only one example of a true diterpenoid compound has so far been reported from sponges. It has been isolated from *Spongia officinalis* (*45*) from which a number of linear furanoterpenes containing 21 and 25 carbon atoms have also been obtained* (see pages 23 and 28).

The new diterpene, named isoagatholactone, proved to be the first naturally occurring compound with the carbon skeleton of isoagathic acid, the acid-catalyzed cyclization product of agathic acid (*30, 159*). Structure (**63**) for isoagatholactone without sterical implications was suggested

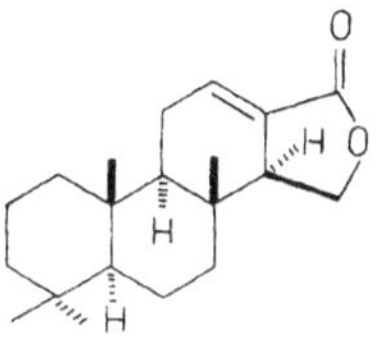

(63)

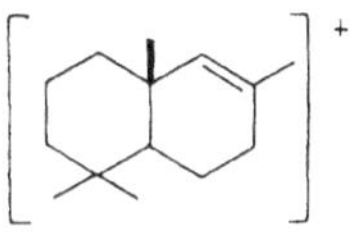

(64)

from spectroscopic data and especially the mass spectrum which gave the base peak at m/e 192 corresponding to the fragment (**64**), originating by a *retro*-Diels-Alder process. Reduction with lithium aluminum hydride of isoagatholactone yielded a diol which was converted by hydrogenolysis over Pd/C to alcohol (**65**). The latter was shown to be identical with a transformation product of grindelic acid (**66**), a known bicyclic diterpene (*149*), thus settling the structure and stereochemistry. Chart 7 outlines the chemical correlation of isoagatholactone (**63**) with grindelic acid (**66**).

---

* The samples of the sponge containing the diterpene proved to be devoid of furanoterpenes, and the specimens giving furanoterpenes did not contain any of the diterpene. Nevertheless, both samples were identified as *Spongia officinalis* and a comparative analysis revealed only slight morphological differences. Probably the two samples represent different subspecies. The identification of sponges is very difficult, particularly those belonging to the family Spongidae.

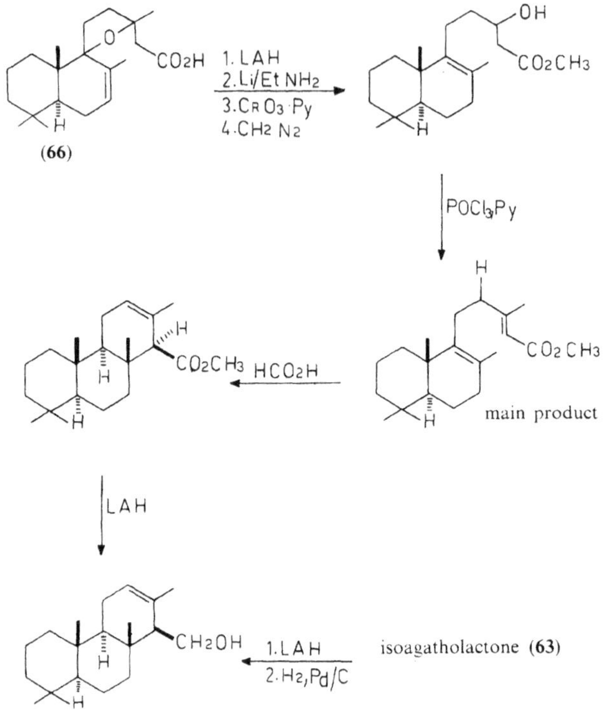

(65)

Chart 7. Chemical correlation of isoagatholactone with grindelic acid

2.3 The C_{21} Furanoterpenes

Linear, closely-related, difuranoterpenes containing 21 carbon atoms have been found to occur in large amount in the sponges of the genus *Spongia*. All of them possess the same carbon skeleton **(67)**, and oxidation in the central chain account for all their differences.

The individual structures are listed in Chart 8.

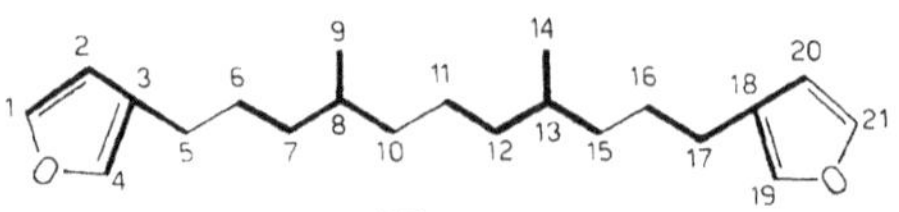

(67)

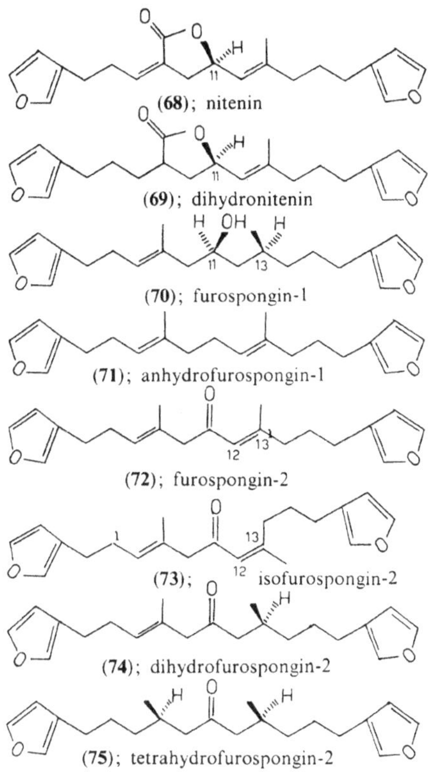

Chart 8. C_{21} linear difurano terpenes from sponges

The first two C_{21} compounds, nitenin and dihydronitenin isolated from *Spongia nitens* (*83*) possess a γ-lactone ring in the central part of the chain. The presence of two β-substituted furan rings, the number, the nature and the stereochemistry of the double bonds and their relative positions with respect to the γ-lactone ring in the major component (0.43% of dry sponge) nitenin (**68**), an oil, $[\alpha]_D$-45.4°, λ_{max} 221 (ε 14000) nm, ν_{max} 1745 and 1670 cm^{-1}, were deduced by conventional spectral techniques (mainly n.m.r.) on nitenin and several of its derivatives. Ozonolysis of niteninic acid, prepared by opening of the lactone ring in alkaline solution, provided decisive proof for the structure of nitenin: it gave succinic acid, 2-oxo-4-hydroxyglutaric acid and 5-oxohexanoic acid. Application of Horeau's partial resolution method (*116*, *117*, *118*) to the diol (**76**), prepared from nitenin by lithium aluminium hydride treatment, resulted in recovery of dextrorotatory 2-phenylbutyric acid, thus showing an R chirality at the sole asymmetric carbon atom (C-11) of nitenin.

Structure (**69**) assigned to the minor constituent dihydronitenin, an oil, $[\alpha]_D$-25.2°, λ_{max} 222 (ε 5230) nm, ν_{max} 1765 cm^{-1}, was deduced on spectral grounds and chemical interrelation with nitenin, which was accomplished by converting the latter to the former using sodium borohydride reduction.

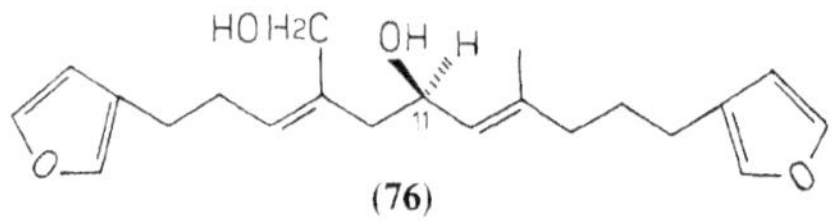

(**76**)

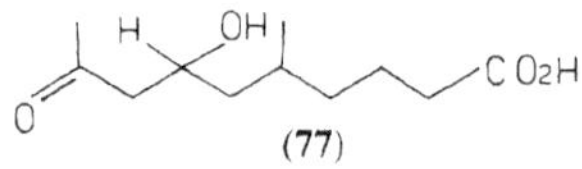

(**77**)

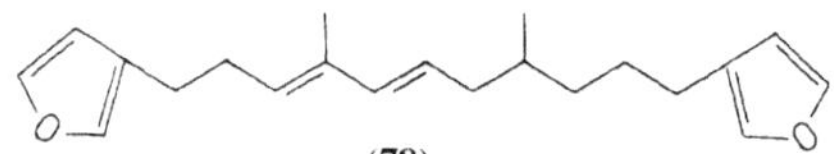

(**78**)

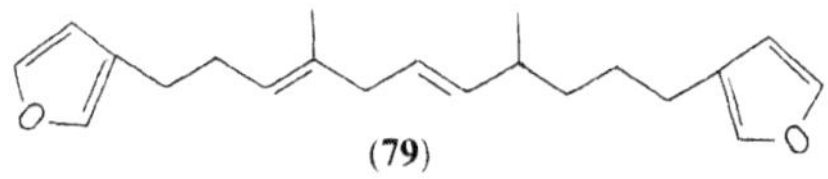

(**79**)

Spongia officinalis and *Hippospongia communis* contain the six other C_{21} furanoterpenes listed in Chart 8 (*58, 59*).

Furospongin-1, $[\alpha]_D$+8.8°, m.p. 35°, is the major compound in both sponges (1% in *S. officinalis* and 0.35% in *H. communis*, dry weight). Structure (**70**) was proposed on spectral grounds and oxidative ozonolysis, which yielded succinic and 9-oxo-7-hydroxy-5-methyldecanoic acid (**77**) (*58*). In confirmation, dehydration of furospongin-1 gave two anhydroderivatives (**78**), with the properties of a conjugated diene, and (**79**). The latter was ozonized to *S* (+)-α-methyladipic acid, thus settling the chirality at C-13. The stereochemistry at C-11 was determined by Horeau's method (*116, 117, 118*): following partial asymmetric esterification of furospongin-1 with excess 2-phenylbutyric acid anhydride, levorotatory 2-phenylbutyric acid was recovered in 10% optical yield.

Structures of the minor components were deduced by spectral evidence and confirmed by chemical correlation with furospongin-1 (*59*).

Difficulty was experienced when attempts were made to carry out selective hydrogenations of both the natural compound (**71**) and the an-hydro compound (**79**) in order to obtain and compare the tetrahydro derivatives. However, metal-catalyzed transfer-hydrogenation (*31*) using decalin and palladium on charcoal under reflux resulted in the formation in high yield of the same tetrahydroderivative from both (**71**) and (**79**).

Both furospongin-2 (**72**) and its Δ^{12}-*cis*-isomer isofurospongin-2 (**73**), distinguishable by the resonances of the 13-methyl signals in their n.m.r. spectra (δ 2.11 in the spectrum of **72**, and 1.83 in the spectrum of **73**), gave, on sodium borohydride-pyridine reduction, an alcohol identical with furospongin-1 (**70**).

The relationship between dihydrofurospongin-2 (**74**) and furospongin-1 (**70**) was established by converting the latter to the former, using chromic anhydride-pyridine complex.

Finally, Pd/C transfer-hydrogenation of dihydrofurospongin-2 (**74**) afforded selectively tetrahydrofurospongin-2 (**75**) which, because it exhibited no optical rotation, is either a racemate or, more probably, the *meso*-diastereoisomer as formulated in (**75**).

Spongia officinalis also yielded, in very small amounts, eight more isomeric C_{21} compounds closely related to furospongin-1 (**70**), the major terpene component of the same sponge. Four of them have one furan ring modified as a γ-hydroxy-α,β-butenolide; the other four being the corresponding β,γ-epoxybutenolides (*53*). Difficulty was experienced in isolating the individual components, particularly the β,γ-epoxybutenolides because of their very facile conversion to the corresponding γ-hydroxy-α,β-butenolides. Structures (**80**)—(**83**) assigned to the γ-hydroxy-α,β-butenolides have been proposed on the basis of the spectral properties determined on mixtures of the pairs (**80**)—(**81**) and (**82**)—(**83**) and conversion of each pair to furospongin-1 (**70**), this was accomplished by sodium borohydride reduction of the lactol rings followed by treatment with di-isobutyl aluminium hydride of the resulting α,β-unsaturated-γ-lactones.

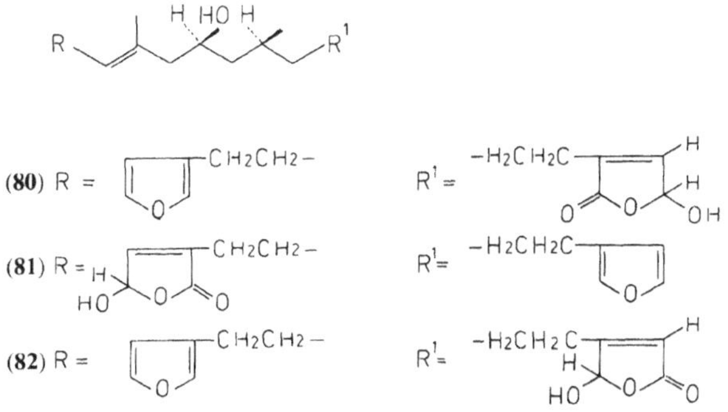

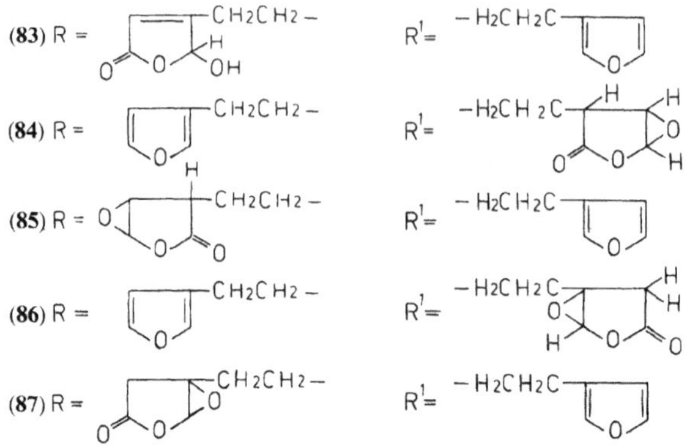

Spectral measurements carried out on the β,γ-epoxybutenolide fraction along with their conversion to the previously mentioned compounds (**80**) to (**83**) suggested that the fraction was a mixture of four compounds with structures (**84**)—(**87**).

The mode of extraction of the sponge (in the dark), and failure to detect any of the above oxidized derivatives in a methanolic solution of furospongin-1 after exposure to light led us to suggest that at least the β,γ-epoxybutenolides are genuine natural products and that possibly oxidation of the furan rings might occur *in vivo* through epoxidation steps in contrast to photooxidation *in vitro*, which is known to proceed *via* intermediate ozonides (*100*). The occurrence of the β,γ-epoxybutenolides, which *inter alia* are compounds of a type not previously encountered, seems also relevant in connection with the problem of the biogenetic derivation of the tyrosine-derived bromometabolites in sponges, which may also be biosynthesized *via* an arene epoxide (see page 4).

Two more linear C_{21} compounds, ircinin-3 and -4, have been isolated in small yield from *Ircinia oros* (*49*). They were shown to have structures (**88**) and (**89**) which are closely related to the isomeric sesterterpenes

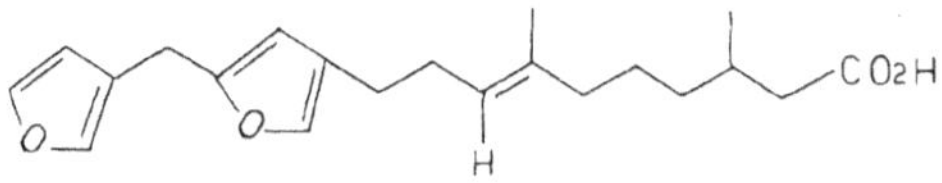

(**88**)

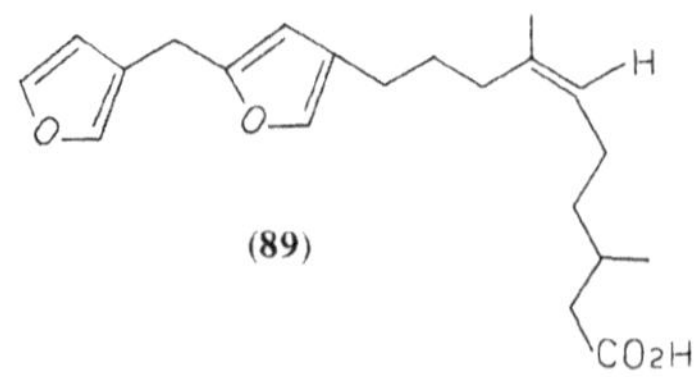

(**89**)

ircinin-1 (**94**) and -2 (**95**), the major terpene components of the same sponge (see page 28). N.m.r. and mass spectrometric investigations of the methyl esters of ircinin-3 (**88**) and -4 (**89**) showed the presence in both substances of the same structural features previously observed in ircinin-1 (**94**) and -2 (**95**). Oxidative ozonolysis of ircinin-3 (**88**) afforded malonic, succinic and 3-methyl-7-oxo-octanoic acids, while similar treatment of ircinin-4 yielded malonic, 5-oxohexanoic and 3-methyladipic acids.

At present the biogenetic origin of these unique C_{21} compounds is a matter of speculation. In view of the occurrence in sponges (genus *Ircinia*) of several furanoid sesterterpenes (see page 28) we prefer the idea that they are derived by degradation of sesterterpenoids to the possibility of biosynthesis by addition of a C-1 unit to a diterpenoid precursor. The occurence in the same organism of the isomeric pair ircinin-1 (**94**) and -2 (**95**) (C_{25}) and ircinin-3 (**88**) and -4 (**89**) (C_{21}) with structures very closely related even in the position and stereochemistry of the central double bonds gives strong support to this hypothesis. In addition, the isolation from *Ircinia spinosula* (*47*) of a C_{35} linear furanoterpene, furospinosulin-3 (**92**), along with the C_{31} difurospinosulin (**93**), which are in the same biogenetic relationship as the above $C_{25}-C_{21}$ compounds, lends further substantial support to this suggestion.

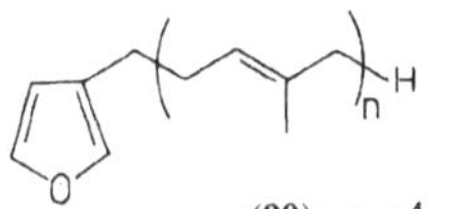

(**90**); n = 4
(**91**); n = 5
(**92**); n = 6

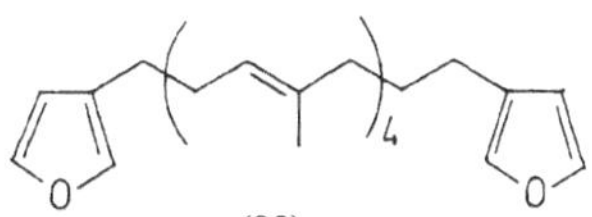

(**93**)

Inter alia, Ircinia spinosula also yielded, in smaller amounts, the lower isoprenologues of furospinosulin-3, with 25 and 30 carbon atoms. The structures of the isoprenologues (**90**), (**91**) and (**92**) as well as the structure of the C_{31} difuranoid compound (**93**) were derived by conventional spectral techniques and ozonolysis (*47*).

2.4 Sesterterpenes

Four related furanoid sesterterpenes have been isolated from sponges of the genus *Ircinia*. Ircinin-1 (**94**) and ircinin-2 (**95**) from *Ircinia oros* were found to be isomeric difurano-sesterterpenes containing an unusual conjugated tetronic acid moiety (*60*). Fasciculatin (**96**) from *Ircinia fasciculata* (*37*) and variabilin (**97**) from *I. variabilis* (*84*) were shown to be closely related monofurano-sesterterpenes.

Two more isomeric linear monofurano-sesterterpenes, furospongin-3 (**98**) and furospongin-4 (**99**), have been isolated in small yield from *Spongia officinalis* (*49*).

Chart 9 lists the structures of the sponge furanoid sesterterpenes.

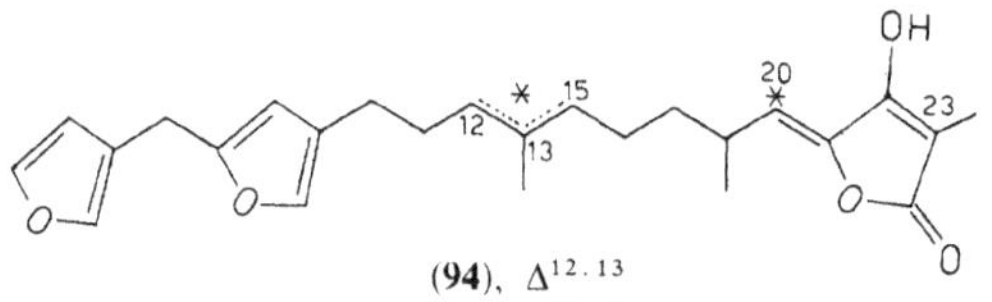

(**94**), $\Delta^{12,13}$
(**95**), $\Delta^{13,15}$

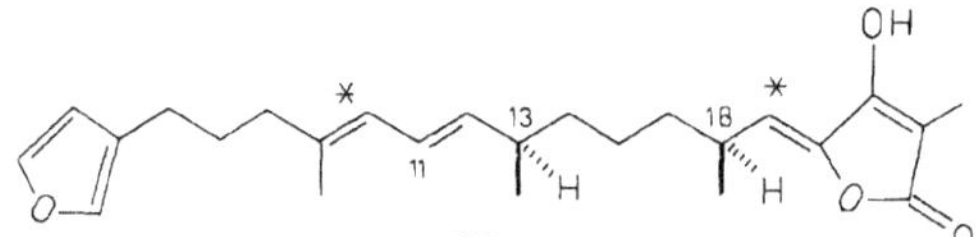

(**96**)

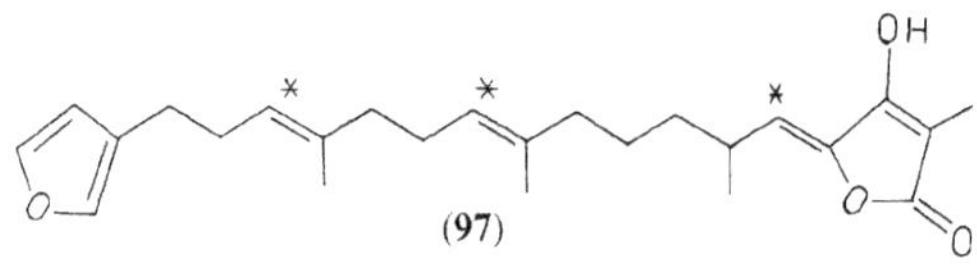

(**97**)

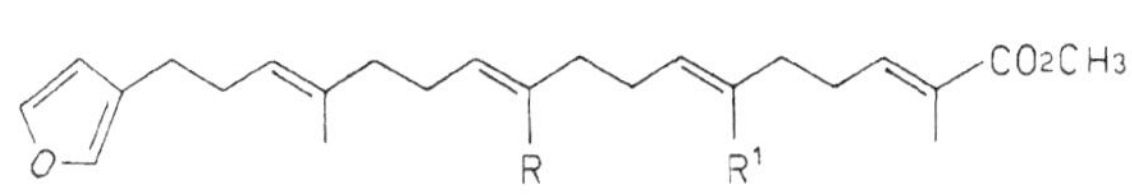

(**98**); R = CO_2H, R^1 = CH_3
(**99**); R = CH_3, R^1 = CO_2H

* Stereochemistry of olefin unknown.

Chart 9. Linear furanoid sesterterpenes in sponges

Ircinin-1 (**94**) and -2 (**95**) proved to be resistant to all attempts at separation and work on their constitution was carried out on the mixture of the two isomers (*60*). The presence of two furan rings linked to each other by a methylene group was indicated by n.m.r. spectroscopy (five broad 1 H singlet at δ 7.29, 7.22, 6.99, 6.25 and 5.78 p.p.m. and 2 H broad singlet at δ 3.65) and confirmed by decoupling experiments. The presence in the n.m.r. spectrum of two singlets at δ 1.55 and 1.66 p.p.m. attributable to methyl groups on double bonds and integrating together for 3 protons suggested that the natural isolate was a mixture of two isomers differing in the position and/or configuration of a double bond. In agreement with this observation, the product of acetylation gave two g.l.c. peaks whose areas were in the same ratio as that of the two vinyl methyl n.m.r. signals. Furthermore vigorous hydrogenation gave a single dodecahydroderivative (**100**); when milder conditions were applied, the selective reduction of the in-chain double bonds occurred to yield a single tetrahydroderivative (**101**).

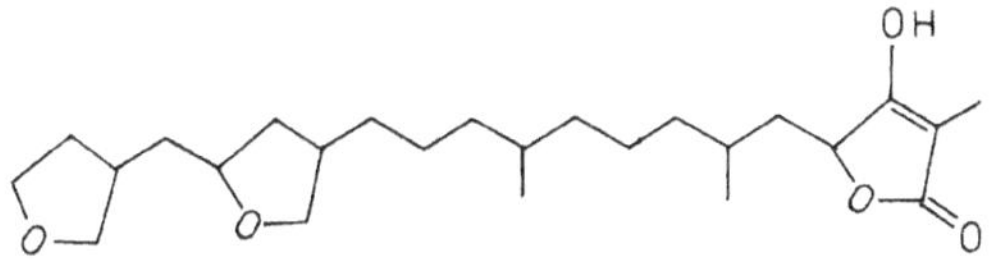

(**100**)

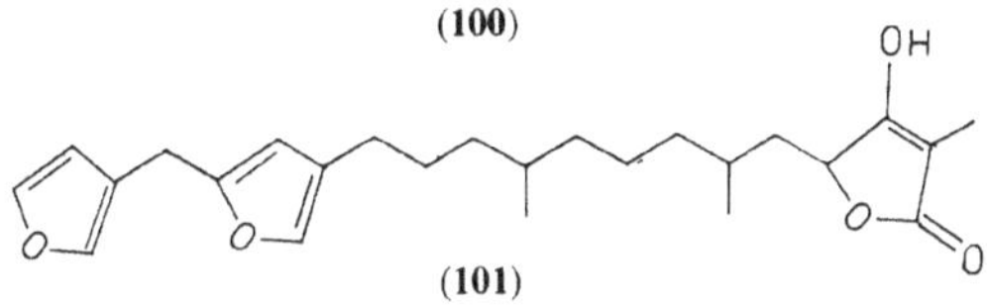

(**101**)

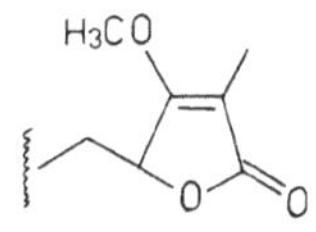

(**102**)

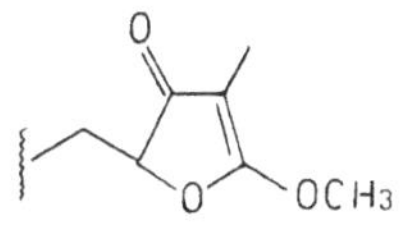

(**103**)

The conjugated tetronic acid moiety in (**94**) and (**95**) was suggested by u.v. [λ_{max} (MeOH) 260 nm; λ_{max} (MeOH/OH^-) 248 and 310 nm] and i.r. (ν_{max} 3150, 1735 and 1635 cm^{-1}) spectra and the n. m. r. signals at δ 8.82 (1 H, exchangeable with D_2O), 5.42 (1 H, d, J 9 Hz; 20-H) and 1.81 (3 H, s, 23-Me) added considerable weight to this assignement. Additional confirmatory evidence was obtained from chemical data: the dodecahydroderivatives (**100**), on methylation with diazomethane, gave, as expected, two O-methyl derivatives [partial structures (**102**) and (**103**)]. Treatment with hot concentrated alkali of the natural mixture of ircinin-1 (**94**) and ircinin-2 (**95**) resulted in the opening of the lactone rings and subsequent benzilic rearrangement of the α-diketones so formed to give the corresponding hydroxydicarboxylic acids (Chart 10). Finally, ozonolysis confirmed that the two isomers differ only in the position of the

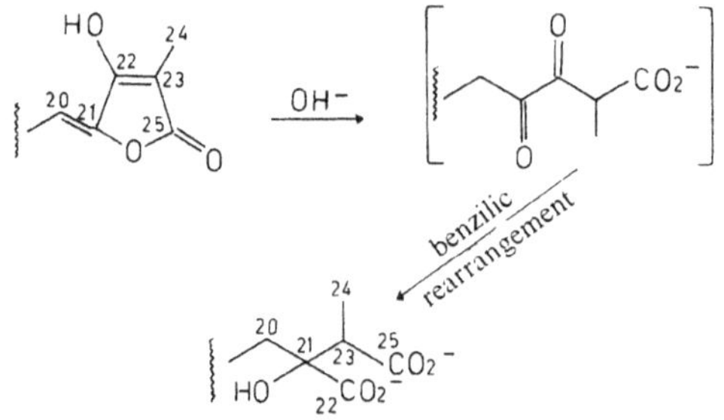

Chart 10. Reaction with alkali of conjugated tetronic acid ring

central double bond; this, together with the foregoing transformations and related spectroscopic properties, allowed ircinin-1 and-2 to be formulated as (**94**) and (**95**).

The structure determination of fasciculatin (**96**) and variabilin (**97**) isolated subsequently (*37*, *84*) followed the same pattern as that used for ircinin-1 and ircinin-2. The stereochemistry of the Δ^{11} double bond in fasciculatin (**96**) was derived from the coupling constants of the two olefinic protons, and the S,S chirality at C-13 and C-18 followed from the production, on ozonolysis, of (2 S, 6 S)-2,6-dimethylpimelic acid.

The last two furanoid sesterterpene listed in Chart 9 which cooccur in *S. officinalis* with the C_{21} terpenes resisted all attempts at separation. Their formulation as halfesters (**98**) and (**99**) rested largely upon the spectroscopic properties of the mixture and ozonolysis (*49*).

Another important group of sesterterpenes with a tetracarbocyclic skeleton has been found in sponges of the genus *Cacospongia*, and in the taxonomically related *Spongia officinalis*. Their structures are shown in Chart 11.

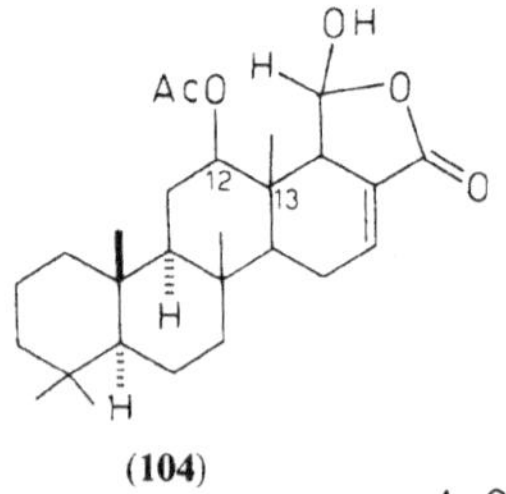

(**104**)

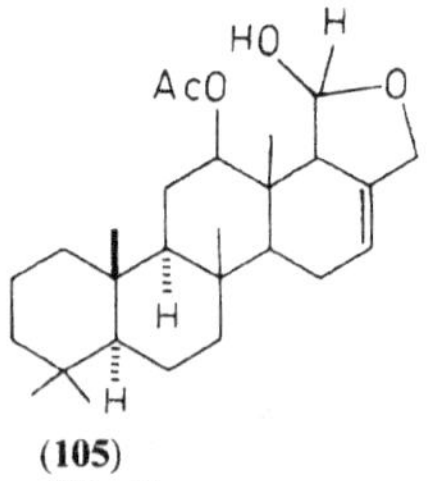

(**105**)

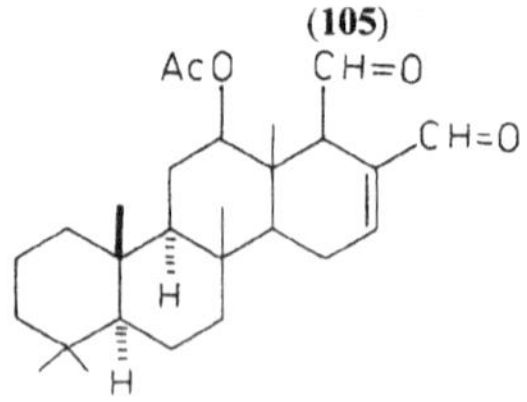

(**106**)

Chart 11. Tetracarbocyclic sesterterpenes in sponges

Scalarin (**104**) from *Cacospongia scalaris* was the first compound of this group to be isolated; its structure determination was reported by FATTORUSSO *et al.* (*78*). The u. v., i. r. and n. m. r. properties [λ_{max} 220 nm; ν_{max} 3350, 3320, 1755 and 1690 cm^{-1}; δ 6.81 (1 H, bm, CH=C) and 5.69 (1 H, bd, C$\underline{H}$OH)] indicated that the double bond, the hydroxyl group and one carbonyl function were included in an unsaturated γ-lactol ring. Supporting evidence for the presence of the lactol ring came from the n. m. r. spectrum of scalarin in pyridine-d_5/NaOD which showed a doublet at δ 9.81 (aldehyde proton) in place of the doublet at δ 5.69. The presence of an acetoxy group (ν_{max} 1733 cm^{-1}; δ 2.09 and 4.91 p. p. m.) accounted for the remaining two oxygen atoms in the molecular formula. Sodium borohydride reduction of scalarin (**104**), followed by removal of the acetyl group and hydrogenation of the product gave the saturated hydroxy-γ-lactone (**107**; ν_{max} 1770 cm^{-1}), chromic acid oxidation of which gave the ketolactone (**108**; ν_{max} 1710, 1780 cm^{-1}). Vigorous nitric acid oxidation of the latter yielded a ring-opened product (**109**), which was degraded (pyrolysis at 350°) to the acid (**110**). The latter was identified by comparison of its ester with an authentic sample. Chart 12 summarizes the degradative reactions of scalarin (**104**).

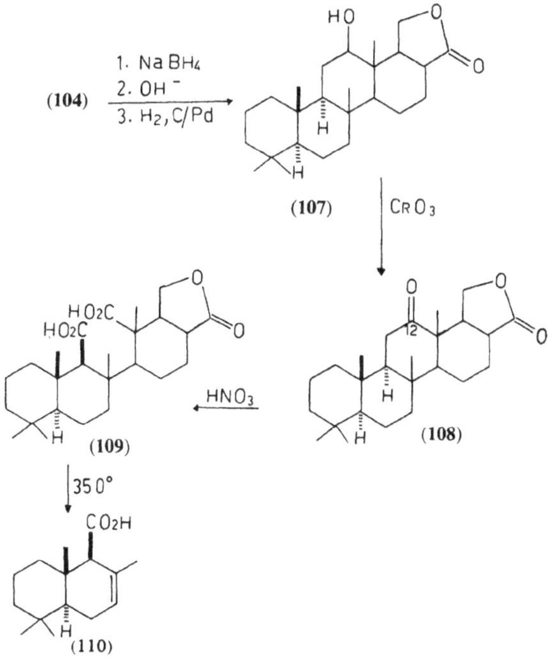

Chart 12. Ring-C opening of scalarin

Base-catalyzed deuterium exchange of the ketolactone (**108**), showed that the acetoxy group in scalarin was in position 12; three deuterium atoms were incorporated, two *alpha* to the ketonic and one *alpha* to the lactone carbonyl. This result also supported the location at C-13 of the fifth methyl group. The fragmentation pattern in the mass spectrum of scalarin fully agreed with the proposed structure (**104**).

Deoxoscalarin (**105**) and scalaradial (**106**) have subsequently been isolated from *Spongia officinalis* (*50*) and *Cacospongia mollior* (*56*), respectively, and their structures assigned on spectral grounds and chemical interrelation with scalarin (**104**), as summarized in Chart 13.

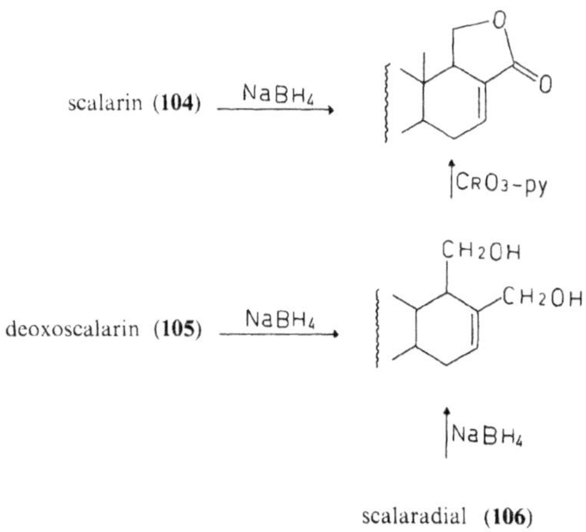

Chart 13. Interrelation of scalarin and its congeners

These closely related cyclic C_{25} terpenes represent a new fundamental skeletal type in sesterterpenes, which seems to be derived from a geranylfarnesyl precursor by cyclization initiated at the isopropylidene group (**111**), in a manner typical of triterpenes. The close biogenetic relationship of the sponge cyclic sesterterpenes to the fern sesterterpene, cheilanthatriol (**112**), described by KHAN *et al.* (*125*), is worthy of note.

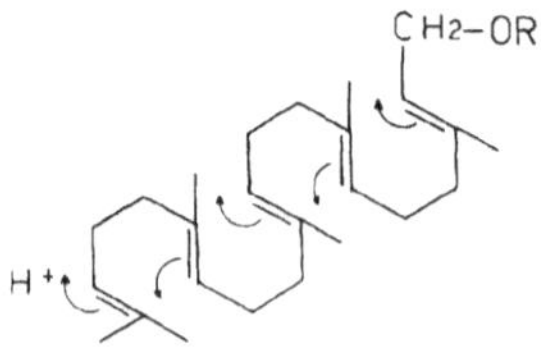

(**111**)

References, pp. 64—72

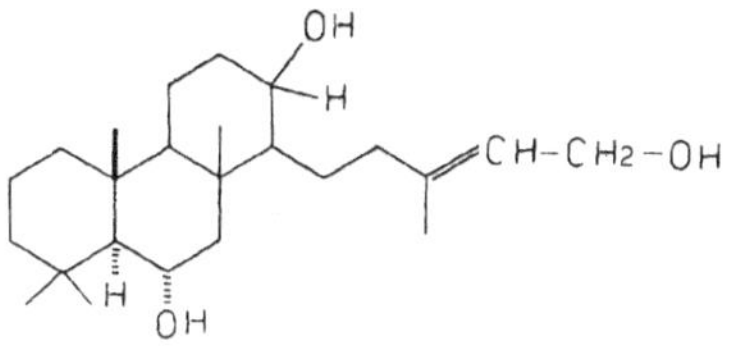

(112)

3. Compounds of Mixed Biogenesis: Mevalonate — Benzenoid Precursor

Compounds of mixed biogenesis originating partly from mevalonate and partly from a benzenoid precursor are widespread in nature: e.g. tocopherol quinones, ubiquinones and plastoquinones, "bioquinones" widely distributed involved in photosynthesis and electron transport (*176*) and the secondary products farnesiferols A – C (*39, 40*) the mould metabolites grifolin (*97*), tauranin (*123*), siccanin (*115*) and its congeners (*148*) and the *Cannabis* consituents derived from geranylolivetol (*136*). Very recently an interesting relative of δ-tocopherol, taondiol (**113**), with a cyclized diterpene chain, has been isolated from the brown alga *Taonia atomaria* by GONZALEZ *et al.* (*101, 104*) and two isomeric C_{21} hydroquinones, zonarol (**114**) and isozonarol, having a bicyclic sesquiterpenoid moiety, have been described from the brown seaweed *Dictyopteris zonarioides* by FENICAL *et al.* (*88*).

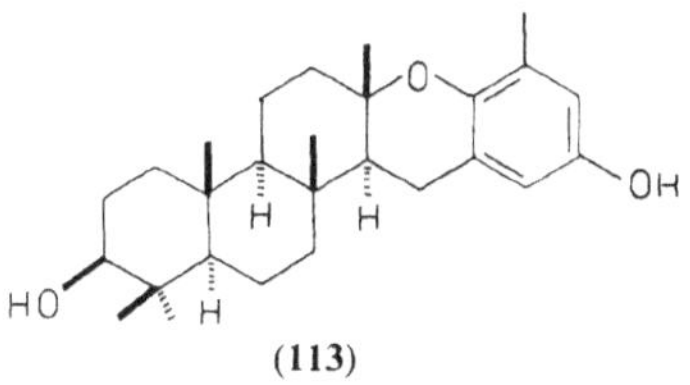

(113)

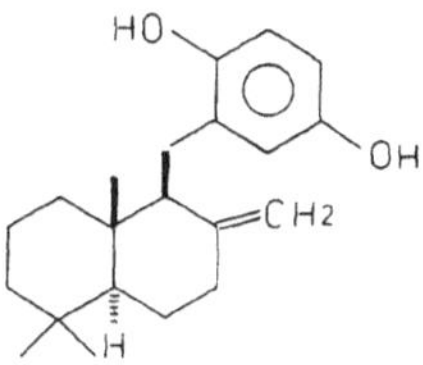

(114)

In recent years a group of related compounds characterized by the presence of an unsubstituted benzoquinone or quinol ring linked to an isoprenoid moiety has been obtained from sponges. Isoprenologous 2-polyprenyl benzoquinones (**115**; n = 6, 7, 8), a novel group of terpenoid quinones, and the corresponding quinols (**116**; n = 6, 7, 8) present in the solvent extracts in much larger amounts have been identified in *Ircinia spinosula* (*47*). Their structures were immediately suggested by spectral properties (u.v., i.r., n.m.r. and ms) of both the quinone and quinol

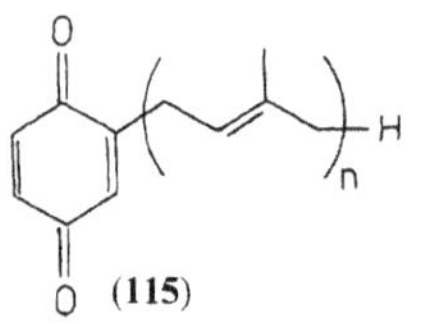

(115)

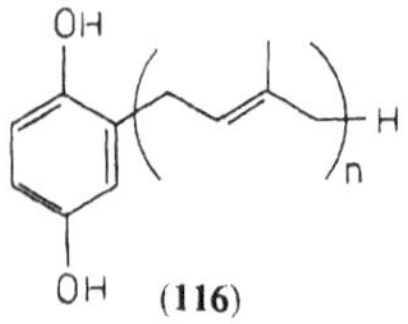

(116)

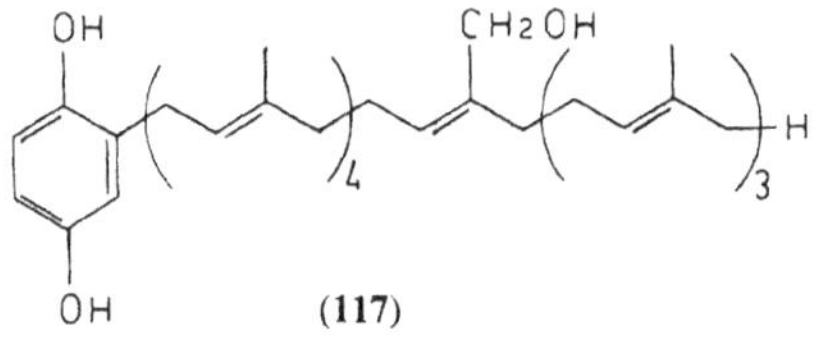

(117)

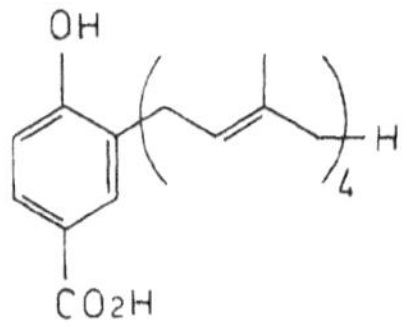

(118)

mixtures and were supported by the formation of acetone, malonic and levulinic acids on ozonolysis. The quinol having a C_{35} side-chain was synthesized by condensation of hydroquinone with farnesyl-geranyl-linalool (*trans*) in the presence of boron trifluoride.

A minor co-metabolite is a hydroxylated 2-octaprenyl quinol (**117**). This was evident from the hydroxyl absorption at 3600—3250 cm^{-1} of the corresponding quinone (obtained on Ag_2O oxidation), the n.m.r. signal at δ 4.11 (s, 2 H; $C\underline{H}_2OH$) which shifted downfield (0.5 p.p.m.) on conversion to the acetate, and the mass spectrum, which confirmed the molecular formula $C_{46}H_{70}O_3$ and showed a peak at M-18. Hydrogenation of the triacetate yielded a perhydroprenyldiacetate, hydrogenolysis of the allylic acetoxyl group occurring simultaneously, with g.l.c. retention time identical to that of the perhydro-diacetate derived from (**116**, n = 8). The fragmentation pattern in the mass spectrum located the hydroxy group in the fifth isoprene unit counting from the quinone ring.

From another *Ircinia* species (*I. muscarum*), 4-hydroxy-3-tetraprenyl-benzoic acid (**118**) has been isolated (*48*), along with 2-tetraprenyl benzoquinone (**115**, n = 4) and the corresponding quinol. This strongly suggests that p-hydroxybenzoic acid is the ring precursor as in ubiquinone biogenesis (*177*). Furthermore the prenylated quinones and quinols could conceivably be the precursors of ubiquinones in sponges. However, we did not detect ubiquinones in *I. muscarum* and could only find ubiquinone-10 (ca. 1 mg from 12 g of extract) in *I. spinosula*, which contains prenylated benzoquinones with 6, 7 and 8 isoprene units (the absence Q-6, Q-7 and Q-8 was established by reverse phase chromatography and comparison with authentic samples).

The acetone extracts of *Halichondria panicea* yielded (*52*) five new related "triprenylphenols", panicein-A, -B_1, -B_2, -B_3 and -C, each containing an aromatic sesquiterpenoid moiety linked to a quinol or quinone system, except for panicein-B_2, which is the corresponding chromenol of panicein-B_3 and probably an artifact of isolation. The structures of the paniceins are shown in Chart 14.

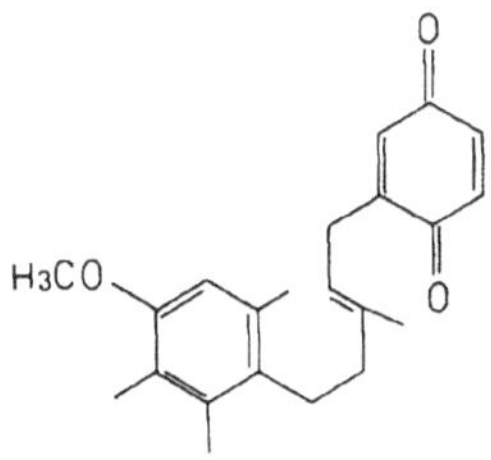

(**119**); panicein-A

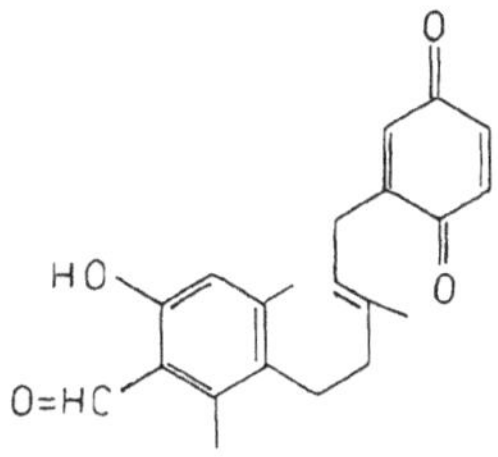

(**120**); panicein-B_1

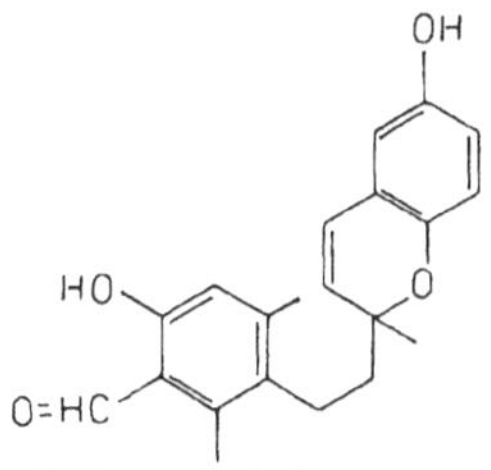

(**121**); panicein-B_2

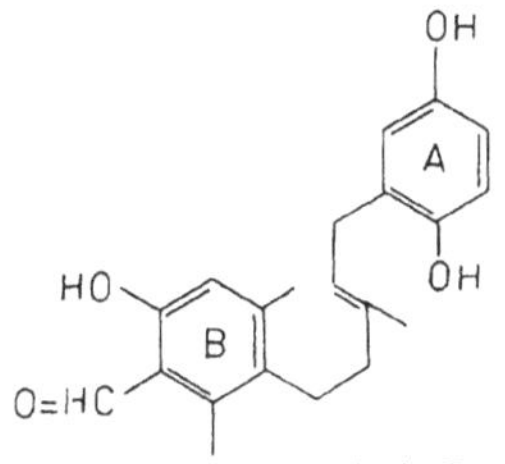

(**122**); panicein-B_3

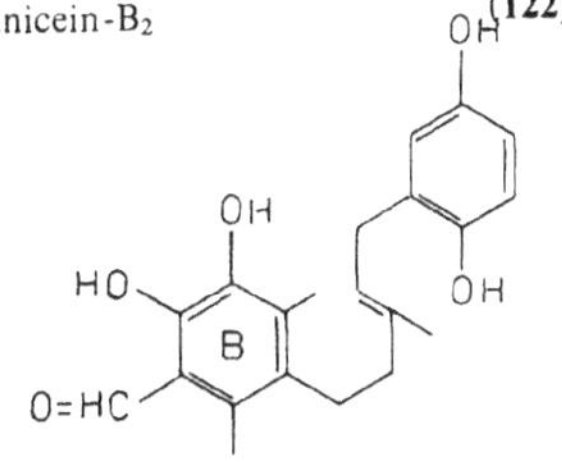

(**123**); panicein-C

Chart 14. Structures of paniceins, triprenylphenols from *Halichondria panicea*

Panicein-B_3, $C_{21}H_{24}O_4$ (**122**), has u.v. (λ_{max} 276 and 346 nm), i.r. (1640 cm^{-1}) and n.m.r. (1 H singlets at δ 10.26 and 12.01, the latter exchangeably with D_2O) spectra indicating the presence of an *o*-hydroxybenzaldehyde chromophore. The remaining two oxygens, indicated by the molecular formula, are incorporated into a monosubstituted 1,4-dihydroxybenzene structure as suggested by a prominent peak in the mass spectrum at m/e 161 corresponding to the ion (**124**), common to all

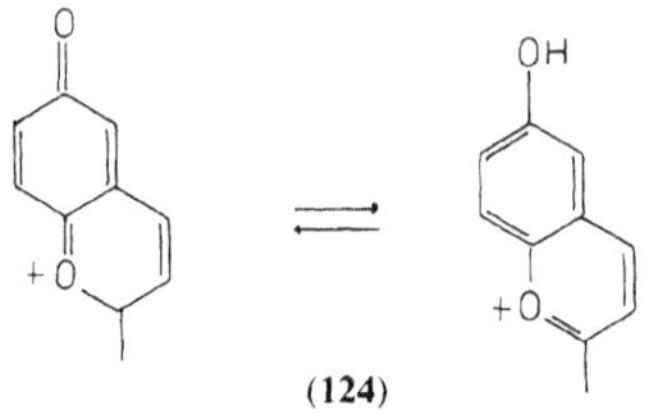

(**124**)

the 2-polyprenylbenzoquinols. Ozonolysis of the triacetate afforded malonic, levulinic, and 2,5-diacetoxyphenylacetic acid. The n.m.r. spectrum of panicein-B_3 also includes signals for two methyl groups (singlets at δ 2.31 and 2.51 p.p.m.) which must be attached to the hydroxybenzaldehyde ring. The arrangement of the substituents on this ring was established by spin decoupling and NOE experiments. From the chemical shift of the ring B proton (δ 6.67 in the spectrum of the triacetate, well separated from the signal of the ring A protons resonating as broad singlet at δ 6.91), it must be either *ortho* or *para* to the phenolic group; irradiation of the aromatic methyl protons at δ 2.31 sharpens the singlet at δ 6.67, while there was no change on irradiation at the downfield aromatic methyl signal. Hence the methyl at δ 2.31 must be *ortho* to the ring proton. Finally, the *ortho* relationship between aldehyde and methyl at δ 2.51 was established by a positive NOE. Of all the possible structures, only (**122**) meets these conditions.

Structures for panicein-B_1 (**120**) and B_2 (**121**), the corresponding quinone and chromenol of panicein-B_3, respectively, and the structure for panicein-A (**119**) were proved by chemical interrelation with panicein-B_3.

Panicein-C (**123**) is also similar to panicein-B_3; the difference resides in the presence of an extra hydroxyl group which is situated in ring *B*. This was suggested by the ultraviolet maxima at 291 and 380 nm and confirmed by ozonolysis of the tetraacetate which yielded 2,5-diacetoxyphenylacetic acid. A green color with $FeCl_3$ and a borate shift in the u.v. spectrum both suggested a catechol structure, and from infrared/n.m.r. spectra one hydroxyl group is chelated with the aldehyde carbonyl group. Since irradiation of the aromatic methyl protons at δ 2.47 resulted in a 13% increase of the intensity of the aldehyde proton signal, it was concluded that the ring B substituents can be arranged either as in (**123**) or as in the alternative structure (**125**). Structure (**123**), with the carbon substitution pattern in ring B identical to those in the other paniceins, is favoured.

References, pp. 64—72

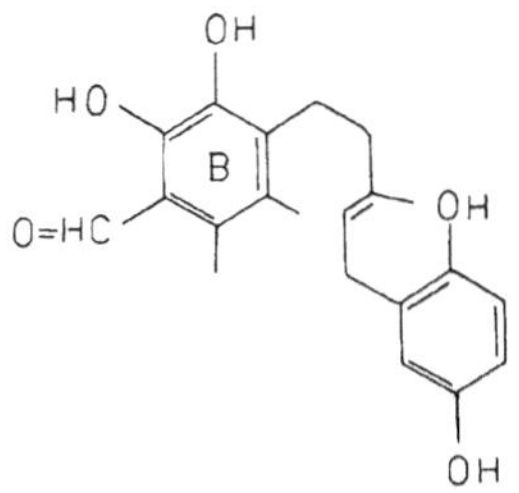

(125)

In the sesquiterpenoid moiety the paniceins have the uncommon feature of an aromatic ring which, notably, has been encountered already in renieratene (**126**) and isorenieratene (**127**). These are arylcarotenoids found in the sponge *Reniera japonica* (syn. *Halichondria panicea*) (*183, 184*), which also yielded a third aromatic carotenoid renierapurpurin (**130**) (*185*).

Phenolic carotenoids, with the hydroxyl groups in the same relative positions as in the paniceins were isolated from *Streptomyces* sp. (*12*), namely 3-hydroxyisorenieratene (**128**) and 3,3′-dihydroxyisorenieratene

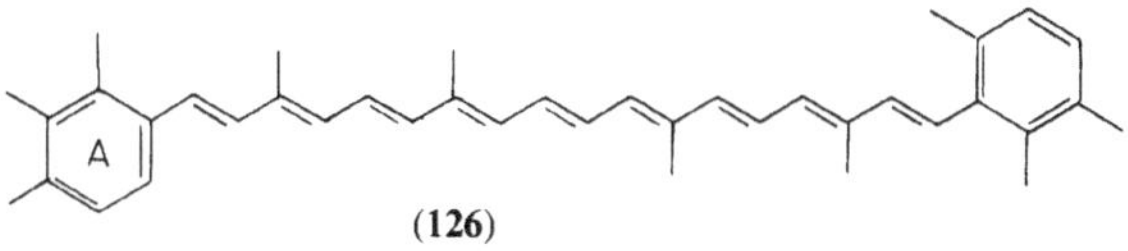

(126)

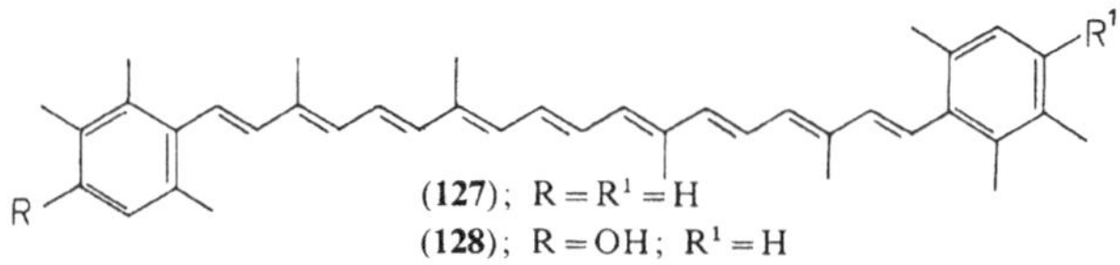

(**127**); $R = R^1 = H$
(**128**); $R = OH$; $R^1 = H$
(**129**); $R = R^1 = OH$

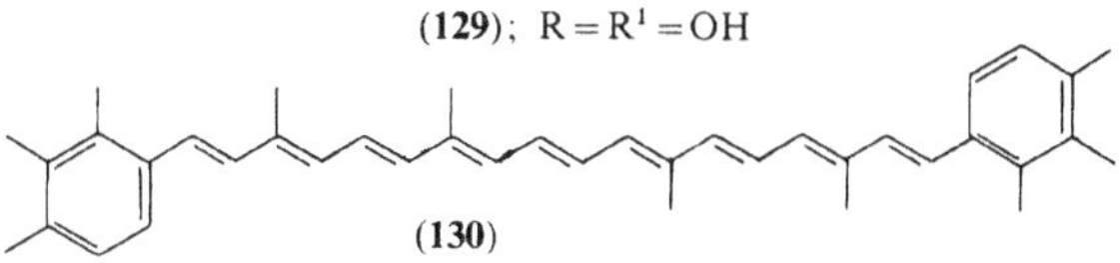

(130)

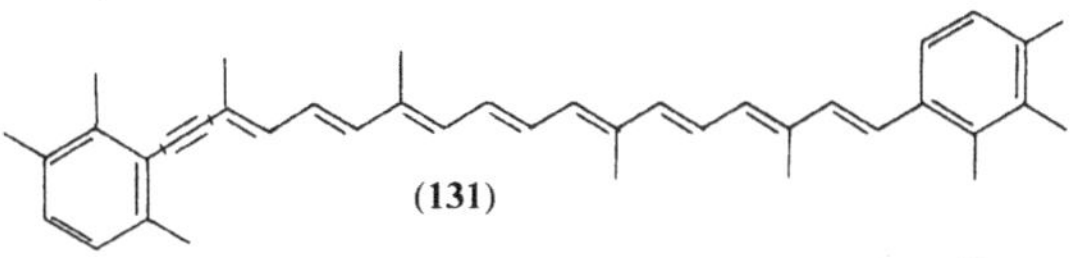

(131)

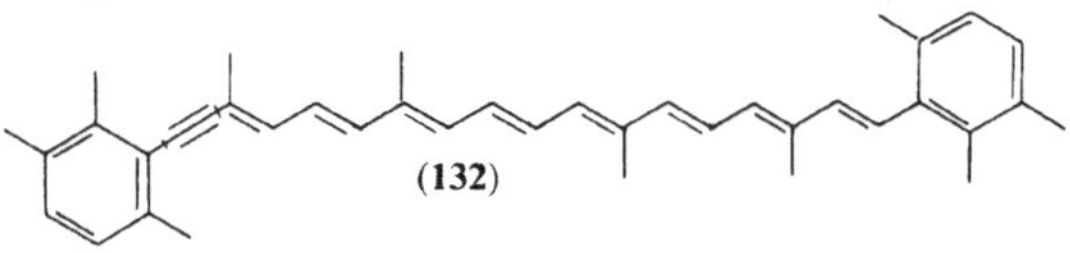

(132)

(**129**). It should be noted that the alternative structure (**125**) for panicein-C has a carbon substitution pattern on ring B identical with that of ring A of renieratene (**126**) and of both the rings of renierapurpurin (**130**). It is of interest that HAMASAKI *et al.* (*109*) have recently found two further aryl-carotenoids (**131**) and (**132**) in *Reniera japonica*, closely related to renieratene and isorenieratene, respectively, and with the unique feature of an acetylenic bond. Other carotenoids having conventional structures were also found in Sponges (*66, 105*).

The biosynthetic derivation of these aromatic groups of terpenoid origin is a matter for conjecture, but it seems likely that 1,2-methyl migration occurs in a monocyclofarnesyl derivative (Chart 15) and in a β-end group of carotenoids as suggested by RAMAGE (*154*), followed by oxidation. The co-occurrence of methyl *trans*-monocyclofarnesate (**62**) and the paniceins in *Halichondria panicea* supports the intermediacy of a monocyclofarnesyl precursor for the latter (*51*).

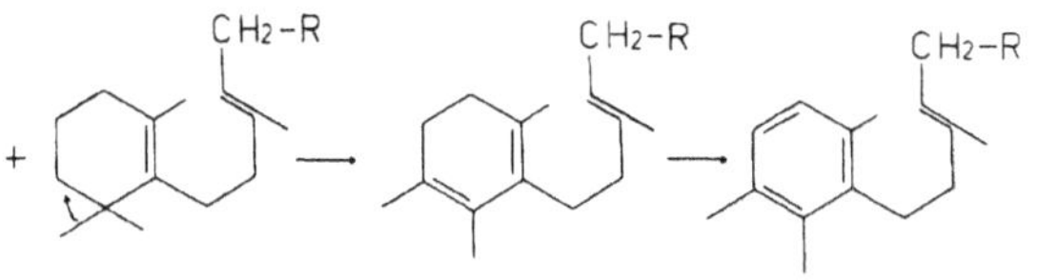

Chart 15. Possible biogenetic route to paniceins

A further sesquiterpenoid hydroquinone, avarol, has been isolated from *Disidea avara* (*139*). It possesses a "friedo" structure (**133**), which could be derived from farnesyl pyrophosphate by cyclization to an intermediate cation (**134**), followed by a "friedo" rearrangement and finally deprotonation. At the time of writing the stereochemistry of avarol was not yet firmly established. The presence of a monosubstituted hydroquinone ring was derived from u.v. (λ_{max} 298 nm), and n.m.r. (three aromatic protons at δ 6.62) data, and conversion to a *p*-benzoquinone (ν_{max} 1660, 1600 cm^{-1}; λ_{max} 245, 315 and 440 nm) by oxidation with silver oxide. The "non-isoprenoid" nature of the bicyclic sesquiterpene moiety of avarol was at first suspected from its n.m.r. spectrum, which includes signals for four methyl group, two tertiary (δ 0.84 and 1.02 p.p.m.), one secondary (δ 1.00, J 6 Hz) and one vinyl (δ 1.50), and a signal for a benzylic methylene group linked to a saturated quaternary carbon (AB system centered at δ 2.64 with J 14 Hz). This eliminated the possibility of a drimane skeleton as for example in zonarol (**114**). Oxidation of avarol dimethyl ether with osmium tetroxide followed by cleavage of the diol with lead tetraacetate gave the ketoaldehyde (**135**). Selective hydrogenation of the aldehyde group afforded the hydroxyketone (**136**). Base-catalyzed deuteriation of this led to the exchange of three protons, thus the acetyl group was

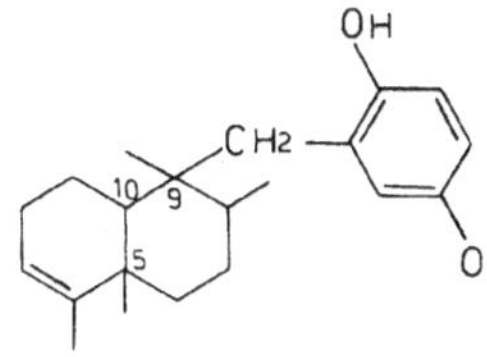

(133)

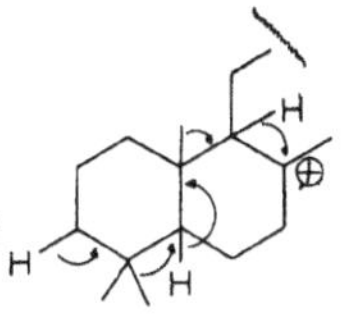

(134)

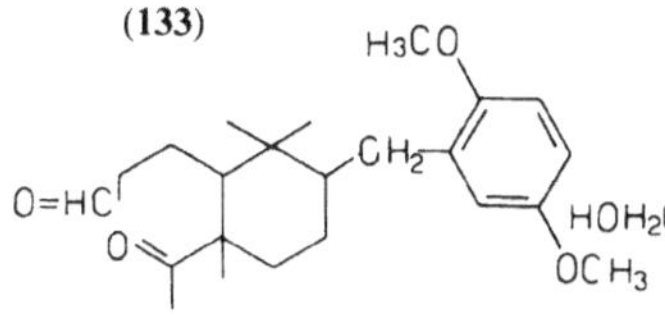

(135)

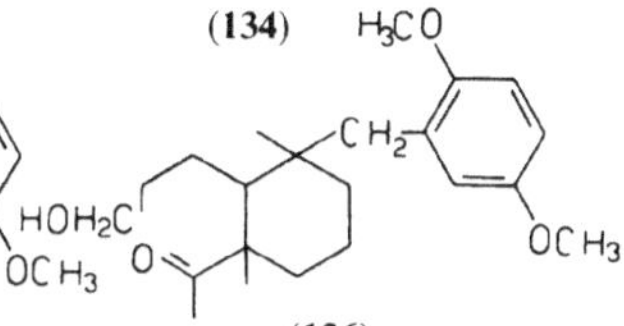

(136)

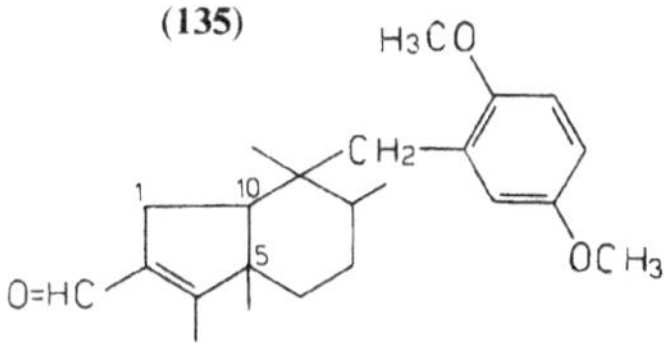

(137)

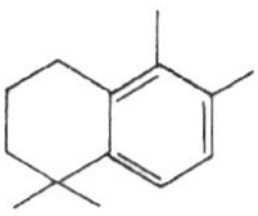

(138)

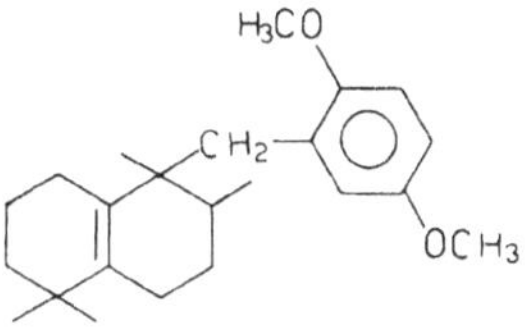

(139)

attached to a quaternary carbon atom. The ketoaldehyde (**135**) was cyclized to the cyclopentenaldehyde (**137**). In its n.m.r. spectrum an AB portion of an ABX system was assigned to protons in position 1 showing that C_{10} is tertiary and C_5 quaternary. Dehydrogenation of avarol (**133**) with palladium-charcoal afforded, in low yield, 1,2,5,6-tetramethylnaphthalene and 1,2,5-trimethylnaphthalene along with major amounts of the tetralin (**138**).

These results can be easily explained by assuming that a preliminary 1,2-methyl migration takes place in the dehydrogenation of avarol. Similar Wagner-Meerwein shifts under dehydrogenation conditions have been reported and they occur, apparently, when an adjacent cationic site is available (*126*). Conversion of avarol by acid into the tetrasubstituted olefin (**139**), which gave, on dehydrogenation, (**138**) in high yield, confirmed the structure of the natural sesquiterpenoid as (**133**). Avarol represents the first "friedo" structure in sesquiterpenoids.

A cyclic sesterterpene linked to a hydroxyhydroquinone ring has recently been isolated as the sodium calcium salt of the disulphate, from *Disidea pallescens* (*44*). For the free phenol, disidein, formula (**140**) can be proposed as the most probable structure, mainly on spectral and biogenetic grounds.

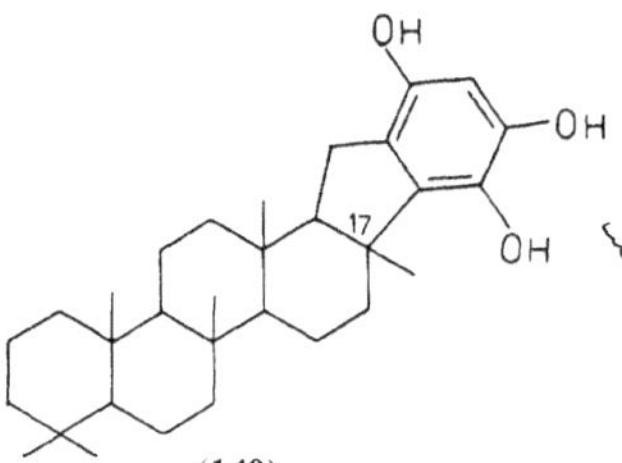

(**140**)

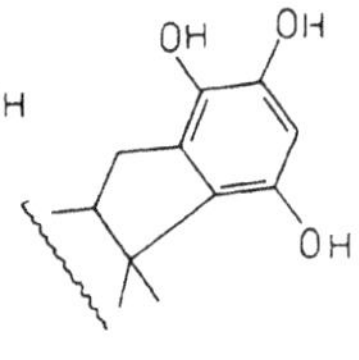

(**141**)

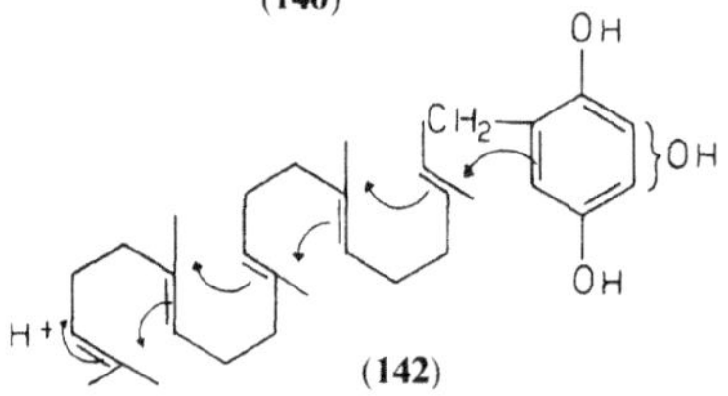

(**142**)

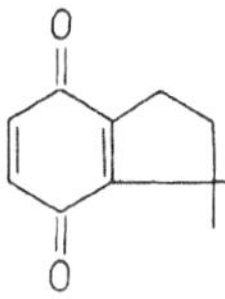

(**143**)

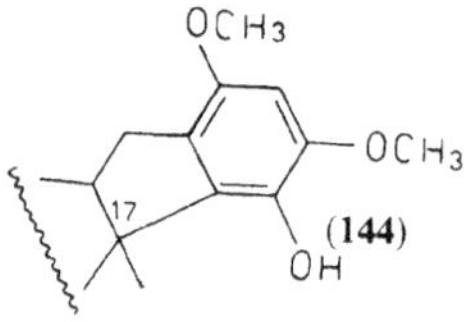

(**144**)

The hydroxyhydroquinone nature of the chromophore in disidein was evident from its conversion on Ag_2O oxidation to an hydroxy-*p*-quinone derivative having ultraviolet absorptions at 280 and 391 nm and i. r. bands at 3500, 1650, 1635, and 1595 cm^{-1}. The molecular formula of disidein, $C_{31}H_{46}O_3$, and the absence of olefinic signals in its n.m.r. spectrum led to consideration of a pentacyclic skeleton in addition to the aromatic ring. Furthermore the n. m. r. spectrum revealed one aromatic proton (δ 6.79) two benzylic hydrogens (δ 2.50 m), and six *tert*-methyl groups [in $CDCl_3$ δ 1.30 (3 H), 1.01 (3 H) 0.85 (12 H)] and this suggested that disidein is represented best by formula (**140**) or the alternative (**141**), which are attractive from the standpoint of biogenesis. In fact, we may imagine that (**142**) undergoes an essentially synchronous process leading to ring formation if H^+ is furnished at C-3. The relatively low-field resonating methyl (δ 1.30 in both the spectra of disidein and the quinone) can be assigned to the 17-methyl group which is deshielded by the neighbouring aromatic ring. In the spectrum of the model compound (**143**) the methyl groups resonate at δ 1.30 and 1.22 p.p.m., respectively; in this compound the cyclopentene ring must be twisted.

A decision in favour of the hydroxylation pattern shown in (**140**) came from n.m.r. spectroscopy of the dimethyl ether (**144**) (obtained from disidein by methanol-hydrochloric acid treatment), using a shift reagent [$Eu(fod_3-d_9)_3$], which induced about the same shift for the 17-methyl resonance and for the benzylic methylene. Evidence for the *meta*-methoxy groups in (**144**) was obtained by DDQ oxidation, which gave a mixture of a methoxy-*p*-quinone and a methoxy-*o*-quinone. Support for the proposed pentacyclic skeleton for the sesterterpene moiety of disidein was also obtained from mass spectral fragmentation studies.

Significantly the carbocyclic system represented by formula (**140**) is reminiscent of those found in scalarin (**104**), deoxoscalarin (**105**) and scalaradial (**106**) isolated from sponges.

Even more significant is the co-occurrence in *Disidea pallescens* of the pentaprenyl-*p*-benzoquinone (**115**, n = 5) (*44*).

4. Sterols

The recognition that sponges contain a variety of sterols which vary from one species to another was first reported in a classical paper by Dorée (*74*). This author isolated a sterol from the sponge *Cliona celata* which he named clionasterol and showed to be different from both cholesterol and spongosterol. The latter, earlier isolated from a mediterranean sponge (*114*), was the first animal sterol to be recognized as different from cholesterol.

Since then, progress on sponge sterol research has been reported by Bergmann who reviewed the subject up to about 1960 (*17*). An earlier review on marine invertebrate sterols was written by the same author in 1949 (*16*).

Austin (*13*), reviewing the work on the sterols of marine invertebrates and plants of the decade up to 1968, was able to deal very briefly with to sterols of sponges because nothing new had appeared in that period. In 1970 Brooks comprehensively reviewed the chemistry of sterols, including marine sterols (*32*).

The extensive researches of Bergmann on sterols were carried out before the introduction of efficient chromatographic methods and he was aware that much of the data accumulated probably referred to sterol mixtures and not to individual compounds. Thus most of the earlier work now requires revision with the aid of more sophisticated techniques of analysis such as t.l.c., g.l.c. and mass spectrometry.

Since 1972 modern reinvestigations of the sterols of sponges have began to appear, which confirm the complexity of sterol composition in this phylum. Sterols of a completely new types have been discovered including unprecedented side-chain alkylation patterns and, more notably, modified tetracyclic nuclei.

We now consider briefly the results reported prior to the application of the modern techniques, while much emphasis will be placed on research published during the past few years mainly on that dealing with the new sterols. A brief summary of the status of our knowledge of the sterol metabolism in sponges is also presented.

4.1 "Conventional" Sterols

Table 1 lists the "conventional" sterols reported in sponges. The term "conventional" is used to cover sterols with the conventional C_{19} tetracyclic nucleus having the usual C_8 side-chain or else modified by the addition of one or two carbon atoms at C-24. The 24-norsterols are also included.

The systematic analyses of BERGMANN dealt with more than fifty species of sponges and his elucidation of sterol structure was limited to the major components.

DOREE's "clionasterol" was found to be a mixture of poriferasterol (**171**) and a monounsaturated sterol (**166**) for which the original name was retained (*178*). These two sterols were subsequently reported as the most common ones in sponges (*16, 17*).

A C_{29}-sterol, chondrillasterol (**172**), which seems to be confined solely to Porifera in the animal kingdom, was reported as the major sterol component of *Chondrilla nucula* (*23*).

Among the C_{28}-sterols, 24-methylenecholesterol (**164**) which is widely distributed in marine organisms was found in *Chalina arbuscula* and *Tetilla laminaris* (*25*). HENZE's "spongosterol" from *Suberites* was separated into neospongosterol (**159**) and cholestanol (**147**), (*21*). Cholestanol (**147**) was found in eleven other sponges (*17*), often as a major sterol constituent, and cholesterol (**149**) was found in *Microciona prolifera* and *Halichondria panicea* (*24*).

BERGMANN also recognized the importance of sterol analysis for the taxonomy and phylogeny of sponges, and attempted to correlate sterol distribution with sponge taxonomy (*17*).

In 1972 ten years after the appearance of the last of BERGMANN's reviews (*17*), ERDMAN and THOMSON reported an investigation on the sterols from two British species. *Cliona celata* and *Hymeniacidon perleve*

Table 1. "*Conventional*" *Sterols Reported in Sponges*

Text no.	Trivial name	Systematic name	References
(145)	—	24-norcholesta-22-en-3β-ol	(*75*)
(146)	—	24-norcholesta-5,22-dien-3β-ol	(*72, 75*)
(147)	cholestanol	5α-cholestan-3β-ol	(*17, 75*)
(148)	22-dehydrocholestanol	5α-cholest-22-en-3β-ol	(*75*)
(149)	cholesterol	cholest-5-en-3β-ol	(*17, 24, 72, 75*)
(150)	lathosterol*	5α-cholest-7-en-3β-ol	(*38*)
(151)	$\Delta^{8(9)}$-cholestenol*	5α-cholest-8 (9)-en-3β-ol	(*38*)
(152)	22-dehydrocholesterol	cholesta-5,22-dien-3β-ol	(*72, 75*)
(153)	$\Delta^{7,22}$-cholestenol*	cholesta-7,22-dien-3β-ol	(*38*)
(154)	7-dehydrocholesterol*	cholesta-5,7-dien-3β-ol	(*38*)
(155)	—	cholesta-5,7,22-trien-3β-ol	(*38*)
(156)	22,23-dihydrobrassicasterol	(24 S)-24-methylcholest-5-en-3β-ol	(*72, 75*)
(157)	Δ^{7}-ergostenol*	24-methylcholest-7-en-3β-ol	(*38*)
(158)	$\Delta^{8(9)}$-ergostenol*	24-methylcholest-8 (9)-en-3β-ol	(*38*)
(159)	neospongosterol	5α-(24 S)-24-methylcholest-22-en-3β-ol	(*17, 24, 75*)
(160)	24-methylenecholestanol	5α-24-methylcholest-24 (28)-en-3β-ol	(*75*)
(161)	22,23-dihydroergosterol*	24-methylcholesta-5,7-dien-3β-ol	(*38*)
(162)	5,6-dihydroergosterol*	24-methylcholesta-7,22-dien-3β-ol	(*38*)
(163)	brassicasterol	(24 R)-24-methylcholesta-5,22-dien-3β-ol	(*72, 75*)
(164)	24-methylenecholesterol	24-methylcholesta-5,24 (28)-dien-3β-ol	(*17,25,70,72,75*)
(165)	ergosterol	(24 R)-24-methylcholesta-5,7,22-trien-3β-ol	(*38*)
(166)	clionasterol	(24 S)-24-ethylcholest-5-en-3β-ol	(*17, 72, 75, 178*)
(167)	β-sitosterol	(24 R)-24-ethylcholest-5-en-3β-ol	(*75*)
(168)	—*	24-ethylcholest-7-en-3β-ol	(*38*)
(169)	—*	24-ethylcholest-8 (9)-en-3β-ol	(*38*)
(170)	—*	24-ethylcholesta-5,7-dien-3β-ol	(*38*)
(171)	poriferasterol	(24 R)-24-ethylcholesta-5,22-dien-3β-ol	(*17, 72, 75, 178*)
(172)	chondrillasterol	(24 R)-24-ethylcholesta-7,22-dien-3β-ol	(*23*)
(173)	fucosterol	24-ethylcholesta-5,24 (28)-dien-3β-ol	(*75*)
(174)	—*	24-ethylcholesta-5,7,22-trien-3β-ol	(*38*)

* Refers to sterols identified by g.l.c.-mass spectrometry and not isolated.

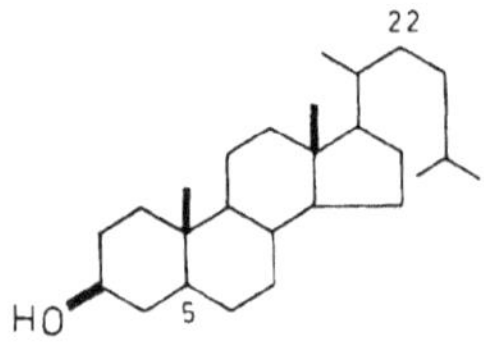

(145); Δ^{22}
(146); $\Delta^{5,22}$

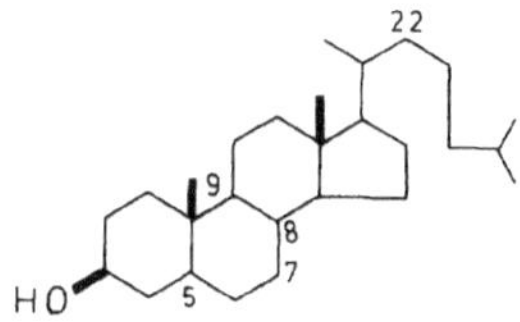

(147)
(148); Δ^{22}
(149); Δ^{5}
(150); Δ^{7}
(151); $\Delta^{8(9)}$
(152); $\Delta^{5,22}$
(153); $\Delta^{7,22}$
(154); $\Delta^{5,7}$
(155); $\Delta^{5,7,22}$

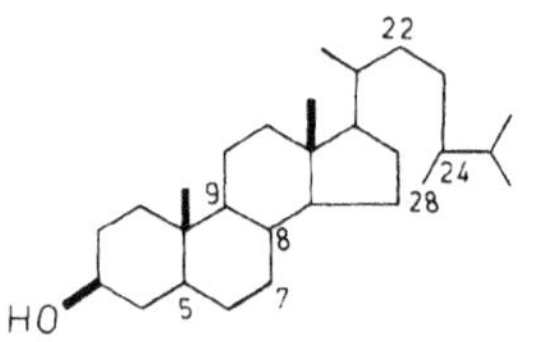

(156); (24 *S*); Δ^{5}
(157); Δ^{7}
(158); $\Delta^{8(9)}$
(159); (24 *S*); Δ^{22}
(160; $\Delta^{24(28)}$
(161); $\Delta^{5,7}$
(162); $\Delta^{7,22}$
(163); (24 *S*); $\Delta^{5,22}$
(164); $\Delta^{5,24(28)}$
(165); (24 *R*); $\Delta^{5,7,22}$

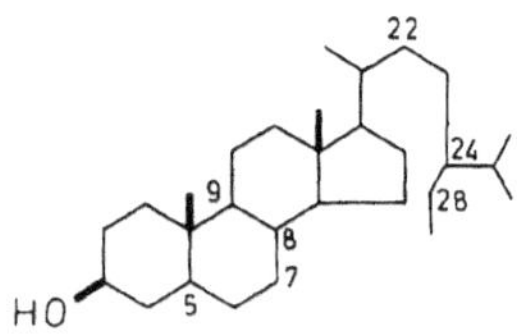

(166); (24 *S*); Δ^{5}
(167); (24 *R*); Δ^{5}
(168); Δ^{7}
(169); $\Delta^{8(9)}$
(170); $\Delta^{5,7}$
(171); (24 *R*); $\Delta^{5,22}$
(172); (24 *R*); $\Delta^{7,22}$
(173); $\Delta^{5,24(28)}$
(174); $\Delta^{5,7,22}$

(*75*). *Cliona celata* was shown to contain a mixture of eight sterols: cholesterol (predominant) (**149**), 24-norcholesta-5,22-dien-3β-ol (**146**), 22-dehydrocholesterol (**152**), 24-methylenecholesterol (**164**), brassicasterol (**163**), 22,23-dihydrobrassicasterol (**156**), poriferasterol (**171**), and clionasterol (**166**). In *Hymeniacidon perleve* cholestanol (**147**) is the predominant sterol accompanied by minor amounts of seven sterols, including three new marine stanols: 24-norcholesta-22-en-3β-ol (**145**), 22-dehydrocholestanol (**148**) and 24-methylenecholestanol (**160**); the remaining sterols were neospongosterol (**159**), brassicasterol (**163**), β-sitosterol (**167**) and fucosterol (**173**). The sterols from both sponges were isolated by

References, pp. 64—72

combined t.l.c. on silica gel/$AgNO_3$ and g.l.c. and identified on spectral grounds (mainly n.m.r. and mass spectrometry). Identification of the known sterols was substantiated by comparison with authentic samples.

It is worth pointing out that this is the first occurrence in sponges of sterols with twenty-six carbon atoms (24-nor); they are widespread in marine invertebrates and to date have been reported in four marine phyla, the Coelenterates (*181*), the Echinoderms (*99*), the Molluscs (*120*), and the Tunicates (*180*).

A detailed g.l.c.-mass spectrometric analysis of the sterols of *Axinella cannabina* (*38*) revealed the presence of fifteen components including Δ^7-, $\Delta^{8(9)}$- and $\Delta^{5/7}$-C_{27}, -C_{28} and -C_{29} sterols (**150**, **151**, **153**, **154**, **155**, **157**, **158**, **161**, **162**, **165**, **168**, **169**, **170**, **174**) along with cholesterol (2%). The major components were ergosterol (**165**) and cholesta -5,7,22-trien-3β-ol (**155**), which were isolated.

In the same sponge Fattorusso *et al.* (*79*) have also found a mixture of 5,8-epidioxy-Δ^6-sterols; the two major components have been isolated and identified as 5,8-epidioxyergosta-6,22-dien-3β-ol (ergosterol peroxide) and 5,8-epidioxycholesta-6,22-dien-3β-ol.

The sterol composition of twenty five sponges has been recently determined by g.l.c. and mass spectrometry of the steryl acetates and the derived stanyl acetates (*72*). This paper confirmed the usefulness of sterol analysis in the classification of sponges. All species were found to contain very complex mixtures, the Δ^5-sterols being the most widely distributed. In Spongidae, $\Delta^{5/7}$-sterols were found almost exclusively in contrast with the earlier reports of Bergmann, who observed in four representative specimens of Spongidae the presence of Δ^5-sterols (*17*). $\Delta^{5/7}$-sterols are also the principal components of two *Axinella* species, and *Disidea avara*. Δ^7-sterols have been found in only one species (*Agelas oroides*) in which stanols have been also found. The latter are the major constituents in *Reniera cratera*, three *Axinella* species and *Chondrilla nucula*.

The stanols of two *Axinella* have subsequently proved to be of a completely new type (see pages 48 and 49).

All the *Verongia* species examined contained as principal components two new sterols designated aplysterol (**179**) and 24,28-didehydroaplysterol (**180**) with a pattern of sidechain alkylation involving "normal" alkylation at C-24 and the unprecedented addition of an "extra" methyl group at C-26 (see page 46; Table 2). Sterols with the aplystane skeleton proved to be absent in all the remaining species examined. C_{26} sterols have been found in very low yield (*ca.* 1% of the total sterol content) in almost all the species examined and their ubiquitous occurrence suggested a common origin from phytoplankton in which they occur (*29*).

A fuller analysis of the minor sterols from *Verongia aerophoba* [70% of its total sterol content is due to aplysterol (**179**) and 24,28-didehydroaplysterol (**180**); Table 2] showed the presence of several known compounds [Δ^5-C_{29}, Δ^5-C_{28}, cholesterol (**149**), porifasterol (**171**), brassicasterol (**163**), 22-dehydrocholesterol (**152**), 24-norcholesta-5,22-dien-3β-ol (**146**) and 24-methylenecholesterol (**164**)] together with a C_{30} di-unsaturated component (*72*).

4.2 "Unusual" Sterols

Marine invertebrates have proved to be sources of unusual sterols having the cholesterol nucleus with side-chains involving new alkylation patterns. Thus gorgosterol (**175**) from gorgonians (*106, 131*), acanthasterol (**176**) from the starfish *Acanthaster planci* (*169*) and 23-demethylgorgosterol (**177**), isolated from two gorgonians (*161*) exemplify the attachment of "extra" carbon atoms at C-22 and C-23; in 29-methylisofucosterol (**178**), isolated from the scallop *Placopecten magellanicus* (*119*), a propylidene group has been elaborated at C-24.

A further structural variant is now represented by aplysterol (**179**) and 24,28-didehydroaplysterol (**180**), the principal sterol components of the sponges of the family Verongidae (Table 2), in which an "extra" carbon atom is attached at C-26 (*70*).

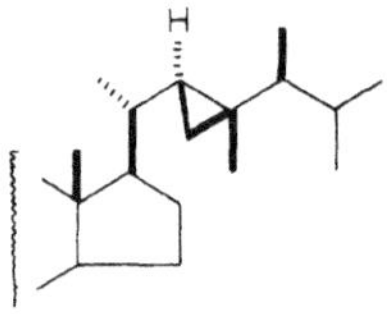

(**175**); Δ^5
(**176**); Δ^7

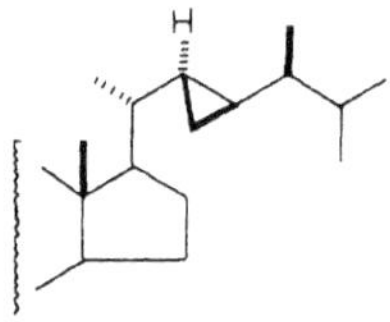

(**177**)

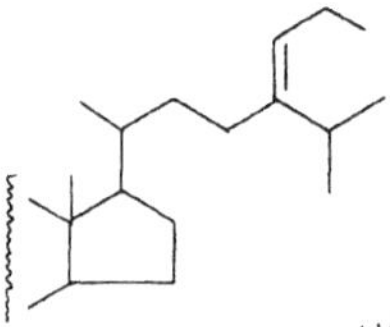

(**178**)

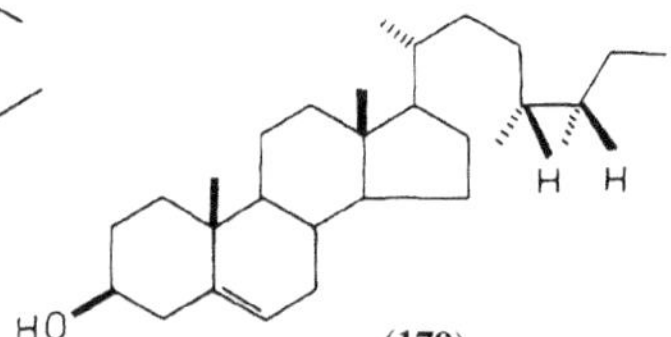

(**179**)

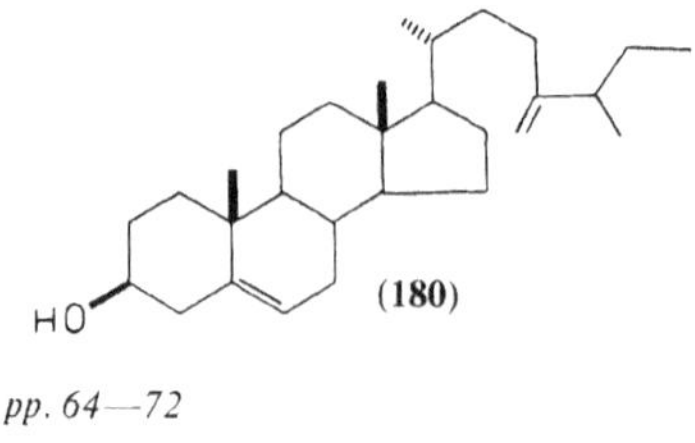

(**180**)

References, pp. 64–72

Table 2. *Sponges Containing Aplysterol* (**179**) *and 24,28-didehydroaplysterol* (**180**) (*72*)

Sponges	Source	% of the total sterol content
Order Dictioceratida		
Family Verongidae		
V. aerophoba	Naples	70%
V. archeri (hard)	Jamaican N. Shore	60%
V. archeri (soft)	British V. Islands	60%
V. fistularis	Bermuda	67%
V. thiona	La Jolla (California)	78%

The presence of a normal 3β-hydroxy-Δ^5-sterol system in both substances which gave, on hydrogenation, the same stanol was established by conventional physical techniques, expecially by using mass spectrometry which showed that the *exo*-methylene group of 24,28-didehydroaplysterol (**180**) was in position 24 (McLafferty rearrangement, peak at m/e 314). The problem concerning the position of the "extra" methyl group was solved by treatment of 24,28-didehydroaplysterol with iodine in benzene to effect migration of the double bond in the side-chain and subsequent exhaustive ozonolysis to give mainly 3-methylpentan-2-one with some butan-2-one. The synthesis of 26-methyl-24-methylenecholesterol definitively established this as the structure of 24,28-didehydroaplysterol (*69*). The synthesis was accomplished *via* a Wittig reaction on 26-methyl-24-oxocholesteryl acetate, the latter being obtained by reaction between 3β-acetoxy-Δ^5-cholenyl chloride and di-*sec*-butyl cadmium. The structure of 24,26-dimethylcholesterol suggested for aplysterol has been confirmed (*69*) by single-crystal X-ray diffraction of its *p*-iodobenzoate, which also revealed the stereochemistry of the side chain as 24 R, 25 S as shown in (**179**).

A new remarkable side chain alkylation pattern has been found in calysterol, the principal sterol component of the sponge *Calyx nicaensis* (*75a*). Structure (**181**), reminiscent of those of gorgosterol (**175**) and its relatives (**176**) and (**177**), has been proposed for this unique sterol on the basis of spectral data (m.s. and n.m.r.) of the acetate, m.p. 105—106°, $[\alpha]_D$-34.3°, and Lemieux oxidation of the corresponding stanol to a β-diketone (**182**). Interestingly, calysterol (**181**) completely decomposes on an alumina-silver nitrate column to give four compounds; the major one has been characterized as (**182a**).

Modifications of the sterol nucleus have also been found in sponges.

The total sterol content of *Axinella polypoides* is a mixture of stanols having a 19-norcholestanol nucleus carrying conventional saturated and mono-unsaturated C_7 (24-nor), C_8, C_9 and C_{10} side-chains (*140*). The major resolved component has been fully characterized (in the form of its acetate) as 19-nor-5α,10β-ergost-*trans*-22-en-3β-ol **(187)**; the related structures **(183)**—**(186)** and **(188)**—**(190)** assigned to the remaining components have been postulated largely on the basis of spectral measurements. The 19-nor-stanol structure was at first suggested from a combination of n.m.r. (only one methyl singlet absorption at δ 0.66, typical for

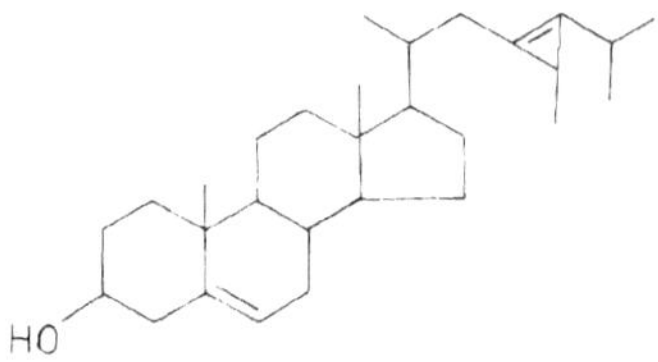

(181)

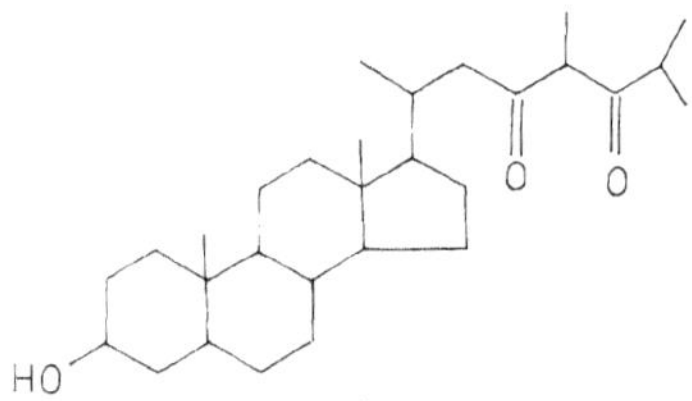

(182)

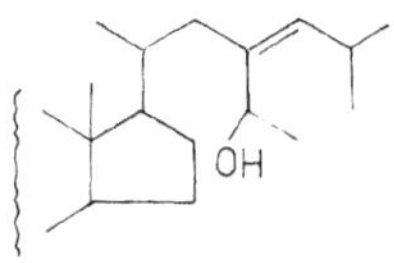

(182a)

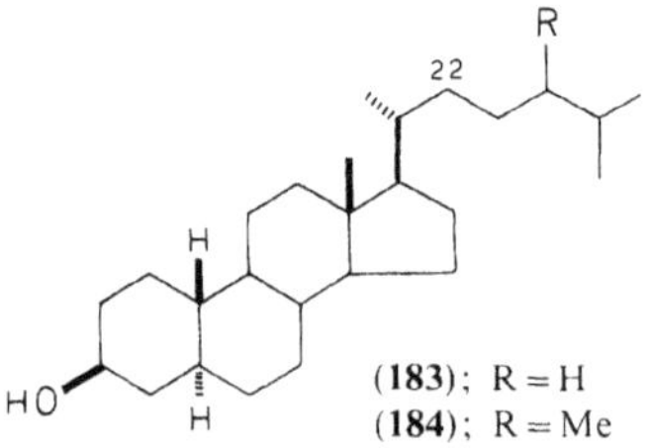

(183); R = H
(184); R = Me
(185); R = Et
(186); R = H, Δ^{22}-*trans*
(187); R = Me, Δ^{22}-*trans*
(188); R = Me, $\Delta^{24(28)}$
(189); R = Et, Δ^{22}-*trans*

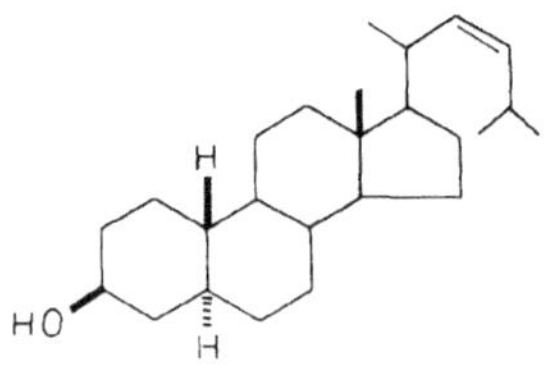

(190)

the 13-Me protons) and mass spectral data. In the mass spectrum of the acetate of **(187)**, all the significant fragments associated with the nucleus are shifted by fourteen mass units in comparison with that of the isomeric 22-dehydrocholestanyl acetate. Thus the peaks at m/e 243 and 241 in the spectrum of the acetate of (**187**) replace the doublet at m/e 257—255. The peak at m/e 201 replaces that at m/e 215 and the vinylic cleavage peak was seen at m/e 330 instead of m/e 344, which also established the position (C-22) of unsaturation in the C_9-side-chain. Confirmation was obtained by converting (**187**) *via* the dihydro derivative into the phenol (**191**) and showing its identity with authentic 19-*nor*-ergosta-1,3,5(10)-trien-3-ol, prepared from ergostan-3-one.

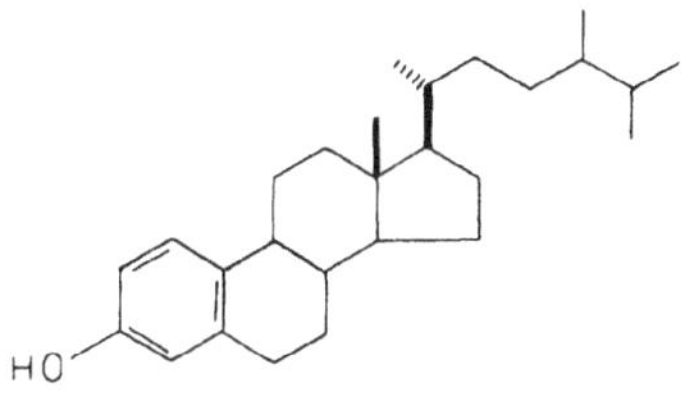

(**191**)

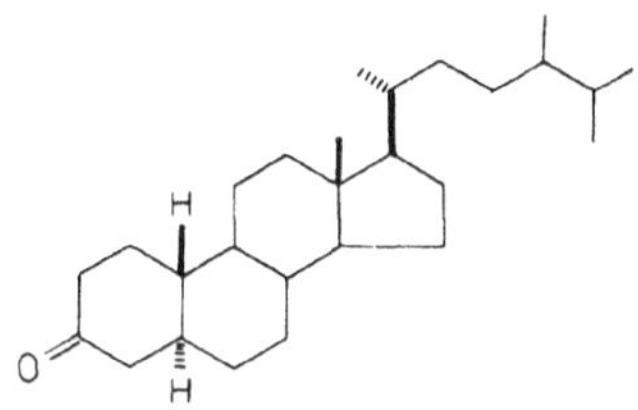

(**192**)

Conversion of the dihydro derivative of (**187**) into the phenol (**191**) was accomplished by oxidation to the corresponding ketone (**192**), followed by bromination and subsequent dehydrobromination of the latter. The stereochemistry of the A/B ring fusion was determined by c.d. measurements on the 3-ketone derivative (**192**), while the hydroxyl configuration was derived from the half-band width (22 Hz) of the C-3 proton signal in the n.m.r. spectrum of the acetate of (**187**), which indicated a degree of coupling consistent with axial orientation of H-3.

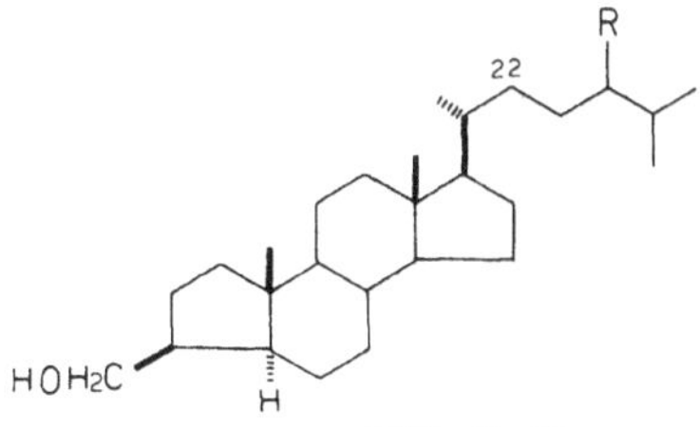

(**193**); R = H
(**194**); R = Me
(**195**); R = Et
(**196**); R = H, Δ^{22}
(**197**); R = Me, Δ^{22}
(**198**); R = Et Δ^{22}

The total sterol content of *Axinella verrucosa* is a mixture of unique stanols with a new 3β-hydroxymethyl-A-*nor*-5α-cholestane nucleus carrying conventional C_8, C_9 and C_{10} side chains (*141*). The major fraction contained the three homologous fully saturated stanols (**193**)—(**195**), of which the C_{27} component makes up *ca.* 70% of the total. All the structural work was done on this fraction. The n.m.r. spectrum includes an 8-line signal centred at δ 3.60 with the characteristic features of the AB portion of an ABX pattern (J 10 Hz, 8 Hz and 6 Hz). Irradiation at δ 2.2 changed the 8-line signal into an AB quartet (J = 10 Hz). This established the presence of a $-CH-CH_2OH$ grouping and, together with the mass spectral fragmentation pattern-almost identical with that of cholestanol, led the authors to suspect a hydroxymethyl-A-nor-sterane structure. Oxidation with dimethyl sulphoxide/dicyclohexylcarbodiimide afforded the corresponding aldehydes which were converted *via* their enol acetates and subsequent ozonolysis to the *nor*-ketones (part-structure **199**). An A-*nor*-cholestan-3-one structure was suggested by carbonyl absorption at 1735 cm^{-1} (5-membered ring ketone) and the exchange of three protons by base-catalyzed deuteriation. On alkaline treatment the ketones (**199**) formed the more stable AB-*cis*-epimers (part-structure **200**) consistent with a *trans*-A/B ring junction. Confirmation was obtained by c.d. measurements on the naturally-derived ketones (**199**) and their epimers (**200**).

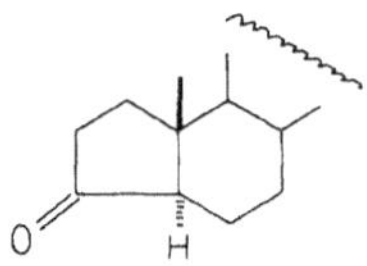

(**199**)

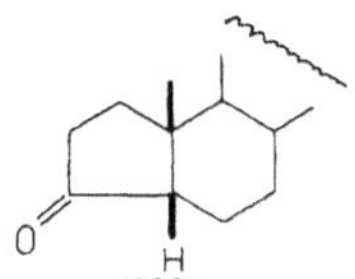

(**200**)

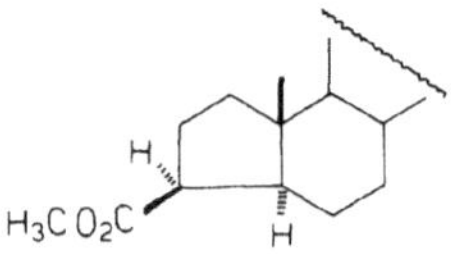

(**201**)

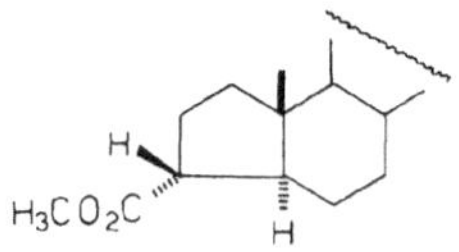

(**202**)

Oxidation of the mixture (**193**)—(**195**) with chromic acid afforded the corresponding carboxylic acids which were converted to the methyl ester (part structure **201**). Inversion at C-3 with sodium methoxide in methanol furnished quantitatively the epimeric esters (part structure **202**) which were shown to be identical with synthetic 3α-carbomethoxy-A-*nor*-5α-cholestane (*91, 150*). The n.m.r. spectra were also identical and the mass spectra were superimposable below m/e 416 (M^+ of the C_{27} component). The major constituent of the epimerised mixture gave a g.l.c. peak with a retention time identical with that of the reference sample 3α-carbomethoxy-A-*nor*-5α-cholestane. The structures of the 22,23-dehydroderivatives (**196**)—(**198**), the remaining members of this series, were assigned mainly on spectral grounds by comparing their n.m.r. and mass spectra with those of the saturated components.

4.3 Sterol Biosynthesis in Sponges

Whereas our knowledge about the sterols present in sponges has rapidly increased in the last few years, our insight into the metabolism of sterols in these animals is still very meagre.

By working with a very small amounts of material WALTON and PENNOCK (*182*) observed that the sponge *Grantia compressa* utilized ^{14}C-mevalonate for the biosynthesis of isoprenoids (squalene, steroids and ubiquinones) in direct contrast with the complete absence of ^{14}C radioactivity in the terpenoids of *Suberites domuncula* incubated with ^{14}C-mevalonate.

The feeding of labelled acetate, mevalonate and methionine to the sponge *Verongia aerophoba* (*71*) yielded no radioactive aplysterol (**179**). Radioactivity from all precursors used was incorporated to some extent into the fatty acids, thus proving that the administered substrates were taken up and metabolized by the animals. A similar situation arose in *Axinella polypoides* and *A. verrucosa*, the sponges containing the 19-nor-stanols and the hydroxymethyl-A-nor-steranes, respectively, when they were fed with labelled acetate. However, when cholesterol-26-^{14}C was administered to these two sponges, the 19-*nor*-C_{27}-stanol and the hydroxymethyl-A-*nor*-C_{27}-sterane fractions were found significantly labelled (*73*). It thus appears that in *A. polypoides* and *A. verrucosa*, the sterols cannot originate from *de novo* biosynthesis but arise by modification of dietary sterols. A similar conclusion is possibly applicable to biosynthesis in the sponge *Verongia aerophoba*.

Clearly, much work is required before any definite conclusions can be drawn about sterol metabolism in sponges. The exogenous sterol requirement is a characteristic of many lower organisms, the majority being invertebrates. Within the invertebrates, the position is still rather confusing as to which members can or cannot synthesize sterols. Literature reports are often conflicting, not only within the phylum, class and order, but even within the species.

A very useful survey of current knowledge of sterol biosynthesis in invertebrates has recently been authored by REES and GOODWIN (*155*).

5. Miscellaneous Compounds

The literature which reports the occurrence of amines and, more generally, products associated with aminoacid metabolism in marine invertebrates results largely from work of ACKERMANN which extended over a number of years (*2*). A brief summary of nitrogen compounds in Porifera by L. S. CIERESZKO (*43*) appeared in 1970.

Histamine (**203**), which is ubiquitous in Nature, was described as occuring in *Geodia gigas* (*3*), *Mycale levis* and *Haliclona sp.* (*137*) and more recently in *Suberites inconstans* (*68*). The latter sponge contains five other histamine-like compounds, three of which seem to possess phenolic groups in their structures, The N,N-dimethyl derivative of histamine (**204**), a potent hypotensive agent, has been isolated from aqueous extracts of *Ianthella* sp. and its structure unambigously established by direct comparison with an authentic sample (*98*). Other products associated with the histidine metabolism were detected by ACKERMANN and MENSSEN (*8*) in *Hippospongia equina*: they were the imidazolyl derivatives of acetic, propionic, lactic and acrylic (uracanic acid) acids. Zöoanemonin (**205**), agmatine (**206**), glycocyamine and a new base, $C_{17}H_{13}N_3$, named hippospongine, were isolated and identified in the extracts of *Hippospongia equina*. Agmatine (**206**), the decarboxylation product of arginine, was also reported from *Hymeniacidon caruncula*, *Thetia lyncurium* and *Halichondria panicea* (*95*) and from *Geodia gigas* (*7*).

ACKERMANN and PANT (*10*) in a search for constituents of the sponge *Calyx nicaensis* isolated trimethylamine, a base $C_4H_9N_3O_2$, and large amounts of taurine (**207**) and monomethyltaurine (**208**). The presence of dimethyltaurine (**209**), taurocyamine (**211**), homarine (**212**) and trigonellin (**213**) was also demonstrated by paper chromatography. The unknown base appeared to be a guanidine derivative but could be distinguished from α- and β-guanylalanine. This is the first report of mono- and dimethyltaurine in the animal kingdom. ROBIN and ROCHE (*156*) described taurine (**207**) from five further species of sponges; in one of them (*Thetia lyncurium*) they also found large amounts of taurocyamine (**211**). Taurine (**207**) was also described by ACKERMANN and LIST (*5*) from *Geodia gigas* in which taurobetaine (trimethyltaurine) (**210**) and zooanemonine (**205**), also occurred. Homarine (**212**), a betaine widely distributed among marine invertebrate phyla, was also identified in a further sponge (*Scypha*) (*96*).

FROENKEL (*90*) in a survey of the carnitine (**214**) levels in a great variety of marine invertebrates found it in all preparations including sponges.

Recently the simple 2-aminoimidazole (**215**), a metabolite possibly associated with arginine metabolism and previously isolated from cultures of *Streptomyces eurocidicus* (*162*), has been obtained in 0.7% yield from *Reniera cratera* (*54*).

The aqueous residue of the extracts of *Verongia aerophoba* yielded large amounts (2.5% of the dry animal) of a strongly fluorescent compound which was shown to be 3,4-dihydroxyquinoline-2-carboxylic acid (**216**) (*76*). This quinoline derivative, probably related to tryptophan metabolism, was not previously known as a natural product.

References, pp. 64—72

A series of N-acylated-2-methylene-β-alanine methyl esters (**217**) has been found in *Faciospongia cavernosa*, in remarkably high concentrations by KASHMAN *et al.* (*121*). The n.m.r. and mass spectra defined the 2-methylene-β-alanine methyl ester structure. Prolonged acid-catalyzed methanolysis cleaved the amine linkage to give 2-methylene-β-alanine methyl ester (**218**) and the free fatty acid methyl esters (normal $C_{16}-C_{20}$). It is interesting to note in this connection that the rare 2-methyl-β-alanine (β-aminoisobutyric acid) (**219**) was found in several sponges by BERGQUIST and HARTMANN (*28*). In this work they analysed 67 species belonging to 50 genera of West Indian sponges for their amino acid composition in order to find a biochemical parameter for sponge systematics.

An unique purine called herbipolin (**220**) was isolated from *Geodia gigas* by ACKERMANN and LIST (*4*). These authors also discovered spongopurine in the same sponge, and identified it (*6*) as 1-methyl-6-iminodihydropurine (1-methyladenine) (**221**). Spongopurine was also found in *Hymeniacidon perleve* (*176a*).

The sponge *Cryptothetia crypta* was the source of three unusual nucleosides: spongothimidine (**222**) and spongouridine (**223**), which are the 1-β-D-arabofuranosides of thymine and uracil (*18*) and spongosine (**224**), which is the 9-β-D-ribofuranosyl-2-methoxyadenine (*19, 26*). These "unnatural" nucleosides are not constituents of sponge nucleic acids (*27*) but are present in the tissue of *C. crypta* in the free state. *C. crypta* also contains free thymine (*20*). Free adenine has been found in *Geodia gigas* (*4*). More recently NIGRELLI and STEMPIEN (*146*) have reported histochemical data on the possible presence of an abnormal nucleic acid in *C. crypta*.

From the unsaponifiable matter of *Spheciospongia vesparia*, BERGMANN and MCALEER (*22*) isolated a crystalline fraction which was shown to be metanethole (**225**), the dimerization product of anethole. From the lack of optical activity the authors suspected that this sponge product was an artifact formed during the saponification procedure.

The isolation of free hydroxyhydroquinone (**226**) has recently been reported from the sponge *Axinella polypoides* (*54*).

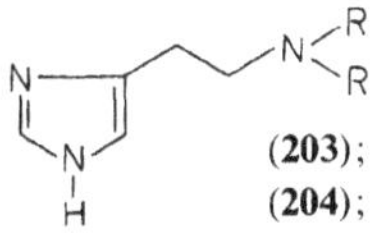

(**203**); R = H
(**204**); R = CH_3

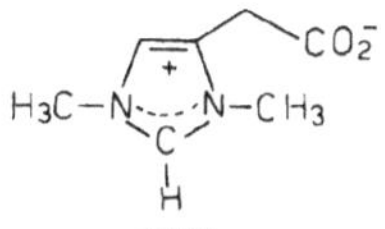

(**205**)

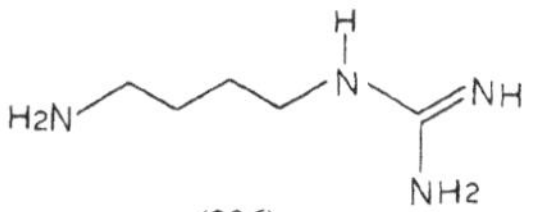

(**206**)

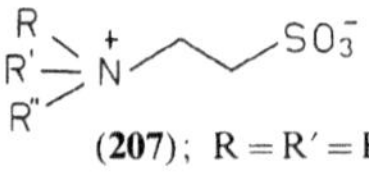

(207); R = R′ = R″ = H

(208); R = R′ = H; R″ = CH_3

(209); R = H; R′ = R″ = CH_3

(210); R = R′ = R″ = CH_3

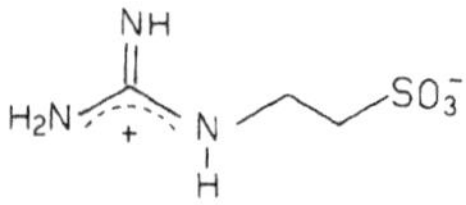

(211)

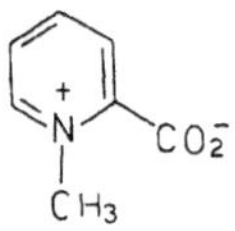

(212)

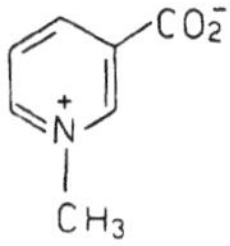

(213)

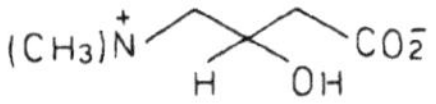

(214)

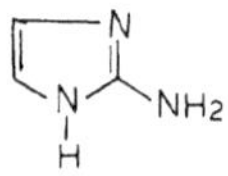

(215)

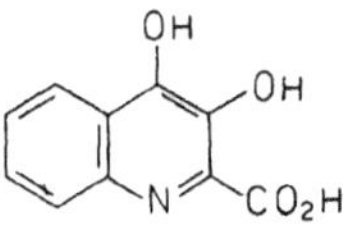

(216)

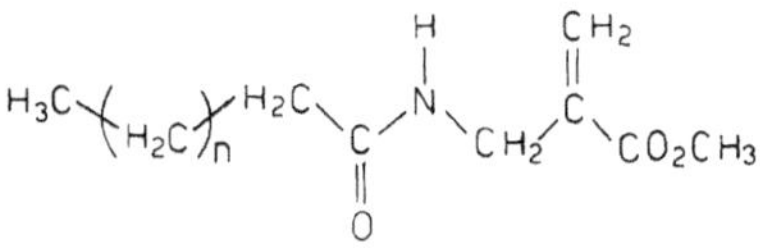

(217)

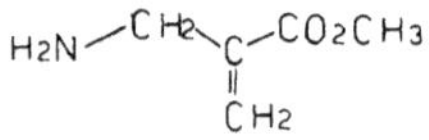

(218)

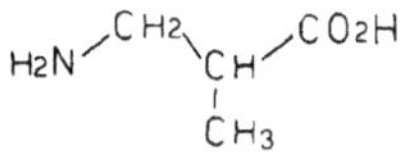

(219)

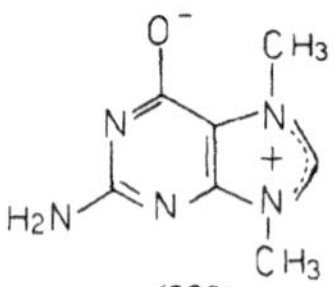

(220)

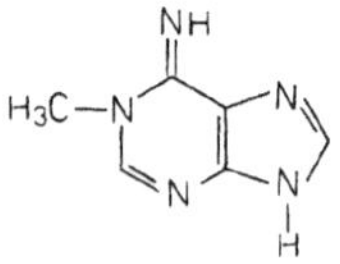

(221)

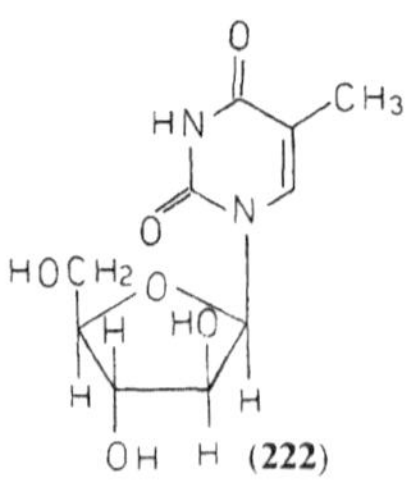

(222)

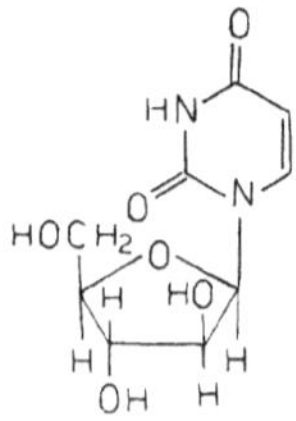

(223)

References, pp. 64—72

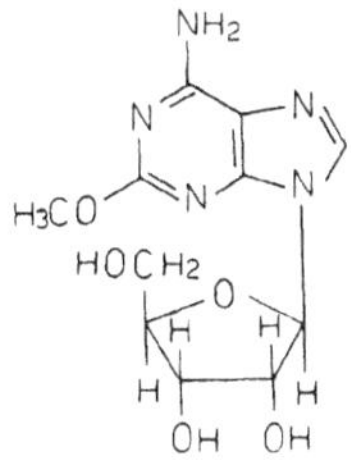

(224)

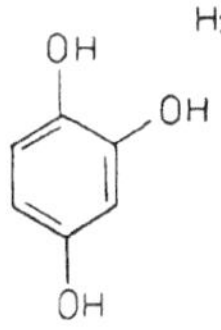

(226)

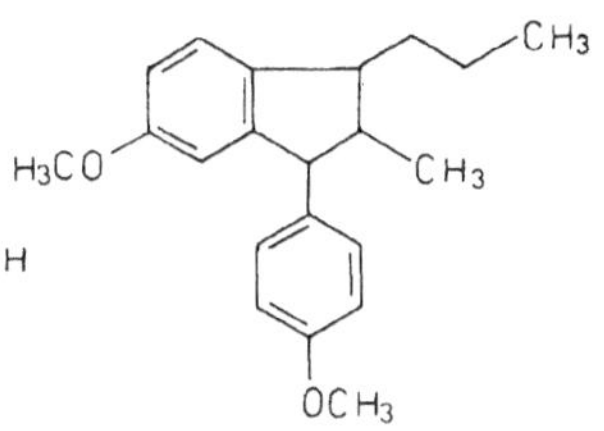

(225)

III. Distribution

In the preceding chapters it has been shown that recent systematic investigations of sponges have led to the isolation of a large number of substances with diverse chemical structures. On the basis of their "molecular characteristics" they can be classified as: A) tyrosine-derived bromo-compounds; B) furanoid terpenes; C) bromopyrrole derivatives and D) prenylated phenols, each group arising along biogenetically very similar lines. Several other miscellaneous substances of sporadic occurrence have also been found. The occurrence of each group of biogenetically related substances, especially the first two, seems to be restricted to taxonomically related species. This was confirmed by a recent survey for the presence of the above groups of compound covering 33 more species representative of 17 families and 9 orders of the class Desmospongiae (*61*), which was undertaken for the purpose of examining this chemical approach to sponge systematics.

Table 3. *Sponges Containing Tyrosine-Derived Bromo-Compounds*

Sponges	References
Order Dictioceratida	
Family Verongidae	
V. aerophoba	(*80, 81, 82, 142, 143*)
V. archeri	(*42*)
V. cauliformis	(*163, 164*)
V. fistularis	(*163, 168*)
V. thiona	(*80, 82, 143*)
V. sp.	(*11*)
Family Disideidae	
Ianthella ardis	(*92*)
Ianthella sp.	(*142*)

The brominated compounds related to 3,5-dibromotyrosine are apparently confined to the Verongidae family and to the closely related *Ianthella* genus (Table 3). It is of interest that Verongidae and *Ianthella* possess two distinctly different sterol patterns, which can be used to distinguish between them (*72*). In fact, after the investigation of 25 species for the presence of aplysterol (**179**) and 24,28-didehydroaplysterol (**180**), Verongidae appear to be the only sponges able to elaborate these unique sterols (Table 2).

Furanoid terpenes have been found only in *Microciona toxystila* (Order Poecilosclerida) and in two families belonging to the Order Dictioceratida (Table 4).

Table 4. *Sponges Containing Furanoid Terpenes*

Sponges	References
Order Dictioceratida	
Family Spongidae	
Spongia nitens	(*83*)
S. officinalis	(*49, 58, 59*)
S. agaricina	(*61*)
Hippospongia communis	(*58, 59*)
Ircinia oros	(*49, 60*)
I. fasciculata	(*37*)
I. variabilis	(*84*)
I. spinosula	(*47*)
Family Aplysillidae	
Pleraplysilla spinifera sample 1	(*55, 62*)
Pleraplysilla spinifera sample 2	(*57*)
Family Disideidae	
Disidea pallescens	(*46*)
Order Poecilosclerida	
Family Clathriidae	
Microciona toxystila	(*46*)

It should be noted that the *Cacospongia* genus belonging to the family *Spongidae* is apparently lacking in furanoid terpenes (*56, 78*), but the two species examined contain the tetracarbocyclic sesterterpenes, scalarin (**104**), which includes in its structure a γ-hydroxy-butenolide ring which could arise *in vivo* by oxidation of a preformed furan ring, and scalaradial (**106**), whose dialdehyde system, could conceivably represent an intermediate stage in the formation of a furan ring. VAN LEAR *et al.* (*179*) have reported the occurrence in *Polyfibrospongia maynardii* (family Spongidae) of bromoindole metabolites, (**34**) and (**35**). but no mention on the presence of furano-compounds has been made.

References, pp. 64—72

In any case, a point arising from the above data is that two distinctly different patterns are recognizable within the Dictioceratida, one found in Verongidae, the other in Spongidae. This strongly supports the wide separation of these two dictyoceratic families, as previously suggested by BERGQUIST and HARTMAN (*28*) on the basis of their free amino-acid composition, and subsequently by DE ROSA *et al.* (*72*) on the basis of sterol patterns.

The bromopyrrole derivatives have been reported from two *Agelas* and three Axinellida (Table 5) which indicate a relationship between *Agelas* and some Axinellida.

Table 5. *Sponges Containing Bromopyrrole Derivatives*

Sponges	References
Order Poecilosclerida	
Family Agelasidae	
Agelas oroides	(*89*)
Agelas sp.	(*173*)
Order Axinellida	
Family Axinellidae	
Phakellia flabellata	(*165*)
Axinella damicornis	(*61*)
A. verrucosa	(*61*)

BERGQUIST and HARTMANN (*28*) observed an anomalous amino acid pattern in the genus *Agelas* with respect to those of other Poecilosclerida and found it difficult to differentiate *Agelas* from typical Axinellidae.

Table 6. *Sponges Containing Prenyl Phenols*

Sponges	References
Order Dictioceratida	
Family Spongidae	
Ircinia muscarum	(*48*)
I. spinosula	(*47*)
Family Disideidae	
Disidea avara	(*139*)
D. pallescens	(*44*)
Order Halichondrida	
Family Halichondridae	
Halichondria panicea	(*52*)

Compounds of mixed biogenesis originating partly from mevalonate and partly from a benzenoid precursor have been found in four Dictioceratida and in one apparently unrelated species (Table 6).

Obviously, there is still much work on more species to be done, but there is no doubt that chemical investigation will make important contributions to many phylogenetic problems within the phylum of Porifera.

IV. Pharmacological Properties

Toxic and antibiotic properties of sponges have been reviewed by NIGRELLI *et al.* (*147*), RUSSELL (*158*) and more recently by BASLOW (*14*). These accounts are mainly concerned with reports of screening studies of crude extracts for antibacterial and antifungal activities (see also *15, 35,*

Table 7. *Isolated Sponge Metabolites Showing some Antimicrobial Activity*

Compounds	Sponge	References
2-(3′,5′-dibromo-1′-hydroxy-4′-oxocyclohexa-2′,5′-dien-1′-yl) acetamide (**1**)	*Verongia fistularis*	(*163*)
	V. cauliformis	(*163*)
	V. aerophoba	(*81, 143*)
aeroplysinin-1 (+) (**4**)	*Verongia aerophoba*	(*81*)
aeroplysinin-1 (−) (**5**)	*Ianthella ardis*	(*92*)
4-bromopyrrol-2-carbonyl guanidine (**31**)	*Agelas* sp.	(*173*)
monobromophakellin (**30**)	*Phakellia flabellata*	(*167*)
dibromophakellin (**28**)	*Phakellia flabellata*	(*167*)
3-(2-aminoethyl)-5,6-dibromoindole (**35**)	*Polyfibrospongia maynardii*	(*179*)
3-(2-methylaminoethyl)-5,6-dibromoindole (**34**)	*Polyfibrospongia maynardii*	(*179*)
dibromo-2-hydroxy-diphenyl ether (**33**)	*Disidea herbacea*	(*166*)
pentabromo-2-hydroxy-diphenyl ether (**32**)	*Disidea herbacea*	(*166*)
4,5-dibromo-2-cyanopyrrole (**23**)	*Agelas oroides*	unpublished results (1)
acanthellin-1 (**49**)	*Acanthella acuta*	unpublished results (2)
furospongin-1 (**70**)	*Spongia officinalis*	(*58*)
nitenin (**68**)	*Spongia nitens*	unpublished results (3)
ircinin-1 and -2 (**94**) (**95**)	*Ircinia oros*	unpublished results (4)
variabilin (**97**)	*Ircinia variabilis*	(*84*)
4,6-dihydroxy-or-6,7-dihydroxy-indole (tentative structure)	*Agelas* sp.	(*172*)

(1) active against *Streptococcus, Diplococcus, Candida albicans, Trichophyton*
(2) active against *Mycobacterium* sp.
(3) active against *Mycobacterium* sp.
(4) active against *S. aureus, Streptococcus, Diplococcus*

145, 163). These investigations of sponges for potentially useful chemotherapeutic agents are of doubtful significance (*14*). In many cases, especially with crude extracts, it is probable that much of the antibiotic activity initially observed on plate tests will prove to be an artifact of preparation and may be a function of pH or concentration of unutilizable foodstuffs for the test organism.

In the last few years, with the isolation of pure compounds and the elucidation of their chemical structures, knowledge of the biological activity of sponge metabolites has increased. Testing of compounds reveals that a number of substances have some antimicrobial activity (Table 7).

The anticancer activity of the bromocompound aeroplysinin-1 (**4**) and the dienone (**1**) is more interesting. Both compounds were active in the KB cells test and aeroplysinin-1 (**4**) revealed L-1210 Leukemia activity in mice.

In this connection it is important to note that the "unusual" arabinosyl derivatives of thymine and uracil (**222** and **223**), so far detected in only one sponge species, *Cryptotethya crypta* (*18*), have served as a model for the synthesis of D-arabinosylcytosine, which has potent anticancer activity; it inhibits the growth of Sarcoma 180, Ehrlich carcinoma and L-1210 Leukemia in mice (*63, 64, 124*). However, human tests of this compound for antitumor activity have been disappointing (*175*).

Although it is disappointing that no active substance from a sponge has ever reached the stage of clinical application, it is apparent from the data presently available that sponges may provide pharmacological agents of interest in medicine.

Addendum

During the few months since completion of this review, several new natural products have been isolated from sponges. Also a review on natural products chemistry of the marine environment has appeared in vol. 5 of "The Sea", which is dedicated to Marine Chemistry (*187*).

The genus *Verongia* continues to yield tyrosine-derived bromocompounds. The latest addition to this group is (**227**) from *V. aurea* (*188*). It represents a major departure from the previously reported compounds listed in Chart 1 in having a rearranged dibromotyrosine skeleton. The hypothesis of an oxime precursor of aerothionin (**7**), homoaerothionin (**8**) and the nitrile aeroplysinin-1 (**4** and **5**) (see **15** and **16**) has received support by isolation of the oximinopyruvic acid (**228**) from a marine sponge *Hymeniacidon sanguinea* (*189*).

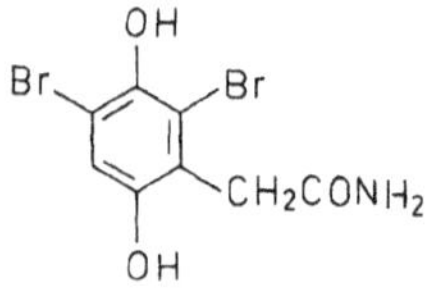

(227)

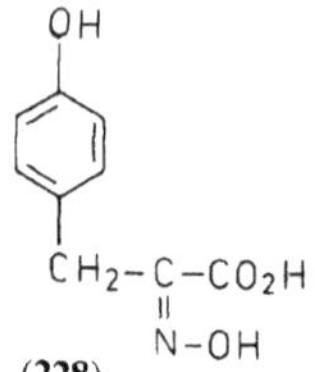

(228)

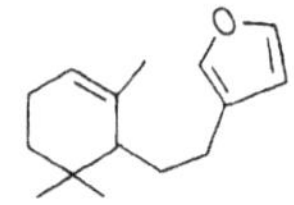

(229)

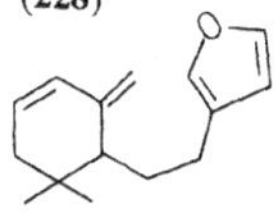

(230)

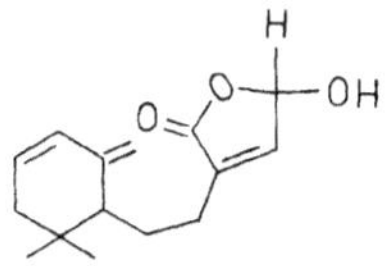

(231)

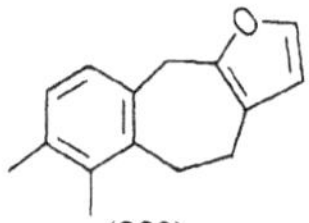

(232)

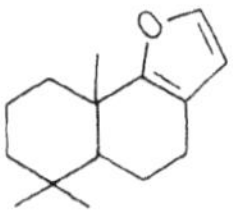

(233)

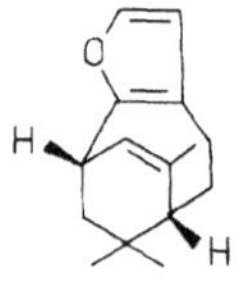

(234)

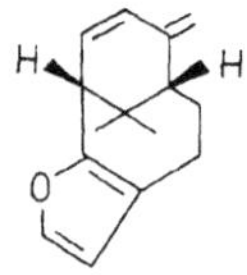

(235)

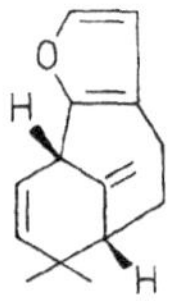

(236)

Examination of the minor sesquiterpene constituents of *Disidea pallescens* has provided a biogenetically interesting assortement of related furanoid sesquiterpenes (*190*). These include three, pallescensin-1 (**229**), pallescensin-2 (**230**) and pallescensin-3 (**231**), of a mono-cyclofarnesane type and five (**232**)—(**236**), having a 2,3-disubstituted furan ring and two additional cycles. The benzenoid component, pallescensine E (**232**), is closely related to the previous pallescensin F (**44**) and G (**43**) from which it could be likely derived by 1,2-methyl migration and subsequent oxidation. The remaining substances pallescensin A–D (**233**) to (**236**) represent further new skeletal variations among sesquiterpenoids.

Pallescensins A–E may be derived from a furanoid mono-cyclofarnesane precursor by cyclization and oxidation as shown in Chart 16.

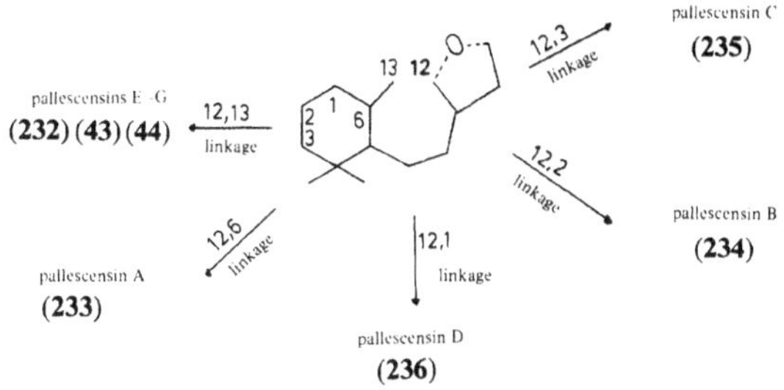

Chart 16. Possible biogenetic route to pallescensins A–G

In addition to being a rich source of furanoid sesquiterpenes, the sponge *Disidea pallescens* has also yielded (*191*) a chromansesquiterpenoid, *ent*-chromazonarol (**237**), which is the antipodal cyclic isomer of the algal hydroquinone zonarol (**114**). The biosynthesis of antipodal terpenoids by different organisms is of considerable interest.

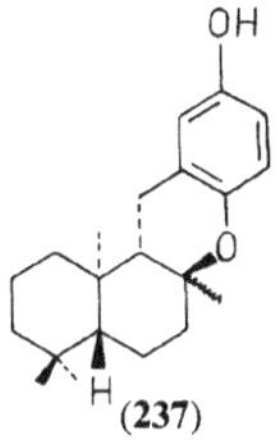

(**237**)

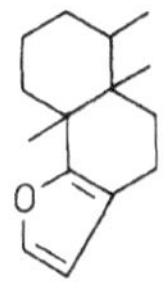

(**238**)

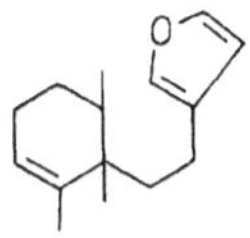

(**239**)

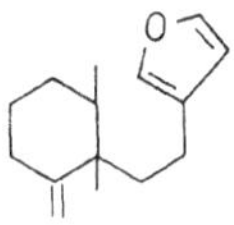

(**240**)

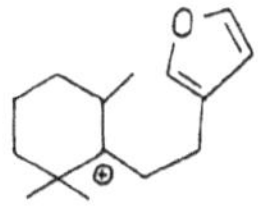

(**241**)

Analysis of the minor sesquiterpene constituents of the sponge *Microciona toxystila* has been completed, leading to the discovery of three more new furanosesquiterpenes (**238**)—(**240**) having a rearranged skeleton (*192*). The coexistence of these three compounds along with the previously described microcionin-1 (**45**) and -2 (**46**) suggests that the whole group is derived from a common biosynthetic precursor, the ion (**241**).

Burreson and Scheuer (*193*) have reported that the sponge *Halichondria sp.*, which previously yielded an amorphane sesquiterpenoid isonitrile (**49a**) together with its corresponding formamide and isothiocyanate, also elaborates a diterpenoid (geranyllinalool) functionalized by the same three groups (**242**)—(**244**).

FATTORUSSO and co-workers (*194*) have discovered two further "unusual" sterols (**245**) and (**246**) in *Calyx nicaensis*. This sponge also

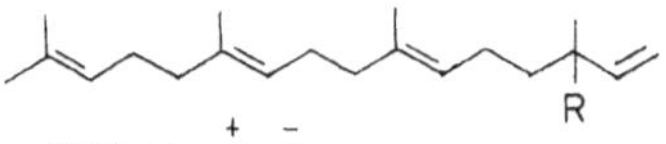

(**242**), R = $\overset{+}{N}\equiv\overset{-}{C}$

(**243**), R = NHCHO

(**244**); R = N=C=S

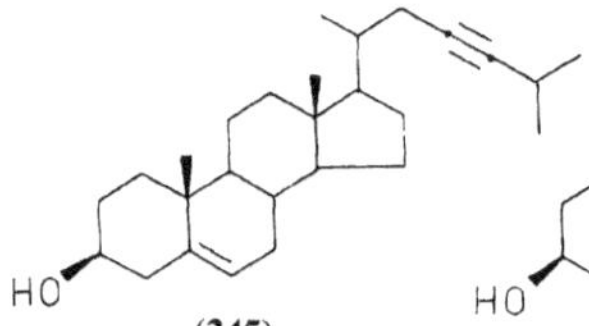

(**245**)

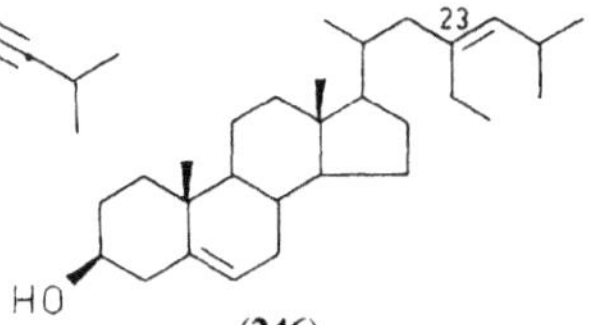

(**246**)

yielded calysterol (**181**), the first sterol found to have a cyclopropene ring in the side chain. Finding (**245**), the first steroid with an acetylenic functionality may have be provided the answer to the question of the biochemical precursors of calysterol (**181**) itself and the gorgonian sterols having a 22,33-cyclopropane ring (**175**)—(**177**).

SHEIKH and DJERASSI (*195*) have examined five sponge species for their steroid content. All sponges contained C_{27}—C_{29}, Δ^5, mono and diunsaturated sterols. Interestingly, in addition to C_{27}—C_{29} sterols, the sponge *Tethya aurantium* also contains Z-24-propylidene-cholest-5-en-3β-ol (**140**), a C_{30} sterol which has previously been encountered in the scallop *Plocopecten magellanicus*. The same sponge has also furnished the 5α, 8α-peroxides of cholesta-5,7-dien-3β-ol, ergosterol, ergosta-5,7,24(28)-trien-3β-ol and 24-ethyl-cholesta-5,7-dien-3β-ol (**247**, **248**, **249** and **250**).

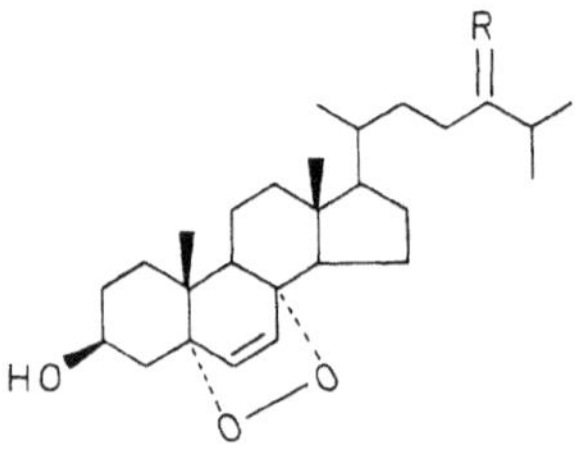

(**247**); R = H_2

(**248**); R = $<^{CH_3}_{H}$; Δ^{22}

(**249**); R = CH_2; $\Delta^{24(28)}$

(**250**); R = CH_2CH_3

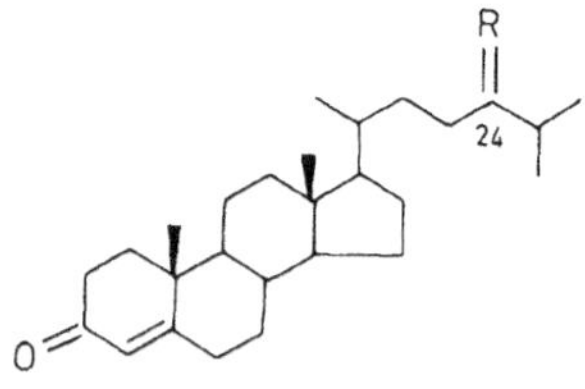

(**251**); R = H_2; 24-nor; Δ^{22}

(**252**); R = H_2; Δ^{22}

(**253**); R = $<^{CH_3}_{H}$; Δ^{22}

(**254**); R = CH_2; $\Delta^{24(28)}$

(**255**); R = $CHCH_3$; $\Delta^{24(28)}$(E)

References, pp. 64—72

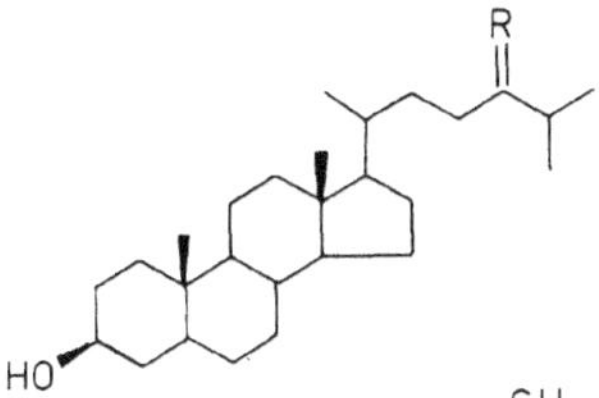

(256); R= 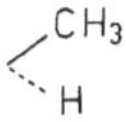

(257); R= 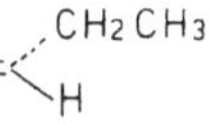

Even more interesting is of the occurrence in the sponge *Stellata clarella* of a series of Δ^4-3-ketones (**251, 252, 253, 254, 255**), the first occasion on which any α,β-unsaturated steran-3-one have been found from marine sources.

In addition to Δ^5-3β-hydroxy-sterols *Stellata clarella* also contains 5α-cholestanol (**147**), 5α-ergostanol (**256**), and 5α-poriferastanol (**257**). The latter two must be counted as new to Porifera.

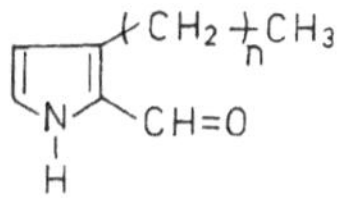

(**258**); n = 18 to 22

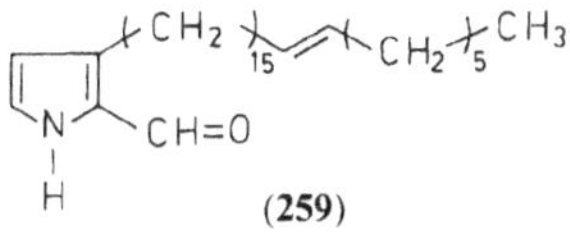

(**259**)

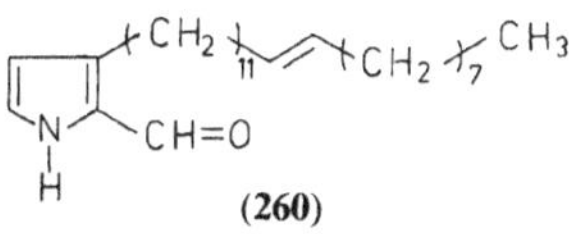

(**260**)

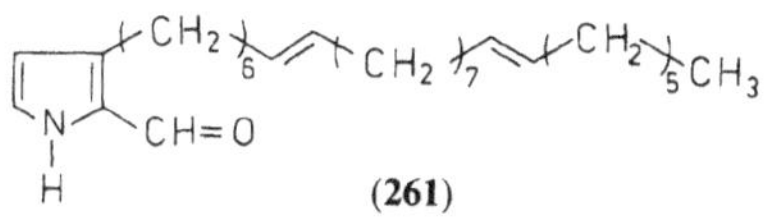

(**261**)

A novel group of compounds characterized by attachment of saturated, mono- and diunsaturated alkyl side chains to position 3 of pyrrol-2-aldehyde (**258**, **259**, **260** and **261**) has been isolated from the sponge *Oscarella lobularis* (*196*), which has also yielded the corresponding carboxylic acids and their methyl esters.

References

1. ACHENBACH, H.: Xanthocillin. Antibiotics (ed. by D. GOTTLIEB and P. D. SHAW), vol. II, p. 26. Berlin-Heidelberg-New York: Springer. 1967.

2. ACKERMANN, D.: Biogene Amine. Ber. phys. med. Ges. Würzb. **70,** 1 (1963).

3. ACKERMANN, D., and P. H. LIST: Über das Vorkommen beträchtlicher Mengen Histamin in der niederen Tierwelt. Z. Physiol. Chem. **308,** 274 (1957).

4. — — Konstitutionsermittlung des Herbipolins, einer neuen tierischen Purinbase. Z. Physiol. Chem **309,** 286 (1957).
— — Zur Konstitution des Zooanemonins und des Herbipolins. Z. Physiol. Chem. **318,** 281 (1960).

5. — — Über das Vorkommen von Taurobetain, Taurin und Inosit in Riesenkieselschwamm. Z. Physiol. Chem. **317,** 78 (1960).

6. — — Konstitution des Spongopurins, einer in der Natur neuen Purinbase. Z. Physiol. Chem. **323,** 192 (1961).

7. ACKERMANN, D., P. H. LIST, and H. G. MENSSEN: Über das Vorkommen einer neuen Purinbase $C_6H_7N_5$ in Riesenkieselschwamm (*Geodia gigas*). Z. Physiol. Chem. **312,** 210 (1958).

8. ACKERMANN, D., and H. G. MENSSEN: N-haltige Inhaltsstoffe des Pferdeschwammes *Hippospongia equina*. Z. Physiol. Chem. **322,** 198 (1960).

9. ACKERMANN, D., and E. MÜLLER: Über das Vorkommen von Dibromtyrosin neben Dijodtyrosin im Spongin. Z. Physiol. Chem. **269,** 146 (1941).

10. ACKERMANN, D., and R. PANT: Inhaltsstoffe des Schwammes *Calyx nicaensis*. Z. Physiol. Chem. **326,** 197 (1971).

11. ANDERSEN, R. J., and D. J. FAULKNER: A novel antibiotic from a sponge of the genus *Verongia*. Tetrahedron Letters **1973,** 1175.

12. ARCAMONE, F., B. CAMERINO, E. COTTA, G. FRANCESCHI, A. GREIN, S. PENCO, and C. SPALLA: New Carotenoids from *Streptomyces mediolani* n. sp. Experientia **25,** 241 (1969).

13. AUSTIN, J.: The Sterols of marine invertebrates and plants, p. 73 in "Advances in Steroid Biochemistry and Pharmacology", ed. by M. H. BRIGGS, vol. I. London and New York: Academic Press. 1970.

14. BASLOW, M. H.: Porifera in "Marine Pharmacology", p. 86. Baltimore: The Williams & Wilkins Co. 1969.

15. BASLOW, M. H., and G. W. READ: Hypotensive and other pharmacologic actions of agents from the sponges *Toxadocia violacea*, *Haliclona viridis* and *Haliclona magnicanulosa*. Proc. West. Pharmacol. Soc. **11,** 117 (1968).

16. BERGMANN, W.: Comparative Biochemical Studies on the Lipids of Marine Invertebrates, with Special Reference to the Sterols. J. Mar. Res. **8,** 137 (1949).

17. — Sterols: Their Structure and Distribution, p. 103 in "Comparative Biochemistry" ed. by M. FLORKIN and S. MASON, vol. 3. New York: Academic Press. 1962.

18. BERGMANN, W., and D. C. BURKE: Contributions to the study of marine products. XXXIV. The nucleosides of sponges III. Spongothymidine and spongouridine. J. Org. Chem. **20,** 1501 (1955).

19. — — Contributions to the study of marine products XL. The nucleosides of sponges IV. Spongosine. J. Org. Chem. **21,** 226 (1956).

20. BERGMANN, W., and R. J. FEENEY: Contributions to the study of marine products XXXII. The nucleosides of sponges. I. J. Org. Chem. **16,** 981 (1951).

21. BERGMANN, W., D. H. GOULD, and E. M. LOW: Contributions to the Study of Marine Products. XVII Spongosterol. J. Org. Chem. **10,** 570 (1945).

22. BERGMANN, W., and W. J. MCALEER: The Isolation of Metanethole from the Sponge *Spheciospongia vesparia*. J. Amer. Chem. Soc. **73,** 4969 (1951).

23. BERGMANN, W., and F. H. MCTIGUE: Contributions to the Study of Marine Products. XVI. Chondrillasterol. J. Org. Chem. **13,** 738 (1948).

24. BERGMANN, W., H. P. SCHELD, and E. M. LOW: Contributions to the Study of Marine Products. XVIII. Microcionasterol and other Sterols of Sponges. J. Org. Chem. **10,** 580 (1945).

25. — — — Contributions to the Study of Marine Products. XIX. Chalinasterol. J. Org. Chem. **10,** 587 (1945).

26. BERGMANN, W., and M. F. STEMPIEN JR.: Contributions to the Study of marine products XLIII. The nucleosides of sponges. V. The synthesis of spongosine. J. Org. Chem. **22,** 1575 (1957).

27. BERGMANN, W., J. C. WATKINS, and M. F. STEMPIEN: Contributions to the Study of Marine Products. XLV. Sponge Nucleic Acids. J. Org. Chem. **22,** 1308 (1957).

28. BERGQUIST, P. R., and W. D. HARTMAN: Free amino acid patterns and the classification of the Desmospongiae. Mar. Biol. **3,** 247 (1969).

28a. BORDERS, D. B., G. O. MORTON, and E. R. WETZEL: Structure of a novel bromine compound isolated from a sponge. Tetrahedron Letters **1974,** 2709.

29. BOUTRY, J. L., A. ALCAIDE, and M. BARBIER: Sur la présence d'un stérol en C_{26} dans un plancton marin végétel. C. R. Acad. Sci., Paris **272,** 1022 (1971).

30. BORY, S., M. FÉTIZON, and P. LASLO: Stéréochemie dans la série de l'acide agathique. Bull. Soc. Fr. *1963,* 2310.

31. BRAUDE, E. A., R. P. LINSTEAD, and P. W. D. MITCHELL: Hydrogen Transfer. Part VI. Metal-catalysed transfer-hydrogenation of Ethylenic Compounds. J. Chem. Soc. **1954,** 3578.

32. BROOKS, C. J. W.: Steroids: Sterols and Bile Acids, p. 1 in "Rodd's Chemistry of Carbon Compounds" (ed. by S. Coffey) vol. IID. Elsevier Publishing Company. 1970.

33. BRYANT, R.: The Sesquiterpenoids in "Rodd's Chemistry of Carbon Compounds" (ed. by S. COFFEY) 2nd ed. vol. IIC, p. 257. Elsevier Publishing Company 1969.

34. BÜCHI, G., W. HOFHEINZ, and J. V. PAUKSTELIS: The Synthesis of (–) Aromadendrene and related Sesquiterpenes. J. Amer. Chem. Soc. **91,** 6473 (1969).

35. BURKHOLDER, P. R., and K. REUTZLER: Antimicrobial activity of some marine sponges. Nature **222,** 983 (1969).

35a. BURRESON, B. J., C. CHRISTOPHERSEN, and P. J. SCHEUER: Cooccurrence of a Terpenoid Isonitrile-Formamide Pair in the Marine Sponge *Halichondria* sp. J. Amer. Chem. Soc. **97,** 201 (1975).

36. CAFIERI, F., E. FATTORUSSO, S. MAGNO, C. SANTACROCE, and D. SICA: Isolation and Structure of Axisonitrile-1 and Axisothiocyanate-1, two unusual Sesquiterpenoids from the marine sponge *Axinella cannabina.* Tetrahedron **29,** 4259 (1973).

37. CAFIERI, F., E. FATTORUSSO, C. SANTACROCE, and L. MINALE: Fasciculatin, a novel sesterterpene from the sponge *Ircinia fasciculata.* Tetrahedron **28,** 1579 (1972).

38. CAFIERI, F., E. FATTORUSSO, C. SANTACROCE, D. SICA, and A. FRIGERIO: Sterols of *Axinella cannabina,* a marine sponge. Steroids (in press).

39. CAGLIOTI, L., H. NAEF, D. ARIGONI, and O. JEGER: Zur Kenntnis der Sesquiterpenes und Azulene 126 Mitteilung. Über die Inhaltsstoffe der *Asa faetida* I. Farnesiferol A. Helv. Chim. Acta **41,** 2278 (1958).

40. — — — — Zur Kenntnis der Sesquiterpene und Azulene 127 Mitteilung. Über die Inhaltsstoffe der *Asa faetida* II. Farnesiferol B und C. Helv. Chim. Acta **42,** 2557 (1959).

41. CHANTRAINE, J. M., G. COMBAUT, and J. TESTE: Phenols Bromes d'une Algue Rouge, *Halopytis incurvus*: Acides carboxyliques. Phytochem. **12,** 1793 (1973).

42. CHIB, J. S.: Personal communication.

43. CIERESZKO, L. S.: Nitrogen Compounds in Porifera and Coelenterata p. 57, in "Comparative Biochemistry of Nitrogen Metabolism" vol. 1 (ed. by J. W. CAMPBELL). London and New York: Academic Press. 1970.

44. CIMINO, G., P. DE LUCA, S. DE STEFANO, and L. MINALE: Disidein, a pentacyclic sesterterpene condensed with an hydroxyquinone moiety, from the sponge *Disidea pallescens*. Tetrahedron **31**, 271 (1975).
45. CIMINO, G., D. DE ROSA, S. DE STEFANO, and L. MINALE: Isoagatholactone, a diterpene of a new structural type from the sponge *Spongia officinalis*. Tetrahedron **30**, 645 (1974).
46. CIMINO, G., S. DE STEFANO, A. GUERRIERO, and L. MINALE: Unpublished results.
47. CIMINO, G., S. DE STEFANO, and L. MINALE: Polyprenyl derivatives from the sponge *Ircinia spinosula*. Tetrahedron **28**, 1315 (1972).
48. — — — Prenylated Quinones in marine sponges: *Ircinia sp.* Experientia **28**, 1401 (1972).
49. — — — Further linear furanoterpenes from marine sponges. Tetrahedron **28**, 5983 (1972).
50. — — — Deoxoscalarin, a further sesterterpene with the unusual tetracyclic carbon skeleton of scalarin, from *Spongia officinalis*. Experientia **29**, 934 (1973).
51. — — — Methyl *Trans*-Monocyclofarnesate from the Sponge *Halichondria panicea*. Experientia **29**, 1063 (1973).
52. — — — Paniceins, unusual aromatic sesquiterpenoids linked to a quinol or quinone system from the marine sponge *Halichondria panicea*. Tetrahedron **29**, 2565 (1973).
53. — — — Oxidized furanoterpenes from the sponge *Spongia officinalis*. Experientia **30**, 18 (1974).
54. — — — Occurrence of hydroxyhydroquinone and 2-aminoimidazole in Sponges. Comp. Biochem. Physiol. **47 B**, 895 (1974).
55. — — — Pleraplysillin-2, a further furanosesquiterpenoid from the sponge *Pleraplysilla spinifera*. Experientia **30**, 846 (1974).
56. — — — Scalaradial, a third Sesterterpene with the tetracarbocyclic skeleton of scalarin, from the sponge *Cacospongia mollior*. Experientia **30**, 846 (1974).
57. — — — Unpublished results.
58. CIMINO, G., S. DE STEFANO, L. MINALE, and E. FATTORUSSO: Furospongin-1, a new C-21 furanoterpene from the sponges *Spongia officinalis* and *Hippospongia communis*. Tetrahedron **27**, 4673 (1971).
59. — — — — Minor C-21 furanoterpenes from the sponges *Spongia officinalis* and *Hippospongia communis*. Tetrahedron **28**, 267 (1972).
60. — — — — Ircinin-1 and -2, linear sesterterpenes from the marine sponge *Ircinia oros*. Tetrahedron **28**, 333 (1972).
61. CIMINO, G., S. DE STEFANO, L. MINALE, and G. SODANO: Metabolism in Porifera-III. Chemical Patterns and the classification of the Desmospongiae. Comp. Biochem. and Physiol. **50 B**, 279 (1975).
62. CIMINO, G., S. DE STEFANO, L. MINALE, and E. TRIVELLONE: New Sesquiterpenes from the marine sponge *Pleraplysilla spinifera*. Tetrahedron **28**, 4761 (1972).
62a. ÇLARDY, J.: Personal communication.
63. COHEN, S. S.: Sponges, cancer chemotherapy and cellular aging. Perspect. Biol. Med. **6**, 215 (1963).
64. — Introduction to the biochemistry of D-arabinosyl nucleosides. In "Progress in Nucleic Acid Research and Molecular Biology" (J. N. DAVIDSON and W. E. COHN eds.) vol. **5**, p. 1. New York: Academic Press. 1966.
65. COSULICH, D. B., and F. M. LOVELL: An X-Ray Determination of the Structure of an Antibacterial Compound from the Sponge, *Ianthella ardis*. Chem. Comm. **1971**, 397.
66. CZECZUGA, B.: Investigations of carotenoids in some fauna of the Adriatic Sea. I. *Verongia aerophoba* (Porifera: Spongidae). Marine Biology **10**, 254 (1971).
67. DALY, J. W., D. M. JERINA, and D. WITKOP: Arene Oxides and the NIH Shift: The Metabolism, Toxicity and Carcinogeneity of Aromatic Compounds. Experientia **28**, 1129 (1972).

68. Daś, N. P., H. S. Lim, and Y. F. Teh: Histamine and Histamine-Like Substances in the marine sponge, *Suberites inconstans*. Comp. gen. Pharmac. **2,** 473 (1971).
69. De Luca, P., M. De Rosa, L. Minale, R. Puliti, G. Sodano, F. Giordano, and L. Mazzarella: Synthesis of 24,28-Didehydroaplysterol and X-Ray Crystal Structure of Aplysterol: Unusual Marine Sterols. J. C. S. Chem. Comm. **(1973),** 825.
70. De Luca, P., M. De Rosa, L. Minale, and G. Sodano: Marine Sterrols with a new Pattern of side-chain alkylation from the Sponge *Aplysina* (= *Verongia*) *aerophoba*. J. C. S. Perkin I, **1972,** 2132.
71. De Rosa, M., L. Minale, and G. Sodano: Metabolism in Porifera. I. Some Studies on the Biosynthesis of Fatty Acids, Sterols and Bromo-Compounds by the Sponge *Verongia aerophoba*. Comp. Biochem. Physiol. **45B,** 883 (1973).
72. — — — Metabolism in Porifera. II. Distribution of sterols. Comp. Biochem. Physiol. **46B,** 823 (1973).
73. — — — Unpublished results.
74. Doree, C.: The occurrence and distribution of cholesterol and allied bodies in the animal Kingdom. Biochem. J. **4,** 72 (1909).
75. Erdman, T. R., and R. H. Thomson: Sterols from the Sponges *Cliona celata* Grant and *Hymeniacidon perleve* Montagu. Tetrahedron **28,** 5163 (1972).
75a. Fattorusso, E.: Personal communication.
76. Fattorusso, E., S. Forenza, L. Minale, and G. Sodano: Isolation of 3,4-dihydroxyquinoline-2-carboxylic acid from the sponge *Aplysina aerophoba*. Gazz. Chim. it. **101,** 104 (1971).
77. Fattorusso, E., S. Magno, L. Mayol, C. Santacroce, and D. Sica: Isolation and Structure of axisonitrile-2, a new sesquiterpenoid isonitrile from the sponge *Axinella cannabina*. Tetrahedron **30,** 3911 (1974).
77a. — — — — — New Sesquiterpenoids from the sponge *Axinella cannabina*. Tetrahedron **31,** 269 (1975).
78. Fattorusso, E., S. Magno, C. Santacroce, and D. Sica: Scalarin, a new pentacyclic C-25 terpenoid from the sponge *Cacospongia scalaris*. Tetrahedron **28,** 5993 (1972).
79. — — — — Sterol Peroxides from the Sponge *Axinella cannabina*. Gazz. Chim. it. **104,** 409 (1974).
80. Fattorusso, E., L. Minale, K. Moody, G. Sodano, and R. H. Thomson: Homoaerothionin, a second tetrabromo-compound from *Aplysina aerophoba* and *Verongia thiona*. Gazz. Chim. it. **101,** 61 (1971).
81. Fattorusso, E., L. Minale, and G. Sodano: Aeroplysinin-1, a New Bromo-Compound from *Aplysina aerophoba*. Chem. Comm. **1970,** 751.
— — — Aeroplysinin-1, an antibacterial bromo-compound from the sponge *Verongia aerophoba*. J. C. S. Perkin I, **1972,** 16.
82. Fattorusso, E., L. Minale, G. Sodano, K. Moody, and R. H. Thomson: Aerothionin, a Tetrabromo-compound from *Aplysina aerophoba* and *Verongia thiona*. Chem. Comm. **1970,** 752.
83. Fattorusso, E., L. Minale, G. Sodano, and E. Trivellone: Isolation and Structure of nitenin and dihydronitenin, new furanoterpenes from *Spongia nitens*. Tetrahedron **27,** 3909 (1971).
84. Faulkner, J. D.: Variabilin, an antibiotic from the sponge, *Ircinia variabilis*. Tetrahedron Letters **1973,** 3821.
85. Faulkner, J, D., and M. O. Stallard: 7-chloro-3,7-dimethyl-1,4,6-tribromo-1-octen-3-ol, a novel monoterpene Alcohol from *Aplysia california*. Tetrahedron Letters **1973,** 1171.
86. Faulkner, J. D., M. O. Stallard, J. Fayos, and J. Clardy: (3 R, 4 S, 7 S)-*trans,trans*-3,7-dimethyl-1,8,8-tribromo-3,4,7- trichloro-1,5-octadiene, a Novel Monoterpene from the Sea Hare, *Aplysia californica*. J. Amer. Chem. Soc. **95,** 3414 (1973).

87. FENICAL, W., J. J. SIMS, and P. RADLICK: Chondriol, a Halogenated Acetylene from the Marine Alga *Chondria oppositiciada*. Tetrahedron Letters **1973**, 313; FENICAL, W., K. B. GIFKINS, and J. CLARDY: X-ray Determination of Chondriol; a Re-assignement of Structure. Tetrahedron Letters **1974**, 1507.
88. FENICAL, W., J. J. SIMS, D. SQUATRITO, R. M. WING, and P. RADLICK: Zonarol and Isozonarol, Fungitoxic Hydroquinones from the Brown Seaweed *Dictyopteris zonarioides*. J. Org. Chem. **38**, 2383 (1973).
89. FORENZA, S., L. MINALE, R. RICCIO, and E. FATTORUSSO: New bromopyrrole derivatives from the sponge *Agelas oroides*. Chem. Comm. **1971**, 1129.
90. FROENKEL, G.: The distribution of Vitamin B_T (Carnitine) throughout the Animal Kingdom. Arch. Biochem. Biophys. **50**, 486 (1954).
91. FUCHS, B., and H. J. E. LOEWENTHAL: The contraction of ring A in 5α-cholestane Derivatives. Tetrahedron **11**, 199 (1960).
92. FULMOR, W., G. E. VAN LEAR, G. O. MORTON, and R. O. MILLS: Isolation and absolute configuration of the aeroplysinin-1 enantiomorphic pair from *Ianthella ardis*. Tetrahedron Letters **1970**, 4551.
93. FURUSAKI, A., E. KUROSAWA, A. FUKUZAWA, and T. IRIE: The Revised Structure and Absolute Configuration of Laurefucin from *Laurencia nipponica* Yamada. Tetrahedron Letters **1973**, 4579.
94. GARCIA, E. E., L. E. BENJAMIN, and R. IAN FRYER: Reinvestigation into the structure of oroidin, a bromopyrrole derivative from a marine sponge. J. C. S. Chem. Comm. **1973**, 78.
95. GARCIA, I., and F. MIRANDA: Sur les dérivés guanidiniques des Spongiaires. Compt. Rend. Soc. Biol. **148**, 1187 (1954).
96. GASTEIGER, E. L., P. C. HAAKE, and J. A. GERGEN: An Investigation of the distribution and Function of homarine (N-methyl picolinic acid). Ann. N. Y. Acad. Sci. **90**, 622 (1960).
97. GOTO, T., H. KAKISAWA, and Y. HIRATA: The structure of Grifolin, an antibiotic from Basidiomycete. Tetrahedron **19**, 2079 (1963).
98. GERMAN, V. F.: Na,Na-Dimethylhistamine, the Hypotensive Principle of the Sponge *Ianthella* sp. J. Pharm. Sci. **60**, 495 (1971).
99. GOAD, L. J., I. RUBINSTEIN, and A. G. SMITH: The Sterols of echinoderms. Proc. Roy. Soc. London B. **180**, 223 (1972).
100. GOLLNIK, K., and G. O. SCHENK: Oxygen as a Dienophile in "1,4-Cycloaddition Reactions" (ed. J. HAMER), p. 288. New York-London: Academic Press. 1967.
101. GONZÁLES, A. G., J. DARIAS, and J. D. MARTIN: Taondiol, a new Component from *Taonia atomaria*. Tetrahedron Letters **1971**, 2729.
102. — — — Caespitol, a New Halogenated Sesquiterpene from *Laurencia caespitosa*. Tetrahedron Letters **1973**, 2381.
103. — — — Furocaespitane, a New Furan from *Laurencia caespitosa*. Tetrahedron Letters **1973**, 3625.
104. GONZÁLES, A. G., and J. D. MARTIN: The Synthesis of a Taondiol Derivative. Tetrahedron Letters **1972**, 2259.
105. GOODWIN, T. W.: Pigments of Porifera. Chemical Zoology, vol. II, p. 37 (ed. by M. FLORKIN and B. T. SCHEER). New York and London: Academic Press. 1968.
106. HALE, R. L., J. LECLERQ, B. TURSCH, C. DJERASSI, R. A. GROSS, A. J. WEINHEIMER, K. GUPTA, and P. J. SCHEUER: Demonstration of a Biogenetically Unprecedent side-chain in the Marine Sterol, Gorgosterol. J. Amer. Chem. Soc. **92**, 2179 (1970).
107. HALL, S. S., D. J. FAULKNER, J. FAYOS, and J. CLARDY: Oppositol, a Brominated Sesquiterpene Alcohol of a New Skeletal Class from the Red Alga, *Laurencia subopposita*. J. Amer. Chem. Soc. **95**, 7187 (1973).
108. HALSTEAD, G. W.: Poisonous and Venomous Marine Animals of the World. U. S. Govt. Printing Office, Washington, D. C., **1**, 271—296 (1965).

109. HAMASAKI, T., N. OKUKADO, and M. YAMAGUCHI: Two Natural Acetylenic Aromatic Carotenoids. Bull. Chem. Soc. Japan **46,** 1884 (1973).

110. HAMMEN, C. S., and M. FLORKIN: Chemical Composition and Intermediary Metabolism-Porifera. Chemical Zoology, vol. II, p. 53, (ed. by M. FLORKIN and B. T. SCHEER). New York and London: Academic Press. 1968.

111. HANESSIAN, S., and J. S. KALTENBRON: Synthesis of a Bromine-Rich Marine Antibiotic. J. Amer. Chem. Soc. **88,** 4509 (1966).

112. HANSON, J. R.: The Di- and Sesterterpene. Part 2. The Sesterterpenes. Chemistry of Terpenes and Terpenoids (ed. by A. A. NEWMAN), p. 200. London and New York: Academic Press. 1972.

113. HAYASHI, H., H. KOMAL, S. EGUCHI, M. NAKAYAMA, S. HAYASCHI, and T. SAKAO: Two New Furanosesquiterpenes from *Actinodaphne longifolia* (Blume) Nakai. Chemistry and Industry **1972,** 572.

114. HENZE, M.: Spongosterin, eine cholestirinartige Substanz aus *Suberites domuncula.* Hoppe-Seyl. Z. **41,** 109 (1904).

115. HIRAI, K., K. T. SUZUKI, and S. NOZOE: The Structure and the Chemistry of Siccanin and Related Compounds. Tetrahedron **27,** 6057 (1971).

116. HOREAU, A.: Principe et Application d'une nouvelle methode de determination des configurations dite Par "Dedoublement Partiel". Tetrahedron Letters **1961,** 506.

117. — Determination des Configurations Par "Dedoublement Partiel". II. Precisions et Complements. Tetrahedron Letters **1962,** 965.

118. HOREAU, A., and H. B. KAGAN: Determination des Configurations Par "Dedoublement Partiel". III. Alcools Steroides. Tetrahedron **20,** 2431 (1964).

119. IDLER, D. R., L. M. SAFE, and E. F. MACDONALD: A New C-30 Sterol (Z)-24-propylidenecholest-5-en-3β-ol-(29-methylisofucosterol). Steroids **18,** 545 (1971).

120. IDLER, D. R., and P. WISEMAN: Molluscan Sterols: A Review. J. Fish. Res. Bull. Canada **28,** 385 (1972).

121. KASHMAN, Y., L. FISHELSON, and I. NE'EMAN: N-Acyl-2-methylene-β-alanine methyl esters from the sponge *Fasciospongia cavernosa.* Tetrahedron **29,** 3655 (1973).

122. KASPEREK, G. J., T. C. BRUICE, H. YAGI, N. KAUBISH, and D. M. JERINA: Solvolytic Chemistry of 1,4-Dimethylbenzene Oxide. A New and Novel Mechanism for the NIH Shift. J. Amer. Chem. Soc. **94,** 7876 (1972).

123. KAWASHIMA, K., K. NAKANISHI, M. TADA, and H. NISHIKAWA: Structure of Tauranin. Tetrahedron Letters **1964,** 1227.

124. KESSEL, D., T. C. HALL, and I. WODINSKY: Transport and phosphorylation as factors in the antitumor action of cytosine arabinoside. Science **156,** 1240 (1967).

125. KHAN, H., A. ZAMEN, G. L. CHETTY, A. S. GUPTA, and S. DEV: Cheilanthatriol — A New Fundamental type in Sesterterpenes. Tetrahedron Letters **1971,** 4443.

126. KING, J. F., and P. DE MAYO: Terpenoid Rearrangements, vol. II, p. 772 in Molecular Rearrangements (ed. by P. DE MAYO). J. Wiley & Sons 1964.

127. KUBOTA, T.: Lower Furanoterpenes in "Cyclopentanoid Terpene Derivatives" p. 279 (ed. by W. I. TAYLOR and A. R. BATTERSBY). New York: Marcel Dekker. 1969.

128. KUROSAWA, E., A FUKUZAWA, and T. IRIE: *trans-* and *cis*-Laurediol, Unsaturated Glycols from *Laurencia nipponica* Yamada. Tetrahedron Letters **1972,** 2121.

129. — — — Isoprelaurefucin, new Bromo Compound from *Laurencia nipponica* Yamada. Tetrahedron Letters **1973,** 4135.

130. KUROSAWA, E., A. FURUSAKI, M. IZAWA, A. FUKUZAWA, and T. IRIE: The Absolute Configurations of Laureatin and Isolaureatin. Tetrahedron Letters **1973,** 3857.

131. LING, N. C., R. L. HALE, and CARL DJERASSI: The Structure and Absolute Configuration of the Marine Sterol Gorgosterol. J. Amer. Chem. Soc. **92,** 5281 (1970).

132. LOVELL, F. M.: The Structure of a Bromine-Rich Marine Antibiotic. J. Amer. Chem. Soc. **88,** 4510 (1966).

133. Low, E. M.: Halogenated Amino Acids of the Bath Sponge. J. Marine Research **10,** 239 (1951).
134. Mazzarella, L., and R. Puliti: Crystal structure and absolute configuration of Aeroplysinin. Gazz. Chim. it. **102,** 391 (1972).
135. McCrindle, R., and K. H. Overton: The Diterpenoids, Sesterterpenoids and Triterpenoids, p. 369, in Rodd's Chemistry of Carbon Compounds (ed. by S. Coffey), vol. II C. Elsevier 1969.
136. Mechoulam, R., and Y. Gaoni: Recent Advances in the Chemistry of Hashish. Vol. 25, p. 175. Progress in the Chemistry of Natural Products. Wien-New York: Springer. 1967.
137. Mettrick, D. F., and J. M. Telford: The histamine content and histidine decarboxylase activity of some marine animals from the West Indies. Comp. Biochem. Physiol. **16,** 547 (1965).
138. Minale, L., R. Riccio, and G. Sodano: Acanthellin-1, a unique isonitrile sesquiterpene from the sponge *Acanthella acuta.* Tetrahedron **30,** 1341 (1974).
139. — — — Avarol, a novel sesquiterpenoid hydroquinone with a rearranged drimane skeleton from the sponge *Disidea Avara.* Tetrahedron Letters **1974,** 3401.
140. Minale, L., and G. Sodano: Marine Sterols: 19-Norstanols from the sponge *Axinella polypoides.* J. C. S. Perkin I, **1974,** 1888.
141. — — Marine Sterols: Unique 3β-Hydroxymethyl-A-nor-5α-steranes from the sponge *Axinella verrucosa.* J. C. S. Perkin I, **1974,** 2380.
142. Minale, L., G. Sodano, W. R. Chan, and A. M. Chen: Aeroplysinin-2, a dibromolactone from marine sponges *Aplysina* (= *Verongia*) *aerophoba* and *Ianthella* sp. Chem. Comm. **1972,** 674.
143. Moody, K., R. H. Thomson, E. Fattorusso, L. Minale, and G. Sodano: Aerothionin and homoaerothionin: two tetrabromo spirocyclohexadienylisoxazoles from *Verongia* sponges. J. C. S. Perkin I, **1972,** 18.
144. Moscowitz, A., E. Charney, V. Weiss, and H. Ziffer: Optical Activity in Skewed Dienes. J. Amer. Chem. Soc. **83,** 4661 (1961).
145. Nigrelli, R. F., S. Jakowska, and I. Calventi: Ectyonin, an antimicrobial agent from the sponge *Microciona prolifera.* Verrill. Zoologica (New York) **44,** 173 (1959).
146. Nigrelli, R. F., and M. F. Stempien: The possible presence of an abnormal nucleic acid in the sponge *Cryptotethya crypta.* J. Histochem. Cytochem. **11,** 395 (1963).
147. Nigrelli, R. F., M. F. Stempien, G. D. Ruggieri, V. R. Liquori, and J. T. Cecil: Substances of potential biomedical importance from marine organism. Fed. Proc. **26,** 1197 (1967).
148. Nozoe, S., and K. T. Suzuki: The Isolation and Structure of Siccanochromenes. Tetrahedron **27,** 6063 (1971).
149. Panizzi, L., L. Mangoni, and M. Belardini: Sui componenti della *Grindelia robusta.* Nota I. Acido grindeico. Gazz. Chim. it. **92,** 523 (1962).
150. Paranjape, B. V., and I. L. Pyle: The Dieckmann Cyclization as a route to A-nor Steroids. Evidence concerning Stereochemistry. J. Org. Chem. **36,** 1009 (1971).
151. Parker, W., J. S. Roberts, and R. Ramage: Sesquiterpene Biogenesis. Quart. Rev. **21,** 331 (1967).
152. Premuzic, E.: Chemistry of Natural Products derived from Marine Sources, p. 417. Progress in the Chemistry of Organic Natural Products, Vol. 29. Wien-New York: Springer. 1971.
153. Quilico, A., F. Piozzi, and M. Pavan: The Structure of Dendrolasin. Tetrahedron **1,** 177 (1957).
154. Ramage, R.: Carotenoid Chemistry in "Chemistry of Terpenes and Terpenoids", p. 287 (ed. by A. A. Newman). London and New York: Academic Press. 1972.

155. Rees, H. H., and T. W. Goodwin: Biosynthesis of Triterpenes, Steroids and Carotenoids in "Biosynthesis" (Specialist Periodical Report), vol. 1, p. 100. London: The Chemical Society. 1972.

156. Robin, Y., and J. Roche: Sur la présence de taurocyamine (guanidotaurine) chez des Coelentérés et des Spongiaires. Compt. Rend. Séanc. Soc. Biol. **148,** 1783 (1954).

157. Rücker, G.: Sesquiterpenes. Angew. Chem. Internat. Edit. **12,** 793 (1973).

158. Russell, F. E.: Comparative pharmacology of some animal toxins. Fed. Proc. **26,** 1206 (1967).

159. Ruzicka, L., and J. R. Hosking: Höhere Terpenverbindungen XLII. Dehydrierung und Isomerisierung der Agathen-Disäure. Helv. Chim. Acta **13,** 1402 (1930).

160. Scheuer, P. J.: Chemistry of marine natural products. Academic Press 1973.

161. Schmitz, F. J., and T. Pattabhiraman: New Marine Sterol Possessing a Side-Chain Cyclopropyl Group: 23-Demethylgorgosterol. J. Amer. Chem. Soc. **92,** 6073 (1970).

162. Seki, Y., T. Nakamura, and Y. Okami: Accumulation of 2-aminoimidazole by *Streptomyces eurocidicus.* J. Biochem. (Tokyo) **67,** 389 (1970).

163. Sharma, G. M., and P. R. Burkholder: Studies on antimicrobial substances of sponges-I. J. Antibiotics, Series A **20,** 200 (1967).

164. — — Studies on the antimicrobial substances of sponges-II. Structure and synthesis of a bromine-containing antibacterial compound from a marine sponge. Tetrahedron Letters **1967,** 4147.

165. — — Structure of dibromophakellin, a new bromine-containing alkaloid from the marine sponge *Phakellia flabellata.* Chem. Comm. **1971,** 151.

166. Sharma, G. M., and B. Vig: Studies on the antimicrobial substances of sponges-VI. Structures of two antibacterial substances isolated from the marine sponge *Disidea herbacea.* Tetrahedron Letters **1972,** 1715.

167. Sharma, G. M., B. Vig, and P. R. Burkholder: Antimicrobial Substances of Marine Sponges-IV. Proceeding of the Conference on Food-Drugs from the Sea, p. 307. New York 1969.

168. — — Studies on the antimicrobial substances-IV. Structure of a bromine containing compound from a marine sponge. J. Org. Chem. **35,** 2823 (1970).

169. Sheikh, Y. M., Carl Djerassi, and B. M. Tursch: Acansterol: A Cyclopropane-containing Marine Sterol from *Acanthaster planci.* Chem. Comm. **1971,** 217.

170. Sims, J. J., W. Fenical, R. M. Wing, and P. Radlick: Marine Natural Products. IV. Prepacifenol, a Halogenated Epoxy Sesquiterpene and Precursor to Pacifenol from the Red Alga, Laurencia filiformis. J. Amer. Chem. Soc. **95,** 972 (1973).

171. Sims, J. J., G. H. Y. Lin, R. M. Wing, and W. Fenical: Marine Natural Products. Concinndiol, A Bromo-Diterpene Alcohol from the Red Alga, *Laurencia concinna.* J. C. S. Chem. Comm. **1973,** 470.

172. Stempien, M. F., Jr.: Antibiotic Substance isolated from a sponge of the genus *Agelas.* Amer. Zool. **6,** 363 (1966).

173. Stempien, M. F., Jr., R. F. Nigrelli, and J. S. Chib: Isolation and synthesis of physiologically active substances from sponges of the genus *Agelas* 164th ACS. National Meeting, New York, 21 MEDI Abstract (1972).

174. Stowe, B. B.: Occurrence and Metabolism of Simple Indoles in Plants. Progress in the Chemistry of Organic Natural Products, vol. 17, p. 248. Wien: Springer. 1959.

175. Talley, R. W., R. M. O'Bryan, W. G. Tucker, and R. V. Loo: Clinical pharmacology and human antitumor activity of cytosine arabinoside. Cancer **20,** 809 (1967).

176. Thomson, R. H.: Naturally occurring Quinones, 2nd Edition, p. 1. London: Academic Press. 1971.

176a. — Private communication.

177. Threlfall, D. R., and G. R. Whistance: Biosynthesis of Isoprenoid Quinones and

Chromanols in "Aspects of Terpenoid Chemistry and Biochemistry" (ed. T. W. GOODWIN), p. 357. London: Academic Press. 1971.

178. VALENTIN, F. R., JR., and W. BERGMANN: The sterols of sponges: Clionasterol and poriferasterol. J. Org. Chem. **6,** 452 (1941).

179. VAN LEAR, G. E., G. O. MORTON, and W. FULMOR: New antibacterial bromoindole metabolites from the marine sponge *Polyfibrospongia maynardii*. Tetrahedron Letters **1973,** 299.

180. VIALA, J., M. DEVYS, and M. BARBIER: Sur la structure des stérols à 26 atomes de carbone du Tunicier *Halocynthia roretzi*. Bull. Soc. Chim. France **1972,** 3626 and references therein.

181. VOOGT, P. A., J. M. VAN DE RUIT, and J. W. A. VAN RHEENEN: On the biosynthesis and composition of sterols and sterol esters in some sea anemones (Anthozoa). Comp. Biochem. Physiol. **48B,** 47 (1974).

182. WALTON, M. J., and J. F. PENNOCK: Some studies on the biosynthesis of ubiquinones and sterols by marine invertebrates. Biochem. J. **127,** 471 (1972).

183. YAMAGUCHI, M.: Chemical Constitution of Renieratene. Bull. Chem. Soc. Japan **30,** 979 (1957).

184. — Chemical Constitution of Isorenieratene. Bull. Chem. Soc. Japan **31,** 51 (1958).

185. — Total Synthesis of Renieratene and renierapurpurin. Bull. Chem. Soc. Japan **33,** 1560 (1960) and references therein.

186. YAMAMURA, S., and Y. HIRATA: A Naturally-Occurring Bromo-Compound, Aplysin-20 from *Aplysia kurodai*. Bull. Chem. Soc. Japan **44,** 2560 (1971).

187. FAULKNER, D. J., and R. J. ANDERSEN: Natural Products Chemistry of the Marine Environment. In "The Sea" (E. D. GOLDBERG, ed.), vol. **5,** p. 679. New York-London-Sydney-Toronto: John Wiley & Sons. 1974.

188. KREJCAREK, G. E., R. H. WHITE, L. P. HAGER, W. O. MCCLURE, R. D. JOHNSON, K. L. RINEHART, Jr., J. A. MCMILLAN, I. C. PAUL, P. D. SHAW and R. C. BRUSCA: A rearranged Dibromotyrosine Metabolite from *Verongia aurea*. Tetrahedron Letters **1975,** 507.

189. CIMINO, G., S. DE STEFANO, and L. MINALE: Occurrence of 4-hydroxyphenylpyruvic acid oxime in the marine sponge *Hymeniacidon sanguinea*. Experientia **31,** 756 (1975).

190. CIMINO, G., S. DE STEFANO, A. GUERRIERO, and L. MINALE: Furanosesquiterpenoids in sponges-I: Pallescensin-1, -2, and -3 from *Disidea pallescens*. Tetrahedron Letters **1975,** 1417.

190a. — — — — Furanosesquiterpenoids in sponges-II. Pallescensins E–G from *Disidea pallescens*: a new skeletal type. Tetrahedron Letters **1975,** 1421.

190b. — — — — Furanosesquiterpenoids in sponges-III. Pallescensins A–D from *Disidea pallescens*: new skeletal types. Tetrahedron Letters **1975,** 1425.

191. CIMINO, G., S. DE STEFANO, and L. MINALE: *ent*-Chromazonarol, a Chroman-Sesquiterpenoid from the sponge *Disidea pallescens*. Experientia **31,** 1117 (1975).

192. CIMINO, G., S. DE STEFANO, A. GUERRIERO, and L. MINALE: Unpublished results.

193. BURRESON, B. J., and P. J. SCHEURER: Isolation of a Diterpenoid Isonitrile from a Marine Sponge. J. C. S. Chem. Comm. **1974,** 1035.

194. FATTORUSSO, E.: Personal Communication.

195. SHEIKH, Y. M., and C. DJERASSI: Steroids from Sponges. Tetrahedron **30,** 4095 (1974).

196. CIMINO, G., S. DE STEFANO, and L. MINALE: Unpublished results.

(Received January, 1975)

Biogenetic-Type Rearrangements of Terpenes

By Robert M. Coates, Department of Chemistry, University of Illinois, Urbana, Illinois, U.S.A.

Contents

Acknowledgement. The author wishes to thank Professor D. ARIGONI (ETH, Zurich) for many enlightening discussions, copies of Ph. D. theses, and unpublished material which assisted in the preparation of this review.

A. Introduction

A remarkable characteristic of naturally occurring terpenes is the bewildering array of carbocyclic structures. As of 1970 more than 200 different carbon skeletons had been identified (excluding nor metabolites), the structures varying from acyclic chains to hexacyclic ring systems and containing almost all ring sizes from three to fourteen members (*1*). Although a considerable number of the 200 are produced by oxidative transformations (*e.g.*, ring cleavage) of pre-existing terpenes, a majority nevertheless represent primary structures formed in nature through multistep cyclization and rearrangement sequences originating from five basic acyclic precursors.

The biogenetic isoprene rule has afforded a theoretical basis with which biogenetic schemes may be derived to correlate both the structure and stereochemistry of the great majority of the primary terpene skeletons (*2*—*8*). A fundamental assumption in the original formulation of the biogenetic isoprene rule and its subsequent applications is that individual steps in the proposed scheme should conform to known mechanisms associated with carbonium ion reactions in solution. The notable success of this version of the preceding isoprene rule has evoked a substantial research effort with the aim of simulating biogenetic schemes, or segments therein, by purely chemical means.

The purpose of the present work is to review the area of biogenetic-type rearrangements of terpenes and relevant model compounds. Since biogenetic-type cyclizations of polyolefins has been rather thoroughly reviewed (*9*—*14*), this closely related topic will, for the most part, be omitted. However, when the rearrangements become intimately connected with a preceding or following cyclization process, as is commonly the case in the mono- and sesquiterpene areas, this aspect has been included. Although parts of this subject have been covered in previous reviews upon rearrangements (*15*) and biogenetic-type synthesis of terpenes (*11, 12*), these have concentrated chiefly upon either the intrinsic chemical changes or the synthetic achievements rather than the mechanistic significance with respect to terpene biogenesis.

The material is organized into four major sections for each isoprenylogous terpene class each of which is further divided into sub-sections containing biogenetically related members. More general discussions of 1,*n*-hydrogen eliminations, 1,*n*-hydrogen rearrangements, and sequential 1,2-rearrangements pertinent to segments of the biogenesis of several terpene classes are treated as a whole in sub-sections C.4 and E.5. In each sub-section one or more biogenetic schemes to representative, related terpenes are presented first for comparison with the corresponding chemically-induced reactions which follow. Dashed arrows are used in the biogenetic pathways in order to stress the hypothetical nature of the schemes. Since most of the schemes shown are considered readily derivable from a knowledge of the structural relationships and the status of biosynthetic investigations*, only occasional literature references are given, *e.g.*, to an appropriate review or a significant alternative biogenetic proposal. Similarly references for the representative terpenes shown are provided only for those compounds which do not appear in the compilation of naturally occurring terpenes by DEVON and SCOTT (*1*). A few key references to biosynthetic studies relevant to the terpene family under consideration are given; however, no attempt to survey all of the rapidly expanding literature in this area has been attempted. The remainder of each section is devoted to presentation and discussion of rearrangements and, in some cases, cyclizations which correspond to one or more of the individual steps in the preceding biogenetic scheme. Since the reaction conditions, the structures of important by-products, and the yield are all germane to the significance of a given reaction, this information, if available, has been provided.

It is appropriate at this stage to discuss certain aspects of terpene biogenesis. The individual steps in biogenetic schemes may be designated as initiating, propagating, and terminating**. There appear to be mainly

* The term "biosynthesis" is used to denote the presumably enzyme-mediated processes by which terpenes are formed in a living organism, or an active preparation therefrom, and the experimental findings obtained with them. The term "biogenesis" refers to the purely hypothetical pathways and mechanisms derived from consideration of the biogenetic isoprene rule and related theoretical proposals. Since biogenetic theory must, if necessary, be modified to accomodate the results of biosynthetic investigations, the distinction between the two terms may ultimately disappear.

** The biogeneses of certain halogen-containing terpenes recently isolated from marine sources presumably involve Br^+ and Cl^+ initiation (*16–19*). Reductive termination, (*i.e.*, hydride transfer from NADPH) is not shown in the scheme since this termination mode appears to be unique to squalene biogenesis (see section E.1).

two types of initiating reactions: protonation (of carbon-carbon double bonds or epoxides), and pyrophosphate ionization. It is reasonable to suppose that the latter step is itself triggered by a proton transfer to a phosphate oxygen. Although the specific incorporation of farnesol into the sesquiterpene-derived alkaloid dendrobine (*20*) and both geranyl-geraniol (*21*) and copalol (*22*) into the fungal diterpene rosenonolactone

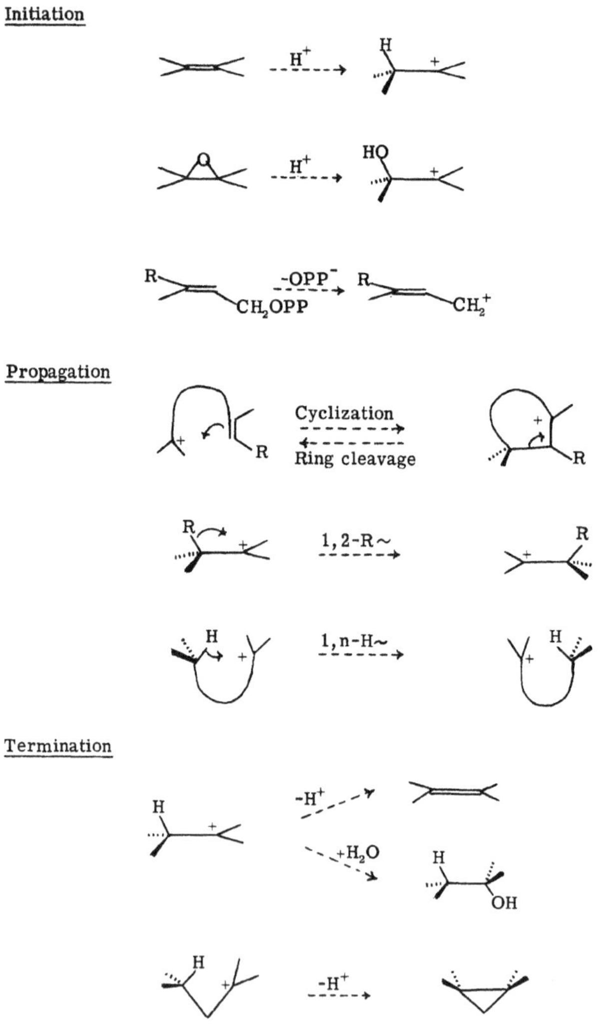

has been observed, it nevertheless seems unlikely that the alcohols are themselves the substrates for these biosynthetic cyclizations. In the biosynthesis of the tetracyclic diterpene kaurene with soluble enzyme preparations geranylgeranyl and copalyl pyrophosphates were efficiently converted to product; however the corresponding alcohols were completely inactive (*23, 24*).

In most biogenetic schemes several propagation steps are required before termination. An extreme example is the pentacyclic triterpene ketone friedelin the biogenesis of which entails fifteen individual propagating stages. One of the original postulates laid down in the application of the biogenetic isoprene rule to triterpene biogenesis is that such multi-step schemes be assumed to proceed uninterrupted to the end product (*i.e.*, termination steps are irreversible) (*3—5*). This "non-stop" hypothesis has been verified in triterpene biosynthesis by the absence of tritium loss (*25—28*) and, albeit negative evidence, in the non-incorporation of various conceivable alcohol intermediates (*29—31*).

However, the biogeneses of certain sesquiterpenes and most tricyclic and tetracyclic diterpenes require one, and in a few cases two, olefin intermediates. For example, the biogenesis of the bicyclic eudesmane, guaiane, and related sesquiterpenes proceeds through a germacradiene intermediate (*cf.*, section C.2). Similarly two deprotonation stopping points (germacradiene and bulnesene) evidently intervene between farnesyl pyrophosphate and the tricyclic patchoulane sesquiterpenes (section C.7). In diterpene biosynthesis, the bicyclic copalyl pyrophosphate is an established intermediate (sections D.1 and D.3) (*24*). Thus, non-stop, single-stop, and evidently douple-stop biogenetic schemes obtain in the sesqui- and diterpene classes.

Another fundamental hypothesis of the biogenetic isoprene rule is concerned with stereochemistry. It was proposed that all propagation and termination steps in multi-stage biogenetic pathways should occur in a synchronized, anti-parallel fashion (*3—6*). Although a series of consecutive, anti-parallel olefin cyclizations may be propagated, and possibly terminated, in a fully concerted mechanism (*i.e.*, one requiring no carbonium ion intermediate, either bridged or localized) by means of through-space, π-π delocalization*, propagation steps involving σ bonds are depicted as proceeding through a stereochemically-linked succession of transient carbonium ion intermediates. While bridged carbonium ion structures are frequently employed to illustrate the stereochemical relationships in biogenetic rearrangements, it is quite irrelevant whether such bridged species themselves be considered as true intermediates or merely

* Convincing evidence in favor of a concerted, double cyclization through a disubstituted double bond and into an aromatic ring has recently been reported (*32*).

transition states between localized, classical carbonium ions. If the intermediates are to be viewed as classical carbonium ions, however, further propagation and the final termination should, according to the biogenetic isoprene rule, be effected stereospecifically anti-parallel with respect to the preceding step. In the biogenetic schemes presented in the following sections, discrete, localized carbonium ions are usually depicted for the sake of clarity. Bridged ions are shown in a few key positions to simplify the derivation of stereochemical relationships.

Although the stereochemistry of the great majority of primary terpenes is in conformity with biogenetic schemes having anti-parallel propagation sequences, a few rare, but significant exceptions are known. One example is the unusual, fungal diterpene pleuromutlin, the biogenesis of which entails, in its early stages, a $C_{10} \rightarrow C_9$ methyl migration *syn* to the preceding $C_9 \rightarrow C_8$ hydrogen shift (**1→3**) (*33*).

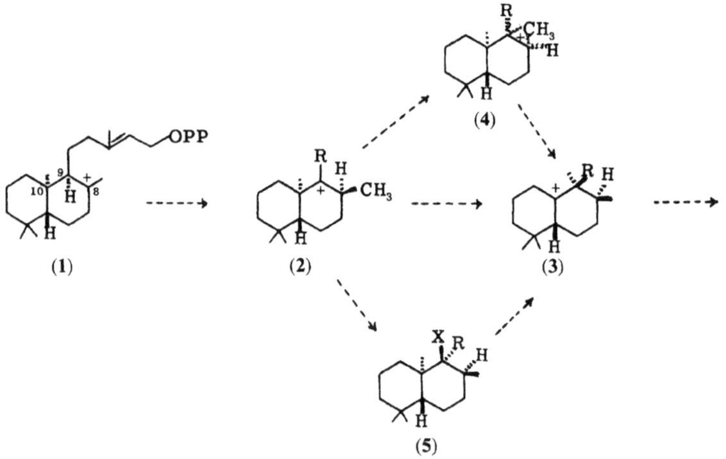

This, and other, apparent violations of the anti-parallel postulate may be rationalized by proposing somewhat more complex mechanisms in which participation of either an appropriately positioned neighboring group within the molecule (*e.g.*, the C_8 methyl group, **2→4**) or a nucleophilic site upon the surface of the surrounding enzyme (X-group) is invoked (**2→5**) (*33, 34*). Alternatively, it does not seem unreasonable to infer that such consecutive *syn* 1,2-rearrangements may occur directly in a non-concerted manner (**1→2→3**) when relatively stable carbonium ions are involved.

A number of other diterpenes [*e.g.*, plathyterpol, (**237**) in section D.1] and triterpenes [neohopene (**408**) and other members of this class, and possibly the cucurbitacins] are known which appear to be derived biogenetically by *syn* backbone rearrangements. The biosynthesis of the trichothecane sesquiterpene verrucarol has been shown to involve consecutive *syn* cyclizations of the two terminal carbons (C_1 and C_{11}) upon the central double bond of farnesyl pyrophosphate (*35*). The biogenesis of cycloartenol (**310**) necessitates *syn*-related 1,2-hydrogen rearrangement and 1,3-hydrogen elimination.

Although the function of the biosynthetic enzymes in the initiation and termination stages of a biogenetic pathway may be readily visualized (*i.e.*, optimization of substrate conformation and maintenance of regiospecificity), it is less clear what role, if any, the enzyme might play in the propagation sequence, particularly rearrangements. A mechanistic variant of the biogenetic isoprene rule is the "X-group" hypothesis in which one or more stabilized intermediates are postulated during propagation (*34*). Although the possibility that the "X-group" could be hydroxyl has been considered, as mentioned above, attempts to realize the further conversion of several hypothetical alcohol intermediates in terpene biosynthesis have been uniformly unsuccessful (*29—31, 36*). However, other nucleophiles (*e.g.*, phosphate) or a strategically-placed nucleophilic site on the enzyme surface could serve as the "X-group". The X-group mechanism is illustrated below with an alternative biogenesis proposed for the E-ring of β-amyrin (**6→10**).

Apart from the function of X-group itself, this mechanism also necessarily differs in stereochemical detail. It will be noted that the lupeol-like intermediate (**7**), in order to set the stage for the following anti-parallel rearrangements (**8→9a** and **9b→10**), has the unnatural stereochemistry at C_{19} (αH). Thus, this proposal suggests that lupeol (**348**) does not lie along the same biogenetic path as the other pentacyclic triterpenes.

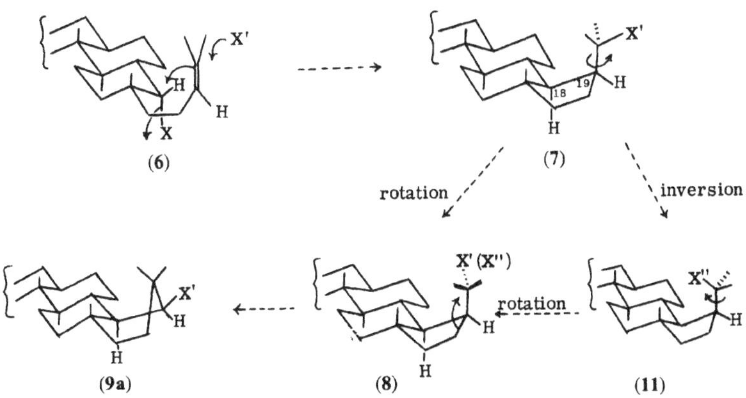

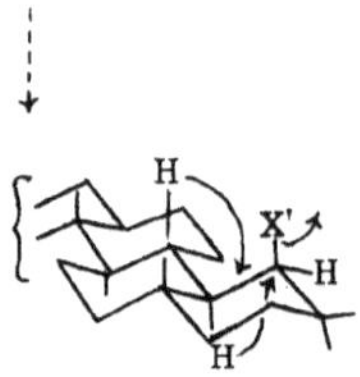
(9b)

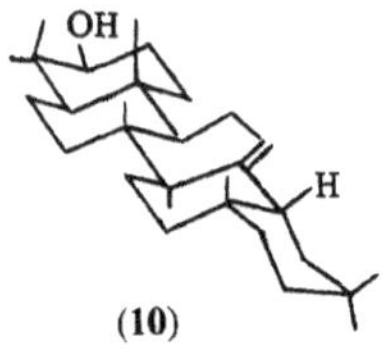
(10)
β-amyrin

Although the above mechanism has in fact been recently disproven by the recent determination of the stereochemistry of the geminal C_{20} methyl groups in β-amyrin biosynthesis (*cf.* section F4), the stereochemical situation may be corrected, albeit with increasing complexity, by participation of another X group (X″) which effects inversion of configuration at the tertiary center prior to rotation (**7**→**11**→**8**–X″). Similar stereochemical difficulties arise when an "X-group" mechanism is considered in connection with the biosynthesis of fusidic acid, a metabolite of the tetracyclic triterpene, $\Delta^{17(20),24}$-protostadien-3β-ol [geometrical isomer of (**307**) about the $\Delta^{17(20)}$ double bond, section E.2] (*37*, *38*). Despite the attractive aspects of X-group involvement in terpene biogenesis, there is as yet no evidence which requires such stabilized intermediates.

Since the evolution of terpene biogenetic theory has been influenced by concurrent advances in the understanding of the mechanisms of carbonium ion reactions, it seems appropriate to consider briefly the matter of carbonium ion structure and its relevance to the stereochemistry of rearrangement. Although the question of the existence of σ-bridged non-classical carbonium ions remains unresolved (*39*, *40*, *41*), it is nevertheless clear that solvolytic rearrangements of certain rigid structures may take place with a high degree of stereospecificity. Illustrative of this phenomenon are the solvolyses of the *endo*- and *exo*-bicyclo[3.2.1]octan-2-yl tosylates (**11**–OTs and **12**–OTs) and the related bicyclo[2.2.2]octan-2-yl tosylate (**13**–OTs) (*42*—*44*). The predominant (95—100%) retention of configuration and structural integrity in the products demonstrates that all three sulfonates undergo S_N1-type heterolysis and, further, that the solvolyses of **11**-OTs and the **12**-OTs/**13**-OTs pair proceed, almost exclusively, through different carbonium ion intermediates. Since the product distribution resulting from a free classical bicyclo[3.2.1]octan-2-yl carbonium ion should be independent of the configuration of its antecedent, the need for an alternative structure(s) for the intermediate(s) is apparent. The high optical purity of the bicyclo[2.2.2]octan-2-ol (**13**-OH) formed from optically active **12**-OTs and **13**-OTs similarly indicates that, at most, a minor amount

(22—27%) of this product arises from a symmetrical, classical carbonium ion intermediate.

		Yield (Optical Purity)[42-44]		
		 (11-OH)	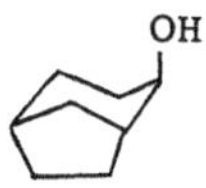 (12-OH)	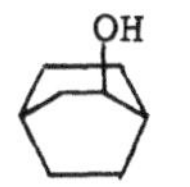 (13-OH)
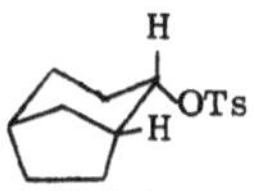 (11-OTs)	80% aq. acetone, pyridine, 49° →	95 (0)	4	1
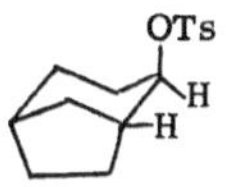 (12-OTs)	" →	0	43 (88)	57 (78)
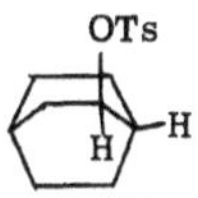 (13-OTs)	" →	0	43 (87)	57 (73)

Three different types of structures have been advanced for the carbonium ion intermediates in these, and other (*45—47*), solvolytic reactions in order to account for stereoselectivity. One is the symmetrically bridged non-classical ion (A), which was favored at the time by GOERING and FICKES (*42—44*). A second proposal suggests that heterolysis leads to non-interconverting ion pair intermediates (B), differing in the location of the counter ion. The substitution reaction takes place with retention of configuration by exchange of the tosylate counterion for a solvent molecule, while Wagner-Meerwein rearrangement occurs only to the opposite carbonium ion face. A third explanation invokes conformationally isomeric carbonium ions (C) which undergo preferential rearrangement of the adjacent σ-bond aligned with the p-orbital and nucleophilic attack fast with respect to conformational equilibration. Carbon-carbon hyperconjugation (vertical stabilization) may have a significant, stabilizing influence upon these conformers and the respective transition states involved in product formation (*48, 49*). One, or more, of these effects could also contribute to the maintenance of stereospecificity in terpene biosynthesis.

A.

B.

C.

The stereospecific migration of the group anti-parallel to the leaving group observed in the rearrangements of the bicyclooctyl tosylates **11**-OTs, **12**-OTs, and **13**-OTs is a characteristic common to many carbonium ion rearrangements, and, as mentioned above, is a basic postulate in the biogenetic isoprene rule. It is, nevertheless, to be expected that, as the stability of the carbonium ion increases, the tendency towards *anti*-parallel migration will be diminished. In fact, many rearrangements with overall *syn* stereochemistry have been reported, some examples of which will be found in the following sections. However, it is appropriate to call attention to this possibility in a more general sense with the two examples shown below (*50, 51*). A number of other *syn* rearrangements of trisubstituted epoxides have been described (*52—54*).

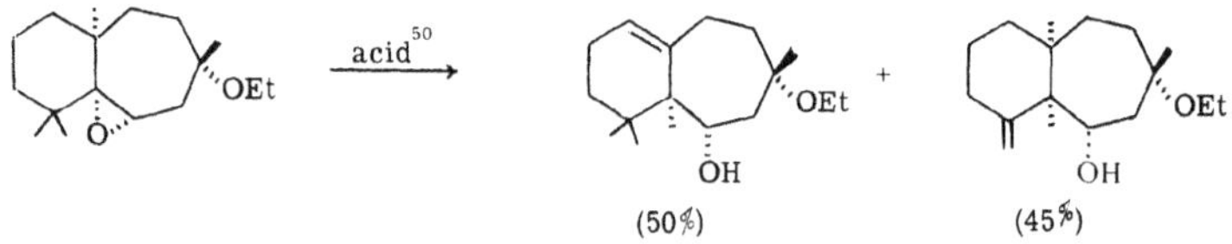

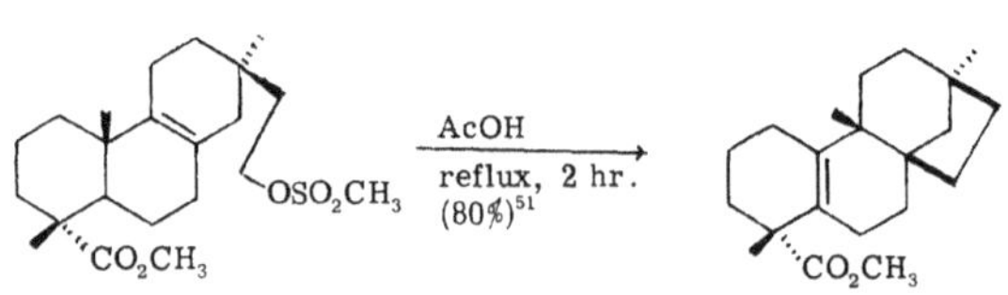

It should be apparent that the realization of a particular rearrangement or sequence of rearrangements, by chemical means cannot be considered proof that the same transformation takes place in the biosynthesis.

However, for the sake of efficiency it seems likely that the enzymes which mediate terpene biosynthesis would frequently capitalize upon the inherent chemical reactivity of their substrates. This expectation is considered sufficient justification to take the optimistic view that the supposed, biogenetic-type rearrangements of terpenes presented in the following sections may reflect kinetic or thermodynamic phenomena which are relevant to the corresponding step or steps in the biosynthetic process. A demonstrably unfavorable chemical reaction may signal a biosynthetic stage in which enzyme action is crucial to the successful outcome of the biological event. The fact that strong acid catalysis and non-aqueous solvents may be required to bring about a carbonium ion rearrangement does not necessarily detract from the biogenetic significance of the result. While such conditions are admittedly quite unlike the typical intracellular medium, the effective pH of a proton donor group protruding into a highly lipophilic region of an enzyme is a matter for conjecture.

B. Monoterpenes

1. Chrysanthemanes, Lavandulanes, Artemisanes, and Santolinanes

The biogenesis of these irregular monoterpenes has been a subject of much interest and speculation since the isolation and structure determination of the first representative, artemisia ketone. Although a number of biogenetic schemes have been proposed (*55, 56, 57*), the suggestion of Bates and Paknikar (*58*) that chrysanthemyl pyrophosphate (**16**-OPP) could serve as a precursor to the artemisane and santolinane skeletons by means of well-precedented cyclopropylcarbinyl→homoallyl rearrangements has the attractive feature of interrelating all the members of this family. The scheme has been expanded (*55*) to encompass lavandulol (**15**) and the more recently discovered yomogi alcohol (**19**). Recent investigations upon the biosynthesis of artemisia ketone and chrysanthemic acid from mevalonate appear to be consistent with this scheme although some alternatives are under consideration (*56, 57, 59—61*).

The first step involves the intermolecular olefin alkylation between two molecules of dimethyl allyl pyrophosphate. A 1,3-proton elimination from the methylene group gives chrysanthemyl pyrophosphate while a 1,2-proton elimination affords the lavandulol skeleton. Of course, these transformations can be equally well represented by formation of either a bridged, non-classical carbonium ion or an intermediate covalently bound to a nucleophile (*e.g.*, X group on an enzyme surface at the same tertiary site) which undergo 1,2- and 1,3-elimination to (**15**) and (**16**-OPP) (*34*, *55*).

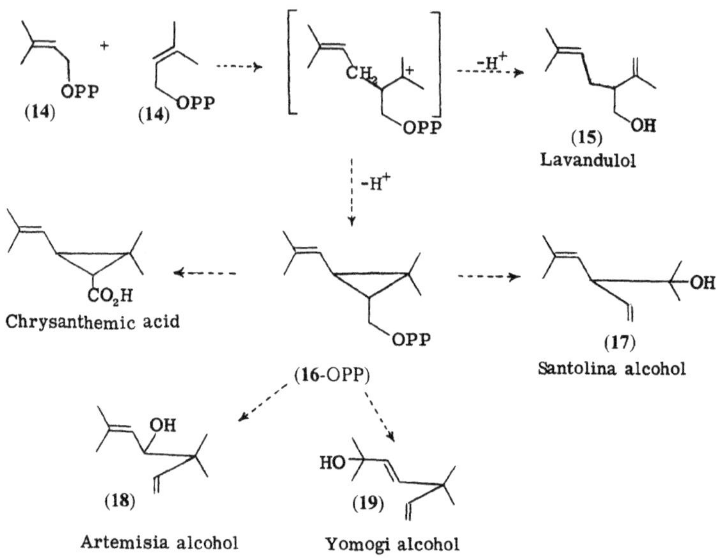

Scheme 1

Although cyclopropane formation will be considered in more detail later (section C.4), it is appropriate to mention a rather striking precedent for the proposed intermolecular cationic condensation of (**14**) to (**15**) and (**16**-OPP). The reaction of phenyl diazomethane with excess olefins (*e.g.*, *trans*-2-butene) at $-70°$ in the presence of trifluoroacetic acid affords cyclopropanes and benzylated products in fairly high yield (40—60%) along with benzyl trifluoroacetate (30—40%) (*62*, *63*). Cyclopropane formation is stereospecific with *cis*- and *trans*-2-butene.

Since incomplete deuterium incorporation into the phenylcyclopropane product (29% d_1) excludes a free carbonium (or diazonium) ion as its precursor, a partially-protonated (partially exchanged) diazo species (**20**) was proposed to react with olefin to give a hexacoordinate, carbenoid-type intermediate (**21**) which gives either cyclopropane or *syn* ion pair (**22**). Intermediate (**20**), although not a normal diazonium ion, is presumably electrophilic and should have diazonium ion character, in support of the biogenetic analogy. Also of interest in the above reaction is the product resulting from the combination of olefin alkylation and

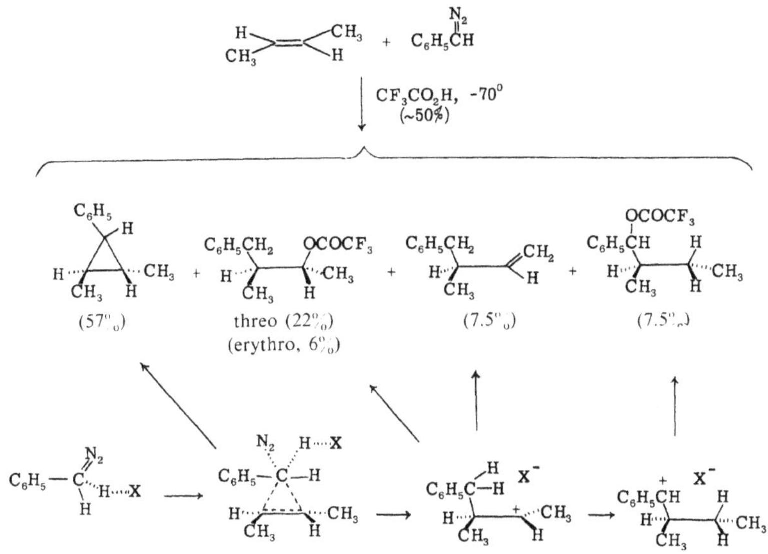

(20) (21) (22) (23)

1,3-hydrogen rearrangement, a process closely paralleling the biogenesis of the cadinanes and himachalanes (sections C.2, C.3, and C.4).

Stimulated by the biogenetic proposal of BATES and PAKNIKAR, several investigators have examined the carbonium ion reactions of chrysanthemyl derivatives (*64—73*). Under acidic conditions cleavage of the bond adjacent to the unsaturated side chain in chrysanthemol (**16**-OH) occurs to give artemisia triene (**24**), a known monoterpene (*64*). Thus the allylic stabilization in the transition state for fragmentation

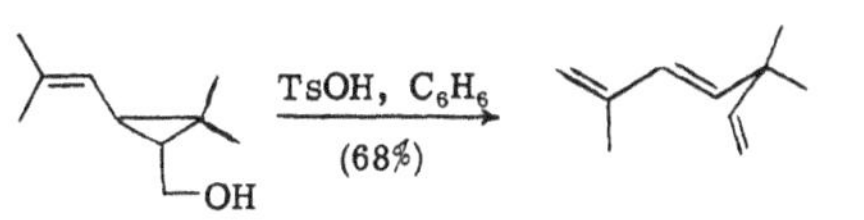

(**16**-OH) (**24**) Artemesia triene

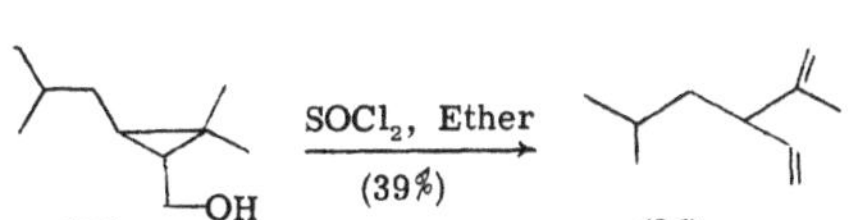

(**25**) (**26**) Santolina diene

of the cyclopropyl-carbinyl carbonium ion dominates over that of the two methyl groups on the adjacent position. In contrast, dihydrochrysanthemol (**25**) cleaves the more highly substituted ring bond, affording santolina diene (**26**) (*64, 73*).

In the buffered hydrolysis of N-methyl-4-(chrysanthemyloxy)-pyridinium iodide (**16**-OPy+I^-), a leaving group which is expected to minimize internal return, the naturally occurring allylic isomers, artemisia and yomogi alcohols (**18** and **19**) are the major products, but a small amount (0.5%) of santolina alcohol (**17**) is present (*66, 67*). Similar results have been reported for amine deamination (*69*) and dinitrobenzoate solvolysis (*66, 67, 70*). The absolute configuration of natural santolina alcohol corresponds to that of chrysanthemic acid, as predicted by the BATES and PAKNIKAR biogenesis (*73*).

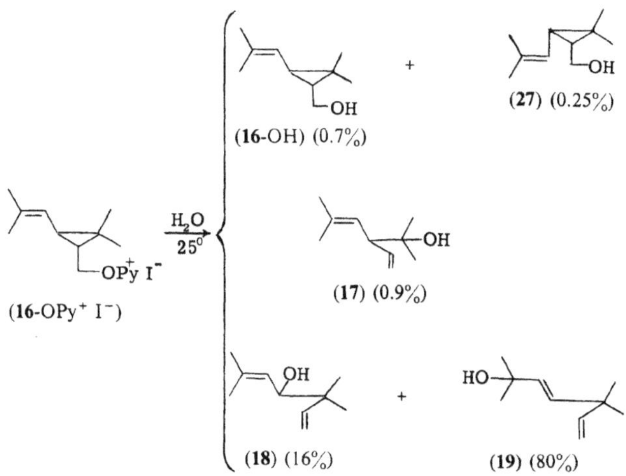

POULTER has demonstrated that chrysanthemyl derivative **16**-OPy^+I^- asymmetrically deuterated at the carbinyl position ionizes with $91 \pm 2\%$ stereoselectivity from a conformation with the leaving group aligned antiparallel to the 1,3-cyclopropyl bond (*68*). Since rotation about the C_1–C_2 bond in the cyclopropylcarbinyl carbonium ion intermediate should be slow relative to nucleophilic capture, this selectivity is reflected in the position of the deuterium atom in the vinyl group of the two major products. The stereoselectivity is attributed to stabilization of the transition state for ionization by charge delocalization into the vinyl group which should be more favorable in the *anti*-parallel alignment. Kinetic data also indicate that the double bond facilitates the ionization process.

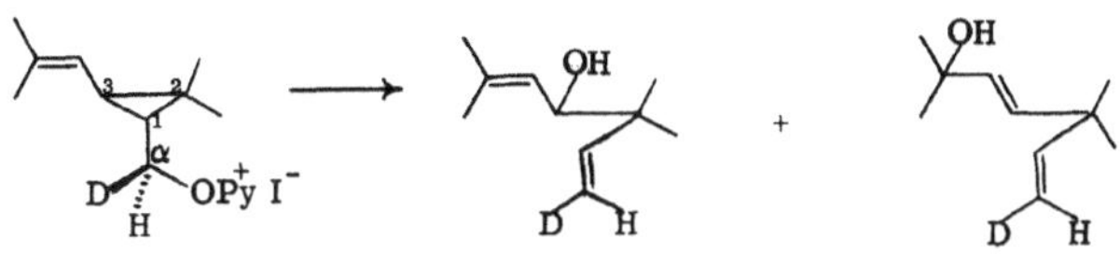

(16-OPy I⁻) (91% stereoselectivity)

It is of interest to note that artemisia alcohol (**18**) produced in the hydrolysis of **16**-OPy$^+$I$^-$ is essentially completely racemic (>98%) (*67*). Apparently (**18**) is formed by nucleophilic capture of the acyclic allylic carbonium ion (**29**) rather than direct attack in the 3 position of the chrysanthemyl carbonium ion (**28**). Nucleophilic substitution upon cyclopropylcarbinyl cations to give homoallyl products occurs with inversion of configuration (*74—76*). In the case of (**28**), however, position 3 is highly hindered by the adjacent gem dimethyl groups; thus collapse to the allylicly stabilized (**29**) is faster than direct substitution. The formation of a small amount of *cis*-chrysanthemol (**27**, 0.25%) is taken to indicate that allylic ion (**29**) recyclizes, at least in part, back to (**28**) and its *cis* isomer (**30**).

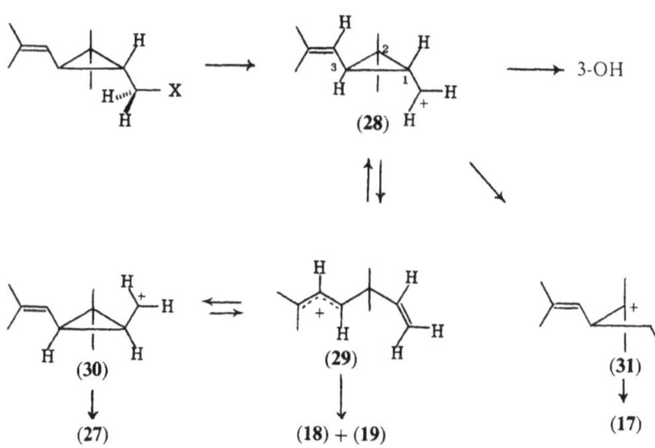

If the S configuration tentatively assigned to the carbinol carbon in natural artemisia alcohol (**18**) (*55*) is sustained, it would indicate that in the actual biosynthesis the nucleophilic substitution occurs with net retention of configuration. Since a direct substitution upon a cyclopropylcarbinyl ion should occur with inversion rather than retention (*74—76*), ring opening to the free allylic cation followed by stereospecific capture is proposed (*55*).

It has been possible to effect the reverse rearrangement, *i.e.*, the conversion of artemisyl derivations to chrysanthemane and/or santolinane products (*71, 72*), as shown below. These homoallyl ⇆ homoallyl rearrangements no doubt proceed through a cyclopropylcarbinyl (chrysanthemyl) intermediate [*e.g.*, (**29**→**28**→**31**)]. The realization of these transformations has led to an alternative proposal that the biogenetic precursor to these irregular terpenes is actually an artemisyl derivative.

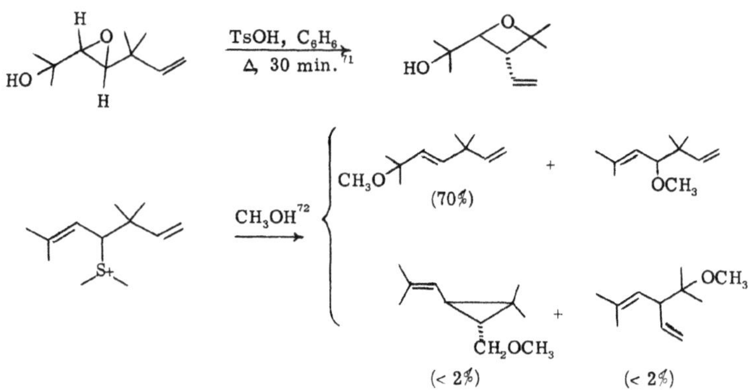

2. Menthanes, Pinanes, Camphanes, Fenchanes, and Thujanes

The biogenesis of this more abundant group of monocyclic and bicyclic monoterpenes proceeds through three main stages: cyclization of an acyclic precursor to the menthane skeleton (Scheme 2), cyclization to the pinanes and rearrangements (Scheme 3) or cyclization to the thujanes (Scheme 4) (*2, 56, 57*). Although biosynthetic investigations have been complicated by unequal incorporation of mevalonate into the two isoprene units, the results thus far are in agreement with these pathways (*56*).

Since both neryl pyrophosphate and linaloyl pyrophosphate (or some equivalent species) (*34*) have been seriously considered as likely acyclic precursors to this family of monoterpenes, it is appropriate to examine first the cationic reactions of these two compounds. A great many investigations have shown that, while geranyl derivatives undergo substitution reactions in a fashion typical of an allylic substrate, both neryl and, less efficiently, linaloyl derivatives lead to cyclic products such as α-terpineol, and its dehydrated relatives limonene, and terpinolene

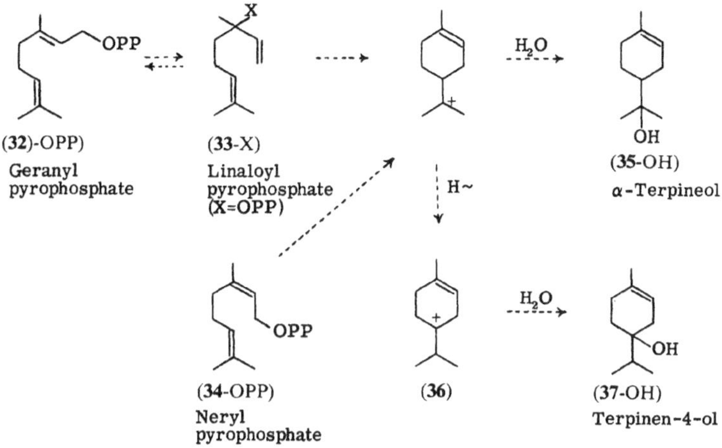

Scheme 2

(*77—84*). A thorough examination of both the rates and products in the hydrolysis of the pyrophosphates established that linaloyl and neryl pyrophosphates react predominantly by distinct pathways (*77—79*). The higher yield of cyclic products from neryl pyrophosphate is attributed to anchimeric assistance [an estimated rate enhancement of 2.5 times in the case of (**34**-OP)] by the 6,7-double bond in the ionization step. Linaloyl pyrophosphate evidently reacts in part from a transoid conformer to give geraniol and from a cisoid conformation to give nerol and cyclic products.

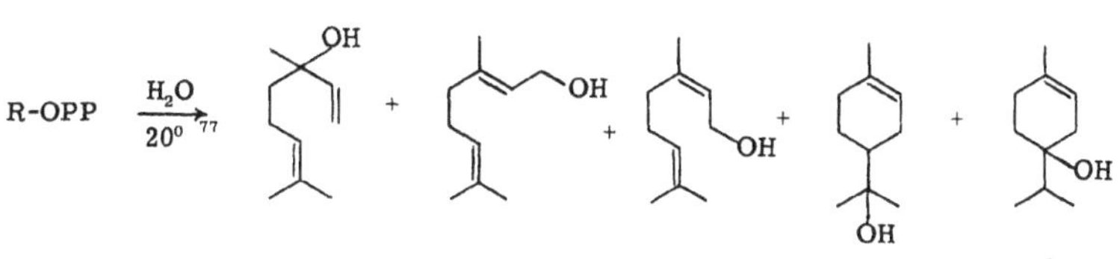

		(33-OH)	(32-OH)	(34-OH)	(35-OH)	(37-OH)
Geranyl-OPP (32-OPP)	→	18%	17%	1%	3%	--
Neryl-OPP (34-OPP)	→	24	0.5	5	67.5	0.4
Linaloyl-OPP (33-OPP)	→	59.7	11	3	23	0.4
α-terpinyl-OPP (35-OPP)	→				80.1	1.0

A remarkable feature of the cyclizations of linaloyl derivatives is the high stereoselectivity in the formation of α-terpineol. Thus, the hydrolysis of (−)-linaloyl phosphate (**33**-OP) affords (+)-α-terpineol with an optical purity of 40% (*79*). The cyclization of linalool to optically active α-terpineol may also be carried out with sulfuric acid in acetic acid (*81*) and aqueous sulfuric acid (*82*). The α-terpinyl *p*-nitrobenzoate produced in the hydrolysis of linaloyl *p*-nitrobenzoate is obtained with an optical purity as high as 87% (*80*). These results require that cyclization occurs, at least in part, from an asymmetric transition state.

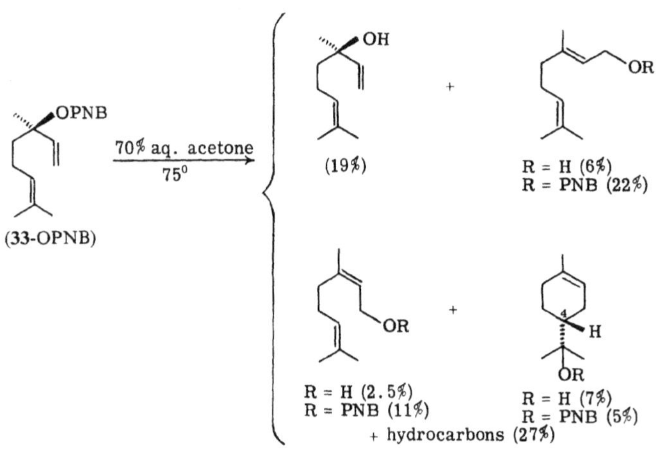

The established stereochemical relationship between linalool and the derived α-terpineol (*79, 80*) is such that the departing hydroxyl group and the proton on the new asymmetric at C_4 bear a *syn* relationship in the cyclization. Since both neryl and α-terpinyl *p*-nitrobenzoates are produced in relatively large amounts from internal return in the hydrolysis of **33**-OPNB, the proposal was made that cyclization in this case occurs from an initially formed cisoid allylic cation *p*-nitrobenzoate ion pair (**38**) (*80*).

In contrast, ion pair return products were not detected in the hydrolysis of (**33**-OP). The persistance of the stereoselectivity in the formation of a α-terpineol in the apparent absence of ion pair intermediates and in the mineral acid-catalyzed isomerization of linalool (conditions in which ion pairs should be unimportant) (*81, 82*) was taken

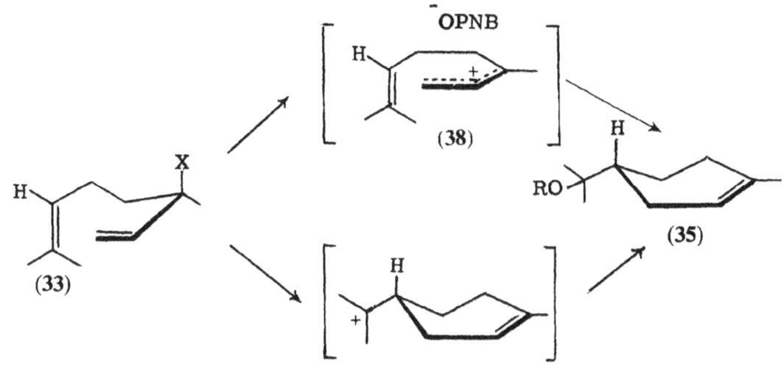

to indicate a concerted S_N2' pathway (*79*). Although a suprafacial S_N2' attack upon C_1 of (**33**-X) [or the derived ion pair (**38**)] in the cyclizations was favored (*79, 80*) in analogy with the stereochemistry of other S_N2' reactions (*85*), it is desirable to have experimental confirmation on this interesting point.

Two other noteworthy features in these solvolyses are the formation of linalool (**33**-OH) with partial (20%) retention of configuration in the hydrolysis of (**33**-OPNB) [complete racemization from (**33**-OP)] and the detection of terpinen-4-ol (**37**-OH) and/or γ-terpinene, presumably *via* hydride shift.

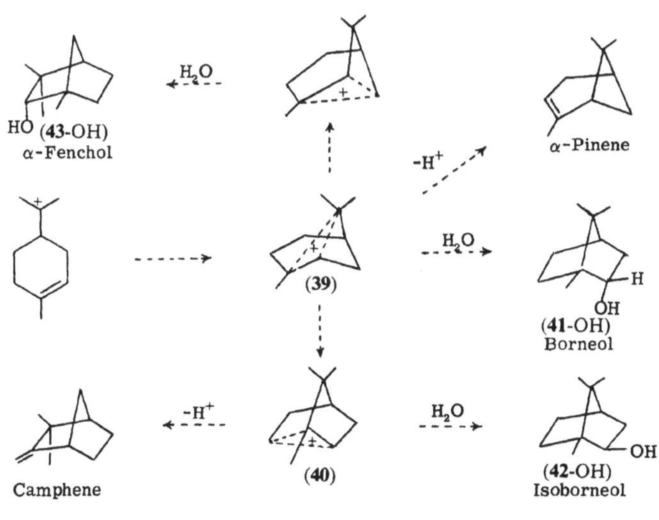

Scheme 3

Bicyclic compounds have apparently never been detected in the products resulting from the carbonium ion reactions of either acyclic precursors [*e.g.*, (**33**-X) or (**34**-X)], or monocyclic menthane reactants [*e.g.*, (**35**-X)] under quite a variety of conditions (*77, 84, 86, 87*). It appears to be necessary to provide additional driving force as in the acid-catalyzed cyclization of carvone enol acetate (**44**) (*87*) or the intramolecular enolate alkylation of keto tosylate (**45**) (*88, 89*) in order to overcome the strain energy of the bicyclo[2.2.1]heptane and bicyclo-[3.1.1]heptane rings. An additional factor in the latter case is, of course, increased electrophilic demand at a primary, as opposed to a tertiary, center.

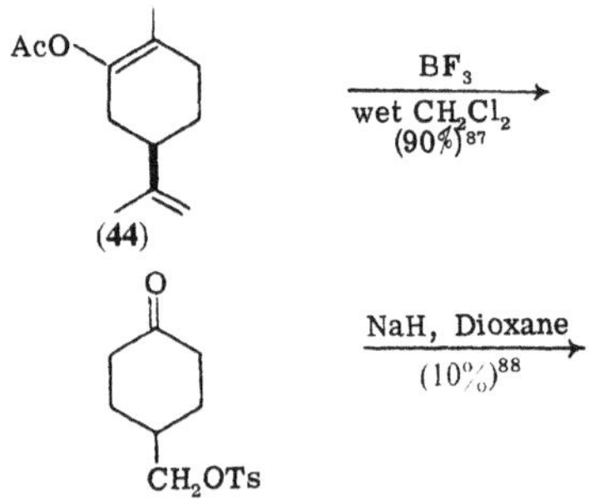

(**44**) (**45**)

The facility and stereospecificity of the interconversions of the camphanyl, bornyl, and pinyl monoterpenes are renowned and provided the initial factual basis for the formulation of non-classical carbonium ion intermediates (*39, 90*). A tremendous number of investigations has been carried out upon the rearrangements of bicyclo[2.2.1]heptanes and the nature of the various carbonium ion intermediates (*39*). However, for the purposes at hand it will be sufficient to illustrate the characteristics of these rearrangements with the classical case of the camphene and pinene hydrochlorides (**46**-Cl and **47**-Cl) rearrangements (*91—93*).

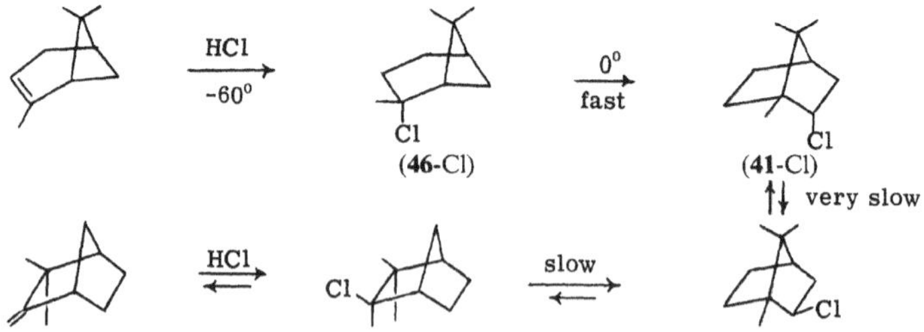

(**46**-Cl) (**41**-Cl) (**47**-Cl) (**42**-Cl)

The stereochemical integrity of these rearrangements (pinene→ bornyl chloride (**41**-Cl) and camphene→isobornyl chloride (**42**-Cl) requires that there be two distinct intermediates capable of generating the isomeric *endo* and *exo* chlorides, respectively. The energy barrier separating the two intermediates is not, however, inacessible as indicated by the slow interconversion of the two bornyl isomers.

The high stereoselectivity of these rearrangements may be explained in terms of discrete non-classical ions (**39** and **40**), or carbonium ion-chloride ion pairs. At this time the evidence seems to weigh against the symmetrical, σ-bridged non-classical structures for the intermediates formed either in the solvolysis of derivatives of 2-alkyl bicyclo[2.2.1]-heptan-2-ols (*39*, *40*), or in the addition of hydrogen chloride to the corresponding olefins (*41*) and for the long-derived carbonium ions observed by nmr spectroscopy (*94*).

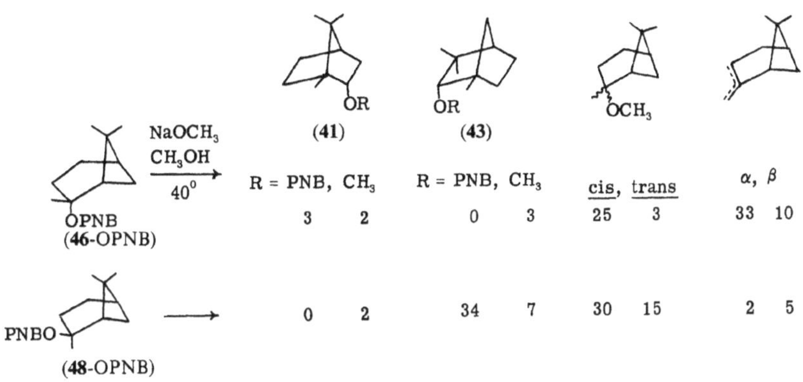

The pinane ring system may rearrange to both bornyl and fenchyl products by migration of either the gem-dimethyl or the methylene bridge (*84*, *95*). Both types of products are usually formed from the addition of acids to the pinenes. Methanolysis of the *cis*- and *trans*-pinyl *p*-nitrobenzoates (**46**-OPNB and **48**-OPNB) gives, in part, separate sets of products, and to a substantial degree a common distribution of products (*96*). Although the energy barrier separating the isomeric ions here is clearly small, it is noteworthy that the bornyl and fenchyl *p*-nitrobenzoates (**41**-OPNB, 3%; **43**-OPNB, 34%) resulting from internal return are evidently formed stereospecifically.

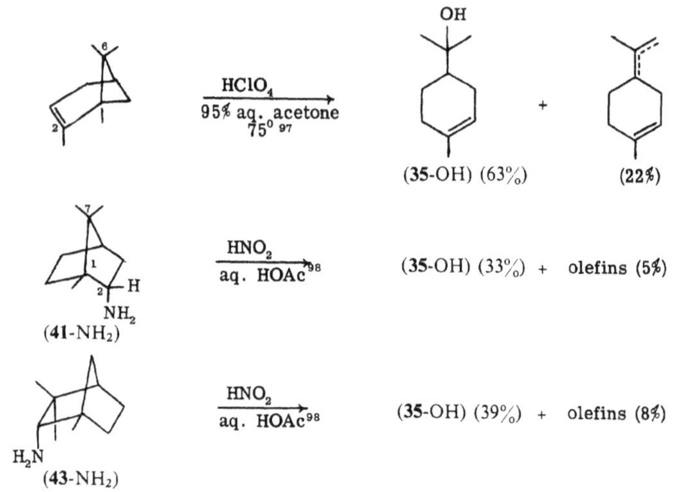

An alternative reaction pathway for both pinyl and bornyl derivatives is cleavage of the gem-dimethyl bridge to give α-terpinyl products (*e.g.*, **35**-OH) or the corresponding elimination products (*95, 97, 98*) the exact reverse of the second cyclization in biogenetic Scheme 3. In view of the strain energy released in these fragmentations, the extent to which the bicyclic structures are retained in many of the rearrangements may seem surprising. The rather poor alignment of the 1,6- and 1,7-C–C bonds with respect to an adjacent p-orbital of a carbonium ion at C_2 may be one reason. Another factor may be that Wagner-Meerwein rearrangements of pinyl derivatives takes place with little rehybridization of the migrating carbon, and hence with the accumulation of very little of the positive charge.

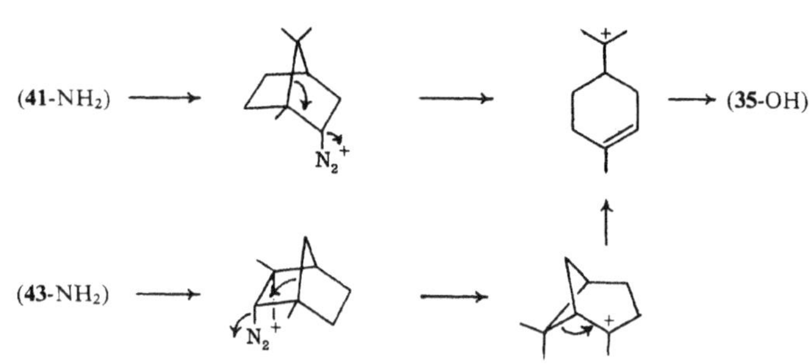

The formation of α-terpineol as the major product from deamination of fenchyl amine (**43**-NH_2) (*98*) is remarkable in that it requires first rearrangement to the more strained pinane skeleton. Small amounts (3—11%) of pinane-type products have been isolated from the bornyl and fenchyl diazonium ions under alkaline conditions (*98 a*). The contra-thermodynamic rearrangement of apobornyl and camphenilyl derivatives to nopinone may be effected in good yield with the driving force of the pinacol-type rearrangement (*98 b*, *99*).

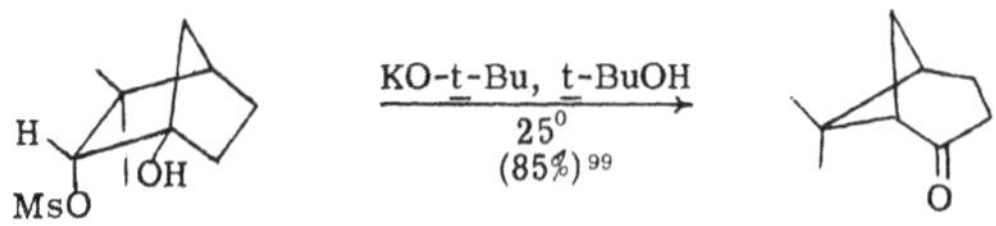

The biogenesis of the thujane family of monoterpenes involves a homoallylic cyclization of the 4-terpinenyl carbonium ion (**36**) to a cyclopropylcarbinyl cation. Tracer experiments seem to be in accord with this scheme (*56*, *100*).

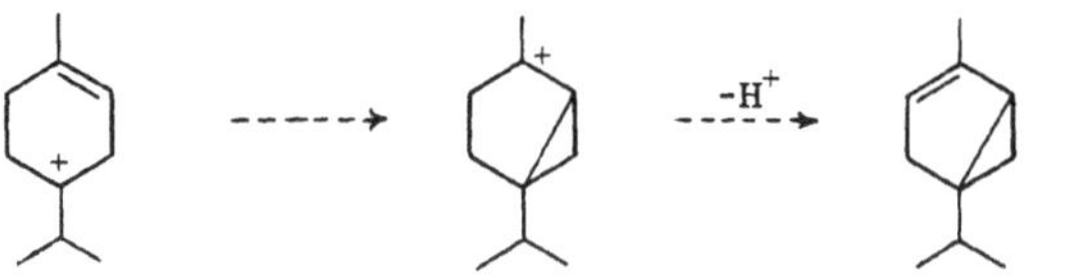

Scheme 4

Although the homoallylic→cyclopropylcarbinyl cyclization is well-precedented in carbonium ion chemistry (*101*, *102*) there seem to be no reports of the direct cyclization of the tertiary 4-terpinenyl carbonium ion. However, deamination of cyclohex-3-enyl amine and solvolysis cyclohex-3-enyl tosylate gives *exo*- and *endo*-bicyclo[3.1.0]hex-2-yl derivatives as 6—43% of the products resulting from nucleophilic capture (*101*, *103*, *104*). The modest yield of bicyclic products in these reactions apparently is the result of competing nucleophilic capture prior to cyclization and hydride shift to the 2-cyclohexenyl cation. More efficient cyclization occurs in the acetolysis of 2-bicyclo[2.2.2]oct-5-enyl tosylate (**49**) owing to the rigid boat-like conformation of the precursor (*105*). The high efficiency of the base-catalyzed cyclization of the epoxide of 4-isopropylidene cyclohexanone may be attributed to the high nucleophilicity of the electron rich double bond of the intermediate enolate anion (*106*).

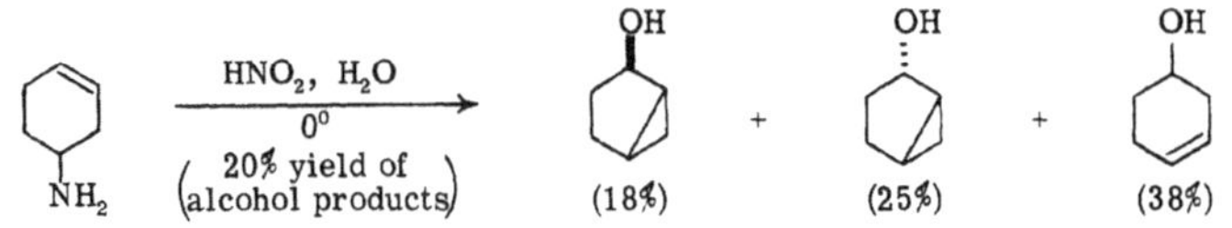

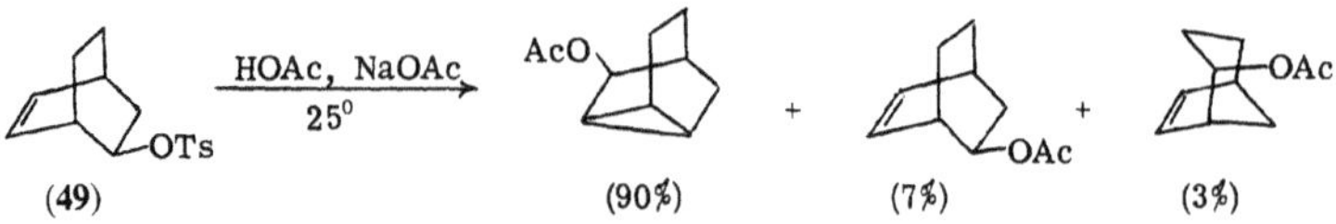

The reaction of sabinene (**50**) and α-thujene with aqueous perchloric acid results in a retro-biogenetic ring opening to terpinen-4-ol (**37**-OH) and the corresponding dienes (*107*). The high optical purity (80—85%) of the terpinen-4-ol indicates that this product is formed mainly by direct nucleophilic attack upon an intermediate cyclopropylcarbinyl carbonium ion.

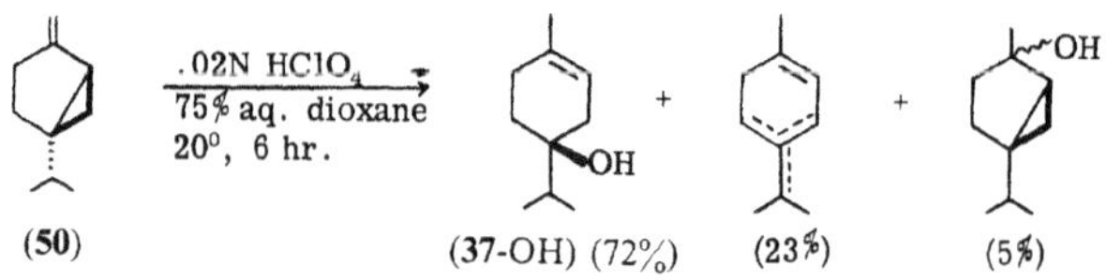

C. Sesquiterpenes

1. Bisabolanes, Sesquicamphanes, Santalanes, and Carotanes

The biogenesis of this group of mono- and bicyclic sesquiterpenes, excepting for the moment the carotanes, begins in a manner analogous to the menthanes, bornanes, and camphanes previously discussed (*108*) (Scheme 5). Closure to a six-membered ring requires a cisoid configuration about C_1–C_4 in the acyclic precursor, *i.e.*, either *cis, trans*-farnesyl pyrophosphate (*108*) or alternatively nerolidyl pyrophosphate (or an equivalent species **53**-X). *Trans, trans*-farnesyl pyrophosphate could serve as predecessor provided it subsequently undergoes allylic isomerization to a nerolidyl derivative (*109, 110*).

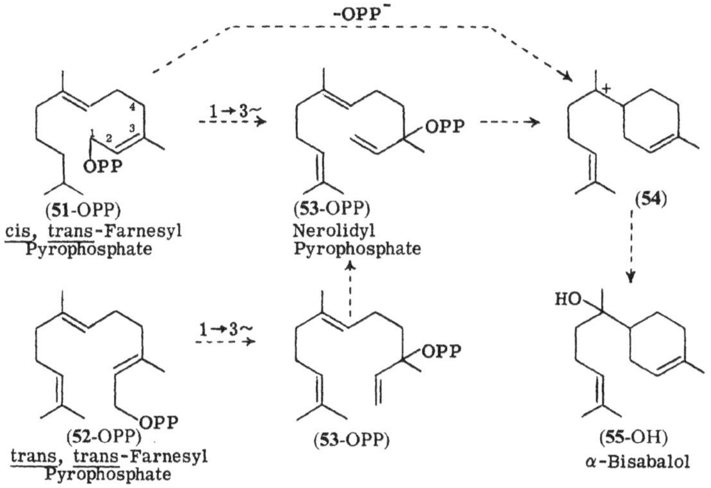

Scheme 5

Chemically-induced cyclization of the farnesols, nerolidol, and their derivatives has been effected under a variety of reaction conditions (*79, 111—114*). Although the earlier investigations by RUZICKA were hampered by the difficulties of separating pure products from the complex reaction mixtures, the formation of the α-bisabalols and bisabolenes was firmly established (*111*). A more recent study (*112*), aided by modern analytical techniques, reports that treatment of either pure *trans, trans*-farnesol (**52**-OH) or a mixture of *cis, trans*- and *trans, trans*-farnesol (**51**-OH + **52**-OH) with 100% formic acid for 5 min at 0° gives (after saponification of the formate esters) α-bisabolol (43%), recovered farnesols (9—12%), diols (17%), and polyenes (17—18%). Very similar product distributions were obtained from *trans*-nerolidol (**53**-OH) in a somewhat faster reaction. By means of catalytic aromatization of the product mixture followed by glc analysis for cadalene (1,6-dimethyl-4-isopropylnaphthalene) it was shown that bicyclic products were present in yields of about 3—12%. Cyclization of (+)-*trans*-nerolidol (**53**-OH, below) with 100% formic acid in pentane affords optically active α- and β-bisabolene (**56** and **57**), the latter having an optical purity of 31% (*113*). As in the analogous conversion of linalool to α-terpineol discussed in section B, the departing hydroxyl group and the proton on the new asymmetric center maintain a *syn* relationship in the cyclization.

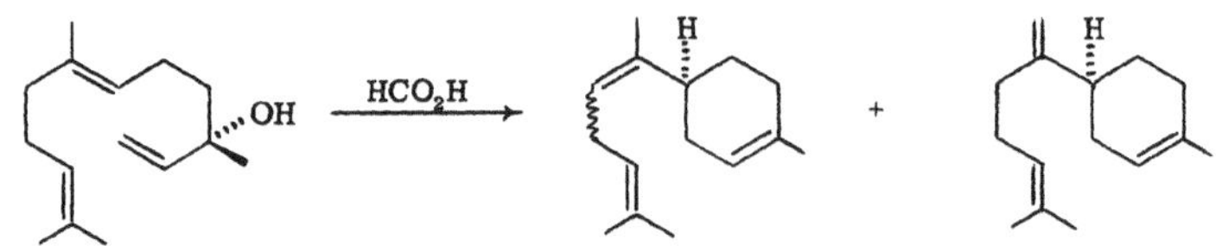

(+)-(**53**-OH) (**56**) (26%) α-Bisabolene (**57**) (35%) (−)-β-Bisabolene + Farnesenes (10%)

Since α-bisabolol, in contrast to α-terpineol, has two asymmetric centers, the interesting question arises as to whether there is stereoselectively in the nucleophilic capture step. If there were a net *anti* addition of the electrophile (C_1) and the nucleophile (H_2O) across the 6,7-double bond (as required for a configurationally stable non-classical carbonium ion intermediate), *cis, trans*-farnesol and *trans*-nerolidol should cyclize to one diastereomeric α-bisabolol (4R, 8R-**55**-OH and its enantiomer). Rittersdorf and Cramer have looked into the matter in a study of hydrolysis of *cis, trans*-farnesyl phosphate (**51**-OP), *trans*-nerolidyl phosphate (**53**-OP), and *cis*-nerolidyl phosphate (**58**-OP) (*79*).

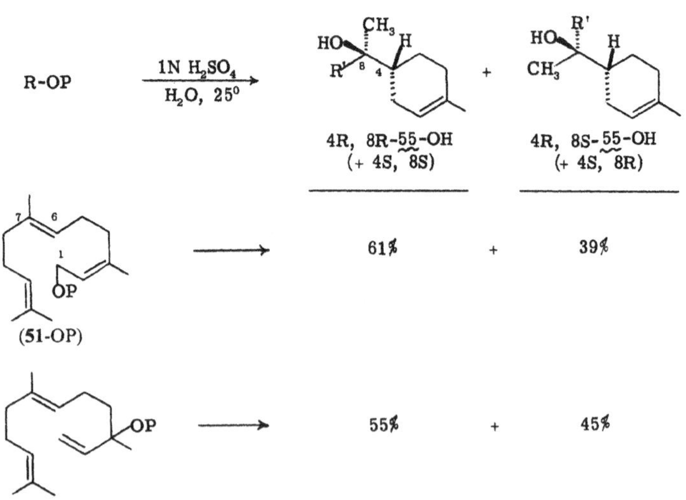

R-OP → (1N H_2SO_4, H_2O, 25°) 4R, 8R-**55**-OH (+ 4S, 8S) + 4R, 8S-**55**-OH (+ 4S, 8R)

(**51**-OP) → 61% + 39%

(**53**-OP) → 55% + 45%

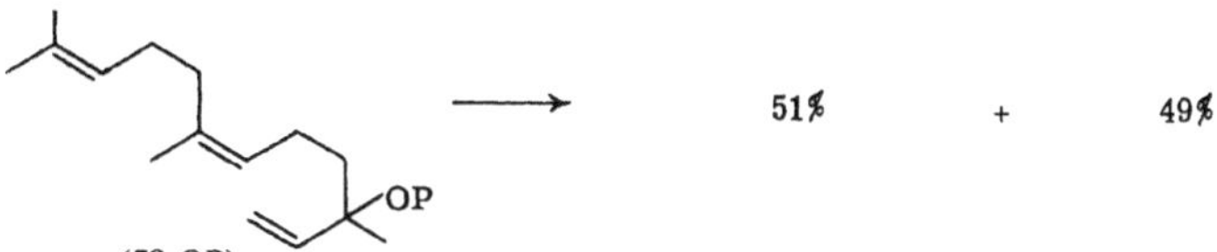

(58-OP)

By means of gas chromatographic analysis they found that both diastereomers were present in comparable, but unequal, amounts. Although no independent evidence was available with which to assign the relative stereochemistries, they assumed that the diastereomer formed in excess would be the one resulting from an *anti* addition. The low stereoselectivity observed in these reactions proves that, if a transient nonclassical carbonium ion is formed at all, it rapidly loses, for the most part, its stereochemical identity. The two diastereomeric bisabolols may then be formed by attack of water with comparable probability upon each of the diastereotopic faces of a planar, tertiary carbonium ion.

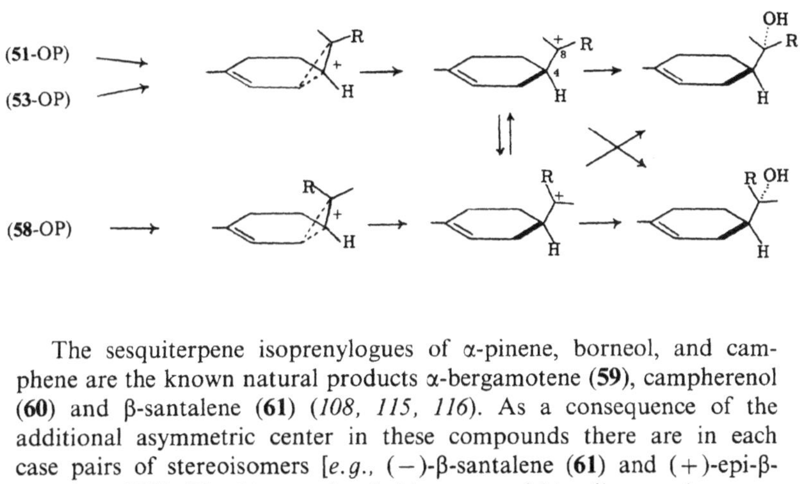

The sesquiterpene isoprenylogues of α-pinene, borneol, and camphene are the known natural products α-bergamotene (**59**), campherenol (**60**) and β-santalene (**61**) (*108, 115, 116*). As a consequence of the additional asymmetric center in these compounds there are in each case pairs of stereoisomers [*e.g.*, (−)-β-santalene (**61**) and (+)-epi-β-santalene (**62**)]. The biogenesis of this group of bicyclic sesquiterpenes is analogous to that discussed for the corresponding monoterpenes (*108*). If it is assumed that (+)-*epi*-β-santalene is formed *via* a 6→2 hydride shift as shown in the scheme, all of the known members of this group may be derived from *cis, trans*-farnesyl pyrophosphate (or nerolidyl pyrophosphate) *via* a bisabolyl carbonium ion which retains the *trans* relationship of the original 6,7-double bond.

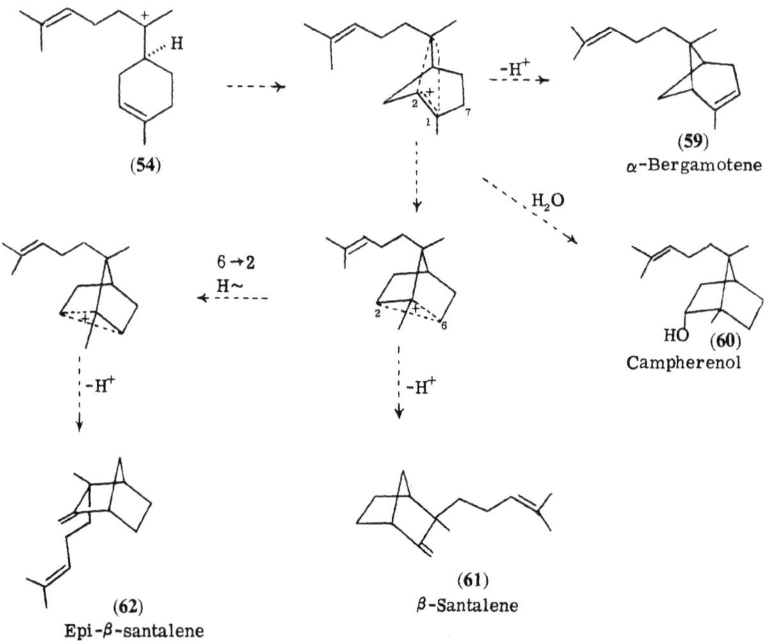

Scheme 6

As in the monoterpenes, no bicyclo[2.2.1]heptanes or bicyclo-[3.1.1]heptanes have as yet been found in the products from the cyclization of farnesol or nerolidol or in the reactions of bisabolyl derivatives. However, with the more electrophilic double bond of an enol acetate, ring closure occurs as in the monoterpene series to give a mixture of the two stereoisomers which were converted onto campherenone and epi-

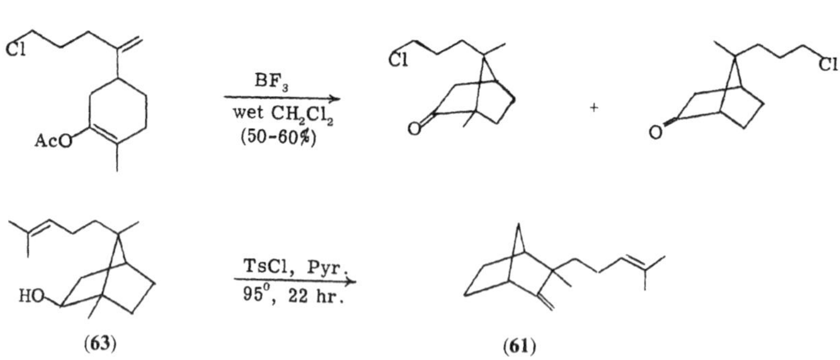

campherenone. Isocampherenol (**63**) and isoepicampherenol upon reaction with tosyl chloride in pyridine rearrange stereospecifically to β-santalene (**61**) and epi-β-santalene (**62**) respectively (*115—117*).

The small group of carotane sesquiterpenes has a perhydroazulene nucleus with an unusual substitution pattern consisting of an angular methyl group and an isopropyl group on the cyclopentane ring (*e.g.*, carotol, **65**). The location of the label from incorporation of radioactive acetate into carotol suggests a biogenesis in which *cis, trans*-farnesyl pyrophosphate (or cisoid nerolidyl pyrophosphate) initially cyclizes onto the central double bond in an anti-Markovnikov fashion (*118*).

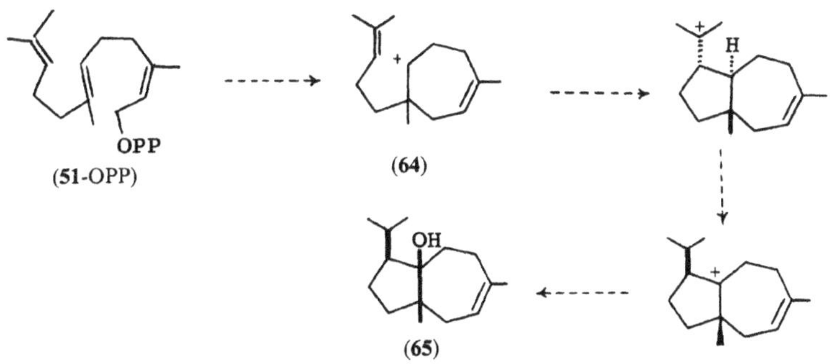

Scheme 7

It is not surprising that there seems to be no appropriate chemical precedent for the *anti*-Markovnikov cyclization to a seven-membered ring. On the contrary, the ring contraction of secondary carbonium ion (**64**) to the tertiary bisabolyl ion should be very rapid. However, the Prins-like conversion of the cycloheptenone (**66**) to the carotol ether (**67**) in the presence of stannic chloride mimics, to some degree, the second cyclization in the biogenesis of carotol (*119*).

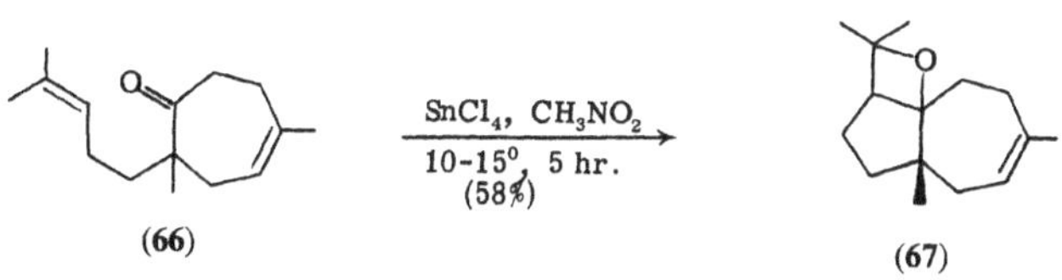

2. Germacranes, Eudesmanes, Guaianes, Cadinanes, and Cubebanes

The biogenesis of the germacrane, eudesmane, and guaiane sesquiterpenes is initiated by a cyclization between the terminal double bond of *trans, trans*-farnesyl pyrophosphate (**52**-OPP) and the electrophilic C_1 position (Scheme 8). The resulting ion (**68**) may then neutralized by reaction with water giving hedycaryol (**69**), by 1,2-proton elimination to the various germacrenes, or by 1,3-proton elimination to bicyclogermacrene (**70**). The *trans-*, *trans*-1,5-cyclodecadiene of the germacranes (*e.g.*, **69**) are proposed to undergo attack of electrophiles upon the 1,2-double bond and Markovnikov cyclization to the decalin-based eudesmane sesquiterpenes (*e.g.*, β-eudesmol). Similarly anti-Markovnikov attack of the electrophile at the 5,6-double bond followed by 2→6 cyclization affords, upon neutralization by proton loss or nucleophilic capture, the hydroazulenic guaiane sesquiterpenes (*e.g.*, bulnesol) (*108*, *120*).

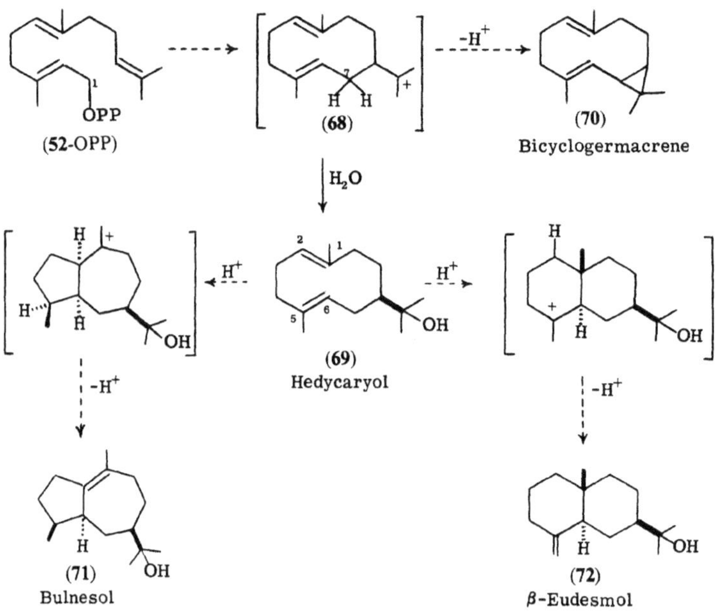

Scheme 8

The electrophilic cyclizations of *trans, trans*-1,5-cyclodecadienes has been extensively investigated, principally with germacrane-type sesquiterpenes (*121*), but also with model substrates (*122*). Electrophiles effect regiospecific addition to the 1,2-double bond and stereospecific cyclization to *trans*-decalins, transformations closely parallel to the biogenesis of the eudesmanes (*123*). Similar cyclizations have been observed with numerous oxygenated germacrane derivatives (*121*).

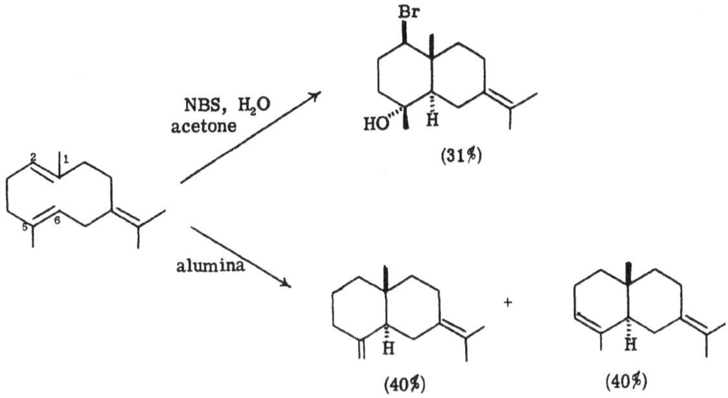

The regiospecificity of these cyclizations may be attributed to Markovnikov participation of the 5,6-double bond in the transition state for Markovnikov protonation of the 1,2-double bond and charge delocalization onto the 5 position. Protonation at the other double bond is less favorable since a Markovnikov orientation would require bonding between carbons 2 and 5 and consequent increase in strain energy. The ability of a remote methyl group to direct the double bond protonation of *trans, trans*-1,5-cyclodecadienes is also evident in the regio-

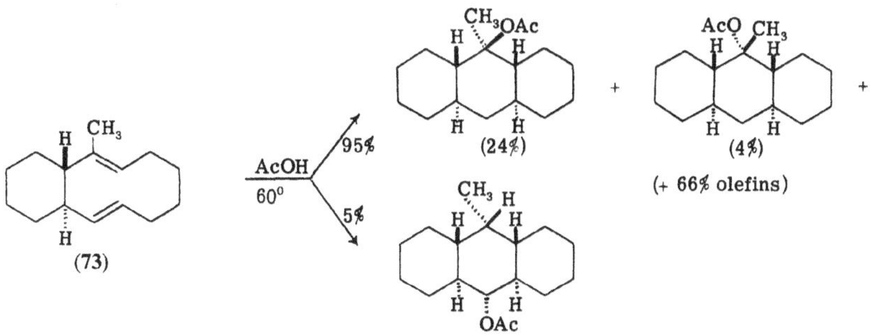

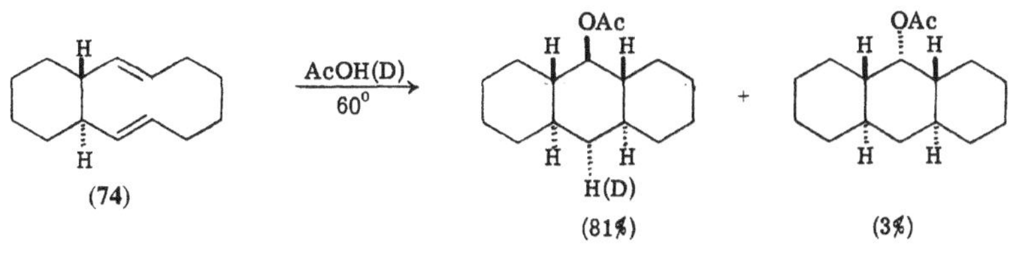

selective addition of acetic acid to model compounds (**73**) and (**74**) (*122*). The observation that *trans*-cyclodecene is completely inert to these reaction conditions attests to the transannular interaction between the double bonds in the dienes.

Another stereochemical characteristic of these cyclizations is the equatorial disposition of both the entering electrophile and nucleophile, as expected for overall *anti* addition to both double bonds (*121—123*). The *anti* mode of addition appears to be followed in the initial electrophilic attack in all cases. However, the stereoselectivity in the approach of the nucleophile is evidently less since minor amounts of axial acetates are formed in the additions to the above model cyclodecadienes (**73**) and (**74**) (*122*).

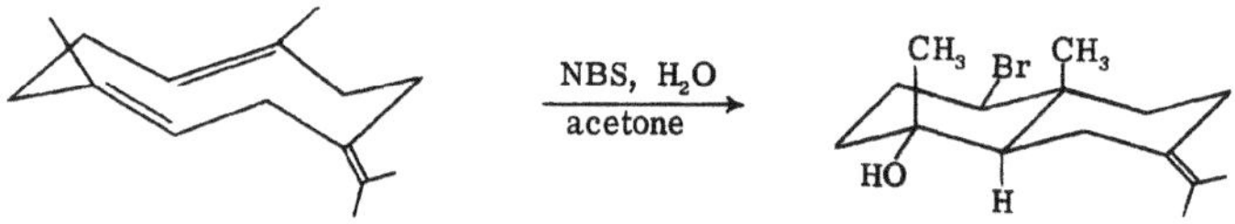

It is possible to effect an anti-Markovnikov cyclization to the perhydroazulene skeleton of the guaianes by the action of acid catalysts upon the 5,6-epoxides of various germacrene sesquiterpenes (*121, 124*). Markovnikov opening of the epoxode with participation of the transannular double bond (cyclobutane-like strain in transition state) does not compete. For contrast, the cyclization of the isomeric epoxide to eudesmanediol is also shown.

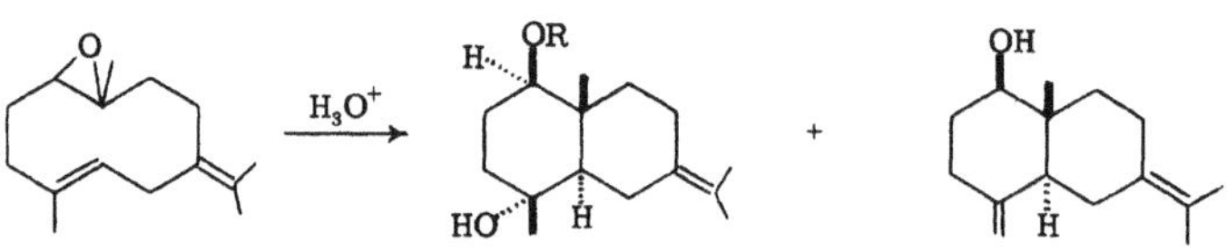

(**75a**) (R = H)
(**75b**) (R = Ts)

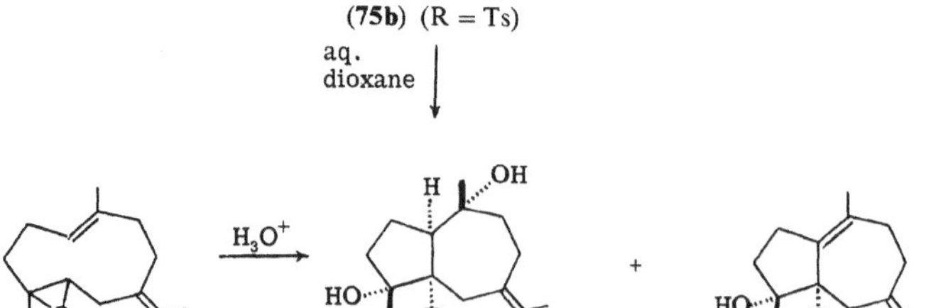

(**76**)

It is of interest to note that eudesmanediol, suitably activated as its mono-tosylate (**75b**), is transformed into the guaianediol (**76**) by hydrolysis, the kinetic driving force being formation of a tertiary carbonium ion (*124*). This type of rearrangement is, in fact, a key step in the total synthesis of bulnesol (**71**) and other guaiane-type sesquiterpenes (*125, 126*). Bulnesol has also been synthesized by way of a stereospecific ring contraction of the bicyclo[4.3.1]decalyl methanesulfonate (**78**) (*127*).

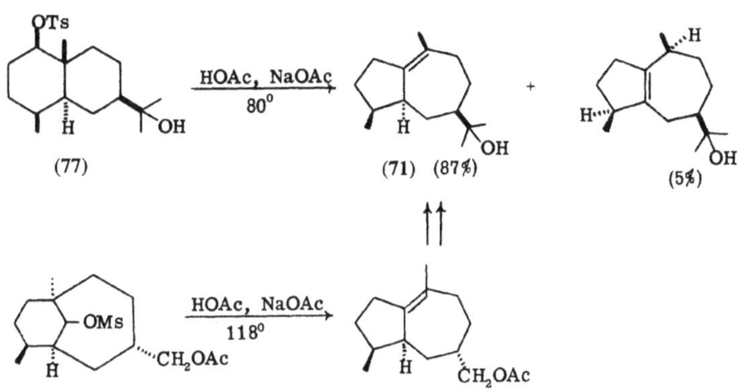

(77) (71) (87%) (5%) (78) (92%)

Since the parent *trans*-decalyl tosylate (**77** lacking methyl and isopropyl substituents) solvolyzed only slightly slower than the compound having the angular methyl group ($k_H/k_{CH_3} = 0.10$), Heathcock, Ratcliffe, and Van concluded that the initial ionization of the tosylate group precedes the rearrangement of the interannular bond (*128*). That is, the solvolytic transition bears little resemblance to the tertiary carbonium which is eventually formed in the rearrangement (see also section D.4).

The cyclization of the various geometrical isomers of the germacrenes (*i.e.*, *cis, trans; trans, cis;* and *cis, cis*) has been suggested as an alternative biogenesis of the guaiane sesquiterpenes. This alternative scheme not only accounts for the usual stereochemistry of normal guaianes (*e.g.*, bulnesol, 71), but also, through cyclization to *trans*-fused perhydroazulenes, provides a stereochemically consistent link-up with the pseudoguaianes (*108*). The isolation of oxygenated derivatives of *cis, trans-* and *trans, cis*-1,5-germacradienes (*129*) and the establishment of the appropriate stereochemical relationships between guaianolides and pseudoguaianolides (*130*), attaches significance to this proposal. In a similar fashion, the stereochemical variations within the aromadendrane family of sesquiterpenes may be rationalized in terms of cyclization of *trans, trans-*, *cis, trans-*, and *cis, cis*-bicyclo germacrenes (double bond isomers of **70**) (*108*). If this proposal is borne out, there arises the interesting question of whether the isomeric germacradienes are formed directly from the corresponding isomer of farnesyl pyrophosphate or by isomerization of a *trans, trans*-germacradiene precursor. The latter possibility is not unreasonable since the known natural germacrane sesquiterpenes with *cis* double bonds (or oxides derived thereform) are highly oxygenated (*129*). In any case, it is appropriate to consider briefly the reactions of *cis, trans*-1,5-cyclodecadienes.

In contrast to the *trans, trans*-1,5-cyclodecadienes and germacradienes, the *cis, trans*-isomers undergo electrophilic cyclization to *cis*-decalins (*121, 131—134*). The parent diene itself reacts with the electrophile at the *cis* double bond initially. As in the case of the *trans, trans*-compounds, the stereochemistry in both addition steps is *anti*.

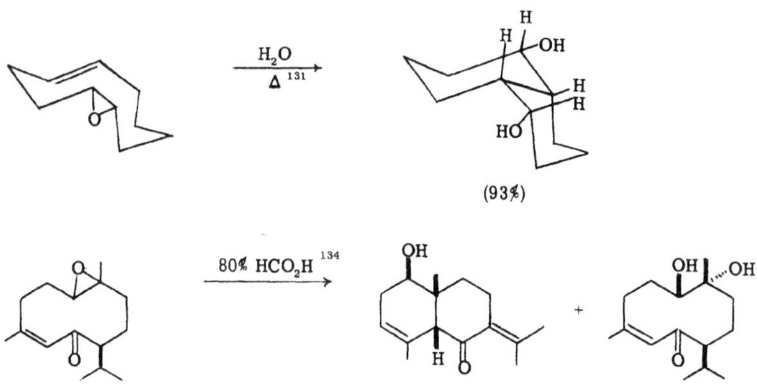

(93%)

(79) (80) (34%) (47%)

The formation of the *cis*-fused eudesmane-type keto alcohol (**80**) from the *trans, cis*-germacrone (**79**) (*134*) suggests that the relatively rare *cis*-eudesmanes (*e.g.*, occidentalol) might arise in nature by such a pathway. There are, however, two viable alternatives. First, it should be recognized that *trans, trans*-germacradienes, under the influence of a biosynthetic enzyme, might cyclize from conformations other than the normal crown conformer, and yield *cis*-eudesmanes (*135*). Another possibility is a facile (non-enzymic?) electrocyclization of a 1,3,5-germacratiene precursor (*136*).

References, pp. 205—230

The cadinane, muurolane, amorphane, and bulgarane sesquiterpenes are stereoisomers of 1,4-dimethyl-6-isopropyldecahydronaphthalene. The biogenesis of these compounds necessitates a *cis*-configuration between C_2 and C_3 at the time of cyclization onto C_6 of the original farnesyl chain. One possibility is a biogenetic scheme originating from either *cis, trans*-farnesyl pyrophosphate (**51**-OPP) or cisoid, *trans*-nerolidyl pyrophosphate (**53**-OPP, equivalent species) (*108, 137—139*). Initial cyclization onto C_1 as in the germacrane biogenesis followed by a 1,3 hydride shift gives rise to a "sickle"-type cyclodecadienyl (**82**) suitable for Markovnikov ring closure to the decalin nucleus.

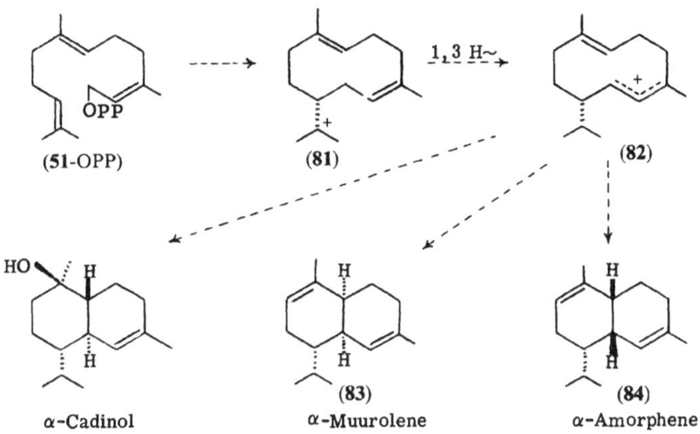

Scheme 9

Alternatively allylic cation (**82**) may be formed from *trans-, trans*-farnesyl pyrophosphate (**52**-OPP) *via* similar cyclization and 1,3-hydride followed by a 2,3-*trans*→*cis* isomerization. This geometrical isomerization migh occur by either proton elimination to the known germacrene D (**85**), rotation, and re-protonation or nucleophilic capture followed by re-ionization (*109*).

Although the timing (*i.e.*, before or after cyclization) and mechanism of the *trans*→*cis* isomerization remain to be established, the results of recent biosynthetic experiments are in line with these biogenetic schemes (*109*, *140*).

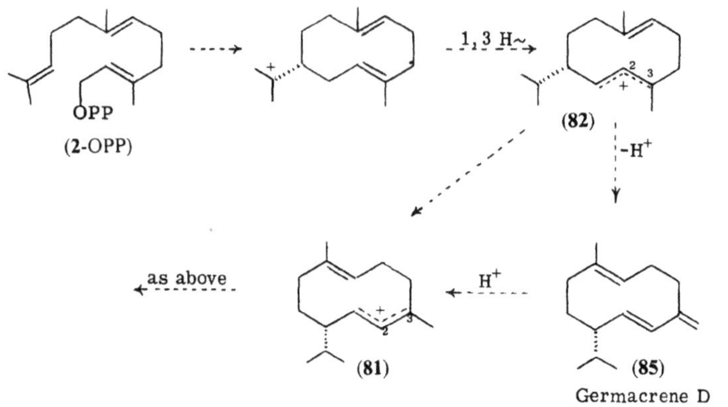

Scheme 10

Cyclizations of cyclodecadienyl carbonium ions have been observed by Hirose and co-workers in the acid-catalyzed reactions of bicyclogermacrene (**70**) and germacrene D (**85**) (*141, 142*). In the former case, protonation presumably occurs on the exocyclic cyclopropane bond adjacent to the double bond affording an allylic carbonium ion resembling (**82**). This W-shaped allylic ion then apparently isomerizes to the "sickle" ion (**81**) (see above) prior to Markovnikov cyclization to δ-cadinene (**86**).

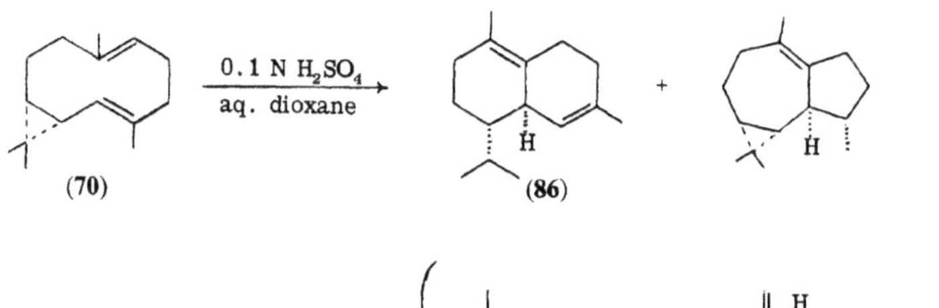

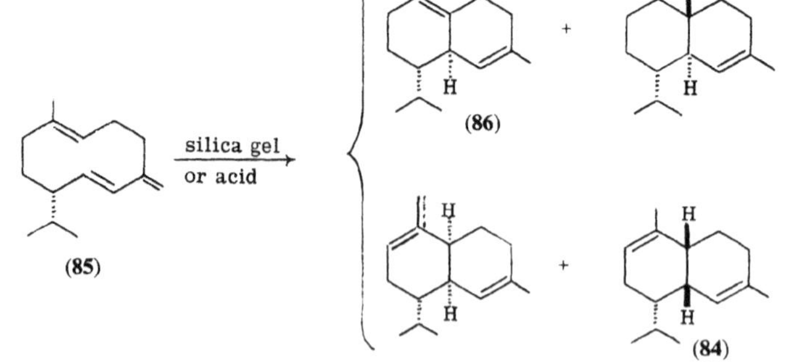

It is interesting that cyclization of germacrene D (**85**) affords cadinenes (γ and δ), muurolenes (α and γ), and amorphene (α) (*142*). Although this might be taken to indicate a non-stereoselective cyclization, the possibility that the products may have equilibrated *via* double bond migrations was not specifically excluded.

The facility of the solvolytic cyclizations of cyclodecadienyl *p*-nitrobenzoates (**87**, **88**, and **89**) is of interest even though the position of the methyl group on the transannular double bond directs the cyclizations to *trans*-fused perhydroazulenic products (*143, 144*). The rate of acetolysis of the 2,6-dimethyl compound (**89**) was estimated to be 10^5-10^6 times faster than *trans*-3-penten-2-yl *p*-nitrobenzoate, indicating a substantial degree of anchimeric assistance by the transannular double bond (*144*).

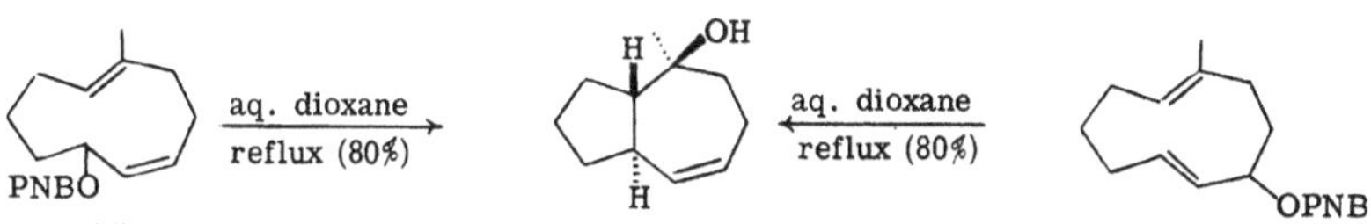

(**87**) (**88**)

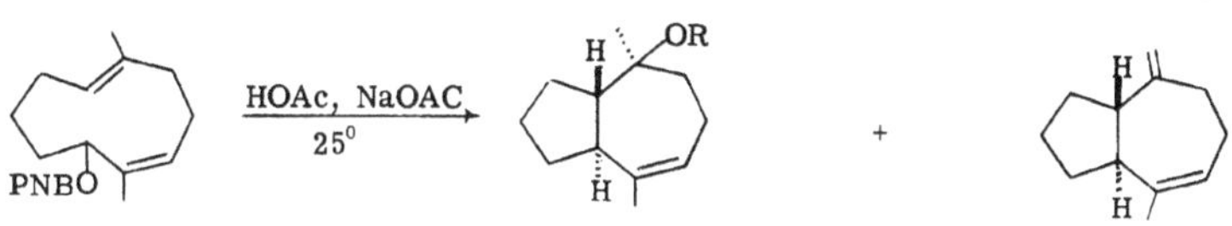

(**89**) (54%) (46%)

The homoallylic alcohol (**90**) is cyclized to ε-cadinene (**91**) by reaction with methanesulfonyl chloride in pyridine (*145*). An intramolecular Prins reaction of the corresponding ketone (preisocalamendiol, **92**) is effected by treatment with aqueous acetic acid, giving isocalamendiol (**93**) (*146*).

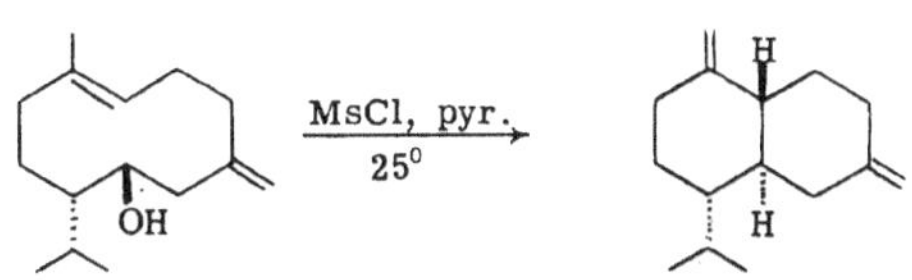

(**90**) (**91**) (35%)

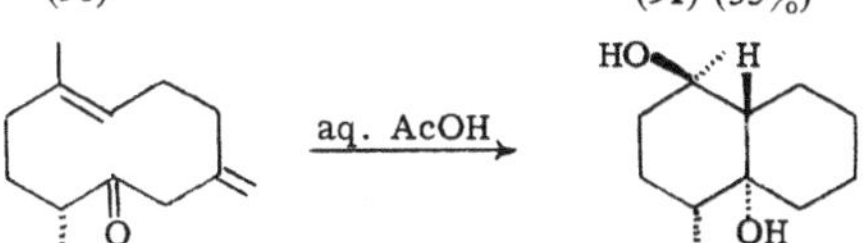

(**92**) (**93**) (47%)

The sesquiterpene α-cubebene (**94**) will be recognized as an isoprenylogue of the monoterpene α-thujene. This tricyclic structure may be biogenetically derived from the carbonium ion precursor to γ-muurolene (see scheme 9) *via* 1,2-hydrogen rearrangement and homoallyl-to-cyclopropylcarbinyl cyclization. Treatment of α-cubebene with aqueous hydrochloric acid effects the reverse of the biogenesis with formation of a mixture of the cadinene isomers and the naturally occurring alcohols cubenol (**95**) and epi-cubenol (**96**) (*137*).

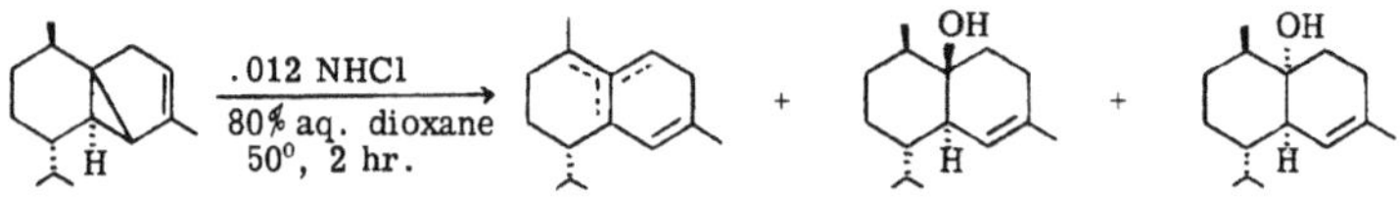

(**94**) (**95**) (**96**)

3. Humulanes, Caryophyllanes, and Himachalanes

If the initial cyclization of *trans, trans-* or *cis, trans*-farnesyl pyrophosphates (**52**-OPP and **51**-OPP) occurs upon the terminal (10, 11) double in an anti-Markovnikov orientation, eleven-membered cyclic dienyl carbonium ions are produced. Deprotonation of the *trans, trans*-intermediate affords humulene (**97**) while bis-homoallyl→cyclobutyl carbinyl cyclization furnishes, after elimination, caryophyllene (**99**). Although the all *trans*-configuration of the three double bonds in humulene would appear to require *trans, trans*-farnesyl pyrophosphate as precursor, the possibility exists that the immediate predecessor of caryophyllene is *cis, trans*-farnesyl (or cisoid nerolidyl) pyrophosphate.

The bicyclic himachalols (**101** and **103**) are derived by cyclization of cisoid allylic carbonium ion (**100**) in a manner analogous to the biogenesis of the cadinenes presented in the previous section. Intermediate (**100**) presumably arises by 1,3-hydrogen shift either directly from the initially formed *cis, trans*-undecadienyl cation, or alternatively from the *trans, trans*-cation with subsequent *trans*→*cis*-isomerization as shown in the scheme. The proposed 1,3-hydrogen rearrangement has recently been experimentally verified in the biosynthesis of longifolene which is formed biogenetically by further cyclization of intermediate (**102**) (see section C.9) (*109*). The stereochemistry at C_6 and C_7 (*cis* CH_3 and H) in the himachalols is unusual in requiring either an overall *syn* addition in the cyclization-hydration processes (**100**→**101**-OH; **100**→**103**) or an *anti* addition from the 6,7-*cis* isomer of ion **100**.

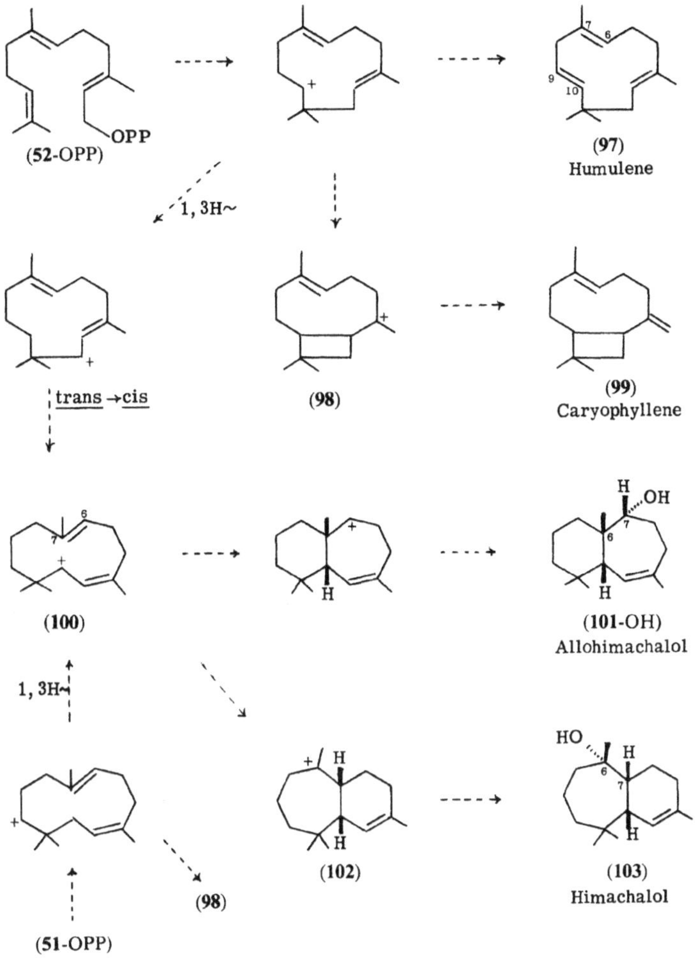

Scheme 11

Humulene reacts with electrophiles initially at the trisubstituted 6,7 double bond in a Markovnikov fashion. With N-bromosuccinimide in aqueous acetone, the attack of bromonium ion triggers a double cyclization to tricyclic bromohydrin (**104**) (*147*). The formation of the *trans*-fused cyclobutane ring by electrophilic cyclization onto the 2,3-double bond is similar to the second ring closure in the caryophyllene biogenesis. By 'dehydration and reductive cleavage of the three-membered ring, it was possible to convert (**104**) into caryophyllene.

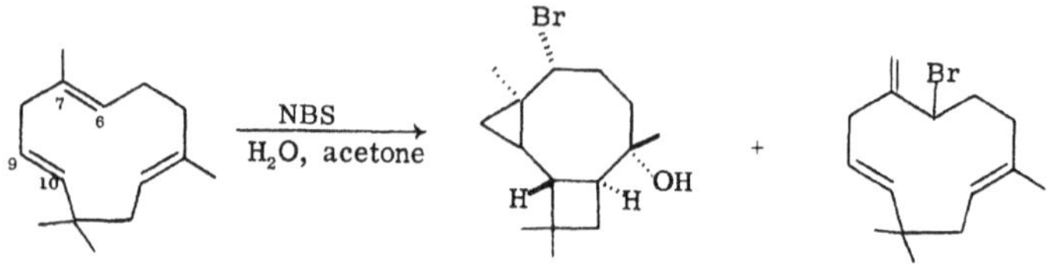

(97) (104) (~25%) (~25%)

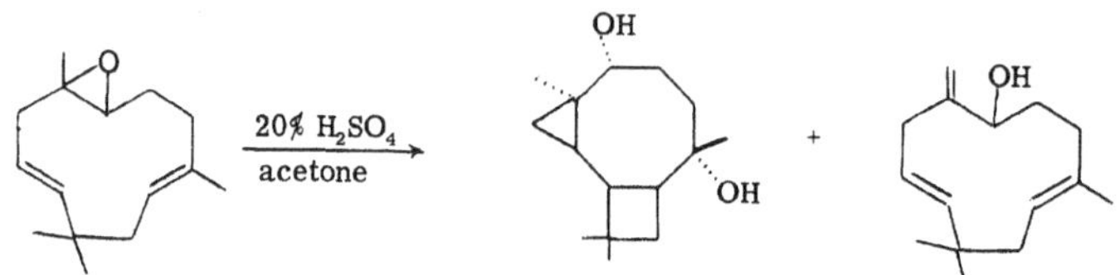

(105) (106) (28%) (107) (15%)

In a very similar manner 6,7-epoxyhumulene (**105**) is converted to tricyclohumuladiol (**106**) and humulenol (**107**), both of which have been isolated from natural sources (*148*, *149*). However, since the starting epoxide (**105**) itself is a known natural product, the close parallelism here raises the possibility that (**106**) and (**107**) may be artefacts of the processing and/or isolation conditions.

Protonation of humulene at position 10 followed by various modes of cyclization and rearrangement has been advanced as a likely biogenetic pathway to a number of unusual tricyclic sesquiterpenes isolated from fungi (hirsuitic acid, marasmic acid, the illudins, illudol, and others) (*150*). In contrast, the initial reaction of humulene with aqueous acid in solution occurs again at the 6,7-double bond (*150*—*154*). The ensuing cyclizations and rearrangements, although of great intrinsic interest, are unrelated to the biogenesis of any known terpenes, and consequently will not be reviewed here (*150*—*154*). Similarly carophyllene (**99**) responds to treatment with various acidic reagents with the production of a remarkable array of tricyclic isomers, transformations which, however, bear no direct relation to terpene biogenesis (*155*, *156*).

Bisarya and Dev have examined the hydrolysis of allohimachalyl tosylate (**101**-OTs) a reaction of obvious biogenetic significance (*157*). The stereochemistry of the nucleophilic capture of the intermediate carbonium ion in this solvolysis is apparently exactly the same as in the biosynthesis since the only alcohol products are himachalol (**101**-OH) and allohimachalol (**103**). The naturally occurring sesquiterpenes α- and β-himachalene (**108** and **109**) are also formed.

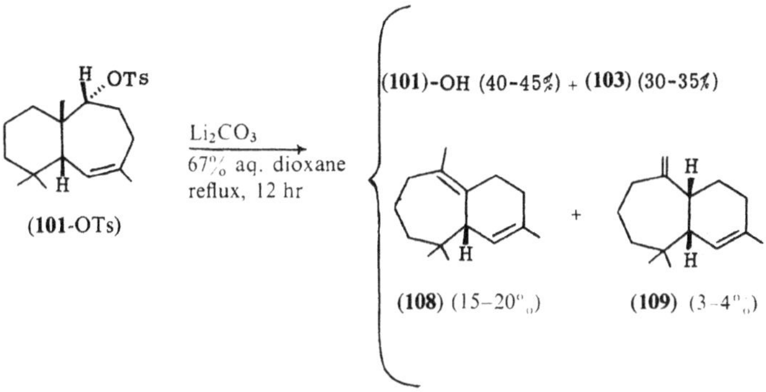

4. 1,n-Hydrogen Eliminations and 1,n-Hydrogen Rearrangements

1,n-hydrogen elimination and 1,n-hydrogen rearrangement reactions are closely related processes which occur in the biogenesis of all classes of terpenes. These two types of reactions will be considered in this section as a whole in order that the mechanistic similarities may be emphasized.

The 1,2-hydrogen elimination reaction (β-elimination) is, as we have seen, a ubiquitous termination step in terpene biogenetic schemes. There is accordingly no need to cite specific examples nor to present precedent from carbonium ion chemistry.

The 1,2-hydrogen shift rearrangement occurs relatively infrequently in the biogenesis of the mono- and sesquiterpenes. One example is the conversion of the 8-terpinyl cation to the 4-terpinyl isomer by 4→8 hydrogen shift (see section B.2 above). In the biogenesis of di- and triterpenes 1,2-hydrogen shifts occur commonly in the many types of backbone rearrangement sequences (see sections D and E below). A possible explanation for the infrequency of 1,2-hydrogen shifts in mono- and sesquiterpene biogenesis is the fact that most of the intermediates are tertiary carbonium ions flanked by methyl, methylene, or bridgehead-methine groups. In this structural environment, a 1,2-hydrogen shift would produce unstable primary, secondary, or bridgehead carbonium ions respectively. Although intermediate secondary carbonium ions are proposed in some biogenetic schemes, they may well be short-lived owing to either relatively rapid nucleophilic capture or further rearrangement to more stable carbonium ions.

Since intramolecular 1,2-hydrogen rearrangements occur commonly in reactions involving carbonium ion intermediates (*157a*), there is no need to present extensive documentation of this process. It is significant, however, that two consecutive 1,2-hydrogen shifts take place in the formation of 2-methylcyclohexanone (**111,** 11%) from deamination of *trans*-2-amino-methylcyclohexanol (**110**) despite the relatively nucleophilic aqueous medium used (*158*). Two 1,2-hydrogen rearrangements occur in the familiar backbone rearrangement leading to the tetracyclic triterpenes. Alternative mechanisms for the formation of **111** (1,3-hydrogen shift and 1,2-hydrogen shift followed by 1,2-hydrogen elimination to the enol) were excluded by deuterium labelling experiments. The yield of 2-methylcyclohexanone drops to 2% in the deamination of the *cis*-isomer of **110,** presumably owing to poor alignment of the now equatorial C_1 proton in the predominant conformer.

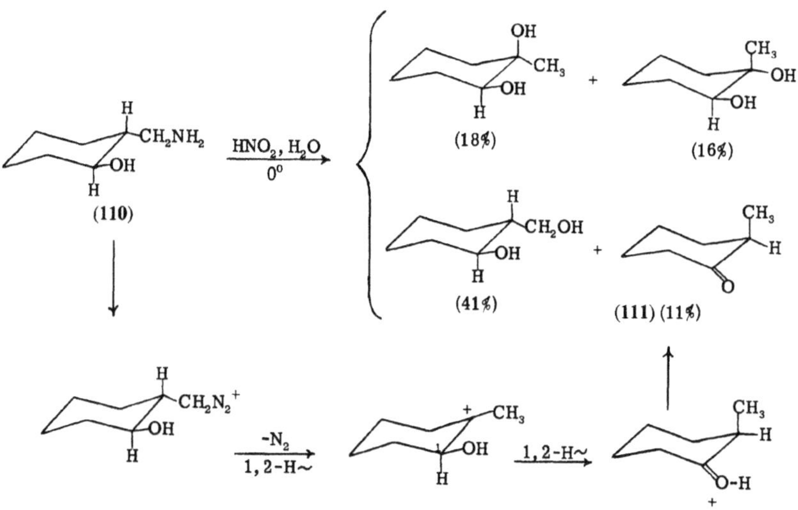

The occurence of 1,2-hydrogen rearrangements is revealed in the nmr spectra of long-lived secondary and tertiary alkyl carbonium ions which have been generated at low temperatures in highly acidic, non-nucleophilic media such as SbF_5–SO_2ClF (*159—166*). Since the 1,2-hydrogen rearrangements were rapid on the nmr time scale at low temperature, it was only possible to place an upper limit on the activation energy for the degenerate rearrangements of *sec*-butyl cation (*163, 164*) and the 2,3-dimethyl-2-butyl cation (*159—162*). Although a static, proton-bridged structure for these carbonium ions was excluded on the basis of the spectral data (*159—162, 164*), this species must, of course, be a transition state in the 1,2-hydrogen rearrangement and is only slightly higher in energy than the classical, localized cations.

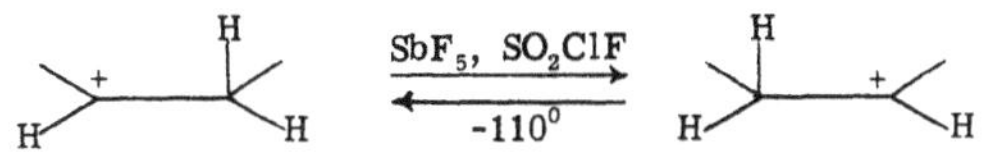

$E_a \leq 6$ kcal/mol

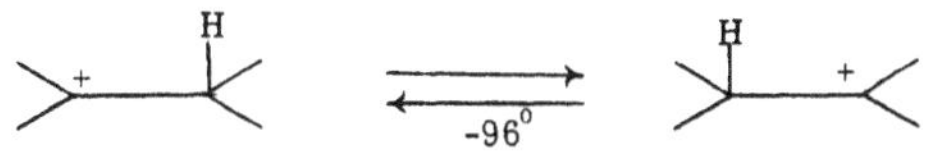

$E_a \leq 6.6$ kcal/mol

The 3,2-hydrogen shift rearrangement in the norbornyl carbonium ion has a substantially higher energy of activation (*165*) which is attributable to the fact that the C_3–H bond cannot become oriented perpendicular to the plane of the carbonium ion. This observation provides evidence for the usual presumption that the most favorable configuration for 1,2-hydrogen rearrangement (and 1,2-hydrogen elimination) has the relevant C–H bond aligned coplanar (dihedral angle = 0°) with the axis of the *p*-orbital of the adjacent carbonium ion (*167, 168*).

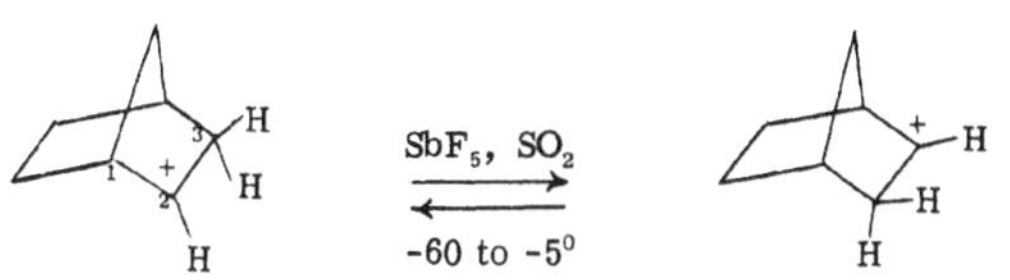

$E_a = 11$ kcal/mol

The cyclopropane ring is a surprisingly common structural component in terpenes of all types. The formation of the three-membered ring is most simply accounted for in biogenetic schemes by either of two processes: homoallyl to cyclopropylcarbinyl carbonium ion cyclizations (*e.g.*, biogenesis of thujene in section B.2) or formal 1,3-hydrogen eliminations. A transformation closely related to the latter is the 1,3-hydrogen rearrangement which has been recently been established in the biosynthesis of certain sesquiterpenes.

The 1,3-hydrogen eliminations may be sub-divided into three classes. First there is the combination of formal α-elimination of pyrophosphoric acid from the $-CH_2OPP$ group of a prenyl pyrophosphate and [2+1]-cycloaddition to form the cyclopropane ring*. This process may be

* Although this mechanistic formalism is a convenient mnemonic device, a pathway involving initial formation of a carbonium ion followed by 1,3-hydrogen elimination is to be regarded as more appropriate for terpene biogenesis. A carbenoid mechanism has, however, been suggested as an alternative (*169*).

intermolecular as in the coupling of two molecules of dimethylallyl, farnesyl, and geranylgeranyl pyrophosphates (**112a**, **112b**, and **112c** respectively) to chrysanthemyl (see section B.1), presqualene (*170, 171*), and prephytoene (*172*) pyrophosphates (**113a**, **113b**, and **113c** respectively). Intramolecular examples of this type of cyclization are found in the biogenesis of 4-carene (**114**), sesquicarene (**115**), bicyclogermacrene (**70**, section C.2), and the diterpene casbene from the appropriate acyclic precursors.

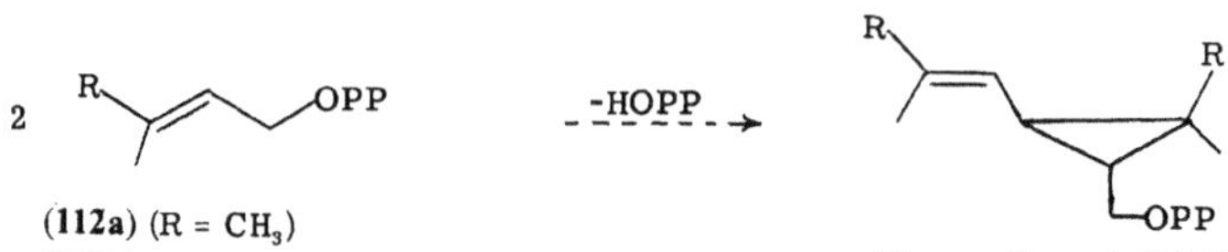

(**112a**) (R = CH_3)
(**112b**) (R = homogeranyl)
(**112c**) (R = homofarnesyl)

Chrysanthemyl-OPP (**113a**)
Presqualene-OPP (**113b**)
Prephytoene-OPP (**113c**)

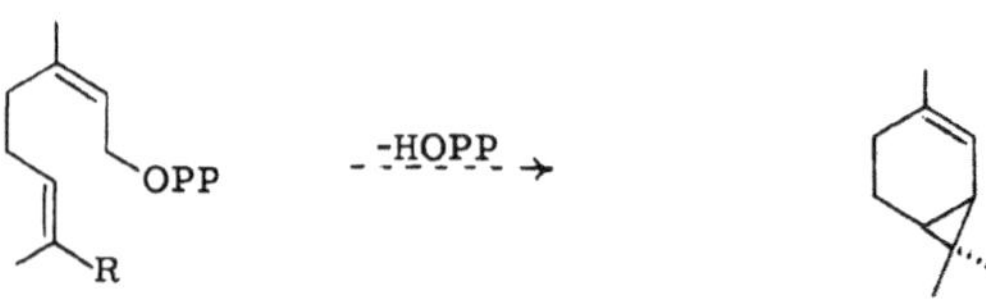

4-Carene (**114** R = CH_3)
Sesquicarene (**115** R = homoprenyl)

Another recurring type of 1,3-hydrogen elimination is the transannular deprotonation of bridged bicyclic ring systems. These include the familiar H_6 elimination from the bornyl and campherenyl carbonium ions to give tricyclene (**116**) and α-santalene (**117**). The same type of 1,3-hydrogen elimination occurs in the biogenesis of the tetracyclic sesquiterpenes ishwarane (**118**) (*173*), cycloseychellene (**187**, section C.7), cyclosativene (**194**, section C.8) and longicyclene (**209**, section C.9) and the pentacyclic diterpene trachylobane (**273**, section D.3).

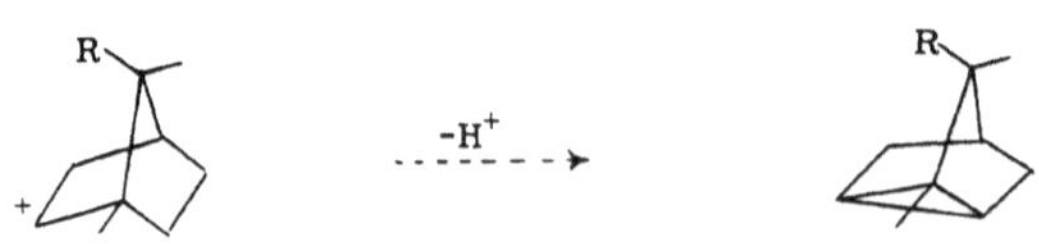

Tricyclene (**116**, R = CH_3)
α-Santalene (**117**, R = homoprenyl)

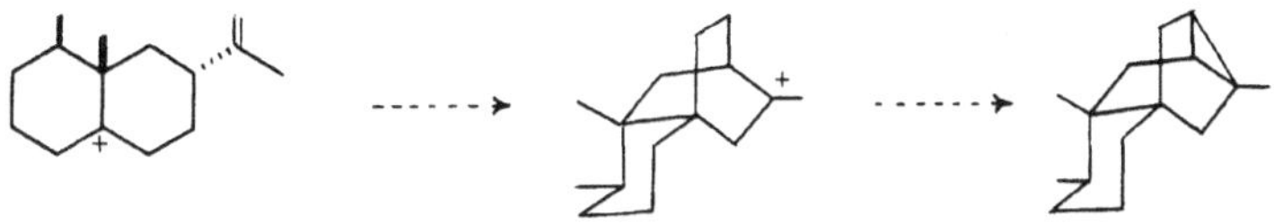

(118)
Ishwarane

The third type of biogenetic three-membered ring formation is a 1,3-proton elimination from a methyl group adjacent to a carbonium ion site. The resulting cyclopropane may be viewed as if a migrating methyl group had been intercepted by deprotonation before reaching its destination. Representatives of this type include the sesquiterpenes debromolaurinterol (*174*) and cycloeudesmol (*175*), the diterpene devadrene, and the triterpenes cycloartenol (**310,** section E.2) and phyllanthol (section E.5).

Cycloeudesmol

Devadrene

1,3-Hydrogen rearrangements, similar to the 1,3-hydrogen eliminations, are proposed in the biogenesis of a number of terpenes. A biogenetic 6,2-hydrogen shift in the bornyl carbonium ion (**119,** R = CH_3) generates its enantiomer and could therefore be one explanation for the low optical purity of many of the bicyclic monoterpenes. Similarly the biogenesis of epi-β-santalene may involve a 6,2-hydrogen shift in the campherenyl carbonium ion (**119,** R = homoprenyl). A related transannular rearrangement may be involved in the biogenesis of the diterpene atiserene (**274,** section D.3).

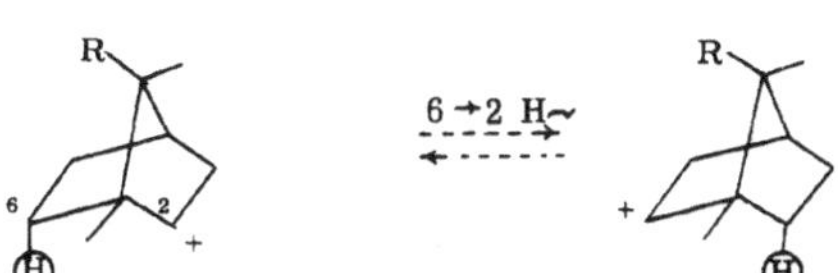

(119)

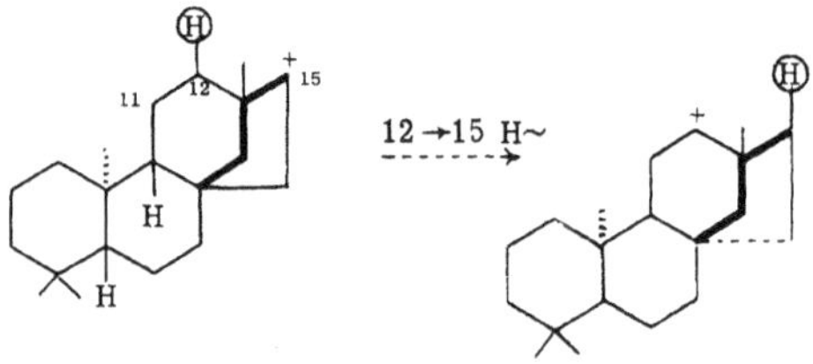

In section C.2 we have seen that a 1,3-hydrogen rearrangement transforms the initially formed cyclodecadienyl ion **81** to allylic carbonium ion **82** which subsequently cyclizes to the decalin ring system of the cadinane group of sesquiterpenes. The tricyclic sesquiterpenes copaborneol, sativene, and various related natural products are derived by a further cyclization (section C.8). A similar 1,3-hydrogen rearrangement takes place in the biogenesis of himachalol and longifolene.

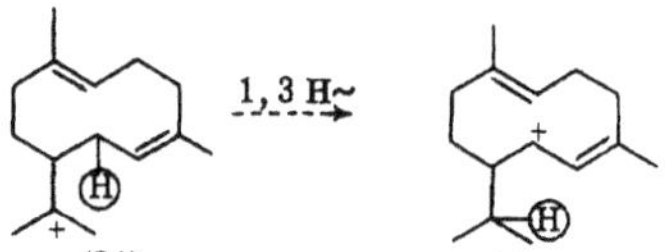

(**81**) (**82**)

Experimental verification for the proposed 1,3-hydrogen rearrangements above has recently been obtained in the biosynthesis of avocettin (cadinene-based sesquiterpene) (*109*), sativene (*109*), longifolene (*109*), dendrobine (derived from copaborneol) (*20, 140*), tutin (derived from copaborneol) (*176, 177*), and culmorin (derived from longiborneol) (*178*).

Although it might be expected that 1,3-hydrogen elimination from carbonium ions to high energy cyclopropanes would be a relatively unfavorable process, numerous examples have been reported in the literature (*98, 157a, 179—190*). For example, deamination of *n*-propyl amine affords cyclopropane in 4—5% yield (*181*). A study of the effect of solvent upon the products resulting from deamination of isobutyl amine indicates that cyclopropane formation is enhanced in non-polar solvents (*182*). 1,3-hydrogen elimination to cyclopropanes has also been found in the solvolysis (*183*) and pyrolysis (*184*) of highly hindered *p*-nitrobenzoates.

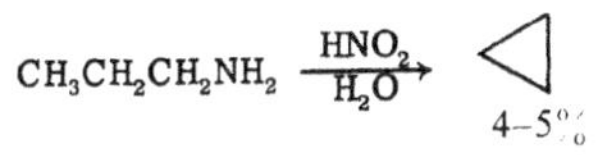

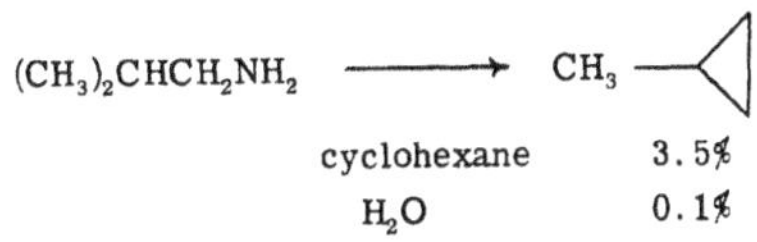

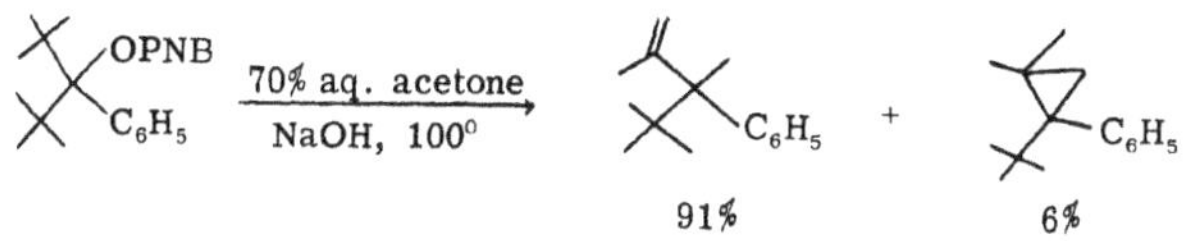

Similarly solvolysis of bornyl derivatives and deamination of bornyl amine produce tricyclene is low yield (*98, 185, 186*). As in the deamination of isobutylamine, the yield of 1,3-hydrogen elimination increases when the carbonium ion is generated in a non-polar solvent (diglyme), in this case from the tosylhydrazone (*187*). That the formation of tricyclene

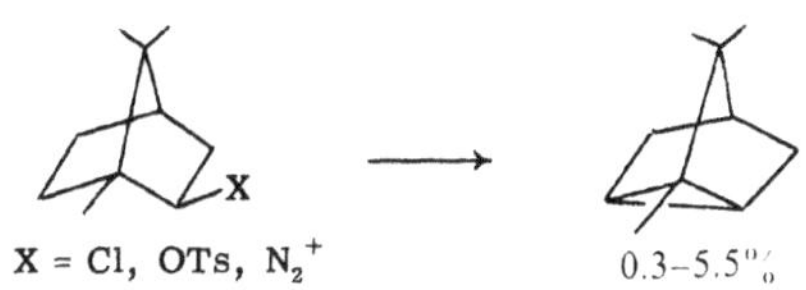

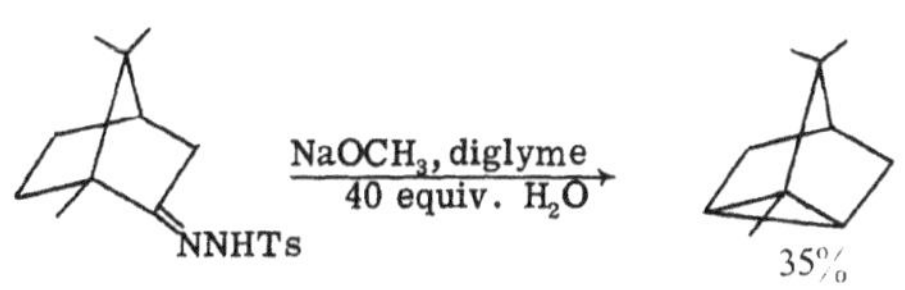

involves mainly a carbonium ion pathway was suggested by deuterium incorporation when the reaction was performed in the presence of deuterium oxide. The increased efficiency of 1,3-hydrogen elimination relative to the acyclic examples above may be attributed to a favorable orientation of the C_6 proton and the fact that the resulting tricyclic ring system is only slightly less stable the strained bicyclo [2.2.1] heptane nucleus of bornane. Other examples of 1,3-hydrogen eliminations are found in the deamination of 2-aminocyclohexanone and its 6,6-dimethyl derivative (**120**) (*188*). The relatively high yield in this case may be a consequence of the increased reactivity of the α-ketodiazonium ion intermediate.

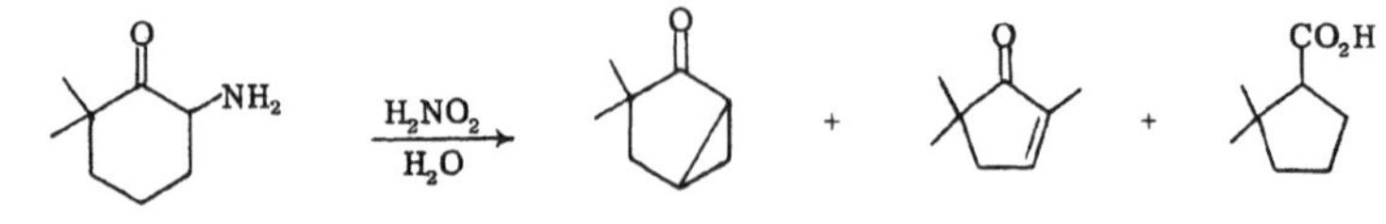

(120) (46%) (23%) (15%)

A remarkably high yield of cyclopropane ring formation results from the deamination of *trans*-9-decalylcarbinyl amine (*180, 189*) and the related *trans*-8-hydrindanylcarbinyl amine (*190*). The further observation that *cis*-9-decalylcarbinylamine, upon deamination, gives no cyclopropane product reveals that 1,3-elimination requires the C–H bond to be aligned co-planar with the three carbons which will eventually comprise the three-membered ring.

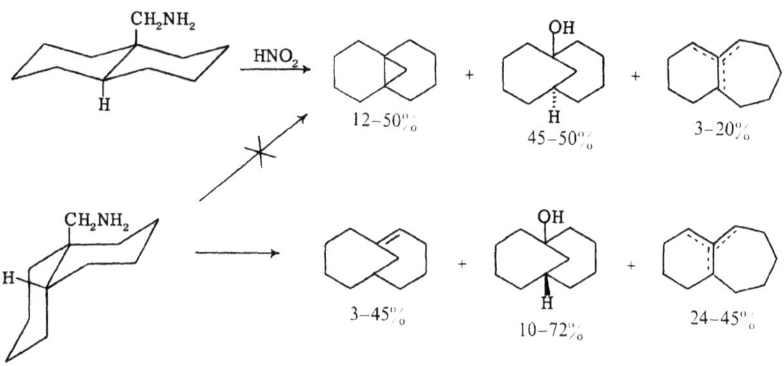

It has been pointed out that there are two distinct stereochemical orientations (referred to here as *syn* and *anti*) of the C–H bond undergoing a 1,3-hydrogen elimination from a carbonium ion (*191*). However, since 1,2-alkyl rearrangement is likely to be more rapid than 1,3-hydrogen elimination, distinguishing between these two modes of elimination becomes difficult. Nevertheless, there does appear to be a stereochemical requirement of co-planarity, *i.e.*, 1,3-elimination is evidently unfavorable from a skew conformation.

Another subtlety in the mechanism of 1,3-hydrogen elimination is the possibility that the proton transfer occurs from a protonated cyclopropane intermediate (*i.e.*, bridged carbonium ion). If 1,3-hydrogen elimination does take place from a protonated cyclopropane, the hydrogen which is eliminated should, for stereoelectronic reasons, lie in the plane of the three-membered ring in the transition state.

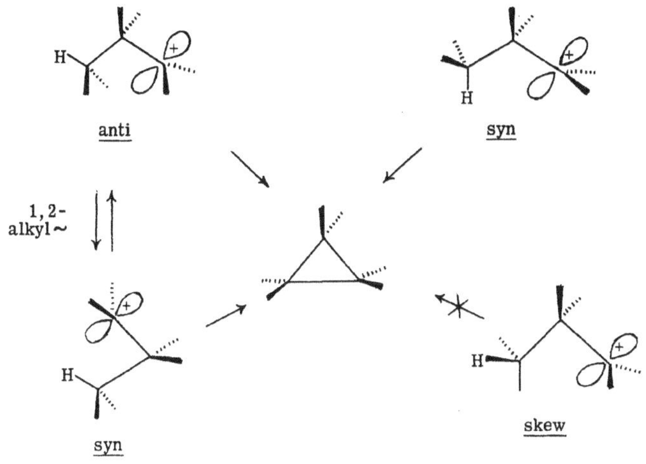

It is appropriate at this point to consider briefly the evidence bearing on the question of protonated cyclopropanes in carbonium ion reactions (*157, 163, 179—181, 192—195*). Deamination of labelled *n*-propyl amine leads to *n*-propyl alcohol 2—3% of which has undergone complete scrambling of the carbon atoms and almost complete scrambling of the hydrogens. The cyclopropane produced in the deamination is also found to have suffered various degrees of hydrogen scrambling prior to the elimination step. SAUNDERS and co-workers have examined the hydrogen exchange processes in several aliphatic and alicyclic cations under stable ion conditions (*163*). In no case was a detectable quantity of a protonated cyclopropane directly observed. From the rates of hydrogen exchange, they have estimated that the protonated cyclopropanes are less stable than the directly observed classical carbonium ions by 2 to 6—7.5 kcal/mol. It is further argued that since a corner-protonated cyclopropane is capable of decomposing in four distinct ways (two modes opening to acyclic carbonium ions and two modes of 1,3-hydrogen rearrangement), this species should be considered to be an intermediate rather than a

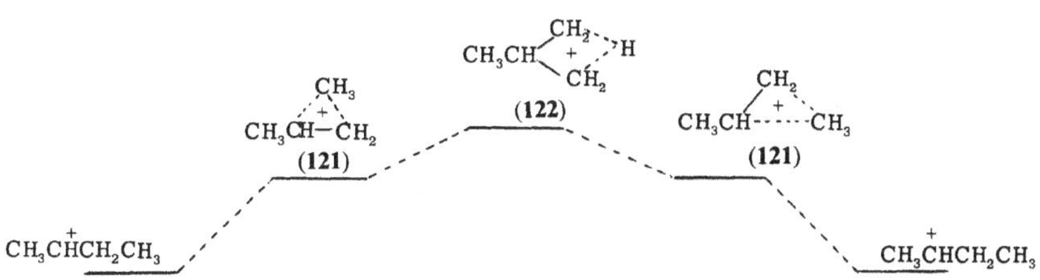

transition state. Hydrogen exchange is then proposed to occur *via* corner-to-corner hydrogen shift through an edge-protonated transition state. The relative energetics for these rearrangements are shown below for the sec-butyl cation.

This diagram illustrates that there are two distinct energy barriers associated with protonated cyclopropanes. The first is the barrier to initial formation of the corner-protonated species **121** (*i.e.*, essentially equivalent to a non-classical bridged ion) and the second is the somewhat higher barrier (edge-protonated transition state **122**) which must be surmounted for hydrogen exchange to occur.

The evidence cited above makes a strong case for the existence of protonated cyclopropane intermediates in carbonium ion reactions but reveals little concerning the structure of such a species. A recent publication by DE PUY, ANDRIST, and FÜNFSCHILLING reports upon the stereochemistry of the electrophilic ring opening of *cis* and *trans* 1,2,3-trimethylcyclopropane with deuterated acids (*196, 197*). While the attack of the entering nucleophile occurred with essentially complete inversion, the overall stereochemistry of protonation took place with both retention (68%) and inversion (32%). The relative amount of retention and inversion was unaffected by changes in both the acid and solvent. Since a symmetrical edge-protonated cyclopropane intermediate offers no ready explanation for the consistent blend of retention and inversion, the authors proposed a trigonal bipyramidal structure (**123**) for the protonated cyclopropane which reacts with the nucleophile at the non-equivalent apical and equatorial positions with slightly different rate constants.

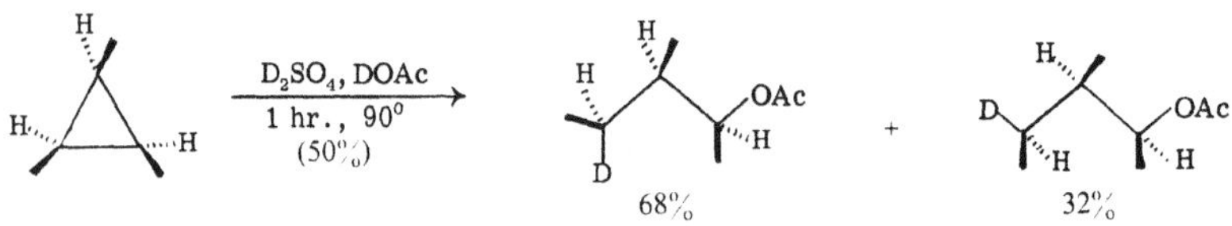

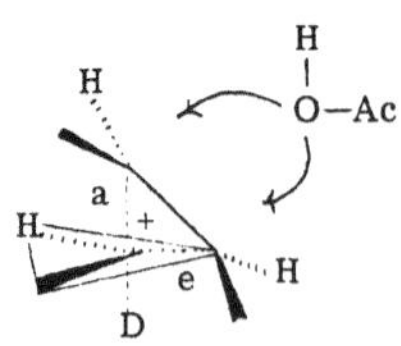

(**123**)

The observation that the *trans* isomer affords a different mixture of products excluded a freely rotating, equilibrating corner-protonated cyclopropane structure. It is interesting to note that in the proposed trigonal bipyramid structure **123** the deuterium atom lies in the plane of the three-membered ring, a position appropriate for either 1,3-elimination or 1,3-hydrogen rearrangement (corner-to-corner shift).

The 1,3-hydrogen rearrangement seems to have been encountered less frequently than the 1,3-elimination in carbonium ion reactions. The small amount of scrambling detected in the deamination of *n*-propyl amine (see above) is in effect a 1,3-hydrogen rearrangement. However, the near equivalence of all of the hydrogens and carbons requires an equilibrating protonated-cyclopropane intermediate rather than simple 1,3-hydrogen shift. Aside from the propyl system, the occurrence of 1,3-hydride shifts in other aliphatic systems seems to be rare. However, in the rigid norbornyl structure the 6,2-hydride shift occurs very readily. For example, acetolysis of either *exo*-camphenilyl brosylate (**124**) or apoisobornyl brosylate (**125**) affords β-fenchoisocamphoryl acetate (**126**) as the major product (*198*). The further observation that the extent of 6→2-hydrogen rearrangement decreases in more nucleophilic solvents (18 to 18.5% in 72.4% aqueous dioxane at 25°) was taken to indicate that the hydrogen rearrangement lags behind the carbon bridging.

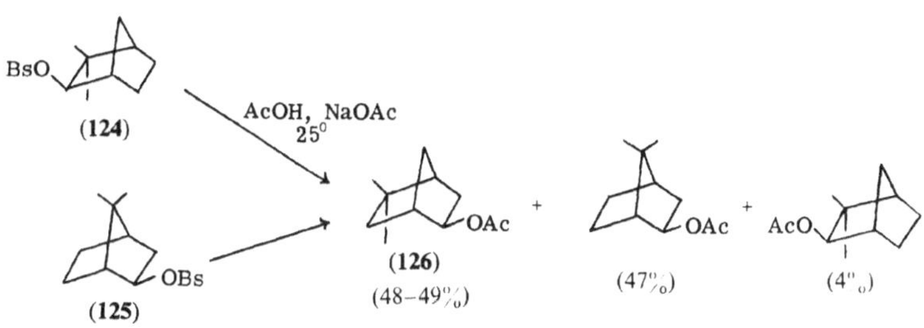

In connection with the sesquiterpene biogenesis involving cyclization of farnesyl pyrophosphate upon the terminal double bond followed by (or concerted with?) 1,3-hydrogen rearrangement (see Schemes 10 and 11), it is interesting that 1,3-hydrogen shifts appear to be favored (relative to other pathways) when carbonium ions intermediates are generated by the so-called π-route (*199—201*). In the acetolysis and formolysis of 2-Δ^3-cyclopentenyl)-2-^{14}C-*p*-nitrobenzene-sulfonate, to norbornyl esters, Lee and co-workers have reported that a fraction (6—25%) of the product is formed by nucleophilic capture either during or immediately after 6→2-hydrogen shift but without further rearrangement (*199*). This finding was explained in terms of a mechanism involving either solvent capture concerted with the 6,2-hydrogen shift or capture of an edge-protonated cyclopropane intermediate.

The unsaturated brosylate (**127**) reacts with exclusive 1,3-hydrogen shift both in formation of acetate products and internal return to brosylates corresponding to the acetate products (*200*). The production of delta-cyclene (**128**) *via* apparent 1,3-hydrogen elimination is analogous to the biogenetic formation of cyclopropanes discussed above (see also section B.1).

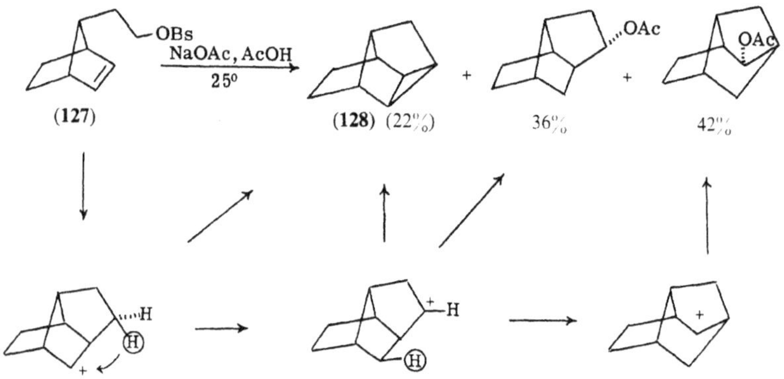

Efficient 1,3-hydrogen rearrangement is also observed in the acetolysis of unsaturated brosylate **129** (*201*). This particular case is significant since acetolysis of the closely related tricyclic brosylate **130**–OBs produces only a small amount of 2-adamantyl acetate (**131**), the product of 1,3-hydrogen rearrangement. It is, therefore, evident that the solvolyses of **129** and **130**–OBs, do not, for the most part, have a common intermediate.

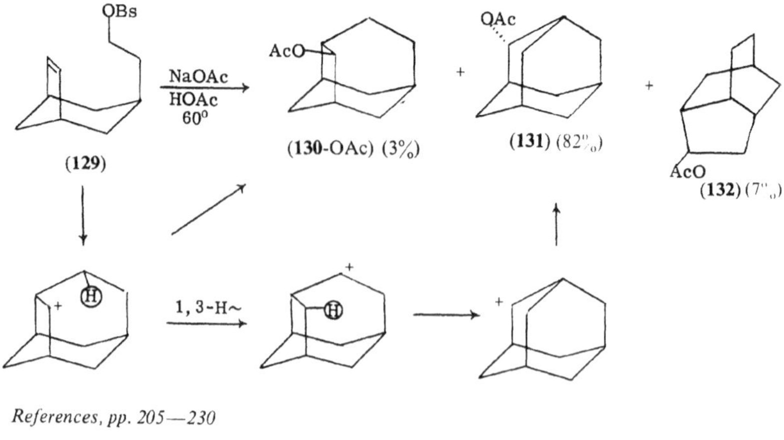

References, pp. 205—230

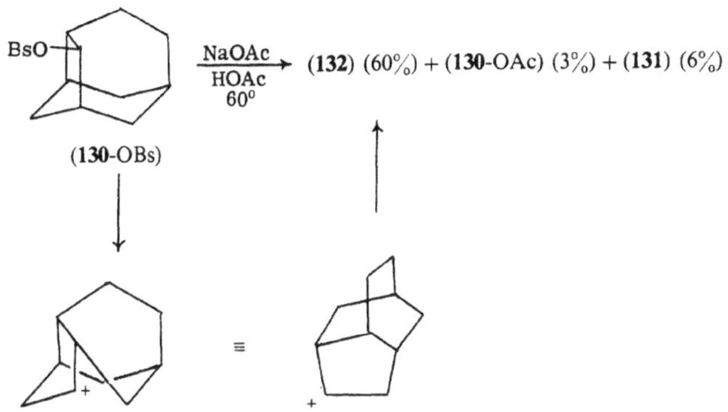

The reasons for the increased tendency for 1,3-hydrogen rearrangement in these olefinic cyclizations are unclear. Possible explanations are (a) facilitation of the rearrangement by interaction of the migrating hydrogen with the sulfonate leaving group within an ion pair, (b) extra electrostatic driving force for hydrogen shift owing to the distance separating the sulfonate from the olefinic carbons bearing most of the positive charge in the initially formed ion pair, or (c) formation of some type of edge-protonated cyclopropane intermediate. The low magnitude of the kinetic α-deuterium isotope effects in the solvolysis of both α,α-dideuterated cyclopentenylethyl *p*-nitrobenzesulfonate and dideuterated **129** ($k_H/k_D = 1.30$) is said to exclude a hydrogen shift in the solvolytic transition state (*199, 201*).

A case of 1,3-hydrogen rearrangement was discovered in the reaction of steroidal epoxide **133** with 60% perchloric acid in chloroform (*202*). The fact that the Δ^8-isomer of the observed product is stable to the reaction conditions demands an intramolecular 14→9 hydrogen transfer prior to elimination.

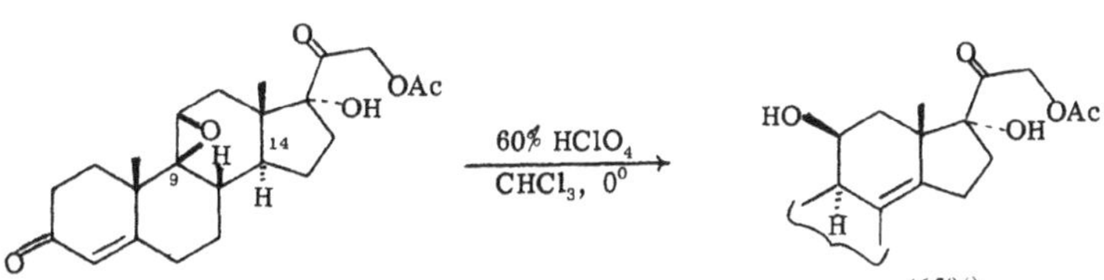

(133) (65%)

1,3-Hydrogen rearrangements have also been detected in the nmr spectra of two long-lived carbonium ions (*203—206*). The degenerative rearrangement of the 2,4-dimethyl pentyl cation occurs without exchange of the hydrogens at position 3, indicating that if a protonated cyclopropane intervenes, it does not undergo rapid corner-to-corner rearrangement (*204—206*). The activation energy for the rapid 6,2-hydrogen rearrangement of the long-lived 2-norbornyl carbonium ion has been determined from its nmr spectrum at $-150°$ (*203*).

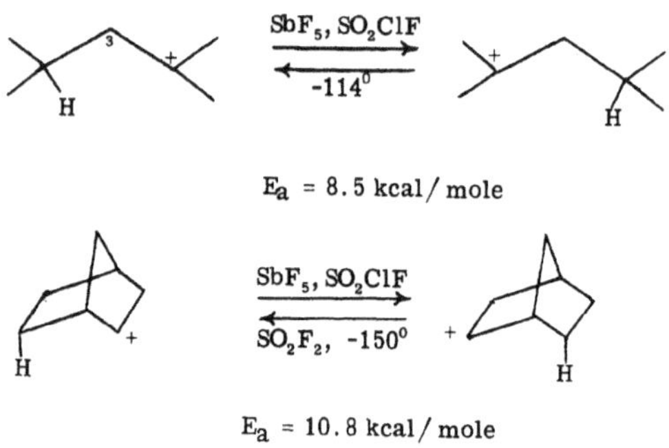

Both 1,4- and 1,5-hydrogen rearrangements have been discovered in the biosynthesis of sesquiterpenes and diterpenes. A 1,4-hydrogen transfer occurs in the formation of the bicarbocyclic nucleus of the trichothecane sesquiterpenes, presumably *via* the parent hydrocarbon trichodiene (*35, 207—209*). The biosynthesis of helicobasidine, a closely related cuparene-type sesquiterpene, also has a 1,4-hydrogen rearrangement step (*210*).

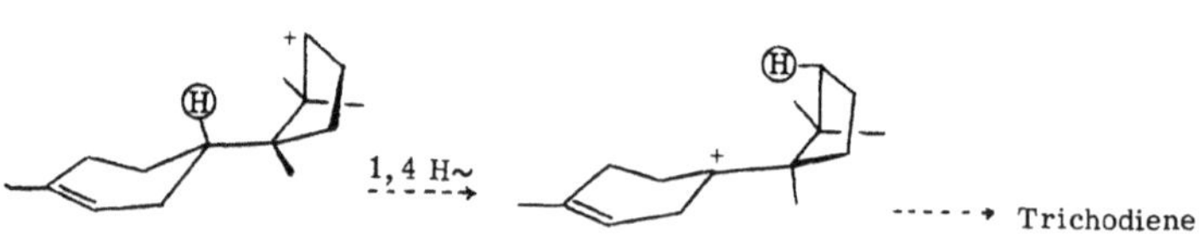

The occurrence of 1,5-hydrogen transfers has been documented in the biosynthesis of the unusual diterpene pleuromutlin (*33*) and the ophiobolin sesterpenes (*211*).

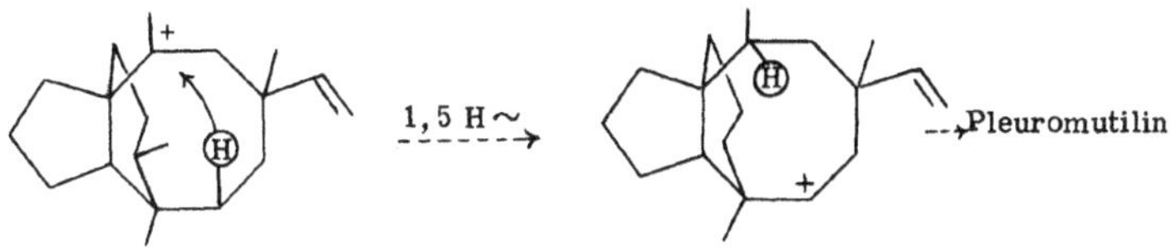

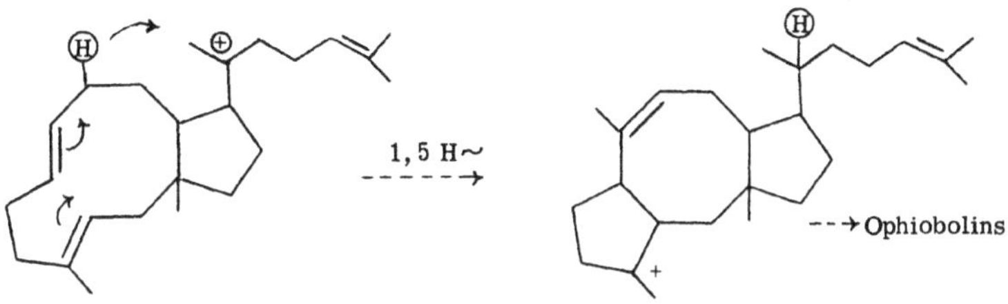

There appear to be few clear-cut examples of 1,4-hydrogen rearrangements in the literature of carbonium ion reactions (*157a, 206, 212, 213*). One case is the phosphoric acid-induced dehydration of dicyclopentadiene derivative **134** to **135** at high dilution (*212*). In contrast, there are numerous reports of 1,5-hydrogen rearrangements in both cyclic and acyclic compounds (*157, 206, 214—222*). In cyclooctane derivatives transannular 1,3- and 1,5-hydrogen rearrangements are frequently the dominant reaction pathway. Thus, the addition of formic acid to *cis*-cyclooctene epoxide affords after saponfication *cis*-1,4-cyclooctane diol as the major product (*214*). Deuterium labelling experiments demonstrated that rearrangement occurs by way of both 1,3- and 1,5-hydrogen transfers.

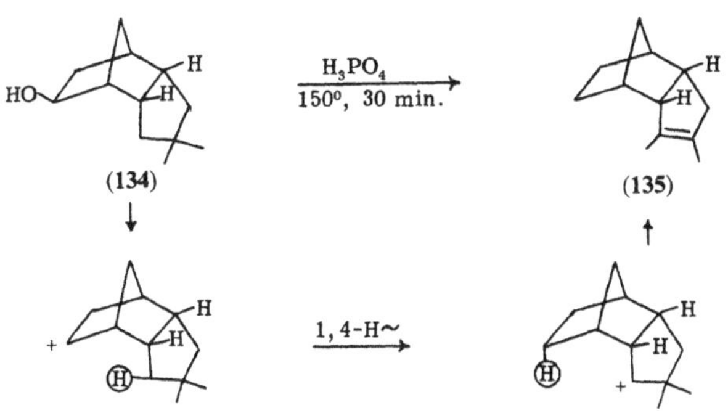

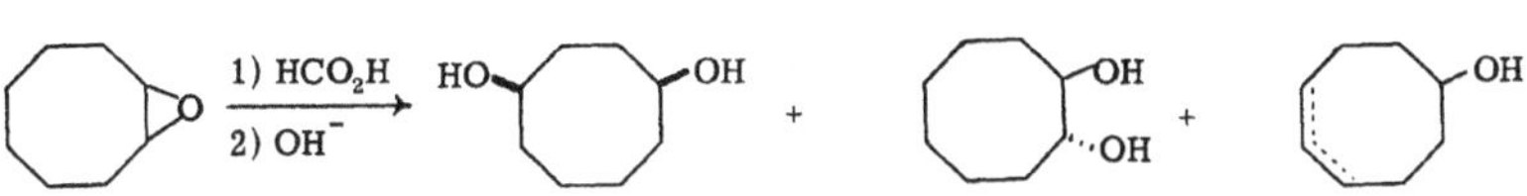

(23–30%) (5–19%) (15%)

Ourisson and co-workers have investigated the remarkably efficient transannular 1,5-hydrogen shift rearrangements of longifolene derivatives and related compounds (*217—219*). The nature of the substituent at C_7 has a profound effect upon both the products of the reaction and the relative rates. With electronegative groups (CN, CO_2CH_3) a ring-contraction occurs. In contrast, the presence of a methyl group diverts the reaction to the 1,5-hydrogen rearrangement pathway.

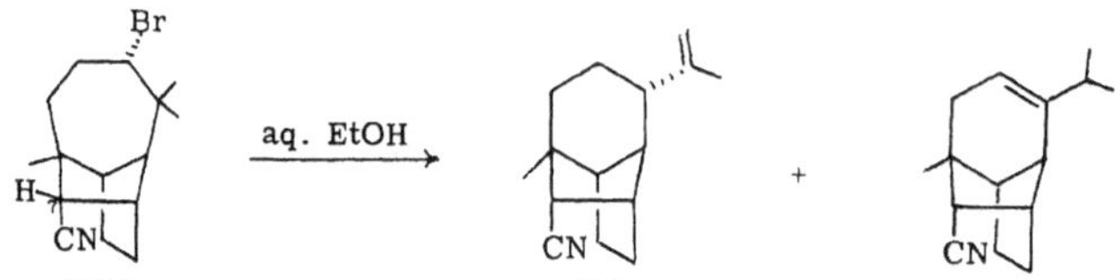

(**136**) 35% 35%

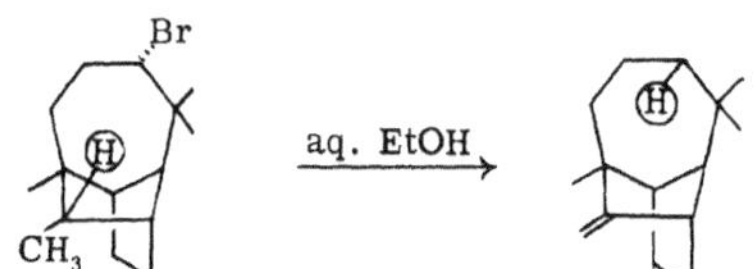

(**137**)

k(**137**)/k(**136**) = 220 (80% aq. EtOH, 60°)

A plot of log k vs. σ^* for a series of nine different substituents gave a curve which could be dissected into two components ($\rho_1 = -0.3$, $\rho_2 = -1.6$). This behavior was interpreted in terms of two competing reactions, the ring contraction and the 1,5-hydrogen shift, the latter being more sensitive to the nature of the substituent. The 220 fold rate enhancement of the solvolysis of **137** over **136** establishes that hydrogen transfer is occurring in the transition state for the reaction.

1,5-Hydrogen rearrangements have also been observed in reactions of acyclic carbonium ions (*157, 204—206, 216, 220—222*). Although most of these involve 1,5-hydride transfer from carbon-bearing-oxygen to form highly stabilized carbonium ions (*e.g.*, **138→139**) (*220*), a few cases of such rearrangements from non-oxygenated carbon have been realized (*216, 221, 222*). Thus, Arigoni and Branca have established, *via* deuterium labelling, that the polyphosphoric acid-induced cyclodehydration of the alcohol **140** proceeds through a 1,5-hydrogen rearrangement (*216, 221*).

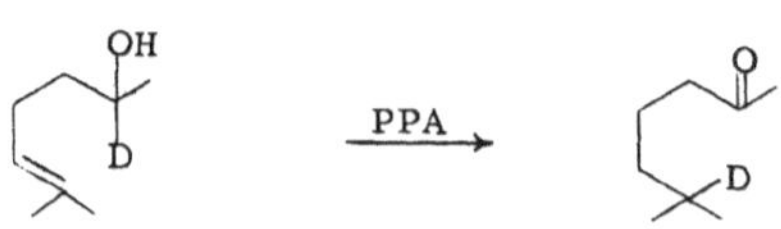

(**138**) (**139**) (47%)

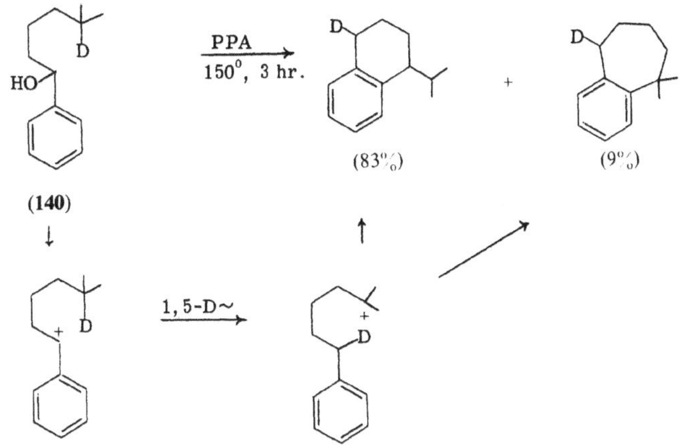

(140)

An interesting example of a 1,5-hydrogen shift combined with a backbone rearrangement is the polyphosphoric acid-induced isomerization of lanosteryl acetate (**141**) to troposteryl acetate (**142**) (*216, 221*). The possibility that this isomerization proceeds *via* a series of 1,2-hydrogen shifts along the side chain was excluded by deuterium labelling at positions 20 (D migrates to C_{25}) and 23 (D_2 unaffected). Since the stereochemistry at C_{20} is inverted in the two hydrogen shift rearrangements, (20→25 and 17→20), the question arises whether the two rearrangements are concerted. That the rearrangement is evidently stepwise, with frequent elimination (including, to some degree, the critical C_{17} proton) enroute to the final product, was revealed from the extent of

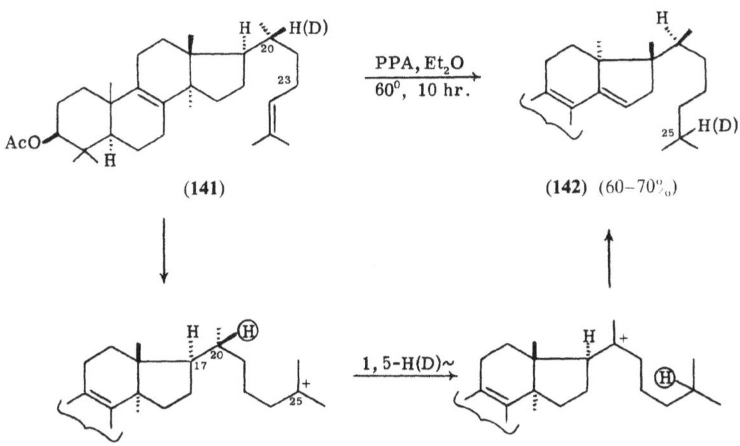

(**141**)

(**142**) (60–70%)

deuterium incorporation in deuterated polyphosphoric acid (major isotopic species were d_{15} and d_{27}; maximally labelled species detected was d_{35}). Thus the stereochemistry at C_{20} established in the rearrangement may be the result of conformational factors and/or stereoselectivity in protonation and deprotonation steps. The generality of these side chain 1,5-hydrogen transfer rearrangements is shown by similar acid-catalyzed isomerizations of euphol acetate and gascardic acid (*216, 221*).

Intramolecular 1,4- and 1,5-hydrogen shifts are revealed in the nmr spectra of 2,5-dimethyl-2-hexyl and 2,6-dimethyl-2-heptyl cations, respectively (*205, 206*). The order of increasing activation energies for the degenerate, non-vicinal hydrogen rearrangements in this series of tertiary cations is 1,5 < 1,3 < 1,4 (E_a for 1,3-hydrogen rearrangement is 8.5 kcal/mole) and appears commensurate with their frequency in carbonium ion reactions.

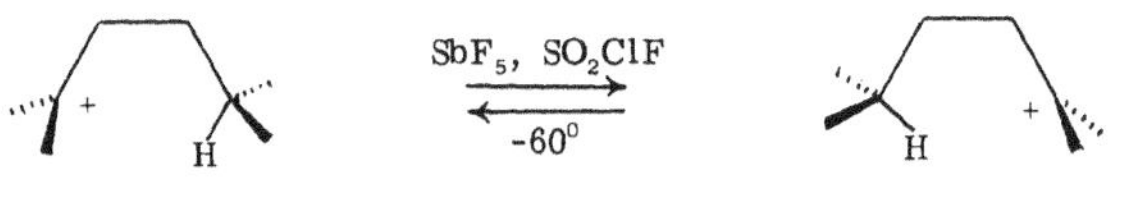

E_a = 12-13 kcal/mol

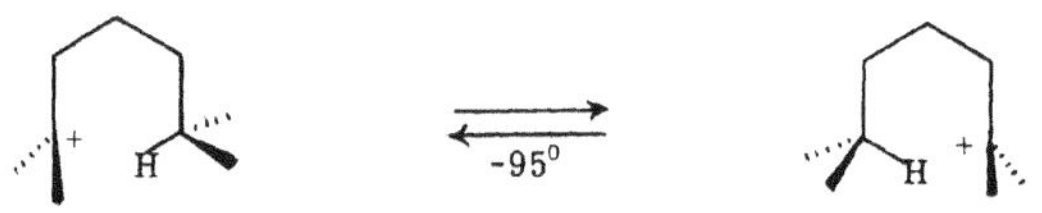

E_a < 6-7 kcal/mol

5. Eremophilanes, Vetispiranes, Valeranes, and Cyperanes

The biogenesis of this group of sesquiterpenes involves rearrangement of a eudesmane precursor by either methyl migration or ring contraction. Thus, the eremophilanes (*e.g.*, eremoligenol) and hinesanes (*e.g.*, hinesol), are considered to arise *via* 1,2-hydrogen shift followed by either angular methyl migration or ring contraction of the six-membered ring bearing the isopropyl substituent (*110, 223*) (Scheme 12). The possibility of angular methyl migration was first proposed by Sir Robert ROBINSON as a means of relating the non-isoprenoid skeleton established for eremophilone with the well-known eudesmanes (*224, 225*). This proposal is of great historical significance in the evolution of the biogenetic isoprene rule since it first called attention to the possibility of skeletal rearrangement in the biosynthesis of terpenes.

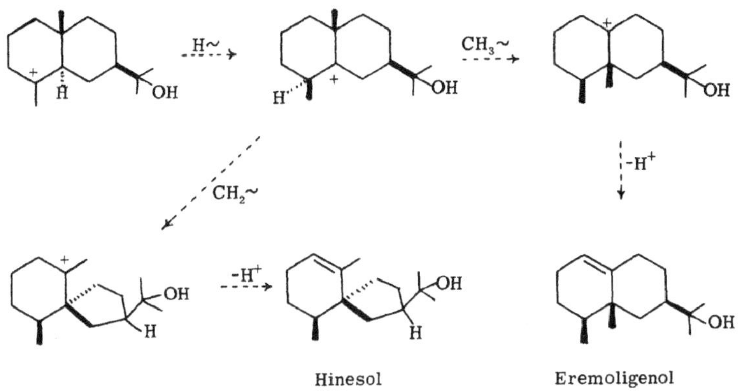

Scheme 12

A second stereochemical sub-group of similarly rearranged bicyclic sesquiterpenes, the nootkatanes (*e.g.*, valeraniol) and the agarospiranes (*e.g.*, agarospirol), may arise from the 5-epi-10-epi-eudesmanes (Scheme 13). This biogenetic scheme is supported by the co-occurrence of the stereochemically related eudesmanes, eremophilanes, and spirovetivanes (*223, 226*). The results of a recent investigation upon the biosynthesis of the eremophilane sesquiterpene petasin (**143**) from [2–^{14}C]-mevalonate and [3R,4R–^{3}H]-mevalonate are also in accord with the above scheme (*227*). The location of one tritium atom at C_4 of petasin is consistent with the proposed 1,2-hydrogen shift from a *trans*-fused eudesmane precursor.

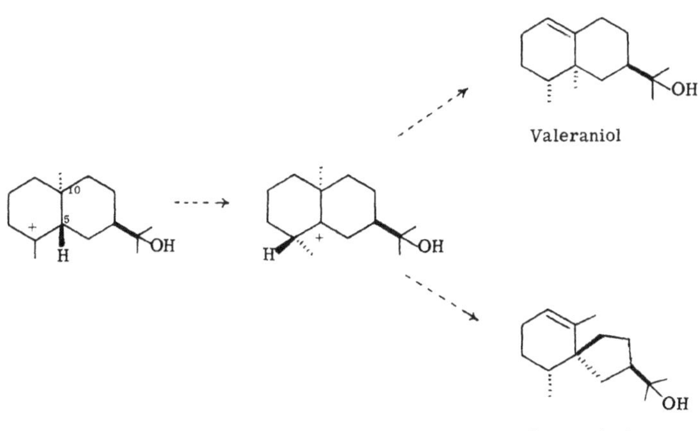

Scheme 13

Capsidiol (**144**), the structure and stereochemistry of which have recently been established by X-ray crystallography (*228*), is the first eremophilane sesquiterpene having the vicinal methyl groups in a *trans* relationship. Although this unusual stereochemistry could arise *via* biogenesis from a *cis*-fused eudesmane (*i.e.*, 5, 10 *cis* as in occidentalol), the location of a hydroxyl group adjacent to the aberrant secondary methyl suggests the alternative possibility that capsidiol is biosynthesized from an eremophilane precursor by oxidation to a 3-keto eremophilane, followed by epimerization and reduction. The biosynthesis of capsidiol from [1,2-$^{13}C_2$]-acetate supports the angular methyl migration proposed in the biogenesis of this modified eremophilane sesquiterpene (*229*).

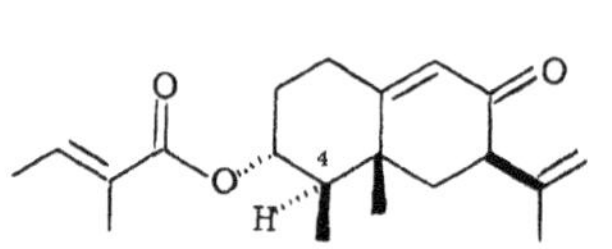

(**143**)

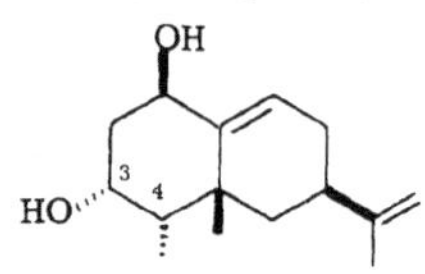

(**144**)

Dihydroalantolactone epoxide (**145**) undergoes acid-catalyzed rearrangement with angular methyl migration, a transformation corresponding to the biogenesis of the eremophilanes (*230*). The structure of the product (**146**) was confirmed by conversion to tetrahydroligularenolide, a derivative of a natural eremophilanolide sesquiterpene lactone. The reverse of this rearrangement has been observed in the formic acid-catalyzed reactions of calarene (**147**) (*231*), calaranol (*231*), calarene epoxide (oxygen *syn* to angular methyl) (*232*), and aristolene epoxide (*233*). A driving force in these rearrangements is the formation of a cyclopropyl carbinyl carbonium ion and subsequent opening of the three-membered ring.

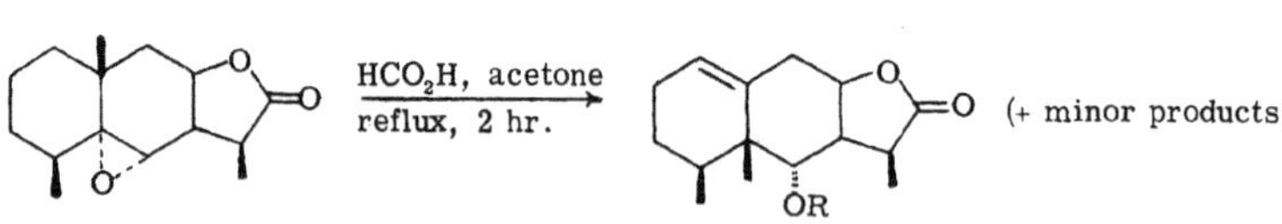

(**145**) (**146**)

R = H (34%)

R = OCHO (10%)

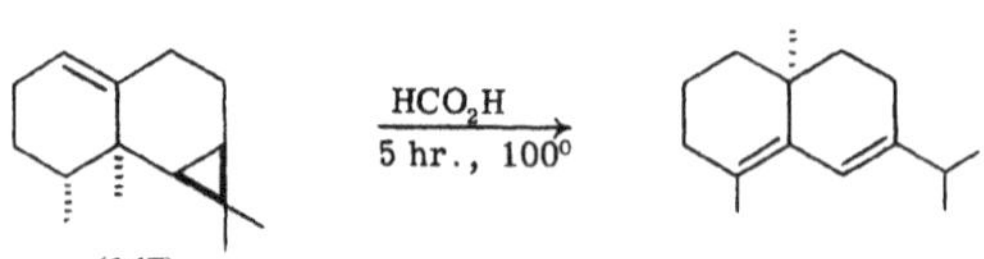

(**147**)

Angular methyl migrations of this type have been observed in the heterolyses of 5α-substituted steroids, a process known as the Westphalen rearrangement (*234*). As an example, the nitrous acid deamination of diacetoxy amine **148** is shown below (*235*). The presence of an electronegative substituent in the 6β-position is apparently usually a prerequisite for the Westphalen rearrangement.

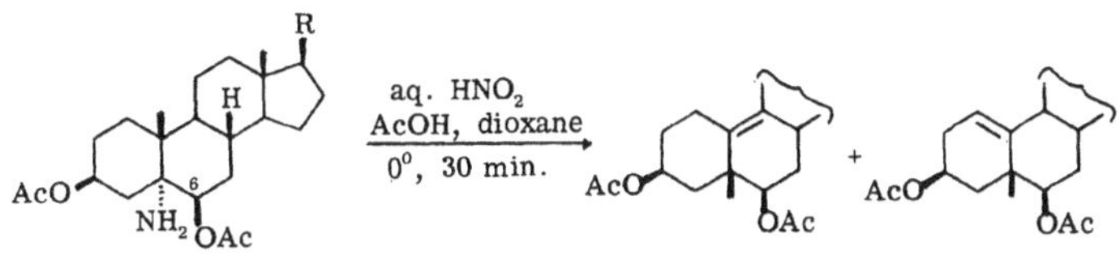

(**148**) (R = C_8H_{17}) (47%) (30%)

Heathcock and co-workers have examined the acid-catalyzed lactonization of 9-methyloctalin carboxylic acids as means to achieve a eudesmane-to-eremophilane rearrangement (*236, 237*). While unsaturated acid **149** (R = H) cyclized to a 2:1 equilibrium mixture of γ-lactone **150** (R = H) and rearranged δ-lactone **151**, no such rearrangement was observed with either the dimethyl analogue (**149,** R = CH_3) or γ-lactone **150** (R = CH_3). That these lactonizations and rearrangements actually involve reversible ring contraction to spiro intermediates instead of angular methyl migration is clear from investigations of Dunham and Lawton (*238*).

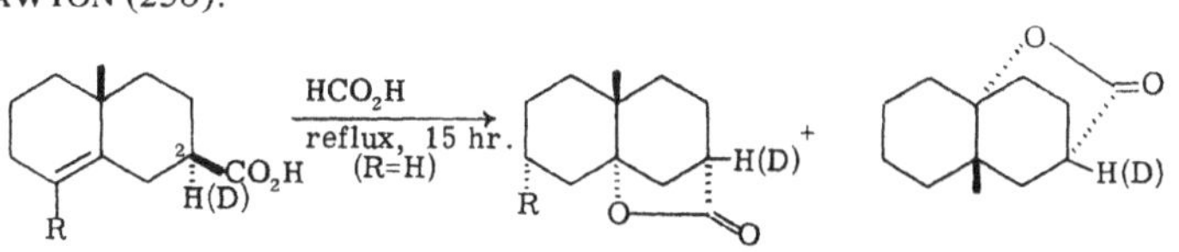

(**149**) (**150**) (2:1) (**151**)

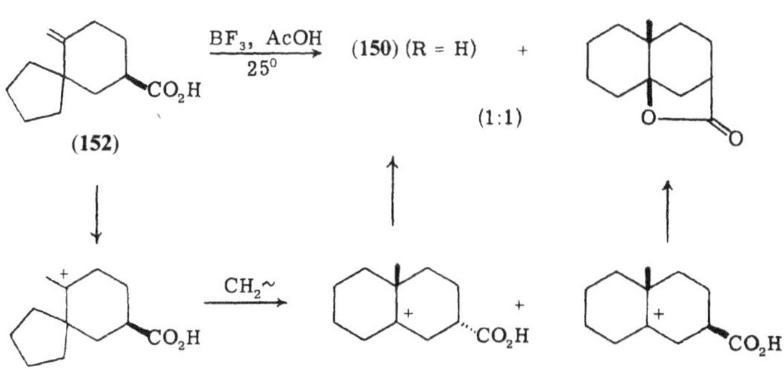

(**152**)

Since lactonization of **149** (R = H) with a deuterium located at position 2 occurred without loss of the label, a mechanism involving simple epimerization of the carboxyl group is excluded. Further the facile lactonization of unsaturated spiro acid **152** to the isomeric decalin γ-lactones tends to support a mechanism for the lactonization of **149** (R = H) *via* a spiro intermediate.

DUNHAM and LAWTON have suggested an alternative biogenesis for the eremophilane sesquiterpenes *via* a series of four decalyl ⇄ spiro rearrangements, a process which effectively shifts the position of the methyl group from one angular position to the other (*238*). This proposal has the disadvantage of requiring a *cis*-fused eudesmane precursor, a known (occidentalol), but rare stereochemical variant, and, at least in the case of capsidiol is excluded by recent biosynthetic results (*229*).

Both spiro and methyl migration rearrangements are proposed to occur when the pentamethyloctalin **153** is subjected to reaction with sulfuric acid in acetic acid (*239*). Isomers **154** and **155** are formed by decalyl ⇄ spiro rearrangement followed by angular methyl shift and elimination. The minor product **156** however, arises by direct methyl migration.

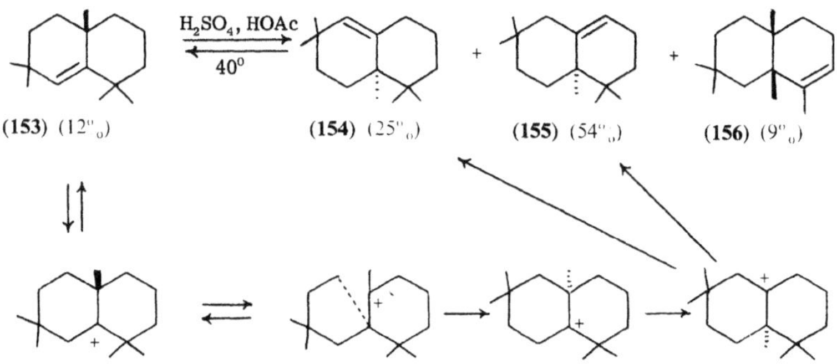

(**153**) (12%) (**154**) (25%) (**155**) (54%) (**156**) (9%)

The reaction of methyloctalin **157** with deuterated polyphosphoric acid causes extensive deuterium incorporation ($d_3 \rightarrow d_{18}$ with d_7 as most abundant species) (*216*). The demonstration of deuterium incorporation into the angular methyl group (average of about 1 deuterium atom by nmr analysis) raises the interesting possibility of a methyl migration rearrangement *via*, 1,3-hydrogen elimination to a cyclopropane followed by opening in the reverse direction. However, a rearrangement through spiro intermediates as discussed above affords an alternative mechanism for deuterium incorporation into the methyl group. With the information available, it is not possible to distinguish between these two alternatives.

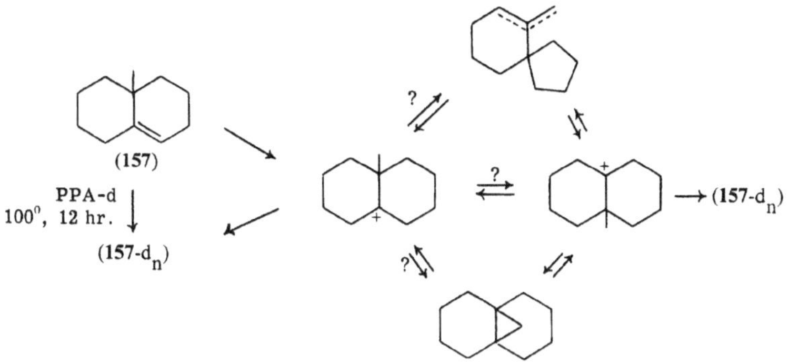

A number of unsuccessful attempts to effect the methyl migration rearrangement of eudesmanes to eremophilanes have been noted (*230, 237, 240*). It would appear that methyl migration does not compete effectively with simple 1,2 proton elimination or, in the case of the lactonization rearrangements discussed above, ring contraction to spiro [5.4] decalyl intermediates. One factor which favors ring contraction is the inherently greater migratory aptitude of methylene over methyl.

The lactonization rearrangements *via* spiro [5.4] decalyl intermediates discussed above provide a precedent for the biogenesis of the spirovetivane sesquiterpenes. The dehydration of β-rotunol (**158**) to spiro dienone **159** under the influence of phosphorous oxychloride in pyridine is apparently the only example of a direct rearrangement of a eudesmane sequiterpene to a spirovetivane (*241*). However, the possibility that β-rotunol might actually have a spirovetivane structure (*i.e.*, hydrated form of **159**) seems not to have been considered. CAINE and DAWSON have reported that spiroenone **160** is transformed into dimethyloctalone **161**, the reverse of the biogenetic rearrangement (*242*).

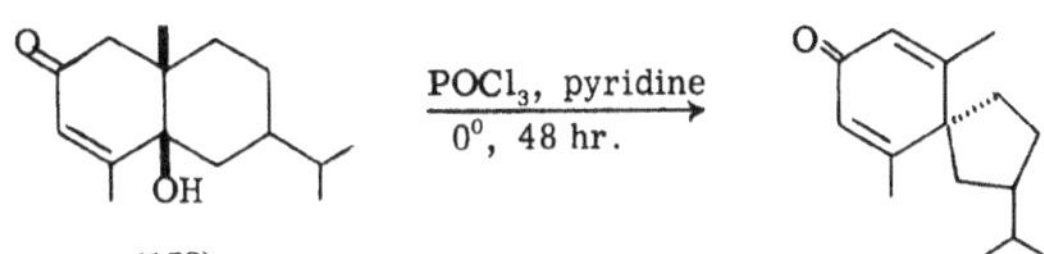

(**158**) (**159**) (39%)

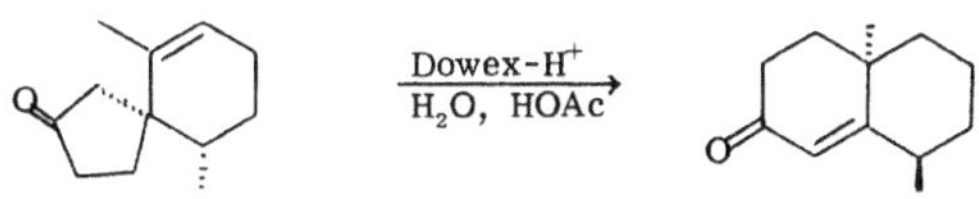

(**160**) (**161**) (>50%)

The biogenesis of the relatively rare valerane and cyperane (*e.g.*, faurinone and cyperolone) sesquiterpenes involves rearrangement of a eudesmane precursor by shift of the C_4 methyl group or ring contraction to a hydrindane nucleus, respectively. Since all of the known natural products bear oxygen functions at the site of rearrangement, an epoxide precursor is usually proposed.

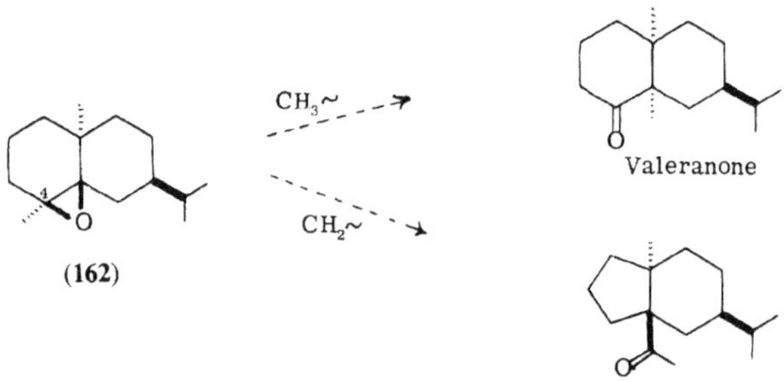

(162) Valeranone Faurinone

Scheme 14

The co-occurrence of cyperolone (**165**) with the well-known eudesmane α-cyperone (**163,** double bond in place of epoxide) suggests that both may be biosynthesized from a common hydroxy selenadiene precursor (*243*). Cyperolone has been synthesized *via* a biogenetically patterned reaction sequence, the key step of which is rearrangement of epoxy α-cyperone **163** to hydrindanone **164** (*243*).

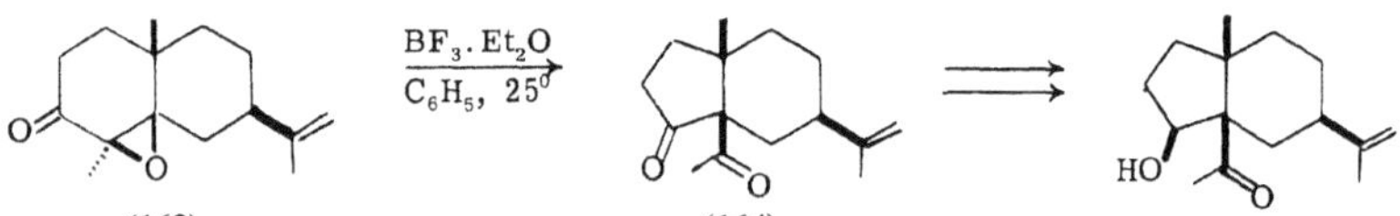

(163) (164) Cyperolone (165)

The rearrangements of three of the four possible diastereomeric 4,5-epoxy-eudesmanes have been examined (*244, 245*). Unfortunately stereoisomer **162,** the proposed precursor to valeranone and faurinone, was not available from the epoxidation reaction used in the synthesis. In the presence of boron trifluoride etherate the epoxy eudesmanes undergo three types of rearrangements: ring contraction to acetyl hydrindanes (C_3 ~) analogous to the biogenesis of the cyperanes, simultaneous ring contraction-ring expansion to *cis*- or *trans*-hydrazulenones *via* migration of the 5,10 bond ($C_{10}\sim$), and ring fragmentation ($C_9\sim$ and C_4–C_5 bond cleavage).

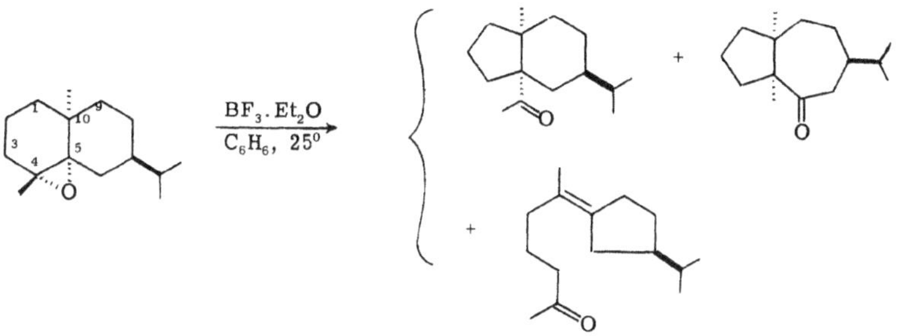

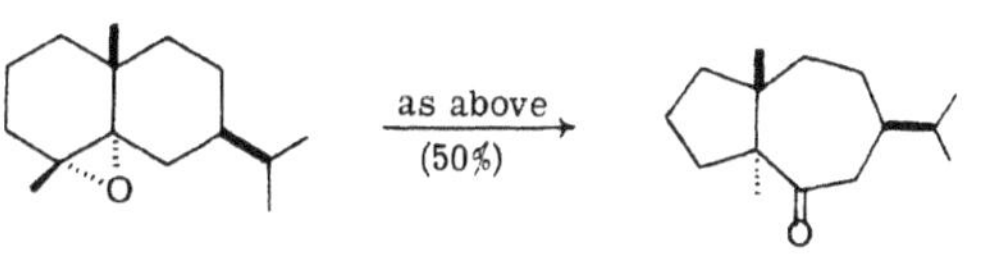

The acid-induced rearrangement of 4β,5β-epoxy-4-methylcholestane affords an A-nor-B-homo ketone in high yield (*246*). Prolonged exposure of the product to boron trifluoride causes a slow rearrangement to 5β-methylcholestan-4-one, very likely *via* the bridged oxycarbonium ion shown. The overall process may be considered as a precedent for the biogenesis of valeranone. The course of the rearrangements of tetra-substituted epoxides may be understood in terms of a preference for initial "axial cleavage" of the epoxide followed by "*trans*-migration" (*234, 247*).

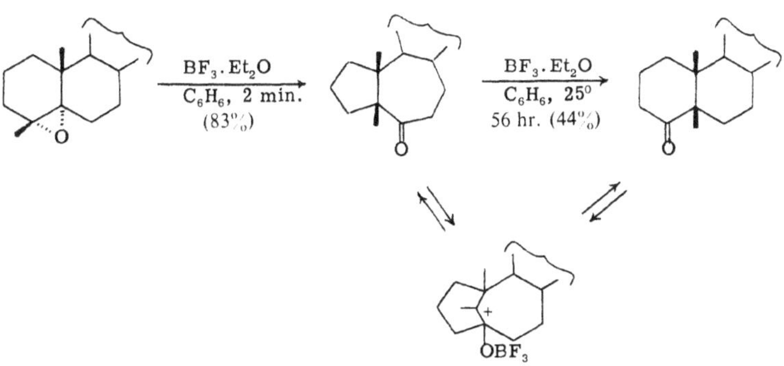

6. Cuparanes, Chamigranes, Thujopsanes, and Widdranes

The biogenesis of this family of bicyclic and tricyclic sesquiterpenes comprises a fascinating and complex combination of rearrangements (*108*, *248*) (Scheme 15). The cuparanes are presumably formed first by double cyclization of *cis*, *trans*-farnesyl pyrophosphate, then 1,4-hydrogen rearrangement. Ring expansion of the cuparenyl cation gives rise to the spiro [5.5] undecane nucleus of the chamigrenes. A second ring expansion followed by angular methyl migration gives the newly isolated α-pseudowiddrene (*249*). Alternatively homoallyl→cyclopropylcarbinyl cyclization furnishes the tricyclic structure of thujopsene. Finally a cyclopropylcarbinyl→cyclopropylcarbinyl→homoallyl rearrangement sequence leads, after hydration, to widdrol. The co-occurrence of all four types of sesquiterpenes bespeaks a close biogenetic relationship (*108*, *249*, *250*).

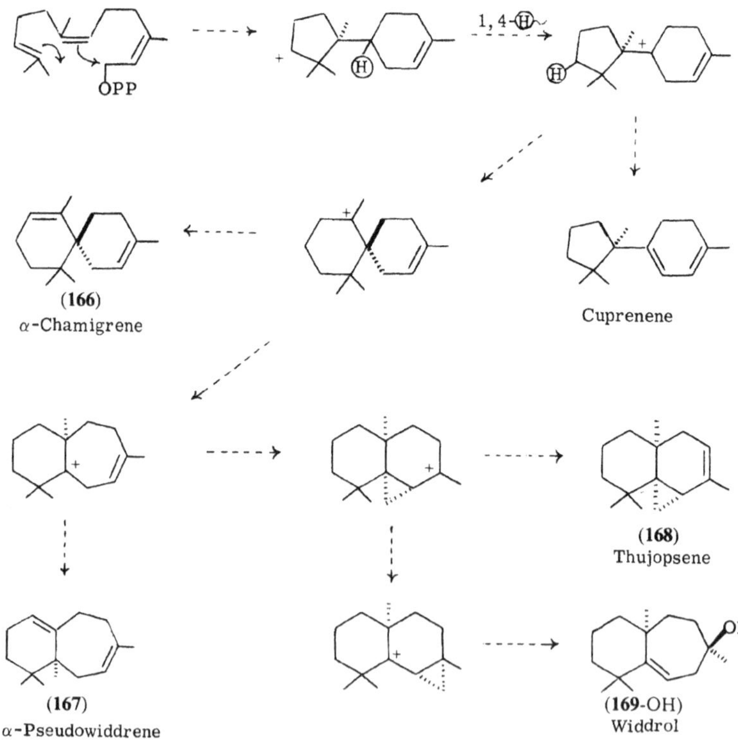

Scheme 15

Upon reaction with aqueous acid thujopsene (**168**) undergoes hydration with rearrangement to widdrol (**169**-OH) (*251*). By means of deuterium labelling, DAUBEN and co-workers have demonstrated that this reaction proceeds *via* a reversible but stereospecific cyclopropylcarbinyl ⇄ cyclopropylcarbinyl rearrangement in which the methylene group migration occurs with retention of configuration (*252—255*). The reversibility of the rearrangement is shown by the observation that widdrol is initially the most rapidly formed product from the reaction of thujopsene with .02 N perchloric acid in 80% aqueous dioxane and, under the same conditions, thujopsene is the initial product from widdrol.

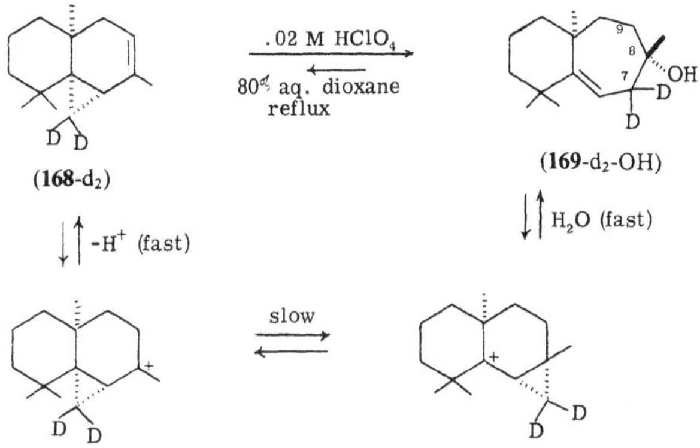

(**168**-d$_2$) (**169**-d$_2$-OH) (**170**) (**171**)

If the rearrangement of thujopsene is carried out in deuterium oxide, the widdrol formed has incorporated more than a single deuterium atom (2% d$_0$, 43% d$_1$, 43% d$_2$ and 11% d$_3$), most which is localized at C$_9$. This result, along with the finding that thujopsanol is rapidly dehydrated (.02 N HCl in 95% methanol) to thujopsene, indicates that the most rapid reaction of ion (**170**) is proton elimination to give thujopsene. Further, it is found that hydrolysis of widdryl *p*-nitrobenzoate in 70% aqueous dioxane gives only widdrol while in anhydrous dioxane containing lithium perchlorate thujopsene is produced. Thus, the most rapid reaction of ion (**171**) in aqueous dioxane is hydration to widdrol. Evidently the cyclopropylcarbinyl-to-cyclopropylcarbinyl rearrangement (**170** ⇄ **171**) has a somewhat higher activation energy than the respective deprotonation and hydration reactions of (**170** and (**171**).

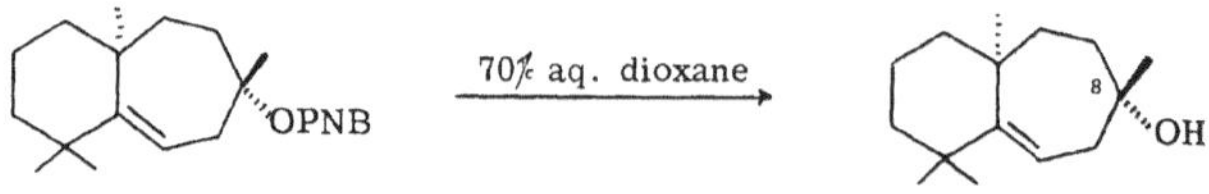

(**169**-OPNB) (**169**-OH)

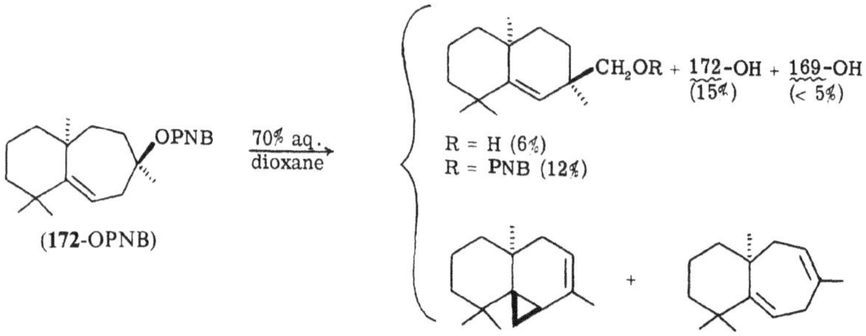

(**172**-OPNB)

(**173**) (14%) (14%)

The capture of ion (**171**) with water to form widdrol occurs with inversion of configuration at C_8. That this stereospecificity is a consequence of a direct reaction with ion (**171**) rather than a steric effect in capture of the corresponding homoallyl cation [*i.e.*, (**169**) ionized] is shown by the hydrolysis of 8-*epi*-widdryl *p*-nitrobenzoate (**172**-OPNB) to *epi*-widdrol (**172**-OH) as the major alcohol product (the small amount of widdrol formed may have been an impurity).

Trans-thujopsene (**173**) is one of the major hydrocarbons produced in this reaction. Thus, there appear to be an isomeric pair of interconvertible cyclopropylcarbinyl cations [(**170**) and (**171**) but with cyclopropane ring *trans* to angular methyl] generated in the solvolysis of (**172**-OPNB), and in the reactions of *trans*-thujopsene. With prolonged reaction times common products begin to be observed, indicating leakage between the *cis*- and *trans*-cyclopropylcarbinyl cations.

In the presence of organic acids, thujopsene undergoes the reverse of its biogenesis, being converted to α- and β-chamigrene (**166** and **174**, respectively) (*250, 256—260*). α-Pseudowiddrene (**167**) is also formed in varying proportion *via* an apparently separate and irreversible competing pathway. After prolonged reaction times bridged tricyclic products become the major products along with small amounts of cuparene (**175**). This last transformation represents the reversal of the cuparenyl→chamigrenyl step in the biogenetic scheme.

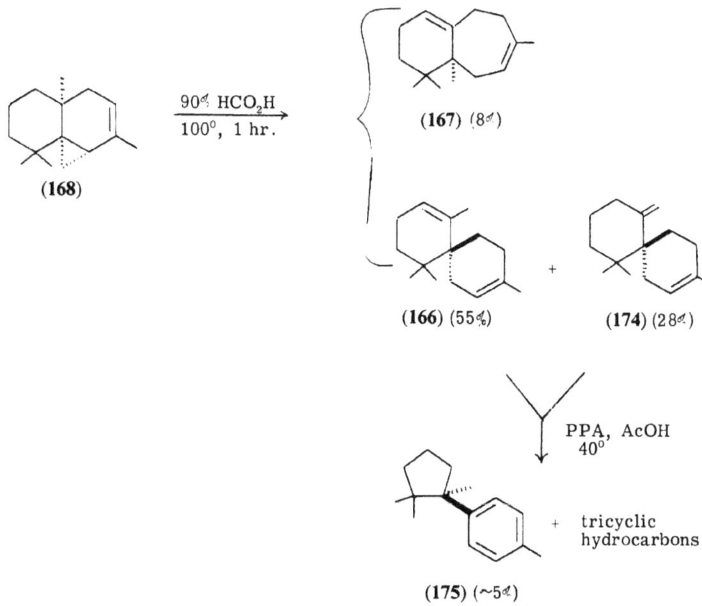

The observation that α- and β-chamigrene are produced in high optical purity (76% and 94—99% optical purities, respectively) in these rearrangements reveals that the ring contraction to the spiro[5.5]undecalyl ring system is essentially irreversible (*257, 260*). It is remarkable that cuparene is also obtained optically active (25—71% optical purity). Apparently the second ring contraction to the cuparenyl system occurs

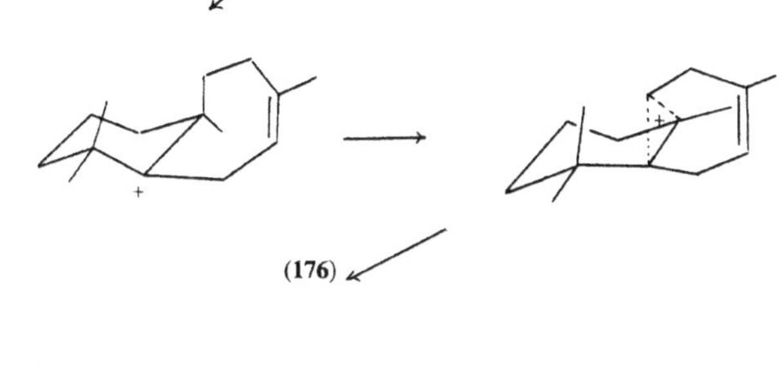

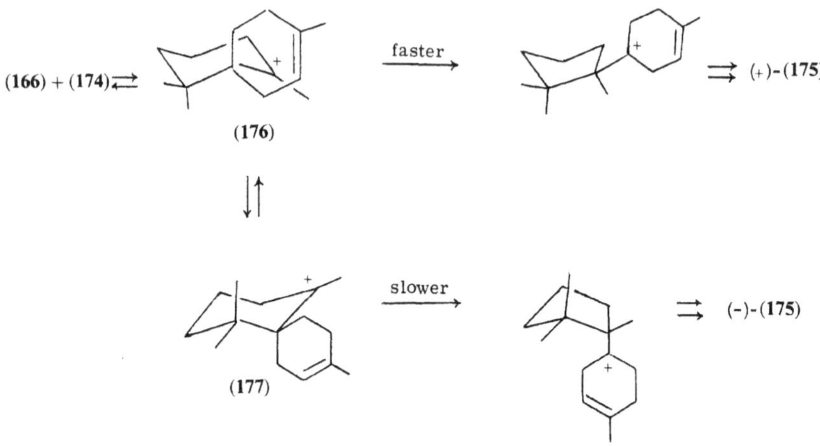

more readily from chamigrenyl cation in conformation (**176**) than the apparently energetically similar conformation (**177**). The stereochemistry of these rearrangements is shown above.

The finding that monocyclo-farnesol undergoes cyclization to α-chamigrene (**166**, 25%) upon reaction with iodine in benzene lead to the suggestion of this pathway as an alternative biogenesis (*261*). However, the co-occurrence patterns mentioned above and the results of biosynthetic experiments with cuparene-based sesquiterpenes (*35*, *207* to *210*) would seem to favor the more comprehensive biogenesis previously presented.

7. Patchoulanes and Seychellanes

The biogenesis of the patchoulane sesquiterpenes requires an initial protonation of the isopropenyl group of α-bulnesene (**178**) which is followed by cyclization and a series of Wagner-Meerwein rearrangements (Scheme 16). Since α-bulnesene is itself regarded as being formed by a proton-induced cyclization of germacratriene (see section C.2 above), the overall conversion of farnesyl pyrophosphate to the patchoulanes is rather unusual in necessarily proceeding *via* two deprotonated intermediates. The closely related isopatchoulane sesquiterpene cyperene (**184**) and its relatives are presumably formed similarly by a protonation-cyclization operation upon 10-epi-α-guaiene having the C_{10} methyl group in the unusual α-configuration. The seychellane sesquiterpenes result from yet another rearrangement, a 1,2-methyl migration in the bridgehead carbonium ion (**181**) and subsequent proton elimination. This biogenetic scheme is supported by the co-occurrence of seychellene, patchouli alcohol, α- and β-patchoulene, β-bulnesene, and β-guaiene in the essential oil of *Pogostemon patchouly* (*262*).

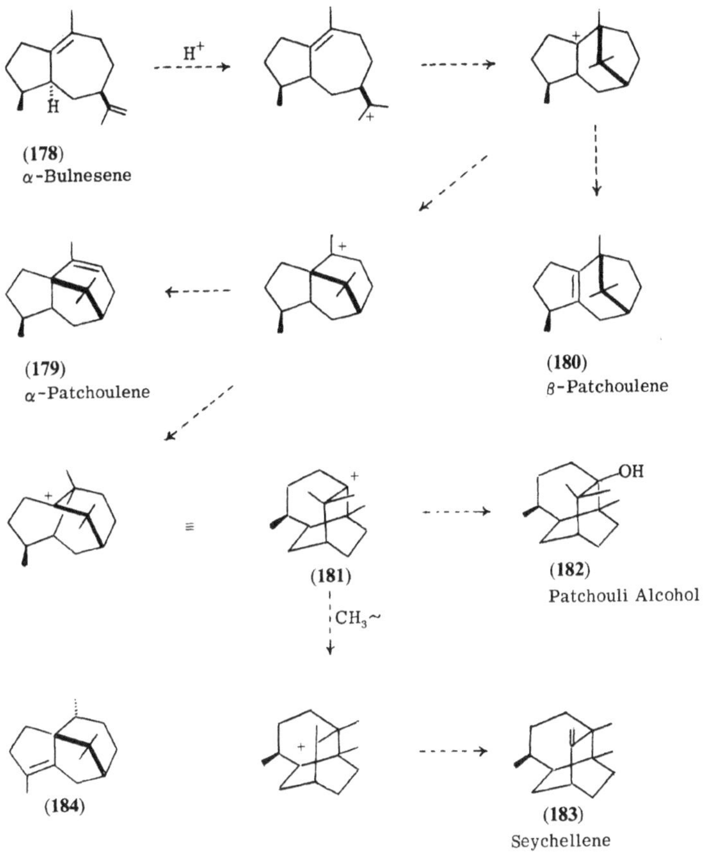

Scheme 16

A biogenetic-like cyclization of α-bulnesol (**71**) to β-patchoulene (**180**) occurs to the extent of ~ 30% in sulfuric acid-acetic acid, along with simple dehydration and subsequent double bond isomerizations to various bulnesenes and guaienes (*263*). A more efficient cyclization to (**180**) (60%) is observed when bulnesol is heated with pyridine and alumina. Evidently β-patchoulene has the most stable structure of the patchoulene group since it is formed from both acid-catalyzed dehydration of patchouli alcohol (**182**) and acid-induced isomerization of α-patchoulene (**179**) (*264*).

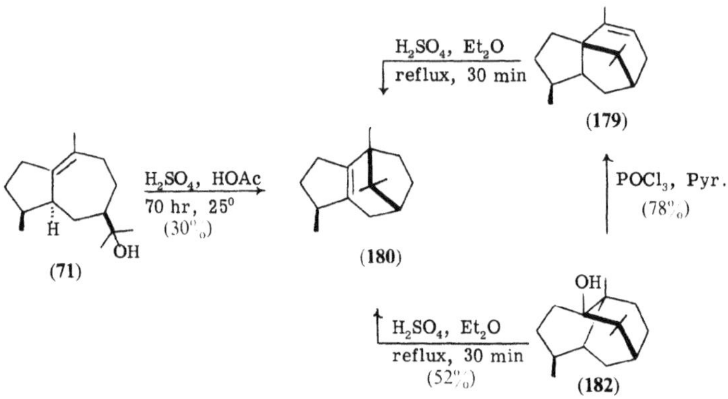

The contra-thermodynamic rearrangement of the dimethyl bridge in β-patchoulene is realized when the corresponding expoxide is treated with boron trifluoride etherate in ether (*265*). A competing migration of the methylene group (C_5), which is *syn* to the epoxide oxygen triggers a sequence of four more rearrangements terminating with the formation of bridged oxide (**185**) (*266*). The second Wagner-Meerwein rearrangement in the biogenetic scheme, the migration of the interannular bond, to give the bridged ring structure of patchouli alcohol occurs when α-patchoulene (**179**) is treated with peroxyacetic acid in chloroform (*264, 265*). It is assumed that patchouli diol (**186**) is formed by rapid rearrangement and hydration of an unstable β-epoxide intermediate since the α-epoxide of α-patchoulene is also isolated and is stable to the reaction

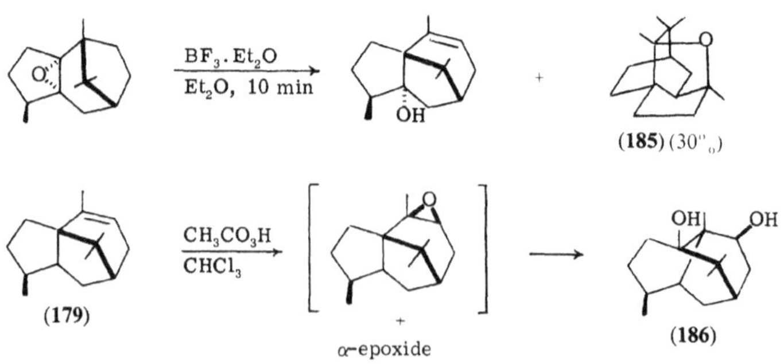

conditions. The reaction of the closely related epoxide of α-cyperene (**184**) with acids leads to a variety of interesting and complex rearrangements (*267, 268*).

Cycloseychellene (**187**) is a new seychellane sesquiterpene which co-occurs with seychellene (**183**) (*269*). An acid catalyzed equilibration between these two compounds has been achieved by reaction of either isomer wich cupric acetate in acetic acid, the equilibrium favoring seychellene (*269*).

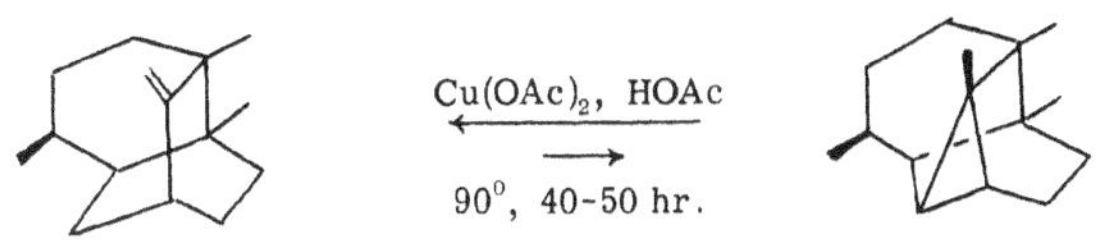

(**183**) (**187**)

8. Copaanes, Copabornanes, Ylanganes, and Sativanes

This group of tricyclic sesquiterpenes may be considered as tricyclic analogues of the bergamotane, sesquicamphane, and santalane sesquiterpenes, in which the bicyclo[3.1.1]heptane and bicyclo[2.2.1]heptane nuclei are bridged with a three carbon chain bearing the isopropyl substituent. The biogenesis (Schemes 17 and 18) proceeds through *cis*-fused decalyl carbonium ions (**188**) and (**191**), precursors to the muurolenes and amorphenes respectively (see scheme 9, section C.2), which undergo cyclizations and rearrangements running closely parallel to the schemes portrayed earlier for their bicyclic relatives. It will be noted that the carbonium ion intermediates bear a close resemblance to the monocyclic bisabolyl cation (**54**).

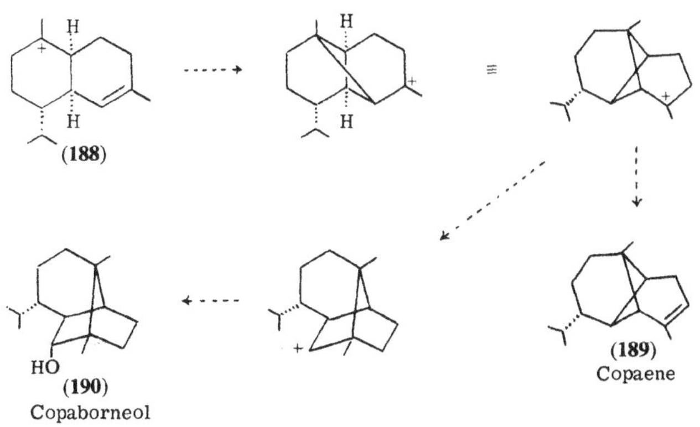

Scheme 17

There are two distinct subgroups of structures depending upon the stereochemical relationship between the ring fusion and the isopropyl substituent. Copaene and copaborneol are accordingly related to the muurolenes (*e.g.*, α-muurolene, **83**) while ylangene and sativene are related to the amorphenes (*e.g.*, α-amorphene, **84**). Sativene and cyclosativene are derived by a further rearrangement analogous to that which leads to the santalenes in the bicyclic series.

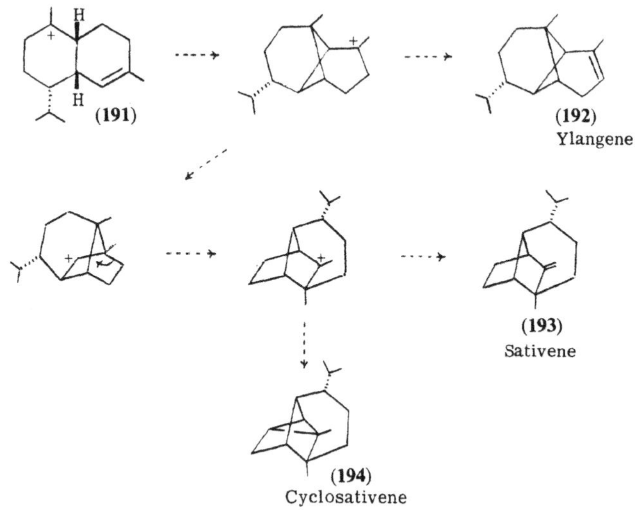

Scheme 18

The stannic chloride-catalyzed cyclization of acyl cyclopropane (**195**) to the tricyclic ketone (**196**) affords a precedent for the biogenesis of the tricyclo[4.4.0.0^{2,7}]nucleus of copaene and ylangene (*270*). The reaction pathway is thought to proceed *via* ring opening, cyclization, hydride shift, and ring closure.

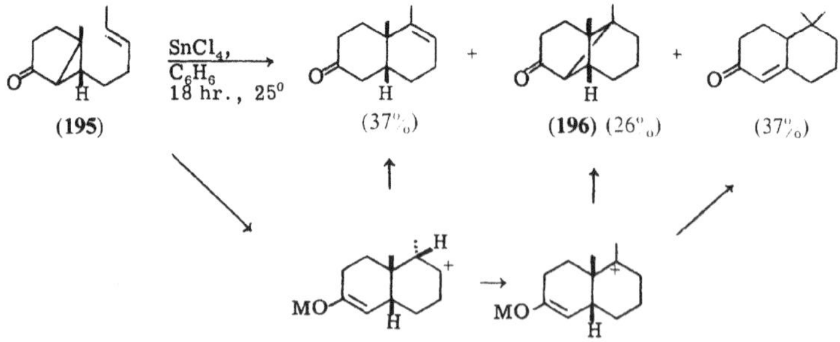

The reaction of either copaene or ylangene, or the corresponding epoxides with protic acids invariably leads to ring opening and formation of products related to α-muurolene (**83**) or α-amorphene (**84**), *i.e.*, the reverse of the biogenetic cyclization (*137, 271—275*). Although the *trans* fused cadinene dihydrochloride is usually isolated from treatment of copaene with anhydrous hydrogen chloride (*271—273*), this product may arise from either addition of hydrogen chloride to δ-cadinene (**86**) or ring juncture isomerization in the subsequent addition of hydrogen chloride to an intermediate muurolene (*276*). No products resulting from Wagner-Meerwein rearrangement (*e.g.*, copaborneol, ylangoborneol, copacomphene, or sativene) have, as yet, been isolated. Evidently the three carbon bridge in copaene and ylangene renders the cyclobutane ring more susceptible to the ring cleavage reaction pathway than that of α-pinene, the monoterpene analogue. Perhaps the additional ring increases the activation energy for the rearrangement as a consequence of increased strain since this process now requires expansion of a six-membered ring to seven members. It is interesting to note that diol (**199**) is a natural product and is formed in the ring-opening of α-copaene epoxide (**198**) with net inversion at the tertiary center (*274*).

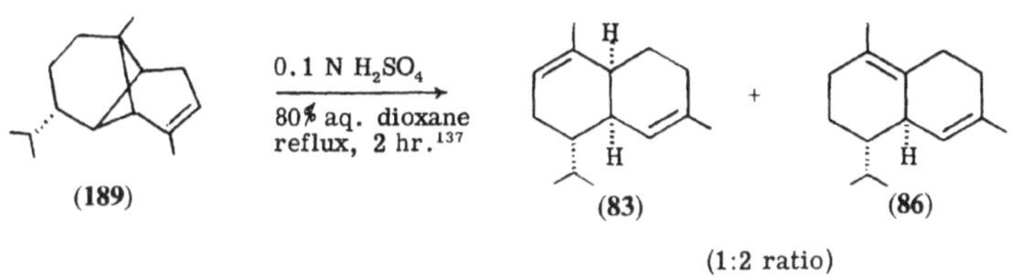

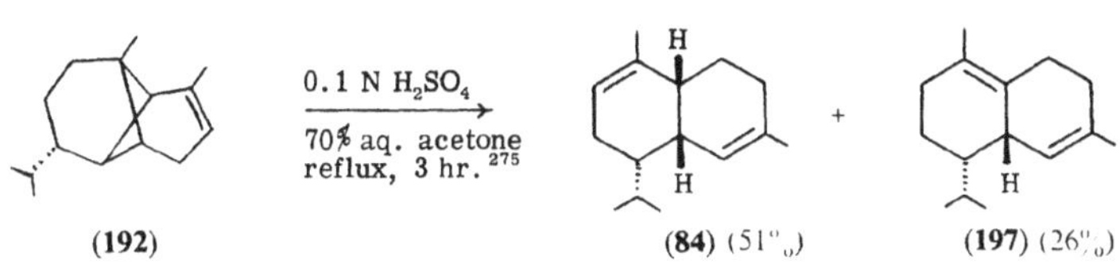

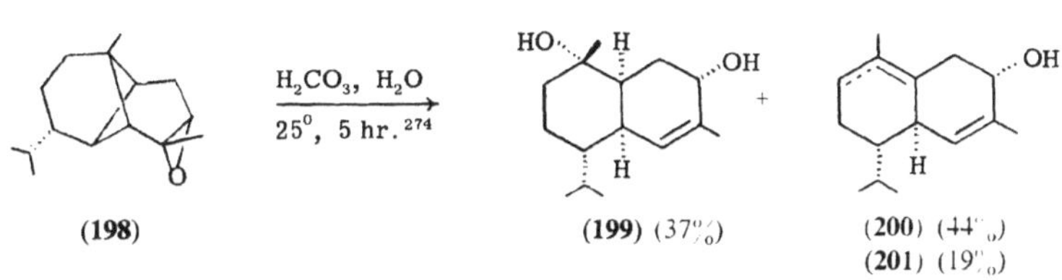

The Wagner-Meerwein rearrangement of copabornanes and ylangobornanes to copacamphanes and sativanes has been observed. Thus, copabornyl brosylate (**190**-OBs) undergoes rearrangement and proton elimination to copacamphene (**202**) (*277*). In a similar fashion sativene (**193**) is formed from isoylangobornyl mesylate (**203**) (*116, 117*). This rearrangement appears to occur with comparable efficiency with both the *endo* and *exo* orientation of the sulfonate leaving group.

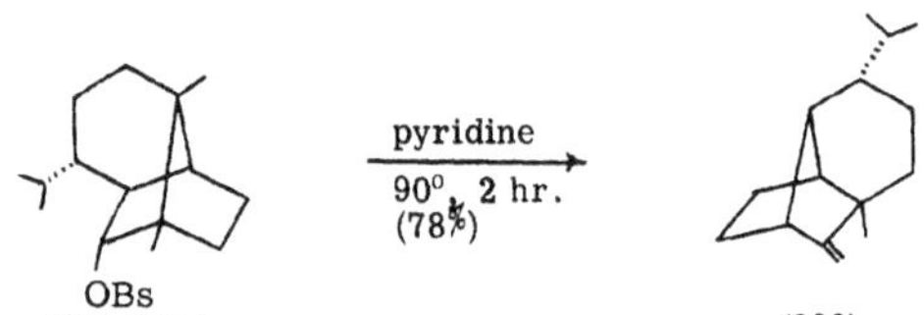

(**190**-OBs) (**202**)

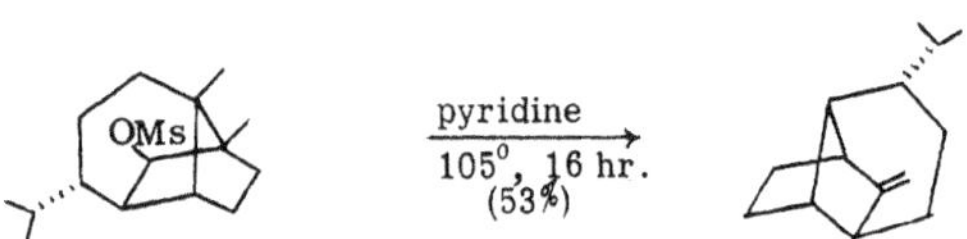

(**203**) (**193**)

Exposure of either sativene (**193**) or copacamphene (**202**) to cupric acetate in refluxing acetic acid produces the same equilibrium mixture of sativene (7—8%), cyclosativene (**194**, 32%), and isosativene (**204**, 60—61%) (*278, 279*). The greater thermodynamic stability of sativene over copacamphene is a consequence of the axial configuration of the isopropyl group on the six-membered ring of the latter. The *endo* bridged isosativene apparently arises from initially formed cyclosativene which undergoes cleavage of the three-membered ring in the opposite sense.

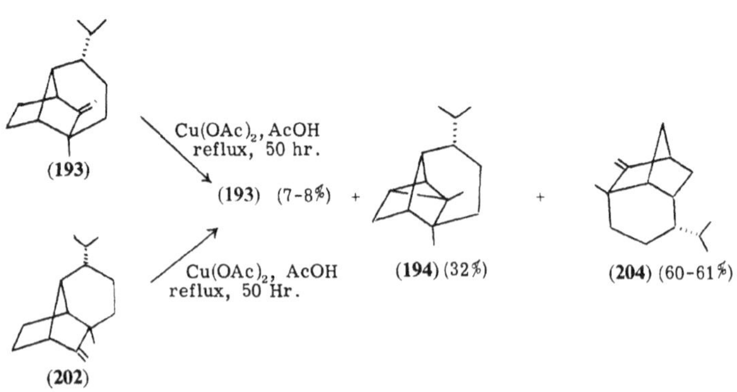

The interconversion of sativene and copacamphene presumably occurs through 1,2-shift of the three carbon bridge. This suggests the possibility that sativene and copacamphene could arise *via* a common biogenetic scheme the last stage of which would be the 1,2-migration of the three carbon bridge (*109*). However, recent investigations upon the biosynthesis of sativene in *Helminthosporium sativuum* and *H. victoriae* exclude the so-called long route passing through a copacamphanyl precursor (*109*). Similarly the picrotoxan sesquiterpenes dendrobine, tutin, and corianyrtin are biosynthesized directly from a copaborneol intermediate (*176*, *177*), *i.e.*, without a rearrangement of the three carbon bridge.

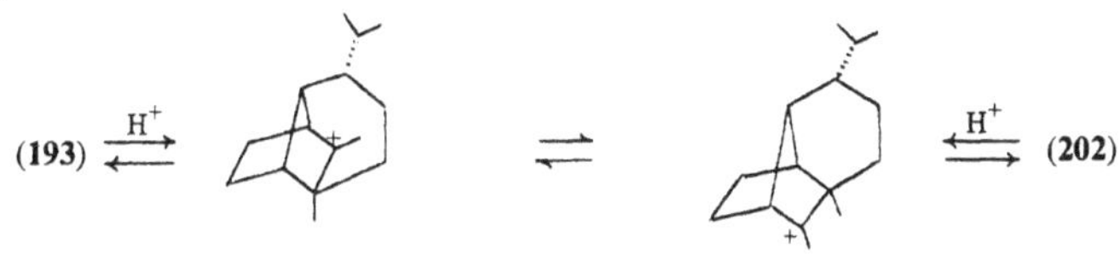

9. Longipinanes, Longibornanes, and Longifolanes

Since the structure of longifolene (**208**) is readily recognized to be similar to sativene with the bridging six-membered ring expanded to seven members, it is not surprising that the biogeneses of these

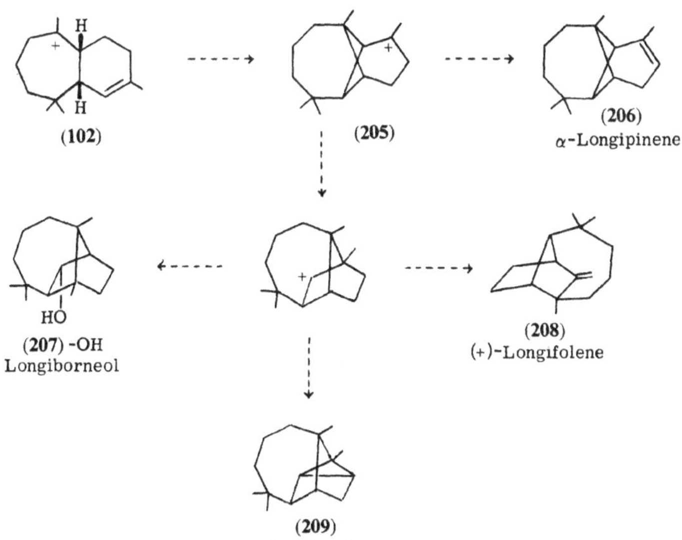

Scheme 19

two groups of tricyclic sesquiterpenes are also similar. Thus, cyclization (Scheme 19) of the himachalyl carbonium ion (**102**) (see section C.3) constructs the ring system of longipinene (**206**). Further rearrangements akin to those presented in the previous section lead to longiborneol (**207**-OH), longifolene (**208**), and longicyclene (**209**). This scheme has been fully verified for the biosynthesis of (+)-longifolene in *Pinus ponderosa* (*109*), (−)-longifolene in *H. sativum* and *H. victoriae* (*109*), and culmorin [a hydroxylated analogue of (−)-longiborneol] in *Fusarium culmorum* (*178*).

Although there seem to be no close precedents for the initial cyclization, all of the subsequent carbonium rearrangements in the scheme have been observed in the chemically induced carbonium ion reactions of the longifolene sesquiterpenes (*233*). Thus, longipinene, longifolene, and longicyclene all undergo addition of hydrogen chloride to form longibornyl chloride (**207**-Cl) (*280—282*). Evidently severe steric hindrance from the four carbon bridge prevents the formation of the *exo* isomer from longifolene in contrast to the initial production of bornyl chloride from camphene in the monoterpene series (*233, 283*). It is also interesting to note that α-longipinene affords longibornyl chloride rather than himachalene hydrochloride, *i.e.*, rearrangement evidently takes precedence over ring cleavage, in contrast to the reaction of copaene and ylangene with hydrogen chloride.

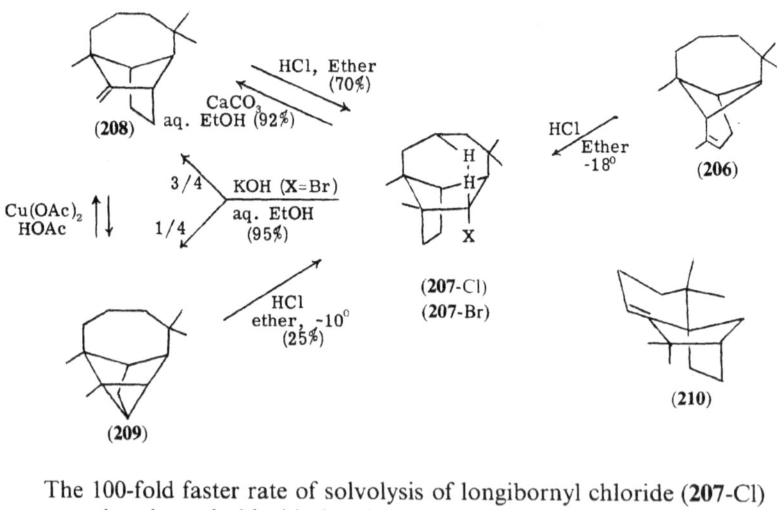

The 100-fold faster rate of solvolysis of longibornyl chloride (**207**-Cl) compared to bornyl chloride has been attributed to relief of an H---H transannular steric interaction in the transition state (*233, 283*). Since longifolene obtained from solvolysis of longibornyl bromide is optically pure (*282*), it is clear that 1,2-migration of the four carbon bridge (analogous to the copacamphene ⇄ sativene rearrangement) does not occur under the reaction conditions. However, the formation of partially racemized (30—80%) isolongifolene (**210**) from exposure of longifolene to strong acid conditions (H_2SO_4/HOAc or $BF_3 \cdot Et_2O$/HOAc) has been taken to indicate that the 1,2-bridge rearrangement of the longifolyl carbonium ion intermediate occurs in competition with the isomerization to (**210**) (*284*). It will be noted that the bridge rearrangement in this case interconverts enantiomers. Isolongifolene (**210**) is formed as a minor product (15—19%) in the cupric acetate/acetic equilibration of longifolene and longicyclene shown above (*282*).

10. Acoranes, Cedranes, and Zizaanes

The simplest biogenesis of the tricyclic cedrane sesquiterpenes (Scheme 20) involves initial cyclization of the β-bisabolyl cation (**211**) to the spiro[4.5]decane nucleus of the acoranes followed by a second cyclization upon the remaining double bond in the cyclohexene ring (*108, 285*). The β-bisabolyl cation may arise less directly from (−)-γ-curcumene *via* additional deprotonation-protonation step, a distinction which is unnecessary for the present purposes (*110*). An alternative biogenesis of the requisite acorenyl cation intermediate has also been proposed in which the cationic precursor to the muurolenes undergoes 1,2-hydride shift, ring contraction and finally 1,4-hydride shift (*110*).

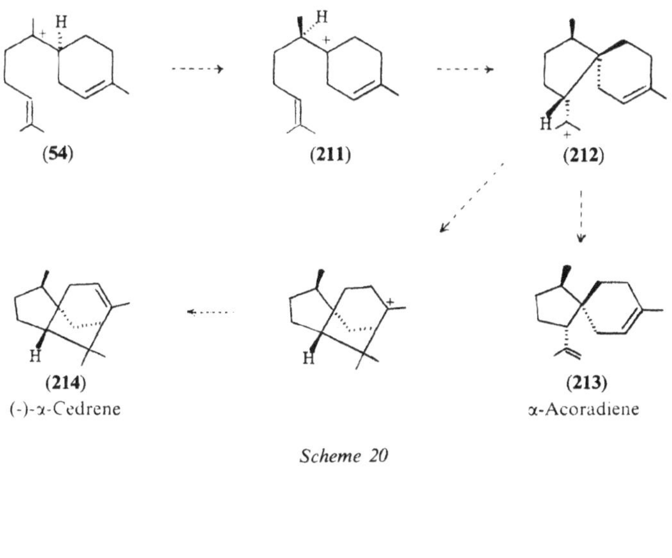

Scheme 20

α-Acoradiene (**213**) and related spiro[4.5]decanes having the proper stereochemistry and functionality undergo remarkably efficient cyclization to cedrene derivatives in the presence of acid catalysts (*110, 285—290*). This transformation is, in fact, a key step in three biogenetic-type syntheses of α-cedrene (**214**) (*286—288*). It has also been possible to achieve a direct tricyclization of nerolidol and farnesol to cedrene and 2-epicedrene in moderate yields (*113, 114*).

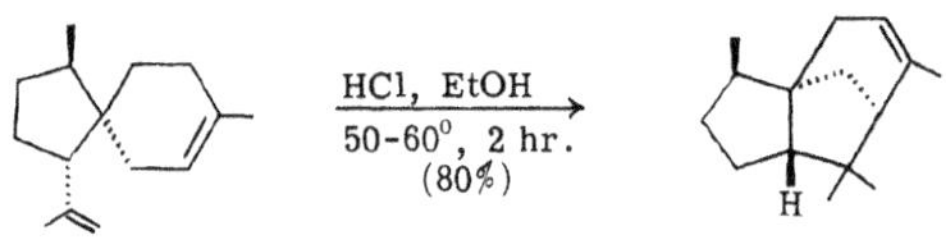

(**213**) (**214**)

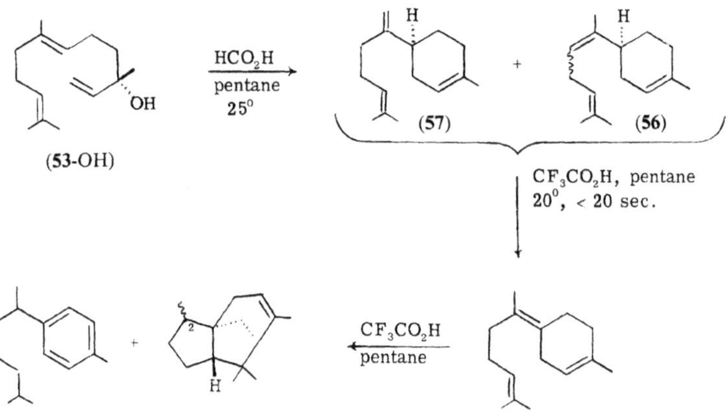

(**53**-OH)

25%   (**214**) (β-CH_3, 30%)   (**215**)
(**216**) (α-CH_3, 20%)

Although the β-bisabolene (**57**) produced is optically active (31% optical purity), isomerization to γ-bisabolene (**215**) prior to further cyclization results in the eventual formation of racemic α-cedrenes (**214** and **216**) (*113*). The absence of detectable amounts of spiro[4.5]decanes indicates that the final cyclization is more rapid than the spiro cyclization step. Since both cedrene (**214**) and 2-epicedrene (**216**) are formed in comparable amounts, it would appear that the bisabolene precursor cyclizes to acorenyl carbonium ion intermediates with the secondary methyl and isopropyl groups both *trans* (as in **212**) and *cis*, the latter affording to 2-epicedrene.

Two distinctly different biogenetic schemes have been entertained in order to derive the similar methanoperhydroazulene ring system of the zizaane sesquiterpenes. One scheme (Scheme 21) proposes a pathway analogous to cedrene biogenesis (*110, 291—293*). However, since both the relative and absolute stereochemistry of zizaene are different from cedrene (*292*), the enantiomeric β-bisabolyl and β-acorenyl cations (**217** and **218**) are required as intermediates. It is perhaps noteworthy that the enantiomeric β-bisabolyl ions (**211**) and (**217**) may be generated from the same enantiomeric α-bisabolyl precursor (**54**) by 1,2-hydride shift to the opposite faces of the carbonium ion in a manner analogous to the 17→20 hydride shifts in tetracyclic triterpene biogenesis (*6*).

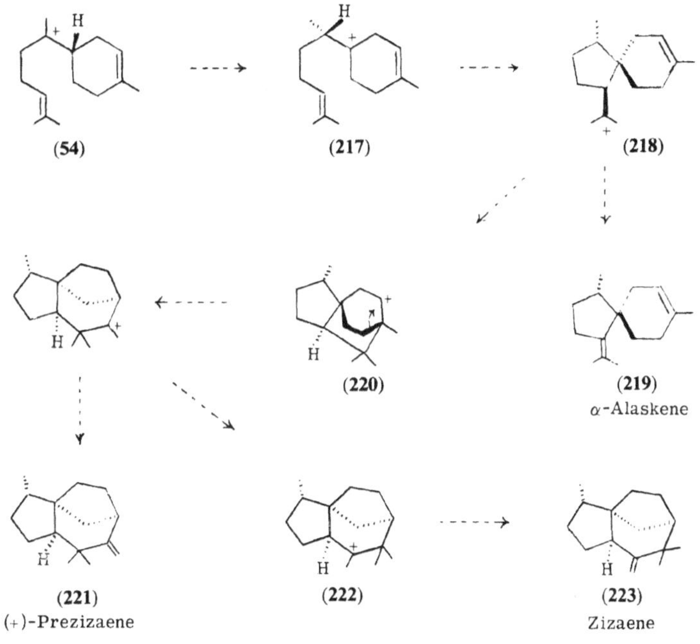

Scheme 21

The alternative proposal for zizaane biogenesis (Scheme 22) involves a final cyclization of a spirovetivane intermediate (*e.g.*, α-isovetivene, **224**) and subsequent ring expansion and elimination to zizaene (**223**) (*294*). The other members of the zizaane group could arise by rearrangements in the reverse direction to that indicated above. Although zizaene (**223**) does co-occur with spirovetivane sesquiterpenes in vetiver oil, the stereochemistry of these spirovetivanes is in fact enantiomeric to that required in this biogenetic scheme. Stereospecific pinacol rearrangements analogous to the ring expansion step (**225**→**222**) in this scheme have been employed in total syntheses of zizaane sesquiterpenes (*294*—*296*).

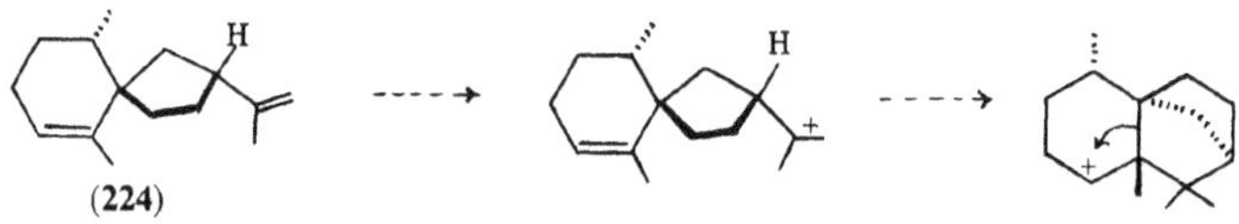

(**224**)
α-Isovetivene

(**225**)

(**222**) --→ (**223**)

Scheme 22

Tomita and Hirose have reported that the reaction of β-acoradiene (**226**) with formic acid affords, after saponification, the newly discovered natural sesquiterpene allocedrol (**227**-OH) (*297*, *298*), along with several other products*. The structure of allocedrol is enantiomeric to that obtained by hydration of the carbonium ion intermediate (**220**) depicted in the zizaene biogenesis. This cyclization is presumably less efficient than the α-acoradiene→cedrene cyclization (**213**→**214**) since ring

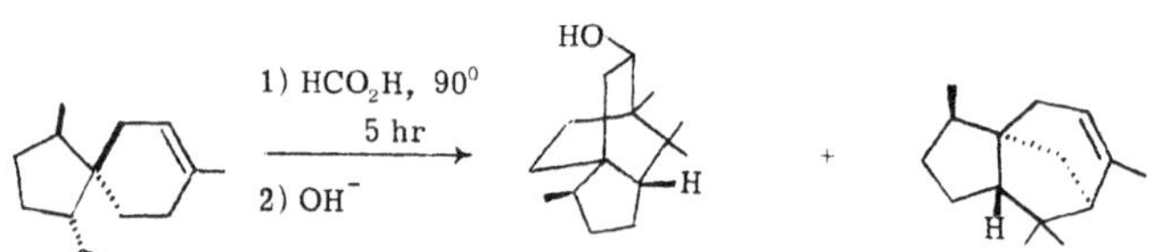

(**226**) (**227**-OH) (20%) (**214** (10%)
Allocedrol

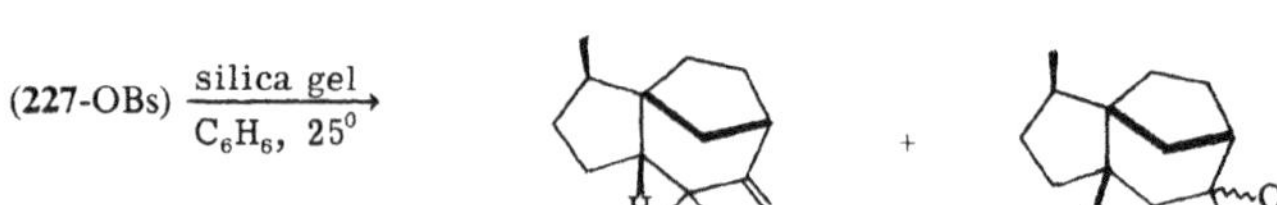

(**228**) (15%) (**229**) (20%)

* Andersen has recently reported the isolation and identification of allocedrol having the absolute configuration corresponding to the zizaane family (*i.e.* enantiomer of **227**-OH) from vetiver oil (*298a*). Pyrolysis of the methanesulfonate at 100° affords (+)-prezizaene (**221**) in 64% yield.

closure proceeds *via* a secondary carbonium ion (or secondary-like bridged ion). Formation of the tertiary carbonium ion is less favorable in this case since the resulting tricyclic structure incorporates a highly strained *trans* bicyclo[3.3.0]octyl moiety. A small amount of α-cedrene (**214**) is also isolated presumably *via* prior isomerization of β-acoradiene (**226**) to α-acoradiene (**213**).

The brosylate of allocedrol (**227**-OBs) upon contact with silica gel, or less efficiently in formic acid at 70°, undergoes rearrangement to (−)-prezizaene [**228**, the enantiomer of natural (+)-prezizaene], the corresponding tertiary alcohol (**229**), and a number of other products (*298*). The low yield in this rearrangement is probably a consequence of the unfavorable *syn* relationship between the brosylate leaving group and the migrating ethane bridge in (**227**-OBs). The sequence of biogenetic-type rearrangements is completed with the observation that prezizaene (**221**) is converted to zizaene (**223**) in formic acid-tetrahydrofuran (*293, 299*). The proportion of zizaene in the product remains low, however, since it rapidly isomerizes to an equilibrium mixture (39 : 7 : 48) of iso-zizaene (**230**), and olefins (**231**) and (**232**). It is of interest to note that (**231**) bears a close structural to the diterpene eremolactone (**233**).

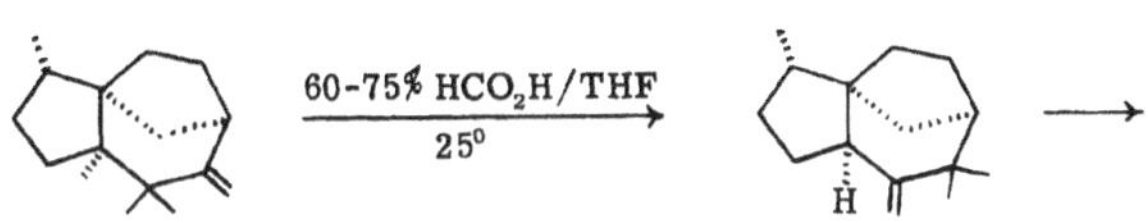

(**221**) (**223**)

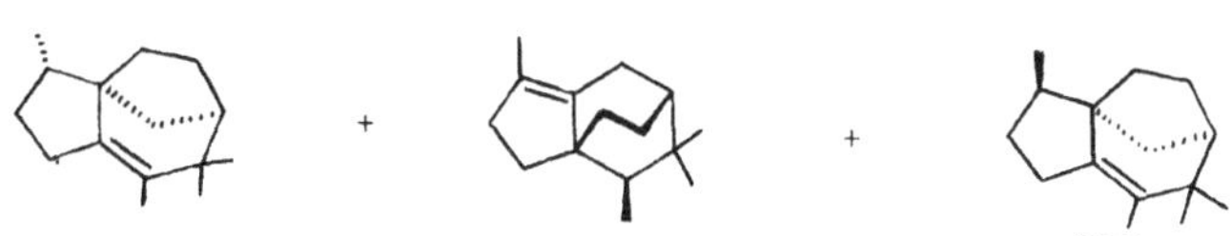

(**230**) (**231**) (**232**)

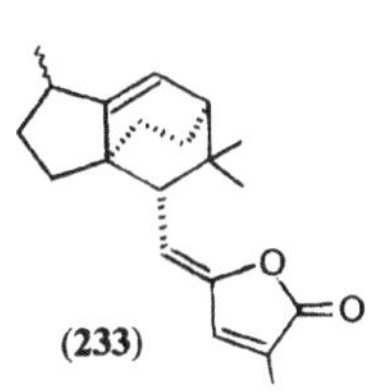

(**233**)

D. Diterpenes

1. Labdanes, Clerodanes, Pimaranes, and Rosanes

The first stage in the biogenesis of the most common types of diterpenes is a proton-initiated bicyclization of geranylgeranyl pyrophosphate (**234**) to the basic labdane or *ent*-labdane nuclei (*e.g.*, copalol, **235**-OH) (*300*) (Scheme 23). A four step backbone rearrangement with sequential hydride and methyl shifts gives rise to the clerodanes (*e.g.*, in the enantiomeric series, kolavenol, **236**). The discovery of such rearranged bicyclic diterpenes having *cis* decalin structures (*e.g.*, plathyterol, **237**) indicates that *syn* backbone rearrangements can occur. It is notable that the stereochemistry of plathyterol requires two *syn* migrations in the biogenesis. A characteristic feature of the structure of diterpenes is the wide-spread occurrence of compounds belonging to both enantiomeric series.

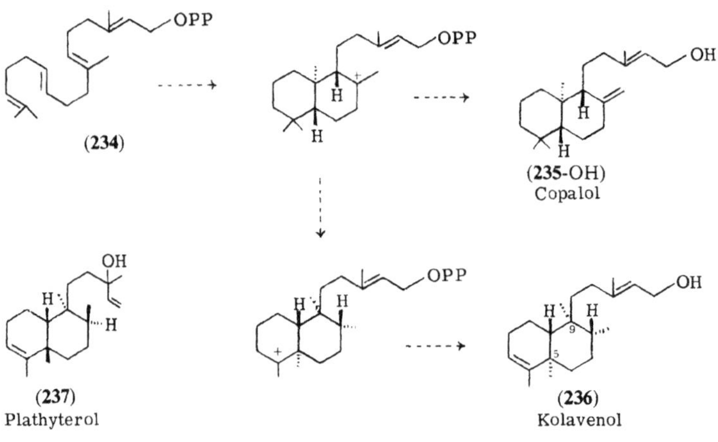

Scheme 23

HADLEY and HALSALL have recently investigated the Lewis acid-catalyzed rearrangements of various labdane epoxides and the bicyclic model compounds (**238**) and (**239**) (*301*). The finding that angular methyl rearrangement occurs with facility in both isomers demonstrates once again that the migrating group need not necessarily be situated *anti* to the leaving group at a tertiary position. Thus, it seems likely that the migration of the methyl group occurs in a second, rapid step following cleavage of the epoxide bond. These rearrangements are considered to have a substantial steric driving force since a 1,3-methyl-methyl diaxial interaction is eliminated in the process.

(238) (85%)

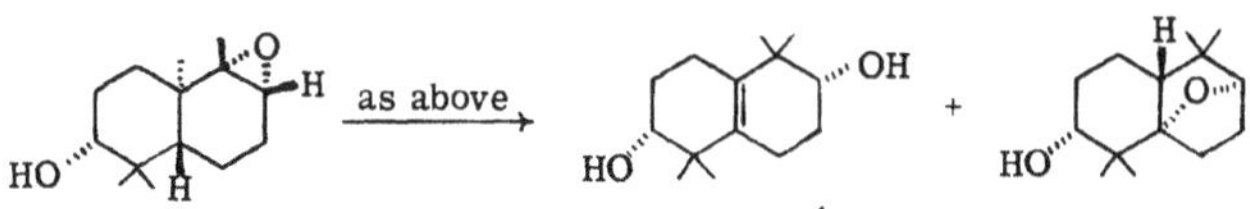

(239) (70%) (15%)

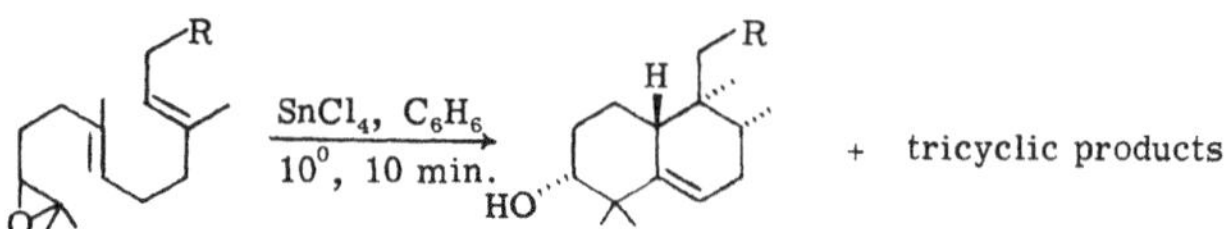

(240) (R = farnesyl) (241) (20-25%)

SHARPLESS and VAN TAMELEN have reported that one of the products obtained from the action of stannic chloride upon squalene 2,3-oxide (**240**) is the hydroxy octalin (**241**) (*302*). In this case, both bicyclization and a three-stage backbone rearrangement have been effected in a single reaction. That no products of the clerodane type (*e.g.*, **236**) have been found in these reactions is attributed to the fact that migration of the C_4 methyl group to C_5 is accompanied by the creation of a 1,3-methyl-methyl diaxial interaction. In support of this explanation is the observation that dolabradiene (**251**) and the acetonides of the erythroxy diols (tricyclic analogues of the clerodanes) rearrange to rosadienes as major products (see below) (*303, 304*). Similarly clerodane-3,4-oxides undergo acid-catalyzed rearrangement to products in which the C_5 methyl group returns to C_4 (*305*). The importance of steric effects is further illustrated by the isomerization of trimethyl octalin (**242**) to (**243**) in which the reverse rearrangement is now favored (*306, 307*).

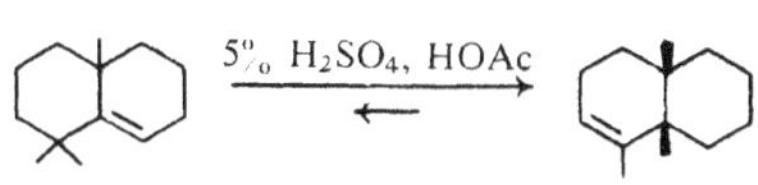

(6%) (242) (94%) (243)

Copalyl pyrophosphate (**235**-OPP) has been demonstrated to be a key branching point in the biosynthesis of tricyclic and tetracyclic diterpenes (*24, 300, 308*). It, therefore, seems likely that $\Delta^{8(20),13}$-labdadienyl pyrophosphate (**244**-OPP) is the precursor to the corresponding enantiomeric diterpenes as shown in the biogenetic scheme below (Scheme 24). Beyond the existence of both enantiomeric series of these tricyclic diterpenes, there is a further stereochemical diversity which results from the cyclization of (**235**-OPP) and (**244**-OPP) to sets of stereoisomers epimeric at the new quaternary carbon at C_{13}, *e.g.*, pimaric acid (**246**) and sandarocopimaradiene (**247**).

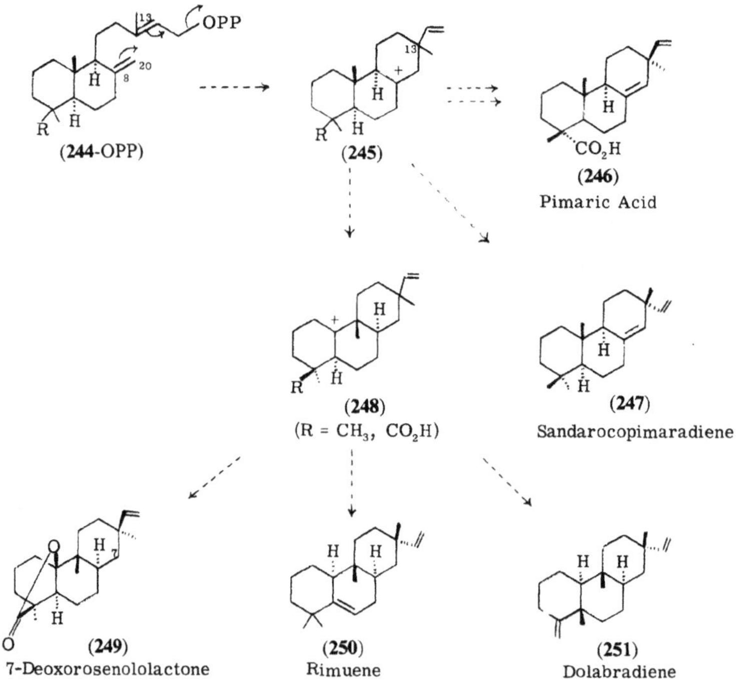

Scheme 24

The occurrence of the rosane lactones (*e.g.*, **249**) raises the interesting question of whether a carbonium ion intermediate (**248**, R = CO_2H) is neutralized by lactonization immediately following the shift of the methyl group from C_{10} to C_9. If so, oxidation of the methyl group would obviously have to precede cyclization. Separate incorporation of mevalonate bearing tritium at positions 4 and 5 into rosenonolactone

(**249** with ketone at C_7) without loss of the label has excluded the intervention of olefinic intermediates having a double bond at either the Δ^5 or $\Delta^{5(10)}$ positions (*21, 300*). An alternative to the combined cyclization-rearrangement-lactonization pathway is hydration of (**248**) ($R = CH_3$) to a 10-hydroxyrosane, with oxidation and lactonization taking place at a later stage.

There have been several investigations upon the cyclization of labdadienol (**244**-OH), manool (**252**), and related compounds to pimaradienes and rosadienes by means of acid catalysis (HCO_2H, HCO_2H/co-solvent, AcOH, and H_2SO_4/aq. AcOH) (*309—313*). Since the same distribution of products is obtained from labdadienol (**244**-OH), manool (**252**), and 13-*epi*-manool, it is clear that cyclization occurs subsequent to ionization of the starting alcohol to the allylic carbonium ion. The C_{13} epimers, $\Delta^{8,15}$-pimaradiene (**253**) and $\Delta^{8,15}$-isopimaradiene (**254**), are formed in approximately equal amounts as the major tricyclic products (43—67%), although small amounts (1.1—1.8%) of pimaradienes with the annular double bond in the Δ^7 and $\Delta^{8(14)}$ positions, as is typically the case with the natural diterpenes, have been detected (*313*). Acid-catalyzed isomerization of these $\Delta^{7,15}$- and $\Delta^{8(14),15}$-pimaradienes and isopimaradienes to the more stable tetrasubstituted double bond isomers (**253**) and (**254**) (to the extent of ~ 70—93%) appears to be a faster process than further rearrangement to the rosadienes (**255** and **256**) (*311, 313*).

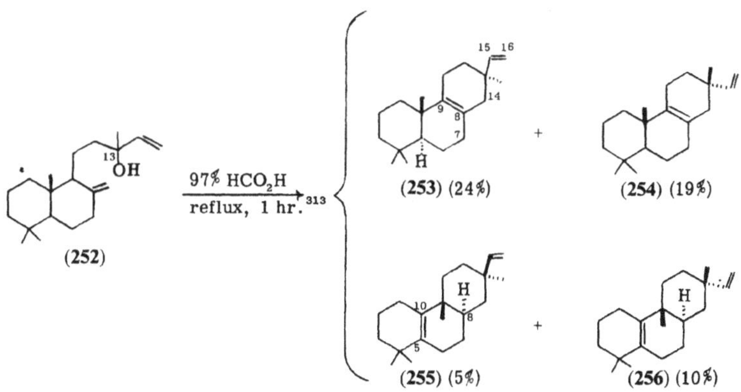

(**252**) (**253**) (24%) (**254**) (19%) (**255**) (5%) (**256**) (10%)

Although relatively small amounts of the $\Delta^{5(10),15}$-rosadienes (**255**) and (**256**) are produced initially, the proportion of these rearranged structures increases to 23—50% with longer reaction times, stronger acids, and/or higher temperatures (*312, 313*). Small amounts (5—11%) of the natural rosadiene, rimuene (**250**), are also formed (*311, 313*). As in the pimaradienes, rimuene is isomerized ($HCl/CHCl_3$) to its more stable tetrasubstituted isomer, $\Delta^{5(10),15}$-rosadiene (**256**) (*314*). Separate exposure of the epimeric rosadienes (**255** and **256**) to refluxing formic acid regenerated some of the corresponding pimaradienes (**253**, 20% after 30 min.; **254**, 3.6% after 4 hr.) establishing that the backbone rearrangement is reversible, despite the strain of the 1,3-methyl diaxial interaction in these structures (*313*). Although protonation of the $\Delta^{8,15}$-pimaradienes (**253** and **254**) occurs from the α direction producing, after methyl shift and elimination, rosadienes having the natural 8α stereochemistry at C_8, with prolonged reaction times the β epimer builds up *via* reversible isomerization to the pimaradienes (*303, 313*). Eventually the vinyl group becomes involved and, after protonation and migration of the C_{13} methyl group to C_{15}, is converted to an isopropyl substituent (*312, 313*).

As mentioned previously, dolabradiene (**251**), and the acetonides of erythroxy-diol X (enantiomer of devadrene (*cf.* section C.4) with dimethyl dioxolane instead of C_{13} vinyl) and erythroxydiol Y (**251** with dimethyl dioxolane instead of C_{13} vinyl group) undergo acid-catalyzed rearrangement of the C_5 methyl to C_4, demonstrating once again the greater stability of the rosadienes (*303, 304*). The formation of $\Delta^{8,15}$-isopimaradiene (**254**, 26%) by complete reversal of the biogenetic backbone rearrangement is noteworthy (*303*).

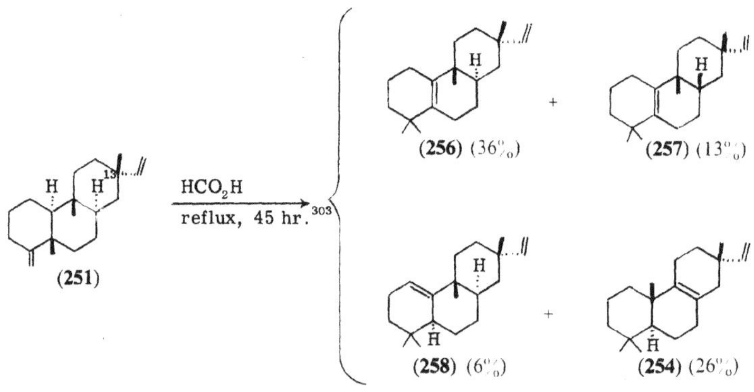

The backbone rearrangement of pimaranes and abietanes has also been brought about through the action of Lewis acids upon 8,9- and 8,14-epoxides (*315—317*). Two biogenetically patterned syntheses of rosenololactone feature a $C_{10} \rightarrow C_9$ methyl group rearrangement in order to construct the rosane skeleton (*317, 318*).

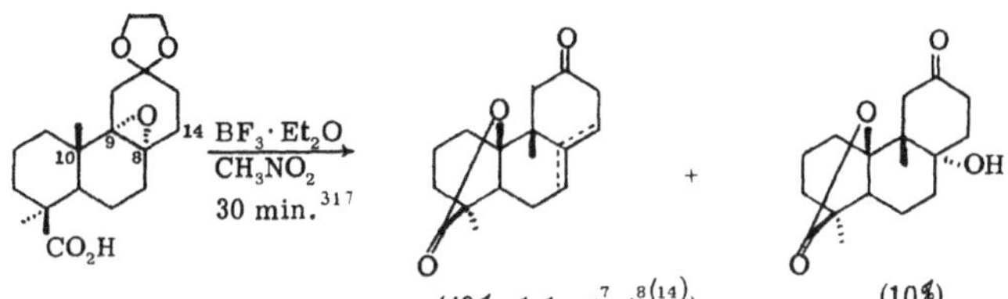

(49%; 1:1, $\Delta^7:\Delta^{8(14)}$) (10%)

2. Abietanes, Strobane, and Cleistanthane

The members of this small family of tricyclic diterpenes have in common an ostensible biogenetic relationship to the pimaradienes. In the usual biogenetic schemes, rearrangement of a substituent (methyl or vinyl) from the C_{13} quaternary position follows protonation of an appropriate pimaradiene precursor. Although the biogenesis of the new diterpene strobic acid (**261**) (*319*) shown below (*320*) fits in nicely with the scheme previously proposed for the abietanes (*2, 5*), an alternative, and quite plausible, origin for the former is a proton-induced cyclization of *cis*-biformene ($\Delta^{8(17),12,14}$-labdatriene), or possibly an oxygenated variant thereof (*320*). The C_{13} stereochemistry of the pimaradiene precursor has been left unspecified since there seems, at present, no basis for selecting one or the other epimer.

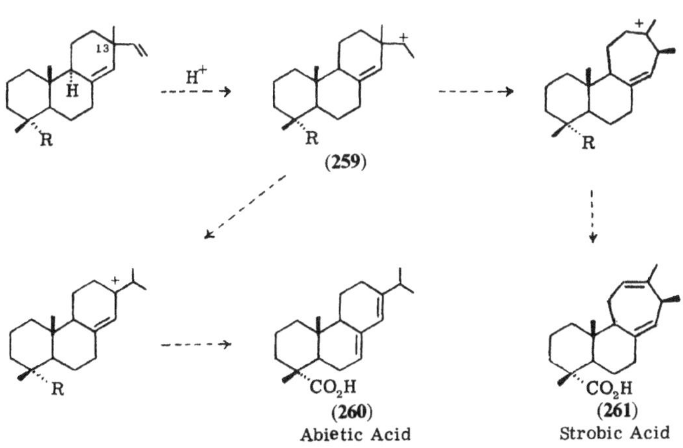

Scheme 25

The acid-catalyzed rearrangement of both pimaric acid (**246**) and isopimaric acid (**262**) to abietic acid (**260**) has been realized by Wenkert and Chamberlain (*321*). Evidently the $C_{13} \rightarrow C_{15}$ methyl migration occurs equally well with either C_{13} stereochemistry. Although both γ and δ-lactones are also formed in this reaction through the previously discussed backbone rearrangement, it is remarkable that protonation at the vinyl group is able to compete with attack on the trisubstituted annular double bond. The pimarane→abietane rearrangement has also been effected through both aprotic deamination of the amino pimarene **263**-NH_2 with isoamyl nitrite in diglyme and perchloric acid-catalyzed elimination of acetic acid from the corresponding acetate (**263**-OAc) in dioxane-acetic acid (*320*). The abietane product obtained from these reactions (20% and 58%, respectively) was dehydroabietic acid (**260** with ring C aromatic), presumably *via* disproportionation of a homoannular diene isomer of (**260**).

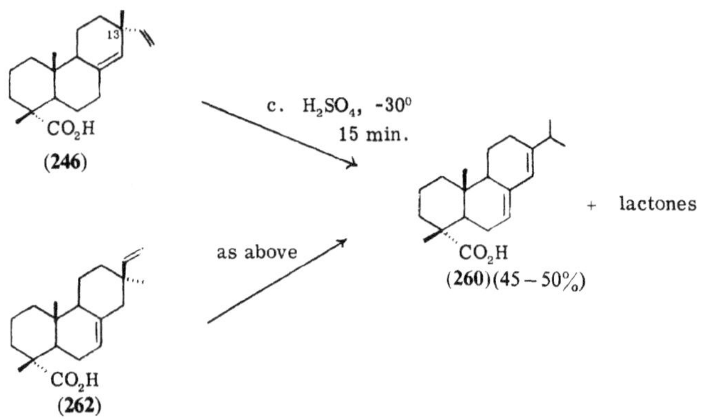

The ring-expansion rearrangement to the strobane nucleus has been observed by Herz and Hall in a study of the reactions of 15-substituted pimarenes (*320*). Thus, hydroxy ester **263**-OH affords the strobane-like isomer (**264**) upon reaction with tosyl chloride in pyridine. The (**263→264**) rearrangement apparently proceeds by way of a stereospecific homoallyl→cyclopropylcarbinyl→homoallyl pathway since substitution reactions at C_{15} occurred with complete retention of configuration in all cases. Hydroxy ester (**264**) reverted stereospecifically to **263**-OAc (78%) in refluxing acetic acid. These transformations resemble those of the sesquiterpenes thujopsene and widdrol (section C.6). Since (**264**) has the C_{14} stereochemistry opposite to that of strobic acid (**261**), the C_{15} epimer of **263**-OH might be expected to furnish products with the proper C_{14} orientation.

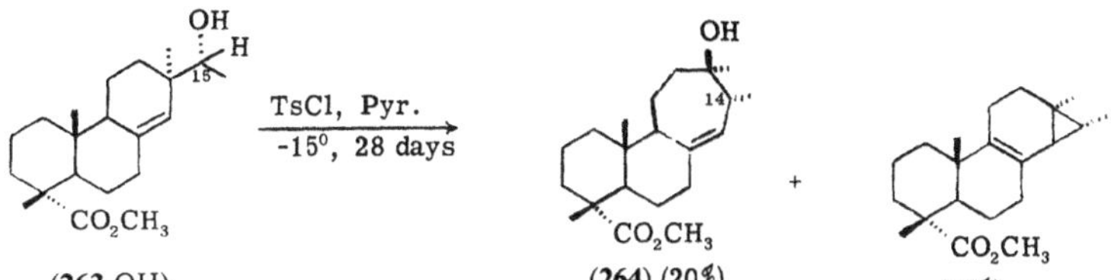

(263-OH) (264) (20%) (22%)

The cassane diterpenes (*e.g.*, cassaic acid, **265**) and the one known cleistanthane, cleistanthanol (**266**), may be derived from an $\Delta^{8(14),15}$-pimaradiene progenitor (possibly bearing additional functionality) through protonation of the nuclear double bond at C_8 and shift of the C_{13} methyl or vinyl group respectively to C_{14} *. Attempts to effect the $C_{13} \rightarrow C_{14}$ methyl rearrangement with dimethyl 7-oxoisopimar-$\Delta^{8(14)}$-ene-16,18-dioate and related compounds having the C_7 oxygen function of the cassanes were not successful (*322*). However, the highly functionalized pimaradiene lactone (**267**) affords products related to cleistanthol under the influence of toluenesulfonic acid in acetic anhydride (*323*).

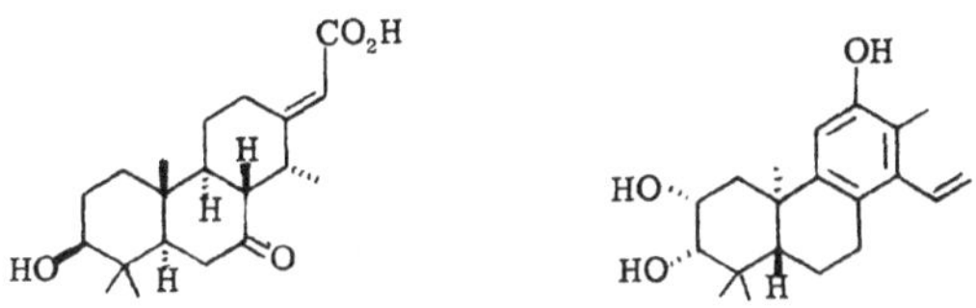

(**265**)
Cassaic Acid

(**266**)
Cleistanthol

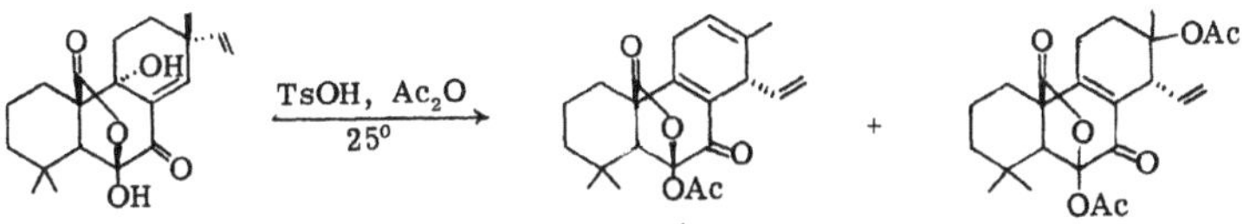

(**267**) (20%) (6%)

* The requisite pimarenyl cation might also be formed more directly from (**245**) (scheme 24) *via* a 14→8 hydride shift.

3. Beyeranes, Kauranes, Atisanes, and Trachylobanes

The biogenetic relationships among the members of this complex group of tetra- and pentacarbocyclic diterpenes were first recognized by WENKERT (*324*). The biogenetic scheme (Scheme 26, updated to include new representatives and depicting classical cations for the sake of clarity*) begins with cyclization of copalyl pyrophosphate (**235**-OPP) to the pimarenyl cation (**268**) (enantiomer of **245**) which further cyclizes upon the remaining double bond in the vinyl group to form the key intermediate (**269**), the immediate precursor of (+)-beyerene (**270**). A Wagner-Meerwein rearrangement gives rise to the tertiary cation isomer (**271**) which upon 1,2- and 1,3-hydrogen elimination yields (−)-kaurene (**272**) and (−)-trachylobane (**273**) (*325*), respectively. The genesis of the bicyclo-[2.2.2]octane moiety of atiserene (**274**), and the related atisine alkaloids, entails a 1,3-hydrogen shift ($C_{12} \rightarrow C_{16}$), Wagner-Meerwein rearrangement, and elimination. The elements of this biogenetic scheme are supported by the co-occurrence of these diterpenes (*325*) and biosynthetic investigations (*300*). Although a complete set of structures stereoisomeric to those depicted below may be similarly generated from the C_{13} epimer of (**268**), the only known natural product of this type is phyllocladene (enantiomer of **272** with ethano bridge *anti* to the C_9 proton) and a few of its derivatives.

In the previously discussed cyclization of manool (**252**) to pimaradienes and the subsequent backbone rearrangement of the latter to rosadienes, the enantiomer of cation (**268**) (*i.e.*, **245**) must certainly be an intermediate. However, despite careful analysis of the reaction mixtures, neither diterpenes of the type shown in the scheme above nor those in the stereoisomeric series (*e.g.*, phyllocladene) could be detected (*312, 313, 326*). Although a tetracyclic formate was actually isolated (8—12%) from treatment of manool with formic acid, labelling experiments have clearly demonstrated that this product arises by initial cyclization to an eight-membered ring and not *via* a pimarenyl cation (*313, 327, 328*). The reluctance of (**268**) to cyclize to the tetracyclic beyeranyl ion (**269**) is attributable to both unfavorable energetics (tertiary→secondary carbonium ion) and poor orbital overlap. The biosynthetic enzymes, therefore, assume a crucial role at this stage in the production of tetracyclic diterpenes *in vivo*.

* A face protonated trachylobane was originally suggested as a key intermediate in the biogenetic scheme (*324*). Since the three-fold symmetric, face-protonated nortricyclene has more recently been discredited as either an intermediate or transition state in the hydrogen shift rearrangements of 2-norbornyl derivatives (*157a*), it is appropriate to discontinue usage of this representation.

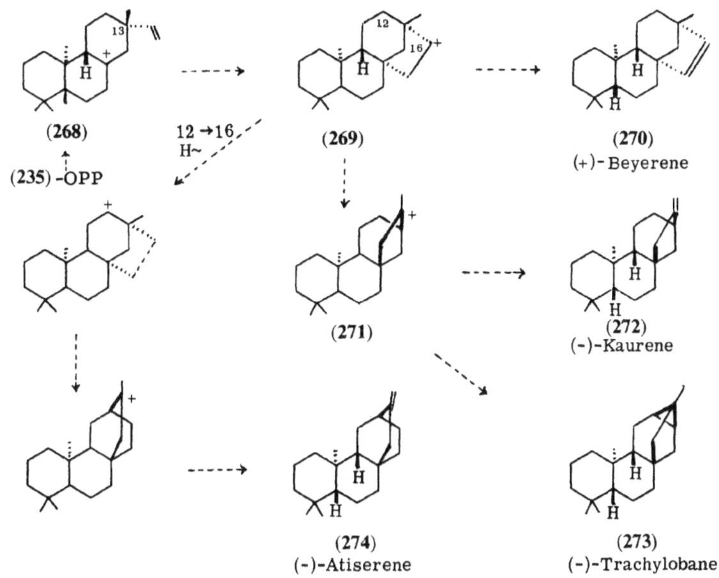

Scheme 26

Considerably greater success has attended the many investigations aimed at simulating the subsequent transformations in the WENKERT biogenetic scheme. Three principal approaches have been employed to generate the beyeranyl and kauranyl cations (**269** and **271**, respectively): reaction of the various olefinic diterpenes and trachylobane with protic acids (*329—331*) or iodine (*332, 333*), reaction of the appropriate epoxy diterpene with acids (*332, 334—340*), and heterolysis of suitably functionalized diterpenes (*341—346*). Similar rearrangements of isohibaene (stereoisomer of **270** with olefin bridge *syn* to C_9 hydrogen) have been reported (*347*). From these investigations it is evident that the Wagner-Meerwein rearrangement, beyeranyl→kauranyl (**269**→**271**), takes place readily. Although this rearrangement has the kinetic driving force resulting from the conversion of a secondary carbonium ion to a tertiary one, it is at the expense of an increase in steric strain. It will be noted that rings B and C are effectively transformed from a *trans*-decalin to a *cis*-decalin, a change which also creates a diaxial alkyl-alkyl steric interaction ($C_{10}-CH_3$ with $C_{12}-H_2$).

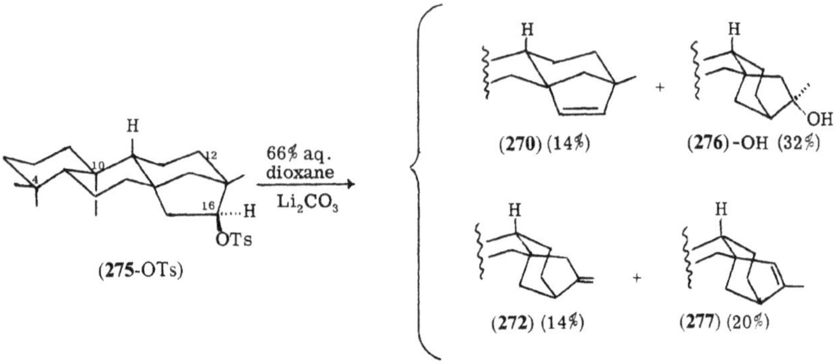

(**275**-OTs) — 66% aq. dioxane, Li_2CO_3 → (**270**) (14%) + (**276**)-OH (32%); (**272**) (14%) + (**277**) (20%)

In the hydrolysis of beyeran-16β-yl tosylate (**275**-OTs) (*341*), solvolyses of oxygenated derivatives of **275**-OTs (*342, 346*), and deamination of the corresponding amine (*346*), mixtures of kaurene (**272**) and isokaurene (**277**) are usually obtained. Similarly in the acid- and iodine-catalyzed rearrangements (*330, 332, 333*) of beyerene and kaurene, mixtures (~2–4:1) of the two kaurene isomers result, the more stable endocyclic isomer predominating. On the other hand, apparently regioselective formation of either isokauren-14-ols (*332, 334*), or kauren-14-ols (*335, 336*) has been reported in the epoxide rearrangements. The reverse rearrangement, i.*e*., kaurene→beyerene has been observed under conditions of iodine (*332, 333*) and acid catalysis (*331, 346*).

The reaction of trachylobane (**273**) with protic acids gives rise to products of both the atiserene and kaurene types, the former usually being favored to a substantial extent (*329, 330*). Thus, cleavage of the two more highly substituted cyclopropane bonds occurs with protonation at the respective less high substituted positions (C_{12} and C_{16}) so as to generate tertiary cations. The apparent preference for C_{16} protonation

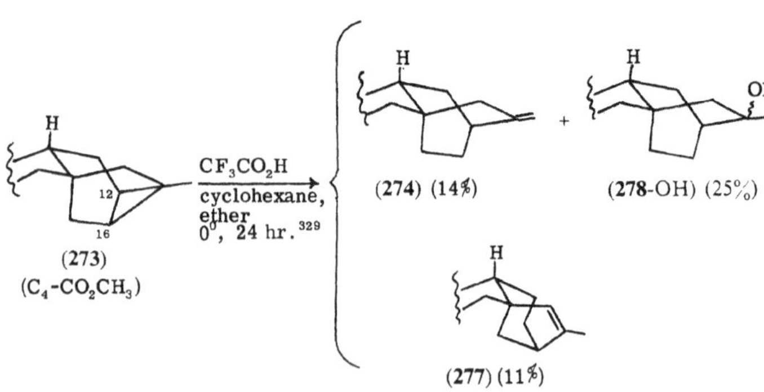

(**273**) (C_4-CO_2CH_3) — CF_3CO_2H, cyclohexane, ether, 0°, 24 hr.[329] → (**274**) (14%) + (**278**-OH) (25%); (**277**) (11%)

may be attributed to incipient steric strain in the transition state leading to the kauranyl cation (the $C_{10}-CH_3 \cdots\cdot C_{12}H_2$ interaction mentioned above) and/or an inherently greater stability of a cyclohexyl carbonium ion (less angle strain) (*330*).

Under the kinetically controlled conditions of the solvolysis reactions and amine deamination, there is no evidence for either 1,3-hydrogen elimination to trachylobane or 1,3-hydrogen shift. Trachylobane formation (12%) has, however, been observed in the decomposition of the tosylhydrazone (**279**) under high temperature, protic conditions (*346*). Since aprotic decomposition (presumably *via* carbene), produces only olefinic products, it would appear that the cyclopropane ring is formed through a cationic pathway.

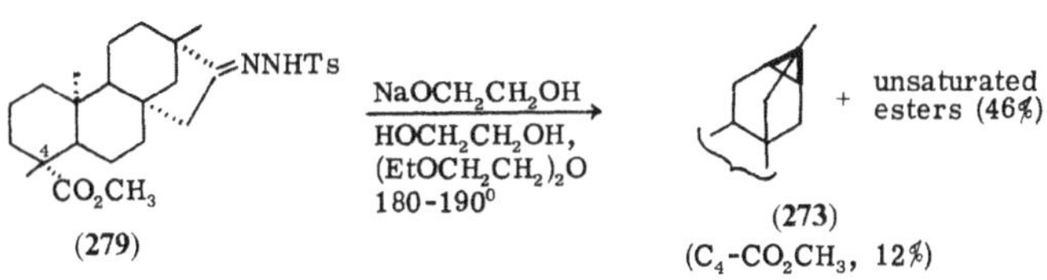

Conversion of kaurene and beyerene to the atiserenes has been observed after prolonged exposure of the diterpene hydrocarbons to hydrogen chloride in chloroform (6–14% after 4–21 days at 25°) (*330*). Similarly prolonged solvolysis (25°, 20 hr.) of the beyerane tosylate **275**-OTs (with $C_4-CO_2CH_3$) (*346*) or kaurene (*331*) in buffered trifluoroacetic acid, a medium of relatively low nucleophilicity, affords (after saponification)

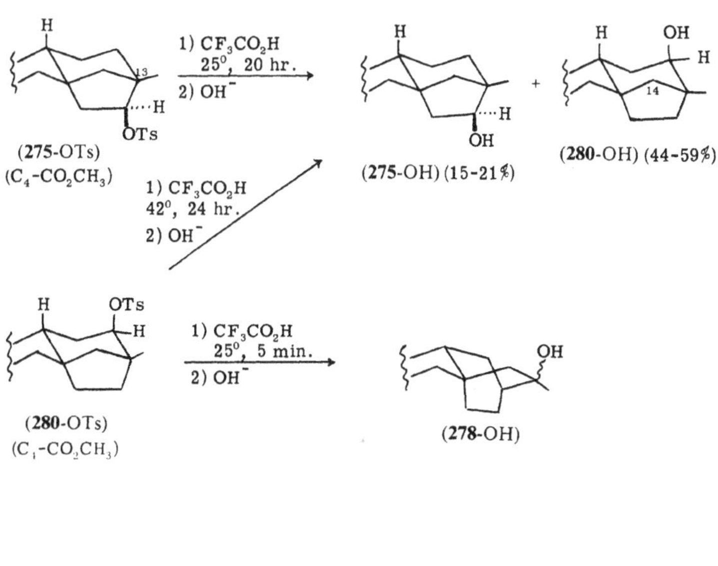

12β-hydroxybeyeranes (**280**-OH, 44–59%). This product is formed by rearrangement of a transient atiserane intermediate which is unstable with respect to **280**-O_2CCF_3 in the thermodynamically-controlled solvolysis. The reversibility of the rearrangement (**275**-OR ⇄ **280**-OR) was shown by the reformation of **275**-OH (15%) from trifluoroacetolysis of **280**-OTs (*346*). The isolation of atiseranol (**278**-OH) from brief exposure of **280**-OTs to trifluoroacetic acid indicates that this tertiary alcohol is the kinetic product (*346*). A closely related, biogenetic-type rearrangement is the acid-catalyzed transformation of jativatriol (Δ^{15}-beyerene-1α,9β,17-triol) to the atiserene-type diterpene sideritol (*348*).

When the prolonged solvolysis of **275**-OTs was carried out in *deuterio*-trifluoroacetic acid, the resulting **280**-OH contained mainly five deuterium atoms (77% d_5), indicating that rapid elimination and reprotonation occurred during the solvolysis (*346*). Nevertheless it was possible to show that the **275**-OTs→**280**-OR rearrangement takes place by way of a 16→12 hydrogen shift rather than through elimination to and reprotonation of a trachylobane intermediate since the latter afforded **280**-OR with incorporation of an additional deuterium atom. It is of interest to note that products resulting from the 16→12 hydrogen shift are produced in the kinetically-controlled acetolysis of 17-norphyllocladan-16β-yl tosylate (enantiomer of **275**-OTs without C_{13}–CH_3) (*344*) and the model compound *exo*-bicyclo[3.2.1]octan-6-yl tosylate (*343*), both of which lack the bridgehead methyl group.

Another example of the kaurane→atiserane rearrangement is the transformation of isokaurene epoxide (**281**) to atisiran-15-one (**282**) (*338, 339*). This remarkable rearrangement apparently proceeds through the four step sequence: Wagner-Meerwein shift (**271**→**269**), 12→16 hydrogen shift, Wagner-Meerwein shift, and 15→16 hydrogen shift. Similar treatment of isophyllocladene epoxide effects rearrangement to neo-atiseran-15-one (*337, 338*).

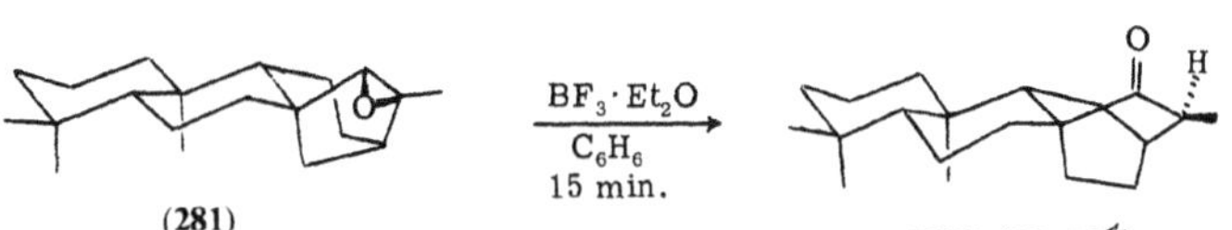

(**281**) (**282**) (29-36%)

4. Gibberellanes, Grayanes, and Aconanes

The kaurane and atisane diterpenes, in a variety of forms, serve as substrates for the elaboration of an array of complex, highly functionalized natural products including the gibberellins, the grayanotoxins, and the aconite alkaloids (*e.g.*, talatizidine, **286**). The biogenesis of each of these three ring systems appears to involve a skeletal rearrangement of a tetracyclic diterpene intermediate subsequent to the cyclizations and rearrangements discussed in the preceding section. Although the mechanism(s) by which such "secondary" biogenetic rearrangements are initiated is uncertain, it seems reasonable to suppose that the transformations resemble carbonium ion rearrangements.

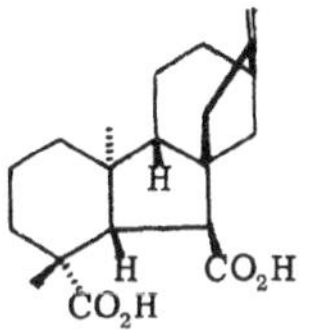

(283)
Gibberellin A_{12}

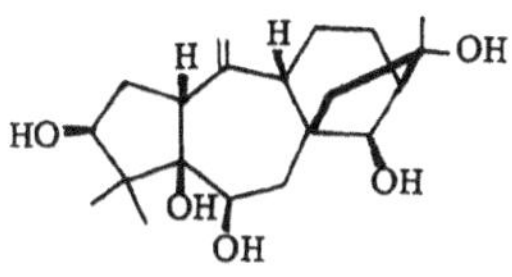

(284)
Grayanotoxin II

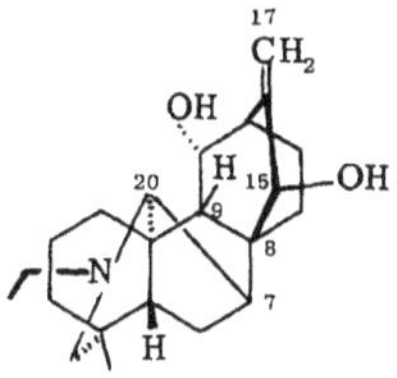

(285)
Denudatin

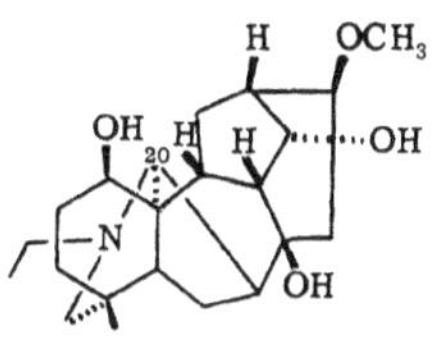

(286)
Talatizidine

The key step in the biogenesis of the gibberellane skeleton is the contraction of the six-membered B ring of a kaurane precursor. Biosynthetic investigations have established that 7β-hydroxykauren-19-oic acid is the immediate predecessor of the gibberellins, undergoing an oxidative, pinacol-like rearrangement to gibberellin A_{12} aldehyde (**288**) with extrusion of C_7 (*300, 349—351*). Chemically induced pinacol rearrangements of kauran-6α,7α-diol monosulfonates have, in fact, been commonly employed (*351—356*) as a means of preparing gibberellin aldehydes (*e.g.*, **287**→**288**) (*355*). It will be noted, however, that the sulfonate leaving group is located at C_7 in all cases reported; thus C_6 rather than C_7 is extruded in these reactions. In a concerted pinacol ring contraction, the rearranging ring bond (C_5–C_6) should migrate to the back-side of the carbon bearing the equatorial leaving group (C_7); hence the aldehyde epimeric to (**288**) (α-CHO) would be expected as the initial product. Perhaps the alkaline conditions of the reaction cause epimerization at C_7 subsequent to the rearrangement.

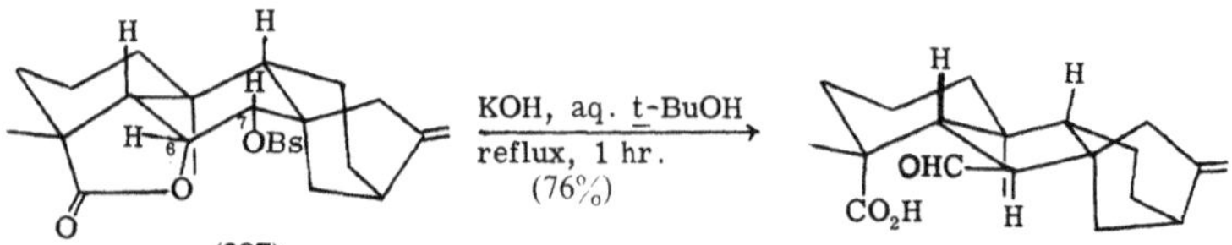

(287) (288)

The biogenesis of the grayane diterpenes entails migration of the C_5–C_{10} interannular bond to C_1 in a kaurane precursor, thus transforming rings A and B from a decalin to a perhydroazulene part-structure. Rearrangements analogous to this process have been reported with C_1-functionalized triterpenes (*357*), steroids (*358, 359*), and a beyerane diterpene (*360*). Solvolysis of 1-tosyloxydecalins (see section C.2) (*125, 126, 128*), and 4-methanesulfonyloxy-1,2,3,4,4a,9,10,10a-octahydrophenanthrenes (*e.g.*, **289**) (*361*) effects rearrangement to perhydroazulene-type products provided the sulfonate leaving group is equatorial and anti-parallel to the interannular bond.

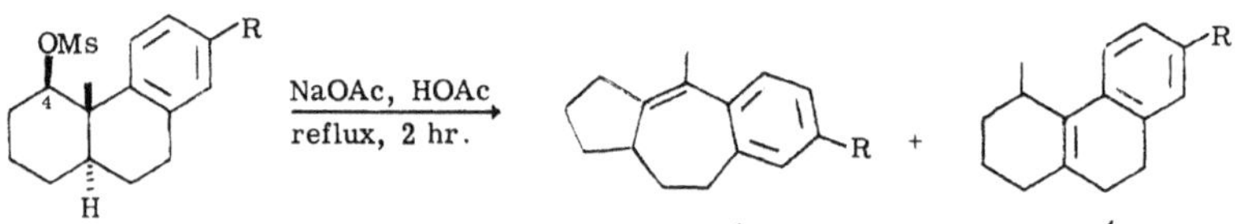

(289)
(R = H, OCH_3)

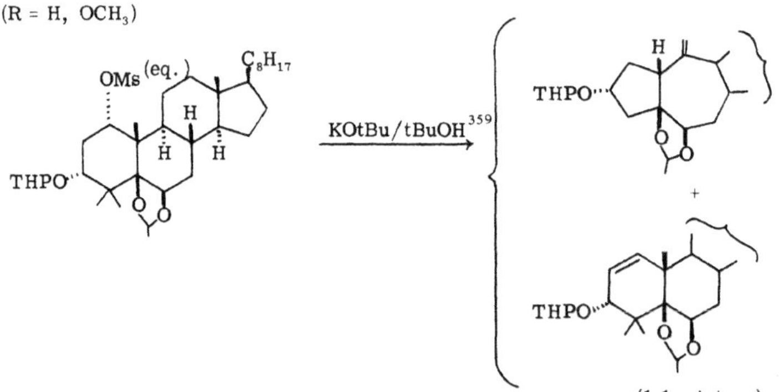

In the epimeric series with the leaving group axial and anti-parallel to the angular methyl group, rearrangement of the methyl group takes place (*358, 361, 362*). Since the rates of acetolysis of (**289**) and its C_4 axial epimer are only slightly affected by the *p*-methoxy substituent (k_H/k_{OCH_3} = 0.75 and 2.2, respectively), the transition state apparently bears little resemblance to the benzylic carbonium ions formed in the rearrangement (*361*). The kinetics and stereoselectivity may be accommodated by either slow ionization to an ion pair (or slowly interconverting, conformationally-isomeric secondary carbonium ions) followed by relatively rapid rearrangement of the requisite C–C bond, or a concerted ionization-rearrangement mechanism with an early transition state having little benzylic character.

The overall transformation of the atisane skeleton (*e.g.*, atiserene, **274**) to the C_{19} aconite alkaloids, apart from oxidative and functionalization steps, requires three principal structural alterations; C_7–C_{20} bond connection, removal of one carbon (C_{17}), and rearrangement of the C_9–C_8 bond to C_{15}. Although the order of these alterations is not known, the isolation of the atisine-type alkaloid, denudatin (**285**), which possesses the C_7–C_{20} link, indicates that this step may occur first.

Since the initial suggestion that the aconane skeleton is biogenetically derived from an atisane precursor (*363–365*), there have been a number of attempts to effect the proposed $C_9 \rightarrow C_{15}$ rearrangement (*366—368*). Pelletier and Ichihara first reported that 15-hydroxy-17-noratisane derivatives undergo rearrangement of the ethanobridge (C_{13}–C_{14}), instead of the C_9–C_8 bond (as in **292**), regardless of the configuration of the hydroxyl group (*366*). Similarly Johnston and Overton found that buffered acetolysis of either the atisine-derived keto tosylate (**290**) or its C_{15} epimer affords only the bridge-rearranged keto olefin (**292**) (*367*). However, gas phase pyrolysis of (**290**) at 600° brings about migration of the C_9–C_8 bond, providing the *seco*-aconane derivative (**291**) in 77% yield (*367*). A cyclic transition state, in which the C_{14},β-proton is transferred to the departing sulfonate group, was proposed for this pyrolytic elimination.

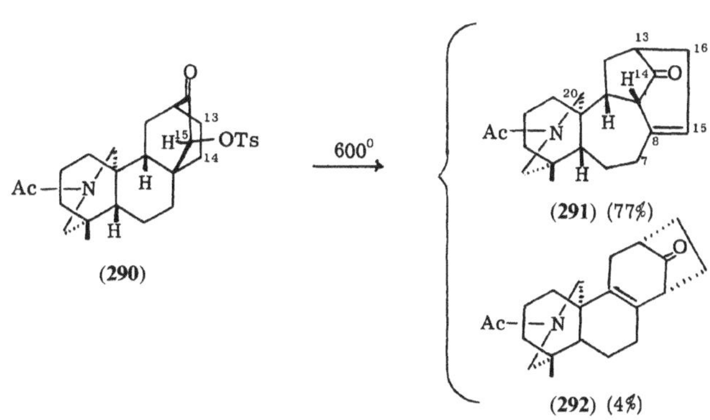

In the enantiomeric series, Ayer and Deshpande have achieved a similar rearrangement of an atisane-like diterpene tosylate in which the two carbon bridge (C_{13}–C_{14}) is unsaturated (*368*). The success of this rearrangement was attributed to charge delocalization by the Δ^{13} double bond in the transition state. Wiesner and co-workers have employed this type of rearrangement in syntheses of talatisamine (*369*) and a compound with the lycoctonine skeleton (*370*). The important discovery was made that elimination of toluenesulfonic acid from ketal tosylates corresponding to (**290**) with tetramethylguanidine-dimethylsulfoxide (180°, 24 hr.) afford 1 : 1 mixtures (80—85%) of the $\Delta^{8,15}$ enone (as in **291**) and its Δ^{7} isomer after ketal hydrolysis. The latter isomer is necessary in order to effect C_7–C_{20} bond formation.

A closely related transformation of possible biogenetic significance is the corresponding rearrangement of a kaurane type diterpene alkaloid derivative having the C_{20}–C_7 bond and functionalized with a 15β-methanesulfonate to a hexacyclic product differing from the aconanes in that C_{14} is attached to C_{16} instead of C_{13} (*cf.* **291**) (*371*). This finding suggests that the aconane skeleton could, in principle, be derived from a kaurane, instead of atisane, precursor, with a final rearrangement of C_{14} (numbering as in **291**) from C_{16} to C_{13}.

E. Triterpenes

1. Squalene

As a consequence of the important role of all *trans*-squalene (**295b**) as a precursor to the triterpenes and steroids, the biosynthesis of squalene has been a subject of great interest (*25, 26*). The key reaction in squalene biosynthesis is the unique head-to-head coupling of the C_{15} intermediate, *trans, trans*-farnesyl pyrophosphate. Although a number of interesting and significant mechanisms have been advanced at one time or another for this unusual reductive coupling transformation, for the present purposes it will be sufficient to discuss the biogenesis as it is now understood.

By means of experiments utilizing stereospecifically deuterated and tritiated precursors, it has been established that two molecules of *trans, trans*-farnesyl pyrophosphate (**112b**) are incorporated into squalene in an unsymmetrical fashion. Thus, a pro (S) hydrogen from C_1 of one farnesyl component is removed in the reaction and replaced with a hydrogen donated by reduced nicotinamide adenine dinucleotide phosphate (NADPH), a cofactor required for squalene biosynthesis. The entering hydrogen is attached to squalene in the (S)-configuration. The other farnesyl unit is incorporated intact, C_1 having undergone inversion of configuration in the coupling reaction.

RILLING discovered that, in the absence of NADPH, a C_{30} pyrophosphate-containing intermediate (presqualene pyrophosphate) may be isolated (*372*). Once the structure and stereochemistry of the derived alcohol were established as **113b**-OH through both chemical degradation (*170, 171, 373*) and total synthesis (*374—376*), the biogenesis of squalene, at least in general terms, became clear (*170, 373, 376—378*) (Scheme 27). As we have seen (Section B.1 and C.4), the structure of presqualene

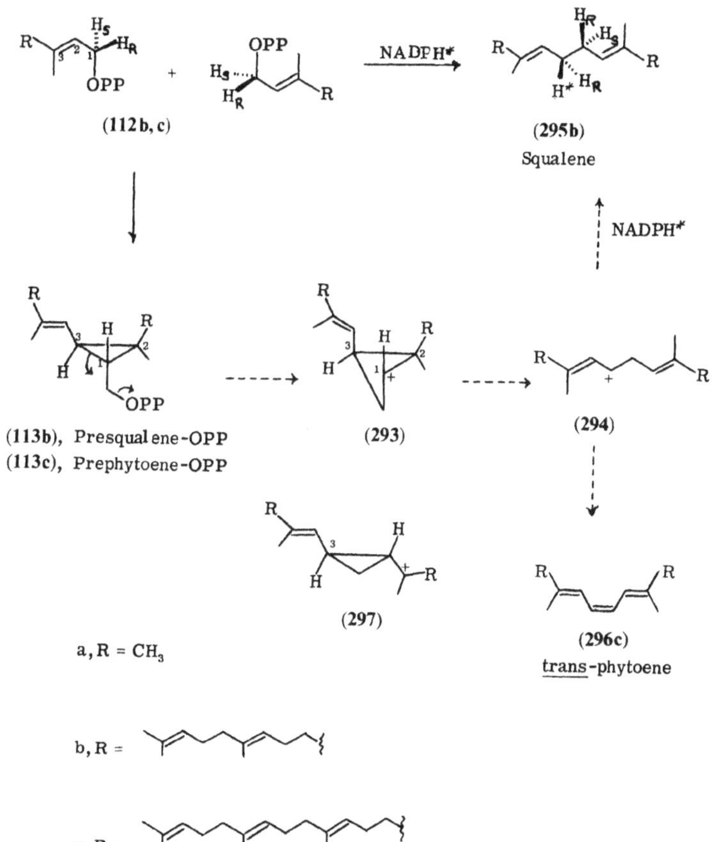

Scheme 27

pyrophosphate (**113b**-OPP) is an intermolecular variant of a familiar type of cyclopropane-containing terpene. The biogenesis of the three-membered ring may be viewed as a stepwise process involving initial bond formation between C_1 of one farnesyl unit and C_2 of the other with loss of pyrophosphate from the former. The resulting tertiary carbonium ion (or an equivalent tetravalent species bound to either pyrophosphate or a nucleophilic group on the enzyme) may subsequently undergo a 1,3-hydrogen elimination from C_1 of the first farnesyl unit to form presqualene pyrophosphate. This pathway is entirely analogous to the biogenesis of chrysanthemyl pyrophosphate presented in an earlier section (B.1).

The second stage in squalene biogenesis entails a cyclopropylcarbinyl-cyclobutyl-homoallyl rearrangement sequence similar to those discussed in connection with the thujopsene-widdrol interconversions (section C.6). Ionization of presqualene pyrophosphate (**113b**) followed by (or concerted with) migration of C_3 to the exocyclic CH_2 group gives rise to a secondary cyclobutyl cation (**293**). Cleavage of the 2,3-bond in the latter generates allylic carbonium ion (**294**), hydride transfer to which affords squalene. Since collapse of the cyclobutyl ion (**293**) to the more stable tertiary cyclopropyl carbinyl isomer (**297**) is expected to be rapid, if not instanteous (see below), the latter may well be a more suitable representation for this intermediate.

The above biogenetic scheme is consistent with the known stereochemistry of squalene biosynthesis. In the ring expansion rearrangement (**113b**→**293**), the configuration at the CH_2 group is inverted. The hydride transfer to C_3 also occurs with inversion and may therefore be concerted with ring opening (*i.e.*, effectively by-passing the open chain allylic carbonium ion **294**). The step-wise pathway is depicted in the scheme for the sake of clarity.

The biosynthesis of phytoene (**296c**), the parent of the carotene family, also proceeds *via* a cyclopropane intermediate referred to as prephytoene pyrophosphate (**113c**) (*172, 379*). Phytoene is derived from (**113c**) by the same rearrangement sequence, the termination step being proton elimination instead of hydride transfer.

The heterolytic reactions of *trans*-chrysanthemol derivatives (**16**-OPP ≡ **113a**-OPP), discussed previously (section B.1) in connection with monoterpene biogenesis, serve as a suitable model for investigation of the chemical behavior of this cyclopropylcarbinyl substrate. The principal products are formed by direct ring opening to compounds having the branched chain structures of the artemisane (C_1–C_3 cleavage) and, to a minor extent, the santolinane (C_1–C_2 cleavage) monoterpenes. Nevertheless, Poulter has been able to detect the presence of rearranged dienols **300**-OH and **301**-OH to the extent of 0.03% in the product mixture from hydrolysis of **16**-OPy^+I^- (*67*).

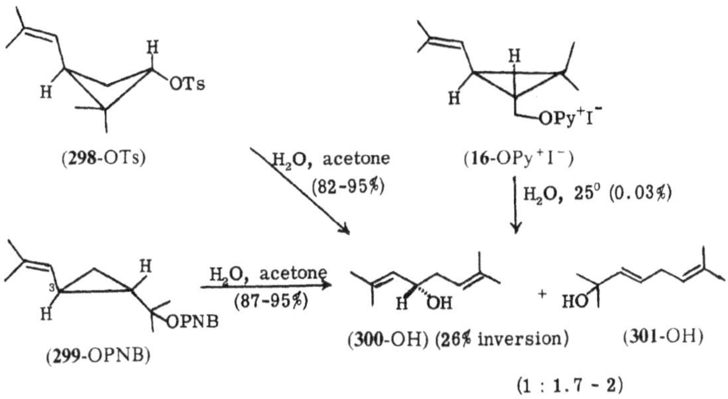

In contrast to the very inefficient rearrangement of **16**-OPy$^+$I$^-$ to squalene-like products, hydrolysis of *trans*-cyclobutyl tosylate (**298**-OTs), a potential progenetor of ion **293** (R = CH_3), affords dienols **300**-OH and **301**-OH in high yield (82—95%) (*67, 380—382*). Since the same product mixture is obtained from hydrolysis of the cyclopropylcarbinyl *p*-nitrobenzoate **299**-OPNB, the two reactions presumably pass through a common cationic intermediate which is very likely the tertiary cyclopropylcarbinyl carbonium ion **297** (R = CH_3). Consistent with this conclusion is the detection of a small amount (0.3%) of the cyclopropylcarbinol (**299**-OH) in the solvolysis product from both **298**-OTs and **299**-OPNB (*382*). The observation that hydrolysis of optically active **298**-OTs and **299**-OPNB affords optically active dienol **300**-OH with net inversion (26% inversion from both **298**-OTs and **299**-OPNB in 80% aqueous acetone; 44—48% inversion from **299**-OPNB in 50% aqueous dioxane) indicates that a large fraction of the product **300**-OH arises from nucleophilic capture prior to ring opening to the presumably racemic open-chain carbonium ion (**294**, R = CH_3) (*382, 383*). The stereochemistry is consistent with a direct attack of water upon C_3 of the cyclopropylcarbinyl carbonium ion intermediate (**297**, R = CH_3). The partial racemization found in these solvolyses is explained by competitive ring opening to the symmetrical allylic carbonium ion.

It is relevant to discuss briefly at this point the stereochemistry and mechanism of the cyclopropylcarbinyl-cyclobutyl-homoallyl rearrangements. Solvolysis of cyclopropylcarbinyl, cyclobutyl, and homoallyl derivatives bearing stereospecific labels or substituents lead to rearranged products in which the stereochemical relationships in the starting material are retained (*74—76*). The stereochemistry is illustrated below with the hydrolysis of specifically deuterated cyclopropylcarbinyl methanesulfonate

302-OMs which affords the d_6-products shown with $97 \pm 4\%$ stereospecificity ($\geq 90\%$ specificity in the case of the d_6-butenols) (*75*)*.

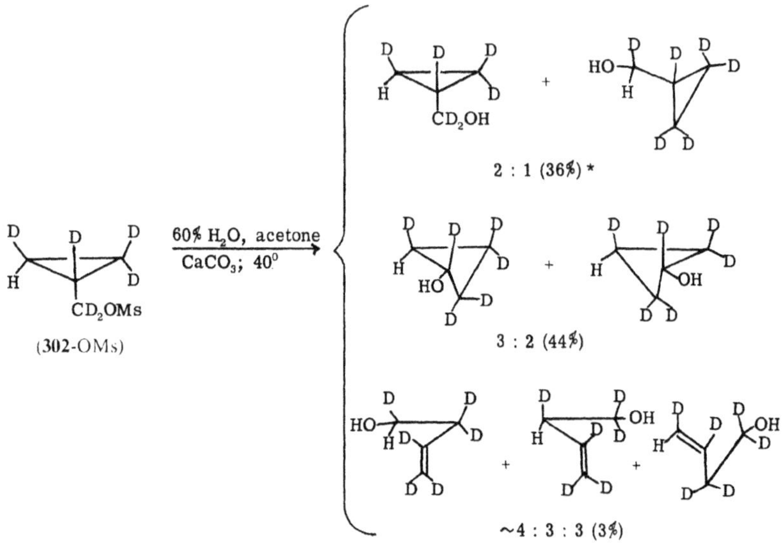

If a cyclobutyl carbonium ion is actually a discrete intermediate in the cyclopropylcarbinyl-cyclopropylcarbinyl rearrangement, the stereospecificity observed requires that the ion be unsymmetrical (puckered). With the assumption that the same stereochemical relationships are maintained in the proposed biogenetic rearrangement of presqualene pyrophosphate (**113b**) to squalene through cyclobutyl ion **293** (or its cyclopropylcarbinyl isomer **297**), it will be seen that the side chains must be *cis* to each other and *cis* to the C_1 hydrogen in the puckered conformation shown (**293**). Thus, the model cyclobutyl tosylate **298**-OTs with the unsaturated side chain and the leaving group in a *trans* (rather than *cis*) relation is the stereochemistry which corresponds to the puckered cyclobutyl carbonium ion (**293**).

It is significant that *trans*-cyclobutyl tosylate **298**-OTs is much more reactive in solvolysis than its *cis* isomer ($k_{trans}/k_{cis} = 250$) (*380, 381*) as is commonly the case with 3-substituted cyclobutyl sulfonates (*76*). The

* A certain amount (17%) of unrearranged cyclopropylcarbinol (**302**–OH) is assumed to have been formed by a direct S_N2 displacement (k_s pathway). The contribution of this pathway is not shown in order to emphasize the near-statistical label distribution in the rearrangement (k_Δ) pathway (*75*).

difference in the solvolytic reactivity between the *trans* and *cis* isomers may be explained by the expectation that both will ionize with concerted and stereospecific ring contraction to *trans* and *cis* cyclopropylcarbinyl carbonium ions respectively. The incipient steric interaction between eclipsing groups in the solvolytic transition state from the *cis* isomer is responsible for its slower rate (*76*).

Hydrolysis of the model cyclobutyl tosylates (**298**-OTs) in the presence of sodium borohydride furnishes the squalene-like diene (**295a**, 34%) and its double isomer **303** (12%) as well as the same two dienols (39%) mentioned previously (*380*). The reaction of **298**-OTs with pyridine in benzene at 70°, conditions conducive to formation of elimination products, provides symmetrical phytoene-like triene **296a** (56%) and its positional isomer **304** (28%). These transformations of cyclobutyl tosylate **298**-OTs mimic closely the later steps proposed in the biogenesis of squalene and phytoene from their respective cyclopropane predecessors.

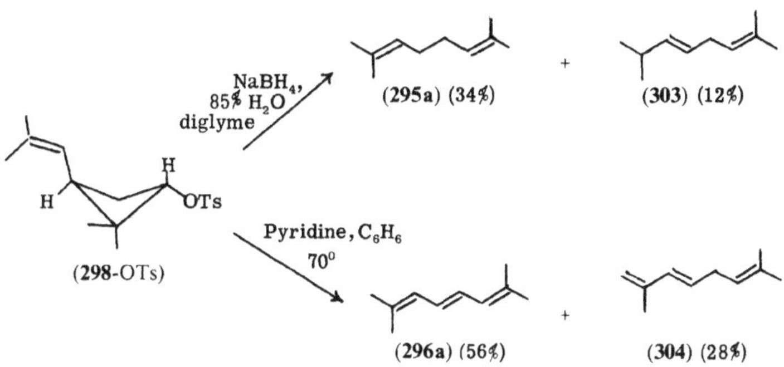

The facile ring opening reactions of **298**-OTs indicate that the steps following formation of cyclobutyl carbonium ion **293** (or equivalent species) in the postulated biogenetic scheme are inherently favorable. It seems clear, however, that the biosynthetic enzyme performs a vital function in suppressing the chemically dominant ring cleavage observed in the heterolytic reactions of presqualene analogues and in optimizing the ring expansion rearrangement. The absence of pertinent experimental information has not inhibited speculation upon means by which enzymatic control might be achieved. Since allylic stabilization provides a driving force for rupture of the C_1–C_3 bond, it has been suggested that immobilization of the adjacent double bond within the active site such that the *p*-orbitals are approximately orthogonal to the C_1–C_3 bond of presqualene pyrophosphate might help to prevent premature ring scission (*380*). Another explanation invokes the influence of the pyrophosphate counter-ion within an initially formed ion pair (*382*). Since ring cleavage results in relatively greater charge separation than bond migration, the ring expansion rearrangement would be rendered more favorable.

2. Lanostanes, Cycloartanes, Euphanes, and Tirucallanes

In contrast to the biogenetic schemes outlined for the mono-, sesqui-, and diterpenes, the biogenesis of most cyclic triterpenes (for some exceptions, see section E.6) is initiated by protonation of 3(S)-squalene-2,3-oxide (*25*, *26*, *384*) (Scheme 28). Multiple cyclization of the substrate folded into a chair, boat, chair-like conformation leads to key carbonium ion intermediate **306** (or its equivalent with an "X" group attached to C_{20}) proton elimination from which gives the protosterols (*e.g.*, $\Delta^{7(20)}$-protosten-3β-ol, **307**). Alternatively a sequence of 1,2-hydrogen and 1,2-methyl migrations along the tertiary, interannular positions terminated by 1,2- or 1,3-proton elimination at the B/C ring juncture gives rise to lanosterol (**309**-OH) and cycloartenol (**310**-OH), respectively. (Elimination of the 9β-hydrogen to form lanosterol could, of course, occur prior to the final hydrogen migration to C_8.) It will be noted that the closure of the cyclopropane ring in the latter (**308**→**310**) takes place *syn* to the hydrogen which has migrated from C_9 to C_8. Attachment of an "X" group to C_9 (α-configuration) of (**308**) may be postulated, in order to restore the presumably more favorable *trans, anti* stereoelectronic relationship. The results of numerous investigations upon the biosynthesis of fusidic acid (a metabolite of the protosterol **307** with the opposite $\Delta^{17(20)}$-double bond

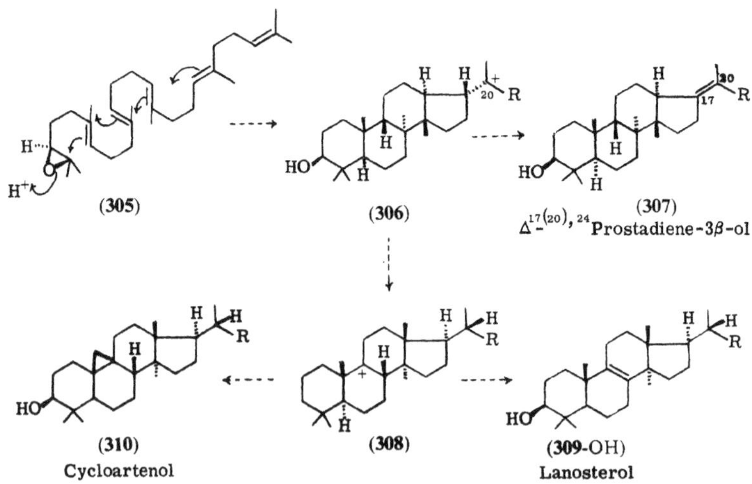

Scheme 28

configuration), lanosterol (or its metabolite, chlosterol), and cycloartenol from specifically labelled mevalonate or squalene are in full accord with the proposed biogenesis (*25, 26*). However, an alternative, somewhat more complex pathway, featuring only tertiary carbonium ions and involving a 1,3-hydrogen rearrangement from a spiro[4.4.0]nonyl-type intermediate, cannot be distinguished since the labelling pattern in the ultimate product is unaffected (*385*).

For the purposes of the present review, the backbone rearrangement protostenyl→lanostenyl (**306**→**308**) and the 1,3-hydrogen elimination to cycloartenol (**308**→**310**) are of primary concern. The acid-catalyzed rearrangements of dihydroprotosterol acetate **311**-OAc (R = 4-methylpentyl) (*386*) and dihydroprotostane-3,20-diol (dihydro **311** with C_{20}–OH) (*387*) to 24,25-dihydrolanosterol acetate (**309**-OAc, R = 4-methylpentyl) demonstrate the inherently favorable nature of the first transformation. Similarly Lewis acid-catalyzed tricyclization of 3(S)-epoxide (**312**)* affords, in addition to 24,25-dihydroisoeuphol (**324**-OH, R = 4-methylpentyl, 3.5%), both dihydroprotosterol **311**-OH (2%) and 24,25-dihydroparkeol (**313**, 3.5%) (*14, 388*). It is of interest to note that parkeol ($\Delta^{9(11),24}$-8β-lanostadien-3β-ol) is a naturally occurring tetracyclic triterpene. The protosterol (**311**-OH) may be an intermediate on the pathway leading to dihydroparkeol (**313**) since the backbone rearrangement (**311**-OH→**313**) could be brought about by separate exposure of **311**-OH to the reaction conditions.

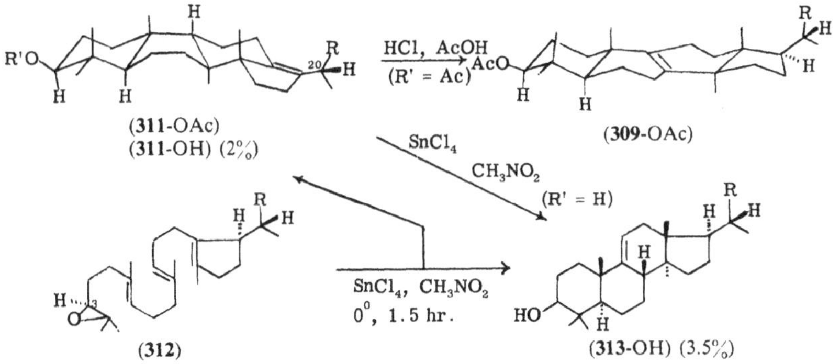

(**311**-OAc)
(**311**-OH) (2%)

(**309**-OAc)

(**312**)

(**313**-OH) (3.5%)

(R = 4-methylpentyl)

The driving force behind this biogenetic-like backbone rearrangement is evidently conformational in nature. In protosterol (**311**) ring B is necessarily constrained to a boat conformation while a more stable all-chair conformation can be adopted by dihydrolanosterol (**309**-OAc).

* This reaction was actually performed with a mixture of the 3(S)-epoxide (**312**) and its 3(R)-diastereomer. The latter is cyclized concurrently to (–)-24,25-dihydroisotirucallol (enantiomeric to **324**–OH at all positions except C_{20}).

Consistent with an inherently greater thermodynamic stability of the lanostane skeleton is the finding that treatment of dihydrolanosterol acetate (**309**-OAc) with hydrogen chloride in chloroform (2 hr.) effects simply partial isomerization of the double bond into the Δ^7 position (*i.e.*, to **314**) (*389*).

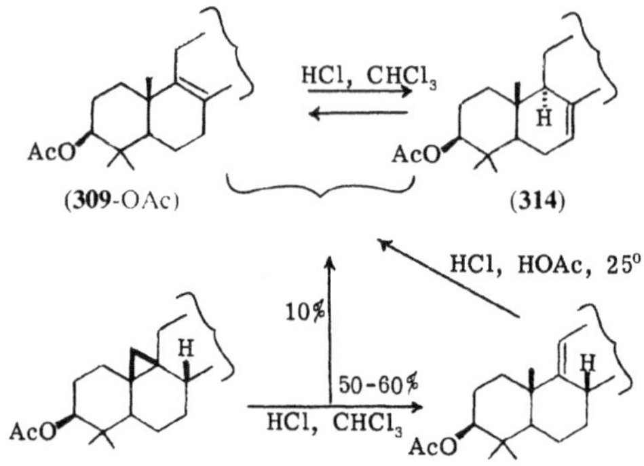

(**309**-OAc) (**314**) (**310**-OAc) (**313**-OAc)

(R = 4-methylpentyl)

The cyclopropane ring of cycloartanol acetate (**310**-OAc) is cleaved regiospecifically at the 9,19 bond, producing dihydroparkeol acetate (**313**-OAc) as the major product and lesser amounts of the equilibrium mixture of Δ^7- and Δ^8-dihydrolanostenol acetates (**309**-OAc and **314**) (*390*—*392*). This transformation is of biogenetic significance since plant sterols are evidently biosynthesized by a similar ring opening reaction of certain cycloartane derivatives (*393, 394*). Parkeol acetate is converted into the equilibrium mixture of **309**-OAc and **314** with hydrogen chloride

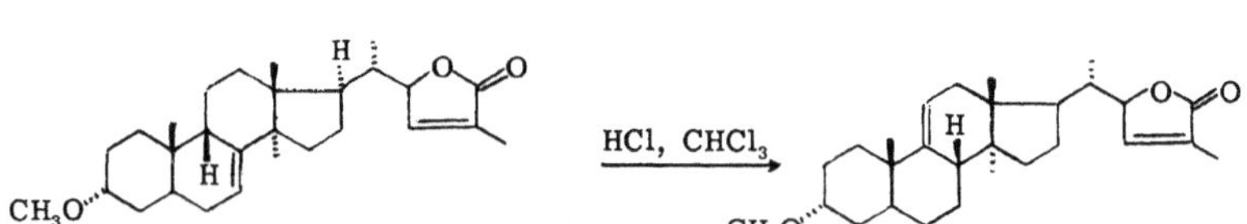

(**315**) (**316**)

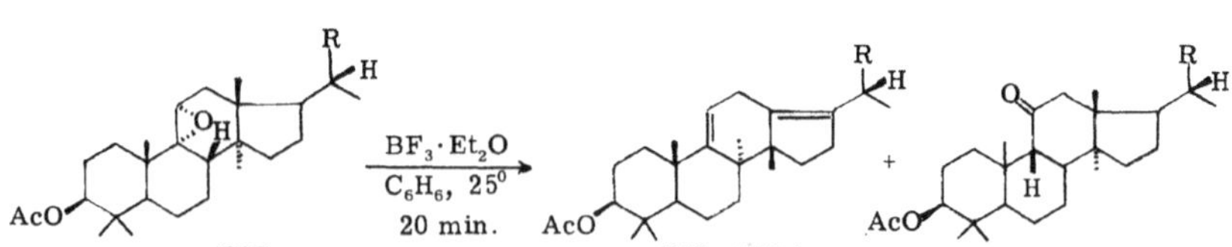

(**317**) (**318**) (47%)

References, pp. 205—230

in acetic acid (*388, 395*). In connection with the formation of **313**-OH (rather than dihydrolanosterol) from treatment of protosterol **311**-OH with stannic chloride (see above), it is interesting the Δ^7-9β-lanostane derivative, abieslactone (**315**), is isomerized to grandisolide (**316**) by the action of hydrogen chloride in chloroform or acetic acid with migration of the 9β hydrogen to C_8 (*396, 397*).

Despite the greather stability of the lanostane nucleus, it is apparently possible to reverse the protostane→lanostane rearrangement. It has been recently reported that the reaction of the epoxide (**317**) of dihydroparkeol acetate with Lewis acids affords, protostadienol acetate **318** (45%), presumably by secondary dehydration of $\Delta^{13(17)}$ protostene-3β,11α-diol 3-monoacetate (*398, 399*).

Although there is reasonable precedent for 1,3-proton elimination from a methyl group to form a three-membered ring under carbonium ion conditions (see section C.4), no 9,19-cyclo steroids were found in the deamination of the 9α-amino-11-keto pregnane derivative (**337**) (*400*). Instead the major products resulted from migration of the C_{10} methyl group to C_9 (see following section for other examples of this rearrangement), even in an aprotic deamination (isoamyl nitrite in benzene with a stoichiometric amount of acetic acid), conditions considered conducive to cyclopropane formation. Consequently it was suggested that efficient 1,3-hydrogen elimination is effected in the biosynthesis of the cycloartane triterpenes

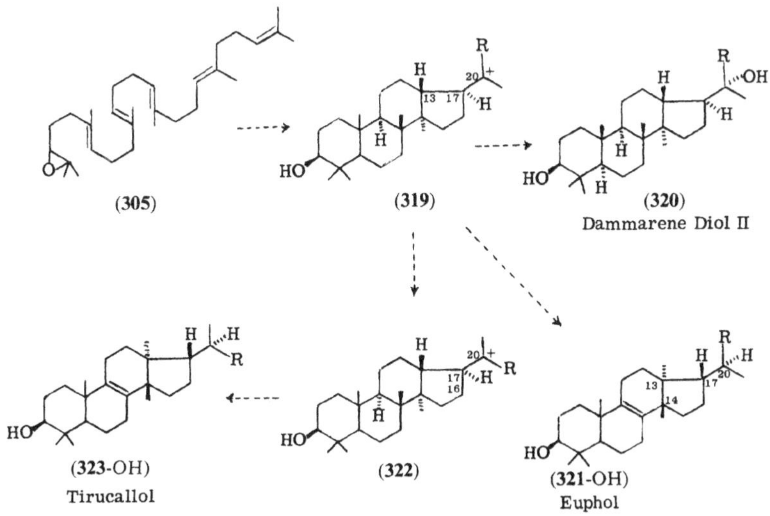

Scheme 29

through proton transfer from the C_{10} methyl group to a suitably-placed basic site on the enzyme surface (*400*).

The euphane and tirucallane triterpenes differ from the lanostanes only in having the opposite stereochemical dispositions at C_{13}, C_{14}, and C_{17} and are themselves epimeric at C_{20} (*384*). The stereochemistry is considered to be a consequence of enzymatic cyclization of squalene epoxide (**305**) from a chair, chair, chair conformation (Scheme 29). The two stereoisomeric forms arise from 17→20 hydrogen shift to the diastereotopic faces of the carbonium ion at C_{20}*.

In contrast to the greater stability of lanosterol with respect to protosterol, the dammarane skeleton is more stable than that of the euphanes or tirucallanes. Thus, 24,25-dihydroeuphol acetate (euphenyl acetate; **321**-OAc, R = 4-methylpentyl) is isomerized efficiently to isodihydroeuphyl acetate (isoeuphenyl acetate, **324**-OAc) having the basic dammarane skeleton (*401*—*404*). Dihydrotirucallol acetate is similarly rearranged to the C_{20} epimer of **324**-OAc (*405, 406*). Since the 13α, 14β stereochemistry at the C/D ring junction of euphol (and tirucallol) forces the C (or B) ring into a half-boat, a conformational driving force is presumed to be responsible for the direction of these rearrangements (*407*).

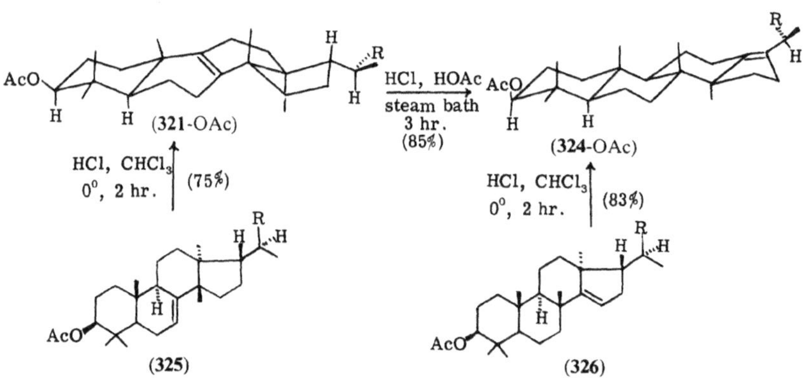

The observation that dihydroapoeuphyl acetate (**326**) is converted to dihydroisoeuphyl acetate (**324**-OAc) under conditions (hydrogen chloride in chloroform, 0°, 2 hr.) which leave dihydroeuphyl acetate (**321**-OAc)

* Carbonium ion (**319**) may also be represented as a bridged species with partial bonding between C_{13} and C_{20}, ring D necessarily adopting a boat-like conformation. A corresponding bridged formulation for (**322**) with partial C_{16}–C_{20} bonding has a chair-like conformation. The stereochemistry implicit in these bridged ions will be considered in more detail in section E.4.

unaffected presumably indicates that, if (**326**) is an intermediate in the **321**-OAc→**324**-OAc rearrangement, it would be difficult to detect (*408*). The same conditions are capable of isomerizing the Δ^7 double bond of dihydrobutyrospermyl acetate (**325**) to the Δ^8 position apparently without further rearrangement (*409—411*).

When the euphenyl→isoeuphenyl rearrangement (**321**-OAc→**324**-OAc) is carried out in deuterated acetic acid at low conversion, the isoeuphenyl acetate obtained (7.5%) is extensively deuterated (8% d_1, 39% d_2, 36% d_3, 12% d_4, 3% d_5) (*412*). The small amount of mono-deuterated product indicates some (8%) rearrangement without concurrent elimination. Since recovered euphenyl acetate (80%) is partially deuterated at C_7 (49%) and the higher deuterated species in the product all contained label at C_7 and C_9, it is clear that equilibration with the Δ^7 isomer (**325**) competes with rearrangement. That some elimination to apoeuphenyl acetate (**326**) occurred during the rearrangement was indicated by the presence of deuterium at C_{15} (6.5% of label). Additional label was located at C_{12} (6.5%), C_{16}, and, by inference, the migrated methyl groups (10%). Methyl groups deuteration could arise *via* reversible 1,3-hydrogen elimination to cyclopropane intermediates* or ring contraction to spiro-type olefin isomers (*216, 238, 414, 415*).

(**321**-OAc)

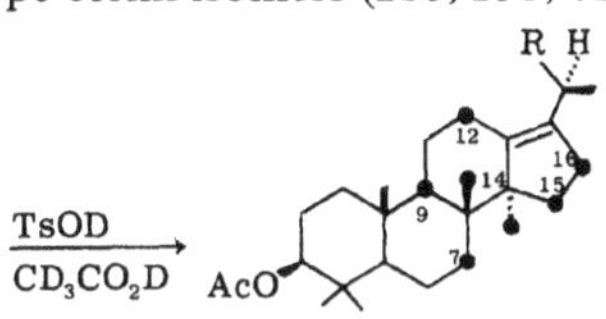

(**324**-OAc) (d_1-d_5, ● = labelled positions)

3. Melianes, Cucurbitacins, and 9(10→19)Abeo-pregnanes

These three groups of highly functionalized, and often partially degraded, natural products have in common a probable biogenesis from a pre-formed tetracyclic triterpene. Although the melianes (also called apotirucallanes, *e.g.*, **327**) could possibly also arise directly by proton loss from C_{15} prior to completion of the normal backbone rearrangement (**322**→**323**), the presence of an oxygen function at C_7 of all known melianes and melicanes (*e.g.*, meldinin, **328**) suggests a likely biogenesis through oxidative rearrangement of a Δ^7-tirucallene progenitor (*416*).

Similar considerations have led to the proposal that the cucurbitacins (*e.g.*, cucurbitacin B, **329**) (*384, 417*) are formed in nature *via* oxidative rearrangement of a $\Delta^{9(11)}$-lanostene precursor (*e.g.*, parkeol, **313**-OH).

* In this connection it is of interest that the first 14,18-cyclo-apotirucallane derivative has recently been isolated (*413*).

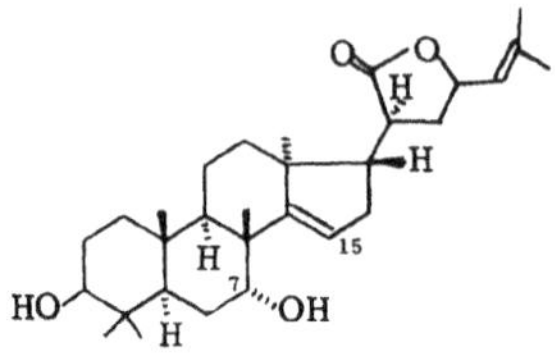

(327)

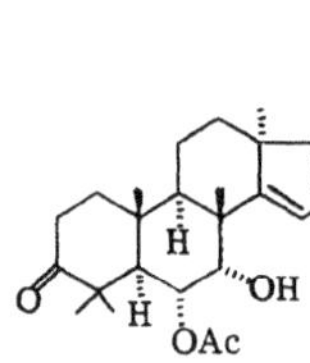

(328)

Meldinin

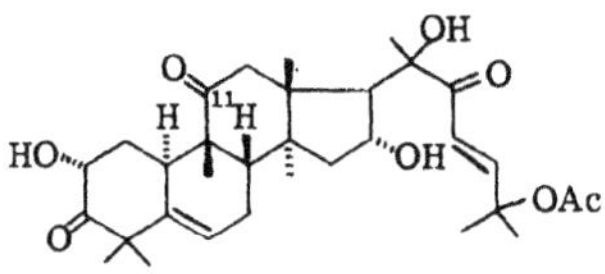

(329)

Cucurbitacin B

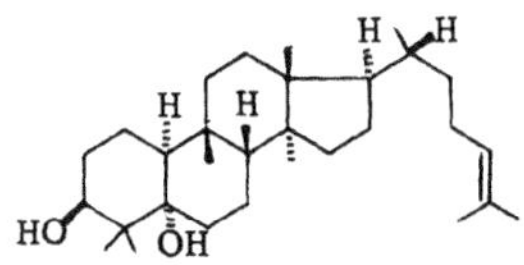

(330)

Litsomentol

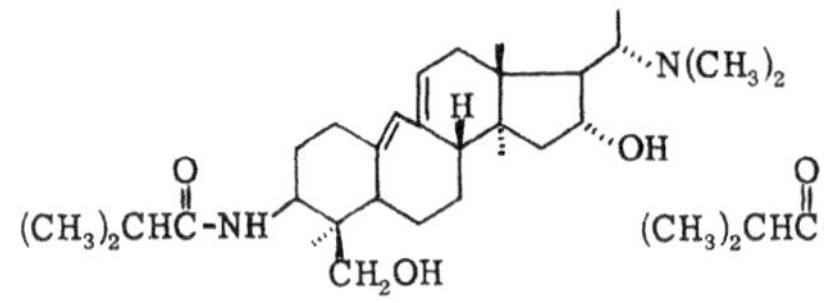

(331)

N-3-Isobutyryl Buxidienine-F

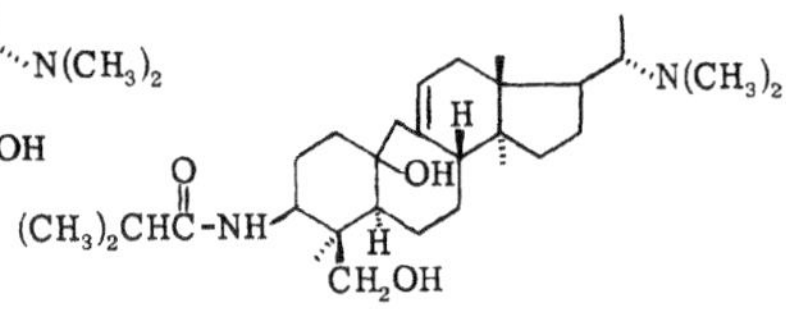

(332)

N-3-Isobutyryl Buxaline F

The recent isolation of the cucurbitacin-like triterpene litsomentol (**330**) lacking a C_{11} oxygen function, however, lays open the possibility of a more direct biogenesis from carbonium ion (**308**) by $C_{10}\rightarrow C_8$ methyl rearrangement (*418*). Yet another alternative is a pathway involving $C_{10}-C_{19}$ bond cleavage of a pre-formed cycloartane precursor (*e.g.*, **310**). Although biosynthetic experiments exclude lanosterol (**309**-OH) as an intermediate (*419*), there appears to be no evidence at hand to distinguish among the other possible pathways.

The 4,4,14-trimethyl-9(10→19)*abeo*-pregnane skeleton is the basic structure of a small group (*e.g.*, N-isobutyryl buxidienine-F and N-isobutyryl buxaline F, **331** and **332**, respectively) within the *Buxus* alkaloid family. The majority of the *Buxus* alkaloids appear to be degraded cycloartane derivatives lacking six carbons from the side chain (C_{22}–C_{27}) and aminated at either C_3 or both C_3 and C_{20}. Since many of these alkaloids have an additional oxygen function at C_{11}, a reasonable biogenesis of the 9(10→19)*abeo*-pregnane skeleton is a cyclopropylcarbinyl-homoallyl ring opening reaction followed by hydration or proton loss (*420*).

Acid-catalyzed biogenetic-type rearrangement of 7,8-epoxytirucallane derivatives has proven to be a viable synthetic approach to the melianes and melicanes (*416, 421*—*423*). It is interesting that these epoxide rearrangements (*e.g.*, **333**→**334**) (*416*) invariably terminate after migration of the C_{14}→C_8 methyl migration, rahter than proceeding further to the fully rearranged dammarenes (see previous section).

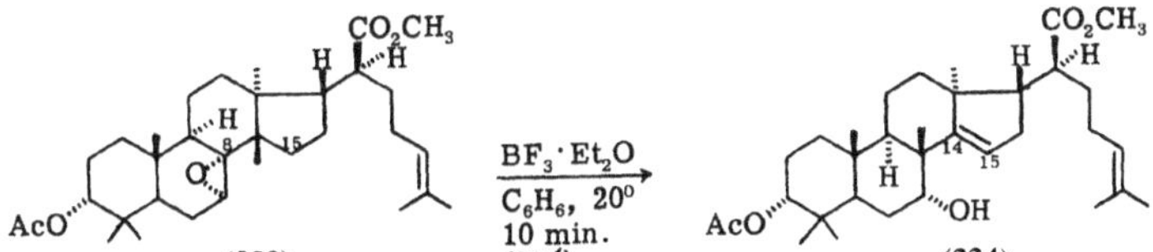

(333) (334)

Rearrangements of this type have also been observed in the reactions of Δ^7-triterpenes of this type with oxidizing electrophiles [dihydrobutyrospermyl acetate (**325**) with chromic acid (*408*) and (**333**) (Δ^7-double bond in place of epoxide) with bromine (*424*)], albeit in rather low yields. 14,15-Epoxymeliacanes (*i.e.*, similar to 14,15-epoxide of **328**) undergo further rearrangement in the presence of acids, the C_{13} methyl group proceeding to C_{14} and forming, after proton loss, a vinyl furan (*425*). However, epoxides of Δ^{14}-unsaturated D-ring δ-lactones (similar to meldinin (**328**) with oxygen inserted between C_{16} and C_{17}) suffer rearrangement in the opposite direction, *i.e.*, C_8→C_{14} methyl migration (*426*).

One possible biogenetic pathway to the cucurbitacins is ring opening of a cycloartane precursor. However, as mentioned in the previous section, the chemically-induced protolysis of cycloartanyl acetate cleaves only the C_9–C_{19} bond of the cyclopropane ring. In this connection it is of interest to note that the cycloartane stereoisomer (**335**) with the three-membered ring in the α orientation undergoes cleavage in the opposite sense (*427*).

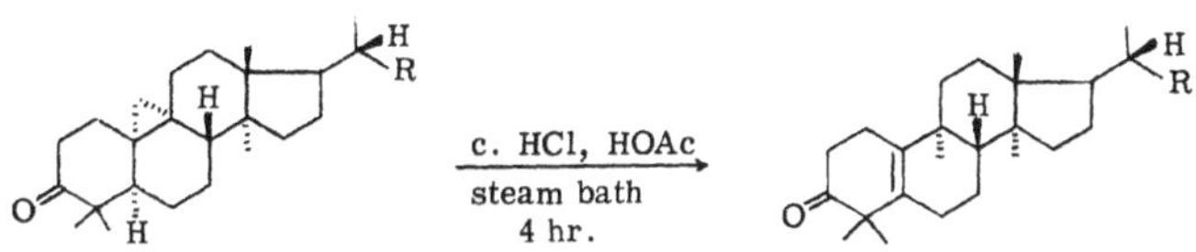

(335)

The acid-catalyzed reactions of 9(11)-epoxides of steroids and triterpenes have been investigated as a means to effect $C_{10} \rightarrow C_9$ methyl rearrangement, and at the same time, generate a C_{11} oxygen function (*398, 399, 428—433*). Three different rearrangement modes have been observed: $C_{11} \rightarrow C_9$ hydrogen shift, $C_8 \rightarrow C_9$ hydrogen shift (followed in some cases by further methyl rearrangements), and $C_{10} \rightarrow C_9$ methyl shift. Although epoxides of $\Delta^{9(11)}$-lanostenes (**317**, previous section) appear to rearrange chiefly *via* the hydrogen shift pathways, efficient methyl rearrangement was observed with 9α,11α-epoxy-4,4-dimethylandrost-5-ene-3, 17-dione (**336**), possibly owing to allylic stabilization (*428*). However, an analogous rearrangement of a closely related epoxide, differing only in the C_{17} substituent, was unsuccessful (*432*). The course of these rearrangements appears to quite sensitive to rather slight changes in structure.

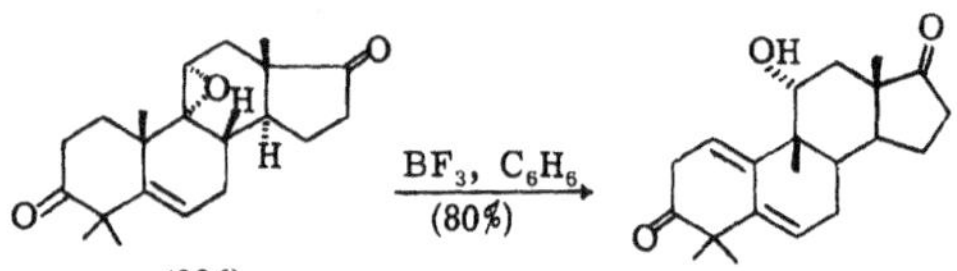

(**336**)

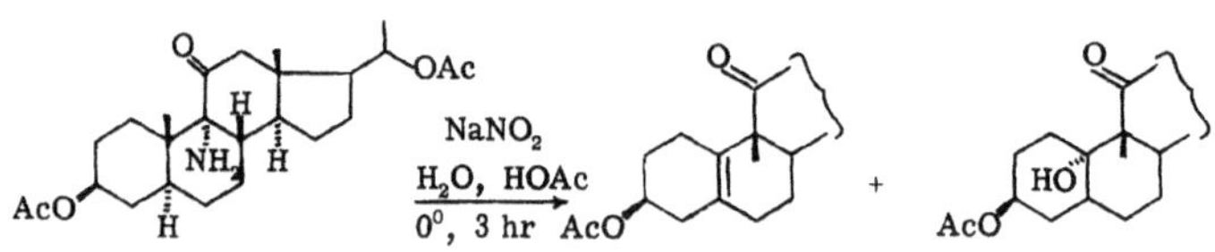

(**337**) (75%) (5%)

Deamination of 9α-amino-11-keto steroids (*e.g.*, **337**) (*400*) is an alternate means to accomplish the $C_{10} \rightarrow C_9$ methyl rearrangement (*400, 432*). In the fluorination of a $\Delta^{9(11)}$ steroid with methyl hypofluorite, a similar methyl migration is observed, ring A being aromatized in the process (*433*).

The availability of 11-hydroxylated buxane alkaloids from natural sources and 11-hydroxy-9β,19-cyclopregnanes of synthetic origin has enabled investigations upon their conversion to 9(10→19)*abeo*-pregnanes through a biogenetic-type rearrangement (*420, 434*). Thus, triol (**338**), prepared from natural cycloxobuxidine-F alkaloids, undergoes selective reaction at the cyclopropylcarbinol moiety, with resultant cleavage of the C_9–C_{10} bond and proton elimination (*420*). The diene diol obtained (buxidienine F, **339**) was identical to a sample obtained by hydrolysis of natural N-3-isobutyryl buxidienine-F (**331**).

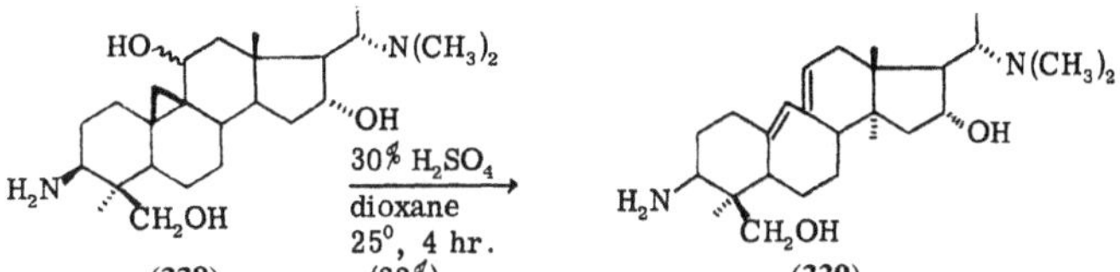

(338) (339)

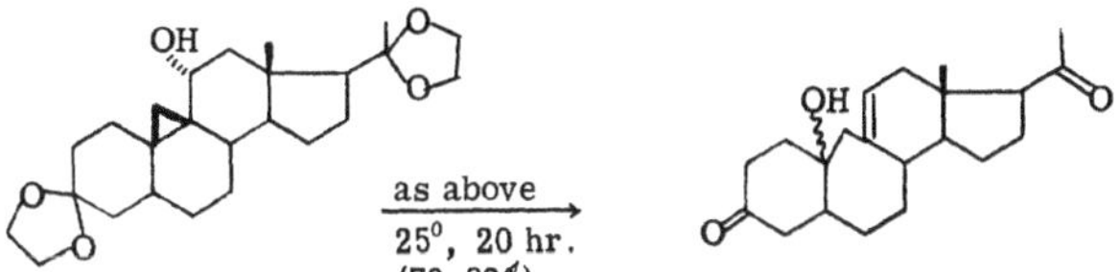

(340) (341)

Exposure of 9β,19-cyclopregnane-11α-ol diketal (**340**) to similar reaction conditions produced homoallyl alcohol (**341**), analogous in structure to N-3-isobutyryl buxaline-F (**332**), in high yield (*434*). Although a small amount (9—14%) of the transoid diene corresponding to (**331**) appeared to have been formed (UV evidence) in the reaction of (**340**) with tosyl chloride in pyridine, this compound could not be isolated. The major product (35%) was the Δ^{11}-olefin with the cyclopropane ring intact.

Deamination of 19-amino-$\Delta^{9(11)}$-lanosten-3β-ol provides a 9(10→19)-*abeo*-lanostene derivative analogous to (**341**) (C_{10} methyl ether in place of C_{10} hydroxyl), presumably by way of a homoallyl-cyclopropylcarbinyl-homoallyl rearrangement (*435*). The initial product was considered to be the 3,10-diol, which underwent methanolysis to the C_{10} methyl ether during recrystallization.

4. Lupanes, Germanicanes, Taraxastanes, Oleananes, and Ursanes

The biogenesis of these pentacyclic triterpenes (*3, 5, 436*) (Scheme 30) involves, in the initial stages, two key cyclopentylcarbinyl→cyclohexyl ring enlargement rearrangements (**343→346,** and **347→349**). The scheme begins with the bridged ion (**342**) produced from tetracyclization of 3-(S)-squalene-2,3-oxide in a chair, chair, chair, boat conformation, which upon fulfilment of the C_{13}–C_{17} bond, affords tertiary cation (**343**). This latter intermediate was depicted previously in the biogenesis of the euphanes and tirucallanes in different rotameric conformations about the C_{17}–C_{20} bond (**319** and **322,** Scheme 29). Migration of the C_{16}

methylene group to C_{20} effects the first ring expansion in preparation for the final E-ring closure. The stereochemistry of these rearrangements is illustrated by the predicted stereospecific hydration of bridged ions (**342** and **343**) to the respective dammarene diols (**320** and **345**).

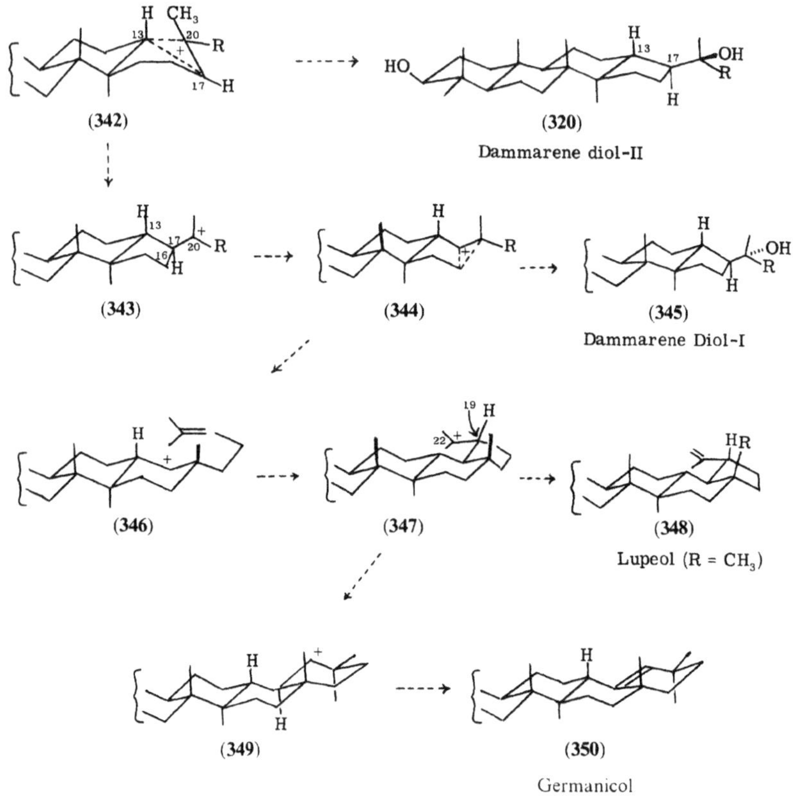

Scheme 30

Secondary carbonium ion (**349**) (or an equivalent bridged ion) marks the point from which the biogenesis of the oleanane and ursane triterpenes diverge (Scheme 31). Thus, two 1,2-hydrogen rearrangements (18α→19α, 13β→18β) and elimination gives β-amyrin (**352**), the parent of the oleanane family. Alternatively 1,2-methyl migration generates the taraxastane skeleton (*e.g.*, Ψ-taraxasterol, **354**), and ultimately, after three 1,2-hydrogen migrations (19β→20β, 18α→19α, 13β→18β) and elimination, α-amyrin (**356**), the ursane prototype.

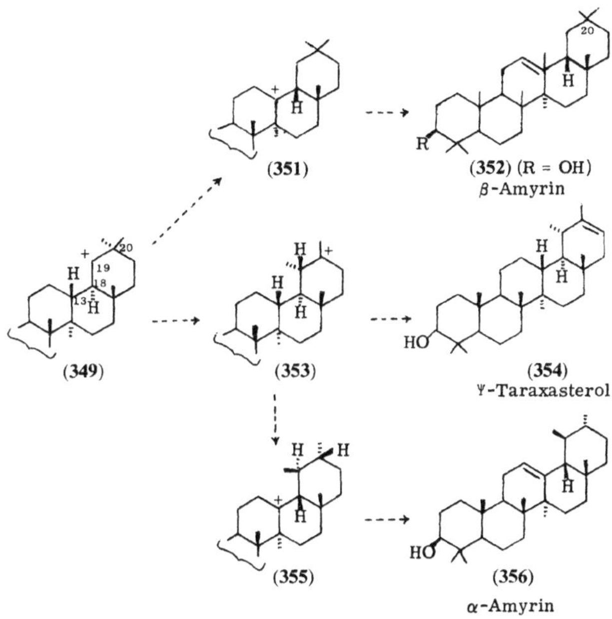

Scheme 31

Alternative biogenetic pathways to these pentacyclic triterpenes (*30, 34, 437*) are excluded by recent investigations upon the biosynthesis of α- and β-amyrin (or metabolites thereof) (*27—30, 438—441*). It is particularly relevant to point out that the stereochemistry of the E-ring geminal methyls at C_{20} (*440, 441*) is inconsistent with that predicted by the X-group biogenetic scheme (see also section A) (*34*). According to this scheme, E-ring closure would be concerted with anti-parallel nucleophilic capture by the X group to give (**7**) (C_{19} epimer of **347**-X). The subsequent ring expansion rearrangement to generate (**349**) must, if concerted with departure of the X group, be preceded by a 180° rotation about the exocyclic C_{19}–C_{22} bond in (**7**). The interchange of the two methyl groups which occurs as a consequence of this rotation is at variance with the results of incorporation of 6-d_3-mevalonate into β-amyrin (*440, 441*).

The rearrangement of pregnane-20β-yl tosylates to D-homo steroids (uranediol rearrangement) (*442—445*) affords an interesting model for the biogenetic ring enlargement (**343→346**). It will be noted that the steroid model differs in the placement of the D-ring quaternary methyl group and the degree of substitution at C_{20}. Formolysis of **357**-OTs brings about stereospecific D-homoannulation to the formate **358**-OCHO (*442*). With short reaction times it was possible to isolate an isomeric tosylate intermediate (**358**-OTs) which undergoes slow formolysis to **358**-OCHO.

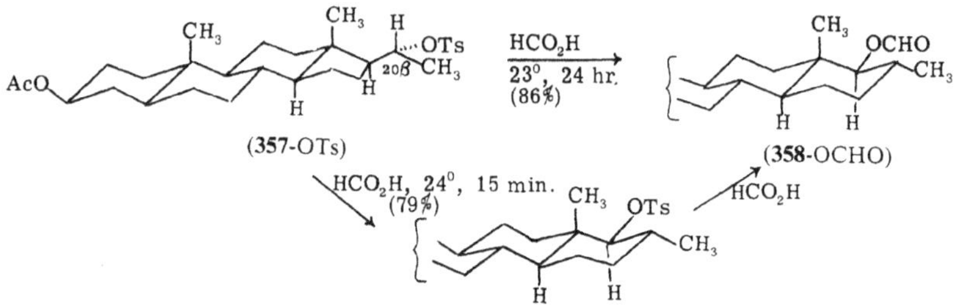

(**357**-OTs) (**358**-OCHO) (**358**-OTs)

This stereospecific substitution reaction (**358**-OTs→**358**-OCHO) with retention of configuration may be explained by formation of a nonclassical carbonium ion intermediate, a carbonium ion-tosylate ion pair, or a conformationally-frozen, localized cation. The observation of a small rate enhancement (1.5 times) in the formolysis of the D-homo tosylate (**358**-OTs) as compared to *trans*-4-*t*-butylcyclohexyl tosylate might be taken to indicate the absence of σ participation in the solvolytic transition state. However, it appears likely that the true ionization rate of **358**-OTs is faster than measured by product formation owing to rapid internal return from an initially formed ion pair. Since the rearrangement (**357**-OTs→**358**-OTs) and the substitution reaction (**358**-OTs→**358**-OCHO) presumably pass through a common intermediate, the formation of **358**-OTs in high yield from the pregnane 20β-tosylate shows that this intermediate, in fact, undergoes internal return faster than formolysis.

The stereospecificity of the uranediol rearrangement is further exemplified by predominant formation of different products in the formolysis of the epimeric 20α-tosylate (**359**-OTs) (*443*, *444*). That olefin (**360**) is formed by 1,2-hydrogen and methyl rearrangements (rather than elimination and secondary acid-catalyzed rearrangement) was demonstrated by introduction of deuterium and tritium label at C_{20}. A substantial

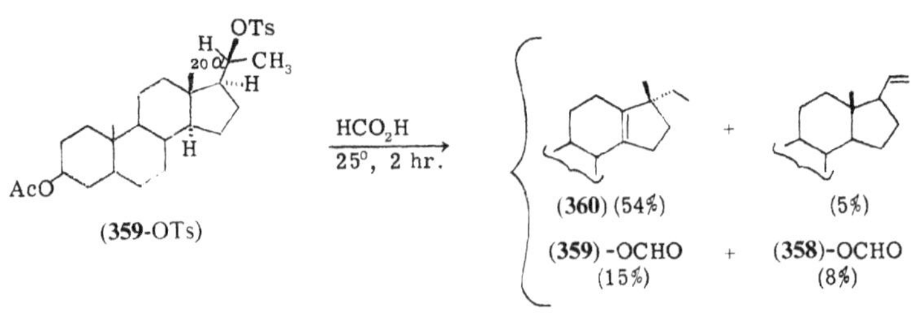

(**359**-OTs)

(**360**) (54%) + (5%)

(**359**)-OCHO (15%) + (**358**)-OCHO (8%)

kinetic isotope effect ($k_H/k_T = 1.7$) was taken to indicate a hydrogen-assisted ionization. The isolation of a small amount (8%) of **358**-OCHO from the product mixture reveals some leakage of the carbonium ion intermediate(s) into the ion(s) from the 20β-tosylate. The relevance of this interconversion to triterpene biogenesis was noted (*444*).

The stereochemistry of the reverse rearrangement (ring contraction) has been examined by means of solvolysis of the bicyclic model methanesulfonate (**361**-OMs) in which the axial C_2 methyl group is labelled with deuterium (*446*). The label proved to be distributed essentially equally between the two diastereotopic methyl groups in the isopropyl substituent of the hydrindanyl alcohol (or acetate) produced in the hydrolysis (or acetolysis). Although this ring contraction rearrangement must pass through a bridged configuration akin to (**344**) (transition state or transient intermediate), this species apparently proceeds on to a more stable, classical tertiary ion (as in **343**) prior to nucleophilic capture. The distinction between the two methyl groups may be lost by either rapid rotation about the exocyclic single bond or non-stereoselective reaction with solvent from the two non-equivalent faces of the carbonium ion intermediate.

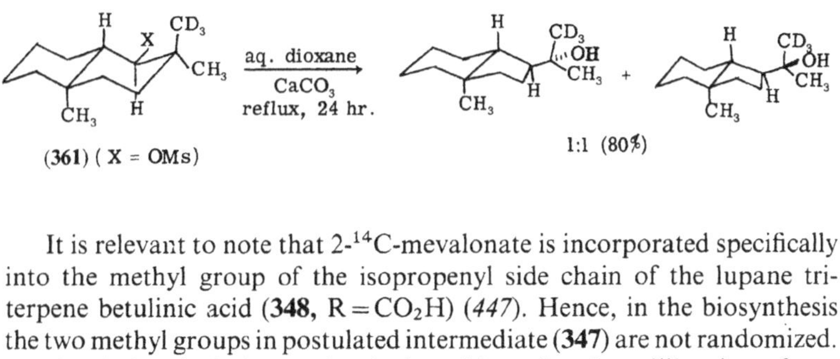

(**361**) (X = OMs) 1:1 (80%)

It is relevant to note that 2-^{14}C-mevalonate is incorporated specifically into the methyl group of the isopropenyl side chain of the lupane triterpene betulinic acid (**348**, R = CO_2H) (*447*). Hence, in the biosynthesis the two methyl groups in postulated intermediate (**347**) are not randomized. A related chemical observation is the acid-catalyzed equilibration of two C_{20} epimeric dammarane diol derivatives (12β-hydroxy-24,25-dihydro version of **320** and **345**) (*448*).

A D-homoannulation rearrangement of possible biogenetic significance has been observed in the acid-catalyzed cyclization of diene acetate (**362**) (*31*). One possible mechanism for the formation of perhydrochrysene-type product (**363**) includes a cyclopentylcarbinyl→cyclohexyl rearrangement analogous to that (**343**→**346**) depicted in biogenetic scheme 30*.

* The authors suggested a slightly different mechanism in which the ring expansion occurs instead by way of C_{13}→C_{20} migration (rather than C_{16}→C_{20} shown).

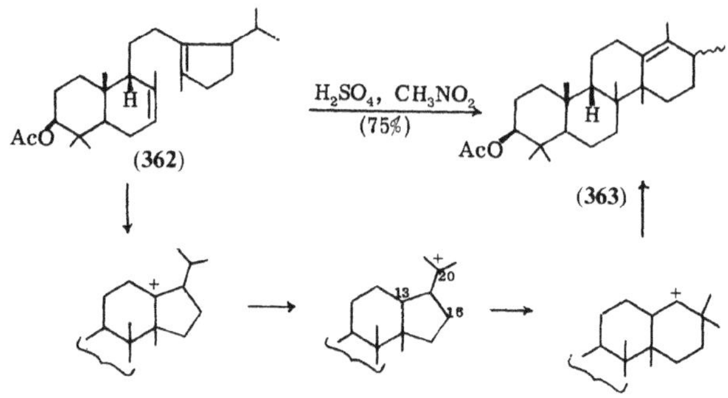

The second cyclopentylcarbinyl→cyclohexyl rearrangement (**347**→**349**), which converts the lupeol skeleton to that of germanicol, has been carried out on lupeol itself. Thus, addition of hydrogen chloride to lupeol (**348**) (or formic acid to lupeone) gives the E-homo chloride **364**-Cl (or formate, **364**-OCHO) (*449, 450*). Ethanolysis in the presence of silver ion effects the reverse rearrangement, whereas dehydrochlorination in refluxing acetic anhydride affords germanicyl acetate (**350**) (*449*). The direct conversion of lupeol to germanicol (**350**) has been accomplished in refluxing formic acid (*451*). A thermodynamically stable equilibrium mixture of $\Delta^{13(18)}$-oleanene and Δ^{13}-18α-oleanene (**365** and **366,** respectively) is reached when germanicene (**350,** H in place of OR) is exposed to concentrated hydrochloric acid in refluxing acetic acid (*452*). This mixture of oleanenes (**365** + **366**) is an end point for rearrangements of the cyclic triterpenes related to β-amyrin.

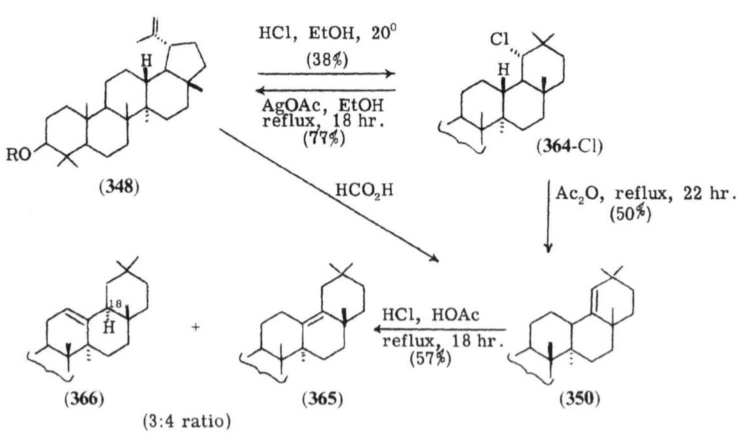

In accord with an *anti*-parallel relationship between the migrating carbon (C_{20}) and the entering nucleophile, the carbon chlorine bond in **364**-Cl is found to be in the equatorial (α) configuration. On the other hand, the acid-catalyzed cyclization of methyl betulonate (**367**), and related diol occurs so as to form an axial carbon-oxygen bond to C_{19}, despite the ostensibly unfavorable stereoelectronics (*453, 454*). It is possible, however, that this transformation is more complex than a simple ring expansion followed by lactonization.

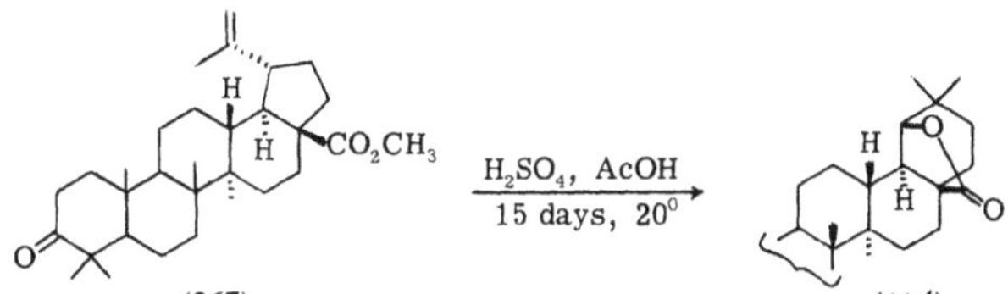

(367) (41%)

The $C_{20} \rightarrow C_{19}$ methyl rearrangement (**349→353,** Scheme 31) to the taraxastane skeleton has also been observed. In refluxing 98—100% formic acid, lupeol gives a mixture of isomers containing Ψ-taraxasterol (**354**), the $\Delta^{13(18)}$ isomer of β-amyrin (**365** with C_3-OH) and possibly (**366**) (C_3-OH) (*452, 455*). Under somewhat stronger acidic conditions, Ψ-taraxasterol (**354** – as benzoate) is epimerized at C_{19} to the more stable lupenol-I (**368** as benzoate) (*451*). The latter reaction is considered to proceed through the Δ^{19}-taraxasterene isomer which is unstable with respect to (**368**) owing to steric interference between the C_{19} methyl and C_{12} methylene groups. The successful isolation of the less stable Ψ-taraxasterol (**354**) from formic acid-treatment is attributed to a combination of the moderate acidity and its insolubility in this medium.

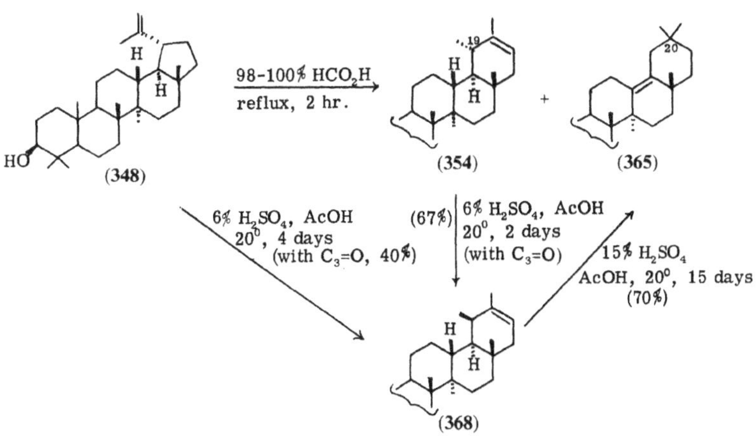

The initial methyl shift rearrangement which takes place in the rearrangement of lupeol to Ψ-taraxasterol (**348→354**) is apparently reversible since the action of stronger acid (15% sulfuric acid in acetic acid) on either lupenone-I (**368,** C_3-ketone) or lupenone (**348,** C_3-ketone) affords $\Delta^{13(18)}$-oleanone (**365**), thereby reinstating C_{20} gem dimethyl group (*451, 456*). Although the end product of the above ring E rearrangements is at least primarily the appropriate derivative of $\Delta^{13(18)}$-oleanene (**365**), and presumably Δ^{12}-18α-oleanene (**366**), the oleanane based dienol (**369**) has been isomerized to the α,β-unsaturated ketone (**370**) having the ursane (α-amyrin) skeleton (*457*).

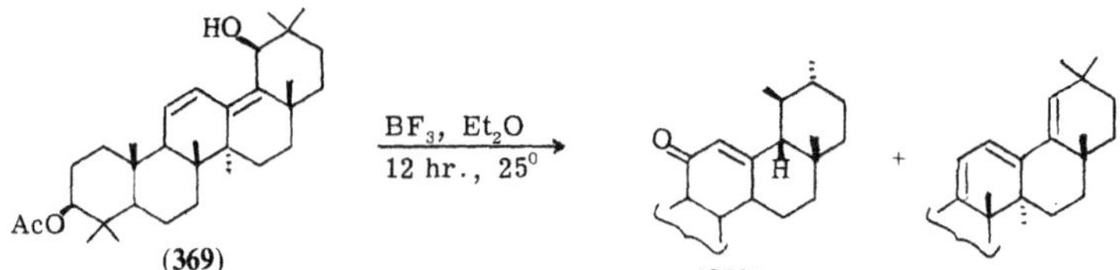

(**369**) (**370**)

5. Taraxeranes, Multiflorane, Glutinanes, Friedelanes, Isoursane, Phyllanthane, Bauerane, and Shionanes

These pentacyclic triterpenes are grouped together in consequence of a common biogenetic origin through sequential backbone rearrangements of either the oleanyl (Scheme 32) or ursyl carbonium ion (**351** and **355,** Scheme 31). Since all adjacent hydrogen and methyl groups which participate are situated in a *trans* (or *anti*) relationship to each other, the rearrangement may be considered to proceed in a concerted fashion through a succession of individual hydrogen- and methyl-bridged nonclassical ions*. However, the occurrence of a number of diterpenes, the biogenesis of which would appear to require *syn* vicinal rearrangements (see section D.1), may be taken to indicate that the *anti* relationship of the groups is an outgrowth of the original folding of the acyclic precursor and not necessarily a prerequisite for the backbone rearrangement.

* A "fully-concerted", multi-step rearrangement implies the existence of a single intermediate (or transition state) in which all migrating groups are simultaneously situated half-way between adjacent carbons. This species appears unreasonable on two counts. Firstly there is no precedent for such extended delocalization through sigma bonds. For example, the rearrangement (**351→374**) if "fully concerted", requires σ-delocation through eight carbon centers. Secondly all carbons along the backbone would have to be flattened into a plane, a conformational change considered energetically prohibitive (*458*).

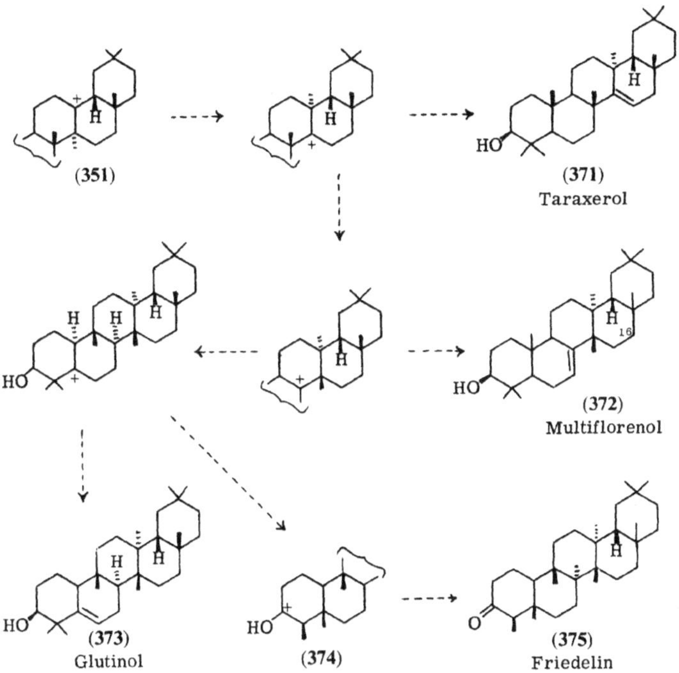

Scheme 32

Owing to the greater thermodynamic stability of the oleanene skeleton, most of the chemically-induced backbone rearrangements in this series proceed in the opposite direction from the biogenesis. Taraxaryl acetate (**371**-OAc) (*459*), various taraxarane derivatives (*460—462*), and multiflorenyl acetate (**372**-OAc) (*463*) are readily isomerized to β-amyrin acetate (**352**-OAc), or the corresponding Δ^{12}-oleanene. The relief of a severe steric interaction between the α-oriented methyl groups at C_{13} and C_{20} is considered to be a major driving force in these rearrangements. An additional factor in the case of multiflorenyl acetate is the resulting conformational change, ring C being converted from a boat to a half-chair.

Under more forcing conditions the parent hydrocarbon Δ^{12}-oleanene (**352,** R = H) undergoes isomerization to the equilibrium mixture of $\Delta^{13(18)}$-oleanene and Δ^{12}-18α-oleanene (**365** + **366**) (*452*). The regiospecific elimination into the Δ^{12} position is attributable to either steric inaccessibility of the tertiary C_{18} hydrogen or unfavorable alignment of

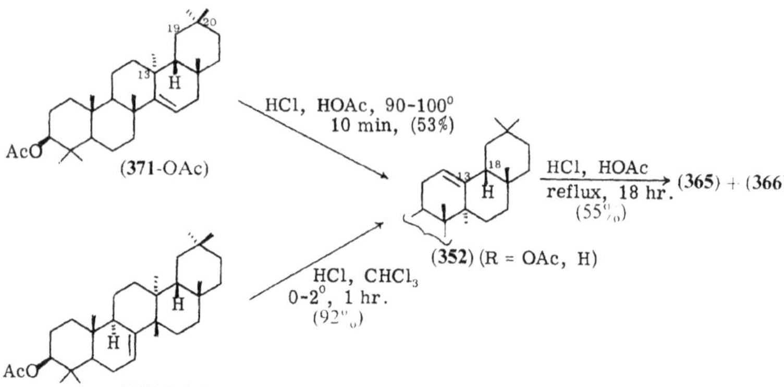

(**371**-OAc) (**372**-OAc) (**352**) (R = OAc, H)

the C_{18} hydrogen (equatorial to ring D in one possible conformation) with respect to the *p* orbital of the carbonium ion center at C_{13}.

Similar rearrangements have been observed in the ursane series. Isoursenyl acetate, the ursane analogue of taraxeryl acetate (**371**, C_{20}-β methyl moved to C_{19}) is converted to α-amyrin acetate (**356**-OAc) with hydrochloric acid in acetic acid (*464*, *465*). Baurenyl acetate (**372** with C_{20}-β methyl moved to C_{19}) is rearranged in like manner to a mixture of α-amyrin acetate and its $\Delta^{13(18)}$-isomer (*466*). This latter rearrangement, however, appears to have less driving force since, under milder conditions (hydrogen chloride in chloroform), sufficient to convert multiflorenyl acetate completely to β-amyrin acetate (**372**-OAc→**352**-OAc), baurenyl acetate is isomerized to its Δ^8 isomer (*463*, *466*).

Despite the inherently greater stability of the oleanene structure, it is possible to effect rearrangement in the biogenetic direction under oxidative conditions. Thus, the reaction of unsaturated ketone (**376**) with selenium dioxide (*467*) or bromine (*467*, *468*) in acetic acid affords the taraxarane-type rearranged dienone (**377**). Allylic alcohol (**378**) in the presence of hydrogen peroxide and acidic catalysts undergoes oxidative cyclization and rearrangement to taraxerene epoxide (**379**) (*469*). Acid-catalyzed heterolysis of the O–O bond of the corresponding allylic hydroperoxide (**378** with C_{11}-OOH), combined with an oxygen analogue of the homoallyl→cyclopropylcarbinyl change and methyl migration, is proposed as a mechanism for this reaction.

It proved possible to effect a second methyl rearrangement (C_8→C_{14}) through chlorination of taxarene derivative (**379**) (as acetate) in the presence of zinc acetate (*469*). The resulting chloro epoxide (**380**) was converted to multiflorenol (**372**) by reductive removal of the extraneous functionality at C_{11}, C_{12}, and C_{15}.

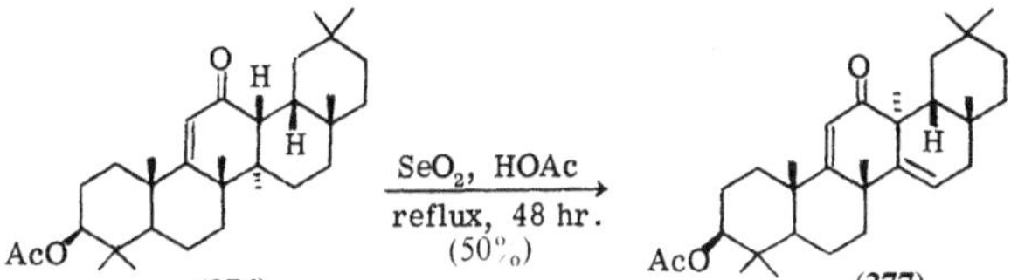

(376) (377)

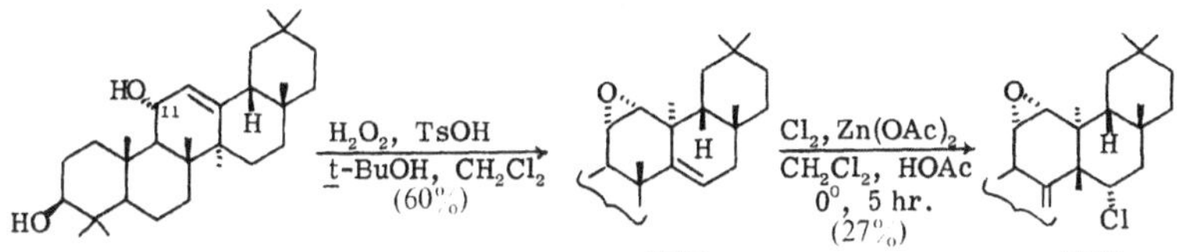

(378) (379) (380)

The biogenesis of the cyclopropane-containing triterpene phyllanthol (**382** without C_{12} ketone) entails 1,3-hydrogen elimination from carbonium ion (**355**) into the methyl group at C_{14}. This methyl group is situated *anti* to the hydrogen which migrates from C_{13} to C_{18} in the preceding rearrangement, in contrast to the *syn* relationship which exists in the analogous cycloartenol (**310**). Although the three-membered ring of phyllanthyl acetate is cleaved by reaction with hydrogen chloride in chloroform with formation of α-amyrin acetate (**356**-OAc) (*470*), it is reported that taraxeren-3β-ol-12-one acetate (**381**) undergoes a biogenetic-type ring closure to the phyllanthane skeleton in the presence of hydrogen chloride in acetic acid (*471*).

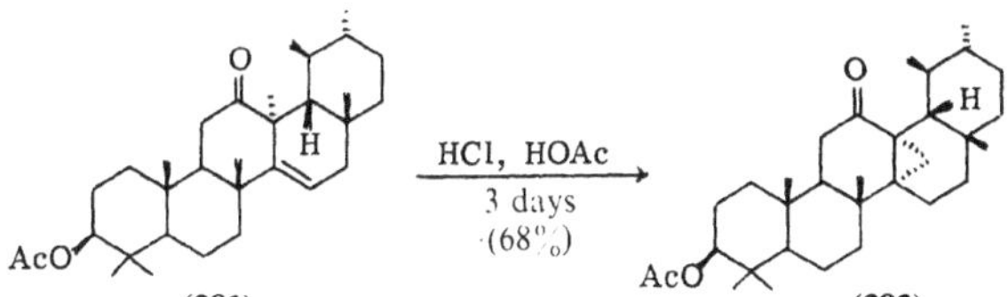

(381) (382)

The retro-biogenetic backbone rearrangement of the glutinene and friedelene triterpenes to oleanenes is remarkable for the number of individual 1,2-migrations involved. Friedelan-3β-ol (*472*), Δ^2-friedelene (*473*), Δ^3-friedelene (**384**) (*474*), Δ^5-glutinene (**383**) (*475*), and $\Delta^{5(10)}$-glutinene (**385**) (*476*) are all converted into the equilibrium mixture of $\Delta^{13(18)}$-oleanene and Δ^{12}-18α-oleanene (**365** + **366**) under a variety of acidic conditions. That the rearrangement of Δ^3-friedelene (**384**) proceeds, at least in part, through $\Delta^{5(10)}$-glutinene (**385**) and Δ^{12}-oleanene

(**352**) was demonstrated by the isolation of these compounds when shorter reaction times were employed (*460*).

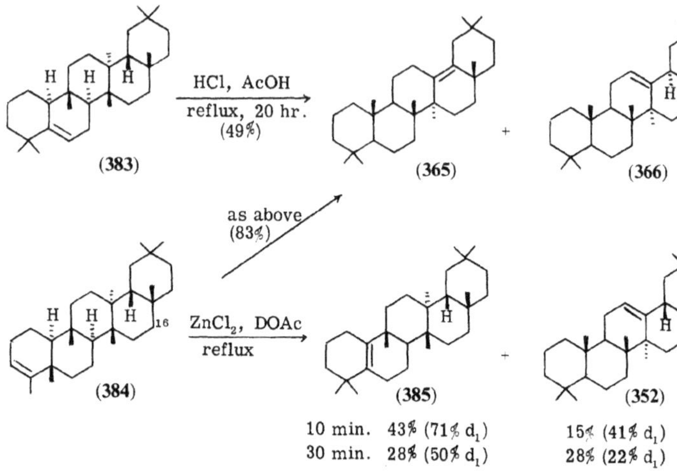

The extent to which other olefinic intermediates might be involved in these rearrangements has been examined by carrying out the reactions in deuterated media ($ZnCl_2$, DOAc or CF_3CO_2D, $CHCl_3$) (*476, 477*). From the amount of deuterium incorporation it was established that $\Delta^{5(10)}$-glutinene (**385**) rearranges to Δ^{12}-oleanene (**352**) largely (75—92%) without elimination-reprotonation occurring along the way. The Δ^3-friedelene$\rightarrow\Delta^{12}$-oleanene rearrangement (**384**$\rightarrow$**352**) occurs both by initial formation of $\Delta^{5(10)}$-glutinene and in part (20—33%) by direct rearrangement with no olefinic intermediates. However, a similar rearrangement of Δ^3-friedelen-16-one (**384** with ketone at C_{16}) afforded, in addition to the corresponding Δ^{12}-olean-16-one and $\Delta^{13(18)}$-olean-16-one, Δ^8-multifloren-16-one (**372** with hydrogen at C_3, Δ^8 double bond, and ketone at C_{16}) (*478, 479*). Evidently the presence of the ketone group at C_{16} alters the relative rates of the competing rearrangement and proton-elimination steps.

A relevant biogenetic-type rearrangement is observed in the acid-catalyzed cyclization of the terminal epoxide (**386**) of 7-farnesyloxy-coumarin (umbelliprenin) (*480*). The friedelin-like, monocyclic ketone (**388**), obtained in 4% yield along with several other products, presumably arises from the monocyclic carbon ion intermediate (**389**) through a sequence of 1,2-shifts. Sesquiterpene derivatives of this type (*e.g.*, ascochlorin, **390**) are, in fact, mold metabolites (*481*). The pattern of incorporation of doubly labelled ^{13}C-acetate into ascochlorin demonstrates that a rearrangement takes place in the biosynthesis (*482*).

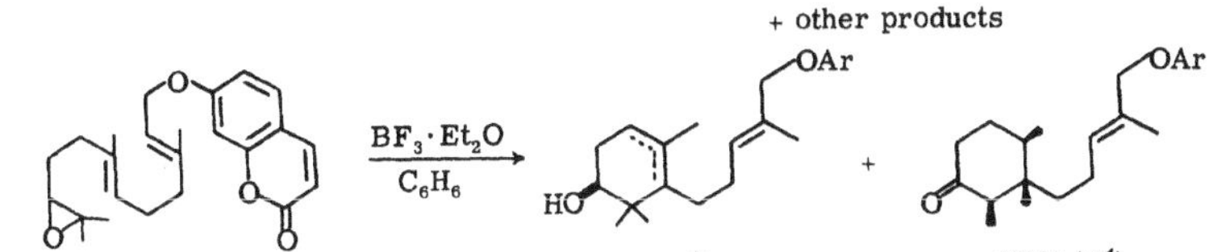

(**386**) (**387**) (9%) (**388**) (4%)

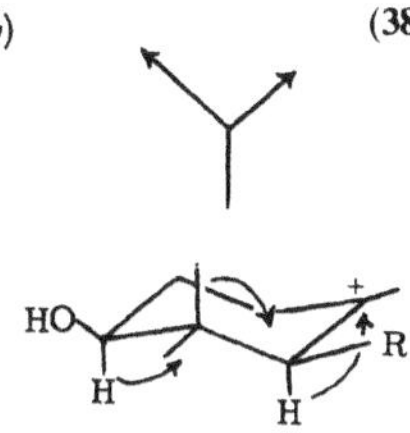

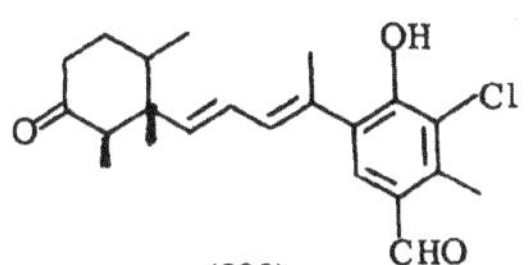

(**390**) (**389**)

The two perhydrochrysene based triterpenes baccharis oxide (**391**) (*483, 484*) and shionone (**392**) are biogenetically derived by backbone rearrangements similar to those portrayed above for glutinol (**373**) and friedelin (**375**), but originating from tetracyclic carbonium ion (**346**) prior to E ring closure (Scheme 30). If cyclization to the bridged oxide moiety of (**391**) immediately follows the $C_{10}{\rightarrow}C_9$ methyl rearrangement, the oxygen nucleophile enters *syn* to the departing methyl group.

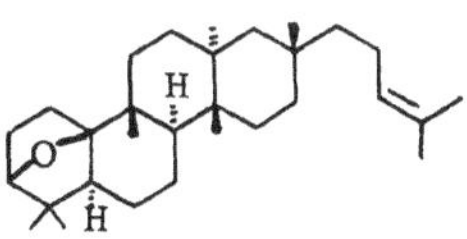

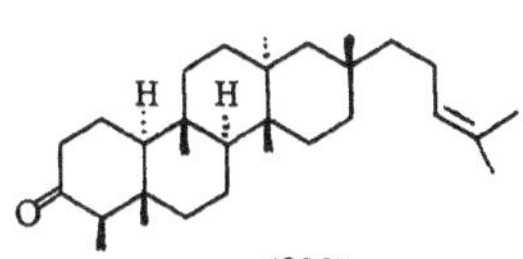

(**391**)
Baccharis oxide

(**392**)
Shionone

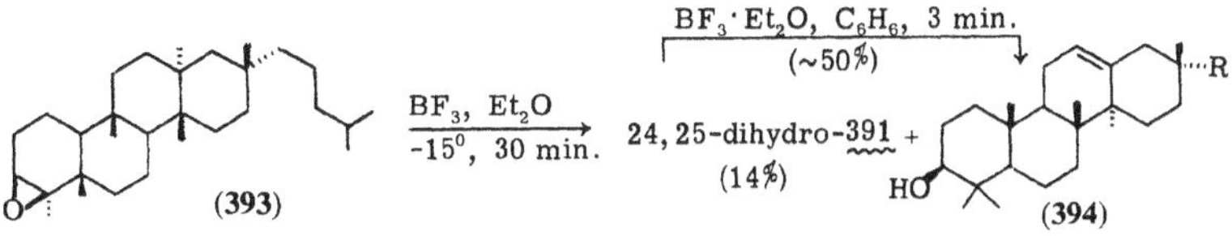

An alternative biogenesis of baccharis oxide is suggested by the boron trifluoride-induced rearrangement of 3β,4β-epoxyshionane (**393**) to 24,25-dihydrobaccharis oxide (*485*). However, the $C_5{\rightarrow}C_4$ methyl migration in this rearrangement occurs *syn* to the epoxide group. The retro-biogenetic backbone rearrangement of dihydrobaccharis oxide to bacchar-12-ene-3β-ol (**394**) may be effected with boron trifluoride in benzene (*483*). The reaction of 3α,4α-epoxy shionane (**393** with α-epoxide) with boron trifluoride in benzene affords both the 3α epimer of (**394**) and a partially rearranged isomer with the double bond in the Δ^8 position (*486*).

A large number of multi-step backbone rearrangements of steroids and related compounds have been recorded in the recent literature (*234, 487—493*). It is appropriate to discuss briefly a few of the more pertinent cases. The generality of these rearrangements is indicated by the apparent lack of sensitivity to various substitutents and changes in ring size. Although in some instances conjugation with a ketone group contributed to the driving force, this is not a requirement. Other contributing factors appear to be movement of double bonds from tri-substituted into tetra-substituted positions, reduction of steric interactions (*e.g.*, 1,3-diaxial methyl-methyl interactions), and lessening of conformational strain (*e.g.*, elimination endo- or exocyclic to the D ring, thereby removing the unstable *trans*-hydrindane moiety). Inductive effects of nearby-substituents evidently also influence these reactions, for cholestane-4β,5α-diol 4-mono acetate (C_4 epimer of **395**) gives only unrearranged olefin under the same conditions which effect the rearrangement of (**395**) shown below (*487*).

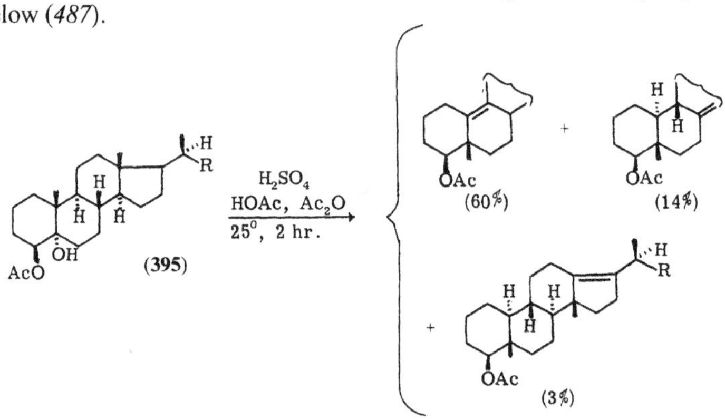

(**395**) (60%) (14%) (3%)

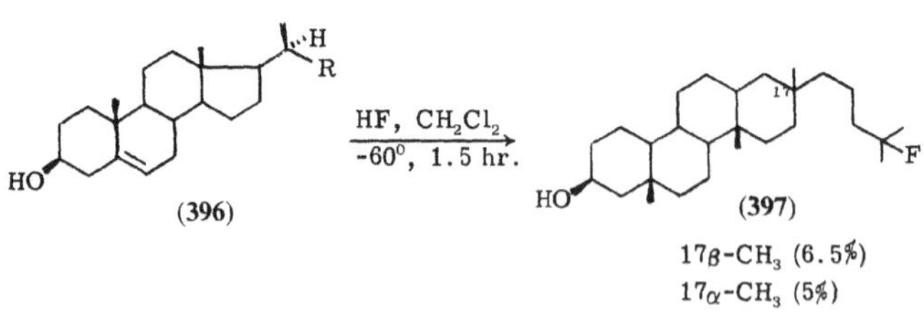

(**396**) (**397**)

17β-CH_3 (6.5%)
17α-CH_3 (5%)

When the dehydration of diol monoacetate (**395**) was carried out in deuterioacetic acid, all rearrangement products were formed free of deuterium, indicating that no elimination-reprotonation takes place under the reaction conditions (*487*). The backbone rearrangement of cholesterol (**396**) in the presence of hydrofluoric acid is combined with expansion of the D ring and 1,5-hydrogen transfer from the side-chain to give the shionane-like product (**397**) (*488*).

The mechanism of such multi-step rearrangements has been investigated with model compounds (**398**) and (**403**-OTs) (*458, 494*). The action of trifluoroacetic acid upon epoxy alcohol (**398**) causes two types of rearrangements; angular methyl migration to (**399**) and (**400**), and geminal methyl shift to friedelin-like ketones (**401**) and (**402**) (*458*). That all *cis* ketone (**401**) is formed as the initial ketonic product by 1,2-hydrogen rearrangement (rather than 1,2-hydrogen elimination to the corresponding enol) is shown by the isolation of (**401** and **401**-d_1) (69% d_1) from (**398** and **398**-d_1), respectively, at short reaction time (10 min.). Thus, ketone (**402**) arises from (**401**) by a secondary acid-catalyzed epimerization.

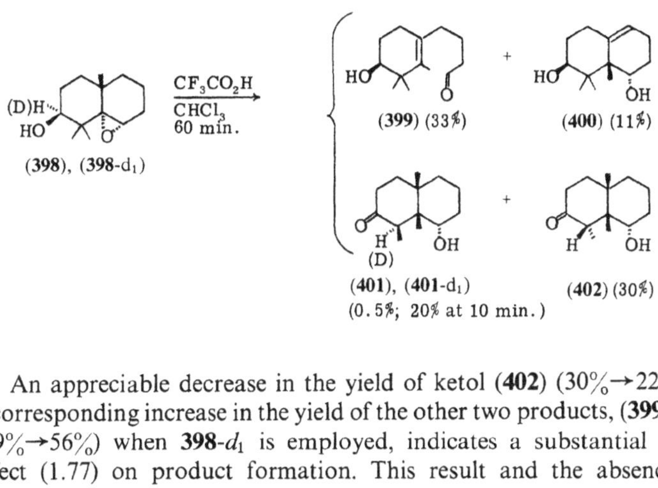

(**398**), (**398**-d_1) (**399**) (33%) (**400**) (11%) (**401**), (**401**-d_1) (0.5%; 20% at 10 min.) (**402**) (30%)

An appreciable decrease in the yield of ketol (**402**) (30%→22%) and a corresponding increase in the yield of the other two products, (**399** + **400**) (49%→56%) when **398**-d_1 is employed, indicates a substantial isotope effect (1.77) on product formation. This result and the absence of a kinetic isotope effect on the relative isomerization rates of (**398**) and (**398**-d_1) ($k_H/k_D = 1.01 \pm .02$), were interpreted in terms of a set of three rapidly equilibrating carbonium ions interconverting *via* 1,2-methyl migrations, which collapse to the observed products through competing 1,2-hydrogen shift (to **401**), ring scission (to **399**), and 1,2-hydrogen elimination (to **400**). The magnitude of the isotope effect on product formation was considered too high for the 1,2-hydrogen shift to be concerted with migration of the geminal methyl.

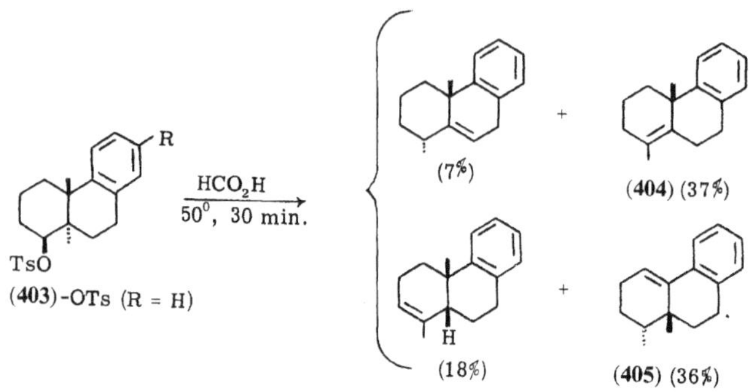

The solvolysis of **403**-OTs (R = H) and its methoxylated analogue (**403**-OTs, R = OCH_3) was examined in ethanol, acetic acid, and formic acid (*494*). The minimal effect of the substituent upon the rate of ethanolysis ($k_{OCH_3}/k_H = 1.6$) and the product distribution in both ethanol and acetic acid was considered in best agreement with a set discreet carbonium ion intermediates undergoing sequential 1,2-shifts. A trend towards an increasing amount of double backbone rearrangement to olefin (**405**) was observed with increasing solvent polarity (EtOH, 8%; AcOH, 14%; HCO_2H, 36%). On the other hand, the acid-catalyzed rearrangement (TsOH, AcOH, 120°) of olefin (**404**) by angular methyl migration to (**405**) was very slow compared to its rapid equilibration with the other two olefins. The greater facility of angular methyl migration from the carbonium ions produced by solvolysis of **403**-OTs was ascribed to a conformational effect. The proposal was made that enzymatic influence upon the rate and specificity of backbone rearrangements in terpene biosynthesis may be effected through control of conformation.

6. Hopanes, Fernanes, Adinanes, and Filicanes

The structural and biogenetic relationships within this class of pentacyclic triterpenes bear a close resemblance to those of the compounds in the preceding section (*495*). One distinctive feature of the hopane and rearranged hopane group is the absence of ring enlargement rearrangements along the biogenetic track (Scheme 33). Instead squalene (or squalene-2,3-oxide) undergoes direct cyclization from mainly an all chair conformation directly to the basic hopane skeleton. The intermediate hopanyl cation (**406**) then serves as a starting point for backbone rearrangements of varying length to neohopene (**408**), $\Delta^{9(11)}$-fernene (**409**), adianene (**410**), or filicene (**411**). The initial two 1,2-hydrogen rearrangements (21β→22β, 17β→21β) take place *syn* to each other.

Another noteworthy characteristic of this triterpene family is the frequent absence of an oxygen function at C_3. Whereas greater than 98% of all other known C_{30} triterpenes possess a C_3 oxygen grouping, about half of the fifty-four members of this particular class lack such functionality (*1*). The non-incorporation of squalene-2-3-oxide into $\Delta^{9(11)}$-fernene (together with a positive incorporation of squalene) supports the expectation that these non-oxygenated triterpenes are biosynthesized by a proton-initiated cyclization of squalene itself (*496*). The incorporation of six tritium atoms into $\Delta^{9(11)}$-fernene from 4-(R)-^{3}H-mevalonate is consistent with the postulated biogenesis (*27*).

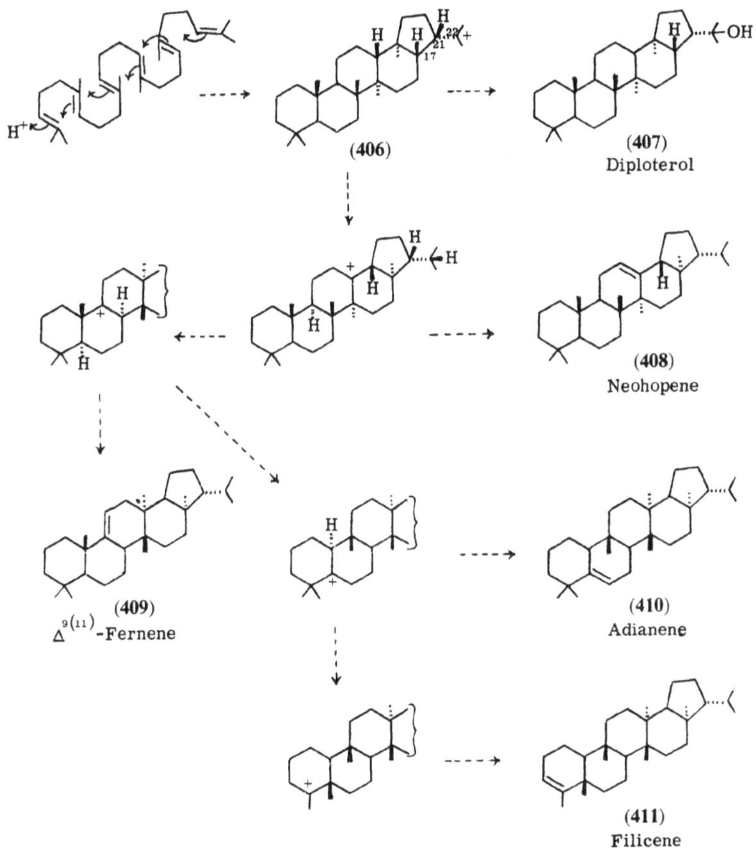

Scheme 33

The acid-catalyzed rearrangements of the triterpenes in this group tend toward the more stable $\Delta^{13(18)}$-neohopene (**412**; also referred to as hopene-II and wallichiene), just as the corresponding pentacycles in the previous sections yielded the equilibrium mixture of $\Delta^{13(18)}$-oleanene and Δ^{13}-18α-oleanene (**365**+**366**). The hopane triterpenes (*e.g.*, diploterol, **407**), under relatively mild conditions, give rise to $\Delta^{17(20)}$-hopenes, the latter undergoing further biogenetic-type rearrangement with $C_{18} \rightarrow C_{17}$ methyl migration to $\Delta^{13(18)}$-neohopene (**412**) (*497—500*). Both neohopene (**408**) (*501—503*) and its 18α-epimer (*502*) are isomerized to the tetra-substituted olefin (**412**).

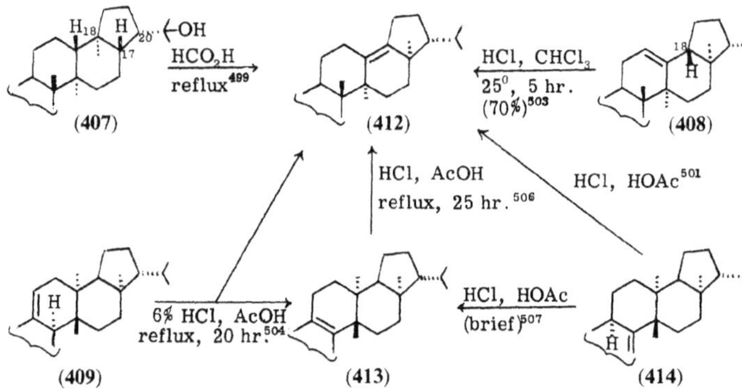

The three fernene isomers (**409**, **413**, and **414**), all of which are natural triterpenes, and various functionalized variants thereof, suffer rearrangement in the retro-biogenetic direction to $\Delta^{13(18)}$-neohopene (**412**) (*501, 504—506*). Evidently Δ^8-fernene (**413**) is relatively resistant to the acidic conditions since mixtures of (**412**) and (**413**) are often obtained. Both Δ^7- and $\Delta^{9(11)}$-fernene (**414** and **409**) are converted to the Δ^8-isomer under relatively mild conditions. The apparent ease of migration of the double bond into the Δ^8 position in this series contrasts with the rapid rearrangement of multiflorenyl acetate (**372**-OAc) to β-amyrin acetate (**352**, R = OAc) mentioned in the preceding section.

Advantage may be taken of the conjugative stabilization afforded by additional unsaturation ($\Delta^{9(11)}$ double bond or 11-ketone) in order to effect rearrangement in the biogenetic direction (*508, 509*). A mixture of $\Delta^{9(11),21}$-hopadiene (**415**) and $\Delta^{12,21}$-hopadiene (not shown) upon heating in the presence of activated fuller's earth gave rise to $\Delta^{7,9(11)}$-fernadiene (**416**, 25%) which was reduced to Δ^8-fernene (**413**) (*508*).

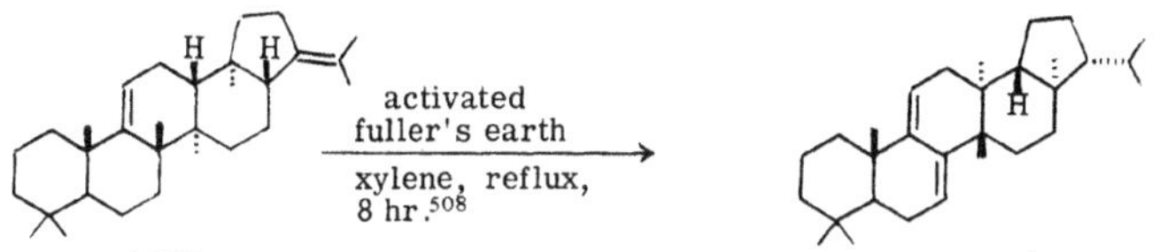

(415) (416) (25%)

Like their counterparts, Δ^5-glutinene (**383**) and Δ^3-friedelene (**384**), in the previous section, adianene (**410**) (*501, 507*), simiarenyl acetate (**417**) (*506, 510*), simiarenone (*510*), and filicene (**411**) (*507, 511*) undergo acid-catalyzed retro-biogenetic rearrangements. Although rearrangement all the way to $\Delta^{13(18)}$-neohopene (**412**) is possible under forcing conditions, a substantial amount of the relatively stable Δ^8-fernene (**413**), has also been isolated from this reaction. The reaction of filicane-3α,4α-oxide (α-epoxide of **411**) with boron trifluoride etherate in benzene gives a mixture of $\Delta^{5(10)}$-adianen-3α-ol and Δ^8-fernene-3α-OH) (*511*).

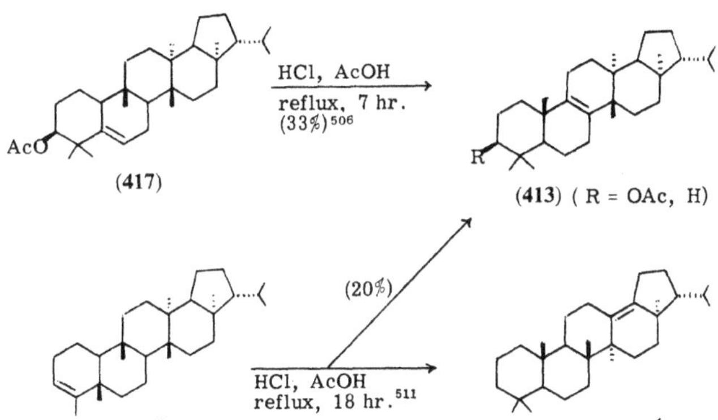

(417) (413) (R = OAc, H)

(411) (412) (14%)

References

1. Devon, T. K., and A. I. Scott: Handbook of Naturally Occurring Compounds, Vol. II Terpenes. New York: Academic Press. 1972.

2. Ružička, L., A. Eschenmoser und H. Heusser: The Isoprene Rule and the Biogenesis of Terpenic Compounds. Experientia **9**, 357 (1953).

3. Eschenmoser, A., L. Ružička, O. Jeger, und D. Arigoni: Zur Kenntnis der Triterpene. Eine stereochemische Interpretation der biogenetischen Isoprenregel bei den Triterpenen. Helv. Chim. Acta **38**, 1890 (1955).

4. Ružička, L.: Bedeutung der theoretischen organischen Chemie für die Chemie der Terpenverbindungen. In: A. Todd (ed.), Perspectives in Organic Chemistry, p. 265. New York: Interscience. 1956.

5. RUŽIČKA, L.: Faraday Lecture — History of the Isoprene Rule. Proc. Chem. Soc. (London) **1959**, 341.
6. RUŽIČKA, L.: Perspektiven der Biogenese und der Chemie der Terpene. Pure Appl. Chem. **6**, 493 (1963).
7. NICHOLAS, H. J.: The Biogenesis of Terpenes in Higher Plants. In: P. BERNFELD (ed.), Biogenesis of Natural Compounds. New York: MacMillan Company. 1963.
8. RICHARDS, J. H., and J. B. HENDRICKSON: The Biosynthesis of Steroids, Terpenes, and Acetogenins. New York: W. A. Benjamin, Inc. 1964.
9. JOHNSON, W. S.: Nonenzymic Biogenetic-like Olefinic Cyclizations. Accounts Chem. Res. **1**, 1 (1968).
10. VAN TAMELEN, E. E.: Bioorganic Chemistry: Sterols and Acyclic Terpene Terminal Epoxides. Accounts Chem. Res. **1**, 111 (1968).
11. GOLDSMITH, D.: Biogenetic-type Synthesis of Terpenoid Systems. In: W. HERZ, H. GRIESEBACH, and G. W. KIRBY (eds.), Progress in the Chemistry of Organic Natural Products, Vol. 29, p. 363. Vienna: Springer. 1971.
12. MONEY, T.: Biogenetic-Type Synthesis of Terpenes. In: W. CARRUTHERS and J. K. SUTHERLAND (eds.), Progress in Organic Chemistry, Vol. 8, p. 29. New York: J. Wiley and Sons. 1973.
13. HARDING, K. E.: On the Stereochemistry of Biogenetic-Like Olefin Cyclizations. Bioorganic Chemistry **2**, 248 (1973).
14. VAN TAMELEN, E. E.: Bioorganic Chemistry: Total Synthesis of Tetra- and Pentacyclic Triterpenoids. Accounts Chem. Res. **8**, 152 (1975).
15. KING, J. F., and P. DE MAYO: Terpenoid Rearrangements. In: P. DE MAYO (ed.), Molecular Rearrangements, Part Two, p. 771. New York: Interscience. 1963.
16. YAMAMURA, S., and Y. HIRATA: A Naturally-occurring Bromo-compound, Aplysin-20 from *Aplysia kurodai*. Bull. Chem. Soc. Japan **44**, 2560 (1971).
17. SIMS, J. J., G. H. Y. LIN, R. M. WING, and W. FENICAL: Marine Natural Products. Concinndiol, A Bromo-diterpene Alcohol from the Red Alga, *Laurencia concinna*. Chem. Commun. **1973**, 470.
18. SUZUKI, M., E. KUROSAWA, and T. IRIE: Olanduliferol, A New Halogenated Sesquiterpenoid from *Laurencia glandulifera* Kützing. Tetrahedron Letters **1974**, 1807.
19. MCMILLAN, J. A., I. C. PAUL, R. H. WHITE, and L. P. HAGER: Molecular Structure of Acetoxyintricatol: A New Bromo Compound from *Laurencia intricata*. Tetrahedron Letters **1974**, 2039.
20. CORBELLA, A., P. GARIBOLDI, G. JOMMI, and M. SISTI: Biosynthesis of the Terpenoid Dendrobine. Early Stages of the Pathway. Chem. Commun. **1975**, 288.
21. ACHILLADELIS, B., and J. R. HANSON: Studies in Terpenoid Biosynthesis. Part V. Biosynthesis of Rosenonolactone. J. Chem. Soc. (C) **1969**, 2010.
22. ACHILLADELIS, B., and J. R. HANSON: Studies in Terpenoid Biosynthesis. I. The Biosynthesis of Metabolites of *Trichothecium Roseum*. Phytochem. **7**, 589 (1968).
23. UPPER, C. D., and C. A. WEST: Biosynthesis of Gibberellins. II. Enzymic Cyclization of Geranylgeranyl Pyrophosphate to Kaurene. J. Biol. Chem. **242**, 3285 (1967).
24. SCHECHTER, I., and C. A. WEST: Biosynthesis of Gibberellins. IV. Biosynthesis of Cyclic Diterpenes from *trans*-Geranylgeranyl Pyrophosphate. J. Biol. Chem. **244**, 3200 (1969).
25. CLAYTON, R. B.: Biosynthesis of Sterols, Steroids, and Terpenoids. Part I. Biogenesis of Cholesterol and the Fundamental Steps in Terpenoid Biosynthesis. Quart. Rev. (Chem. Soc. London) **19**, 168 (1965).
26. MULHEIRN, L. J., and P. J. RAMM: The Biosynthesis of Stereols. Chem. Soc. Rev. **1**, 259 (1972).
27. GOODWIN, T. W.: Biosynthesis of Carotenoids and Plant Triterpenes (Ciba Lecture). Biochem. J. **123**, 293 (1971).

28. GIGER, R.: Analogien und Unterschiede in der Biosynthese von α- und β-Amyrin. Ph. D. Thesis, ETH, Zürich, 1973.

29. IMMER, H., und K. HUBER: Synthese und Biologische Auswertung von 3β,20(R)-Dihydroxy-Protost-24-en. Helv. Chim. Acta **54,** 1346 (1971).

30. BARTON, D. H. R., G. MELLOWS, D. A. WIDDOWSON, and J. J. WRIGHT: Biosynthesis of Terpenes and Steroids. Part IV. Specific Hydride Shifts in the Biosynthesis of Lanosterol and β-Amyrin. J. Chem. Soc. (C) **1971,** 1142.

31. VAN TAMELEN, E. E., R. G. LEES, and A. GRIEDER: Cyclization of a Terpenoid Diene with Preformed A-B-D Rings and Its Significance for the Mechanism of Terpenoid Terminal Epoxide Cyclizations. J. Amer. Chem. Soc. **96,** 2255 (1974).

32. BARTLETT, P. A., J. I. BRAUMAN, W. S. JOHNSON, and R. A. VOLKMANN: Concerning the Mechanism of a Nonenzymic Biogenetic-Like Olefinic Cyclization. J. Amer. Chem. Soc. **95,** 7502 (1973).

33. ARIGONI, D.: Some Studies in the Biosynthesis of Terpenes and Related Compounds. Pure Appl. Chem. **17,** 331 (1968).

34. CORNFORTH, J. W.: Olefin Alkylation in Biosynthesis. Angew. Chem. Intern. Ed. **7,** 903 (1968).

35. ARIGONI, D., D. E. CANE, B. MULLER, and C. TAMM: The Mode of Incorporation of Farnesyl Pyrophosphate into Verrucarol. Helv. Chim. Acta **56,** 2946 (1973).

36. FORRESTER, J. M., and T. MONEY: Sequence Studies in Biosynthesis: Trichothecin. Canad. J. Chem. **50,** 3310 (1972).

37. MULHEIRN, T. J., and E. CASPI: Mechanism of Squalene Cyclization. The Biosynthesis of Fusidic Acid. J. Biol. Chem. *246,* 2494 (1971).

38. EBERSOLE, R. C., W. O. GODTFREDSEN, S. VANGEDAL, and E. CASPI: Mechanism of Oxidative Cyclization of Squalene. Concerning the Mode of Formation of the 17(20) Double Bond in the Biosynthesis of Fusidic Acid by *Fusidium coccineum.* J. Amer. Chem. Soc. **96,** 6499 (1974).

39. SARGENT, G. D.: The 2-Norbornyl Cation. In: G. A. OLAH and P. VON R. SCHLEYER (eds.), Carbonium Ions, Vol. III, pp. 1099—1200, especially p. 1177. New York: Wiley-Interscience. 1972.

40. BROWN, H. C.: The Question of σ Bridging in the Solvolysis of 2-Norbornyl Derivatives. Accounts Chem. Res. **6,** 377 (1973).

41. BROWN, H. C., and K.-T. LIU: Additions to Bicyclic Olefins. VII. Electrophilic Addition of Hydrogen Chloride and Deuterium Chloride to Norbornene, 2-Methylnorbornene, and Related Bicyclic Olefins. Evidence for a Carbonium Ion Process and the Capture of Unsymmetrical (Classical) 2-Norbornyl Cations. J. Amer. Chem. Soc. **97,** 600 (1975).

42. GOERING, H. L., and G. N. FICKES: Ionic Reactions in Bicyclic Systems. VI. Solvolytic Studies of Bicyclo[3.2.1]octan-2-yl and Bicyclo[2.2.2]octan-2-yl Systems. J. Amer. Chem. Soc. **90,** 2856 (1968).

43. — — Ionic Reactions in Bicyclic Systems. V. Solvolysis of *endo*-Bicyclo[3.2.1]octan-2-yl (Equatorial) *p*-Toluenesulfonate. J. Amer. Chem. Soc. **90,** 2848 (1968).

44. — — Ionic Reactions in Bicyclic Systems. VII. Solvolysis of Optically Active *exo*-Bicyclo[3.2.1]octan-2-yl and Bicyclo[2.2.2]octan-2-yl *p*-Toluenesulfonate. J. Amer. Chem. Soc. **90,** 2862 (1968).

45. BERSON, J. A.: Memory Effects and Stereochemistry in Multiple Carbonium Ion Rearrangements. Angew. Chem. Intern. Ed. **7,** 779 (1968).

46. BERSON, J. A., M. S. POONIAN, and W. J. LIBBEY: Memory Effects in Multiple Carbonium Ion Rearrangements. II. Behavior of the Potentially Symmetrical Cation in the Ring-Expansion Reactions of 7-Norbornylcarbinyl Systems. J. Amer. Chem. Soc. **91,** 5567 (1969).

47. BECKER, K. B., and C. A. GROB: Nucleophilic Reactions at Tertiary Carbon. Part I.

The 1,2-Dimethyl-1-Cyclohexyl Cation. Helv. Chim. Acta **56**, 2723 (1973) and pertinent references cited therein.

48. TRAYLOR, T. G., W. HANSTEIN, H. J. BERWIN, N. A. CLINTON, and R. S. BROWN: Vertical Stabilization of Cations by Neighboring σ Bonds. General Considerations. J. Amer. Chem. Soc. **93**, 5715 (1971).
49. SARGENT, G. D., and T. J. MASON: Solvolysis of 8-Vinyl-*exo*-8-bicyclo[3.2.1]octyl-3,5-Dinitrobenzoate. Evidence for Stabilization by Carbon-Carbon-Hyperconjugation in the Transition State for Formation of a Tertiary Allylic Cation. J. Amer. Chem. Soc. **96**, 1063 (1974).
50. ITÔ, S., K. ENDO, and T. NOZOE: Stereochemistry of Widdrol. Tetrahedron Letters **1964**, 3375.
51. HERZ, W., A. K. PINDER, and R. N. MIRRINGTON: Resin Acids. IX. Cationic Cyclization of Pimaric Acid Derivatives. Partial Synthesis of (−)-Hibaene. J. Organ. Chem. (USA) **31**, 2257 (1966).
52. HALSALL, T. G., E. R. H. JONES, E. L. TAN, and G. R. CHAUDHRY: The Chemistry of Triterpenes and Related Compounds. Part XLV. The Acid-catalysed Rearrangements of the 5α,6α-Diols and 5,6-Epoxides of Some 4,4-Dimethylsteroids. J. Chem. Soc. (C) **1966**, 1374.
53. BLUNT, J. W., M. P. HARTSHORN, and D. N. KIRK: Reactions of Epoxides. Part XV. Boron Trifluoride-catalysed Rearrangements of 5,6-Epoxides in the 4,4-Dimethylcholestane Series. J. Chem. Soc. (C) **1968**, 635.
54. COXON, J. M., M. P. HARTSHORN, and C. N. MUIR: Reactions of Epoxides - XXI. Boron Trifluoride Catalysed Rearrangements of Some 3α-Substituted-5,6-Oxidocholestanes. Tetrahedron **25**, 3925 (1969).
55. EPSTEIN, W. W., and C. D. POULTER: A Survey of Some Irregular Monoterpenes and Their Biogenetic Analogies to Presqualene Alcohol. Phytochem. **12**, 737 (1973).
56. BANTHORPE, D. V., B. V. CHARLWOOD, and M. J. O. FRANCIS: The Biosynthesis of Monoterpenes. Chem. Rev. **72**, 115 (1972).
57. BANTHORPE, D. V., and B. V. CHARLWOOD: Biogenesis of Terpenes. In: A. A. NEWMAN (ed.), Chemistry of Terpenes and Terpenoids, p. 337. London: Academic Press. 1972.
58. BATES, R. B., and S. K. PAKNIKAR: Terpenoids. IX. Biogenesis of Some Monoterpenoids Not Derived from a Geranyl Precursor. Tetrahedron Letters **1965**, 1453.
59. BANTHORPE, D. V., and B. V. CHARLWOOD: Biosynthesis of Artemisia Ketone. Nature New Biology **231**, 285 (1971).
60. PATTENDEN, G., and R. STORER: Studies on the Biosynthesis of Chrysanthemum Monocarboxylic Acid. Tetrahedron Letters **1973**, 3473.
61. SUGA, T., T. SHISHIBORI, K. KOTERA, and R. FUJII: The Biosynthesis of a Non-Head to Tail Monoterpene, Artemisia Ketone. Chemistry Letters **1972**, 533.
62. CLOSS, G. L., R. A. MOSS, and S. H. GOH: Formation of Cyclopropanes in Acid-Catalyzed Decomposition of Phenyldiazomethane in Olefins. J. Amer. Chem. Soc. **88**, 364 (1966).
63. CLOSS, G. L., and S. H. GOH: The Acid-Induced Reaction of Aryldiazomethanes with Olefins. Mechanism of Reaction. J. Organ. Chem. (USA) **39**, 1717 (1974).
64. CROMBIE, L., P. A. FIRTH, R. P. HOUGHTON, D. A. WHITING, and D. K. WOODS: Cyclopropane Cleavage of Chrysanthemic Acid Relatives to Santolinyl, Artemisyl, and Lavandulyl Structures: Acid-catalyzed and Biosynthetic Experiments. J. Chem. Soc. Perkin I **1972**, 642.
65. BATES, R. B., and D. FELD: Terpenoids. XIII. Conversion of Chrysanthemyl Alcohol to a Triene with the Artemesia Ketone Skeleton. Tetrahedron Letters **1967**, 4875.
66. POULTER, C. D., S. G. MOESINGER, and W. W. EPSTEIN: Solvolysis of N-methyl-

4-(chrysanthemyloxy)pyridinium Iodide – A Model for Non-head-to-tail Monoterpene Biosynthesis. Tetrahedron Letters **1972,** 67.

67. POULTER, C. D.: Model Studies in Terpene Biosynthesis. J. Agr. Food Sci. **22,** 167 (1974).

68. — Model Studies of Terpene Biosynthesis. Stereoselective Ionization of *N*-Methyl-4-[(α*S*,1*R*,3*R*)-chrysanthemyloxy]pyridinium-d_1 Iodide. J. Amer. Chem. Soc. **94,** 5515 (1972).

69. SASAKI, T., S. EGUCHI, M. OHNO, and T. UMEMURA: Studies on Chrysanthemic Acid VII. Thermal Decomposition of Chrysanthemyl Oxalate and Deamination of Chrysanthemylamine. J. Organ. Chem. (USA) **36,** 1968 (1971).

70. — — — — Solvolysis of *cis*- and *trans*-Chrysanthemyl 3,5-Dinitrobenzoates. The Conjugative Transmission Ability of Cyclopropane Ring. Chemistry Letters **1972,** 503.

71. THOMAS, A. F., and W. PAWLAK: The Synthesis and Reactions of Yomogi Alcohol. Conversion of the Artemisyl Skeleton to the Santolinyl Skeleton by a 1,2-Shift of a Vinyl Group. Synthesis of Santolinatriene. Helv. Chim. Acta **54,** 1822 (1971).

72. TROST, B. M., P. CONWAY, and J. STANTON: Some Aspects of Terpene Biosynthesis – A Model. Chem. Commun. **1971,** 1639.

73. POULTER, C. D., R. J. GOODFELLOW, and W. W. EPSTEIN: The Absolute Configuration of Santolina Alcohol from *Ormenis Multicaulus*. Tetrahedron Letters **1972.** 71.

74. WIBERG, K. B., and G. SZEIMIES: Acid-Catalyzed Solvolyses of Bicyclobutane Derivatives. Stereochemistry of the Cyclopropylcarbinyl-Cyclopropylcarbinyl and Related Rearrangements. J. Amer. Chem. Soc. **92,** 571 (1970).

75. MAJERSKI, Z., and P. VON R. SCHLEYER: The Stereochemistry of Cyclopropylcarbinyl Rearrangements. Synthesis and Solvolysis of Cyclopropylcarbinyl-1,1′,1′-*trans*-2,3,3-d_6 Methanesulfonate. J. Amer. Chem. Soc. **93,** 665 (1971).

76. WIBERG, K. B., B. A. HESS, JR., and A. J. ASHE III: Cyclopropylcarbinyl and Cyclobutyl Cations. In: G. A. OLAH and P. von R. SCHLEYER (eds.), Carbonium Ions Vol. III, p. 1295. New York: Wiley-Interscience. 1972.

77. CRAMER, F., und W. RITTERSDORF: Die Hydrolyse von Phosphaten und Pyrophosphaten einiger Monoterpenalkohole. Modellreaktionen zur Biosynthese der Monoterpene. Tetrahedron **23,** 3015 (1967).

78. RITTERSDORF, W., und F. CRAMER: Die Hydrolyse von 2,3-Dihydroterpenyl-Phosphaten und -Pyrophosphaten. Tetrahedron **23,** 3023 (1967).

79. — — Cyclization of Nerol and Linalool on Solvolysis of Their Phosphate Esters. Tetrahedron **24,** 43 (1968).

80. WINSTEIN, S., A. VALKANAS, and C. F. WILCOX: The Solvolysis of Linalyl *p*-Nitrobenzoate and the Stereochemical Aspects of the Resulting 1–3 and 1–5 Rearrangements. J. Amer. Chem. Soc. **94,** 2286 (1972).

81. STEPHAN, K.: Über eine Umwandlung von Linalool in Terpineol vom Schmelp. 35°. J. prakt. Chem. **58,** 109 (1898).

82. ZEITSCHEL, O.: Über das Nerol und seine Darstellung aus Linalool. Ber. dtsch. chem. Ges. **39,** 1780 (1906).

83. BUNTON, C. A., D. L. HACHEY, and J. P. LERESCHE: Deamination of Nerylamine and Geranylamine. J. Organ. Chem. (USA) **37,** 4036 (1972).

84. WHITTAKER, D.: The Monoterpenes. In: A. A. NEWMAN (ed.), Chemistry of Terpenes and Terpenoids, Chapter 2. London: Academic Press. 1972.

85. STORK, G., and W. N. WHITE: The Stereochemistry of the S_N2' Reaction II. J. Amer. Chem. Soc. **78,** 4609 (1956).

86. WILCOX, C. F., and S. S. CHIBBER: Solvolysis of 3-Cyclohexenylcarbinyl Derivatives. J. Organ. Chem. (USA) **27,** 2332 (1962).

87. FAIRLIE, J. C., G. L. HODGSON, and T. MONEY: Synthesis of (±)-Camphor. J. Chem. Soc. Perkin I **1973,** 2109.

88. MUSSO, H., K. NAUMANN, und K. GRYCHTOL: Asterane, IV. Synthesen des Norpinans (Bicyclo[3.1.1]heptan). Chem. Ber. **100,** 3614 (1967).
89. THOMAS, M. T., and A. G. FALLIS: The Total Synthesis of (±)α- and (±)β-Pinene: A General Route to the Bicyclo[3.1.1]heptanes. Tetrahedron Letters **1973,** 4687.
90. BERSON, J. A.: Carbonium Ion Rearrangements in Bridge Bicyclic Systems. In: P. DE MAYO (ed.): Molecular Rearrangements, Part One, p. 111. New York: Interscience. 1963.
91. MEERWEIN, H., und K. VAN EMSTER: Untersuchungen in der Camphen-Reihe, I: Über den Reaktionsmechanismus der Isoborneol ⇌ Camphen-Umlagerung. Chem. Ber. **53,** 1815 (1920).
92. — — Über die Gleichgewichts-Isomerie zwischen Bornylchlorid, Isobornylchlorid und Camphen-Chlorhydrat. Chem. Ber. **55,** 2500 (1922).
93. MEERWEIN, H., und J. VORSTER: Über das Pinenchlorhydrat. J. prakt. Chem. **147,** 83 (1936).
94. OLAH, G. A., J. R. DEMEMBER, C. Y. LUI, and R. D. PORTER: Stable Carbonium Ions. CX. The 1,2-Dimethylnorbornyl Cation. J. Amer. Chem. Soc. **93,** 1442 (1971).
95. BANTHORPE, D. V., and D. WHITTAKER: Rearrangements of Pinane Derivatives. Quart. Rev. (Chem. Soc. London) **20,** 373 (1966).
96. SALMON, J. R., and D. WHITTAKER: Rearrangements of Pinane Derivatives. Part III. Solvolysis of the 2-Pinanyl p-Nitrobenzoates. J. Chem. Soc. (B) **1971,** 1249.
97. VALKANAS, G., und N. ICONOMOU: Reaktionen in der Terpen-Reihe. 1. Mitteilung. Hydratisierung von α-Pinen. Helv. Chim. Acta **46,** 1089 (1963).
98. HÜCKEL, W., und H.-J. KERN: Änderung des Molekülbaus bei chemischen Reaktionen, XXI. Desaminierung in der Bicyclo[2.2.1]heptan-Reihe. Liebigs Ann. Chem. **728,** 49 (1969).
98a. KIRMSE, W., und G. AREND: Bornyl- und Fenchyldiazonium-Ionen in alkalischen Medien. Chem. Ber. **105,** 2738 (1972).
98b. — — Umlagerung 1-substituierter Apobornyldiazonium-Ionen. Chem. Ber. **105,** 2746 (1972).
99. PAUKSTELIS, J. V., and B. W. MACHARIA: *Trans*-Coplanar Rearrangements. The Conversion of Camphor to Nopinone. Tetrahedron **29,** 1955 (1973).
100. BANTHORPE, D. V., J. MANN, and K. W. TURNBULL: Terpene Biosynthesis. Part II. Biosynthesis of Thujane Derivatives in *Thuja, Tanacetum* and *Juniperus* Species. J. Chem. Soc. (C) **1970,** 2689.
101. HANACK, M., and H.-J. SCHNEIDER: Neighboring-Group Effects and Rearrangements in Reactions of Cyclopropylmethyl, Cyclobutyl, and Homoallyl Systems. Angew. Chem. Intern. Ed. **6,** 666 (1967).
102. STORY, P. R., and B. C. CLARK, JR.: Homoallylic and Homoaromatic Cations. In: G. A. OLAH and P. VON R. SCHLEYER (eds.), Carbonium Ions Vol. III, p. 1007. New York: Wiley-Interscience. 1972.
103. BLOSS, A. S., P. R. BROOK, and R. M. ELLAM: The Formation of the 2-Bicyclo-[3.1.0]hexyl Cation by Deamination and Solvolysis, and the Effect of Methyl Substitution at C-5. J. Chem. Soc. Perkin II **1973,** 2165.
104. HANACK, M., und W. KEBERLE: Über die Bildung von Bicyclo[3.1.0]hexanol-(2) aus Δ^3-Cyclohexenyl-Derivaten. Chem. Ber. **96,** 2937 (1963).
105. LEBEL, N. A., and J. E. HUBER: The Tricyclo[2.2.2.$0^{2,6}$]octan-3-ols and Derivatives. Preparation, Structure and Reactivity Studies. J. Amer. Chem. Soc. **85,** 3193 (1963).
106. GAONI, Y.: Base Induced Isomerizations of γ,δ-Epoxyketones – II. Syntheses in the Thujane Series. *D,L*-Sabina Ketone and *D,L-Cis* Sabinene Hydrate. Tetrahedron **28,** 5525 (1972).
107. NORIN, T., and L.-A. SMEDMAN: The Cyclopropylcarbinyl Rearrangements in Sabinene and α-Thujene. Acta Chem. Scand. **25,** 2010 (1971).

108. Parker, W., J. S. Roberts, and R. Ramage: Sesquiterpene Biogenesis. Quart. Rev. (Chem. Soc. London) **21,** 331 (1967).

109. Arigoni, D.: Stereochemical Aspects of Sesquiterpene Biosynthesis. 9th IUPAC Conference on Natural Products, Ottawa, Canada, June 24–28, 1974.

110. Marshall, J. A., St. F. Brady, and N. H. Andersen: The Chemistry of Spiro[4.5]Decane Sesquiterpenes. In: Herz, Grisebach, Kirby (eds.) Progress in the Chemistry of Organic Natural Products, Vol. 31, p. 283, 1974.

111. L. Ružička, und E. Capato: Höhere Terpenverbindungen XXIV. Ringbildungen bei Sesquiterpenen. Totalsynthese des Bisabolens und eines Hexahydro-Cadalins. Helv. Chim. Acta **8,** 259 (1925).

112. Gutsche, C. D., J. R. Maycock, and C. T. Chang: Acid-catalyzed Cyclization of Farnesol and Nerolidol. Tetrahedron **24,** 859 (1968).

113. Andersen, N. H., and D. D. Syrdal: Chemical Simulation of the Biogenesis of Cedrene. Tetrahedron Letters **1972,** 2455.

114. Ohta, Y., and Y. Hirose: Electrophile Induced Cyclization of Farnesol. Chemistry Letters **1972,** 263.

115. Hikino, H., N. Suzuki, and T. Takemoto: Structure and Absolute Configuration of Campherenone and Campherenol, Sesquiterpenoids of *Cinnamomum camphora.* Chem. Pharm. Bull. (Japan) **19,** 87 (1971).

116. Hodgson, G. L., D. F. MacSweeney, and T. Money: Synthesis of (±)-Campherenone, (±)-Epicampherenone, (±)-β-Santalene, (±)-epi-β-Santalene, (±)-α-Santalene, (±)-Ylangocamphor, (±)-Copacamphor, and (±)-Sativene. J. Chem. Soc. Perkin I **1973,** 2113.

117. Eck, C. R., G. L. Hodgson, D. F. MacSweeney, R. W. Mills, and T. Money: A Stereospecific Synthetic Route to Campherenone, Campherenol, Epicampherenone, β-Santalene, Epi-β-santelene, Ylangocamphor, Copacamphor, Sativene, and Copacamphene. J. Chem. Soc. Perkin I **1974,** 1938.

118. Souček, M.: On Terpenes. CXLVIII. Biosynthesis of Carotol in *Daucus carota* L. A Contribution to Configuration of Carotol and Daucol. Collect. Czech. Chem. Comm. **27,** 2929 (1962).

119. Demole, E., P. Enggist et C. Borer: Application synthétiques de la cyclisation d'alcools tertiaires γ-éthyléniques en α-bromotétrahydrofurannes sous l'action du N-bromosuccinimide. II. Cyclisation du (±)-nérolidol en diméthyl-2,5-(méthyl-4-pentène-3-yl)-2-cycloheptène-4-one, tétraméthyl-3,3,7,10-oxa-2-tricyclo[5.5.0.0^{1,4}]-dodécène-9, β-acoratiene, cédradiène-2,8-*epi*-2-α-cédrène et α-cédrène. Helv. Chim. Acta **54,** 1845 (1971).

120. Hendrickson, J. B.: Stereochemical Implications in Sesquiterpene Biogenesis. Tetrahedron **7,** 82 (1959).

121. Sutherland, J. K.: Regio- and Stereo-Specificity in the Cyclisation of Medium Ring 1,5-Dienes. Tetrahedron **30,** 1651 (1974).

122. Wharton, P. S., R. A. Kretchmer, R. J. Kilian, and T. Oda: Ten-Membered Rings. Transannular Double-Bond Participation in Acid-Promoted Cyclizations. J. Organ. Chem. (USA) **39,** 3755 (1974).

123. Brown, E. D., M. D. Solomon, J. K. Sutherland, and A. Torre: A Possible Intermediate in Sesquiterpene Biosynthesis. Chem. Commun. **1967,** 111.

124. Brown, E. D., and J. K. Sutherland: The Conversion of a Germacrane into Guaiane Derivatives. Chem. Commun. **1968,** 1060.

125. Heathcock, C. H., and R. Ratcliffe: A Stereoselective Total Synthesis of the Guaiazulenic Sesquiterpenoids α-Bulnesene and Bulnesol. J. Amer. Chem. Soc. **93,** 1746 (1971).

126. Kato, M., H. Kosugi, and A. Yoshikoshi: Solvolytic Rearrangement of 1β-Tosyloxy-4α, 8aβ-dimethyldecalin Derivatives; A Synthesis of (±) Bulnesol. Chem. Commun. **1970,** 185.

127. MARSHALL, J. A., and J. J. PARTRIDGE: The Total Synthesis of (±)-Bulnesol and Related Studies. Tetrahedron **25,** 2159 (1969).

128. HEATHCOCK, C. H., R. RATCLIFFE, and J. VAN: Synthesis of Hydroazulenes by Solvolytic Rearrangement of 9-Methyl-1-decalyl Tosylates. J. Organ. Chem. (USA) **37,** 1796 (1972).

129. WATKINS, S. F., N. H. FISCHER, and I. BERNAL: Neutron Diffraction Structure of Melampodin. Its Role in the Reclassification of the Germacranolides. Proc. Nat. Acad. Sci. (USA) **70,** 2434 (1973).

130. ANDERSON, G. D., R. GITANY, R. S. MCEWEN, and W. HERZ: Relative and Absolute Configuration of Pseudoivalin. Tetrahedron Letters **1973,** 2409.

131. DITTMANN, W., und F. STÜRZENHOFECKER: Synthese von *racem.* (1S:4S:9R) (1R:4R:9S)-*cis*-Decalin-diol-(1.4). Liebigs Ann. Chem. **688,** 57 (1965).

132. TRAYNHAM, J. G., and H. H. HSIEH: Addition Reactions of *cis,trans*-1,5-Cyclodecadiene. J. Organ. Chem. (USA) **38,** 868 (1973).

133. TRAYHNHAM, J. G., G. R. FRANZEN, G. A. KNESEL, and D. J. NORTHINGTON, Jr.: Addition Reactions of *cis, trans*-1,5-Cyclodecadiene. J. Organ. Chem. (USA) **32,** 3285 (1967).

134. IGUCHI, M., M. NIWA, and S. YAMAMURA: Biogenetic Model Reactions of Epoxygermacrone. Chem. Commun. **1972,** 689.

135. NOZOE, S., and C. KANEKO: Personal communication cited in reference *108.*

136. HORTMANN, A. G., D. S. DANIEL, and J. E. MARTINELLI: Biogenetically Patterned Total Synthesis of (±)-Occidentalol and 7-Epi-(−)-occidentalol. J. Organ. Chem. (USA) **38,** 728 (1973).

137. OHTA, Y., K. OHARA, and Y. HIROSE: Acid Catalyzed Isomerization of α-Cubebene, α-Copaene, and α-Ylangene. Tetrahedron Letters **1968,** 4181.

138. ANDERSEN, N. H.: Biogenetic Implications of the Antipodal Sesquiterpenes in Vetiver Oil. Phytochem. **9,** 145 (1970).

139. ANDERSEN, N. H., and D. D. SYRDAL: Terpenes and Sesquiterpenes of *Chamaecyparis nootkatensis* Leaf Oil. Phytochem. **9,** 1325 (1970).

140. CORBELLA, A., P. GARIBOLDI, and G. JOMMI: Aspects of the Biosynthesis of the Terpenoid Dendrobine. Chem. Comm. **1973,** 729.

141. NISHIMURA, K., N. SHINODA, and Y. HIROSE: A New Sesquiterpene Bicyclogermacrene. Tetrahedron Letters **1969,** 3097.

142. YOSHIHARA, K., Y. OHTA, T. SAKAI, and Y. HIROSE: Germacrene D, A Key Intermediate of Cadinene Group Compounds and Bourbonenes. Tetrahedron Letters **1969,** 2263.

143. MARSHALL, J. A., W. F. HUFFMAN, and J. A. RUTH: Stereoselective Synthesis of Hydroazulenes from Cyclodecadienols. J. Amer. Chem. Soc. **94,** 4691 (1972).

144. WHARTON, P. S., and M. D. BAIRD: Conformation and Reactivity in the *cis, trans*--2,6-Cyclodecadienyl System. J. Organ. Chem. (USA) **36,** 2932 (1971).

145. IGUCHI, M., M. NIWA, and S. YAMAMURA: Biogenetic Model Reactions of Germacrone-type Sesquiterpenes. Chem. Commun. **1971,** 974.

146. — — — Biogenetic-type Reactions of Acoragermacrone. Tetrahedron Letters **1973,** 4367.

147. GREENWOOD, J. M., M. D. SOLOMON, J. K. SUTHERLAND, and A. TORRE: A Cyclisation of Humulene. J. Chem. Soc. (C) **1968,** 3004.

148. MCKERVEY, M. A., and J. R. WRIGHT: The Acid-catalysed Cyclisation of 1,2-Humulene Epoxide, a Possible Intermediate in the Biosynthesis of Caryophyllene. Chem. Commun. **1970,** 117.

149. NAYA, Y., and M. KOTAKE: Natural Occurrence of Humulol and Tricyclohumuladiol. Bull. Chem. Soc. Japan **42,** 2405 (1969).

150. BAINES, D., J. FORRESTER, and W. PARKER: Acid-catalysed Cyclisation of Humulene. J. Chem. Soc. Perkin I **1974,** 1598.

151. NICKON, A., T. IWADARE, F. J. MCGUIRE, J. R. MAHAJAN, S. A. NARANG, and B. UMEZAWA: The Structure, Stereochemistry, and Genesis of α-Caryophyllene Alcohol (Apollan-11-ol). J. Amer. Chem. Soc. **92,** 1688 (1970).

152. NAYA, Y., and Y. HIROSE: Acid Catalyzed Isomerization of Humulene (Part 1). Chemistry Letters **1973,** 133.

153. — — Acid Catalyzed Isomerization of Humulene (Part 2). Stereochemistry and the Reaction Pathways. Chemistry Letters **1973,** 727.

154. DAUBEN, W. G., J. P. HUBBELL, and N. D. VIETMEYER: Acid-Catalyzed Rearrangements of Humulene. J. Organ. Chem. (USA) **40,** 479 (1975).

155. MCKILLOP, T. F. W., J. MARTIN, W. PARKER, J. S. ROBERTS, and J. R. STEVENSON: The Synthesis of Neoclovene. J. Chem. Soc. (C) **1971,** 3375.

156. BAINES, D., C. ECK, and W. PARKER: Epi-Clovene – An Acid Catalysed Rearrangement Product of Caryolan-1-ol. Tetrahedron Letters **1973,** 3933.

157. BISARYA, S. C., and S. DEV: Studies in Sesquiterpenes – XXXIV. Structure of Allohimachalol. Tetrahedron **24,** 3869 (1968).

157a. FRY, J. L., and G. J. KARABATSOS: Intramolecular Hydride Shifts in Carbonium Ions. In: G. A. OLAH and P. VON R. SCHLEYER (eds.) Carbonium Ions Vol. II, p. 521. New York: Wiley-Interscience. 1970.

158. FĂRCASIU, D., C. KASCHERES, and L. H. SCHWARTZ: Nitrous Acid Deamination of 2-(Aminomethyl)cyclohexanol. The Question of a 1,3-Hydride Shift or Two Consecutive 1,2-Hydride Shifts. J. Amer. Chem. Soc. **94,** 180 (1972).

159. OLAH, G. A., and J. LUKAS: Stable Carbonium Ions. XLVII. Alkylcarbonium Ion Formation from Alkanes *via* Hydride (Alkide) Ion Abstraction in Fluorosulfonic Acid-Antimony Pentafluoride-Sulfuryl Chlorofluoride Solution. J. Amer. Chem. Soc. **89,** 4739 (1967).

160. OLAH, G. A., J. R. DEMEMBER, A. COMMEYRAS, and J. L. BRIBES: Stable Carbonium Ions. LXXXV. Laser Raman and Infrared Spectroscopic Study of Alkylcarbonium Ions. J. Amer. Chem. Soc. **93,** 459 (1971).

161. OLAH, G. A., and J. A. OLAH: Alkylcarbonium Ions: In: G. A. OLAH and P. VON R. SCHLEYER (eds.), Carbonium Ions Vol. II, p. 715. New York: Wiley-Interscience. 1970.

162. SAUNDERS, M., and P. VOGEL: Equilibrium Deuterium Isotope Effects in Systems Undergoing Rapid Rearrangements. Methyl Interchange in Dimethylisopropylcarbonium Ion. J. Amer. Chem. Soc. **93,** 2561 (1971).

163. SAUNDERS, M., P. VOGEL, E. L. HAGEN, and J. ROSENFELD: Evidence for Protonated Cyclopropane Intermediates from Studies of Stable Solutions of Carbonium Ions. Accounts Chem. Res. **6,** 53 (1973).

164. SAUNDERS, M., E. L. HAGEN, and J. ROSENFELD: Rearrangement Reactions of Secondary Carbonium Ions. Protonated Cyclopropane Intermediates Formed from *sec*-Butyl Cation. J. Amer. Chem. Soc. **90,** 6882 (1968).

165. SAUNDERS, M., P. VON R. SCHLEYER, and G. A. OLAH: Stable Carbonium Ions. XI. The Rate of Hydride Shifts in the 2-Norbornyl Cation. J. Amer. Chem. Soc. **86,** 5680 (1964).

166. OLAH, G. A.: Carbocations and Electrophilic Reactions. Angew. Chem. Intern. Ed. **12,** 173 (1973).

167. BROUWER, D. M., and H. HOGEVEEN: The Importance of Orbital Orientation as a Rate-Controlling Factor in Intramolecular Reactions of Carbonium Ions. Rec. trav. chim. Pays-Bas **89,** 211 (1970).

168. VOGEL, P., M. SAUNDERS, W. THIELECKE, and P. VON R. SCHLEYER: Exceptionally High Barriers to 1,2-Hydride Shifts in the 1-Adamantyl Cation. Tetrahedron Letters **1971,** 1429.

169. COHEN, T., G. HERMAN, T. M. CHAPMAN, and D. KUHN: A Laboratory Model for the Biosynthesis of Cyclopropane Rings. Copper-Catalyzed Cyclopropanation of Olefins by Sulfur Ylides. J. Amer. Chem. Soc. **96,** 5627 (1974).

170. EPSTEIN, W. W., and H. C. RILLING: Studies on the Mechanism of Squalene Biosynthesis. The Structure of Presqualene Pyrophosphate. J. Biol. Chem. **245,** 4597 (1970).

171. POPJÁK, G., J. EDMOND, and S.-M. WONG: Absolute Configuration of Presqualene Alcohol. J. Amer. Chem. Soc. **95,** 2713 (1973).

172. ALTMAN, L. J., L. ASH, R. C. KOWERSKI, W. W. EPSTEIN, B. R. LARSEN, H. C. RILLING, F. MUSCIO, and D. E. GREGONIS: Prephytoene Pyrophosphate. A New Intermediate in the Biosynthesis of Carotenoids. J. Amer. Chem. Soc. **94,** 3257 (1972).

173. GOVINDACHARI, T. R., P. A. MOHAMED, and P. C. PARTHASARATHY: Ishwarane and Aristolochene, Two New Sesquiterpene Hydrocarbons from *Aristolochia Indica.* Tetrahedron **26,** 615 (1970).

174. IRIE, T., M. SUZUKI, E. KUROSAWA, and T. MASAMUNE: Laurinterol, Debromolaurinterol and Isolaurinterol, Constituents of *Laurencia Intermedia* Yamada. Tetrahedron **26,** 3271 (1970).

175. FENICAL, W., and J. J. SIMS: Cycloeudesmol, An Antibiotic Cyclopropane Containing Sesquiterpene from the Marine Alga, *Chondria Oppositiclada* Dawson. Tetrahedron Letters **1974,** 1137.

176. TURNBULL, K. W., W. ACKLIN, D. ARIGONI, A. CORBELLA, P. GARIBOLDI, and G. JOMMI: Biological Conversion of Copaborneol into Tutin. Chem. Commun. **1972,** 598.

177. CORBELLA, A., P. GARIBOLDI, and G. JOMMI: Biosynthesis of Tutin from (4R)-[4-$^3H_1$]Mevalonic Acid. Chem. Commun. **1972,** 600.

178. HANSON, J. R., and R. NYFELER: The Mevalonoid Origin of the Hydrogen Atoms of Culmorin. Chem. Commun. **1975,** 171.

179. KEATING, J. T., and P. S. SKELL: Free Carbonium Ions. In: G. A. OLAH and P. VON R. SCHLEYER (eds.), Carbonium Ions Vol. II, p. 573. New York: Wiley-Interscience. 1970.

180. FRIEDMAN, L.: Carbonium Ion Formation from Diazonium Ions. In: G. A. OLAH and P. VON R. SCHLEYER (eds.), Carbonium Ions Vol. II, p. 655. New York: Wiley-Interscience. 1970.

181. KARABATSOS, G. J., C. E. ORZECH, J. L. FRY, and S. MEYERSON: Carbonium Ions. XI. Deamination of 1-Aminopropane and the Question of Protonated Cyclopropane *vs.* the 1-Propyl Cation. J. Amer. Chem. Soc. **92,** 606 (1970).

182. FRIEDMAN, L., and J. H. BAYLESS: Aprotic Diazotization of Aliphatic Amines. Hydrocarbon Products and Reaction Parameters. J. Amer. Chem. Soc. **91,** 1790 (1969).

183. TANIDA, H., and H. MATSUMURA: Solvolyses of Aryldineopentylcarbinyl, Aryl-*tert*-butylneopentylcarbinyl, and Aryldi-*tert*-butylcarbinyl *p*-Nitrobenzoates. Effects of the Bulky Alkyl Groups at the Reaction Center and Substituents in the Aryl Ring. J. Amer. Chem. Soc. **95,** 1586 (1973).

184. ABRUSCATO, G. J., and T. T. TIDWELL: Steric Crowding in Organic Chemistry. VI. Reactivity of Tri-*tert*-butylethylene and Related Compounds. J. Organ. Chem. (USA) **37,** 4151 (1972).

185. GREAM, G. E., D. WEGE, and M. MULAR: The Camphenehydro (Methylcamphenilyl) and Isobornyl (Bornyl) Cations. I. Generation of the Enantiomeric Cations by the π- and σ-Routes of Solvolysis. Austral. J. Chem. **27,** 567 (1974).

186. BUNTON, C. A., and C. O'CONNOR: Structure of the Trimethylnorbornyl Carbonium Ion. Chem. and Ind. **1965,** 1182.

187. SHAPIRO, R. H., J. H. DUNCAN, and J. C. CLOPTON: The Effect of Base Concentration and Solvent Polarity on the Base Catalyzed Decomposition of Camphor Tosylhydrozone under Aprotic Conditions. J. Amer. Chem. Soc. **89**, 1442 (1967).

188. EDWARDS, O. E., and M. LESAGE: The Deamination of Alicyclic α-Aminoketones. Canad. J. Chem. **41**, 1592 (1963).

189. WESTMAN, T. L.: Studies in the 9-Substituted Decalins. Ph. D. Thesis, University of California, Berkeley, California, 1961.

190. DAUBEN, W. G., and P. LANG: The Preparation of 14β,18-Cyclosteroids from N-Desmethyl-N(20)-Conene Derivatives and 18-Aminosteroids. Tetrahedron **20**, 1259 (1964).

191. NICKON, A., and N. H. WERSTIUK: 1,3-Eliminations. I. Stereochemical Considerations and Terminology. J. Amer. Chem. Soc. **89**, 3914 (1967).

192. COLLINS, C. J.: Protonated Cyclopropanes. Chem. Rev. **69**, 543 (1969).

193. LEE, C. C.: Protonated Cyclopropanes. In: S. G. COHEN, A. STREITWIESER, and R. W. TAFT, Progress in Physical Organic Chemistry, Vol. 7, p. 129. New York: Interscience. 1970.

194. FRIEDMAN, L., and A. T. JUREWICZ: Factors Affecting the Formation and Stability of the Equilibrating Protonated Cyclopropane. J. Amer. Chem. Soc. **91**, 1800 (1969).

195. — — Hydrocarbon Forming Pathways from Amine Diazotizations. J. Amer. Chem. Soc. **91**, 1803 (1969).

196. DEPUY, C. H., A. H. ANDRIST, and P. C. FÜNFSCHILLING: Stereochemistry of the Electrophilic Ring Opening of Cyclopropanes by D^+. Evidence for an Unsymmetrical, Nonrotating, Corner-Protonated Cyclopropane. J. Amer. Chem. Soc. **96**, 948 (1974).

197. DEPUY, C. H.: Stereochemistry and Reactivity in Cyclopropane Ring-Cleavage by Electrophiles. Fortschr. chem. Forsch. **40**, 73 (1973).

198. COLTER, A., E. C. FRIEDRICH, N. J. HOLNESS, and S. WINSTEIN: The Apoisobornyl-*exo*-Camphenilyl Nonclassical Cation. J. Amer. Chem. Soc. **87**, 378 (1965).

199. LEE, C. C., and B.-S. HAHN: Rearrangement Studies with Carbon-14. XXXIV. The π Route to the Norbornyl Cation from Solvolyses of 2-(Δ^3-Cyclopentenyl)-2-^{14}C-*p*-Nitrobenzenesulfonate. J. Amer. Chem. Soc. **91**, 6420 (1969).

200. BLY, R. S., R. K. BLY, A. O. BEDENBAUGH, and O. R. VAIL: π-Electron Participation in the Acetolysis of β-(*syn*-7-Norbornenyl) ethyl *p*-Bromobenzenesulfonate. J. Amer. Chem. Soc. **89**, 880 (1967).

201. SPURLOCK, L. A., and K. P. CLARK: Nature of the Carbonium Ion. X. The 2-Protoadamantyl Cation. J. Amer. Chem. Soc. **94**, 5349 (1972).

202. WENDLER, N. L., R. P. GRABER, C. S. SNODDY, JR., and F. W. BOLLINGER: Transannular Hydrogen Transfer in the Steroid Series. J. Amer. Chem. Soc. **79**, 4476 (1957).

203. OLAH, G. A., G. LIANG, G. D. MATEESCU, and J. L. RIEMENSCHNEIDER: Stable Carbocations. CL. Fourier Transform ^{13}C Nuclear Magnetic Resonance and X-Ray Photoelectron Spectroscopic Study of the 2-Norbornyl Cation. J. Amer. Chem. Soc. **95**, 8698 (1973).

204. BROUWER, D. M., and J. A. VAN DOORN: Spectroscopic Observation of a 1,3-Hydrogen Shift in 2,4-Dimethylpentyl Cation. Rec. trav. chim. Pays-Bas **88**, 573 (1969).

205. SAUNDERS, M., and J. J. STOFKO, JR.: Intramolecular Hydride Shifts in Carbonium Ions. J. Amer. Chem. Soc. **95**, 252 (1973).

206. STOFKO, J. J., Jr.: NMR Studies I. Hydride Shifts in Alkyl Carbonium Ions. II. Nitrogen Inversion in Tertiary Amines. Ph. D. Thesis, Yale University, New Haven, Conn., 1972.

207. ACHILLADELIS, B. A., P. M. ADAMS, and J. R. HANSON: Studies in Terpenoid Biosynthesis. Part VIII. The Formation of the Trichothecane Nucleus. J. Chem. Soc. Perkin I **1972**, 1425.

208. MÜLLER, B., and CH. TAMM: Biosynthesis of the Verrucarins and Roridins. Part 4. The Mode of Incorporation of (3R)-[(5R)-5-^{3}H]-Mevalonate into Verrucarol. Evidence for the Identity of the C(11)-Hydrogen Atom of the Trichothecane Skeleton with the (5R)-Hydrogen Atom of (3R)-Mevalonic Acid. Helv. Chim. Acta **58**, 483 (1975).
209. TAMM, CH.: The Antibiotic Complex of the Verrucarins and Riordins. In: W. HERZ, H. GRISEBACH, and G. W. KIRBY (eds.), Progress in the Chemistry of Organic Natural Products, Vol. 31, p. 63, Vienna: Springer-Verlag. 1974.
210. ADAMS, P. M., and J. R. HANSON: Studies in Terpenoid Biosynthesis. Part VII. The Biosynthesis of Helicobasidin. J. Chem. Soc. Perkin I **1972**, 586.
211. CORDELL, G. A.: The Occurrence, Structure Elucidation, and Biosynthesis of the Sesterpenes. Phytochem. **13**, 2343 (1974).
212. WILDER, P., Jr., D. J. CASH, R. C. WHELAND, and G. W. WRIGHT: Rearrangements of Dihydrodicyclopentadiene Derivatives in Phosphoric Acid. J. Amer. Chem. Soc. **93**, 791 (1971).
213. COREY, E. J., and R. D. BALANSON: A Simple Synthesis of (±)-Cedrene and (±)-Cedrol Using a Synchronous Double Annulation Process. Tetrahedron Letters **1973**, 3153.
214. COPE, A. C., M. M. MARTIN, and M. A. MCKERVEY: Transannular Reactions in Medium-sized Rings. Quart. Rev. (Chem. Soc. London) **20**, 119 (1966).
215. PRELOG, V., and J. G. TRAYNHAM: Transannular Hydride Shifts. In: P. DE MAYO (ed.), Molecular Rearrangements, Part One, p. 593. New York: Interscience. 1963.
216. BRANCA, Q.: Säurekatalysierte 1,5-Hydridverschiebungen in Substraten mit flexibler Konformation: Modellverbindungen und Tetracyclische Triterpene. Ph. D. Dissertation (Nr. 4575) Eidgenössische Technische Hochschule Zürich, 1970.
217. STÉHELIN, L., J. LHOMME, and G. OURISSON: Transannular Hydride Shifts. A Mechanistic Study in the Longifolene Series. J. Amer. Chem. Soc. **93**, 1650 (1971).
218. STÉHELIN, L., L. KANELLIAS, and G. OURISSON: Solvolysis of 7-Substituted Bicyclo-[3.3.1]non-3-yl Tosylates. A Kinetic Proof of σ(C–H) Participation. J. Organ. Chem. (USA) **38**, 847 (1973).
219. — — — Solvolysis of 9,9-Dimethylbicyclo[3.3.1]non-3-yl Tosylate. Enhancement of σ(C–H)Participation by Steric Blocking. J. Organ. Chem. (USA) **38**, 851 (1973).
220. HILL, R. K., and R. M. CARLSON: Acid-catalyzed 1,5-Hydride Transfer in Acyclic Molecules. Mechanism and Stereochemistry. J. Amer. Chem. Soc. **87**, 2772 (1965).
221. BRANCA, Q., und D. ARIGONI: Säurekatalysierte 1,5-Hydridverschiebungen in Systemen mit flexibler Konformation. Chimia **23**, 189 (1969).
222. ATKINSON, R. S., and R. H. GREEN: 1,5-Hydride Shifts in Acyclic Systems Containing α,β-Unsaturated Ketones and *p*-Methoxyphenyl Groups. J. Chem. Soc. Perkin I **1974**, 394.
223. ANDERSEN, N. H., M. S. FALCONE, and D. D. SYRDAL: Structures of Vetivenenes and Vetispirenes. Tetrahedron Letters **1970**, 1759.
224. ROBINSON, R.: The Structural Relations of Natural Products, p. 12. Oxford: Clarendon Press. 1955.
225. PENFOLD, A. R., and J. L. SIMONSEN: The Constitutions of Eremophilone, Hydroxyeremophilone, and Hydroxydihydroeremophilone. Part III. J. Chem. Soc. **1939**, 87.
226. SULSER, H., J. R. SCHERER, and K. L. STEVENS: The Structure of Paradisiol, a New Sesquiterpene Alcohol from Grapefruit Oil. J. Organ. Chem. (USA) **36**, 2422 (1971).
227. BROOKS, C. J. W., and R. A. B. KEATES: Biosynthesis of Petasin in *Petasites hybridus*. Phytochem. **11**, 3235 (1972).
228. BIRNBAUM, G. I., A. STOESSL, S. H. GROVER, and J. B. STOTHERS: The Complete Stereostructure of Capsidiol. X-Ray Analysis and ^{13}C Nuclear Magnetic Resonance

of Eremophilane Derivatives Having *trans*-Vicinal Methyl Groups. Canad. J. Chem. **52,** 993 (1974).

229. BAKER, F. C., C. J. W. BROOKS, and S. A. HUTCHINSON: Biosynthesis of Capsidiol in Sweet Peppers *(Capsium frutescens)* Infected with Fungi: Evidence for Methyl Group Migration from ^{13}C Nuclear Magnetic Resonance Spectroscopy. Chem. Commun. **1975,** 293.

230. KITAGAWA, I., H. SHIBUYA, Y. YAMAZOE, H. TAKENO, and I. YOSIOKA: Conversion of Dihydroalantolactone to Tetrahydroligularenolide. A Biogenetic-type Transformation of Eudesmanolide to Eremophilanolide. Tetrahedron Letters **1974,** 111.

231. BÜCHI, G., F. GREUTER, and T. TOKOROYAMA: Terpenes – XVII. Structure of Calarene and Stereochemistry of Aristolone. Tetrahedron Letters **1962,** 827.

232. STREITH, J., P. PESNELLE, et G. OURISSON: Le β-gurjunène. Identification avec la calarene. Bull. soc. chim. France **1963,** 518.

233. OURISSON, G.: Molecular Rearrangements of Terpenes. Proc. Chem. Soc. (London) **1964,** 274.

234. KIRK, D. N., and M. P. HARTSHORN: Steroid Reaction Mechanisms. Chapter 8. Amsterdam: Elsevier Publishing Company. 1968.

235. SNATZKE, G., und A. VEITHEN: 19-Nor-5β-methyl-steroide, IV. Die HNO_2-Desaminierung von 5α-Amino-Steroidén. Liebigs Ann. Chem. **703,** 159 (1967).

236. HEATHCOCK, C. H., and T. R. KELLY: Sesquiterpenoids – IV. Acid-Catalyzed Methyl Migration in the 9-Methyl Decalins. Tetrahedron **24,** 3753 (1968).

237. HEATHCOCK, C. H., and Y. AMANO: Sesquiterpenoids – V. The Synthesis and Formolysis of 4aβ,8β-Dimethyl-8aα-Hydroxydecahydronaphth-2β-oic Acid and 4aβ,8β-Dimethyl-8aα-Hydroxydecahydronaphth-2α-oic Acid Lactone. Tetrahedron **24,** 4917 (1968).

238. DUNHAM, D. J., and R. G. LAWTON: Spiro Intermediates in Sesquiterpene Rearrangements and Synthesis. J. Amer. Chem. Soc. **93,** 2075 (1970).

239. HOCHSTETLER, A. R.: Acid-Catalyzed Angular Methyl Migration in a Substituted Octalin. J. Organ. Chem. (USA) **39,** 1400 (1974).

240. HUFFMAN, J. W.: Attempted Duplication of the Methyl Shift in Eremophilane Biosynthesis. J. Organ. Chem. (USA) **37,** 2736 (1972).

241. HIKINO, H., K. AOTA, D. KUWANO, and T. TAKEMOTO: Sesquiterpenoids. XLI. Structure and Absolute Configuration of α-Rotunol and β-Rotinol, Sesquiterpenoids of *Cyperus rotundus*. Tetrahedron **27,** 4831 (1971).

242. CAINE, D., and J. B. DAWSON: The Synthesis and Acid-catalysed Rearrangement of a Spiro[4.5]dec-6-en-2-one. Chem. Commun. **1970,** 1232.

243. HIKINO, H., N. SUZUKI, and T. TAKEMOTO: Synthesis of Cyperolone and 3-*epi*-Cyperolone. Chem. Pharm. Bull. (Japan) **15,** 1395 (1967).

244. HIKINO, H., T. KOHAMA, and T. TAKEMOTO: Rearrangement of 4,5-Epoxy-Eudesmanes with Boron Trifluoride. Tetrahedron **25,** 1037 (1969).

245. MEHTA, G., G. L. CHETTY, U. R. NAYAK, and S. DEV: Studies in Sesquiterpenes – XXVIII. BF_3-Induced Rearrangement of Eudesmane- and 7-Epieudesmane-based 1,2-Epoxides. Tetrahedron **24,** 3775 (1968).

246. HARTSHORN, M. P., and D. N. KIRK: Reactions of Epoxides – IV. The Rearrangements of 4α,5α- and 4β,5β-Epoxy-4-Methylcholestanes with Boron Trifluoride, and a Novel Ketone Rearrangement. Tetrahedron **20,** 2943 (1964).

247. HARTSHORN, M. P., and D. N. KIRK: Reactions of Epoxides – VI. A Conformational Analysis of Rearrangements of Tetra-substituted Epoxides. Tetrahedron **21,** 1547 (1965).

248. DAUBEN, W. G., and P. OBERHÄNSLI: Constituents of Hiba Wood Oil. The Isolation and Synthesis of Two Isomeric Cuprenenes. J. Organ. Chem. (USA) **31,** 315 (1966).

249. ITÔ, S., K. ENDO, and H. NARITA: α-Pseudowiddrene, A New Sesquiterpene Hydrocarbon from *Thujopsis dolabrata* Sieb. et Zucc. Tetrahedron Letters **1974,** 1041.

250. ITÔ, S., K. ENDO, T. YOSHIDA, M. YOTAGAI, and M. KODAMA: Chamigrene, a Sesquiterpene Hydrocarbon of a Novel Carbon Skeleton. Chem. Commun. **1967,** 186.
251. NAGAHAMA, S.: Hydration of Thujopsene to Widdrol. Bull. Chem. Soc. Japan **33,** 1467 (1960).
252. DAUBEN, W. G., and L. E. FRIEDRICH: The Mechanism of the Transformation of Thujopsene to Widdrol. Tetrahedron Letters **1964,** 2675.
253. — — Cyclopropylcarbinyl Rearrangements in the Thujopsene Series. Tetrahedron Letters **1967,** 1735.
254. DAUBEN, W. G., and E. I. AOYAGI: Stereochemistry of Cyclopropylcarbinyl Rearrangements Thujopsene-Widdrol Interconversion. Tetrahedron **26,** 1249 (1970).
255. DAUBEN, W. G., L. E. FRIEDRICH, P. OBERHÄNSLI, and E. I. AOYAGI: Thujopsene Rearrangements. The Cyclopropylcarbinyl System. J. Organ. Chem. (USA) **37,** 9 (1972).
256. ITÔ, S., M. YATAGAI, and K. ENDO: Acid-catalyzed Rearrangement of Thujopsene. I. Structure Determination of New Compounds. Tetrahedron Letters **1971,** 1149.
257. ITÔ, S., M. YATAGAI, K. ENDO, and M. KODAMA: Acid-catalyzed Rearrangement of Thujopsene II. The Reaction Pathway. Tetrahedron Letters **1971,** 1153.
258. DAUBEN, W. G., and L. E. FRIEDRICH: Thujopsene Rearrangements. The Ring System *via* Methyl Group Rearrangements. J. Organ. Chem. (USA) **37,** 241 (1972).
259. DAUBEN, W. G., and E. I. AOYAGI: Thujopsene Rearrangements. The Ring System *via* Ring Contraction. J. Organ. Chem. (USA) **37,** 251 (1972).
260. DAENIKER, H. U., A. R. HOCHSTETLER, K. KAISER, and G. C. KITCHENS: The Acid-Catalyzed Isomerization of Thujopsene. J. Organ. Chem. (USA) **37,** 1 (1972).
261. KATO, T., S. KANNO, and Y. KITAHARA: Cyclization of Polyenes – V. Synthesis of α-Chamigrene by the Cyclization of *cis-* and *trans*-Monocyclofarnesols. Tetrahedron **26,** 4287 (1970).
262. WOLFF, G., et G. OURISSON: Le Seychellene Isolement et Structure. Tetrahedron **25,** 4903 (1969).
263. BATES, R. B., and R. C. SLAGEL: Conversion of Bulnesol to Patchouli Alcohol, Guaiol, and "δ-Guaiene". J. Amer. Chem. Soc. **84,** 1307 (1962).
264. BÜCHI, G., R. E. ERICKSON, and N. WAKABAYASHI: Terpenes. XVI. Constitution of Patchouli Alcohol and Absolute Configuration of Cedrene. J. Amer. Chem. Soc. **83,** 927 (1961).
265. BÜCHI, G., W. D. MACLEOD, and J. PADILLA O.: Terpenes. XIX. Synthesis of Patchouli Alcohol. J. Amer. Chem. Soc. **86,** 4438 (1964).
266. BANG, L., I. G. GUEST, et G. OURISSON: Transpositions a Etapes Multiples de L'Epoxide du β-Patchoulene. Tetrahedron Letters **1972,** 2089.
267. BANG, L., M. A. DIAZ-PARRA, et G. OURISSON: Transpositions a Etapes Multiples d'Epoxydes Sesquiterpeniquies – I. Action de L'acide Formique sur L'Epoxyde de Cyperene. Correlation entre le Cyperene et L'α-Cedrene. Tetrahedron **29,** 2087 (1973).
268. BANG, L., et G. OURISSON: Transpositions a Etapes Multiples D'Epoxydes Sesquiterpenique – II. Action D'Acides de Lewis sur L'Epoxyde de Cyperene. Tetrahedron **29,** 2097 (1973).
269. TERHUNE, S. J., J. W. HOGG, and B. M. LAWRENCE: Cycloseychellene, A New Tetracyclic Sesquiterpene from *Pogostemon Cablin.* Tetrahedron Letters **1973,** 4705.
270. STORK, G., and P. A. GRIECO: Olefin Participation in the Acid-Catalyzed Opening of Acylcyclopropanes. IV. Cyclization of 5-Methyl-6-*endo*-(*trans*-3-pentenyl)bicyclo-[3.1.0]hexan-2-one. Tetrahedron Letters **1971,** 1807.
271. MAYO, P. DE, R. E. WILLIAMS, G. BÜCHI, and S. H. FEAIRHELLER: The Absolute Stereostructure of Copaene. Tetrahedron **21,** 619 (1965).

272. KAPADIA, V. H., B. A. NAGASAMPAGI, V. G. NAIK, and S. DEV: Studies in Sesquiterpenes - XXII. Structure of Mustakone and Copaene. Tetrahedron **21,** 607 (1965).

273. WESTFELT, L.: β-Copaene and β-Ylangene, Minor Sesquiterpenes of the Wood of *Pinus silvestris* L. and of Swedish Sulphate Turpentine. Acta Chem. Scand. **21,** 152 (1967).

274. OHLOFF, G., und M. PAWLAK: Säurekatalysierte Isomerisierung von (−)-*a*-Copaenepoxide. Helv. Chim. Acta **53,** 245 (1970).

275. OHTA, Y., and Y. HIROSE: The Acid-catalyzed Isomerization of α-Ylangene and α-Ylangene Epoxide. Bull. Chem. Soc. Japan **46,** 1535 (1973).

276. WESTFELT, L.: α,γ, and ε-Muurolene, Major Sesquiterpenes of the Wood of *Pinus silvestris* L. and of Swedish Sulphate Turpentine. Acta Chem. Scand. **20,** 2852 (1966).

277. KOLBE-HAUGWITZ, M., and L. WESTFELT: Copaborneol, Constitution and Synthesis. Acta Chem. Scand. **24,** 1623 (1970).

278. MCMURRY, J. E.: The Total Synthesis of Copacamphene and Its Acid-catalyzed Interconversion with Sativene. J. Organ. Chem. (USA) **36,** 2826 (1971).

279. SMEDMAN, L. A., E. ZAVARIN, and R. TERANISHI: Composition of Oxygenated Monoterpenoids and Sesquiterpenoid Hydrocarbons from the Cortical Oleoresin of *Abies Magnifica* A. Murr. Phytochem. **8,** 1457 (1969).

280. WESTFELT, L.: The Structure of α-Longipinene, a Minor Sesquiterpene of the Wood of *Pinus silvestris* L. and of Swedish Sulphate Turpentine. Acta Chem. Scand. **21,** 159 (1967).

281. NAFFA, P., et G. OURISSON: Le Longifolène. (III.) - Addition des hydracides halogénés sur le longifolène. Les halogénures de longibornyle. Bull. Soc. Chim. France **1954,** 1410.

282. NAYAK, U. R., and S. DEV: Studies in Sesquiterpenes - XXXV. Longicyclene, the First Tetracyclic Sesquiterpene. Tetrahedron **24,** 4099 (1968).

283. OURISSON, P., et. G. OURISSON: Le Longifolène (IV). Mécanisme de l'addition des hydracides halogénés sur le longifolène. Mécanisme de la solvolyse des halogénures de longibornyle. Bull. Soc. Chim. France **1954,** 1415.

284. RANGANATHAN, R., U. R. NAYAK, T. S. SANTHANAKRISHNAN, and S. DEV: Studies in Sesquiterpenes - XL Isolongifolene (Part 1): Structure. Tetrahedron **26,** 621 (1970).

285. TOMITA, B., and Y. HIROSE: Terpenoids. XXVI. Acoradiene and Acorenol, Key Intermediates of Cedrane Group Sesquiterpenoids, and Their Transformation into (−)-α-Cedrene. Tetrahedron Letters **1970,** 143.

286. COREY, E. J., N. N. GIROTRA, and C. T. MATHEW: Total Synthesis of *dl*-Cedrene and *dl*-Cedrol. J. Amer. Chem. Soc. **91,** 1557 (1969).

287. CRANDALL, T. G., and R. G. LAWTON: A Biogenetic-Type Synthesis of Cedrene. J. Amer. Chem. Soc. **91,** 2127 (1969).

288. LANSBURY, P. T., V. R. HADDON, and R. C. STEWART: Latent Homoallylic Ions in Carbocyclic Ring Construction. α-Cedrene. J. Amer. Chem. Soc. **96,** 896 (1974).

289. ANDERSEN, N. H., and D. D. SYRDAL: The Alaskenes - Precursors of Tricyclic Sesquiterpenes. Tetrahedron Letters **1970,** 2277.

290. — — The Absolute Stereochemistry of the Alaskenes and Acorone-related Sesquiterpenes. Tetrahedron Letters **1972,** 899.

291. KIDO, F., H. UDA, and A. YOSHIKOSHI: The Structure of Zizanoic Acid, A Novel Sesquiterpene in Vetiver Oil. Tetrahedron Letters **1967,** 2815.

292. — — — The Stereochemistry of Zizanoic Acid. Tetrahedron Letters **1968,** 1247.

293. ANDERSEN, N. H., and M. S. FALCONE: Prezizaene and the Biogenesis of Zizaene. Chem. and Ind. **1971,** 62.

294. MACSWEENEY, D. F., R. RAMAGE, and A. SATTAR: Biogenetic Relationships of the Vetiver Sesquiterpenes. Tetrahedron Letters **1970,** 557.

295. Kido, F., H. Uda, and A. Yoshikoshi: Synthetic Study of Zizaane-type Sesquiterpenoids. J. Chem. Soc. Perkin I **1972,** 1755.
296. MacSweeney, D. F., and R. Ramage: A Stereospecific Total Synthesis of Zizanoic and Isozizanoic Acids. Tetrahedron **27,** 1481 (1971).
297. Tomita, B., T. Isono, and Y. Hirose: Terpenoids. XXVIII. Acorane Type Sesquiterpenoids from *Juniperus Rigida* and Hypothesis for the Formation of New Tricarbocyclic Sesquiterpenoids. Tetrahedron Letters **1970,** 1371.
298. Tomita, B., and Y. Hirose: *Allo*-cedrol: A New Tricarbocyclic Sesquiterpene Alcohol. Phytochem. **12,** 1409 (1973).
298a. Andersen, N. H.: From Nerdidol to Cedrene and Vetiver Tricyclics: Chemical Realization of the Step-wise Processes. Paper No. 159, VI International Congress of Essential Oils, San Francisco, California, September 11, 1974.
299. Andersen, N. H., S. E. Smith, and Y. Ohta: Rearrangement of Zizaene-related Sesquiterpenes. Chem. Commun. **1973,** 447.
300. Hanson, J. R.: The Biosynthesis of the Diterpenes. In: W. Herz. H. Griesebach, and G. W. Kirby (eds.), Progress in the Chemistry of Organic Natural Products, Vol. 29, p. 395. Vienna: Springer. 1971.
301. Hadley, M. S., and T. G. Halsall: Rearrangement of Epoxides. Part I. Preparation and Rearrangement of the α- und β-Epoxides of 14,15-Bisnorlabd-8-en-13(R and S)-yl Acetates and of Related Epoxides. J. Chem. Soc. Perkin I **1974,** 1334.
302. Sharpless, K. B., and E. E. van Tamelen: Terpene Terminal Epoxides. Skeletal Rearrangement Accompanying Bicyclization of Squalene 2,3-Oxide. J. Amer. Chem. Soc. **91,** 1848 (1969).
303. Kitadani, M., C. Kabuto, K. Sakai, A. Yoshikoshi, and Y. Kitahara: Backbone Rearrangement of Dolabradiene. Chemistry Letters **1974,** 963.
304. Connolly, J. D., R. McCrindle, R. D. H. Murray, A. J. Renfrew, K. H. Overton, and A. Malera: Constituents of *Erythroxylon monogynum* Roxb. Part II. Erythroxydiols X, Y, and Z; Two Novel Skeletal Types of Diterpenoids. J. Chem. Soc. (C) **1966,** 268.
305. McCrindle, R., and E. Nakamura: Constituents of *Solidago* Species. Part VI. The Constitution of Diterpenoids from a Chemically Distinct Variety of *Solidago serotina*. Canad. J. Chem. **52,** 2029 (1974).
306. Marshall, J. A., and A. R. Hochstetler: Photosensitized Isomerizations of 10-Methyl-1(9)-octalins. J. Amer. Chem. Soc. **91,** 648 (1969).
307. Saucy, G., R. E. Ireland, J. Bordner, and R. E. Dickerson: Acid-Catalyzed Cyclization of 4-(2,6,6-Trimethylcyclohexenyl)-2-methylbutanal. X-Ray Structure Analysis of the Major Product. J. Organ. Chem. (USA) **36,** 1195 (1971).
308. Fall, R. R., and C. A. West: Purification and Properties of Kaurene Synthetase from *Fusarium moniliforme*. J. Biol. Chem. **246,** 6913 (1971).
309. Dolby, L. J., and R. H. Iwamoto: Studies of Terpene Chemistry. II. Model Studies of the Synthesis of the C-D Ring System of Gibberellic Acid. J. Organ. Chem. (USA) **30,** 2420 (1965).
310. Edwards, O. E., and R. S. Rosich: Cationic Cyclization of Bicyclic Diterpenes. Canad. J. Chem. **46,** 1113 (1968).
311. Wenkert, E., and Z. Kumazawa: Manool-derived Diterpenic Hydrocarbons and Related Products. Chem. Commun. **1968,** 140.
312. McCreadie, T., and K. H. Overton: The Conversion of Labdadienols into Pimara- and Rosa-dienes. J. Chem. Soc. (C) **1971,** 312.
313. Hall, S. F., and A. C. Oehlschlager: Cationic Rearrangements and Cyclizations of Diterpenoid Olefins. Tetrahedron **28,** 3155 (1972).
314. Connolly, J. D., R. McCrindle, R. D. H. Murray, and K. H. Overton: The Constitution and Stereochemistry of Rimuene. J. Chem. Soc. (C) **1966,** 273.

315. HERZ, W., and H. J. WAHLBORG: Resin Acids. III. 9-Hydroxyabietic Acid and Its Transformation Products. J. Organ. Chem. (USA) **30**, 1881 (1965).
316. APSIMON, J. W., and H. KREHM: Diterpene Chemistry III. Chemistry of Some Epoxides Derived from Methyl Pimarate. Canad. J. Chem. **47**, 2859 (1969).
317. HANCOCK, W. S., L. N. MANDER, and R. A. MASSY-WESTROPP: A Synthesis of Rosenonolactone from Podocarpic Acid. J. Organ. Chem. (USA) **38**, 4090 (1973).
318. MCCREADIE, T., K. H. OVERTON, and A. J. ALLISON: A Synthesis of Rosenonolactone and Deoxyrosenonolactone. J. Chem. Soc. (C) **1971**, 317.
319. ZINKEL, D. F., and B. P. SPALDING: Structure of Strobic Acid. Tetrahedron **29**, 1441 (1973).
320. HERZ, W., and A. L. HALL: Resin Acids XXVI. Biogenetic-Type Rearrangements of the Homoallylic Cation from Methyl 15(R)-Hydroxypimar-8(14)-en-18-oate. J. Organ. Chem. (USA) **39**, 14 (1974).
321. WENKERT, E., and J. W. CHAMBERLIN: The Stereochemistry of Some Resin Acid Derivatives. J. Amer. Chem. Soc. **81**, 688 (1959).
322. JOHNSTON, J. P., and K. H. OVERTON: The Pimarane→Cassane Rearrangement. Synthesis of a Potential Intermediate from Isopimaric Acid. J. Chem. Soc. Perkin I **1973**, 853.
323. ELLESTAD, G. A., M. P. KUNSTMANN, and G. O. MORTON: Conversion of a Pimarane Diterpene into the Cleistanthane Ring System. Chem. Commun. **1973**, 312.
324. WENKERT, E.: Structural and Biogenetic Relationships in the Diterpene Series. Chem. and Ind. **1955**, 282.
325. BRIGGS, L. H., and G. W. WHITE: Constituents of the Essential Oil *Araucaria araucana*. Tetrahedron **31**, 1311 (1975).
326. CHIN, W. J., R. E. CORBETT, D. R. LAUREN, and R. A. J. SMITH: The Dehydration Products of Sandaracopimar-15-en-8β-ol. Austral. J. Chem. **22**, 2033 (1969).
327. FOURREY, J.-L., J. POLONSKY, and E. WENKERT: The Mechanism of the Transformation of Manool into 14α-Hibyl Formate. Chem. Commun. **1969**, 714.
328. EDWARDS, O. E., and B. S. MOOTOO: Cationic Cyclization of Deuterated Manool. Can. J. Chem. **47**, 1189 (1969).
329. HUGEL, G., L. LODS, J. M. MELLOR, et G. OURISSON: Diterpènes de trachylobium. III. Réactions des dérivés trachylobaniques. Bull. soc. chim. France **1965**, 2894.
330. APPLETON, R., A. J. MCALEES, A. MCCORMICK, R. MCCRINDLE, and R. D. H. MURRAY: The Acid-catalysed Rearrangement of Diterpene Hydrocarbons. Part I. Kaurene, Isoatisirene, Stachene, and Trachylobane. J. Chem. Soc. (C) **1966**, 2319.
331. FAIRLIE, J. C., A. J. MCALEES, R. MCCRINDLE, and E. NEIDERT: Rinc C Functionalized Diterpenoids. Part III. 12-Substituted *ent*-Beyeranes. Canad. J. Chem. **52**, 706 (1974).
332. YOSHIKOSHI, A., M. KITADANI, and Y. KITAHARA: Interconversion Between Hibaene and Kaurene. Tetrahedron **23**, 1175 (1967).
333. MORI, K., and M. MATSUI: Diterpenoid Total Synthesis-VIII (±)-Kaur-16-en-19-oic Acid, (±)-Kaur-16-en-19-ol, (±)-Monogynol, and Some Oxygenated Kauranes. Tetrahedron **24**, 3095 (1968).
334. KAPADI, A. H., and S. DEV: Chemical Transformation of (+)-Hibaene into (−)-Kaurene. Tetrahedron Letters **1965**, 1255.
335. HANSON, J. R.: The Chemistry of the Tetracyclic Diterpenoids – V, Stereochemical Studies in the Stachene Series. Tetrahedron **23**, 793 (1967).
336. — The Chemistry of the Tetracyclic Diterpenoids – X, Some Beyerene 2 and 3-Alcohols. Tetrahedron **26**, 2711 (1970).
337. GUNN, P. A., R. MCCRINDLE, and R. G. ROY: Synthesis of a Neoatiseranone, the Enantiomer of a Ketone Obtained by the Acid-catalyzed Rearrangement of Isophyllocladene Epoxide. J. Chem. Soc. (C) **1971**, 1018.
338. BAKER, K. M., L. H. BRIGGS, J. G. ST. C. BUCHANAN, R. C. CAMBIE, B. R. DAVIS,

R. C. HAYWARD, G. A. S. LONG, and P. S. RUTLEDGE: Diterpenes. Part X. Some Transformations of Phyllocladene and Isophyllocladene. J. Chem. Soc. Perkin I **1972,** 190.

339. MACMILLAN, J., and E. R. H. WALKER: Terpenoids. Part V. Rearrangement of *ent*-Kaurane 15β,16β-Epoxide to *ent*-(16R)-Atisan-15-one. J. Chem. Soc. Perkin I **1972,** 1274.

340. MURRAY, R. D. H., R. W. MILLS, A. J. MCALEES, and R. MCCRINDLE: The Acid Catalysed Rearrangement of a Diterpenoid Epoxide. Tetrahedron **30,** 3399 (1974).

341. SOBTI, R. R., and S. DEV: Biogenetic-type Transformations in Diterpenoids. Tetrahedron Letters **1966,** 3939.

342. GHISALBERTI, E. L., and P. R. JEFFERIES: The Chemistry of the Euphorbiaceae. XVI. A Rearrangement of the Beyerane Skeleton. Austral. J. Chem. **19,** 1759 (1966).

343. APPLETON, R. A., J. C. FAIRLIE, R. MCCRINDLE, and W. PARKER: The Solvolytic Behavior of *exo*- and *endo*-Bicyclo[3.2.1]octane-6-toluene-*p*-sulphonates. J. Chem. Soc. (C) **1968,** 1716.

344. APPLETON, R. A., P. A. GUNN, and R. MCCRINDLE: Buffered Acetolyses of the Toluene-*p*-Sulphonates of *exo*- and *endo*-17-Norkauran- and 17-Norphyllocladan-16-ols. J. Chem. Soc. (C) **1970,** 1148.

345. COATES, R. M., and E. F. BERTRAM: Structural Modifications of Isosteviol. Partial Synthesis of Atiserene and Isoatiserene. J. Organ. Chem. (USA) **36,** 2625 (1971).

346. — — Biogenetic-Like Rearrangements of Tetracyclic Diterpenes. J. Organ. Chem. (USA) **36,** 3722 (1971).

347. KITADANI, M., K. ITÔ, and A. YOSHIKOSHI: Synthesis and the Iodine-Catalyzed Rearrangement of Isohibaene. Bull. Chem. Soc. Japan **44,** 3431 (1971).

348. VON CARSTENN-LICHTERFELDE, C., F. M. PANIZO, T. G. DE QUESADA, B. RODRIGUEZ, S. VALVERDE, W. A. AYER, and J.-A. H. BALL: Correlation of the Diterpenoids Sideritol and Jativatriol. Canad. J. Chem. **53,** 1172 (1975).

349. HANSON, J. R., J. HAWKER, and A. F. WHITE: Studies in Terpenoid Biosynthesis. Part IX. The Sequence of Oxidation on Ring B in Kaurene-Gibberellin Biosynthesis. J. Chem. Soc. Perkin I **1972,** 1892.

350. GRAEBE, J. E., P. HEDDEN, and J. MACMILLAN: The Ring Contraction Step in Gibberellin Biosynthesis. Chem. Commun. **1975,** 161.

351. HEDDEN, P., J. MACMILLAN, and B. O. PHINNEY: Fungal Products. Part XII. Gibberellin A_{14}-aldehyde, an Intermediate in Gibberellin Biosynthesis in *Gibberella fujikuroi.* J. Chem. Soc. Perkin I **1974,** 587.

352. GALT, R. H. B., and J. R. HANSON: New Metabolites of Gibberella fujikuroi. Part VII. The Preparation of Some Ring-B Norderivatives. J. Chem. Soc. **1965,** 1565.

353. CROSS, B. E., NORTON, K., and J. C. STEWART: The Biosynthesis of the Gibberellins. Part III. J. Chem. Soc. (C) **1968,** 1054.

354. CROSS, B. E., and I. L. GATFIELD: New Metabolites of *Gibberella fujikuroi.* Part XVII. The Partial Synthesis of Gibberellin A_{15} Norketone from 7-Hydroxykaurenolide. J. Chem. Soc. (C) **1971,** 1539.

355. HANSON, J. R., and J. HAWKER: Preparation of [^{14}C]-Gibberellic Acid. Phytochemistry **12,** 1073 (1973).

356. BEARDER, J. R., J. MACMILLAN, and B. O. PH'NNEY: 3-Hydroxylation of Gibberellin A_{12}-Aldehyde in *Gibberella Fujikuroi* Strain Rec-193 A. Phytochemistry **12,** 2173 (1973).

357. HUNECK, S.: Triterpene VIII. Die Umlagerung von 1-substituierten Triterpenen in A-Nor-B-Homo-Triterpene. Tetrahedron Letters **1963,** 1977.

358. SHOPPEE, C. W., R. E. LACK, S. C. SHARMA, and L. R. SMITH: Steroids and Walden Inversion. Part LX. Some Reactions of the Epimeric 5α-Cholestan-1-ols and the Solvolysis of their Toluene-*p*-Sulphonates. J. Chem. Soc. (C) **1967,** 1155.

359. OKUNO, T., and T. MATSUMOTO: Synthesis of a *cis*-Perhydroazulene Derivative Related to Grayanotoxin. Tetrahedron Letters **1969,** 4077.

360. VON CARSTENN-LICHTERFELDE, C., S. VALVERDE, and B. RODRÍGUEZ: Studies on Diterpenes from Genus *Sideritis*. XIII. Two New Stachene Derivatives from *Sideritis angustifolia* (Labiatae). Austral. J. Chem. **27,** 517 (1974).

361. WHITLOCK, H. W., P. B. REICHARDT, and F. M. SILVER: Synthesis and Solvolytic Rearrangement of Epimeric 4-Methanesulfonyloxy-4a-methyl-*trans*-1,2,3,4,4a,9,10,10a-octahydrophenanthrene and Their 7-Methoxy and 1,1-Dimethyl Derivatives. Conformationally Rigid Homobenzylic Systems. J. Amer. Chem. Soc. **93,** 485 (1971).

362. SMITH, L. L., T. J. FOELL, and D. M. TELLER: Retropinacol Rearrangement of 1α-Hydroxy Steroids. A New Route to 1β-Methyl 19-Norsteroids. J. Organ. Chem. (USA) **30,** 3781 (1965).

363. COOKSON, R. C., and M. E. TREVETT: Aconitum and Delphinium Alkaloids. Part II. Interrelation of the Functional Groups of Delpheline. J. Chem. Soc. **1956,** 3121.

364. VALENTA, Z., and K. WIESNER: Biogenetic Interrelationships of Diterpene Alkaloids. Chem. and Ind. **1956,** 354.

365. WIESNER, K., and Z. VALENTA: Recent Progress in the Chemistry of the Aconite-Garrya Alkaloids. In: Progress in the Chemistry of Organic Natural Products, Vol. 26, pp. 26–89. Wien: Springer. 1958.

366. PELLETIER, S. W., and A. ICHIHARA: Skeletal Rearrangements of Atisine Derivatives. Chem. and Ind. **1967,** 2149.

367. JOHNSTON, J. P., and K. H. OVERTON: A Laboratory Model for the Atisane→Aconane Conversion. J. Chem. Soc. Perkin I **1972,** 1490.

368. AYER, W. A., and P. D. DESHPANDE: Rearrangements in the Diterpenoid Series. I. The Solvolysis of Methyl 15β-tosyloxy-13-isopropyl-17-noratis-13-en-18-oate. Canad. J. Chem. **51,** 77 (1973).

369. WIESNER, K., T. Y. R. TSAI, K. HUBER, S. E. BOLTON, and R. VLAHOV: Total Synthesis of Talatisamine, a Delphinine Type Alkaloid. J. Amer. Chem. Soc. **96,** 4990 (1974).

370. PRZYBYLSKA, M., T. Y. R. TSAI, and K. WIESNER: Conversion of the Alkaloid Atisine into a Compound with the Lycoctonine Skeleton. Chem. Commun. **1975,** 297.

371. KODAMA, M., H. KURIHARA, and S. ITÔ: A Novel Rearrangement in the C-20 Diterpene Alkaloids. Formation of Bridged Bicyclo[4.3.1]dec-1-ene System. Tetrahedron Letters **1975,** 1301.

372. RILLING, H. C.: A New Intermediate in the Biosynthesis of Squalene. J. Biol. Chem. **241,** 3233 (1966).

373. EDMOND, J., G. POPJÁK, S.-M. WONG, and V. P. WILLIAMS: Presqualene Alcohol. Further Evidence on the Structure of a C_{30} Precursor of Squalene. J. Biol. Chem. **246,** 6254 (1971).

374. CAMPBELL, R. V. M., T. CROMBIE, and G. PATTENDEN: Synthesis of Presqualene Alcohol. Chem. Commun. **1971,** 218.

375. COATES, R. M., and W. H. ROBINSON: Stereoselective Total Synthesis of (±)-Presqualene Alcohol. J. Amer. Chem. Soc. **93,** 1785 (1971).

376. ALTMAN, L. J., R. C. KOWERSKI, and H. C. RILLING: Synthesis and Conversion of Presqualene Alcohol to Squalene. J. Amer. Chem. Soc. **93,** 1782 (1971).

377. VAN TAMELEN, E. E., and M. A. SCHWARTZ: Mechanism of Presqualene Pyrophosphate-Squalene Biosynthesis. J. Amer. Chem. Soc. **93,** 1780 (1971).

378. RILLING, H. C., C. D. POULTER, W. W. EPSTEIN, and B. LARSEN: Studies on the Mechanism of Squalene Biosynthesis. Presqualene Pyrophosphate, Stereochemistry and a Mechanism for Its Conversion to Squalene. J. Amer. Chem. Soc. **93,** 1783 (1971).

379. GREGONIS, D. E., and H. C. RILLING: The Stereochemistry of *trans*-Phytoene Synthesis. Some Observations on Lycopersene as a Carotene Precursor and a Mechanism of the Synthesis for *cis*- and *trans*-Phytoene. Biochemistry **13,** 1538 (1974).

380. COATES, R. M., and W. H. ROBINSON: Solvolysis of *trans*-2,2-Dimethyl-3-(2′-methylpropenyl)-cyclobutyl Tosylate. Model Reactions Relevant to Squalene Biosynthesis. J. Amer. Chem. Soc. **94,** 5920 (1972).

381. POULTER, C. D., O. J. MUSCIO, C. J. SPILLNER, and R. G. GOODFELLOW: Model Studies of Terpene Biosynthesis. Cationic Rearrangements Leading to Head-to-Head Terpenes. J. Amer. Chem. Soc. **94,** 5921 (1972).

382. POULTER, C. D., O. J. MUSCIO, and R. J. GOODFELLOW: Biosynthesis of Head-to-Head Terpenes. Carbonium Ion Rearrangements Which Lead to Head-toHead Terpenes. Biochemistry **13,** 1530 (1974).

383. MUSCIO, O. J., JR., and C. D. POULTER: Model Studies of Terpene Biosynthesis. Synthesis and Absolute Configuration of (+)-*trans*-2,2-Dimethyl-3-(2′-methylpropenyl)cyclobutanol. J. Organ. Chem. (USA) **39,** 3288 (1974).

384. OURISSON, G., P. CRABBÉ, and O. R. RODIG: Tetracyclic Triterpenes. San Francisco: Holden-Day. 1964.

385. VAN TAMELEN, E. E., J. D. WILLETT, and R. B. CLAYTON: On the Mechanism of Lanosterol Biosynthesis from Squalene 2,3-Oxide. J. Amer. Chem. Soc. **89,** 3371 (1967).

386. VODTFREDSEN, W. O., S. VANGEDAL, R. DÄTWYLER, G. VISCONTI, and D. ARIGONI: unpublished work cited in reference 33. Also Ph. D. Thesis, G. VISCONTI DI MODRONE, ETH, Zurich, 1968.

387. COREY, E. J., and H. YAMAMOTO: Correlation of a Protosterol from 20,21-Dehydro-2,3-Oxidosqualene and 2,3-Oxidosqualene-Sterol Cyclase with Dihydrolanosterol. Tetrahedron Letters **1970,** 2385.

388. VAN TAMELEN, E. E., and R. J. ANDERSON: Biogenetic-Type Total Synthesis. 24,25-Dihydrolanosterol, 24,25-Dihydro-$\Delta^{13(17)}$-protosterol, Isoeuphenol, (−)-Isotirucallol, and Parkeol. J. Amer. Chem. Soc. **94,** 8225 (1972).

389. MARKER, R. E., E. L. WITTLE, and L. W. MIXON: Sterols. XVI. Lanosterol and Agnosterol. J. Amer. Chem. Soc. **59,** 1368 (1937).

390. BARTON, D. H. R.: Triterpenoids. Part III. *cyclo*-Artenone, a Triterpenoid Ketone. J. Chem. Soc. **1951,** 1444.

391. BENTLEY, H. R., J. A. HENRY, D. S. IRVINE, and F. S. SPRING: Triterpene Resinols and Related Acids. Part XXVIII. The Non-saponifiable Fraction from *Strynos nux-vomica* Seed Fat: The Structure of *cyclo*-Artenol. J. Chem. Soc. **1953,** 3673.

392. BARTON, D. H. R., J. E. PAGE, and E. W. WARNHOFF: Triterpenoids. Part XVIII. The Constitutions of Phyllanthol and *cyclo*-Artenol. J. Chem. Soc. **1954,** 2715.

393. REES, H. H., and T. W. GOODWIN: Biosynthesis of Triterpenes, Steroids, and Carotenoids. In: Biosynthesis. A Specialist Periodical Report of the Chemical Society, Vol. 1, p. 59, 1972.

394. HEINTZ, R., and P. BENVENISTE: Plant Sterol Metabolism. Enzymatic Cleavage of the 9β,19β-Cyclopropane Ring of Cyclopropyl Sterols in Bramble Tissue Cultures. J. Biol. Chem. **249,** 4267 (1974).

395. VAN TAMELEN, E. E., and J. W. MURPHY: Formation of the Lanosterol System through Biogenetic-Type Cyclization. J. Amer. Chem. Soc. **92,** 7204 (1970).

396. KUTNEY, J. P., and N. D. WESTCOTT: The Structure of Abieslactone. Tetrahedron Letters **1971,** 3463.

397. IRIE, H., S. UYEO, and K. KURIYAMA: Acid-catalyzed Rearrangement of Abieslactone. Tetrahedron Letters **1971,** 3467.

398. GUEST, I. G., and B. A. MARPLES: Steroids. Part XIV. Transformations of 8α,9α- and 9α,11α-Epoxylanostanes and Related Compounds. J. Chem. Soc. (C) **1971,** 1468.

399. KAZLAUSKAS, R., J. T. PINHEY, and J. J. H. SIMES: Conversion of Lanosterol into a Compound with the Carbon Skeleton of Fusidic Acid. J. Chem. Soc. Perkin I **1972,** 1243.

400. EDWARDS, O. E., and T. SANO: Deamination of Two Steroidal α-Aminoketones. Canad. J. Chem. **47,** 3489 (1969).

401. Vilkas, M., G. Dupont, et R. Dulou: Contribution à l'étude des résines d'euphorbiacées. V. Sur l'α-euphol (suite). Bull. soc. chim. France **1949,** 813.

402. Christen, K., M. Dünnenberger, C. B. Roth, H. Heusser, und O. Jeger: Zur Kenntnis der Triterpene. Über weitere Beiträge zur Konstitution des Euphadienols und seine nahe Verwandtschaft mit dem Lanostadienol. Helv. Chim. Acta **35,** 1756 (1952).

403. Dawson, M. C., T. G. Halsall, and R. E. H. Swayne: The Chemistry of the Triterpenes. Part XVII. Some Aspects of the Chemistry of Tetracyclic Triterpenes. J. Chem. Soc. **1953,** 590.

404. Arigoni, D., R. Viterbo, M. Dünnenberger, O. Jeger, und L. Ružička: Zur Kenntnis der Triterpene. Konstitution und Konfiguration von Euphol und *iso*-Euphenol. Helv. Chim. Acta **37,** 2306 (1954).

405. Arigoni, D., O. Jeger, und L. Ružička: Zur Kenntnis der Triterpene. Über die Konstitution und Konfiguration von Tirucallol, Euphorbol und Elemadienolsäure. Helv. Chim. Acta **38,** 222 (1955).

406. Barbour, J. B., W. A. Lourens, F. L. Warren, and K. H. Watling: The Euphorbia Resins. Part IX. Corresponding Isomerizations in Tirucallol and Euphorbol. J. Chem. Soc. **1955,** 2194.

407. Barton, D. H. R., J. F. McGhie, M. K. Pradhan, and S. A. Knight: The Constitution and Stereochemistry of Euphol. J. Chem. Soc. **1955,** 876.

408. Lawrie, W., W. Hamilton, F. S. Spring, and H. S. Watson: Triterpenoids. Part LIII. The Constitution and Stereochemistry of Butyrospermol. J. Chem. Soc. **1956,** 3272.

409. Irvine, D. S., W. Lawrie, A. S. McNab, and F. S. Spring: Triterpenoids. Part L. The Constitution of Butyrospermol. J. Chem. Soc. **1956,** 2029.

410. Dawson, M. C., T. G. Halsall, E. R. H. Jones, G. D. Meakins, and P. C. Phillips: The Chemistry of the Triterpenes and Related Compounds. Part XXIX. The Chemistry of Butyrospermol. J. Chem. Soc. **1956,** 3172.

411. Dawson, M. C., T. G. Halsall, E. R. H. Jones, and P. A. Robins: The Chemistry of the Triterpenes. Part XVI. The Action of Hydrogen Chloride on Butyrospermol. J. Chem. Soc. **1953,** 586.

412. Nakatani, Y., G. Ponsinet, G. Wolff, J. L. Zundel, et G. Ourisson: La Transposition Spinale Euphénol-Isoeuphénol Catalysée par D^+. Tetrahedron **28,** 4249 (1972).

413. Ferguson, G., P. A. Gunn, W. C. Marsh, R. McCrindle, R. Restivo, J. D. Connolly, J. W. B. Fulke, and M. S. Henderson: Tetranortriterpenoids and Related Substances. Part XVII. A New Skeletal Class of Triterpenoids from *Guarea glabra* (Meliaceae). J. Chem. Soc. Perkin I **1975,** 491.

414. Izawa, H. Y. Katada, Y. Sakamoto, and Y. Sato: Rearrangement of Steroid-14-ene. Tetrahedron Letters **1969,** 2947.

415. Anastasia, M., M. Bolognesi, A. Fiecchi, G. Rossi, and A. Scala: A Ready Synthesis of 17α Steroids. J. Organ. Chem. (USA) **40,** 2007 (1975).

416. Cotterrell, G. P., T. G. Halsall, and M. J. Wriglesworth: A Chemical Model for a Possible Oxidative Rearrangement in the Biosynthesis of Triterpenes: The Rearrangement of 7α,8α- and 8,9-Epoxytirucallanes. J. Chem. Soc. (C) **1970,** 1503.

417. Lavie, D., and E. Glotter: The Cucurbitanes, a Group of Tetracyclic Triterpenes. In: W. Herz, H. Grisebach, and G. W. Kirby (eds.), Progress in the Chemistry of Organic Natural Products, Vol. 29, p. 307. Wien: Springer. 1971.

418. Govindachari, T. R., N. Viswanathan, and P. A. Mohamed: Litsomentol, A New Tetracyclic Triterpene. Chem. Commun. **1971,** 665.

419. Zander, J. M., and D. C. Wigfield: The Biosynthesis of Cucurbitacin B. Chem. Commun. **1970,** 1599.

420. Herlem-Gaulier, D., F. Khuong-Huu-Lainé, et R. Goutarel: Alcaloïdes Stéroidiques LXXII. Famille des Buxacées (10e communication). Alcaloïdes du groupe des

cyclobuxidines retirés du *Buxus balearica* Willd: N-3-isobutyryl cyclobuxidine-F (baléabuxidine), N-3-isobutyryl cycloxobuxidine-H, N-3-benzoyl cycloxobuxidine-F et cycloxobuxoxazine-C (baléabuxoxazine). Bull. soc. chim. France **1968,** 763.

421. BUCHANAN, J. G. ST. C., and T. G. HALSALL: The Conversion of Turraeanthin and Turraeanthin A into Simple Meliacins by a Route Involving an Oxidative Rearrangement of Probable Biogenetic Importance. J. Chem. Soc. (C) **1970,** 2280.

422. LAVIE, D., and E. C. LEVY: Meliane-Meliacin Relationship. Tetrahedron **27,** 3941 (1971).

423. MERRIEN, A., and J. POLONSKY: The Natural Occurence of Melianodiol and its Diacetate in *Samadera madagascariensis* (Simaroubaceae): Model Experiments on Melianodiol Directed towards Simarolide. Chem. Commun. **1971,** 261.

424. HALSALL, T. G., and R. J. WESTON: A Tirucall-7-ene to Apotirucall-14-ene Rearrangement Initiated by Bromine. Chem. Commun. **1972,** 1212.

425. ADESOGAN, E. K., D. A. OKORIE, and D. A. H. TAYLOR: Limonoids from *Khaya anthotheca* (Welw.) C. D. C. J. Chem. Soc. (C) **1970,** 205.

426. EKONG, D. E. U., and M. D. SELEMA: The Meliacins (Liminoids): Boron Trifluoride-catalysed Rearrangements of 14,15-Epoxides in the Meliacins. Chem. Commun. **1970,** 783.

427. FRACHEBOUD, M.: Etudes sur le Comportement du Cycle Propanique de Certains Dérivés du Cycloartenol. Ph. D. Thesis, ETH, Zurich, 1966.

428. APSIMON, J. W., and J. M. ROSENFIELD: Methyl Migration by Epoxide Cleavage. The Effect of Carbonium Ion Stabilisation by a Neighbouring Double Bond on the Direction of Migration on Cleavage of 9β,11β-Epoxy-4,4-dimethylandrost-5-ene-3,17-dione. Chem. Commun. **1970,** 1271.

429. APSIMON, J. W., R. R. KING, and J. J. ROSENFELD: Approaches to the Synthesis of Triterpenoids. I. Methyl Group Migration to the 9β-Position in Steroids as a Route to the Cucurbitacins. A Possible Structural Feature Directing Epoxide Cleavage Rearrangements. Canad. J. Chem. **47,** 1989 (1969).

430. CAMPBELL, A. C., C. L. HEWETT, M. S. MAIDMENT, and G. F. WOODS: Formation of Some Novel 9β-Methyl-19-norpregnane Derivatives. J. Chem. Soc. Perkin I **1974,** 1799.

431. LEVY, E. C., and D. LAVIE: Attempted Skeletal Rearrangements in the Lanostane Series. Israel J. Chem. **8,** 677 (1970).

432. MAKHUBU, L. P., Z. G. HAJOS, and G. R. DUNCAN: Selective Hydrolysis and Oxidation of the Bismethylenedioxy Protective Group During Angular Methyl Migration in Corticosteroid Analogs. Canad. J. Chem. **52,** 1744 (1974).

433. BARTON, D. H. R., L. J. DANKS, A. K. GANGULY, R. H. HESSE, G. TARZIA, and M. M. PECHET: Organic Reactions of Fluoroxy-compounds: Addition Reactions of Unactivated and Deactivated Unsaturated Linkages of Steroids. Chem. Commun. **1969,** 227.

434. KUPCHAN, M. S., J. W. A. FINDLAY, P. HACKETT, and R. M. KENNEDY: The Chemistry of 9β,19-Cyclo Steroid Derivatives. J. Organ. Chem. (USA) **37,** 2523 (1972).

435. BARTON, D. H. R., D. KUMARI, P. WELZEL, L. J. DANKS, and J. F. MCGHIE: Photochemical Transformations. Part XXV. The Synthesis of Cycloartenol. J. Chem. Soc. (C) **1969,** 332.

436. HALSALL, T. G., and R. T. APLIN: A Pattern of Development in the Chemistry of Pentacyclic Triterpenes. In: L. ZECHMEISTER (ed.), Progress in the Chemistry of Organic Natural Products, Vol. 22, p. 153. Vienna: Springer. 1964.

437. COREY, E. J., and S. K. GROSS: Direct Biosynthesis of the 29,30-Bisnoramyrin System from 29,30-Bisnor-2,3-oxidosqualene in Pea Seedlings. J. Amer. Chem. Soc. **90,** 5045 (1968).

438. SEO, S., Y. TOMITA, and K. TORI: Biosynthesis of Oleanene- and Ursene-type Triterpenes from [4-^{13}C]Mevalonic Acid in Tissue Cultures of *Isodon japonicus* Hara. Chem. Commun. **1975,** 270.

439. SUGA, T., T. SHISHIBORI, and S. KIMOTO: Biosynthesis of Triterpenoids. The Stereochemistry of the Squalene Formation and its Cyclization to β-Amyrin, Chemistry Letters **1972,** 129.

440. — — — Triterpenoid Biosynthesis. The Stereospecificity in the Enzymatic Cyclization of Squalene to β-Amyrin. Chemistry Letters **1972,** 313.

441. ARIGONI, D., and H. BRUNNER: unpublished results cited in reference 28.

442. HIRSCHMANN, H., F. B. HIRSCHMANN, and A. P. ZALA: The Uranediol Rearrangement. J. Organ. Chem. (USA) **31,** 375 (1966).

443. LEE, H., and M. E. WOLFF: C-18 Functional Steroids and D-Homo Steroids. J. Organ. Chem. (USA) **32,** 192 (1967).

444. HIRSCHMANN, F. B., D. M. KAUTZ, S. S. DESHMANE, and H. HIRSCHMANN: Formolysis of 3β-Acetoxy-5α-pregnan-20α-yl *p*-toluene-sulfonate. Tetrahedron **27,** 2041 (1971).

445. HIRSCHMANN, F. B., and H. HIRSCHMANN: Inversions of Both Adjacent Centers in the Formolysis of a 2,2,6-Trialkylcyclohexyl Tosylate. Formation of a 13α-D-Homo Steroid. J. Organ. Chem. (USA) **38,** 1270 (1973).

446. COATES, R. M., and S. K. CHUNG: Stereochemistry in the Solvolytic Ring Contraction of 2,2,4aα-Trimethyl-1-decalyl Methanesulfonate. A Model Reaction Pertaining to Triterpene Biogenesis. J. Organ. Chem (USA) **38,** 3677 (1973).

447. BOTTA, L.: Zur Biogenese von Verbindungen der Lupanreihe. Ph.D. Dissertation, ETH, Zürich, 1968.

448. TANAKA, O., M. NAGAI, T. OHSAWA, N. TANAKA, K.-I. KAWAI, and S. SHIBATA: Chemical Studies on the Oriental Plant Drugs. XXVII. The Acid Catalyzed Reactions and the Absolute Configuration at C_{20} of Dammarane Type Triterpenes. Chem. Pharm. Bull. (Japan) **20,** 1204 (1972).

449. HALSALL, T. G., E. R. H. JONES, and G. D. MEAKINS: The Chemistry of the Triterpenes. Part XI. The Conversion of Lupeol into Germanicol (*iso*-Lupeol). The Structure of Lupeol Hydrochloride. J. Chem. Soc. **1952,** 2862.

450. AMES, T. R., G. S. DAVY, T. G. HALSALL, and E. R. H. JONES: The Chemistry of the Triterpenes. Part XII. The Action of Formic Acid on Lupeol. J. Chem. Soc. **1952,** 2868.

451. AMES, T. R., J. L. BETON, A. BOWERS, T. G. HALSALL, and E. R. H. JONES: The Chemistry of the Triterpenes and Related Compounds. Part XXIII. The Structure of Taraxasterol, Ψ-Taraxasterol (Heterolupeol), and Lupenol-I. J. Chem. Soc. **1954,** 1905, and references cited therein.

452. BROWNLIE, G., M. B. E. FAYEZ, F. S. SPRING, R. STEVENSON, and W. S. STRACHAN: Triterpenoids. Part XLVIII. Olean-13(18)-ene: Isomerisation of Olean-12-ene and Related Hydrocarbons with Mineral Acid. J. Chem. Soc. **1956,** 1377.

453. DAVY, G. S., T. G. HALSALL, and E. R. H. JONES: The Chemistry of the Triterpenes. Part IX. Elucidation of the Betulin-Oleanolic Acid Relationship. J. Chem. Soc. **1951,** 2696.

454. DAVY, G. S., T. G. HALSALL, E. R. H. JONES, and G. D. MEAKINS: The Chemistry of the Triterpenes. Part X. The Structures of Some Isomerisation Products from Betulin and Betulinic Acid. J. Chem. Soc. **1951,** 2702.

455. HALSALL, T. G., E. R. H. JONES, and R. E. H. SWAYNE: The Chemistry of the Triterpenes and Related Compounds. Part XXII. The Conversion of Lupeol into Ψ-Taraxasterol (Heterolupeol). J. Chem. Soc. **1954,** 1902.

456. AMES, T. R., T. G. HALSALL, and E. R. H. JONES: The Chemistry of the Triterpenes. Part VII. An Interrelationship between the Lupeol and the β-Amyrin Series. Elucidation of the Structure of Lupeol. J. Chem. Soc. **1951,** 450.

457. LEONARD, R., and J. B. THOMSON: Isomerisation of a β-Amyrin Derivative to an α-Amyrin Derivative. Chem. Commun. **1972,** 1281.

458. WHITLOCK, H. W., JR., and A. H. OLSON: The Concertedness, or Lack Thereof, of a Multiple Carbonium Ion Rearrangement. J. Amer. Chem. Soc. **92,** 5383 (1970).

459. BEATON, J. M., F. S. SPRING, R. STEVENSON, and J. L. STEWART: Triterpenoids. Part XXXVII. The Constitution of Taraxerol. J. Chem. Soc. **1955,** 2131.
460. COURTNEY, J. L., R. M. GASCIOGNE, and A. Z. SZUMER: Triterpenes of the Friedelane Series. Part III. The Course of the Friedelene-Oleanene Rearrangement. J. Chem. Soc. **1958,** 881.
461. BROOKS, C. J. W.: Observations on Taraxerol (Skimmiol). J. Chem. Soc. **1955,** 1675.
462. KOLLER, E., A. HIESTAND, P. DIETRICH, und O. JEGER: Zur Kenntnis der Triterpene. Überführung von Taraxerol in $\Delta^{13,18}$-Oleanen. Helv. Chim. Acta **33,** 1050 (1950).
463. SENGUPTA, P., and N. H. KHASTGIR: Terpenoids and Related Compounds – III. Bauerenol and Multiflorenol from *Gelonium multiflorum* A. Juss. The Structure of Multiflorenol. Tetrahedron **19,** 123 (1963).
464. LAIRD, W., F. S. SPRING, and R. STEVENSON: Triterpenoids. Part LVIII. The Synthesis of Isoursenol, the Ursane Analogue of Taraxerol. J. Chem. Soc. **1961,** 2638.
465. CHIVERS, H., R. E. CORBETT, and R. E. M. MITCHELL: Extractives from the Leaves of *Olearia paniculata*. J. Chem. Soc. (C) **1966,** 1814.
466. LAHEY, F. N., and M. V. LEEDING: A New Triterpene Alcohol, Bauerenol. Proc. Chem. Soc. (London) **1958,** 342.
467. BUDZIAREK, R., J. D. JOHNSTON, W. MANSON, and F. S. SPRING: Triterpene Resinols and Related Acids. Part XXII. *iso*-β-Amyrenonol and *iso*-β-Amyradienonol. J. Chem. Soc. **1951,** 3019.
468. JEGER, O., und L. RUŽIČKA: Zur Kenntnis der Triterpene. Überführung des β-Amyrins in ein neues Dien-dion-Derivat. Helv. Chim. Acta **28,** 209 (1945).
469. AGATA, I., E. J. COREY, A. G. HORTMANN, J. KLEIN, S. PROSKOW, and J. J. URSPRUNG: Oxidative Rearrangements of Pentacyclic Triterpenes. A Method for the Synthesis of Certain Naturally Occurring Triterpenes from α- and β-Amyrin. J. Organ. Chem. (USA) **30,** 1698 (1965).
470. BARTON, D. H. R., and P. DE MAYO: Triterpenoids. Part XIII. Phyllanthol, the First Hexacarbocyclic Triterpenoid. J. Chem. Soc. **1953,** 2178.
471. BEATON, J. M., J. D. EASTON, M. M. MACARTHUR, F. S. SPRING, and R. STEVENSON: Triterpenoids. Part XLV. The Conversion of α-Amyrin into Phyllanthol. The Constitution of "*iso*-α-Amyradienonyl-II Acetate. J. Chem. Soc. **1955,** 3992.
472. COREY, E. J., and J. J. URSPRUNG: The Structures of the Triterpenes Friedelin and Cerin. J. Amer. Chem. Soc. **78,** 5041 (1956).
473. DUTLER, H., O. JEGER, und L. RUŽIČKA: Zur Kenntnis der Triterpene. Zur Konstitution und Konfiguration von Friedelin und Cerin; ein Beitrag zur Biogenese pentacyclischer Triterpene. Helv. Chim. Acta **38,** 1268 (1955).
474. BROWNLIE, G., F. S. SPRING, R. STEVENSON, and W. S. STRACHAN: Triterpenoids. Part LII. The Constitution and Stereochemistry of Friedelin and Cerin. J. Chem. Soc. **1956,** 2419.
475. BEATON, J. M., F. S. SPRING, R. STEVENSON, and J. L. STEWART: Triterpenoids. Part LIV. The Constitution of Alnusenone. Tetrahedron **2,** 246 (1958).
476. COATES, R. M.: On the Friedelene-Oleanene Rearrangement. Tetrahedron Letters **1967,** 4143.
477. WHITLOCK, H. W., JR., and M. C. SMITH: The Enzymatic Requirement for "Concerted" Backbone Rearrangement. Tetrahedron Letters **1968,** 821.
478. KIKUCHI, T., M. NIWA, M. TAKAYAMA, T. YOKOI, and T. SHINGU: Application of Homonuclear INDOR Technique. Structure of an Acid-Induced Rearrangement Product of 16-Keto-friedel-3-ene. Tetrahedron Letters **1973,** 1987.
479. KIKUCHI, T., M. TAKAYAMA, T. TOYODA, M. ARIMOTO, and M. NIWA: Studies on the Neutral Constituents of *Pachysandra terminalis* Sieb. et Zucc. V. Structures of Pachysandiol-B and Pachysonol, New Friedelin Type Triterpenes. Chem. Pharm. Bull. (Japan) **21,** 2243 (1973).

480. VAN TAMELEN, E. E., and R. M. COATES: Biogenetic-type Synthesis of (±)-Farnesiferol A and (±)-Farnesiferol C. Chem. Commun. **1966,** 413.

481. ALDRIDGE, D. C., A. BORROW, R. G. FOSTER, M. S. LARGE, H. SPENCER, and W. B. TURNER: Metabolites of *Nectria coccinea*. J. Chem. Soc. Perkin I **1972,** 2136.

482. TANABE, M., and K. T. SUZUKI: Detection of C–C Bond Fission during the Biosynthesis of the Fungal Triprenylphenol Ascochlorin using [1,2-^{13}C]-Acetate. Chem. Commun. **1974,** 445.

483. ANTHONSEN, T., T. BRUUN, E. HEMMER, D. HOLME, A. LAMVIK, E. SUNDE, and N. A. SORENSEN: Baccharis Oxide, a new Triterpenoid from *Baccharis halimifolia* L. Acta Chem. Scand. **24,** 2479 (1970).

484. MO, F., T. ANTHONSEN, and T. BRUUN: Revised Structure of the Triterpenoid Baccharis Oxide. Acta Chem. Scand. **26,** 1287 (1972).

485. TACHIBANA, K., and T. TAKAHASHI: The Conversion of Shionone into Dihydrobaccharis Oxide. Tetrahedron Letters **1975,** 1857.

486. YAMADA, S., S. YAMADA, Y. MORIYAMA, Y. TANAHASHI, and T. TAKAHASHI: A BF_3-Et_2O Catalyzed Rearrangement of 3α,4α-Epoxyshionane. Tetrahedron Letters **1972,** 5043.

487. BATHURST, E. T. J., J. M. COXON, and M. P. HARTSHORN: Reactions of 5-Hydroxy Steroids. XIII. The Acid-Catalysed Reactions of Cholestane-4,5-diol 4-Acetates and Cholestane-4,5-diol Diacetates. Austral. J. Chem. **27,** 1505 (1974).

488. BOURGUIGNON, P., J.-C. JACQUESY, R. JACQUESY, J. LEVISALLES, et J. WAGNON: Stéréochimie. XXXVIII. Stéroides fluorés–VII. Réarrangement spinal du cholestérol. Bull. soc. chim. France **1971,** 269.

489. BLUNT, J. W., M. P. HARTSHORN, and D. N. KIRK: Reactions of Epoxides – XVII. "Backbone Rearrangements" of Cholest-5-ene and 5,6α-epoxy-5α-cholestane. Tetrahedron **25,** 149 (1969).

490. KIRK, D. N., and P. M. SHAW: Backbone Rearrangements of Androst-5-ene and D-Homoandrose-5-ene: A Novel Racemisation. Chem. Commun. **1971,** 948.

491. MONNERET, C., Q. KHUONG-HUU, et R. GOUTAREL: Transpositions spinales de D-homo androstène-13 dione-3,17. Bull. soc. chim. France **1972,** 291.

492. BASCOUL, J., E. NOYER, et A. CRASTES DE PAULET: Transpositions spinales in vitro en milieu acide. IV. Influence de la distance ≪tension-fonction déclenchante≫ sur le cours des transpositions spinales. Bull. soc. chim. France **1972,** 2744.

493. ZUNDEL, J. L., G. WOLFF, et G. OURISSON: La transposition spinale de l'A-nor euphénone. Bull. soc. chim. France **1973,** 3206.

494. WHITLOCK, H. W., JR., and L. W. OVERMAN: Solvolytic Rearrangements Accompanied by Multiple Alkyl Shifts. J. Amer. Chem. Soc. **93,** 2247 (1971).

495. BERTI, G., and F. BOTTARI: Constituents of Ferns. In: L. REINHOLD and Y. LIWSCHITZ (eds.), Progress in Phytochemistry, Vol. 1, p. 589. London: Interscience Publishers. 1968.

496. BARTON, D. H. R., G. MELLOWS, and D. A. WIDDOWSON: Biosynthesis of Terpenes and Steroids. Part III. Squalene Cyclisation in the Biosynthesis of Triterpenoids; the Biosynthesis of Fern-9-ene in *Polypodium vulgare* Linn. J. Chem. Soc. (C) **1971,** 110.

497. BARTON, D. H. R., P. DE MAYO, and J. C. ORR: Triterpenoids. Part XXIV. Further Investigations on the Constitution of Zeorin. J. Chem. Soc. **1958,** 2239.

498. FAZAKERLEY, H., T. G. HALSALL, and E. R. H. JONES: The Chemistry of Triterpenes and Related Compounds. Part XXXIV. The Structure of Hydroxyhopanone. J. Chem. Soc. **1959,** 1877.

499. AGETA, H., K. IWATA, and Y. ÔTAKE: A Fern Constituent: Diplopterol, a Triterpenoid Isolated from *Diplopterygium glaucum* Nakai. Chem. Pharm. Bull. (Japan) **11,** 407 (1963).

500. AGETA, H., K. IWATA, and K. YONEZAWA: Fern Constituents: Fernene and Diploptene, Triterpenoid Hydrocarbons Isolated from *Dryopteris crassirhizoma* Nakai. Chem. Pharm. Bull. (Japan) **11,** 408 (1963).

501. NAKAMURA, S., T. YAMADA, H. WADA, Y. INOUE, T. GOTO, and Y. HIRATA: The Structures of Five New Triterpenoids Obtained from *Rhododendron linearifolium*. Tetrahedron Letters **1965,** 2017.

502. AGETA, H., K. SHIOJIMA, and Y. ARAI: Fern Constituents: Neohopene, Hopene-II, Neohopadiene, and Fernadiene isolated from *Adiantum* Species. Chem. Commun. **1968,** 1105.

503. BERTI, G., A. MARSILI, J. MORELLI, and A. MANDELBAUM: Boron Trifluoride-catalyzed Rearrangements of Some Tetrasubstituted Neotriterpene Epoxides – II. Hopene-II Oxide and Its Analogue in the A: B-Neoallobetulin Series. Tetrahedron **27,** 2217 (1971).

504. AGETA, H., K. IWATA, and S. NATORI: A Fern Constituent, Fernene. A Triterpenoid Hydrocarbon of a New Type. Tetrahedron Letters **1963,** 1447.

505. LIN, Y.-Y., H. KAKISAWA, Y. SHIOBARA, and K. NAKANISHI: The Structure of Davallic Acid. Chem. Pharm. Bull. (Japan) **13,** 986 (1965).

506. NISHIMOTO, K., M. ITO, S. NATORI, and T. OHMOTO: The Structures of Arundoin, Cylindrin, and Fernenol Triterpenoids of Fernane and Arborane Groups of *Imperata Cylindrica* var. *Koenigii*. Tetrahedron **24,** 735 (1968).

507. AGETA, H., K. IWATA, and S. NATORI: Fern Constituents: Adianene, Filicene, 7-Fernene, Isofernene and Diploptene. Triterpenoid Hydrocarbons Isolated from *Adiantum monochlamys*. Tetrahedron Letters **1964,** 3413.

508. IGUCHI, K., and H. KAKISAWA: Synthesis of the Triterpenes Fernadiene and Fern-8-ene. Chem. Commun. **1970,** 1486.

509. KAKISAWA, H., and K. IGUCHI: Synthesis of 18α-Fernane Derivatives. A Backbone Rearrangement of Pentacyclic Triterpenes. Chem. Commun. **1970,** 1488.

510. APLIN, R. T., H. R. ARTHUR, and W. H. HUI: The Structure of the Triterpene Simiarenol (a E: B-*friedo*-Hop-5-ene) from the Hong Kong species of *Rhododendron simiarum*. J. Chem. Soc. (C) **1966,** 1251.

511. BERTI, G., F. BOTTARI, and A. MARSILI: Structure and Stereochemistry of a Triterpenoid Epoxide from *Adiantum Capillus-Veneris*. Tetrahedron **25,** 2939 (1969).

(Received August 18, 1975)

Chemistry of the Ansamycin Antibiotics

By K. L. Rinehart, Jr., and L. S. Shield,
Roger Adams Laboratory, University of Illinois, Urbana, Illinois, U.S.A.

With 35 Figures

Contents

I. Introduction

The ansamycins constitute a class of antibiotics characterized by an aliphatic bridge linking two non-adjacent positions of an aromatic nucleus. The name, originally suggested by PRELOG, is derived from the term "ansa compounds" coined by LÜTTRINGHAUS in 1942 (*71*). It is the structures of the ansamycins which have provided the name for these antibiotics and their structures are remarkably complex, as we shall see in the next section, but it is the bioactivities which have attracted the greatest attention. A derivative of one of the rifamycins, rifampicin, is marketed widely for treatment of tuberculosis and other infections caused by gram-positive organisms (*69*), while other derivatives of rifamycin and streptovaricin are biological probes because of their binding to DNA-dependent RNA polymerase (*42a*, *103*, *162*) and their inhibition of reverse transcriptases (*3*, *8*, *15*, *33*, *40*, *92*, *103*, *135*). Maytansine and related compounds are powerful anti-tumor agents (*61*, *62*, *63*, *150*). Bioactivities of the ansamycins have been reviewed extensively elsewhere (*4*, *69*, *101*, *111*, *119*, *153*) and need not be exhaustively described here, although they will be noted in the section of this review dealing with reactions of individual antibiotics. Instead, the present review will stress the chemistry of the compounds, attempting to correlate the properties of related members of the class. The literature coverage of this review includes articles appearing through April, 1975. No attempt has been made to review patents in any comprehensive way.

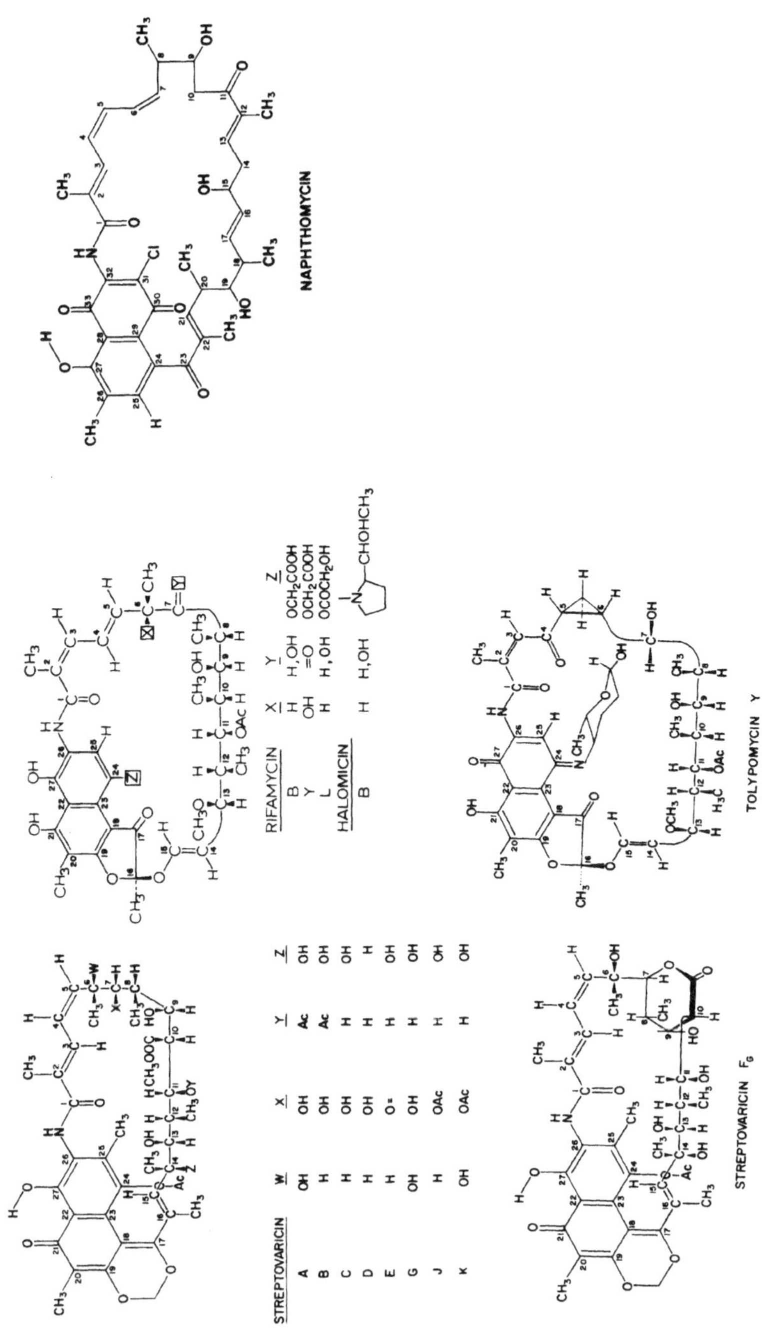

Fig. 1. Structures of naturally occurring naphthoquinonoid and naphthalenoid ansamycins

II. Structures of the Naturally Occurring Ansamycins

Nearly all of the structures assigned to the ansamycins rest, ultimately, on x-ray investigations. Consequently, little attention will be paid in this section to those degradative reactions which allowed the assignment of the structures of individual antibiotics by chemical means. Instead, the degradations will be covered in Section III, which describes the reactions of the compounds.

One can divide the ansamycins into two sub-classes, those whose ansa bridges are attached to nominal naphthoquinone or naphthalene nuclei and those whose ansa bridges are attached to benzoquinone or benzene nuclei. The former sub-class includes the streptovaricins, the rifamycins, the halomicins, the tolypomycins, and naphthomycin (Fig. 1). The latter sub-class includes geldanamycin and the maytansinoid compounds (Fig. 2). Properties of the naturally occurring ansamycins are shown in Table 1.

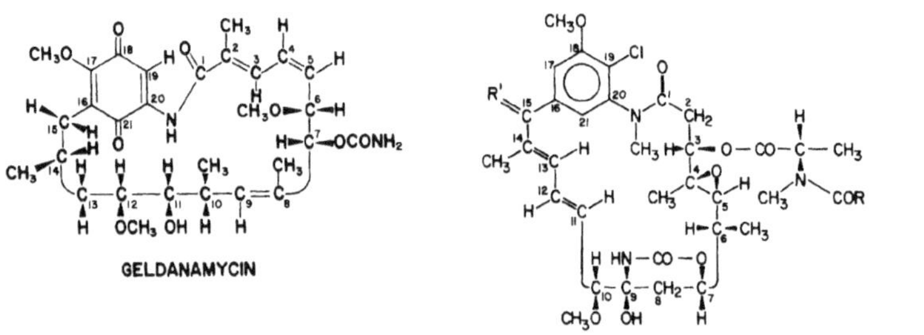

MAYTANSINE : R = CH_3 R' = H, H
MAYTANPRINE : R = CH_2CH_3 R' = H, H
MAYTANBUTINE : R = $CH(CH_3)_2$ R' = H, H
COLUBRINOL : R = $CH(CH_3)_2$ R' = H, OH
COLUBRINOL ACETATE : R = $CH(CH_3)_2$ R' = H, $OCOCH_3$

Fig. 2. Structures of naturally occurring benzoquinonoid and benzenoid ansamycins

A. Streptovaricins

The streptovaricins were the first ansamycins reported (*100, 130, 158*) and they also provide the most numerous group of antibiotics within the naphthoquinonoid ansamycins. Structures have been assigned to nine members of the family (Fig. 1) – streptovaricins A–G (*107*), plus J (*103*), and K (*47*). Streptovaricins H and I were reported early as chromatographic entities, but have not been described again until recently, when atropisostreptovaricin F_C was shown to have the same chromatographic

Table 1. *The Ansamycins*

Name	Producing Organism	Molecular Formula	Mol. wt.	M. p. (°C)	$[\alpha]_D$,°[d]	Ref.
Naphthoquinonoid						
Streptovaricin A	*Streptomyces spectabilis*	$C_{42}H_{53}NO_{16}$	827	233—243	+610	(*103*)
Streptovaricin B	*Streptomyces spectabilis*	$C_{42}H_{53}NO_{15}$	811	190—193	+576	(*103*)
Streptovaricin C	*Streptomyces spectabilis*	$C_{40}H_{51}NO_{14}$	769	189—191	+602	(*103*)
Streptovaricin D	*Streptomyces spectabilis*	$C_{40}H_{51}NO_{13}$	753	172—175	+590.3	(*103*)
Streptovaricin E	*Streptomyces spectabilis*	$C_{40}H_{49}NO_{14}$	767	198—202	+412.3	(*103*)
Streptovaricin F_G[a]	*Streptomyces spectabilis*	$C_{39}H_{47}NO_{14}$	753	222—224	+550	(*103*)
Streptovaricin G	*Streptomyces spectabilis*	$C_{40}H_{51}NO_{15}$	785	193—194	+473	(*103*)
Streptovaricin J	*Streptomyces spectabilis*	$C_{42}H_{53}NO_{15}$	811	177—180	+436	(*103*)
Streptovaricin K	*Streptomyces spectabilis*	$C_{42}H_{53}NO_{16}$	827	162—165		(*47*)
Rifamycin B	*Nocardia mediterranei* n. sp.[c] *Streptomyces tolypophorus*	$C_{39}H_{49}NO_{14}$	755	dec 160—164	−11[e]	(*123, 127*)
Rifamycin Y	*Nocardia mediterranei* n. sp.	$C_{39}H_{47}NO_{15}$	769		+325[e]	(*68*)
Rifamycin L	*Nocardia mediterranei* n. sp.	$C_{39}H_{49}NO_{14}$	755	152—153 dec		(*64*)
Rifamycin O[b]	*Nocardia mediterranei* mutant	$C_{39}H_{47}NO_{14}$	753	dec 160	+71.5[f]	(*121*)
Rifamycin S[b]	*Nocardia mediterranei* mutant	$C_{37}H_{45}NO_{12}$	695	180—182 dec	+476[e]	(*122, 128*)
Rifamycin SV[b]	*Nocardia mediterranei mutant*	$C_{37}H_{47}NO_{12}$	697	dec 140	−4[e]	(*122, 128*)
Halomicin B	*Micromonospora halophytica* sp. n. var. n. *nigra*	$C_{43}H_{58}N_2O_{12}$	794	178—182	+73.1	(*36*)
Tolypomycin Y	*Streptomyces tolypophorus*	$C_{43}H_{54}N_2O_{14}$	822	120 sinter	+325	(*44, 53*)
Naphthomycin	*Streptomyces collinus* Lindenbein	$C_{40}H_{46}ClNO_9$	683	dec 200	+432	(*2, 159*)
Benzoquinonoid						
Geldanamycin	*Streptomyces hygroscopicus* var. *geldanus* var. *nova*	$C_{29}H_{40}N_2O_9$	560	252—255	+55	(*22*)
Maytansine	*Maytenus ovatus* Loes	$C_{34}H_{46}ClN_3O_{10}$	691	171—172	−145	(*62*)
Maytanprine	*Maytenus buchananii* (Loes) R. Wilczek	$C_{35}H_{48}ClN_3O_{10}$	705	169—170	−125	(*63*)
Maytanbutine	*Maytenus buchananii* (Loes) R. Wilczek *Colubrina texensis* Gray (Rhamnaceae)	$C_{36}H_{50}ClN_3O_{10}$	719	178—180	−126	(*63*)
Colubrinol	*Colubrina texensis* Gray (Rhamnaceae)	$C_{36}H_{50}ClN_3O_{11}$	735	194—196	−94	(*150*)
Colubrinol acetate	*Colubrina texensis* Gray (Rhamnaceae)	$C_{38}H_{52}ClN_3O_{12}$	777	179—182	−127	(*150*)

[a] Streptovaricin F_G was originally referred to as streptovaricin F. [b] Rifamycins O, S and SV were originally isolated from chemical degradation of rifamycin B but have recently been isolated from a mutant of *Nocardia mediterranei* (*65, 134*). [c] This organism was originally referred to as *Streptomyces mediterranei*, but was reclassified in 1969 (*136*). The rifamycins were reported to be the same as the nancimycins (*118*). [d] Determined in $CHCl_3$ solution, except as noted. [e] CH_3OH solution. [f] Dioxane solution.

properties as the reported streptovaricin H (*47, 84*). The three-dimensional structure of the streptovaricins rests on the x-ray analysis of a heavy atom derivative (**1**) (*149*) of the most abundant component, streptovaricin C, prepared by acetylation followed by treatment with *p*-bromobenzeneboronic acid, as shown here.

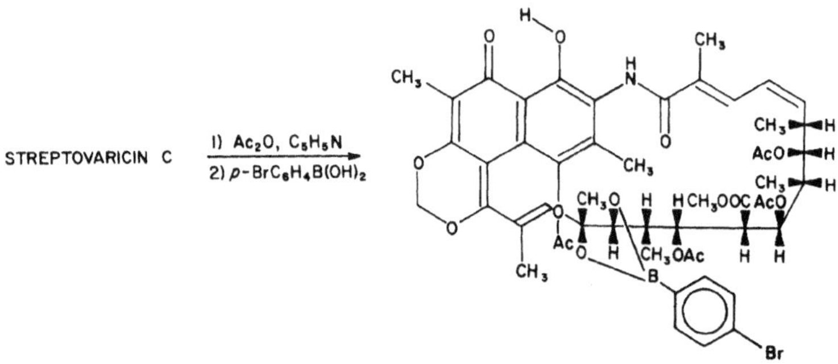

ATROPISOSTREPTOVARICIN C TRIACETATE CYCLIC *p*-BROMOBENZENEBORONATE

(**1**)

The derivative examined by x-ray was, in fact, the atropisomer of the natural streptovaricin C derivative; the subject of atropisomerism will be discussed in detail in Section III.A.1. Structures of the other streptovaricins rest mainly on comparison of their spectral properties to those of streptovaricin C. Streptovaricins B and J have also been related to streptovaricin C by acetylation (*103*), streptovaricin A to streptovaricins G and K by acetylation (*103*), streptovaricin E to streptovaricin C by reduction (*103*), streptovaricin G to streptovaricin F_G by lactonization (*56*), and streptovaricin C to streptovaricins A, D, and G by biosynthetic conversion. (The latter topic will be discussed in Section V.A, dealing with Biosynthesis.)

Streptovaricin C, Streptovaricin B, Streptovaricin J $\xrightarrow{Ac_2O,\ C_5H_5N}$ Streptovaricin C Triacetate + Streptovaricin C Tetraacetate

Streptovaricin G, Streptovaricin A, Streptovaricin K $\xrightarrow{Ac_2O,\ C_5H_5N}$ Streptovaricin G Triacetate

Streptovaricin E $\xrightarrow{NaBH_4}$ Streptovaricin C

Streptovaricin G $\xrightarrow{OH^-,\ r.t.}$ Streptovaricin F_G

B. Rifamycins

The initial reports from Sensi *et al.*, described rifamycins (originally spelled rifomycins) A, B, C, D, and E, and spectral properties of these compounds were given (*123*, *127*). However, in subsequent experiments, the Lepetit group was able by addition of diethylbarbituric acid to the

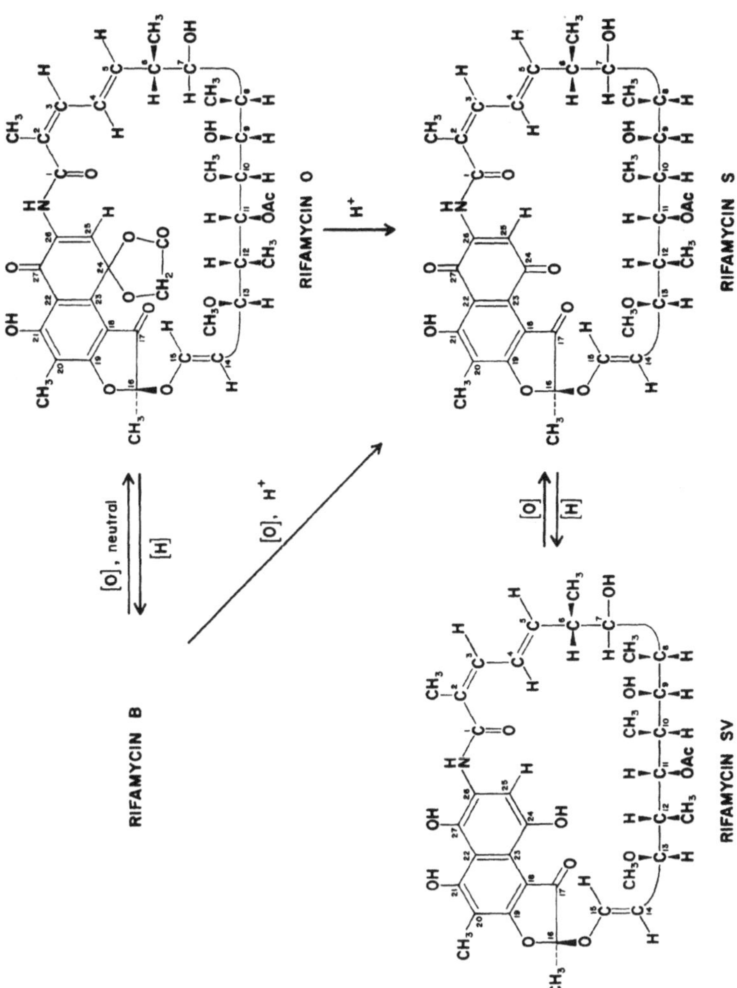

Fig. 3. Conversion of rifamycin B to rifamycins O, S and SV

fermentation medium to alter the composition of the antibiotic mixture produced (*55, 82, 83, 123*); this procedure stimulated the production of rifamycin B and inhibited that of the other components so that only rifamycin B was obtained. Rifamycin B (together with its conversion products rifamycins O, S and SV, whose structures are shown in Fig. 3) was the only component of the originally reported rifamycin mixture to be used for further work. Upon standing at room temperature or upon aeration at room temperature, rifamycin B solutions were observed to become "activated" (*121, 123, 140*), apparently by conversion to a more potent antibiotic, rifamycin S, as shown in Fig. 3. The same sequence can also be accomplished chemically by oxidation of rifamycin B to rifamycin O (containing a lactide at C-24), followed by hydrolysis of rifamycin O to rifamycin S (which removes the glycolic acid). Reduction of rifamycin S reforms the hydroquinone and gives rifamycin SV, which is also an antibiotic. Rifamycins Y (*68*) and L (*64*), whose structures are shown in Fig. 1, are related to rifamycin B and are the only other naturally occurring rifamycins reported thus far, except for 13-O-demethylrifamycin B, 13-O-demethylrifamycin SV, and 13-O-demethyl-11-O-deacetylrifamycin SV (*120*). (In a sense, rifamycins O, S and SV can also be regarded as naturally occurring rifamycins, since mutant strains of the usual rifamycin producers have been developed which accumulate these compounds.) The numbering system used in Fig. 1 is different from that usually employed for the rifamycins and is designed to facilitate comparison to the streptovaricins. A correspondence chart is provided in Table 2.

Table 2. *Numbering Scheme for Streptovaricins and Rifamycins*

Scheme used here	Usual scheme for rifamycins	Scheme used here	Usual scheme for rifamycins
C-1	C-15	C-15	C-29
2	16	16	12
3	17	17	11
4	18	18	5
5	19	19	6
6	20	20	7
7	21	21	8
8	22	22	9
9	23	23	10
10	24	24	4
11	25	25	3
12	26	26	2
13	27	27	1
14	28		

For x-ray crystallographic investigation by the heavy atom method rifamycins B and Y (*9—12*) were converted to their respective *p*-iodo-anilides (**2, 3**). Rifamycins L and Y were compared to rifamycin B by their spectral properties and by demonstrating that rifamycins B and L are both formed biosynthetically from rifamycin SV by reaction with glycolic acid and that rifamycin B is then converted to rifamycin Y (*64, 66*). These biosynthetic studies will be described in Section V.B.

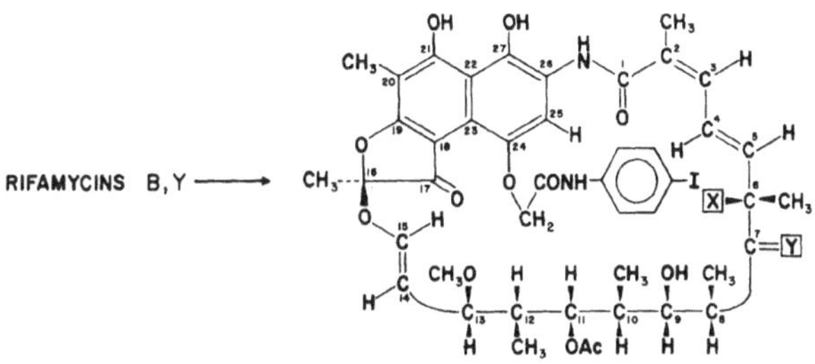

RIFAMYCIN B *p*-IODOANILIDE (2)
RIFAMYCIN Y *p*-IODOANILIDE (3)

(2): X=H, Y=H,OH
(3): X=OH, Y=O

The rifamycins differ from the streptovaricins in several ways. Most fundamental is the presence of an oxygen atom separating C-15 from C-16 in the rifamycins whereas the streptovaricins contain an uninterrupted carbon chain in the ansa bridge. Second, the rifamycins contain a fully aromatic ring (left-hand ring in Fig. 1) which is a quinone methide ring in the streptovaricins. Third, the rifamycins lack one aromatic methyl group of the streptovaricins. Finally, different ansa ring methyl groups have been oxidized in the two antibiotics. The C-10 methyl of the rifamycins has been oxidized to a carbomethoxyl group in the streptovaricins, while a methyl substituent present in the streptovaricins at C-14 is lacking in the rifamycins. Recent biosynthetic work to be discussed in Section V.B. indicates the C-14 methyl group has been removed by oxidation.

C. Halomicins

According to an initial report, the halomicins constitute a complex of antibiotics, from which halomicins A, B, C and D were described (*154*). However, only the structure of halomicin B (Fig. 1) has been reported (*36*). This compound is like rifamycin B, but differs in the group attached to the naphthoquinone nucleus. The structure of halomicin B was established by its conversion to rifamycin S and a substituted pyrrolidine (**4**) on treatment with nitrous acid. The structure of the pyrrolidine and its relative and absolute stereochemistry were based on analysis of proton magnetic resonance (pmr) and optical rotatory dispersion (ORD) spectra of derivatives, to be discussed in Section III.C.

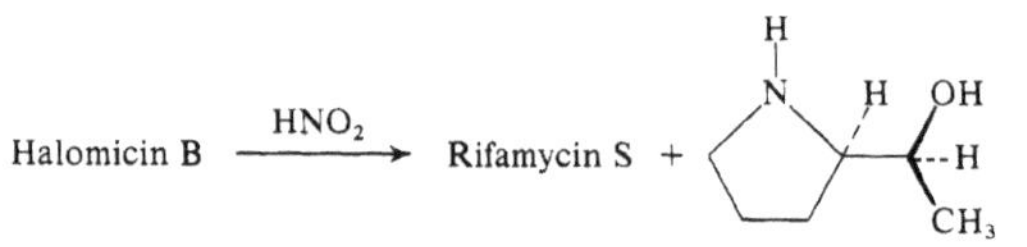

(**4**)

D. Tolypomycins

The tolypomycins have also been described as a complex of antibiotics (*44, 129*). Here, too, only one member has been described, tolypomycin Y (Fig. 1) (*43, 53, 54*). This compound was converted by mild

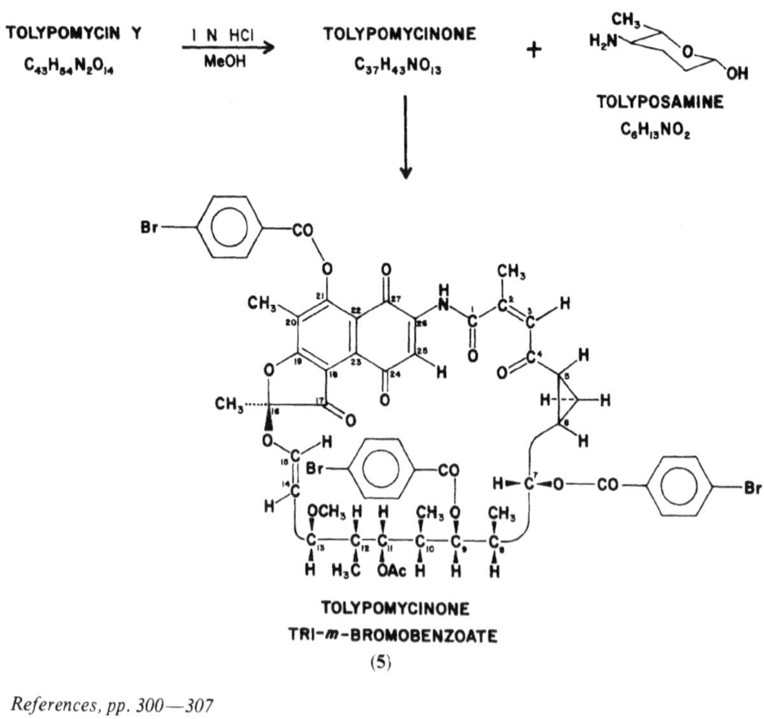

(5)

acidic hydrolysis to tolypomycinone (*52*), whose tri-*m*-bromobenzoyl derivative (**5**) was used for structure determination by x-ray crystallography (*50*). Tolypomycin Y differs from rifamycin B in two ways. First, a C-4 keto group and a cyclopropane ring (containing C-5 and C-6) replace the 4,5-double bond, C-6 and the allylic methyl of rifamycin B. Second, the dihydronaphthoquinone ring of rifamycin B is a naphthoquinone ring in tolypomycin Y and the glycolic acid group of rifamycin B is replaced by a 2,3,4,6-tetradeoxy-4-amino sugar (tolyposamine) attached as a quinone imide. The tolypomycins and the rifamycins are clearly related biosynthetically and, in fact, the microorganism which produces tolypomycin also produces the rifamycins (*54*).

E. Naphthomycin

The most recent addition to the naphthoquinonoid ansamycins is naphthomycin, whose structure was assigned (Fig. 1) largely on the basis of its proton magnetic resonance spectrum (*2, 159*). This structure is most nearly like that of the streptovaricins, but naphthomycin contains a naphthoquinone nucleus and a chloro (rather than a methyl) substituent *ortho* to the amide nitrogen. It also lacks the methylenedioxy and enol acetate groups of the streptovaricins and its ansa bridge is longer by six carbon atoms than those of the other naphthoquinonoid ansamycins. The phenolic hydroxyl of naphthomycin was originally assigned to C-27 (*159*), but a more recent report (*23, 48*) suggests that it is at C-25 instead.

F. Geldanamycin

The only benzoquinonoid ansamycin of undisputably microbial origin is geldanamycin (Fig. 2) (*22*). The streptomycete which produces geldanamycin makes no other ansamycin but does produce several other antibiotics (*96*). The structure of geldanamycin has been assigned by x-ray crystallography, employing the direct method on the underivatized antibiotic (*25*). In addition to containing a smaller aromatic system than the naphthoquinonoid ansamycins geldanamycin also contains a shorter ansa bridge with fewer methyl groups. Its relationship to the naphthoquinonoid ansamycins can be seen most clearly in the dienamide portion of the molecule as well as in the general nature of the ansa system.

G. Maytansinoids

A number of compounds which appear to be related to the ansamycins have been isolated from plants in recent years. This group, the maytansinoids, includes maytansine itself (*62*), its homologs maytanprine and maytanbutine (*63*), plus colubrinol, an oxygenated derivative of maytanbutine, and its acetate (*150*). Although the maytansinoids have been obtained from plant extracts, their close structural similarity to geldanamycin (including an ansa bridge of the same chain length) has led to speculation that they are, in fact, produced by a microorganism, perhaps a fungus, associated with the plants (*14*). As yet, no convincing positive or negative evidence is available. The structure of maytansine was assigned by an x-ray crystallographic investigation of its 3-bromopropyl ether (**6**) (*14*). The other structures are largely based on spectral evidence and their relationship to maytansine.

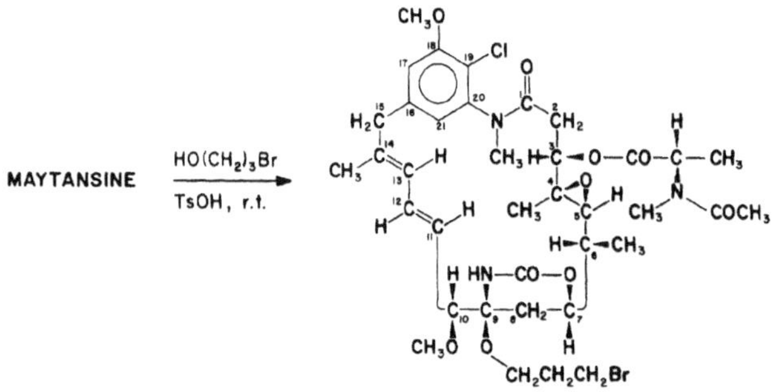

MAYTANSINE 3-BROMOPROPYL ETHER

(6)

H. Three-dimensional Structures

Those ansamycins whose structures have been assigned by x-ray crystallography provide intriguing three-dimensional models (Fig. 4). In each case the ansa ring lies above the aromatic system (*9, 11, 12, 14, 25, 50, 149*) and this can lead to special magnetic resonance shielding of the protons and carbon atoms which reside in this region (*98*), as we shall see in Section IV.B., dealing with nuclear magnetic resonance. The coiling of the bridge above the ring also leads to an atropisomerism observed so far only for the streptovaricins (see Section III.A.1. below). The rigidity of the quaternary carbon (C-16) of the rifamycins, halomicin, and tolypomycin presumably hinders the formation of atropisomers of these compounds, while the aromatic rings of the benzoquinonoid compounds may be too small to prevent their passing through the center of the ansa ring.

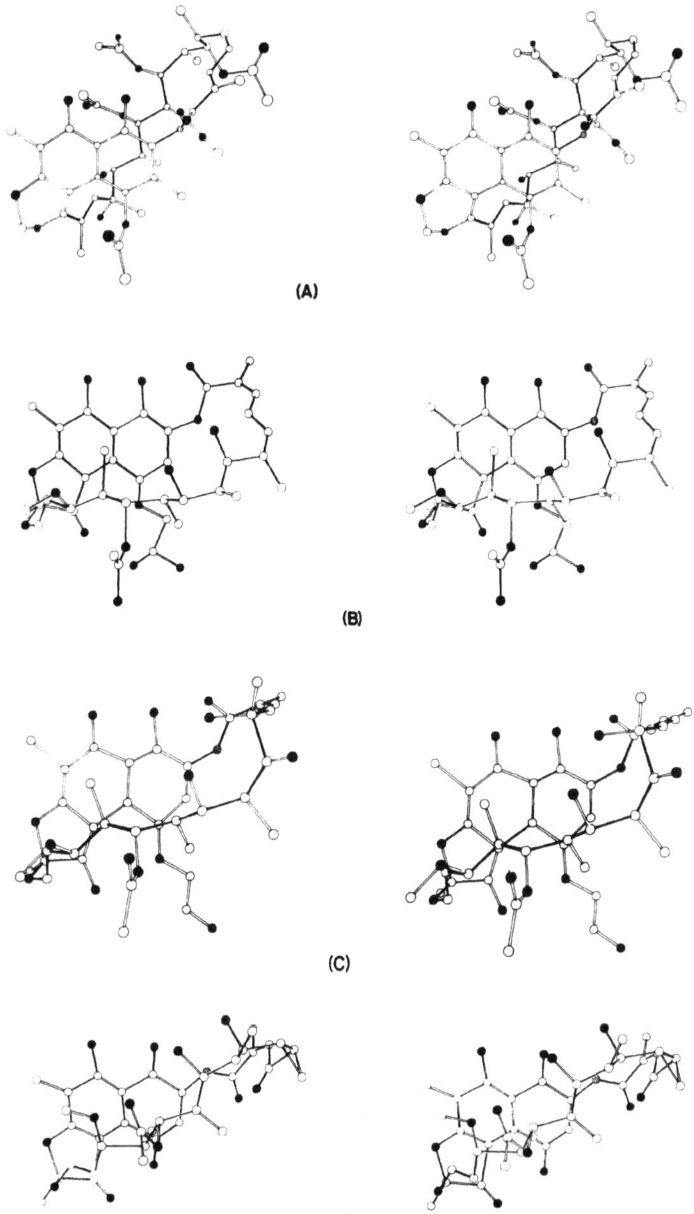

Fig. 4 (continued on p. 244)

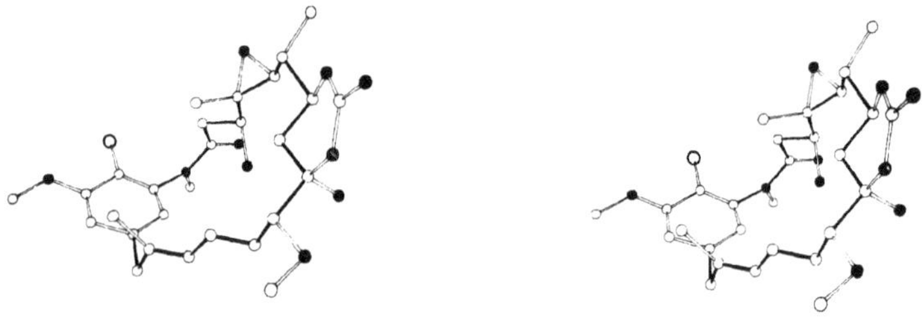

(E)

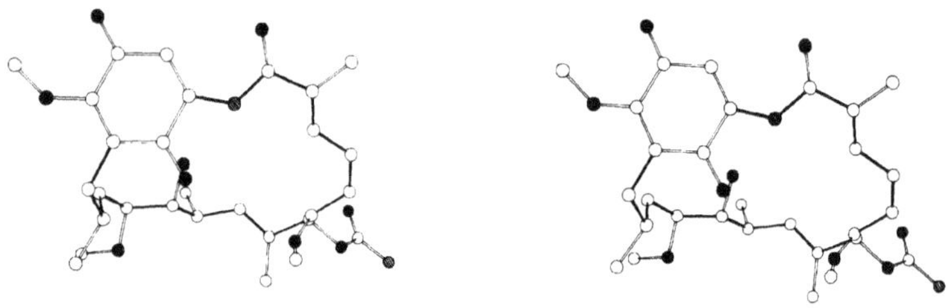

(F)

Fig. 4. Three-dimensional stereo models of ansamycins and their derivatives: A) atropisostreptovaricin C triacetate cyclic *p*-bromobenzeneboronate (**1**), B) rifamycin B *p*-iodoanilide (**2**), C) rifamycin Y *p*-iodoanilide (**3**), D) tolypomycinone tri-*m*-bromobenzoate (**5**), E) maytansine 3-bromopropyl ether (**6**), F) geldanamycin. All drawings except that for geldanamycin are from WANG and PAUL (*148*); that for geldanamycin was obtained from Dr. D. J. DUCHAMP, The Upjohn Company (*25*)

III. Reactions of the Ansamycins

In the nearly twenty years since the ansamycins were first reported as antibacterial agents, well over one thousand compounds have been prepared from them. The largest group of these are derivatives made from rifamycins B, S or SV in attempts to improve their antibacterial properties, and a second large group of compounds are those prepared from the streptovaricins and rifamycins in the course of structural studies of those antibiotics. At any rate it seems appropriate to divide arbitrarily the reactions of the ansamycins into those that were important in the structural assignment, mainly those resulting in carbon-carbon or carbon-oxygen bond cleavages, and those that were primarily directed toward studying the bioactivities of the compounds, which in the main did not result in carbon-carbon or carbon-oxygen bond cleavages. The former are labeled as degradation reactions and the latter as derivatization reactions. We shall treat the degradation reactions and then the derivatization reactions for each of the antibiotics, in the same order as in Section II.

A. Streptovaricins

1. Degradations

Intensive efforts have been directed through the years toward isolation of the individual streptovaricins from the crude antibiotic complex and these have given the components shown in Fig. 1 (*100, 103, 107, 108, 130, 133, 158*). Each was characterized by high resolution mass spectrometry to establish the formula of its molecular ion. Streptovaricin C is the most abundant component and the principal effort was directed first toward assigning its structure (*20*), as detailed below. Structures of the other streptovaricins were then established by comparison of their nmr and mass spectra with those of streptovaricin C (*103, 110*). All of the streptovaricins except streptovaricin D fragment under electron impact to give intense ions at lower masses – *m/e* 390, 324, 297, 269 – whose compositions are $C_{23}H_{20}NO_5$, $C_{18}H_{14}NO_5$, $C_{17}H_{15}NO_4$, and $C_{16}H_{13}O_4$, respectively (*107, 115*). These ions, which are accounted for by a fragmentation sequence discussed in Section IV.C., established a unit common to all streptovaricins consisting of C-16 through C-27, the amide nitrogen and C-1 through C-6, together with their substituents. Very extensive comparisons of the proton magnetic resonance spectra (*1, 103, 107*) of the individual streptovaricins then established the nature of the differences in the substitution patterns in the region C-6 through C-16. As shown in Section II.A., the conclusions reached as to the differences were then confirmed in part by chemical conversion of one streptovaricin to another (streptovaricin E→streptovaricin C, streptovaricin G→streptovaricin F_G) or by conversion to common derivatives (streptovaricins A, G and K yielding the same oligoacetate and streptovaricins B, C and J also yielding a common acetate).

It required a prolonged search to find the heavy-atom derivative (see Section II.A.) of streptovaricin C suitable for x-ray analysis and during that time extensive chemical degradation was carried out. Hydrolysis and alcoholysis attempts in acid and base proved singularly unsuccessful and no large fragments of any streptovaricin could be obtained. The most successful reaction involved hydrolysis of the methylenedioxy group and enol acetate in methanolic ammonia (*91*). This reaction, illustrated in Fig. 5 for streptovaricin C, was accompanied by keto-enol isomerization and oxidation of the resulting hydroquinone to give the product, here labeled damavaricin C (from *dea*cetyl*dem*ethylenestrepto*varicin* C). The cleavage reaction is accompanied by lactonization to give damavaricin F_C, as shown. Analogous products (damavaricins) were obtained from streptovaricins A, B, D and G.

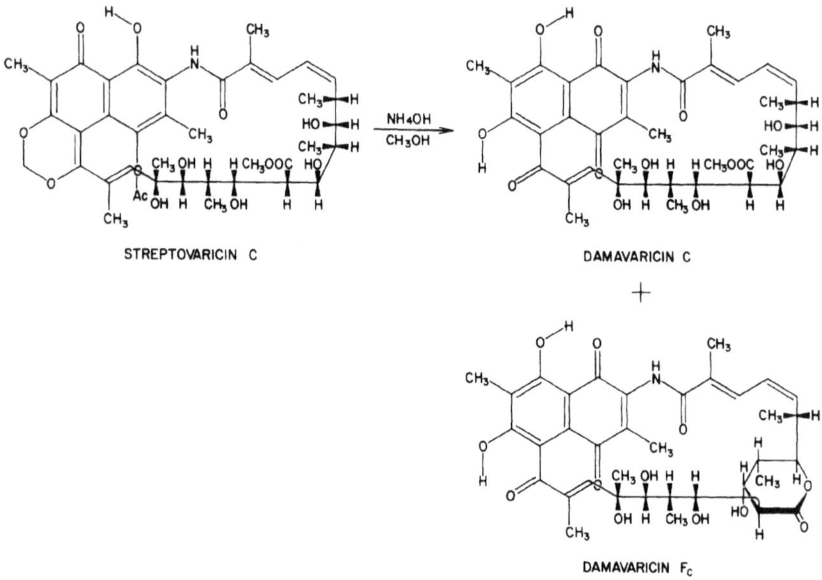

Fig. 5. Products of basic hydrolysis of streptovaricin C

Instead of hydrolysis, the principal degradative method proved to be oxidation, especially periodate oxidation (*104, 108, 113*). Nearly all streptovaricins (all but streptovaricin D) react with one mole of periodate to give products with the aliphatic portion of the molecule cleaved but still attached to the aromatic nucleus. With streptovaricins B, C, E, F_G, J and K, where there is only one vicinal glycol unit, cleavage occurs between C-13 and C-14 to give cyclic hemiacetals called streptovals (*1, 112*). The structure of streptoval C is shown in Fig. 6. If streptovaricin C is first triacetylated then treated with periodate, cleavage still occurs, but the product, *aldehydo*-streptoval C triacetate, cannot form a hemiacetal and on chromatography loses two moles of acetic acid to give streptovadienal.

Streptovaricins A and G contain two vicinal glycol units and cleavage could occur at either unit, though cleavage between C-6 and C-7 to give isostreptovals appears to be favored. This is illustrated for streptovaricin A in Fig. 6. If the latter streptovaricins (*e.g.*, streptovaricin A) are treated instead with excess periodate, a double cleavage occurs to give the orange prestreptovarone ($C_{29}H_{29}NO_9$) (*104, 108*), which contains the chromophore of streptovaricin. In addition, a water-soluble dialdehyde, varicinal A (*109*), was isolated from streptovaricin A (Fig. 6). The field ionization mass spectrum of varicinal A assigned its molecular formula while its nmr spectrum identified contiguous protons by decoupling and largely assigned its structure, together with relative stereochemistry at C-9 through C-12.

The aromatic fragment, prestreptovarone, contained the dienamide unit of the streptovaricins and the proton absorptions of that unit were used to assign its substitution pattern. Further oxidation of prestreptovarone with osmium tetroxide-periodate, which cleaves alkenes, gave the smaller orange ketone streptovarone ($C_{24}H_{23}NO_9$) (*102, 104, 108*), which still contains the streptovaricin chromophore but in which the dienamide system is replaced by a pyruvamide (Fig. 6). Streptovarone proved to be the key oxidative degradation product since it could also be obtained directly from all the streptovaricins (except streptovaricin D) or from the streptovaricin complex by oxidation with osmium tetroxide-periodate. It thus served as an accessible intermediate for further degradations.

Streptovarone clearly contained pyruvamide, methylenedioxy, and aryl acetate groups. The first was detected from hydrolysis of streptovarone to give methyl pyruvate, the second from the pmr spectrum of streptovarone and its hydrolysis to formaldehyde, and the third from the infrared spectrum of streptovarone (*104, 117*). All three groups were lost when streptovarone was treated with mild acid (Fig. 7), which yielded the deep red dapmavarone (*dea*cetyl*dep*yruvyl*dem*ethylenedioxystrepto*varone*, $C_{18}H_{16}O_7$), which lacked proton magnetic resonance absorptions for those groups (*104*). The latter compound was then treated with dilute hydrogen peroxide in base, which removed the acyl side chain in a Dakin reaction. Both the side chain and the aromatic nucleus were isolated as 2-methyl-4-oxo-2-pentenoic acid and 3,7-dimethyl-2,5,6,8-tetrahydroxynaphthoquinone. The former was a known compound and the latter's structure was assigned from the symmetry of its nmr spectrum and the similarity of its ultraviolet spectrum to that of the known 2,6-dimethoxy-5,8-dihydroxynaphthoquinone.

Due to the symmetry of the dimethyltetrahydroxynaphthoquinone there are only two positions for the acyl group of dapmavarone – at C-5 or C-6. The pmr spectrum of dapmavarone shows that its acyl chain actually contains a cyclic ether, with an ether oxygen substituted at the carbon

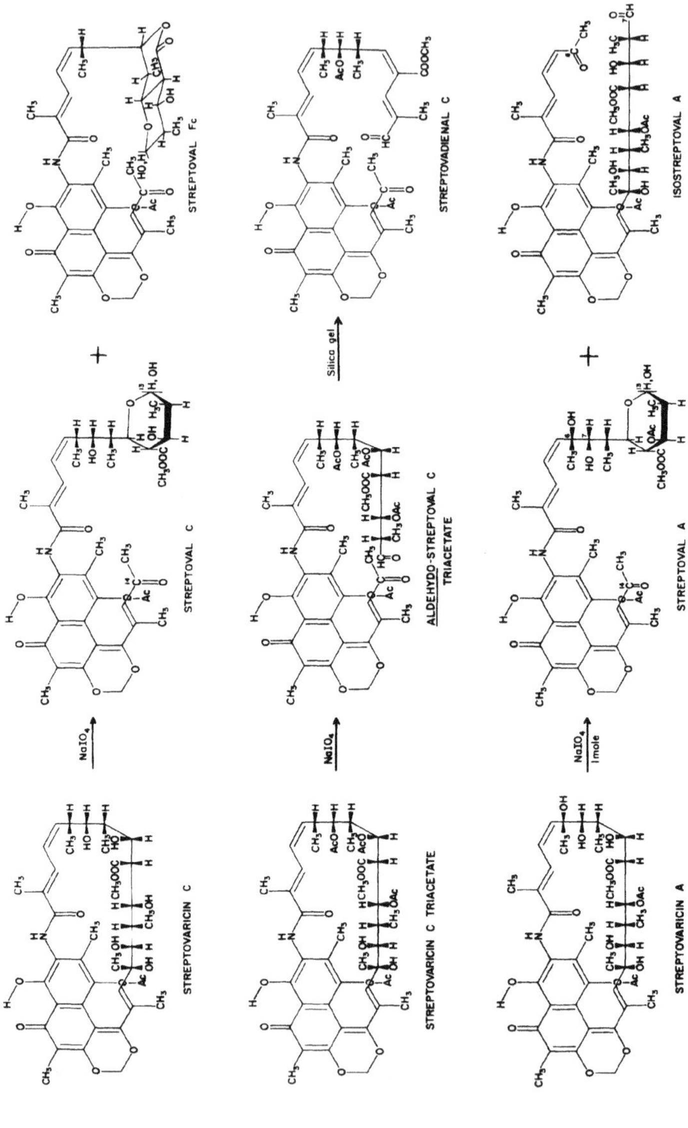

bearing the α-methyl group. Thus, two possibilities existed for dapmavarone, cyclic ethers involving either O-4 or O-6 of the tetrahydroxynaphthoquinone. The hydrogen-bonded keto carbonyl of dapmavarone decided in favor of the latter linkage. Thus, the partial formula (**a**) can be

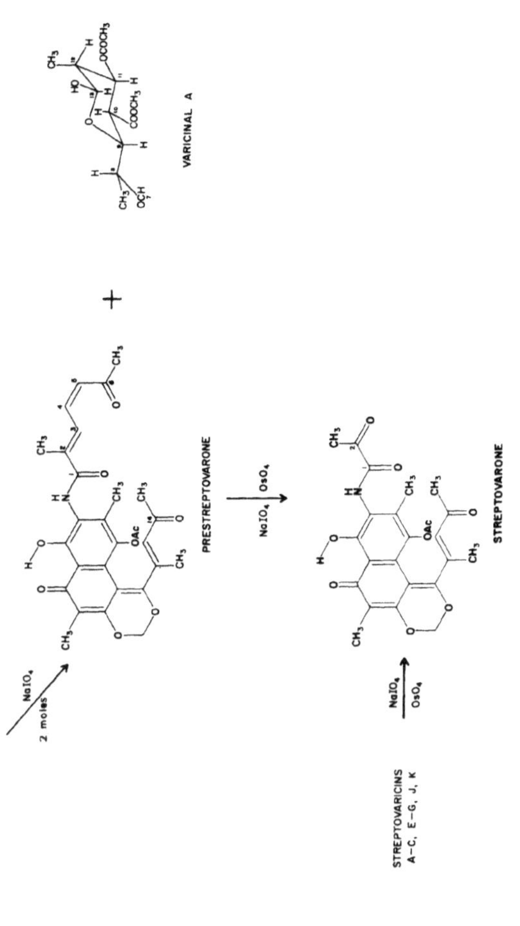

Fig. 6. Streptovals and related degradation products from the streptovaricins

written for streptovarone, in which one of the oxygen atoms must be replaced by a nitrogen.

An 8-hydroxy-1,4-naphthoquinone nucleus was established in (**a**) by spectroscopic properties and pK_a of streptovarone, while the pyruvamide

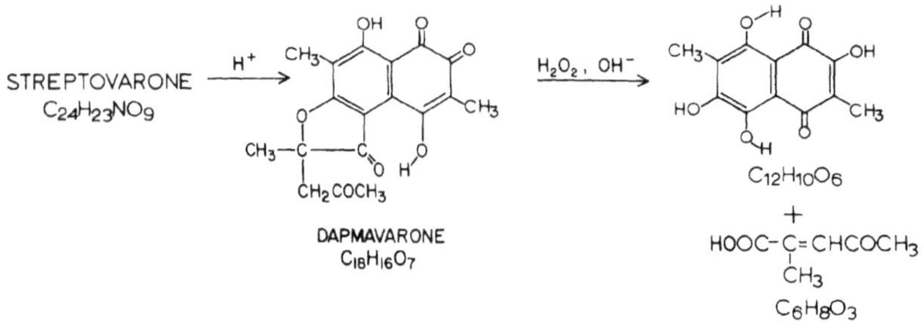

Fig. 7. Hydrolysis and oxidation products derived from streptovarone

lost in converting streptovarone to dapmavarone was located at C-2 of (**a**) by analogy to the behavior of known 2-amino-1,4-naphthoquinones. A ketone was located at C-4′ (in addition to the ketone in the pyruvamide

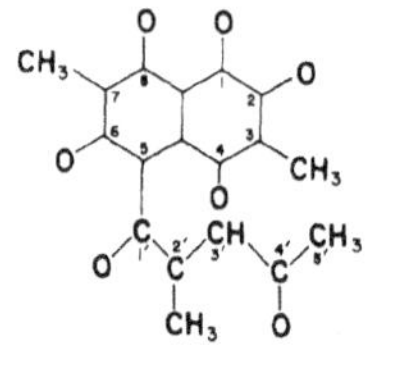

a

group) by the nmr properties of a dioxime (Fig. 8). The enol acetate was located at C-4 of (**a**) by a curious reaction, the loss of molecular acetic anhydride (identified by infrared spectroscopy and gas chromatography)

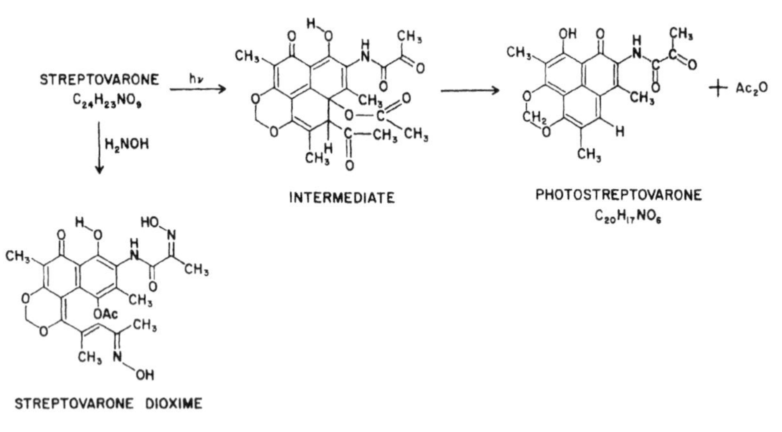

Fig. 8. Products derived from streptovarone

References, pp. 300—307

on photolysis of streptovarone to give photostreptovarone (Fig. 8) (*116*), a yellow 8-hydroxyphenalenone, identified by its spectroscopic properties. The methylenedioxy group, which is retained in photostreptovarone, must then link O-6 and O-1′ of (**a**), the only two remaining oxygenated positions available. Prestreptovarone and streptoval C are converted to similar photolysis products.

With the structure of streptovarone assigned, that of prestreptovarone follows automatically from the structure of the dienamide as that shown in Fig. 6. Linkage of the remaining aliphatic portion of the streptovaricins (that corresponding to varicinal A) must be to the two aliphatic ketone carbons (C-6 and C-14 in the streptovaricins). Extensive spin decoupling carried out on streptovaricins C, D and G demonstrated the symmetrical nature of C-7 through C-13 in those components, while shifts of H-7 or H-11 on acetylation or lactonization, or loss of H-7 on oxidation, completed the structural assignments of streptovaricins A, B, E, F, J and K. Additional structural evidence regarding the aliphatic side chain was derived from osmium tetroxide-sodium periodate oxidation of streptovaricin C triacetate, which gave the dialdehydes (**7**) and (**8**), and of streptoval C triacetate, which gave (**9**), as well as from ruthenium tetroxide oxidation of streptovaricin C tetraacetate, which gave (**10**) (*113*).

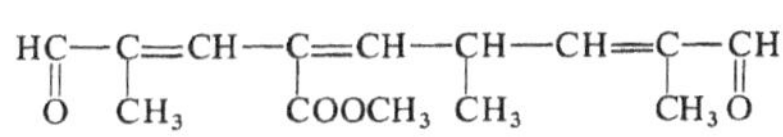

(**7**)

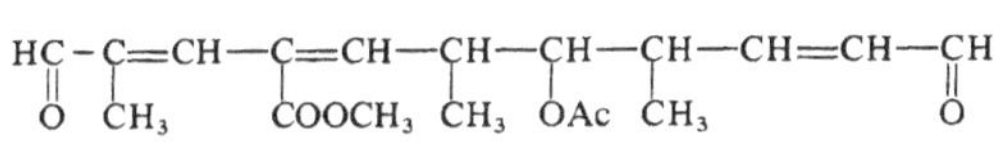

(**8**)

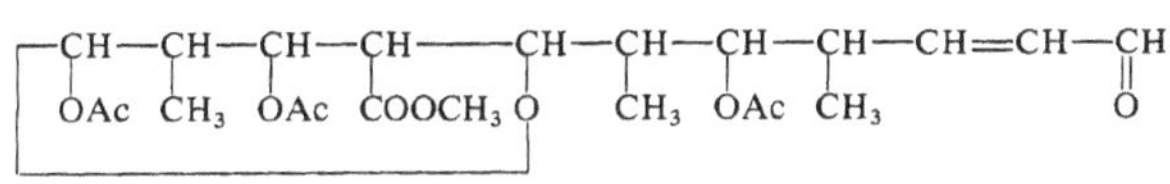

(**9**)

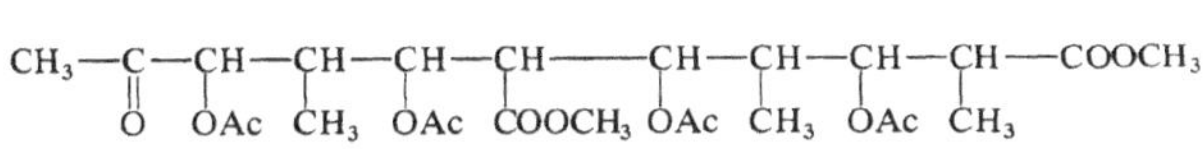

(**10**)

2. *Derivatizations*

Surprisingly little effort has been expended in attempts to prepare derivatives of streptovaricins with improved chemotherapeutic properties, compared to the enormous (and successful) effort directed toward rifamycin B (*7, 16, 17, 45*). The simplest reaction employed has been thermal isomerization, a reaction in which individual streptovaricins are converted to their atropisomers by heating in an inert solvent (*105*). Streptovaricin C, in which the ansa bridge lies above the aromatic nucleus, is converted in refluxing xylene to atropisostreptovaricin C, in which the ansa bridge lies below the aromatic nucleus. Similar atropisomers can be prepared from other streptovaricins. The equilibrium, which can be approached from either isomer, favors the natural isomer, except for lactonized streptovaricins (F_C, F_G), where the atropisomer is greatly favored. A slightly more complicated reaction occurs if the reaction is carried out in pyridine, since the refluxing base promotes cyclization (*e.g.* of streptovaricin C to streptovaricin F_C), as well as atropisomerization, as shown in Fig. 9.

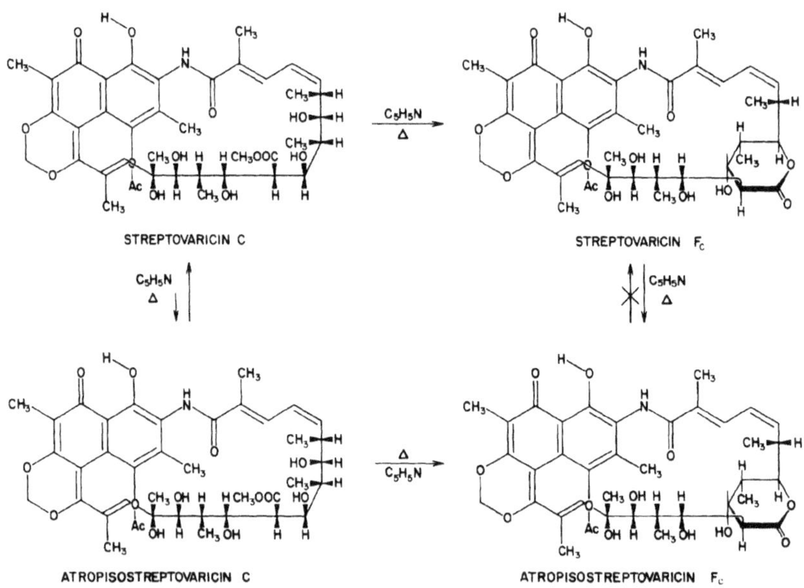

Fig. 9. Atropisomers of streptovaricins

Most spectral properties (uv, ir, pmr, mass spectral) are nearly identical for the isomers. The natural isomers show intense positive rotations at the *D* line of sodium, while the atropisomers show equally intense negative rotations (streptovaricin C, $[\alpha]_D$ +602°; atropisostreptovaricin C, $[\alpha]_D$ −551°), and the circular dichroism curves of the atropisomers are nearly mirror images, as shown in Fig. 10. Surprisingly, the biological properties

References, pp. 300—307

of the atropisomers are very similar to those of the streptovaricins. Atropisostreptovaricin C is slightly less antibacterial than streptovaricin C, but somewhat more effective in inhibiting reverse transcriptase (*105*).

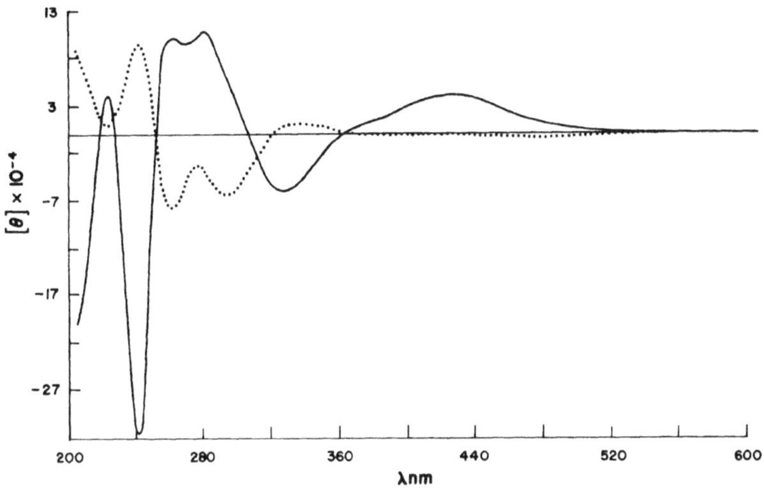

Fig. 10. Circular dichroism curves for streptovaricin C (——) and atropisostreptovaricin C (·······)

Derivatives of streptovaricins also can be converted to their atropisomers, sometimes at lower temperatures than the streptovaricins themselves. In refluxing benzene an equilibrium mixture of streptovaricin C triacetate and its atropisomer is formed. This reaction led to a complication in assigning the structure of streptovaricin C by x-ray crystallography (*149*), since the x-ray study was carried out on a cyclic *p*-bromobenzeneboronate formed by heating streptovaricin C triacetate in benzene with *p*-bromobenzeneboronic acid and the atropisomeric cyclic *p*-bromobenzeneboronate was formed along with the natural isomer. The compound chosen for x-ray analysis was the cyclic *p*-bromobenzeneboronate (**1**) of atropisostreptovaricin C triacetate; thus, the structure of natural streptovaricin C is that shown in Fig. 1. Like the atropisomers of the streptovaricins themselves the atropisomers of their derivatives have biological properties resembling those of the natural isomer derivatives (*47, 56*).

Lactonization is another simple reaction of the streptovaricins, by which they are converted to the corresponding streptovaricin F. This reaction was first noted in Section II.A. in the conversion of streptovaricin G to streptovaricin F_G and encountered again as a side reaction in Fig. 9 where heating streptovaricin C in pyridine gives streptovaricin F_C. Lactonization reduces greatly whatever biological activity any streptovaricin possesses.

The most extensively studied streptovaricin derivatives are the streptovaricin acetates (*103*). As noted in Section II.A., acetylation was a key reaction in interrelating the structures of the streptovaricins. Since streptovaricins A, G and K all give the same compound, streptovaricin G triacetate (streptovaricins A and K diacetate), on acetylation, and streptovaricins A and K have one more acetate group than streptovaricin G, streptovaricins A and K are monoacetates of streptovaricin G. Similarly, streptovaricins C, B and J give the same compounds, streptovaricin C tri- and tetraacetates (streptovaricins B and J di- and triacetates), and streptovaricins B and J have one more acetate group than streptovaricin C. The positions of the acetates in streptovaricins A, B, J and K were located by shifts of appropriate protons in magnetic resonance spectra.

Eight streptovaricins (A—G and J) and three synthetic streptovaricin acetates, streptovaricin G triacetate (streptovaricins A and K diacetates) and streptovaricin C tri- and tetraacetates (streptovaricins B and J di- and triacetates), have been assayed for their antibacterial spectra and activity against *E. coli* DNA-dependent RNA polymerase. The relative *in vitro* antibacterial activities of the individual streptovaricins parallel their relative potencies as inhibitors of bacterial DNA-dependent polymerase. There are relatively minor differences between the biological activities of the unacetylated streptovaricins (C and G) and their monoacetylated analogs (streptovaricins A, B, J and K). A C-7 hydroxyl group appears to be a desirable, if unnecessary, feature for antibacterial activity (since streptovaricins J and K are less active than the other four) and a hydroxyl group at C-14 seems advantageous (since streptovaricin D is less active than the others). It is possible that the C-9 hydroxyl is the most important ansa chain feature for antibacterial activity and RNA polymerase inhibition, since the more fully acetylated streptovaricins showed no antibacterial activity and no activity against *E. coli* RNA polymerase.

Although multi-acetylation reduced antibacterial activity, when streptovaricin C tri- and tetraacetates and streptovaricin G triacetate were tested against the reverse transcriptase of Rauscher leukemia virus (RLV), the acetylated compounds proved to be more active than the unacetylated or mono-acetylated streptovaricins. The atropisomers of the streptovaricin oligoacetates are also inactive as antibacterial agents or inhibitors of *E. coli* RNA polymerase but more active in inhibiting reverse transcriptase. The cyclic benzeneboronate and *p*-bromobenzeneboronate derivatives of streptovaricin C (**11**, **13**, respectively) and atropisostreptovaricin C (**12**, **1**, respectively) triacetates, shown in Fig. 11, also inhibit reverse transcriptase but are inactive as antibacterial agents.

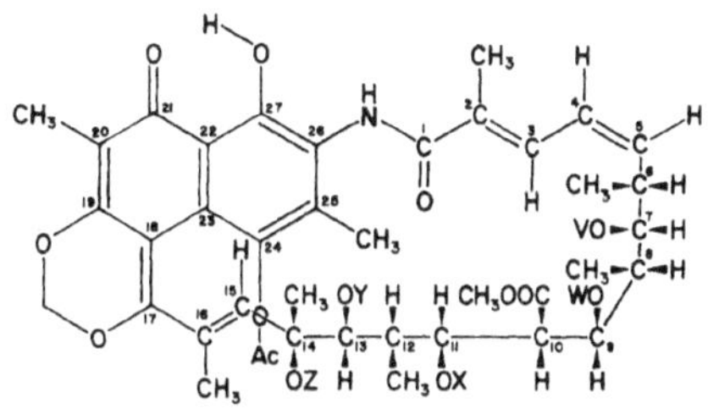

(**11**): V = W = X = Ac, Y + Z = $>BC_6H_5$

(**12**): atropisomer of (**11**)

(**13**): V = W = X = Ac, Y + Z = $>BC_6H_4Br(p)$

(**1**): atropisomer of (**13**)

Streptovaricin C 7,9-acetonide: V + W = $C(CH_3)_2$, X = Y = Z = H

Streptovaricin C 9,11-acetonide: W + X = $C(CH_3)_2$, V = Y = Z = H

Fig. 11. Derivatives involving the hydroxyl groups of streptovaricin C and atropisostreptovaricin C

Acetonides of streptovaricin C have also been prepared (Fig. 11) (*47*). Their structures were assigned by acetylation and comparison with the acetonides prepared from streptovaricins B and J. The acetonides are biologically active as inhibitors of both bacterial RNA polymerase and reverse transcriptase (RT).

Hydrogenation of streptovaricin C over a palladium-charcoal catalyst yielded 2,3-dihydrostreptovaricin C and 2,3,4,5-tetrahydrostreptovaricin C (*47*). These compounds retain biological activity in inhibiting bacterial RNA polymerase and reverse transcriptase.

Recently, Sasaki, *et al.* (*112a*), have prepared a series of 19-O-alkyl derivatives (**14**) of damavaricin C (Fig. 5) by reaction of damavaricin C with alkyl halides in the presence of silver oxide. Most of the derivatives have greater activity than damavaricin C against *Bacillus subtilis* and *Mycobacterium smegmatis* and many have enhanced activity against *Staphylococcus aureus* as well. Activity was also observed against membrane mutants of *E. coli* but not against the wild strain of *E. coli*.

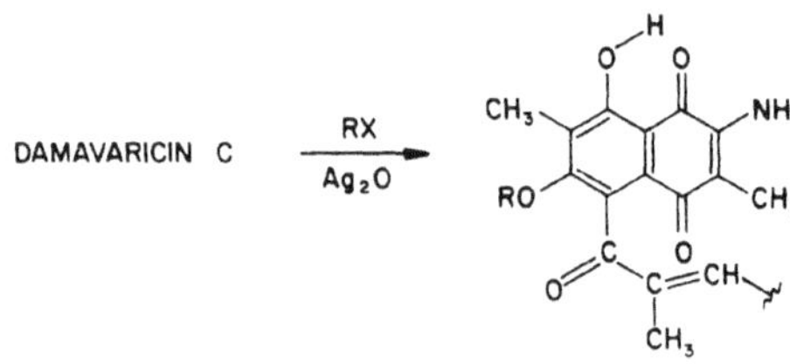

(14)

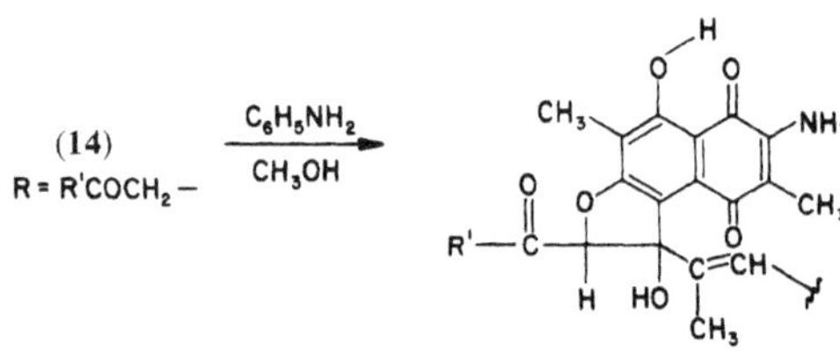

(15)

Some of the 19-O-alkyl derivatives, containing keto groups, undergo aldol cyclization with the 17-keto group, giving products of the general structure (**15**) (*112b*). The latter compounds are reported to be antibacterial, with activities approximating those found for the non-cyclized compounds.

In addition to the derivatives above, the streptovals (Fig. 6) and their acetates have been extensively investigated for their biological activity (*70*). They are mostly inactive as antibacterial agents or inhibitors of *E. coli* RNA polymerase, but show dramatically improved inhibition of reverse transcriptase and are also active in inhibiting tumor viral plaque formation. Streptoval C, the most readily obtained, is quite active and *aldehydo*-streptoval C triacetate and streptovadienal are especially active.

B. Rifamycins

1. Degradations

As noted in Section II.B., the structures of rifamycins B and Y rest on x-ray analyses of their *p*-iodoanilides (*9, 11, 12*). However, a parallel investigation of rifamycin B by chemical degradation and spectroscopic

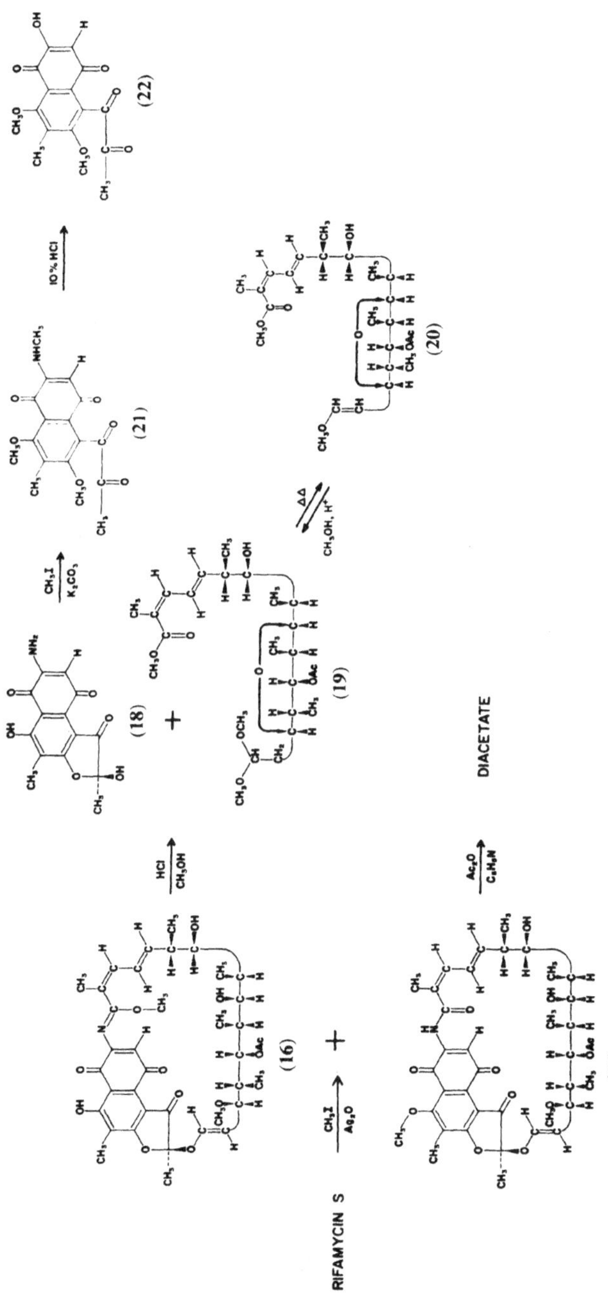

Fig. 12. Degradation of rifamycin S *via* its imino methyl ether (**16**)

characterization assigned the same structure (*67, 93, 94, 97*). Key reactions and products involved in the chemical investigation will now be described.

The chemical study was carried out on rifamycin S, derived from rifamycin B by the series of simple reactions shown earlier in Fig. 3. The key degradation of rifamycin S began with its methylation (Fig. 12) with methyl iodide and silver oxide to yield two isomeric derivatives, the imino methyl ether of the amide function (**16**) and the 21-O-methyl ether (**17**). The diacetyl derivative formed from the monomethyl ether (**17**) showed no infrared –OH bands, indicating a total of three hydroxyl groups for rifamycin S. The imino methyl ether (**16**), unlike rifamycin S, could be cleaved by mild methanolysis (Fig. 12), which yielded mainly two products, the aromatic naphthoquinone derivative (**18**), and the aliphatic dimethyl acetal methyl ester (**19**), which could be interconverted with its enol methyl ether (**20**). Molecular formulas of (**18**) and (**19**) showed that these two compounds contained the entire carbon skeleton of the rifamycin molecule.

The structure of the aromatic portion was deduced in part by methylation of (**18**), which yielded two di-O-methyl derivatives (the mono-N-methylated **21** and the corresponding di-N-methylated compound), thus establishing two hydroxyl groups in (**18**). Replacement of the methylated amino group of (**21**) by hydroxide ion gave (**22**), establishing that the amino group was on a quinone ring.

Further information regarding the aromatic chromophore of (**18**) came from vigorous ozonolysis of rifamycin S (Fig. 13), which gave acids (**23**) and (**24**). The structure of the tricarboxylic acid (**24**) was established by comparison of its dimethyl ether trimethyl ester (**25**) with

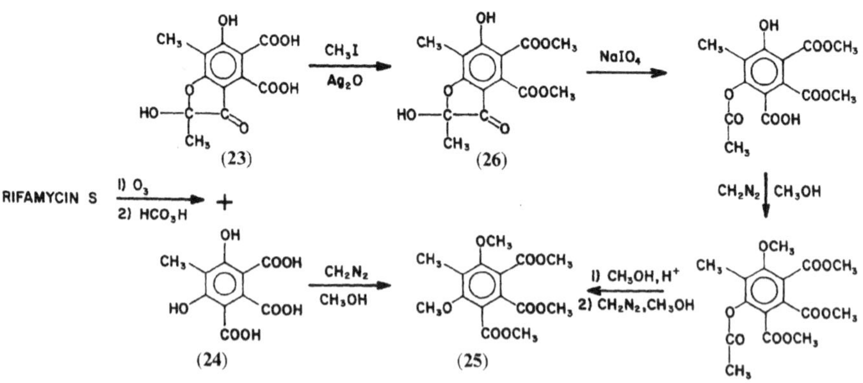

Fig. 13. Degradation of rifamycin S *via* oxidative ozonolysis

the same compound prepared from the known dimethyl ether tricarboxylic acid, while the dicarboxylic acid (**23**) could be converted to (**25**) by the series of reactions shown in Fig. 13. Cleavage of (**26**) by periodate requires an α-hydroxy ketone in (**23**), while the hemiketal structure of (**26**) requires attachment of its acyl side chain to a carbon *ortho* to one of the aromatic hydroxyls.

Assignment of structures to the ozonolysis acids (**23**) and (**24**) (Fig. 13) was nearly sufficient to establish the constitution of (**18**). A *p*-quinonoid system must be present in (**18**), because rifamycin S did not react with lead tetraacetate as it would if an *o*-quinonoid system were present, and, as noted above, the amino group must be on the quinone ring. The only remaining point is the precise location of the nitrogen atom, whether it be at C-2 or C-3 of the quinone ring of (**18**) (C-26 or C-25, respectively, of the original rifamycin S). This was settled by the reaction sequence of Fig. 14.

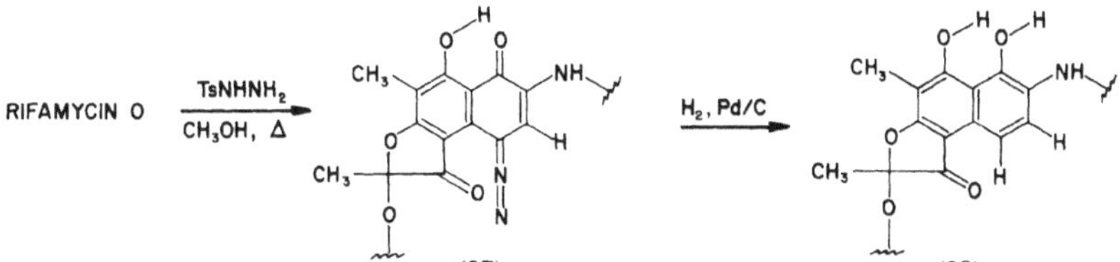

(**27**) (**28**)

Fig. 14. Removal of a quinone oxygen from rifamycin O

Catalytic hydrogenation (Fig. 14) of the diazo product (**27**) of the reaction of rifamycin O with *p*-toluenesulfonylhydrazine yielded the dihydroxynaphthalene derivative (**28**), whose spectral and chemical properties indicated the *peri*-relationship of the C-21 and C-27 hydroxyl groups and the presence of the two *ortho* aromatic protons. The latter structural feature required the amide group to be placed at C-26.

Study of the aliphatic portion of rifamycin S commenced with the observation that it contains an α-methyldienamide. This follows from comparison of the ultraviolet spectra of rifamycin S and tetrahydrorifamycin S (Fig. 15), which showed that the amide carbonyl was conjugated with two double bonds, and from characterization of (**19**), whose ultraviolet spectrum was that of a methyl dienoate and whose proton magnetic resonance spectrum showed olefinic protons at C-3, C-4 and C-5. Products isolated from the vigorous oxidation of tetrahydrorifamycin S (Fig. 15) indicated the substitution pattern of the aliphatic bridge

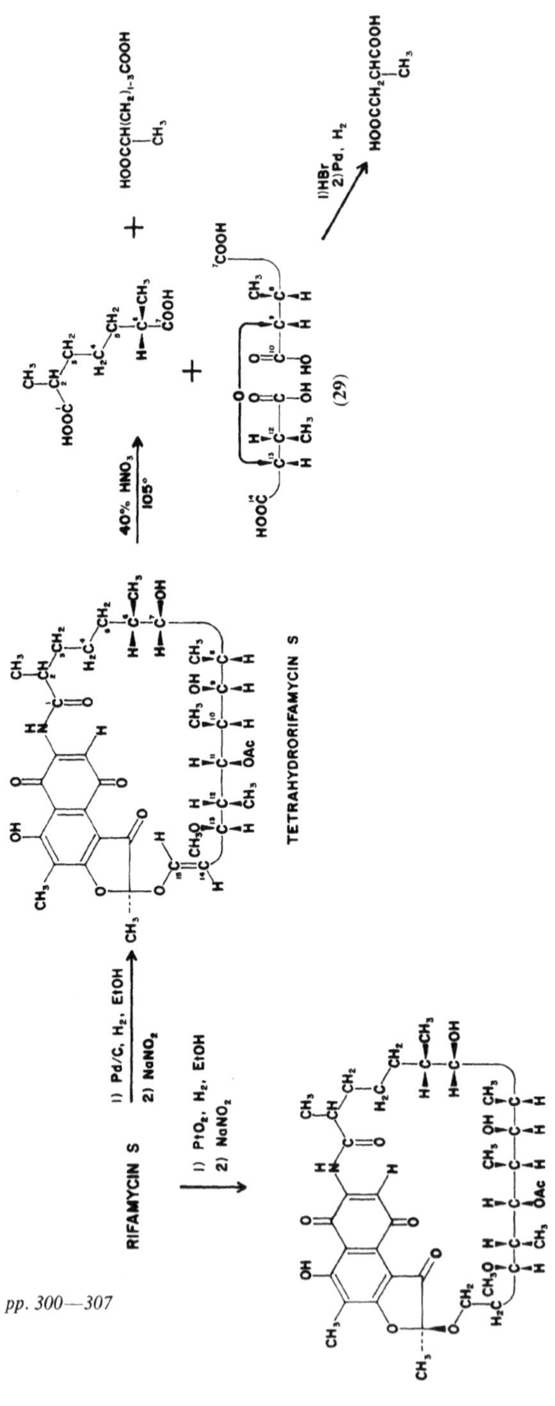

Fig. 15. Hydrogenation of rifamycin S and oxidation of tetrahydrorifamycin S

between C-1 and C-14. A series of α-methyl aliphatic diacids was obtained, including *meso*- and (+)-forms of α,α′-dimethylpimelic acid. These diacids established the substitution pattern at C-1 through C-6 of the chain and indicated the absolute configuration at C-6, with the implication being that tetrahydrorifamycin S was a mixture of two products, epimeric at C-2, which could not be separated. Assignment of the absolute configuration at C-6 was especially important, since it allowed the complete absolute configuration of rifamycin B to be assigned from the x-ray structure determination of (**2**).

Assignment of substitution at C-7 through C-14 involved the structurally symmetrical tetracarboxylic acid (**29**) (Fig. 15), whose structure was assigned from its nuclear magnetic resonance spectrum and by conversion to methylsuccinic acid by treatment with hydrogen bromide and subsequent hydrogenation. The tetracarboxylic acid could best be explained as arising from oxidative cleavage between C-10 and C-11 of a tetrahydropyran ring formed between C-9 and C-13 in acidic solution. The structure of (**29**) established methyl groups at C-8 and C-12 and oxygen atoms at C-9 and C-13.

These conclusions were supported by additional degradative experiments carried out on the methanolysis product (**19**) (Fig. 16). Oxidative ozonolysis of (**19**), followed by diazomethane treatment, gave (**30**) with loss of C-1 through C-5 and the C-2 methyl group, thus confirming the dienoate structure. Compound (**30**) was converted to its propionate, which was pyrolyzed to the α,β-unsaturated ester (**31**), establishing a hydroxyl in the β-position of (**30**). Oxidative ozonolysis of (**31**) followed by diazomethane treatment gave (**32**), which was methanolyzed to remove the acetate and oxidized to the ketone (**33**). The pmr spectrum of (**33**) located the keto group at the central position of the pyran ring, hence the original acetate group at C-11 of (**30**). When considered together with the structure of the tetracarboxylic acid (**29**), the location of the acetate completed the structural assignment of (**32**), hence of (**31**), (**30**) and (**19**). It should be noted that the terminal carbomethoxyl group of (**30**) appears as a dimethyl acetal in (**19**), which was identified both in the pmr spectrum of (**19**) and by its interconversion with an enol ether group in (**20**) (Fig. 12).

One point of attachment in rifamycin of the aliphatic chain (**19**) to the aromatic chromophore (**18**) was determined directly, in that the dienoate of (**19**) is the dienamide of rifamycin S. The second point of attachment was assigned less directly, as follows.

Rifamycin S contains three hydroxyl groups, a phenolic hydroxyl which can be methylated and two aliphatic hydroxyls which can be acetylated, as noted above. Only one aliphatic hydroxyl is present in (**19**); thus, the other has cyclized to the tetrahydropyran ring and is either on C-9 or C-13. Comparison of the pmr spectra of rifamycin S and hexa-

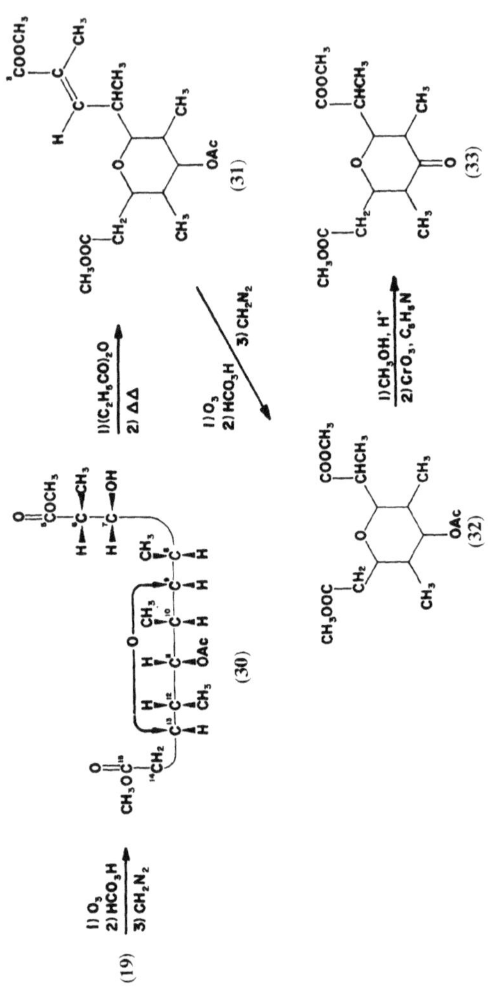

Fig. 16. Degradation products from the aliphatic portion (**19**) of rifamycin S

hydrorifamycin S, which was obtained by hydrogenation of rifamycin S over platinum oxide (Fig. 15), shows that the terminal two carbons of the aliphatic chain (**19**) are an enol ether in rifamycin S ($-O-CH=CH-$). The vinyl ether oxygen could be attached either to a methyl group or to the aromatic nucleus, but the position of the methoxyl protons in the pmr spectrum of rifamycin S corresponds to that of a saturated ether and does not change in the spectrum of hexahydrorifamycin S. Consequently, the vinyl ether is attached to the aromatic nucleus and the methoxyl is at C-9 or C-13, with a hydroxyl on the other carbon, as noted above. A decision is reached in favor of C-13 for the methoxyl, since that position is allylic, which would explain the ready loss of methoxyl by cyclization, *e.g.*, to give (**19**). Location of the methoxyl completed the assignment of structure to rifamycin S by chemical means.

2. Derivatizations

As noted at the outset of this review, hundreds of compounds derived from rifamycins B, S and SV have been prepared (in the laboratories of Lepetit S.p.A. and CIBA-Geigy AG) in attempts to optimize the activity of the rifamycins against Gram-positive bacteria, Gram-negative bacteria, and *M. tuberculosis*, to improve absorption and retention behavior, and to lower toxicity (*5, 28, 32, 58—60, 78, 119, 137*). Some of the more promising derivatives have been tested *in vivo* and several have been put to therapeutic use. Many of the same rifamycin derivatives have also been tested as inhibitors of tumor virus reverse transcriptases (RNA-dependent DNA polymerases) and as potential antitumor agents (*38—40, 131, 132, 138, 139, 143, 145, 146, 163*). In general these compounds do not inhibit mammalian polymerases. To list all these derivatives would require a separate review and most of them have, in fact, been included in reviews elsewhere (*4, 92, 111, 119, 120, 126, 151—153*). What we shall do here is to show the general types of derivatization reactions of the rifamycins, giving some indication of whether bioactivity was enhanced or diminished in the products.

Before turning to a detailed discussion of the rifamycin derivatives, however, we must first note that three distinct rifamycins – rifamycins B, S and SV (Table 1, Figs. 1, 3) – have been employed as starting materials for the derivatizations. It had been noted that the antibacterial activity of the original rifamycin mixture produced by *N. mediterranei* increased in aqueous solution in the presence of atmospheric oxygen (*122*). It was determined that the transformation in aqueous solution of rifamycin B to rifamycins O and S (Fig. 3) caused this "activation" and that rifamycin B itself has little *in vivo* antibacterial activity (*30, 32, 75, 126, 128, 140*). Rifamycin S is not suitable for therapeutic use (*126*), but its reduction product, rifamycin SV (Rifocin) (*4, 31, 72, 74, 141, 142*), has good physicochemical and biological properties for therapeutic purposes and has been used for the parenteral treatment of tuberculosis, of infections due to Gram-positive bacteria and of infections of the biliary tract.

The simplest large class of derivatives was that derived from rifamycin B by conversion of its carboxyl group to various amides, hydrazides and esters (*99, 119, 125*). Many of these derivatives possess high antibacterial activity, thus supporting the hypothesis (*125*) that the free carboxyl group causes the inactivity of rifamycin B *in vivo*. Very high *in vivo* antibacterial activity and low toxicity were shown by the N,N-diethyl amide of rifamycin B (Fig. 17), Rifamide (*24, 29, 73, 77, 95, 126*), which has been used clinically. The tertiary amides are more active than the esters or the primary or secondary amides, but diminished antibacterial activity was noted for tertiary amides substituted with hydroxyalkyl, chloroalkyl and related groups (*126*).

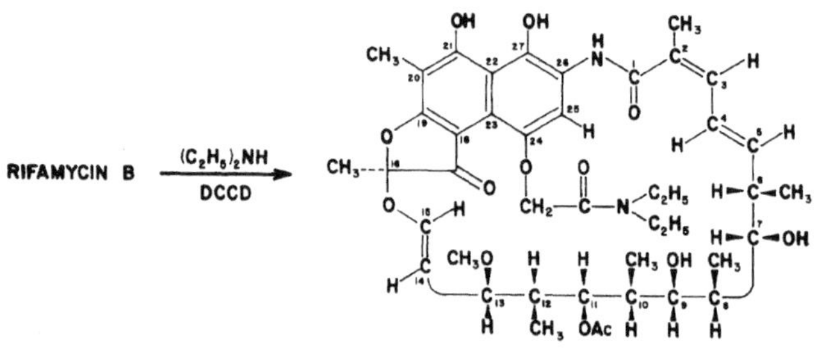

Fig. 17. Conversion of rifamycin B to rifamide, a clinically useful derivative

A second large group of rifamycin derivatives has been obtained from rifamycin S. The C-24 carbonyl group can be selectively condensed with some types of substituted amines under conditions where the hydrogen-bonded C-27 carbonyl group is inert. Rifamycin O can also be used as starting material in some of these reactions due to its facile hydrolysis to rifamycin S.

Derivatives of rifamycins S and SV of the quinonimine type (Fig. 18), formed by condensation of rifamycin O with aromatic amines (**34**), hydrazides (**35**), amidrazones (**36**) and aminoguanidines (**37**) (*37*), showed high *in vitro* activity against Gram-positive bacteria but poor *in vivo* absorption (*37, 126*). The unsubstituted aminoguanidine derivative (**38**, Fig. 18) displayed especially good broad-spectrum antibiotic properties but was not suitable for oral or parenteral use.

Simple aliphatic primary and secondary amines and ammonia undergo conjugate addition to the quinone ring instead of direct addition to the quinone carbonyl, giving C-25 substituted derivatives of rifamycin SV (**39**)* (*5, 60*). Some of these derivatives have excellent antibacterial

* Two recent reviews of derivatives of rifamycin employed as inhibitors of reverse transcriptase have shown structures of 24-dialkylamino substituted rifamycins. However, no references were given to their preparation and it seems possible they may have been 25-dialkylamino-substituted hydroquinones, instead (*40, 163*).

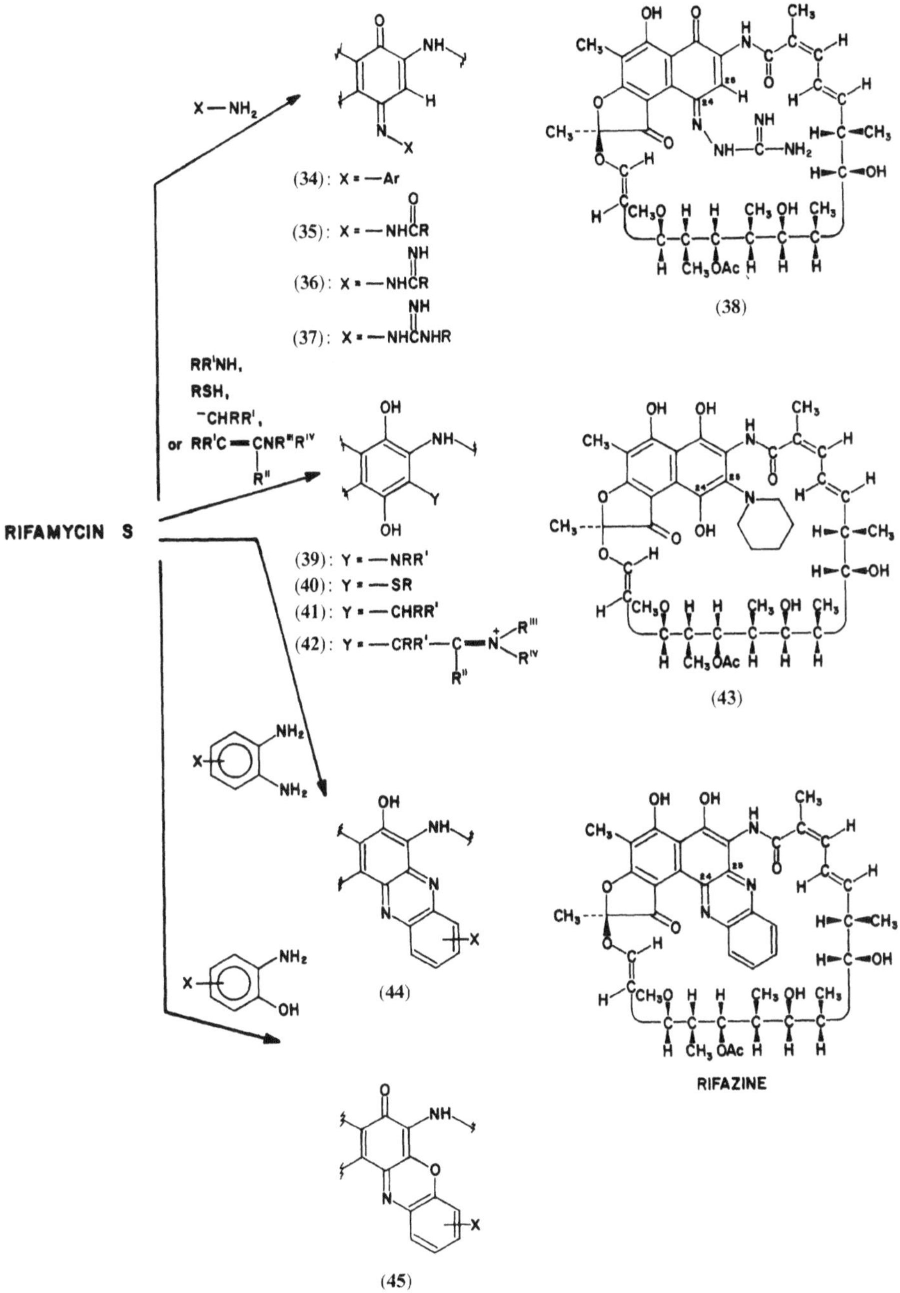

Fig. 18. Rifamycin S and SV derivatives produced by quinonimine formation, conjugate addition or both

properties (*5, 58*) and other 25-alkylamino and C-25 heterocyclic derivatives of rifamycin SV showed strong inhibition of RNA tumor virus polymerases (*38*). Compounds with simple 25-alkylamino substituents and those with C-25 heterocycles containing more than one heteroatom in their rings were inactive, but the 25-piperidyl derivative (**43**) was inhibitory. Addition of polar substituents to the piperidyl ring destroyed its activity, but nonpolar substituents, especially cyclohexyl and cyclohexylalkyl groups, greatly enhanced the polymerase inhibition.

Other nucleophiles, including mercaptans, carbanions, and enamines (Fig. 18) (*60*), can also add in a conjugate fashion to the quinone ring but the derivatives obtained have not proved to be useful as antibacterial agents. The thioalkyl derivatives (**40**) were quite active *in vitro* but were not orally absorbed, while the carbanion addition products (**41**) were good inhibitors of bacterial DNA-directed RNA polymerase but were relatively inactive on bacteria. The enamine products (derived by cyclization of **42**) were quite complex and their activity does not appear to have been reported.

A combination of the two reactions (quinonimine formation, conjugate addition) is observed in the reaction of rifamycin S (Fig. 18) with aromatic 1,2-diamines, which gives phenazines (**44,** X = alkyl, halo, or condensed aromatic ring) (*5*), or with *o*-aminophenols, which gives phenoxazinones (**45,** X = alkyl, chloro, nitro) (*5*). Rifazine (*35*), the simplest of the phenazine series, showed high antibacterial activity but disrupted liver function. Other derivatives with this phenazine or phenoxazinone substitution pattern at C-24 and C-25 showed enhanced antibacterial activity in both the quinonoid and hydroquinonoid forms, and two of them (like **44,** but derived from 1,2-diaminonaphthalene, and like **45,** but derived from 1-amino-2-naphthol) gave weak inhibition of RNA-directed DNA polymerase and DNA-directed DNA polymerase activity of RNA tumor viruses of murine, feline, and avian origin (*38*).

The most numerous C-25 substitution products are those derived from the 25-aminomethyl derivatives (**46**) (*76*) of rifamycin SV obtained by the Mannich reaction of rifamycin S with secondary amines in the presence of formaldehyde (Fig. 19)*. The aminomethyl derivatives upon oxidation in acidic medium give 25-formylrifamycin SV (rifaldehyde), the starting point for preparation of many of the best rifamycin derivatives (**47—52**) (*28, 78, 126*). The most active rifaldehyde derivatives are generally N,N-disubstituted hydrazones (**48**) and O-alkyl oximes (**49**), such as those shown in Fig. 19 (*21*). The O-methyl oxime and the hydrazone formed from N-amino-N′-methylpiperazine were especially potent and

* Although rifamycin S is employed, rifamycin SV may be assumed to be an intermediate (at least in an equilibrium mixture) in the electrophilic substitution reaction.

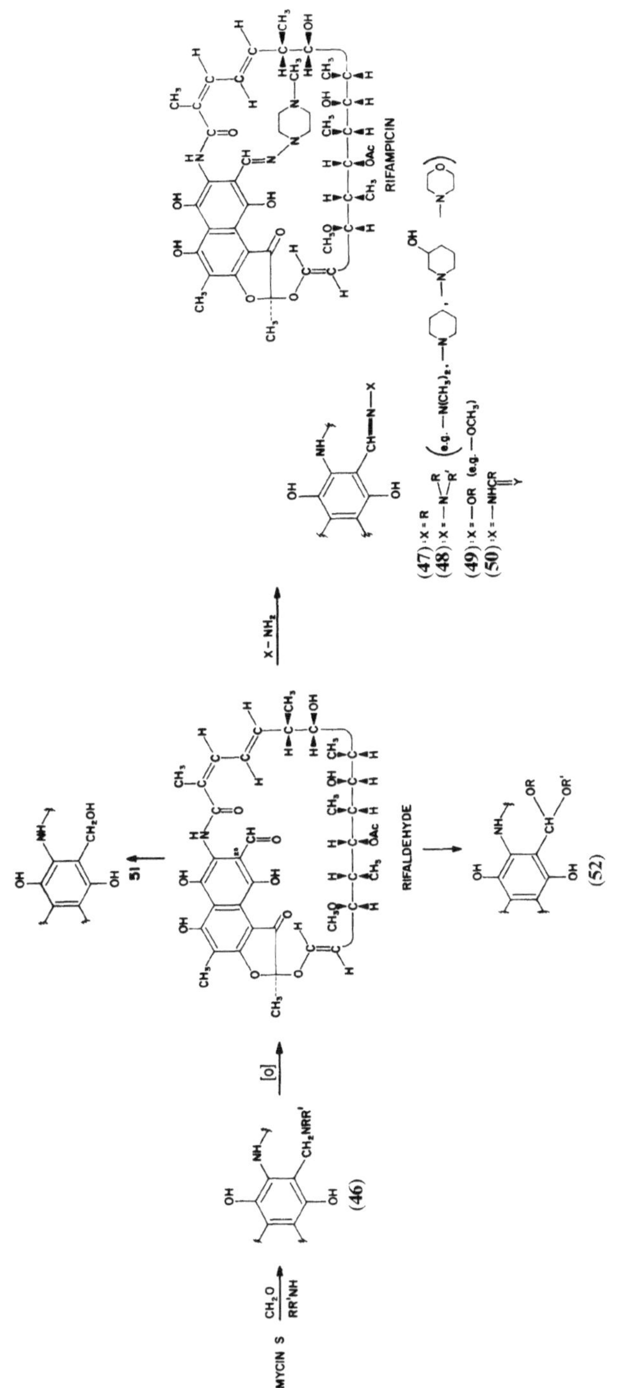

Fig. 19. Compounds derived from rifamycin S or SV *via* aminomethylation of C-25

could be administered orally *in vivo*. The latter compound, called rifampin in the United States and rifampicin elsewhere, has been marketed widely under the trademarks Rifadin (Lepetit) and Rimactane (CIBA-Geigy) (*69, 79, 80*). It is this derivative which accounts for the commercial success of rifamycin.

Many of the rifaldehyde derivatives have also been screened for inhibition of the RNA-directed DNA polymerases (reverse transcriptases) of murine sarcoma virus (MSV) and Rauscher murine leukemia virus (RLV) and the purified DNA polymerases from human normal and leukemic lymphoblasts (*40, 163*). Surprisingly perhaps, the best compounds shown in Fig. 19 (**48, 49,** rifampicin) were inactive. Further testing of derivatives of rifaldehyde indicated that a longer, bulkier side chain is needed at C-25 for inhibition of viral polymerases.

The alkyl substituted hydrazones of rifaldehyde, which have good antibacterial activity, are not effective against MSV reverse transcriptase but their analogs with large heterocyclic rings and bulky side chains are effective (*40*). When the methyl group of the inactive compound rifampicin is replaced by benzyl (or 2-phenylethyl or 3-phenylpropyl or hydrogen), viral polymerases are inhibited. Diphenyl, dinitrophenyl, cycloheptyl, and cyclooctyl substituted hydrazones (**48**) are also excellent inhibitors. Among the homologs of the O-methyl oxime of rifaldehyde (**49**), those with ethyl and propyl side chains are also inactive, but the *n*-butyl derivative is a good inhibitor of RNA-directed DNA polymerase, the *n*-pentyl and *n*-octyl compounds are better, and benzyl, 2-phenylethyl, 3-phenylpropyl and benzyhydryl O-alkyl oximes are especially potent. Among some 180 rifamycin derivatives tested against MSV polymerase, all except one of the ten most active inhibitors (active at 5—20 μg/ml) were rifamycin SV derivatives with long, bulky C-25 chains and 15 of the 16 next most active derivatives (inhibitory at 20 μg/ml) had large C-25 side chains. On the other hand, 155 rifamycins which did not inhibit the MSV polymerase even at the 100 μg/ml level included 20 rifamycin SV derivatives with relatively short C-25 side chains, 18 24-deoxy-24,25-substituted rifamycin SV derivatives, 16 24-deoxy-24-substituted rifamycin SV derivatives and 44 24-substituted rifamycin B derivatives (*40*).

Condensation of rifaldehyde with larger hydrazines produces the rifazacyclo (*146*) and the rifazone (*6, 146*) series of derivatives. Of these, rifazacyclo-16 (**48,** $R+R' = -(CH_2)_{15}-$) is one of the most potent and rifazone-8_2 (**48,** $R = R' = n-C_8H_{17}-$) one of the most selective inhibitors of the viral polymerase *in vitro*.

Other compounds of interest have been obtained by acylating N-demethylrifampicin (see Fig. 19) and rifaldehyde hydrazone with diacid derivatives to give a series of rifamycin dimers (*26, 145*). A simpler dimeric material, the azine of rifaldehyde (from two moles of rifaldehyde and one of hydrazine) – called rifamazine (*26*), has been extensively studied recently and shown to inhibit both DNA-dependent RNA polymerase and reverse transcriptase.

Aside from the derivatives described thus far, involving substitution at C-24 and C-25, almost no derivatives of the rifamycins have been found to have superior biological properties. Mono-acyl derivatives of rifamycins B, O, S and SV derived by acylating the phenolic hydroxyl at C-21 showed only very slight changes in antibacterial activities from those of the parent compounds (*119, 122, 124, 126*). On the other hand, acetylation of both the hydroxyl groups on C-7 and C-9 gave an inactive derivative and the C-7 hydroxyl also appears to be required for activity from the results with rifamycin Y (Fig. 1), which is inactive (*9, 152*).

Alkali has been used to deacetylate rifamycin S and various rifamycin derivatives at C-11 (*5, 81*). Both the quinone and hydroquinone forms of rifamycin S and the derivatives were tested. In every case, antibacterial activity was retained, though reduced relative to the acetates. Other modifications of the aliphatic ansa bridge of the rifamycins also tend to destroy or decrease antibacterial activity (*5, 80, 126, 152*). Preparation of many of these compounds was already described in Section III.A., since they were involved in the structural studies on rifamycin. Thus, tetrahydrorifamycin S (Fig. 15) and tetrahydrorifamycin SV had slightly reduced antibacterial activity relative to the parent compounds, while hexahydrorifamycin S (Fig. 15) and hexahydrorifamycin SV showed greatly reduced activity (*5*). Epoxidation of the diene system by monoperphthalic acid gave the 2,3- and 4,5-monoepoxides plus the 2,3,4,5-diepoxide of rifamycin S, all of which have strongly reduced antibacterial activity (*5*). The iminomethyl ether (**16**) was completely inactive (*126*).

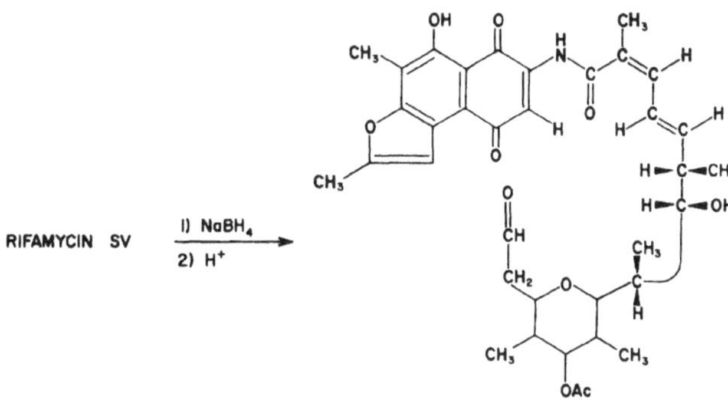

Fig. 20. Cleavage of the ansa ring to give rifarubin S

Mild acidic hydrolysis of various rifamycins brought about replacement of the allylic methoxyl group at C-13 by the hydroxyl oxygen at C-9 to give compounds analogous to the cyclic ether (**19**), but retaining the aromatic nucleus. These cyclic compounds were inactive in both their quinone and hydroquinone forms (*5*).

Opening of the ansa ring appears to destroy the antibacterial activity, as judged by rifarubin S (Fig. 20) (*5, 152*), which was practically inactive. From the ease of formation of rifarubin S from borohydride-reduced rifamycin SV it appears that stability of the ketal function at C-23 depends on the presence of the C-17 keto group.

C. Halomicins

Very little of the chemistry of the halomicins has been reported. As shown earlier, in Section II.C., halomicin B was degraded with nitrous acid to rifamycin S and 2-(1-hydroxyethyl)pyrrolidine (**4**) (*36*). The structure of the substituted pyrrolidine was assigned from spectral data by comparison to corresponding properties of L-prolinol. Two reaction sequences carried out on the pyrrolidine yielded (**53**) and (**54**), as shown below. The relative stereochemistry of the pyrrolidine was established by

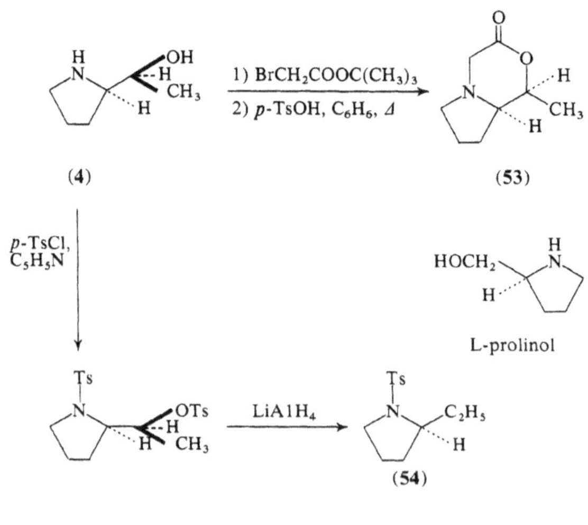

comparison of nmr proton coupling data for the cyclic compound (**53**) (in which the methine protons on the six-membered ring could be shown to be *cis*) and for the compound formed by a similar reaction carried out on L-prolinol. The absolute stereochemistry of the pyrrolidine was determined by comparison of the Cotton effects (which were opposite in sign) in the circular dichroism spectra of (**54**) and the 1-tosyl-2-methylpyrrolidine obtained from L-prolinol by reactions identical with those which gave (**54**). The attachment of (**4**) to the rifamycin unit in halomicin B was shown to involve the pyrrolidine nitrogen since the hydroxyl of (**4**) could be acetylated and the pK_a of halomicin B 21,27-di-O-methyl ether was like that of *p*-methoxyaniline. Attachment of the pyrrolidine to C-24 or C-27 would accord with the nitrous acid cleavage of halomicin B to rifamycin S and the pyrrolidine. Attachment at C-24 (rather than C-27) appears to have been assigned largely on the basis of analogy, although halomicin B was regenerated by heating rifamycin O (which is substituted at C-24) with the pyrrolidine in tetrahydrofuran.

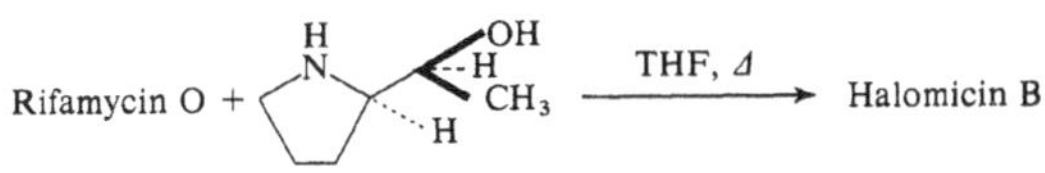

(**4**)

A few halomicin B derivatives (from O-methylation and O-acetylation) were prepared in the course of its structure determination (*36*), but bioactivities of these derivatives have not been reported. Nor have the bioactivities of the individual members of the halomicin complex (halomicins A, B, C, D) been reported. Structure-activity relationships are, therefore, impossible to discern.

D. Tolypomycin

1. Degradations

As was described in Section II, the chemical study of tolypomycin Y began with cleavage of the antibiotic to give tolypomycinone and tolyposamine (p. 240) (*52, 53, 54*). The structure of tolyposamine was assigned as a 4-amino-2,3,4,6-tetradeoxyhexose by its oxidation with permanganate to ammonia, acetic acid, and succinic acid. Its relative stereochemistry at C-4 and C-5 was assigned from the coupling constants of the protons at those positions in the pmr spectrum and the absolute configuration was assigned as L- by application of Hudson's rule to the axial and equatorial anomeric methyl glycosides of the N-benzoyl derivative. Attachment of tolypomycinone *via* the amino group of tolyposamine was deduced since the pmr spectrum of tolypomycin contains both α-

and β-anomeric protons and the antibiotic can be converted to a mixture of methyl glycosides.

As noted in Section II, the structure of tolypomycinone rests firmly on the x-ray study of its tri-*m*-bromobenzoate (**5**) (*50*). However, the structure was also assigned by chemical means (*52*). Chemical degradation (Fig. 21) commenced with the hydrolysis of tolypomycinone to the

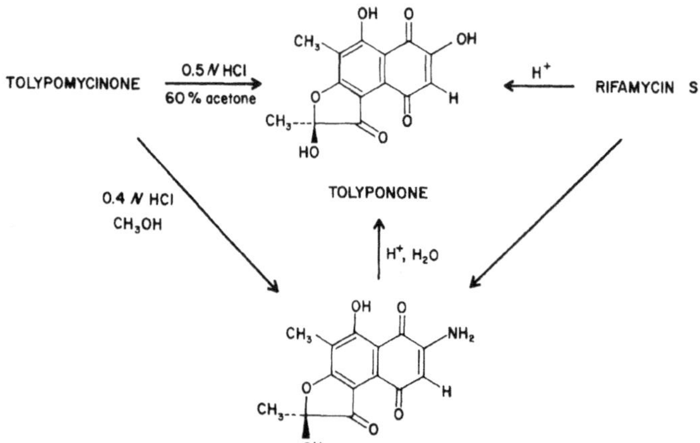

(18)

Fig. 21. Aromatic degradation products derived from tolypomycinone

same hydroxynaphthoquinone (tolyponone) and aminonaphthoquinone (**18**) obtainable from rifamycin S (Section III.B.) under like conditions. The other compounds isolated were two aliphatic aldehydes, (**55**) and (**56**), one the acetate of the other (Fig. 22). Spin decoupling in its

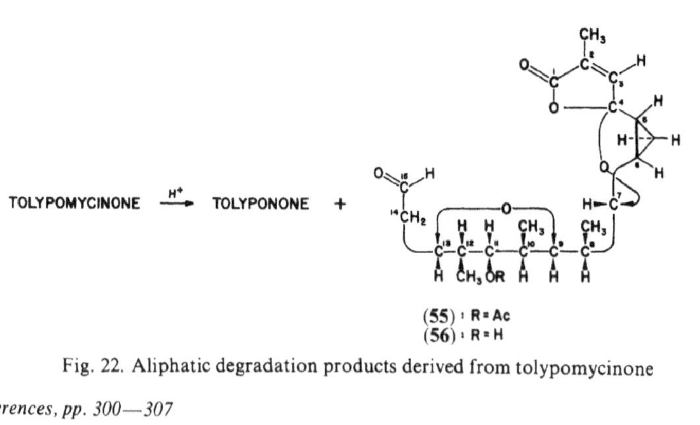

(**55**) : R = Ac
(**56**) : R = H

Fig. 22. Aliphatic degradation products derived from tolypomycinone

References, pp. 300—307

pmr spectrum assigned the structure of (**55**) from C-7 through C-15 and indicated the group $-CH{=}\underset{|}{C}CH_3$ as well.

A cyclopropane ring was also apparent in the pmr spectrum of (**55**). Oxidation of (**55**) with osmium tetroxide followed by lead tetraacetate gave a new γ-lactone (**57**) (Fig. 23) with the same structure as (**55**) at C-7 through C-15. Inclusion of the cyclopropane ring completes the assignment of structure to (**57**), while consideration of the groups lost in the oxidation allows assignment of (**55**).

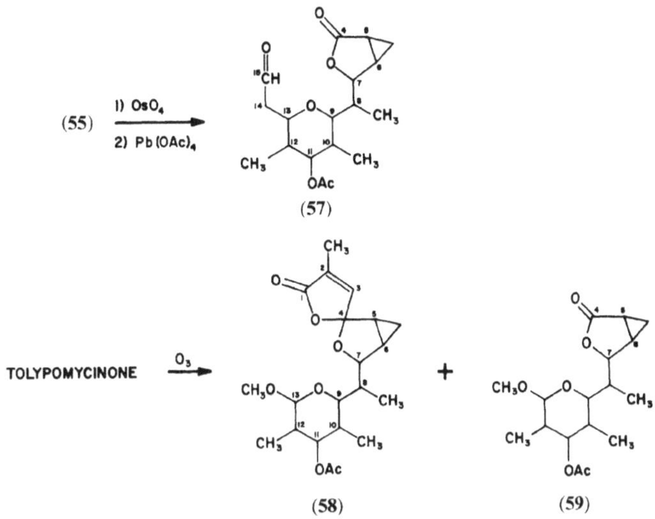

Fig. 23. Degradation products derived by oxidation of aliphatic portion of tolypomycinone

Compounds related to (**55**) and (**57**) were obtained by ozonolysis of tolypomycinone, which gave (**58**) and (**59**), with the acetaldehyde group replaced by methoxyl (Fig. 23). This serves to locate the methoxyl group in tolypomycinone at C-13.

The nature of the aliphatic portion of tolypomycinone at C-2 and C-3 and from C-7 through C-15 was then confirmed by spin decoupling of its pmr spectrum. The cyclopropane was shown to be present by the pmr spectra of selected derivatives, and the keto group at C-4 by comparison of the ultraviolet spectra of tolypomycinone and 2,3-dihydrotolypomycinone. Attachment of C-1 (the carbonyl group) of (**55**) to the amino group of (**18**) and of C-15 of (**55**) through oxygen to C-16 of (**18**) rests mainly on pnalogy to rifamycin.

The position of tolypomycinone involved in attachment to tolypos-amine was assigned by a process of elimination. The functional groups of tolypomycinone appeared in the infrared and nmr spectra of tolypo-mycin Y and four hydroxyl groups were established by acetylation. Carbonyl groups were established at C-4 by spectral considerations and at C-27 by the hydrogen-bonded *peri*-hydroxyl at C-21. The C-17 carbonyl was not specifically eliminated as a binding site, but the difference in chemical shift of the C-25 proton between nmr spectra of tolypomycin Y and tolypomycinone was felt to favor C-24 as the point of attachment.

2. *Derivatizations*

In vitro and *in vivo* antimicrobial activities have been reported for tolypomycin Y (*43, 57*), but few derivatives have been tested. The most important of these is tolypomycin R (the hydroquinone form of tolypo-mycin Y), which was obtained as one of the products of catalytic hydro-genation of tolypomycin Y.

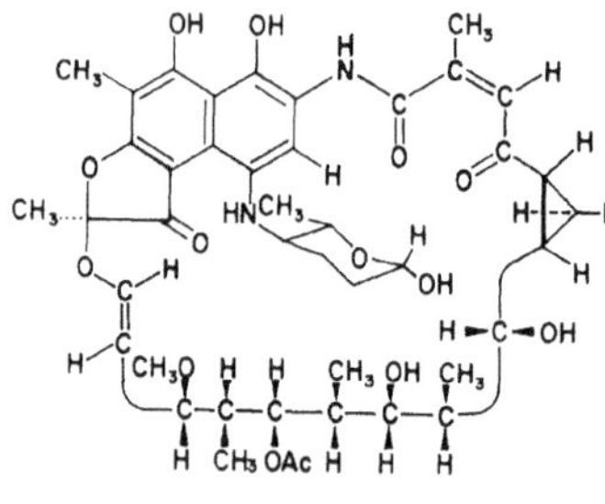

TOLYPOMYCIN R

Tolypomycins Y and R (*57*) show outstanding *in vitro* growth-inhibi-tion against Gram-positive bacteria and to some extent against some Gram-negative bacteria, approximately equivalent to the activity of rifampicin. Tolypomycins Y and R are very effective therapeutically in mice against Gram-positive bacterial infection by subcutaneous, intra-peritoneal and intravenous administration.

E. Naphthomycin

The reported assignment of structure to naphthomycin (*159*) was based entirely on spectral data for the antibiotic itself, with lanthanide shift reagents providing the bulk of the argument thus far presented. A more detailed discussion is needed to provide a convincing structural argument. The original study of naphthomycin (*2*) described a product obtained by hydrogenation over palladium in acetic anhydride but its structure is uncertain and no biological data have been reported for the derivative. The biological activity (*2*) of naphthomycin itself is interesting in that it inhibits Gram-positive bacteria and fungi but its antibacterial activity is reversed by addition of cysteine. It is also a vitamin K antagonist.

F. Geldanamycin

1. Degradations

Only a few degradation reactions have been described for geldanamycin (Fig. 2). From the standpoint of the structure proof, the most important was that which produced the aminobenzoquinone methyl geldanamycinate (*114*) by treating geldanamycin with potassium car-

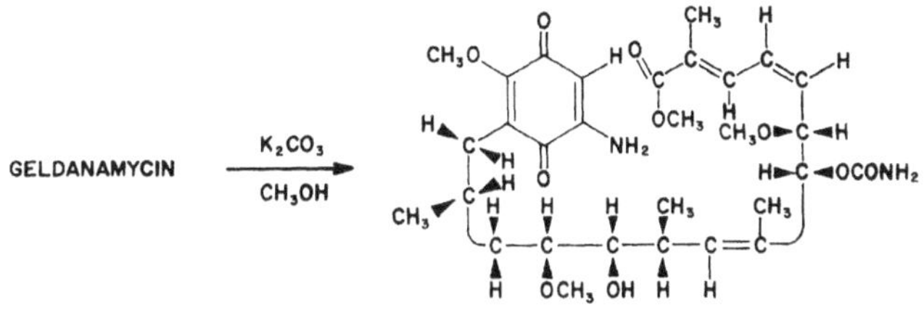

METHYL GELDANAMYCINATE

bonate in refluxing methanol-chloroform. Analysis of the nmr spectrum of methyl geldanamycinate was instrumental in assigning the ring proton and the amide group to the same side of the benzoquinone ring in geldanamycin.

2. Derivatizations

Extensive sets of derivatives have been described for geldanamycin (*88, 160*). Although smaller than rifamycin, geldanamycin also has a quinone ring with an unsubstituted carbon adjacent to a quinone carbonyl. By taking advantage of this structural similarity, compounds were prepared analogous to the rifazine and rifampicin classes of derivatives (Figs. 24 and 25).

Reaction of geldanamycin with an *o*-phenylenediamine directly, or reaction first with barium hydroxide to give 17-des-O-methylgeldanamycin (**60**) followed by reaction with an *o*-phenylenediamine, gave products of the type illustrated by geldanazine in Fig. 24. Geldanazine and

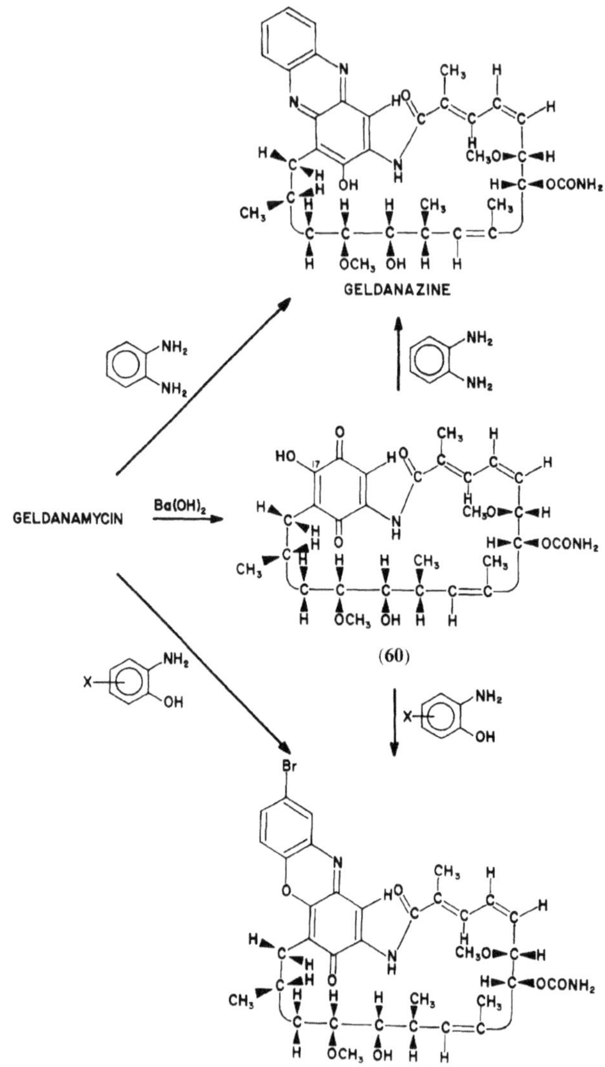

(61)

Fig. 24. Phenazines and phenoxazinones derived from geldanamycin

methyl-substituted geldanazines are active as reverse transcriptase inhibitors.

Similarly, reaction of geldanamycin or 17-des-O-methylgeldanamycin (**60**) with an *o*-aminophenol gave phenoxazinones like the bromo deriva-

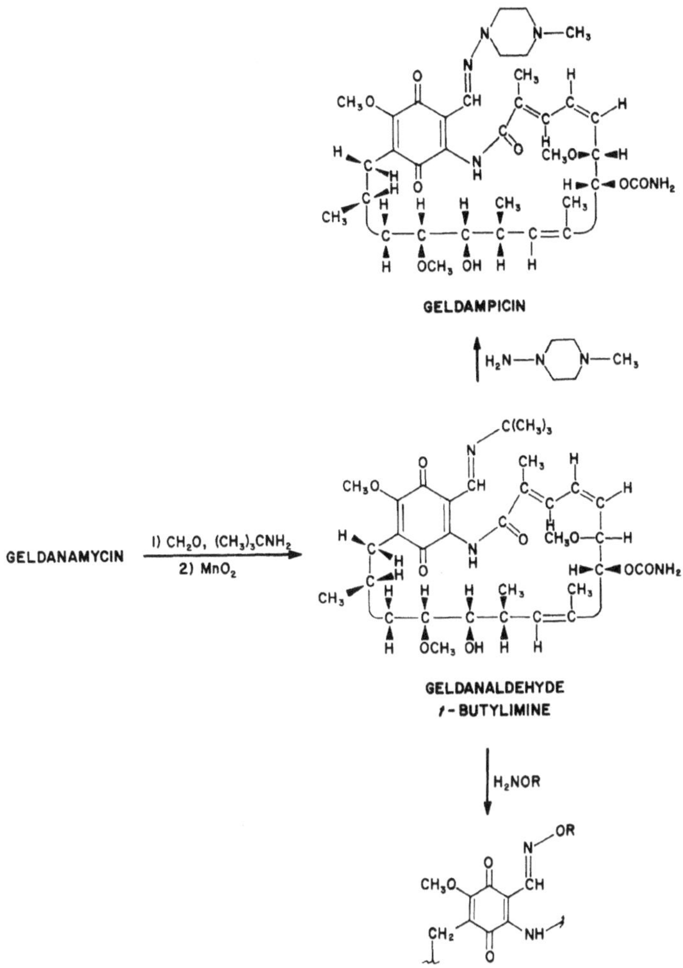

(62)

Fig. 25. Derivatives of geldanaldehyde

tive (**61**). Other substituents included chloro, alkyl, carbomethoxy, and carboxymethyl groups. Like the phenazines, most of the substituted phenoxazinones have reverse transcriptase inhibitory activity. Halo-substituted derivatives such as (**61**) appear to be the most promising.

Finally, the reaction of geldanamycin with formaldehyde and *t*-butylamine followed by oxidation gave geldanaldehyde *t*-butylimine (Fig. 25). The imine inhibits reverse transcriptase in its own right and can also be exchanged with hydrazines or O-alkylhydroxylamines to give hydrazones (like geldampicin, the analog of rifampicin) or oximes (**62**) of geldanaldehyde analogous to the derivatives formed from rifaldehyde (Fig. 19).

Several geldanamycin derivatives have shown low *in vitro* antimicrobial activities, while the substituted phenazines related to geldanazine gave minimal inhibition of *E. coli* RNA polymerase. Although geldanamycin itself, with no substituents at C-18 or C-19, did not inhibit MSV RNA-dependent DNA polymerase, a number of phenoxazinone and phenazine derivatives of geldanamycin (Fig. 24) inhibit RLV reverse transcriptase, as noted above.

The C-17 methoxyl group (quinone substituent) of geldanamycin which was replaced by a hydroxyl on treatment with barium hydroxide (Fig. 24), can also be replaced by amino groups to give (**63**) by treatment with ammonia or primary or secondary amines (Fig. 26). The only other geldanamycin derivatives reported involve modification of the ansa bridge by hydrogenation of the C-4, C-5 double bond and by acylation of the C-11 hydroxyl (see Fig. 2). Biological properties of these derivatives have not yet been described.

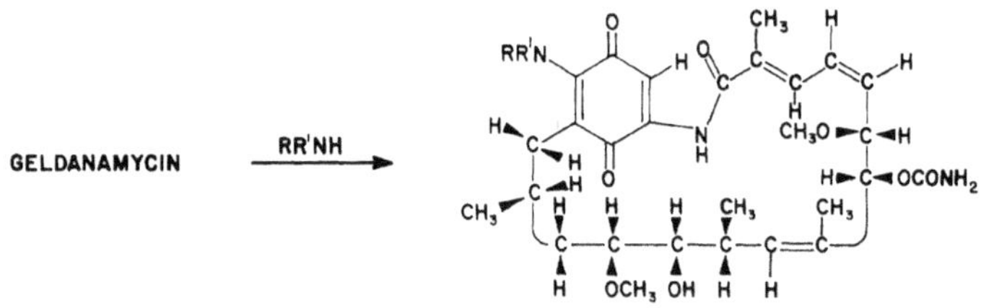

(63): R,R' = H, or CH$_3$

Fig. 26. Derivatives of geldanamycin derived by substitution at C-17

G. Maytansinoids

Very few degradative reactions have been described for the maytansinoids, since their structures were assigned from x-ray analysis of a readily prepared derivative (*14*). Moreover, the amounts of material available have been small, discouraging chemical investigations. The methyl and ethyl ethers (replacing the hydroxyl at C-9) of maytansine were obtained in the course of the initial extractions from *Maytenus ovatus* (*62*), and their ready formation suggested the preparation of the 3-bromopropyl ether (**6**) for structure determination by x-ray analysis. The ethyl ether was reported to lack antileukemic activity, possibly because the carbinolamide function is blocked (*61*).

Additional information about structure-activity relationships of the maytansinoids has come from the naturally occurring compounds maysine, normaysine and maysenine (Fig. 27) (*61*). Maytansine and maytanvaline have both been converted to maysine by treatment with mild base and normaysine can be converted to maysenine by removal of the epoxide group. Compounds containing the macrocyclic ring system of maytansine but not the ester moiety (maysine, normaysine, maysenine)

$$\text{Maytansine, Maytanvaline} \xrightarrow[\text{aq. } CH_3OH]{Na_2CO_3} \text{Maysine}$$

$$\text{Normaysine} \xrightarrow[\text{HOAc}]{CrCl_3} \text{Maysenine}$$

lack antileukemic activity. Since maytanside esters such as maytansine (*62*) itself, maytanprine (*63*), maytanbutine (*63*), maytanvaline (*61*), colubrinol (*150*), and colubrinol acetate (Fig. 2) (*150*) are all potent antileukemic agents, it appears that the ester group is required for bioactivity (*161*).

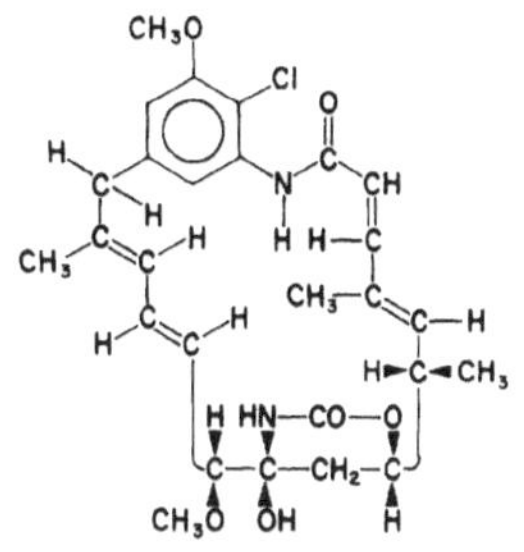

MAYSENINE

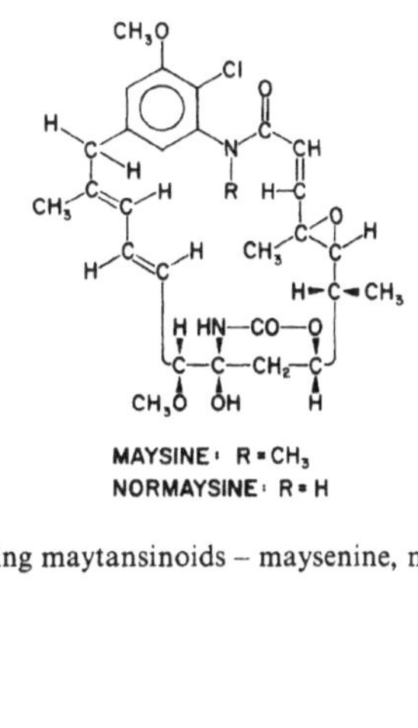

MAYSINE: R = CH_3
NORMAYSINE: R = H

Fig. 27. Naturally occurring maytansinoids – maysenine, maysine, and normaysine

H. Conclusions Regarding Bioactivity

So much has been written on the derivatives of the ansamycins, especially rifamycin derivatives, that little more can be added here. Two activities have been widely investigated for streptovaricins and rifamycins. The first of these is antibacterial activity. Activity on the enzymatic level in inhibiting bacterial RNA polymerase has been used as a model for antibacterial activity, since *in vitro* antibacterial activity is rather well predicted by inhibition of DNA-dependent RNA polymerase. Unfortunately, the absorption required for *in vivo* activity is more difficult to achieve or correlate.

The basic ansamycin structure consisting of a naphthoquinone or naphthohydroquinone ring and its spanning chain with its various substituents is required for antibacterial activity and *E. coli* RNA polymerase inhibition. Neither the rifamycin aromatic chromophore alone (**18**, Fig. 12) (*126*) nor streptovarone (Fig. 6) (*104*) has much activity, nor does the aliphatic ansa chain of rifamycin by itself (**19**, Fig. 12) or varicinal A (Fig. 6) (*109*). Cleavage of the ansa chain, as in the streptovals (Fig. 6) or rifarubin S (Fig. 20) (*5*), also leads to loss of activity. Benzoquinonoid and benzenoid ansamycins are not potent antibacterial agents.

Inhibition of the isolated *E. coli* RNA polymerase does not seem to be affected as dramatically by changes within the aromatic part of the molecule, since the chromophore is somewhat different for the streptovaricins, rifamycins and tolypomycins. Moreover, streptovaricin contains a 25-methyl group, while rifamycins and tolypomycins contain no C-25 substituents. All inhibit bacterial RNA polymerase. On the other hand, derivatives with differing substitution in the aromatic region of the molecule differ considerably in their effects on bacterial growth *in vitro* and on chemotherapeutic activity (*5*). One impressive example is the lack of *in vivo* activity of rifamycin B itself compared to the activity of rifamycin SV; another is the great dependence of *in vivo* activity on the nature of C-25 substituents in rifamycin SV and the similar dependence on the 19-O-alkyl substituent in damavaricins. In the aliphatic portion of the ansamycins, the shape of the ansa ring (which is altered by lactone or ether formation) and the nature of the C-7 (and possibly the C-9) substituents both affect RNA polymerase inhibition and antibacterial activity (*103, 111, 153*). The ansamycins' capacity to inhibit enzyme activity and to affect bacterial growth is usually diminished by other modifications – acetylation, hydrogenation, oxidation – of the aliphatic bridge (*9, 103, 152*).

The second major activity investigated for the ansamycins – reverse transcriptase inhibition – does not follow closely that of bacterial inhibition. Thus, the most active antibacterial rifamycin derivatives are rather

References, pp. 300—307

inactive as reverse transcriptase inhibitors (*153*). This seems to be mainly a matter of polarity, with longer, bulkier side-chains at C-24 or C-25 required for reverse transcriptase inhibition (*40*). Similarly, enhanced lipophilicity achieved by acetylating the streptovaricins destroys antibacterial activity but enhances reverse transcriptase inhibition. Enhanced lipophilicity does not seem the only criterion, however, since cleavage of the ansa chain to give streptovals also enhances RT activity at the expense of antibacterial activity. Thus, the intact ansa ring is not a requirement for RT inhibition. For that matter, much smaller sub-units, including pre-streptovarone and streptovarone, are effective inhibitors of reverse transcriptase (*70*).

A third activity, anti-leukemic activity, has only been demonstrated by the maytansinoids (*61—63, 150*). An insufficient number of compounds has been reported for extensive conclusions to be drawn, but it does appear that an ester at C-3 and a carbinolamide at C-9 are both required for anti-leukemic activity.

IV. Physical Properties of the Ansamycins

A. Ultraviolet, Rotatory, Infrared Spectra

With the exception of the maytansinoids (which are colorless) (*61* to *63, 150*), the ansamycins are all yellow to orange crystalline compounds, soluble in polar organic solvents and insoluble in water and non-polar organic solvents (*2, 22, 36, 54, 122, 123, 127, 128, 130, 158, 159*). The streptovaricins tend to form firmly bound solvates and this tendency gave some difficulty in obtaining accurate microanalyses (*108*).

The electronic spectra of the ansamycins are complex, but most show high extinction coefficients ($>25{,}000$) at wave lengths below 230 nm, with maxima at longer wave lengths for the naphthoquinonoid antibiotics than for the benzoquinonoid compounds, as would be expected. Ultraviolet spectra of streptovaricin complex (*84*), rifamycins B (*123*), S (*122*) and SV (*122*), and geldanamycin (*22*) are shown in Fig. 28. That of maytansine (*62*) has not been published, but it is reported to have maxima at 223 (ε 29,800), 243 (27,100 sh), 254 (27,200), 282 (5,690), and 290 nm (5,520). It is of some interest that the spectra of the streptovaricins resemble that of rifamycin S but not that of rifamycin B; this hampered early recognition of the antibiotics' similarity.

Specific rotations of the natural streptovaricins are all strongly positive at the sodium *D* line, with $[\alpha]_D = +400°$ to $+700°$ (*47, 103*), and this has been associated with the helicity of the chromophore, since the atrop-

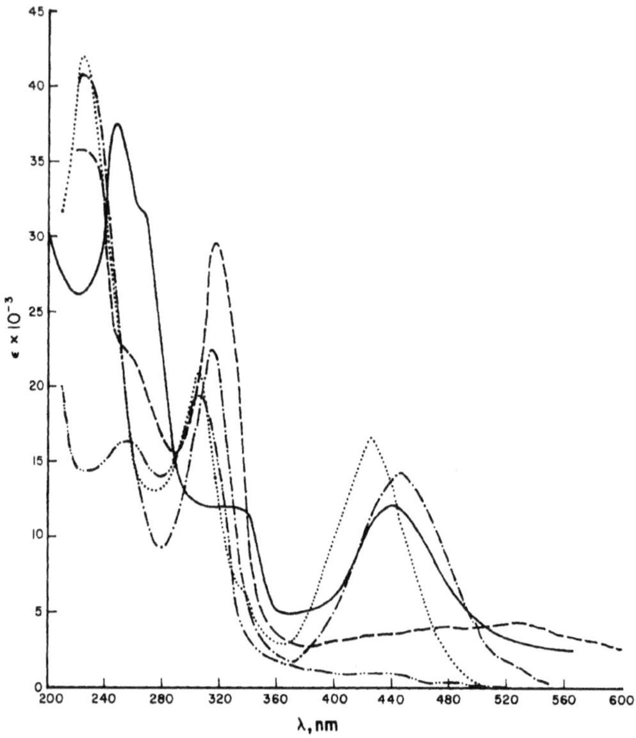

Fig. 28. Electronic spectra of streptovaricin complex (———), rifamycins B (··········), S (---------) and SV (-·-·-·-·-·), and geldanamycin (-··-··-··-··)

isomeric streptovaricins have nearly equally strongly negative rotations, $[\alpha]_D = -400°$ to $-600°$ (*105*). This opposite helicity is even more apparent in the circular dichroism spectra, where natural and atropisomeric streptovaricins have nearly mirror image spectra (Fig. 10) (*105*). Lowering the rotational barrier giving rise to atropisomerism, by converting streptovaricin C to damavaricin C, does not diminish the specific rotation; damavaricin C has $[\alpha]_D$ $+806°$ (*91*). However, the atropisomeric damavaricin F_C ($[\alpha]_D$ $-734°$) is also formed as the favored isomer. On the other hand, removing the barrier entirely by cleaving the ansa bridge, as in the conversion of streptovaricin C to streptoval C ($[\alpha]_D$ $-92.3°$) (*105*) gives much reduced rotations.

Specific rotations of the other ansamycins are less exceptional (Table 1). Rifamycin B has a small negative rotation ($-11°$) (*123*) while rifamycin Y is, surprisingly, strongly positive ($+325°$) (*68*). Other rifamycins, O (*121*), S (*128*) and SV (*128*), have varying rotations, with only rifamycin S having a strong rotation, which is positive (+476°). Tolypomycin Y has a large negative specific rotation (−325°) (*53*), and naphthomycin a strong positive rotation (+432°) (*2*). The benzoquinonoid ansamycins have generally lower specific rotations, some positive (geldanamycin, $[\alpha]_D$ +55°) (*22*), some negative (maytansine, $[\alpha]_D$ −145°; colubrinol, $[\alpha]_D$ −94°) (*62, 63, 150*).

Infrared spectra of the ansamycins are rather unexceptional, with the expected absorptions being observed for quinone, hydroxyl, ester, and amide groups. Among the more distinctive bands are those for the enol acetate of the streptovaricins (sometimes observed as a shoulder) near 1760 cm^{-1} (*84*) and the lactide band of rifamycin O at 1822 cm^{-1} (*121*).

B. Magnetic Resonance Spectra

Considerable attention has been directed toward the magnetic resonance spectra of the ansamycins, since these spectra have played a major role in the assignments of structures to the ansamycins and their degradation products. Table 3 shows proton magnetic resonance assignments for streptovaricins A, C, D, F and J (partial) (*1, 47, 56*), for rifamycin S (complete) (*34*), and for naphthomycin (complete assignment but structure uncertain) (*159*). Other streptovaricin spectra have also been assigned, but complete spectral analyses have not been reported for other naphthoquinonoid ansamycins. Table 4 shows the proton magnetic resonance assignments for geldanamycin (*114, 144*) and maytansine (*62*). Table 5 shows the carbon magnetic resonance assignments for streptovaricins A, C, D, F and J (*49*) and for rifamycin S (*27, 86, 87*).

A particular point of interest in the spectra of the ansamycins is the effect of the aromatic ring system on the protons or carbons of the ansa bridge. Two effects have been described. The first derives from a comparison of the spectra of the streptovaricins with their corresponding atropisomers, which shows that only the carbons and protons in the immediate region of the overlap causing the atropisomerism differ significantly in the two spectra (*105*); *e.g.*, C-15, C-16 and the acetate carbonyl carbon move to 149.3, 126.1, 170.6 ppm from 153.9, 130.3 and 169.0, respectively. A second effect is the appearance of certain protons, those found directly above the ring, at abnormally high field. For instance, the spectrum of rifamycin S contains methyl absorptions at 0.21 and 0.66 ppm, ascribed to the 12-methyl and 10-methyl, respectively. The former remarkable position was, in fact, the first signal that the rifamycins did in fact contain an ansa bridge.

Table 3. *Proton Magnetic Resonance Spectra of Naphthoquinonoid Ansamycins*

Proton	δ, ppm[a]						
	Streptovaricins (*1, 47, 56*)					Rifa-mycin S[b] (*34*)	Naphtho-mycin[d,e] (*159*)
	A[b]	C[b]	D[b]	F[c]	J[b]		
2 -CH_3	2.04	2.23	2.21	2.01	2.22	2.01	2.15s
3 -H	7.82	7.76	7.61	8.39	7.91		6.15m
4 -H	6.29	6.49	6.46	6.48	6.43		6.12
5 -H	5.80	5.77	5.75	5.80	5.63	5.90	6.12
6 -H	—	3.13	2.95	—	3.45	2.30	
6 -OH	2.50*	—	—	5.52	—		
6 -CH_3	1.31	1.21	1.24	1.28	1.24	0.84	
7 -H	3.36	3.45	3.50	4.56	4.76	3.52	
7 -OH	2.86*	3.09*	3.55*	—	—	3.62	
7 -$OCOCH_3$	—	—	—	—	2.08†		
8 -H	2.25	2.10	2.30	2.87	2.22	1.76	
8 -CH_3	0.95	0.97†	0.97	1.44	0.97	1.00	
9 -H	4.00	4.16	4.14	4.36	4.17	2.97	
9 -OH	3.35*	3.86*	4.14*	5.52	3.50*	3.62	
10 -H	3.06	2.89	2.87	3.36	2.83	1.50	
10 -CH_3	—	—	—	—	—	0.66	
10 -$COOCH_3$	3.73	3.71	3.70	—	3.71	—	
11 -H	5.18	4.07	3.91	4.36	4.09	4.59	
11 -OH	—	4.18*	4.31*	5.52	3.63*	—	
11 -$OCOCH_3$	1.95†	—	—	—	—	2.01	
12 -H	2.00	2.10	1.95	2.06	1.95	1.91	2.25 (H–18)
12 -CH_3	0.76	0.95†	0.78	1.29	1.01	0.21	0.94 (18–CH_3)
13 -H	2.86	3.58	3.37	4.17	3.63	3.33	3.08 (H–19)
13 -OH	4.35*	4.18*	4.80*	5.52	4.26*	—	
13 -OCH_3	—	—	—	—	—	3.08	
14 -H	—	—	2.58	—	—	5.02	2.43 (H–20)
14 -OH	4.68*	4.18*	—	5.52	3.22*	—	—
14 -CH_3	1.19	1.15	0.69	1.57	1.17	—	0.80 (20–CH_3)
15 -H	5.60	5.68	5.42	5.99	5.66	6.16	5.95 (H–21)
16 -CH_3	2.28	2.28	2.03	2.63	2.31	1.70	2.01 (22–CH_3)
17,19 -O-CH_2-O-	5.19, 5.77	5.20, 5.84	5.20, 5.83	5.24, 5.98	5.19, 5.84	—	7.98 (H–25)
20 -CH_3	1.95†	1.97‡	1.97†	1.97†	1.99†	2.30	2.39 (26–CH_3)
21 -OH	—	—	—	—	—	12.51	9.63 (21–OH)
24 -$OCOCH_3$	2.22	2.28	2.27	2.43†	2.31	—	
25 -H	—	—	—	—	—	7.77	
25 -CH_3	1.99†	2.02‡	1.99†	2.25†	2.05†	—	
26 -NH-	7.68–[f] 7.74	7.89–[f] 7.94	7.93	9.41	7.80–[f] 7.82	8.38	
27 -OH	14.18	14.21	14.10	14.76	14.25	—	

[a] Signals labeled *, †, ‡ may be interchanged with other signals of a like designation. [b] $CDCl_3$ solvent. [c] Pyridine-d_5 solvent. [d] CD_2Cl_2 solvent. [e] The naphthomycin *ansa*-chain is longer by six carbon atoms than those of streptovaricin or rifamycin and is somewhat

differently substituted. Consequently, while C–1 through C–5 of naphthomycin are like C–1 through C–5 of the others, C–18 through C–33 of naphthomycin correspond to C–12 through C–27 of the streptovaricins and rifamycin S and are so aligned in the table. As noted in the text, the 19–H and 21–OH of naphthomycin should probably be interchanged. Other proton absorptions assigned naphthomycin are as follows: 6–H, 6.53 ppm; 7–H, 5.47; 8–H, 2.25; 8–CH_3, 1.16; 9–H, 3.53; 10–H, 2.53; 10–H, 3.21; 12–CH_3, 1.71; 13–H, 6.76; 14–H(2), 2.30; 15–H, 3.98; 16–H, 5.59; 17–H, 5.44. [f] Concentration dependent.

Table 4. *Proton Magnetic Resonance Spectra of Geldanamycin and Maytansine*

Geldanamycin (*114, 144*)		Maytansine (*62*)	
Proton	δ, ppm[a, b]	Proton	δ, ppm[b, c]
–NHCO–	9.14 sb	–CONCH$_3$	3.22 s
2–CH_3	1.91	2–H (2)	2.21 d of d
			2.65 d of d
3–H	6.95 d	3–H	4.79 d of d
4–H	6.56 t	4–CH_3	0.87 s
5–H	5.80 t	5–H	3.04 d
6–H	4.34 d	6–H	0.80*
6–OCH_3	3.22 s**	6–CH_3	1.34 d
7–H	4.86 sb	7–H	4.28 m
7–$OCONH_2$	6.45 sb		
8–CH_3	1.61 sb	8–H (2)	2.50*
9–H	5.51 d	9–NH	6.24 sb
		9–OH	3.53 s
10–H	3.61	10–H	3.50 d
10–CH_3	0.97 d†	10–OCH_3	3.38 s
11–H	3.29*	11–H	5.66 d of d
12–H	3.07*	12–H	6.42 d of d
12–OCH_3	3.23 s**		
13–H (2)	1.45 sb	13–H	6.70 db
14–H	1.91		
14–CH_3	0.76†	14–CH_3	1.69 sb
15–H (2)	2.42 m	15–H (2)	3.13 d
			3.67 d
17–OCH_3	3.93 s	17–H	6.75 d**
		18–OCH_3	3.99 s
19–H	7.02 s		
		21–H	6.84 d**
		2′–H[d]	5.35 q
		3′–H (3)[d]	1.37 d
		2′–N–CH_3[d]	2.89 s
		2′–N–$COCH_3$[d]	2.15 s

[a] DMSO–d_6 solvent. [b] Assignments marked *, †, ** may be interchanged in the column in which they appear. [c] $CDCl_3$ solvent. [d] Numbered as follows:

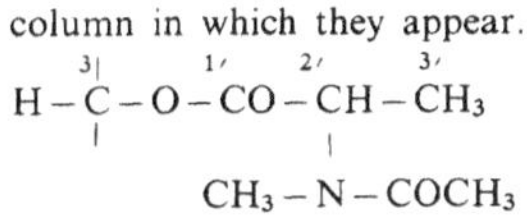

Table 5. *Carbon Magnetic Resonance Spectra of Naphthoquinonoid Ansamycins*

Carbon		Streptovaricins (*49*), δ, ppm					Rifamycin S (*27, 86, 87*), δ, ppm[b,d]
		A[a,b]	C[a,b]	D[a,b]	F[b,c]	J[a,b]	
C-1	$-C(=O)-$	168.7**	169.0#	169.4	170.2‡	169.3‡	169.0
C-2	$=C<$	130.8	126.7	127.5‡	131.2	127.1	131.0
C-2	$-CH_3$	13.3	13.1	12.9	13.4*	13.2	20.0
C-3	$=CH-$	132.1	135.0	134.7	132.1	134.8	133.2
C-4	$=CH-$	124.6	124.2	124.0	126.2	124.5	124.4
C-5	$=CH-$	142.4	143.9	144.1	140.0	141.5	142.4
C-6	$>CO-$	75.0			78.1		
C-6	$>CH-$		41.6	41.9		41.6	39.2
C-6	$-CH_3$	25.0	22.1*	22.2	27.8	22.3*	16.8
C-7	$>CHO-$	82.1	82.7	83.6	88.5	81.9	73.6
C-8	$>CH-$	36.9	38.9	38.8	40.7	38.4	33.0
C-8	$-CH_3$	19.0†	15.8	15.7	16.5	15.9	11.4
C-9	$>CHO-$	77.1	77.6	77.6	75.8†	75.1	77.7
C-10	$>CH-$	46.8	47.6	47.4	54.9	48.8	37.4
C-10	$-C(=O)-$	173.4	173.4	173.0	173.1	173.7	
C-10	$-OCH_3$	52.2	51.9	51.9		51.8	
C-10	$-CH_3$						8.8
C-11	$>CHO-$	75.0‡	74.0‡	73.4†	75.3†	74.4†	73.6
C-11	$-C(=O)-$						172.6
C-11	$-CH_3$						20.9
C-12	$>CH-$	34.9	38.7	38.5*	30.4	37.5	37.4

References, pp. 300—307

Table 5 (continued)

Carbon		Streptovaricins (*49*), δ, ppm					Rifamycin S (*27, 86, 87*), δ, ppm[b,d]
		A[a,b]	C[a,b]	D[a,b]	F[b,c]	J[a,b]	
C-12	$-CH_3$	10.8	10.4	9.1	12.3	10.3	11.4
C-13	–CHO–	71.6‡	70.6‡	70.4†	71.7†	70.7†	81.9
C-13	$-OCH_3$						56.8
C-14	–CO–	77.6‡	77.3		77.3	77.4	
C-14	–CH–			37.5*			115.7
C-14	$-CH_3$	23.4†	21.9*	15.1	25.1	21.3*	
C-15	=CH–	154.8	153.9	153.6	148.7	154.0	145.3
C-16	=C–	132.1	130.3	127.9‡	134.0	130.4	108.6
C-16	$-CH_3$	12.6	12.7	12.7	12.9*	12.5	22.2
C-17	=C–	168.5**	168.8#	169.0	171.1‡	168.4	191.1
C-18	=C–	102.3	102.0	101.9	102.6	102.2	111.2
C-19	=C–	159.3	159.4	159.6	161.2	159.7	172.4
C-20	=C–	113.4	113.3	113.4	114.3	113.4	115.7
C-20	$-CH_3$	7.5	7.4	7.4	7.5	7.4	7.4
C-21	–C(=O)–	188.7	188.5	188.7	189.7	188.8	166.7
C-22	=C–	107.3	107.2	106.9	107.5	107.3	111.0
C-23	=C–	125.4	125.4	125.6	126.7	125.4	131.3
C-24	=C–	138.3	136.6	136.9	137.3	136.8	184.5
C-25	=C–	122.3	121.7	121.5	123.1	122.1	117.4
C-25	$-CH_3$	13.6	13.9	14.2	14.1	14.1	
C-26	=C–	135.9	135.0	134.5	137.3	135.1	139.4
C-27	=C–	150.5	153.5	153.1	156.4	153.6	181.6
	$-OCH_2O-$	89.7	89.7	89.8	90.9	89.8	

Table 5 (continued)

Carbon	Streptovaricins (*49*), δ, ppm					Rifamycin S (*27, 86, 87*), δ, ppm[b,d]
	A[a,b]	C[a,b]	D[a,b]	F[b,c]	J[a,b]	
−C=O	169.8	169.0‡	169.0	172.7	169.0	
CH$_3$	21.3	21.2	21.3	21.0	21.3	
−C=O	171.7				170.9	172.6
CH$_3$	21.3				21.3	20.9

[a] CD_2Cl_2 solvent. [b] Assignments marked *, **, †, ‡, ‡ may be interchanged in columns in which they appear. [c] CD_3OD solvent. [d] $CDCl_3$ solvent.

C. Mass Spectra

Although mass spectrometry has been used in characterizing individual ansamycins, relatively few systematic studies have appeared. Field desorption mass spectrometry appears to be the method of choice for characterizing these compounds at early stages, since this ionization method gives predominantly molecular ions. This has been demonstrated for individual streptovaricins and for the streptovaricin complex as well (*103a*). In the latter case it was possible to estimate the amounts of individual streptovaricins in the mixture. Rifamycins have also been studied by the field desorption method (*165*). Initial studies of the electron impact mass spectral fragmentations of the streptovaricins and rifamycins mainly listed ions resulting from the expected losses of small fragments – methanol, water, acetic acid, etc. (*107, 115*).

The mass spectrum of geldanamycin (*114*) is interesting in that its molecular ion is a doublet, at m/e 560 ($C_{29}H_{40}N_2O_9$) and 562 ($C_{29}H_{42}N_2O_9$), arising from the well known tendency of quinones to undergo reduction in the ion source of a mass spectrometer. In addition, the base peak in the geldanamycin spectrum is observed at m/e 517, resulting from loss of HNCO from the quinone molecular ion. The loss of HNCO is prominent even in the field desorption spectrum of geldanamycin and is a feature of the mass spectra of geldanamycin derivatives as well.

Extensive studies have been carried out on the electron impact mass spectrometric fragmentation of the streptovaricins (*107, 115*). These demonstrated that all the streptovaricins (except streptovaricin D) give very intense ions at lower masses – at m/e 390, 324, 297, and 269 – whose compositions were assigned as $C_{23}H_{20}NO_5$, $C_{18}H_{14}NO_5$, $C_{17}H_{15}NO_4$, and $C_{16}H_{13}O_4$, respectively, by high resolution. Even after the structures of the streptovaricins were assigned, these ions were difficult to relate to the structures since their molecular formulas did not correspond to any set of contiguous atoms in the molecules. However, it was ultimately recognized that the compositions of the ions correspond to a fragmentation in the mass spectrometer analogous to the photochemical reaction of streptovarone and prestreptovarone (Section III.A.). Indeed, the mass spectra of prestreptovarone and photoprestreptovarone both contain the ions at *m/e* 390, 297, and 269 and the mass spectra of streptovarone and photostreptovarone both contain the ions at *m/e* 324, 297 and 269, all of the same elemental composition as those in the spectra of the streptovaricins themselves. A possible mechanism for this fragmentation for streptovaricin C is outlined in Fig. 29, with the intermediate (**64**) corresponding to a similar intermediate in Fig. 8. Fragmentation of one or more ions corresponding to (**65**) accounts for the *m/e* 390, 324, and 297 ions; additional loss of C-26 and the nitrogen atom accounts for the *m/e* 269 ion.

Another extensive study of ansamycin fragmentation patterns has been reported by ZERILLI, *et al.* (*164, 166*). Although only rifamycin S and rifamycin W (which is more closely related to the streptovaricins than to the other rifamycins) gave molecular ions by electron impact mass spectrometry, methylated derivatives of rifamycin B, rifamycin SV and rifamycin (as well as underivatized rifamycin S) gave characteristic cleavages between the enol ether oxygen and C-16 as well as successive cleavages between the nitrogen atom and C-1, between C-6 and C-7, between C-7 and C-8, and between C-9 and C-10 of the ansa chain.

V. Biosynthesis of the Ansamycins

The ansamycin antibiotics represent a new class of natural products, combining an aromatic nucleus with a branching aliphatic side chain, and considerable interest has centered on their formation in nature. Their common structural features suggest immediately a similar biosynthetic pathway. Moreover, the streptovaricins, rifamycins, tolypomicin B and halomicin B all have the same number of carbon atoms in the ansa ring (C_{17}), while geldanamycin and the maytansinoids also have the same number of carbon atoms (C_{15}). Thus far, biosynthetic

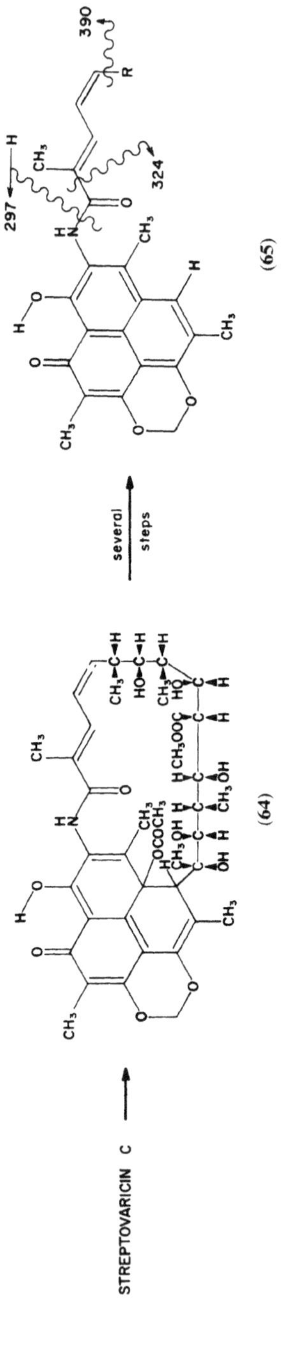

Fig. 29. Origin of major peaks in mass spectral fragmentation of streptovaricin C

experiments have been reported for rifamycin (*13, 51, 66, 85, 155—157*), streptovaricin (*23, 89—91*) and geldanamycin (*42, 46*); these studies, indeed, support the hypothesis of a common biosynthetic pathway for rifamycin and streptovaricin, with minor variations apparent for geldanamycin. In succeeding sections we shall discuss in turn biosynthetic results for these three ansamycins, presenting first for each an indication of what primary precursors are incorporated into the antibiotic and then describing intermediates identified.

Extensive and often difficult chemical degradations were required to determine the distribution of isotopes when ^{14}C-labeled radiotracers were used in early biosynthetic studies (*13, 19, 56, 89*). The more recent use of cmr spectroscopy employing the pulsed Fourier transform method has made the study of the ansamycins easier, especially that of their aromatic chromophores, which are not easily degraded. Carbon-13 occurs in 1.1% natural abundance and natural abundance cmr spectra of many ansamycins have been reported, as described in Section IV. When highly enriched ^{13}C-labeled compounds are used as precursors in ansamycin-producing cultures, cmr data reveal the extent and site of incorporation of ^{13}C-labeled precursors (*106*).

A. Streptovaricins

A priori, it seemed likely that the primary precursors of the streptovaricin *ansa* bridge should be repeating propionate and acetate units (Fig. 30). This presumption was based on the known biosynthetic pathway of macrolide antibiotics lacking the aromatic nucleus (*18, 147*). However, as pointed out by Corcoran (*19*), propionate and acetate could in principle be incorporated in two directions, an amide-head direction (Fig. 30, Scheme **a**) and an amide-tail direction (Scheme **b**). These two directions can be distinguished from one another by observing whether C-1 is labeled by propionate or by a one-carbon unit, whether C-3, C-5, C-17, and C-19 are labeled by propionate or by acetate, and whether C-21 is labeled by propionate or by some other precursor. Preliminary studies were directed at establishing this point by chemical degradation to products containing specific carbon atoms (*56*), but this approach was abandoned when cmr techniques became available (*vide supra*). It also seemed likely that methionine would be the source of the methylenedioxy and methoxy carbons, while the origin of the ring carbons C-22 through C-27 and the attached C-25 methyl could have been acetate, propionate, or other primary metabolites such as glucose. In any event, studies (*90*) with ^{14}C-labeled compounds established that [*methyl*-^{14}C]-methionine (0.11%), and sodium [*carboxy*-^{14}C]malonate (0.11%) were

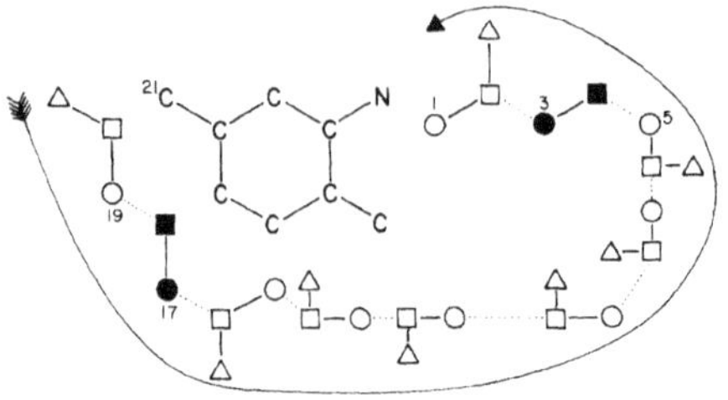

a

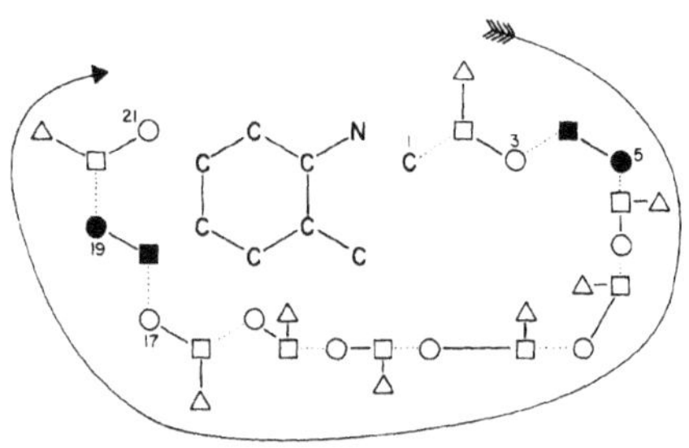

b

Fig. 30. Possible biosynthetic pathways for streptovaricins: (**a**) amide-head direction; (**b**) amide-tail direction; △ – □ – ○ = CH_3–CH_2–COOH derived, ■ – ● = CH_3–COOH derived

well incorporated, [1-^{14}C]mevalonolactone (0.068%), [6-^{14}C]glucose (0.05%) and [*carboxy*-^{14}C]propionate (0.032%) were incorporated to an intermediate extent, and sodium [α-^{14}C]pyruvate, [U-^{14}C]tryptophan, [U-^{14}C]phenylalanine, and [4-^{14}C]erythrose hardly at all. [*Carboxy*-^{13}C]-propionate, [*methylene*-^{13}C]malonate, and [*methyl*-^{13}C]methionine were then administered and the carbons labeled were identified, employing the carbon assignments made previously (see Table 5), in the cmr spectra of the labeled streptovaricin D (the most abundant streptovaricin found in the biosynthetic studies). Results showed that [*carboxy*-^{13}C]propionate labels eight carbon atoms to approximately the same extent, C-1, C-5, C-7, C-9, C-11, C-13, C-15 and C-19, thus establishing the amide-head direction of biosynthesis (Scheme **a**) (*90*). By implication, carbons C-2, C-6, C-8, C-10, C-12, C-14, C-16, and C-20, plus the 2-, 6-, 8-, 10-, 12-, 14-, 16- and 20-methyl carbons were also derived from propionate. Sodium [2-^{13}C]malonate labeled equally C-4 and C-18 (by implication, carbons C-3 and C-17 were also derived from malonate), which further confirmed the amide-head path (*89*). The labeling of C-19, one of the ring carbons, by propionate implied a continuous sequence of propionate and acetate units from C-20 through C-1. Neither propionate nor malonate labeled C-21 through C-27.

[*Methyl*-^{13}C]methionine labeled, as expected, the methylenedioxy and methoxy carbons and, in addition, labeled the 25-methyl carbon (*90*). Neither [*methyl*-^{13}C]acetate nor [6-^{13}C]glucose gave enrichment adequate for an interpretable cmr spectrum. The origin of the remaining "C_7N unit" (C-21 through C-27, plus the amide nitrogen) cannot be determined from the data reported thus far on streptovaricin itself. However, results with rifamycin and geldanamycin indicate a shikimate-related pathway, as will be described presently. The origin of the carbons of streptovaricin D is summarized in Fig. 31.

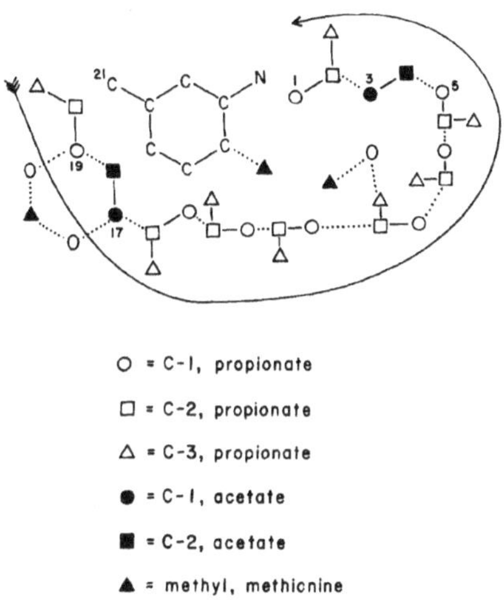

Fig. 31. Proposed biosynthesis of streptovaricin D

No intermediates have been observed thus far for streptovaricin (or any other ansamycin) between the primary metabolite and an assembled unit containing all of the skeletal carbon atoms of the antibiotic. However, considerable evidence is accumulating on the sequence of events which transpire following assemblage of the carbon skeleton. This evidence is derived from the identification of numerous compounds related to the streptovaricins but differing from them in details of oxygenation pattern or methylation (*23*, *91*), and from experiments interconnecting radioactively labeled streptovaricins (*89*).

Damavaricin D (*91*) (*de*acetyl*de*methylenedioxystrepto*varicin* D, Fig. 32), a biologically active precursor which can also be derived by chemical degradation (by treatment with methanolic ammonia) of streptovaricin D (see Section III.A.), was isolated by repeated chromatography of the crude complex from *S. spectabilis*. It was identified largely by comparison of its nmr spectrum to that of streptovaricin D. Damavaricin D lacks three carbons and one oxygen of streptovaricin D. Carbon magnetic resonance, ir, and pmr data indicate that the lost atoms result from replacing the enol acetate by an enol group and the methylenedioxy by phenol and enol groups; in addition, the resulting hydroquinone is oxidized to a quinone.

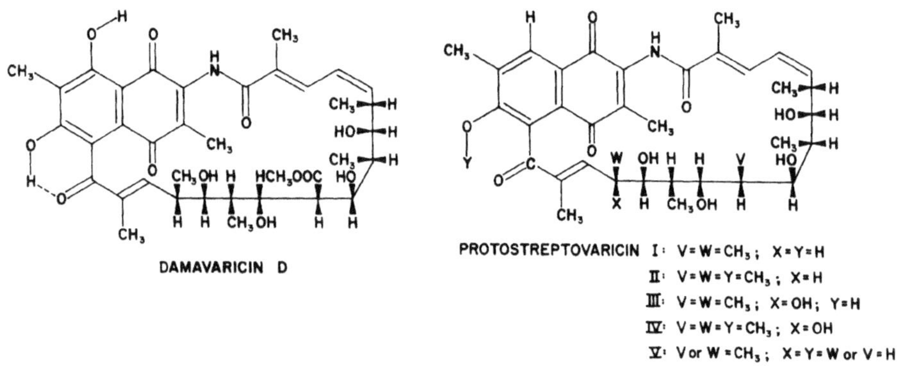

Fig. 32. Biosynthetic precursors to streptovaricins

Protostreptovaricins I—V (*23*) (Fig. 32) are still earlier precursors, isolated in small amounts by repeated chromatography of fractions from the streptovaricin complex. They contain the complete carbon skeleton of the streptovaricins but lack much of their oxygenation. Protostreptovaricins I—IV differ from the streptovaricins in retaining the C-10 methyl (which appears as a carboxyl group in the streptovaricins) and in lacking a hydroxyl group at C-21. In addition, protostreptovaricins I and III lack the methylenedioxy carbon, while protostreptovaricins II and IV contain a methoxyl substituent (instead of a methylenedioxy group) at C-19. The methoxyl group can be converted to a methylenedioxy group by oxidation.

A final intermediate isolated, protostreptovaricin V, is like the others but lacks one aliphatic C-methyl group. The missing methyl presumably has been lost by decarboxylation and thus seems likely to be the C-10 methyl, which appears as a carboxyl in the streptovaricins, or the C-14 methyl, which is lacking in the rifamycins (Fig. 1). If the latter is the case then protostreptovaricin V may well be at the branch point between streptovaricin and rifamycin biosynthesis (*vide infra*). The proposed role of individual protostreptovaricins in the biosynthesis of streptovaricins is shown in Fig. 33. It seems clear that at least some of the steps can be carried out on a variety of substrates.

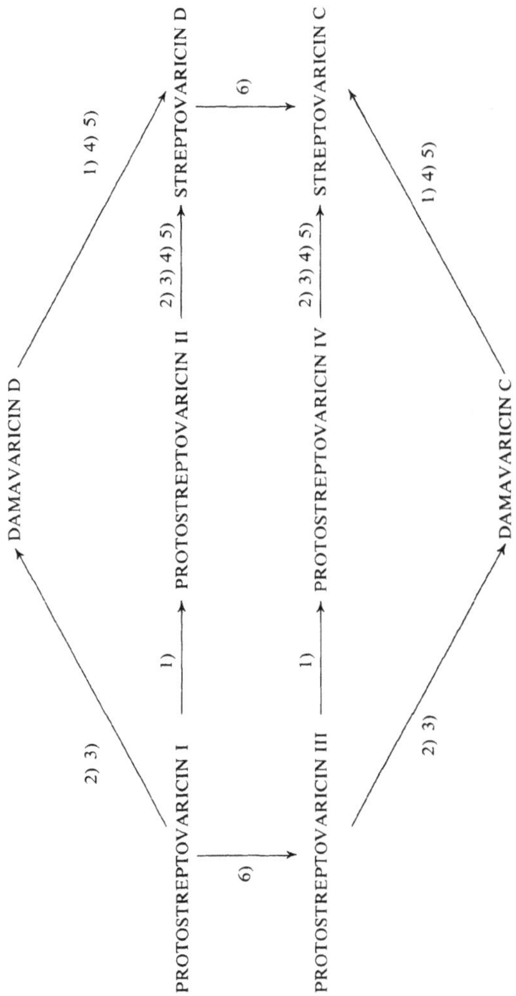

Fig. 33. Proposed routes from protostreptovaricins to streptovaricins. Numbered steps are: 1) methylation of 19-hydroxyl; 2) conversion of 10-methyl to 10-carbomethoxyl group; 3) hydroxylation at C-21; 4) enolization at C-17, oxidation of 17-hydroxyl and 19-methoxyl to methylenedioxy group; 5) reduction at C-24 and C-27 to hydroquinone, acetylation of 24-hydroxyl; 6) hydroxylation at C-14

Radioactive streptovaricin D was prepared biosynthetically by administering [*methyl*-^{14}C]methionine to a growing culture of *S. spectabilis* (*89*). The streptovaricin D isolated was then incubated in a growing culture; after two hours, labeling was found in streptovaricin C as well

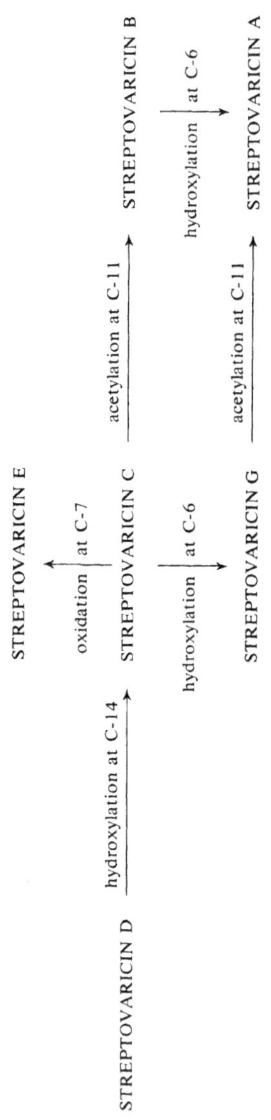

Fig. 34. Probable biosynthetic relationship of the streptovaricins

as D. In a separate experiment [7-^{3}H]streptovaricin C, prepared by reduction of streptovaricin E with sodium borotritide, was administered to a cell-free enzyme preparation. Radioactivity in the streptovaricin complex was then found in streptovaricins B and A as well as C. The probable biosynthetic relationship of the streptovaricins is shown in Fig. 34.

B. Rifamycins

An early report (*19*), based on degradation of radioactive rifamycins derived from ^{14}C-labeled precursors, suggested the amide-tail direction of biosynthesis for rifamycin. This was reversed later in favor of the amide-head direction when sodium [*carboxy*-^{13}C]propionate, sodium [*methylene*-^{13}C]propionate, sodium [*methyl*-^{13}C]propionate sodium [*carboxy*-^{13}C]acetate and sodium [*methyl*-^{13}C]acetate were fed tc *Nocardia mediterranei* cultures (*156*). The rifamycin B isolated was converted for spectroscopic study to rifamycin S, through rifamycin O (see Figs. 1 and 3). Eight propionate units were shown to take part in the formation of rifamycin S, with one of them losing a methyl (the C-14 methyl) in the process. Carbon magnetic resonance data for the propionate-labeled samples showed that C-1, C-5, C-7, C-9, C-11, C-13, C-15, and C-19 were derived from the carboxyl of propionate, C-2, C-6, C-8, C-10, C-12, C-14, C-16, and C-20 were derived from the methylene carbon of propionate, and the 2-, 6-, 8-, 10-, 12-, 16-, and 20-methyl carbons were derived from the methyl group of propionate. This labeling pattern corresponded to the amide-head direction.

The origin of the methoxyl group at C-13 was not demonstrated by a carbon-13 precursor, but the earlier carbon-14 study (*13*) using [*methyl*-^{14}C]methionine showed its origin to be methionine. The remaining C_7N unit was felt to be derived from a metabolite in the glucose→

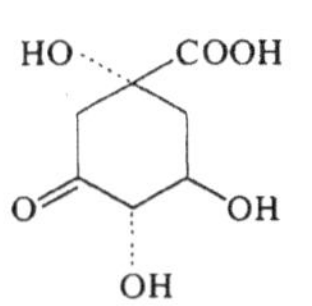

5-dehydroquinic acid

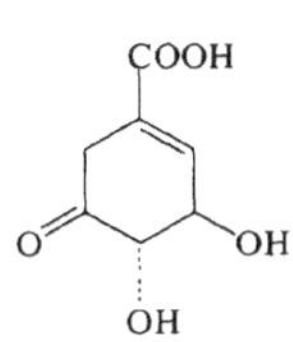

5-dehydroshikimic acid

shikimate pathway, since [1-^{13}C]glucose labeled C-1 and C-10 (*51, 155*) and [1-^{13}C]glycerate labeled C-8 and C-3 (*155*). On the other hand, [U-^{14}C]shikimic acid was not incorporated (*51*), suggesting a branch point before shikimate itself. Positioning of the amino group *meta* to the carboxyl function could occur by transamination if the intermediate metabolite were 5-dehydroshikimate or 5-dehydroquinate.

A mixture of new rifamycins was obtained from a mutant of *Nocardia mediterannei*, which normally produces rifamycin B (*157*). The main component was a new metabolite named rifamycin W (*85*), which lacks the ether linkage between C-15 and C-16 and has an extra methyl at C-14. Incorporation of ^{13}C-enriched precursors led to the conclusion that there are eight propionate units and two acetate units in

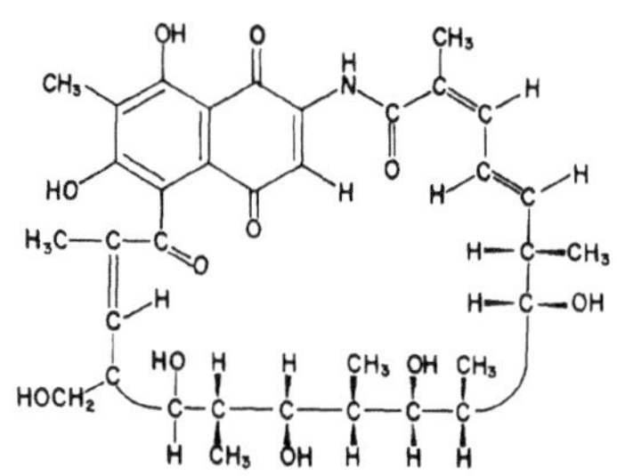

RIFAMYCIN W

rifamycin W, as in rifamycin B and streptovaricin D. The Lepetit workers have also indicated that rifamycin W is converted biologically to rifamycin B, which would require the loss of the propionate-derived methyl group from C-14 and the insertion of an oxygen between two carbons of the same propionate unit (C-15 and C-16).

C. Geldanamycin

Preliminary biosynthetic results for geldanamycin (*46*), a benzoquinonoid ansamycin, showed that it followed the same general pathways as the naphthoquinonoid ansamycins, the streptovaricins and rifamycins. A number of ^{14}C-labeled precursors were fed to fermentation cultures and the geldanamycin produced showed that [*methyl*-^{14}C]methionine and [*carboxy*-^{14}C]propionate were incorporated very well, [*carboxy*-^{14}C]ace-

References, pp. 300—307

tate, [*carboxy*-^{14}C]malonate, *L*-[*guanido*-^{14}C]arginine, and *L*-[*ureido*-^{14}C]-citrulline less well, and formate very little, which suggested that the ansa chain was formed from propionate and acetate, that methionine should label the three O-methyl groups, and that the carbamate group was derived from arginine or citrulline. Conversion to 17-des-O-methylgeldanamycin (Fig. 24) showed that one-third of the label was lost with the 17-methoxyl carbon.

Carbon magnetic resonance spectroscopy was used to demonstrate that sodium [*carboxy*-^{13}C]propionate labeled C-1, C-7, C-9, and C-13 (and, by implication, that C-2, C-8, C-10, C-14 and the 2-, 8-, 10- and 14-methyl carbons were also derived from propionate), and that [*methyl*-^{13}C]methionine was the source of the three methoxyl groups (*42, 46*). This confirmed the amide-head path of biosynthesis (Fig. 35).

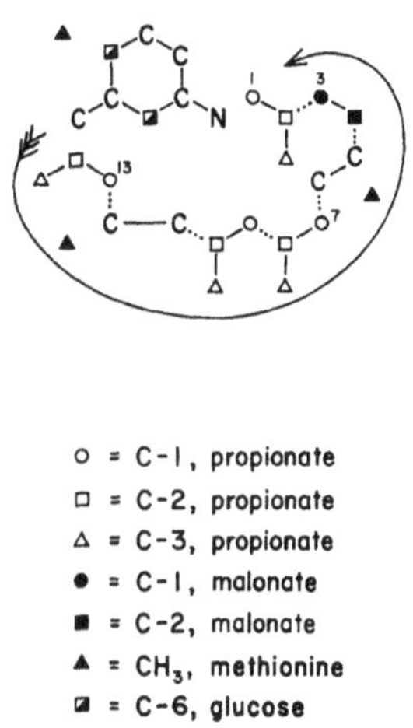

Fig. 35. Proposed biosynthesis of geldanamycin

Further study (*42*) using [*methylene*-^{13}C]malonate has shown that, while the ansa chain of geldanamycin contains four propionate units, only one acetate (malonate) unit is incorporated, at C-3 and C-4. Since four ansa-chain carbon atoms (C-5, C-6, C-10, C-11) remain biosynthetically undefined, geldanamycin is biogenetically distinct from rifamycin S and streptovaricin D. The C_7N unit of geldanamycin has, like that in rifamycin S (*155*), been shown to be derived from glucose; *D*-[6-^{13}C]-

glucose labeled C-17 and C-21 (*42*), which would be the expected positions from a shikimate-type intermediate.

Thus, the three ansamycins whose biosyntheses have been investigated are derived from closely related biosynthetic pathways. The streptovaricin and rifamycin ansa chains are derived from seven propionate and two acetate (malonate) units, the geldanamycin ansa chain from four propionate, one malonate and two still unidentified 2-carbon units. In two of the three antibiotics (and presumably in streptovaricin as well) a C_7N unit is derived from glucose *via* a shikimate-type pathway. Although a number of late intermediates in the biosynthesis of rifamycin and streptovaricins have been identified, much remains to be done.

References

1. ANTOSZ, F. J.: University of Illinois, unpublished results.

2. BALERNA, M., W. KELLER-SCHIERLEIN, C. MARTIUS, H. WOLF, and H. ZÄHNER: Stoffwechselprodukte von Mikroorganismen: Naphthomycin, ein Antimetabolit von Vitamin K. Arch. Mikrobiol. **65,** 303 (1969).

3. BALTIMORE, D.: RNA-Dependent DNA Polymerase in Virions of RNA Tumour Viruses. Nature (London) **226,** 1209 (1970).

4. BERGAMINI, N., and G. FOWST: Rifamycin SV. A Review. Arzneim. Forsch. **15,** 951 (1965).

5. BICKEL, H., F. KNÜSEL, W. KUMP, and L. NEIPP: New Derivatives of Rifamycin Antibiotics. Antimicrob. Agents Chemother. **1966,** 352 (1967).

6. BISSEL, J., C. HATIE, A. N. TISCHLER, and M. CALVIN: Preferential Inhibition of the Growth of Virus-Transformed Cells in Culture by Rifazone-8_2, a New Rifamycin Derivative. Proc. Nat. Acad. Sci. USA **71,** 2520 (1974).

7. BORDEN, E. C., W. W. BROCKMAN, and W. A. CARTER: Selective Inhibition by Streptovaricin of Splenomegaly Induced by Rauscher Leukaemia Virus. Nature **232,** 214 (1971).

8. BROCKMAN, W. W., W. A. CARTER, L.-H. LI, F. REUSSER, and F. R. NICHOL: Streptovaricins Inhibit RNA Dependent DNA Polymerase Present in an Oncogenic RNA Virus. Nature (London) **230,** 249 (1971).

9. BRUFANI, M., S. CERRINI, W. FEDELI, and A. VACIAGO: Rifamycins: An Insight into Biological Activity Based on Structural Investigations. J. Mol. Biol. **87,** 409 (1974).

10. BRUFANI, M., W. FEDELI, G. GIACOMELLO, and A. VACIAGO: Struttura cristallina e molecolare della rifamicine: dati preliminari. Chemotherapia **7,** 145 (1963).

11. — — — — The X-Ray Analysis of the Structure of Rifamycin B. Experienta **20,** 339 (1964).

12. — — — — The X-Ray Analysis of the Structure of Rifamycin Y. Experientia **23,** 508 (1967).

13. BRUFANI, M., D. KLUEPFEL, G. C. LANCINI, J. LEITICH, A. S. MESENTSEV, V. PRELOG, F. P. SCHMOOK, and P. SENSI: Über die Biogenese des Rifamycins S. Helv. Chim. Acta **56,** 2315 (1973).

14. BRYAN, R. F., C. J. GILMORE, and R. C. HALTIWANGER: Crystal and Molecular Structure and Absolute Configuration of Maytansine (3-Bromopropyl) Ether. J. Chem. Soc., Perkin Trans. 2, 7, 897 (1973).
15. BYRD, D. M., and W. A. CARTER: Ansamycins. (B) Streptovaricins. In: W. A. CARTER (Edit.), Selective Inhibitors of Viral Functions. Cleveland: CRC Press, Inc. 1973.
16. CARTER, W. A., W. W. BROCKMAN, and E. C. BORDEN: Streptovaricins Inhibit Focus Formation by MSV (MLV) Complex. Nature **232,** 212 (1971).
17. CARTER, W. A., P. M. PITHA, W. W. BROCKMAN, E. C. BORDEN, and L. W. MARSHALL: Selective Inhibition of Viral Functions: The Antiviral Action of Interferon and Streptovaricin. Medicine **51,** 181 (1972).
18. CORCORAN, J. W.: Speculation on the Origin of Macrolide Antibiotics. In: Z. VANEK and Z. HOSTALEK (Edit.), Biogenesis of Antibiotic Substances, p. 131. Prague: Czechoslovak Academy of Sciences. 1965.
19. CORCORAN, J. W., and M. CHICK: Biochemistry of the Macrolide Antibiotics. In: J. F. SNELL (Edit.), Biosynthesis of Antibiotics, Vol. I, p. 159. New York: Academic Press. 1966.
20. COVERDALE, C. E.: Structural Studies on Streptovaricin C. Ph. D. Thesis, University of Illinois, Urbana, 1965.
21. CRICCHIO, R., G. LANCINI, G. TAMBORINI, and P. SENSI: Oximes of 3-Formylrifamycin SV. Synthesis, Antibacterial Activity, and Other Biological Properties. J. Med. Chem. **17,** 396 (1974).
22. DEBOER, C., P. A. MEULMAN, R. J. WNUK, and D. H. PETERSON: Geldanamycin, A New Antibiotic. J. Antibiot. **23,** 442 (1970).
23. DESHMUKH, P. V., K. KAKINUMA, J. J. AMEEL, K. L. RINEHART, JR., P. F. WILEY, and L. H. LI: Protostreptovaricins I–V. J. Am. Chem. Soc. **98,** 870 (1976).
24. DEZULIAN, V., M. G. SERRALUNGA, and G. MAFFII: Pharmacology and Toxicology of Rifamide. Appl. Toxicol. Pharmacol. **8,** 126 (1966).
25. DUCHAMP, D. J., and K. L. RINEHART, JR.: Unpublished results.
26. FIETTA, A. M., and L. G. SILVESTRI: Mechanism of Action of Rifamazine, a Member of a New Class of (Dimeric) Rifamycins. Eur. J. Biochem. **52,** 391 (1975).
27. FUHRER, H.: Zuordnung des ^{13}C-NMR-Spektrums von Rifamycin-S aufgrund der selektiven Protonen-Entkopplung. Helv. Chim. Acta **56,** 2377 (1973).
28. FÜRESZ, S., V. ARIOLI, and R. PALLANZA: Antimicrobial Properties of New Derivatives of Rifamycin SV. Antimicrob. Agents Chemother. **1965,** 770 (1966).
29. FÜRESZ, S., V. ARIOLI, and R. SCOTTI: Rifamycin B Diethylamide (Rifamide): Activity in Experimental Infections, Absorption and Elimination in Man. Arzneim. Forsch. **15,** 802 (1965).
30. FÜRESZ, S., and R. SCOTTI: Some Laboratory and Clinical Experiences with Rifamycin B. Antibiot. Annu. **1959—1960,** 285 (1960).
31. — — Further Studies on Rifamicin SV: *In Vitro* Activity and Elimination in Man. Farmaco Ed. Sci. **16,** 262 (1961).
32. FÜRESZ, S., and M. T. TIMBAL: Antibacterial Activity of Rifamycins. Chemotherapia **7,** 200 (1963).
33. GALLO, R. C.: Reverse Transcriptase, the DNA Polymerase of Oncogenic RNA Viruses. Nature (London) **234,** 194 (1971).
34. GALLO, G. G., E. MARTINELLI, V. PAGANI, and P. SENSI: The Conformation of Rifamycin S in Solution by ^{1}H NMR Spectroscopy. Tetrahedron **30,** 3093 (1974).
35. GALLO, G. G., C. R. PASQUALUCCI, N. MAGGI, R. BALLOTTA, and P. SENSI: Rifamycins. XLII. Rifazine, a New Phenazinic Derivative of Rifamycins. Farmaco Ed. Sci. **21,** 68 (1966).
36. GANGULY, A. K., S. SZMULEWICZ, O. Z. SARRE, D. GREEVES, J. MORTON, and J. MCGLOTTEN: Structure of Halomicin B. Chem. Commun. 395 (1974).

37. GRECO, A. M., R. BALLOTTA, and P. SENSI: Rifomicina. XXI. Su una serie di prodotti derivanti dalla condensazione della rifomicina O con amine, idrazidi, amidrazoni e aminoguanidine. Farmaco Ed. Sci. **16,** 755 (1961).

38. GREEN, M., J. BRAGDON, and A. RANKIN: 3-Cyclic Amine Derivatives of Rifamycin: Strong Inhibitors of the DNA Polymerase Activity of RNA Tumor Viruses. Proc. Nat. Acad. Sci. USA **59,** 1294 (1972).

39. GURGO, C., D. P. GRANDGENETT, G. F. GERARD, and M. GREEN: Interaction of Purified Ribonucleic Acid Directed Deoxyribonucleic Acid Polymerase of Avian Myeloblastosis Virus and Murine Sarcoma-Leukemia Virus with a Rifamycin SV Derivative. Biochem. **13,** 708 (1974).

40. GURGO, C., R. RAY, and M. GREEN: Inhibitors of Tumor Virus RNA-DNA Polymerase. J. Nat. Cancer Inst. **49,** 61 (1972).

41. HABER, A., R. D. JOHNSON, and K. L. RINEHART, JR.: Geldanamycin. III. Carbon-13 Magnetic Resonance Study. To be submitted for publication.

42. — — — Geldanamycin. IV. Carbon-13 and Carbon-14 Biosynthetic Study. To be submitted for publication.

42a. HARTMANN, G., W. BEHR, K.-A. BEISSNER, K. HONIKEL, and A. SIPPEL: Antibiotica als Hemmstoffe der Nucleinsäure- und Proteinsynthese. Angew. Chem. **80,** 710 (1968).

43. HASEGAWA, T.: Tolypomycin, A New Antibiotic. VII. Assay Method for Tolypomycin Y. J. Antibiot. **25,** 25 (1972).

44. HASEGAWA, T., E. HIGASHIDE, and M. SHIBATA: Tolypomycin, A New Antibiotic. II. Production and Preliminary Identification of Tolypomycin Y. J. Antibiot. **24,** 817 (1971).

45. HOROSZEWICZ, J. S., K. L. RINEHART, JR., S. S. LEONG, and W. A. CARTER: Activity of Pure Streptovaricins and Fractionated Streptovaricin Complex Against Friend Virus. Antimicrob. Agents Chemother. **7,** 281 (1975).

46. JOHNSON, R. D., A. HABER, and K. L. RINEHART, JR.: Geldanamycin Biosynthesis and Carbon Magnetic Resonance. J. Amer. Chem. Soc. **96,** 3316 (1974).

47. KAKINUMA, K.: University of Illinois, unpublished results.

48. KAKINUMA, K., and P. V. DESHMUKH: University of Illinois, unpublished results.

49. KAKINUMA, K., B. I. MILAVETZ, and K. L. RINEHART, JR.: Carbon-13 Nuclear Magnetic Resonance Spectra of the Streptovaricins and Related Compounds. J. Org. Chem., in press.

50. KAMIYA, K., T. SUGINO, Y. WADA, M. NISHIKAWA, and T. KISHI: The X-Ray Analysis of Tolypomycinone Tri-*m*-bromobenzoate. Experientia **25,** 901 (1969).

51. KARLSSON, A., G. SARTORI, and R. J. WHITE: Rifamycin Biosynthesis: Further Studies on Origin of the Ansa Chain and Chromophore. Eur. J. Biochem. **47,** 251 (1974).

52. KISHI, T., M. ASAI, M. MUROI, S. HARADA, E. MIZUTA, S. TERAO, T. MIKI, and K. MIZUNO: Tolypomycin. I. Structure of Tolypomycinone. Tetrahedron Lett. **2,** 91 (1969).

53. KISHI, T., S. HARADA, M. ASAI, M. MUROI, and K. MIZUNO: Tolypomycin. II. Structures of Tolyposamine and Tolypomycin Y. Tetrahedron Lett. **2,** 97 (1969).

54. KISHI, T., H. YAMANA, M. MUROI, S. HARADA, M. ASAI, T. HASEGAWA, and K. MIZUNO: Tolypomycin, a New Antibiotic. III. Isolation and Characterization of Tolypomycin Y. J. Antibiot. **25,** 11 (1972).

55. KLUEPFEL, D., G. C. LANCINI, and G. SARTORI: Metabolism of Barbital by *Streptomyces mediterranei.* Appl. Microbiol. **13,** 600 (1965).

56. KNÖLL, W. M. J.: University of Illinois, unpublished results.

57. KONDO, M., T. OISHI, and K. TSUCHIYA: Tolypomycin, A New Antibiotic. V. *In Vitro* and *In Vivo* Antimicrobial Activity. J. Antibiot. **25,** 16 (1972).

58. KRADOLFER, F., L. NEIPP, and W. SACKMANN: Chemotherapeutic Activity of New Derivatives of Rifamycin. Antimicrob. Agents Chemother. **1966,** 359 (1967).

59. KUMP, W., and H. BICKEL: Zur Kenntnis von Rifamycin-S. Reaktionen des Ansa-ringes. Helv. Chim. Acta **56**, 2323 (1973).
60. — — Zur Kenntnis von Rifamycin-S. Reaktionen des chinoiden Nucleus. Helv. Chim. Acta **56**, 2348 (1973).
61. KUPCHAN, S. M., Y. KOMODA, A. R. BRANFMAN, R. G. DAILEY, JR., and V. A. ZIMMERLY: Novel Maytansinoids. Structural Interrelations and Requirements for Antileukemic Activity. J. Amer. Chem. Soc. **96**, 3706 (1974).
62. KUPCHAN, S. M., Y. KOMODA, W. A. COURT, G. J. THOMAS, R. M. SMITH, A. KARIM, C. J. GILMORE, R. C. HALTIWANGER, and R. F. BRYAN: Maytansine, a Novel Antileukemic Ansa Macrolide from *Maytenus ovatus*. J. Amer. Chem. Soc. **94**, 1354 (1972).
63. KUPCHAN, S. M., Y. KOMODA, G. J. THOMAS, and H. P. J. HINTZ: Maytanprine and Maytanbutine, New Antileukemic Ansa Macrolides from *Maytenus buchananii*. Chem. Commun. 1065 (1972).
64. LANCINI, G. C., G. G. GALLO, G. SARTORI, and P. SENSI: Isolation and Structure of Rifamycin L and Its Biogenetic Relationship with Other Rifamycins. J. Antibiot. **22**, 369 (1969).
65. LANCINI, G., and C. HENGELLER: Isolation of Rifamycin SV from a Mutant *Streptomyces mediterranei* Strain. J. Antibiot. **22**, 637 (1969).
66. LANCINI, G. C., J. E. THIEMANN, G. SARTORI, and P. SENSI: Biogenesis of Rifamycins. The Conversion of Rifamycin B into Rifamycin Y. Experientia **23**, 899 (1967).
67. LEITICH, J., W. OPPOLZER, and V. PRELOG: Über die Konfiguration des Rifamycins B und verwandter Rifamycine. Experientia **20**, 343 (1964).
68. LEITICH, J., V. PRELOG, and P. SENSI: Rifamycin Y und seine Umwandlungsprodukte. Experientia **23**, 505 (1967).
69. LESTER, W.: Rifampin: A Semisynthetic Derivative of Rifamycin – A Prototype for the Future. Annu. Rev. Microbiol. **26**, 85 (1972).
70. LI, L. H., C. H. COWIE, L. G. GRAY, D. M. MORAN, T. D. CLARK, and K. L. RINEHART, JR.: Effects of Streptovaricins and Their Degradation Products on RNA-Directed DNA Polymerase of Rauscher Leukemia Virus. Submitted for publication.
71. LÜTTRINGHAUS, A., and H. GRALHEER: Über eine neue Art atropisomerer Verbindungen. Justus Liebigs Ann. Chem. **550**, 67 (1942).
72. MAFFII, G., G. BIANCHI, P. SCHIATTI, and G. G. GALLO: Absorption, Biliary Excretion and Distribution of Rifomycin SV. Farmaco Ed. Sci. **16**, 246 (1961).
73. MAFFII, G., and P. SCHIATTI: Distribution and Excretion of Rifamide in Rats and Dogs. Toxicol. Appl. Pharmacol. **8**, 138 (1966).
74. MAFFII, G., P. SCHIATTI, G. BIANCHI, and M. G. SERRALUNGA: Rifomycin. XVIII. Pharmacological Studies with Rifomycin SV. Farmaco Ed. Sci. **16**, 235 (1961).
75. MAFFII, G., and M. T. TIMBAL: Pharmacology of Rifamycin B. Antibiot. Annu. **1959—1960**, 277 (1960).
76. MAGGI, N., V. ARIOLI, and P. SENSI: Rifamycins. XLI. A New Class of Active Semisynthetic Rifamycins. N-Substituted Aminomethyl Derivatives of Rifamycin SV. J. Med. Chem. **8**, 790 (1965).
77. MAGGI, N., G. G. GALLO, and P. SENSI: Rifamicine. XXXVI. Proprietà chimiche e chimico-fisiche della dietilamide della rifamicina B. Farmaco Ed. Prat. **20**, 147 (1965).
78. MAGGI, N., R. PALLANZA, and P. SENSI: New Derivatives of Rifamycin SV. Antimicrob. Agents Chemother. **1965**, 765 (1966).
79. MAGGI, N., C. R. PASQUALUCCI, R. BALLOTTA, and P. SENSI: Rifampicin: A New Orally Active Rifamycin. Chemotherapia **11**, 285 (1966).
80. MAGGI, N., A. VIGEVANI, G. G. GALLO, and C. R. PASQUALUCCI: Acetyl Migration in Rifampicin. J. Med. Chem. **11**, 936 (1968).
81. MAGGI, N., A. VIGEVANI, and R. PALLANZA: Desacetyl-Rifamycins: Preparation and Antibacterial Properties. Experientia **24**, 209 (1968).

82. MARGALITH, P., and H. PAGANI: Rifomycin. XIII. Fermentation and Production of Rifomycin Complex. Appl. Microbiol. **9**, 320 (1961).
83. — — Rifomycin. XIV. Production of Rifomycin B. Appl. Microbiol. **9**, 325 (1961).
84. MARTIN, P. K.: Structural Studies on the Streptovaricin Antibiotics. Ph. D. Thesis, University of Illinois, Urbana, Illinois, 1965.
85. MARTINELLI, E., G. G. GALLO, P. ANTONINI, and R. J. WHITE: Structure of Rifamycin W. A Novel Ansamycin from a Mutant of *Nocardia mediterranei*. Tetrahedron **30**, 3087 (1974).
86. MARTINELLI, E., R. J. WHITE, G. G. GALLO, and P. J. BEYNON: Carbon-13 NMR Spectra of Rifamycins. Tetrahedron **29**, 3441 (1973).
87. — — — — Carbon-13 NMR Spectrum of Rifamycin S: A Re-examination of the Assignments with Special Reference to their Biogenetic Implication. Tetrahedron Lett. **15**, 1367 (1974).
88. McMILLAN, M. W.: Studies on the Antibiotic Geldanamycin and Biologically Active Derivatives and Analogues. Ph. D. Thesis, University of Illinois, Urbana, Illinois, 1974.
89. MILAVETZ, B. I.: Biosynthesis of the Streptovaricins. Ph. D. Thesis, University of Illinois, Urbana, Illinois, 1975.
90. MILAVETZ, B., K. KAKINUMA, K. L. RINEHART, JR., J. P. ROLLS, and W. J. HAAK: Carbon-13 Magnetic Resonance Spectroscopy and the Biosynthesis of Streptovaricin. J. Amer. Chem. Soc. **95**, 5793 (1973).
91. MILAVETZ, B. I., K. L. RINEHART, JR., F. J. ANTOSZ, P. V. DESMUKH, K. KAKINUMA, P. K. MARTIN, K. SASAKI, T. R. WITTY, L. H. LI, and F. REUSSER: Identification and Preparation of Damavaricins, Biologically Active Precursors of Streptovaricins. J. Antibiot. **29**, 201 (1976).
92. MOSS, B.: Ansamycins: (A) Rifamycin SV Derivatives. In: W. A. CARTER (Edit.), Selective Inhibitors of Viral Functions, p. 313. Cleveland: CRC Press, Inc. 1973.
93. OPPOLZER, W., and V. PRELOG: Über die Konstitution und die Konfiguration der Rifamycine B, O, S und SV. Helv. Chim. Acta **56**, 2287 (1973).
94. OPPOLZER, W., V. PRELOG, and P. SENSI: Konstitution des Rifamycins B und Verwandter Rifamycine. Experientia **20**, 336 (1964).
95. PALLANZA, R., S. FÜRÉSZ, M. T. TIMBAL, and G. CARNITI: *In Vitro* Bacteriological Studies on Rifamycin B Diethylamide (Rifamide). Arzneim. Forsch. **15**, 800 (1965).
96. PETERSON, D. H., K. L. RINEHART, JR., and R. PANDEY: Unpublished results. The Upjohn Company and the University of Illinois.
97. PRELOG, V.: Constitution of Rifamycins. Pure Appl. Chem. **7**, 551 (1963).
98. PRELOG, V., and W. OPPOLZER: Ansamycine, eine neuartige Klasse von mikrobiellen Stoffwechselprodukten. Helv. Chim. Acta **56**, 2279 (1973).
99. QUINN, F. R., and J. S. DRISCOLL: Structure-Activity Correlations among Rifamycin B Amides and Hydrazides. J. Med. Chem. **18**, 4 (1975).
100. RHULAND, L. E., K. F. STERN, and H. R. REAMES: Streptovaricin. III. *In vivo* Studies in the Tuberculous Mouse. Am. Rev. Tuberc. Pulm. Dis. **75**, 588 (1957).
101. RINEHART, K. L., JR.: Antibiotics with Ansa Rings. Acc. Chem. Res. **5**, 57 (1972).
102. RINEHART, K. L., JR., and F. J. ANTOSZ: Chemistry of the Streptovaricins. VII. Revised Structures for Streptovarone and Streptovaricin C. J. Antibiot. **25**, 71 (1972).
103. RINEHART, K. L., JR., F. J. ANTOSZ, K. SASAKI, P. K. MARTIN, M. L. MAHESHWARI, F. REUSSER, L. H. LI, D. MORAN, and P. F. WILEY: Relative Biological Activities of Individual Streptovaricins and Streptovaricin Acetates. Biochemistry **13**, 861 (1974).
103a. RINEHART, K. L., JR., J. C. COOK, JR., K. H. MAURER, and U. RAPP: Field Desorption Mass Spectra of Antibiotics. J. Antibiot. **27**, 1 (1974).
104. RINEHART, K. L., JR., C. E. COVERDALE, and P. K. MARTIN: Chemistry of the Streptovaricins. II. Streptovarone and Prestreptovarone. J. Amer. Chem. Soc. **88**, 3150 (1966).

105. RINEHART, K. L., JR., W. M. J. KNÖLL, K. KAKINUMA, F. J. ANTOSZ, I. C. PAUL, A. H.-J. WANG, F. REUSSER, L. H. LI, and W. C. KRUEGER: Atropisomeric Streptovaricins. J. Amer. Chem. Soc. **97,** 196 (1975).

106. RINEHART, K. L., JR., W. M. J. KNÖLL, B. MILAVETZ, and K. KAKINUMA: Carbon-13 in Biosynthetic Studies of Antibiotics. In: P. D. KLEIN and S. V. PETERSON (Edit.), Proceedings of the First International Conference on Stable Isotopes in Chemistry, Biology, and Medicine (CONF-730525), p. 20. Springfield, Va.: National Technical Information Service, U.S. Department of Commerce. 1973.

107. RINEHART, K. L., JR., M. L. MAHESHWARI, F. J. ANTOSZ, H. H. MATHUR, K. SASAKI, and R. J. SCHACHT: Chemistry of the Streptovaricins. VIII. Structures of Streptovaricins A, B, D, E, F, and G. J. Amer. Chem. Soc. **93,** 6273 (1971).

108. RINEHART, K. L., JR., P. K. MARTIN, and C. E. COVERDALE: Chemistry of the Streptovaricins. I. Characterization of Streptovaricins A, B, C, D, E, and G. J. Amer. Chem. Soc. **88,** 3149 (1966).

109. RINEHART, K. L., JR., and H. H. MATHUR: Chemistry of the Streptovaricins. IV. Structure of Varicinal A. J. Amer. Chem. Soc. **90,** 6241 (1968).

110. RINEHART, K. L., JR., H. H. MATHUR, K. SASAKI, P. K. MARTIN, and C. E. COVERDALE: Chemistry of the Streptovaricins. V. Structures of Streptovaricins A and C. J. Amer. Chem. Soc. **90,** 6241 (1968).

111. RIVA, S., and L. G. SILVESTRI: Rifamycins: A General View. Annu. Rev. Microbiol. **62,** 199 (1972).

112. SASAKI, K.: University of Illinois, unpublished results.

112a. SASAKI, K., T. NAITO, T. SATOMI, and K. ONODERA: Chemical Modification of Streptovaricin C. I. 19-Substituted Damavaricin C. J. Antibiot., in press.

112b. SASAKI, K., T. NAITO, T. SATOMI, and K. MOMOKI: Chemical Modification of Streptovaricin C. II. The Intramolecular Aldol Condensation of 19-O-Acetonyldamavaricin C and its Analogs. J. Antibiot., in press.

113. SASAKI, K., K. L. RINEHART, JR., and F. J. ANTOSZ: Chemistry of the Streptovaricins. VI. Oxidation Products from Streptovaricin C. J. Antibiot. **25,** 68 (1972).

114. SASAKI, K., K. L. RINEHART, JR., G. SLOMP, M. F. GROSTIC, and E. C. OLSON: Geldanamycin. I. Structure Assignment. J. Amer. Chem. Soc. **92,** 7591 (1970).

115. SCHACHT, R. J.: Studies in the Photochemical Derivatives of the Streptovaricin Antibiotics. Ph. D. Thesis, University of Illinois, Urbana, Illinois. 1969.

116. SCHACHT, R. J., and K. L. RINEHART, JR.: Chemistry of the Streptovaricins. III. Photostreptovarone, a Phenalenone from Photolytic Loss of Acetic Anhydride. J. Amer. Chem. Soc. **89,** 2239 (1967).

117. SCHLEGEL, D. C.: Part One. Synthetic and Degradative Studies of Two Streptovaricin Degradation Products, Streptovarone and Dapmavarone. Part Two. Synthetic and Spectral Studies of Highly Substituted 1,4-Naphthoquinones. Ph.D. Thesis, University of Illinois, Urbana, Illinois. 1969.

118. SCHWARZ, J. S. P.: Identification of Rifamycin B as Nancimycin. J. Antibiot. **20,** 238 (1967).

119. SENSI, P.: A Family of New Antibiotics, the Rifamycins. Prog. Org. Biol. Med. Chem. **1,** 337 (1964).

120. — Inhibitors of the Transcribing Enzymes. Pure Appl. Chem. **35,** 383 (1973).

121. SENSI, P., R. BALLOTTA, and A. M. GRECO: Rifomycin. V. Rifomycin O, a New Antibiotic of the Rifomycin Family. Farmaco Ed. Sci. **15,** 228 (1960).

122. SENSI, P., R. BALLOTTA, A. M. GRECO, and G. G. GALLO: Rifomycin. XV. Activation of Rifomycin B and Rifomycin O. Production and Properties of Rifomycin S and Rifomycin SV. Farmaco Ed. Sci. **16,** 165 (1961).

123. SENSI, P., A. M. GRECO, and R. BALLOTTA: Rifomycin. I. Isolation and Properties of Rifomycin B and Rifomycin Complex. Antibiot. Annu. **1959—1960,** 262 (1960).

124. Sensi, P., A. M. Greco, G. G. Gallo, and M. T. Timbal: Rifomycin. VI. Chemical and Biological Correlations between some Derivatives of Rifomycin B and Rifomycin O. Farmaco Ed. Sci. **15,** 235 (1960).

125. Sensi, P., N. Maggi, R. Ballotta, S. Fürész, R. Pallanza, and V. Arioli: Rifamycins. XXXV. Amides and Hydrazides of Rifamycin B. J. Med. Chem. **7,** 596 (1964).

126. Sensi, P., N. Maggi, S. Fürész, and G. Maffii: Chemical Modifications and Biological Properties of Rifamycins. Antimicrob. Agents Chemother. **1966,** 699 (1967).

127. Sensi, P., P. Margalith, and M. T. Timbal: Rifomycin, A New Antibiotic – Preliminary Report. Farmaco Ed. Sci. **14,** 146 (1959).

128. Sensi, P., M. T. Timbal, and G. Maffii: Rifomycin. IX. Two New Antibiotics of Rifomycin Family: Rifomycin S and Rifomycin SV. Experientia **16,** 412 (1960).

129. Shibata, M., T. Hasegawa, and E. Higashide: Tolypomycin, A New Antibiotic. I. *Streptomyces tolypophorus* nov. sp., a New Antibiotic, Tolypomycin-Producer. J. Antibiot. **24,** 810 (1971).

130. Siminoff, P., R. M. Smith, W. T. Sokolski, and G. M. Savage: Streptovaricin. I. Discovery and Biologic Activity. Am. Rev. Tuberc. Pulm. Dis. **75,** 576 (1957).

131. Smith, R. G., and R. C. Gallo: Minireview. Agents Which Inhibit Reverse Transcriptases. Life Sci. **15,** 1711 (1974).

132. Smith, R. G., J. Whang-Peng, R. C. Gallo, P. Levine, and R. C. Ting: Selective Toxicity of Rifamycin Derivatives for Leukaemic Human Leucocytes. Nature (London) New Biol. **236,** 166 (1972).

133. Sokolski, W. T., N. J. Eilers, and P. Siminoff: Paper Chromatography and Assay of Components of Streptovaricin. Antibiot. Annu. **1957—1958,** 119 (1958).

134. Sugarawa, S., K. Karasawa, M. Watanabe, and T. Hidaka: Production of Rifamycin O by Streptomyces 4107A_2. J. Antibiot. Ser. A, **17,** 29 (1964).

135. Temin, H. M., and S. Mizutani: RNA-Dependent DNA Polymerase in Virions of Rous Sarcoma Virus. Nature (London) **226,** 1211 (1970).

136. Thiemann, J. E., G. Zucco, and G. Pelizza: A Proposal for the Transfer of *Streptomyces mediterranei* Margalith and Beretta 1960 to the Genus *Nocardia* as *Nocardia mediterranei* (Margalith and Beretta) Comb. Nov. Arch. Mikrobiol. **67,** 147 (1969).

137. Thiry, L., and G. Lancini: Mode of Action of Rifamycin and Aminopiperazine Derivatives on Animal Viruses and Cells. FEBS Sym. **22,** 177 (1972).

138. Thompson, F. M., L. J. Libertini, U. R. Joss, and M. Calvin: Detergent Effects on a Reverse Transcriptase Activity and on Inhibition by Rifamycin Derivatives. Science **178,** 505 (1972).

139. Thompson, F. M., A. N. Tischler, J. Adams, and M. Calvin: Inhibition of Three Nucleotide Polymerases by Rifamycin Derivatives. Proc. Nat. Acad. Sci. USA, **71,** 107 (1974).

140. Timbal, M. T.: Rifomycin. II. Antibacterial Activity of Rifomycin B. Antibiot. Annu. **1959—1960,** 271 (1960).

141. Timbal, M. T., and A. Brega: Rifomycin. XVII. Rifomycin SV, Treatment of Experimental Infections. Farmaco Ed. Sci. **16,** 191 (1961).

142. Timbal, M. T., R. Pallanza, and G. Carniti: Rifomycin. XVI. Bacteriological Studies of Rifomycin SV *in vitro*. Farmaco Ed. Sci. **16,** 181 (1961).

143. Ting, R. C., S. S. Yang, and R. C. Gallo: Reverse Transcriptase, RNA Tumour Virus Transformation and Derivatives of Rifamycin SV. Nature (London) New Biol. **236,** 163 (1972).

144. Tipton, C. D.: Structural and Synthetic Studies of the Antibiotics Lomofungin and Geldanamycin. Ph. D. Thesis. University of Illinois, Urbana. 1971.

145. Tischler, A. N., U. R. Joss, F. M. Thompson, and M. Calvin: Synthesis of Some Rifamycin Derivatives as Inhibitors of an RNA-Instructed DNA Polymerase Function. J. Med. Chem. **16,** 1071 (1973).

146. TISCHLER, A. N., F. M. THOMPSON, L. J. LIBERTINI, and M. CALVIN: Rifamycin Derivatives as Inhibitors of a Ribonucleic Acid Instructed Deoxyribonucleic Acid Polymerase Function. Effect of Lipophilicity. J. Med. Chem. **17,** 948 (1974).

147. VANEK, Z., and J. MAJER: Macrolide Antibiotics. In: D. GOTTLIEB and P. D. SHAW (Edit.), Antibiotics. Vol. II. Biosynthesis, p. 154. New York: Springer-Verlag. 1967.

148. WANG, A. H.-J., and I. C. PAUL: Structure of a Derivative of Streptovaricin C Triacetate: Crystal and Molecular Structure of the Atropisomer of the Cyclic *p*-Bromobenzeneboronate Ester of Streptovaricin C Triacetate. To be submitted for publication.

149. WANG, A. H.-J., I. C. PAUL, K. L. RINEHART, JR., and F. J. ANTOSZ: Chemistry of the Streptovaricins. IX. X-Ray Crystallographic Structure of a Streptovaricin C Derivative. J. Amer. Chem. Soc. **93,** 6275 (1971).

150. WANI, M. C., H. L. TAYLOR, and M. E. WALL: Plant Antitumour Agents: Colubrinol Acetate and Colubrinol, Antileukaemic Ansa Macrolides from *Colubrina texensis.* Chem. Commun. 390 (1973).

151. WEHRLI, W., F. KNÜSEL, K. SCHMID, and M. STAEHELIN: Interaction of Rifamycin with Bacterial RNA Polymerase. Proc. Nat. Acad. Sci. USA **61,** 667 (1968).

152. WEHRLI, W., and M. STAEHELIN: The Rifamycins – Relation of Chemical Structure and Action on RNA Polymerase. Biochim. Biophys. Acta **182,** 24 (1969).

153. — — Actions of the Rifamycins. Bacteriol. Rev. **35,** 290 (1971).

154. WEINSTEIN, M. J., G. M. LUEDEMANN, E. M. ODEN, and G. H. WAGMAN: Halomicin, A New *Micromonospora*-produced Antibiotic. Antimicrob. Agents Chemother. **1967,** 435 (1968).

155. WHITE, R. J., and E. MARTINELLI: Ansamycin Biogenesis: Incorporation of [1-^{13}C]-Glucose and [1-^{13}C]Glycerate into the Chromophore of Rifamycin S. FEBS Lett. **49,** 233 (1974).

156. WHITE, R. J., E. MARTINELLI, G. G. GALLO, G. LANCINI, and P. BEYNON: Rifamycin Biosynthesis Studied with ^{13}C Enriched Precursors and Carbon Magnetic Resonance. Nature (London) **243,** 273 (1973).

157. WHITE, R. J., E. MARTINELLI, and G. LANCINI: Ansamycin Biogenesis: Studies on a Novel Rifamycin Isolated from a Mutant Strain of *Nocardia mediterranei.* Proc. Nat. Acad. Sci. USA **71,** 3260 (1974).

158. WHITFIELD, G. B., E. C. OLSON, R. R. HERR, J. A. FOX, M. E. BERGY, and G. A. BOYACK: Streptovaricin. II. Isolation and Properties. Am. Rev. Tuberc. Pulm. Dis. **75,** 584 (1957).

159. WILLIAMS, T. H.: Naphthomycin, A Novel Ansa Macrocyclic Antimetabolite. Proton NMR Spectra and Structure Elucidation Using Lanthanide Shift Reagent. J. Antibiot. **28,** 85 (1975).

160. WITTY, T. R.: Unpublished results. University of Illinois.

161. WOLPERT-DEFILIPPES, M. K., R. H. ADAMSON, R. L. CYSYK, and D. G. JOHNS: Initial Studies on Cytotoxic Action of Maytansine, a Novel Ansa Macrolide. Biochem. Pharmacol. **24,** 751 (1975).

162. YAMAZAKI, H., S. MIZUNO, K. NITTA, R. UTAHARA, and H. UMEZAWA: Biochemical Mechanism of Action of Substance B44P. J. Antibiot. **21,** 63 (1968).

163. YANG, S. S., F. M. HERRERA, R. G. SMITH, M. S. REITZ, G. LANCINI, R. TING, and R. C. GALLO: Rifamycin Antibiotics: Inhibitors of Rauscher Murine Leukemia Virus Reverse Transcriptase and of Purified DNA Polymerases from Human Normal and Leukemic Lymphoblasts. J. Nat. Cancer Inst. **49,** 7 (1972).

164. ZERILLI, L. F., and G. G. GALLO: EUR. Report **4765,** f.i.e., 217 (1971).

165. ZERILLI, L. F., M. LANDI, G. G. GALLO, K. H. MAURER, and U. RAPP: EI and FD Mass Spectra of Rifamycins. 23rd Annual Conference on Mass Spectrometry and Allied Topics, Houston, 1975.

166. ZERILLI, L. F., M. LANDI, N. RIMORINI, and G. G. GALLO: Spettri di Massa delle Rifamicine. Ann. Chim. (Rome) **64,** 199 (1974).

(Received August 25, 1975)

The Chemistry of Tryptophan in Peptides and Proteins

By A. Fontana and C. Toniolo,
Istituto di Chimica Organica dell'Università, Padova, Italy

With 10 Figures

Contents

Acknowledgements. – The authors wish to express their gratitude and admiration to their distinguished teacher in peptide and protein chemistry, the late Professor ERNESTO SCOFFONE. Much of the scientific work of one of us (A. F.) here reviewed has been carried out under this guidance.

Special thanks are due to Dr. W. E. SAVIGE (CSIRO, Division of Protein Chemistry, Parkville, Australia) and to Dr. T. F. SPANDE (National Institutes of Health, Bethesda, U.S.A.) for making unpublished material available, and in particular to Dr. SAVIGE for help in the preparation of the manuscript. The valuable aid of Mrs. M. PERNECHELE-GIACON in typing the manuscript is also acknowledged.

This work was carried out with Italian C.N.R. aid.

I. Introduction

The chemistry of tryptophan (**1**) has probably been the object of more intense investigation than that of any other amino acids. This is

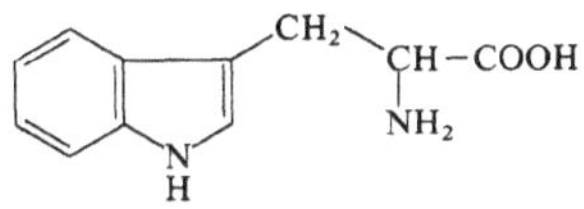

(1)

undoubtedly due to the particular reactivity of the indole nucleus and to the extensive chemistry developed over the years regarding this important and ubiquitous functionality. Only the sulfur amino acids have received comparable attention from chemists and biochemists.

Tryptophan is a widely distributed naturally occurring amino acid, essential to man and many living organisms. In addition to being incorporated into proteins, tryptophan is metabolized along several different pathways, which lead to compounds of great biological importance.

The impetus for continued research into the chemistry of tryptophan in proteins during the last period has come largely from interest in the chemical modification of this amino acid residue for structure-activity studies on enzymes. In addition, many studies have been prompted by a desire to elucidate the structure of the numerous metabolic breakdown products of tryptophan itself and also by an interest in the synthesis of this essential amino acid.

The aims of research on the chemistry and biochemistry of tryptophan are so varied that complete coverage of the subject warrants a book rather than a chapter. The reader may find herein a somewhat personal selection of subject matter, which includes brief accounts of important topics as well as omissions. The availability of reviews has allowed the authors to make only brief summaries in some instances and to discuss in more detail only the more recent contributions.

The topics selected for coverage are generally related to protein chemistry and progress from synthesis of tryptophan to a study of the role of tryptophan residues in the biological function of enzymes. For a better comprehension of the material reviewed in this chapter, a short presentation of some aspects of the chemical synthesis, biosynthesis and metabolism of tryptophan is also included.

Although the main part of the chapter covers the chemistry of the tryptophan residue, a section dealing with physico-chemical techniques used in assessing structural characteristics of tryptophan-containing peptides and proteins has been included. Tryptophan residues contribute significantly to the optical, stereooptical and fluorescent properties of proteins and their spectral characteristics are considered to be important gauges of protein conformation.

II. General

1. Structure and Reactivity

The resonance energy of the indole nucleus has been evaluated as 47–49 kcal/mole from its heat of combustion. Of the resonance structures which can be written, structure (**2**) has to be considered the most important since it involves full conjugation in both rings. Of the

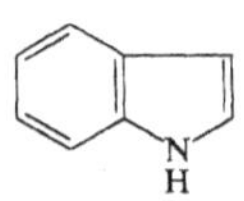

(**2**)

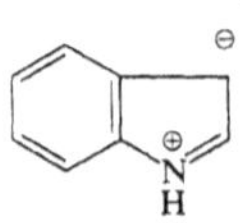

(**3**)

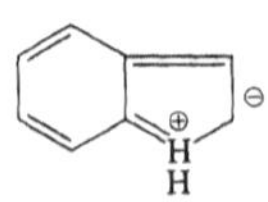

(**4**)

References, pp. 426—449

ionic structures (**3**) and (**4**), only (**3**) with a negative charge at the 3-position permits all the uncharged resonance forms possible for the benzene ring, whereas (**4**) involves an *o*-quinoid benzene ring. Formula (**3**) explains the higher reactivity of position 3 toward electrophilic reagents (*1, 398*).

Theoretical calculations to predict the reactivity of the indole nucleus give indefinite results (*61, 322, 399*). Several approaches have been used, all successfully predicting high reactivity at C-3, but none is able to give a satisfactory order of reactivity for the other positions. These methods do not take into account possible steric effects which might be responsible for the decrease in the experimentally observed reactivity; differences in the structures of attacking electrophiles are also neglected (*137*).

Calculations of π-electron densities of indole using the HÜCKEL method, when appropriate parameters are introduced into the calculation to take into account the electronegativity of the nitrogen atom, give a π-density order of $3 > 2 > 4 \sim 5 \sim 6 \sim 7$ (*61*). The method employed predicts correctly the order of positional reactivity of the pyrrole ring, whereas it has not predictive value for positions in the benzene ring.

With 3-substituted indoles, two general possibilities for the experimentally observed 2-substitution of electrophilic reagents have been considered (*207, 208*). The electrophile can attack directly at C-2 or it can attack the more nucleophilic 3-position and subsequently rearrange intramolecularly to C-2 as shown below.

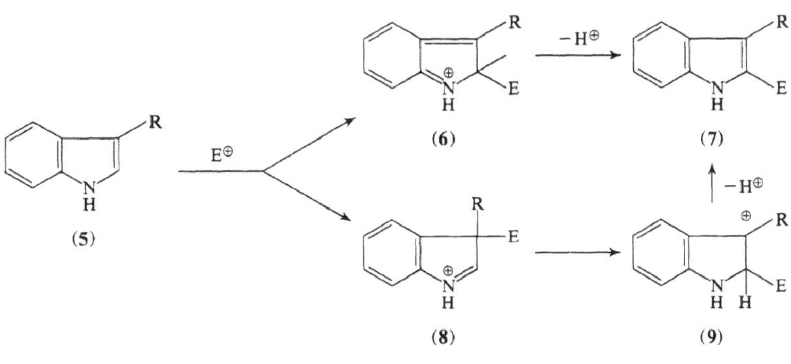

The intermediate (**8**) must be considered the more stable in the absence of steric effects. On the other hand the formation of (**6**) could be kinetically favored with bulky substituents.

SUNDBERG (*398*) has reviewed electrophilic substitution in 3-substituted indoles and, on the basis of the bulk of evidence available, concluded that electrophilic substitution seems to proceed by initial attack at C-3 followed by rearrangement to the 2-substituted indole. However, evidence has now been found for possible direct substitution at C-2, without initial attack at the 3-position followed by migration of the substituent.

Recent studies carried out by CASNATI and coworkers (*69, 70*) seem to indicate that revision of the mechanism outlined above is necessary. These authors studied the reactivity of 3-alkylated indoles towards alkylation with allyl bromides and clearly showed that direct attack at position 2 of the indole nucleus is competitive with alkylation at position 3. In particular, with reagents of enhanced electrophilic character direct attack at the 2-position increases (*69*). It was suggested that the positional selectivity is determined by the electronic density of 3-alkyl-indoles. In fact, an alkyl group in an aromatic system decreases the electron density in the α-position and increases it in the β-position (*272*).

Another hitherto unobserved type of rearrangement which may have synthetic and biogenetic implications has been found by CASNATI *et al.* (*71*). 3-Alkyl-1-allyl-indoles rearrange under acid catalysis by trifluoroacetic acid to give 2-allyl derivatives. A 3-protonated species seems to be the most probable intermediate for this new type of rearrangement, which was clearly shown to occur intramolecularly.

2. Chemical Synthesis

Tryptophan has been prepared by various synthetic routes since the first successful synthesis by ELLINGER and FLAMAND in 1907 (*107*). Chemical approaches are numerous, because of the efforts given over the years to accomplish economical syntheses of this important amino acid. L-Tryptophan has been proved to be essential for animal nutrition, and it is likely that it will be used as a feed additive, as well as in medicine, if the production costs fall significantly in the near future (*291*). Recent results leading to very economical microbial production of L-tryptophan (see following Section II.3) indicate that biosynthetic procedures could perhaps supplement the chemical methods.

The extensive literature dealing with the chemical synthesis of tryptophan has been summarized several times; the reader is referred to the reviews of JULIAN *et al.* (*217*), SUMPTER and MILLER (*393*),

GREENSTEIN and WINITZ (*156*), ACHESON (*1*) and SUNDBERG (*401*). Some new synthetic results are considered in the following paragraphs.

Synthesis of tryptophan *via* Fischer cyclization is one of the classical procedures and consists in heating, with acid, phenylhydrazones, which cyclize to the indole compounds. Several 5-derivatives of tryptophan have been prepared by this route using the appropriate derivative of phenylhydrazine. The mechanism of the Fischer cyclization has been extensively studied and reviewed (*400*).

A general approach to the synthesis of tryptophan involves gramine (**10**) as an alkylating agent for carbanions derived from diethyl-

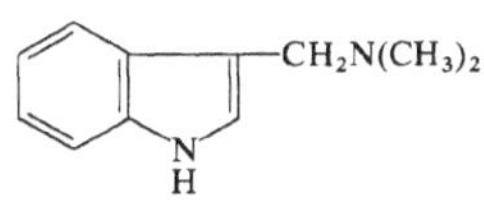

(**10**)

acetamidomalonate, α-nitrocarboxylate esters or diethylnitromalonate and similar compounds. Hydrolysis, decarboxylation, and in the case of the nitro-derivatives reduction, affords D,L-tryptophan (*217, 401*).

Reduction of 3-(2-nitrovinyl-)indoles constitutes one of the most general synthetic routes to tryptamine and the corresponding carboxyl compound (**11**) has been used to synthesize D,L-tryptophan. Reduction

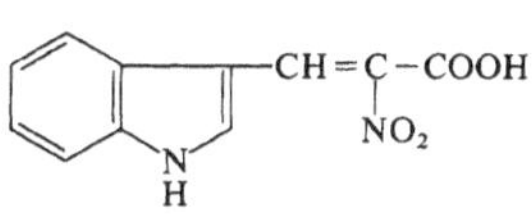

(**11**)

of (**11**) can be achieved by lithium aluminium hydride or by catalytic reduction. Recently electrolytic reduction has been thoroughly investigated (*18, 457, 458*).

Acrolein can be used as starting material for tryptophan synthesis (*227, 238, 239*). The Michael-type condensation between acrolein (**12**) and ethyl acetamidomalonate (**13**) gives γ-acetamido-γ,γ-dicarbethoxy-butyraldehyde (**14**), whose phenyhydrazone by the Fischer indole synthesis gives D,L-tryptophan. Of the hitherto reported methods, the synthetic route employed by KOMACHIYA *et al.* (*227*) appears most suitable for the industrial synthesis of D,L-tryptophan.

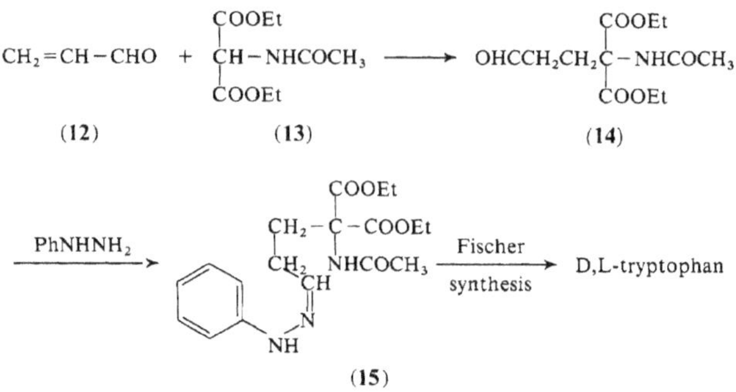

MAEDA and YOSHIDA (*238, 239*) by hydroformylation of acrolein obtained 1,1-diacetoxy-4,4-dialkoxybutane (**17**), which by Buecherer's reaction gave β-(hydanto-5-yl-)propionaldehyde acetal (**18**). The phenylhydrazone of (**18**), by the Fischer indole synthesis, gave tryptophan hydantoin (**20**). Attempts to obtain crystalline (**20**) were unsuccessful,

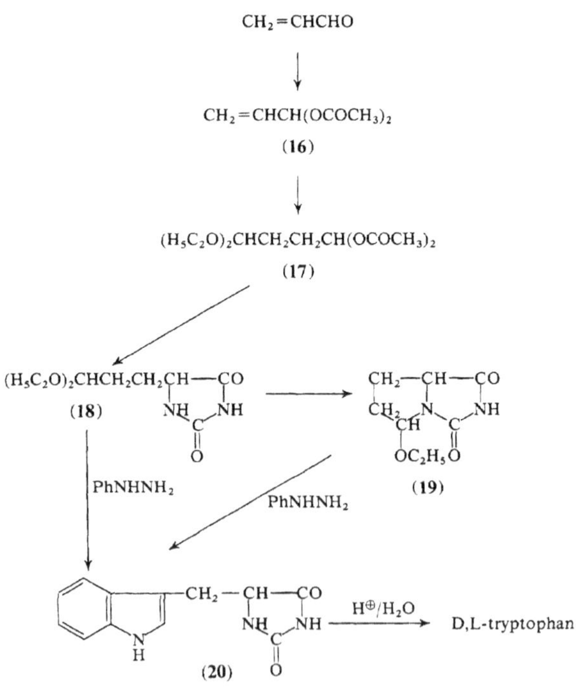

but 5-alkoxy-1,3-dioxoperhydropyrro[1,2-c]imidazole (**19**) could be isolated in crystalline form and was analogously converted to tryptophan hydantoin (**20**).

A general synthesis of tryptophan involves condensation of 3-formylindole with oxazolinones derived from N-acyl-glycine. Recently, Kirby and Varley (*225*) have used this method to synthesize tryptophan stereoselectively labelled with tritium and deuterium in the β-methylene group. The synthesis employed is depicted in the following scheme (**21**) to (**24**) and (**25**). The hydrogenation step leading to the racemate

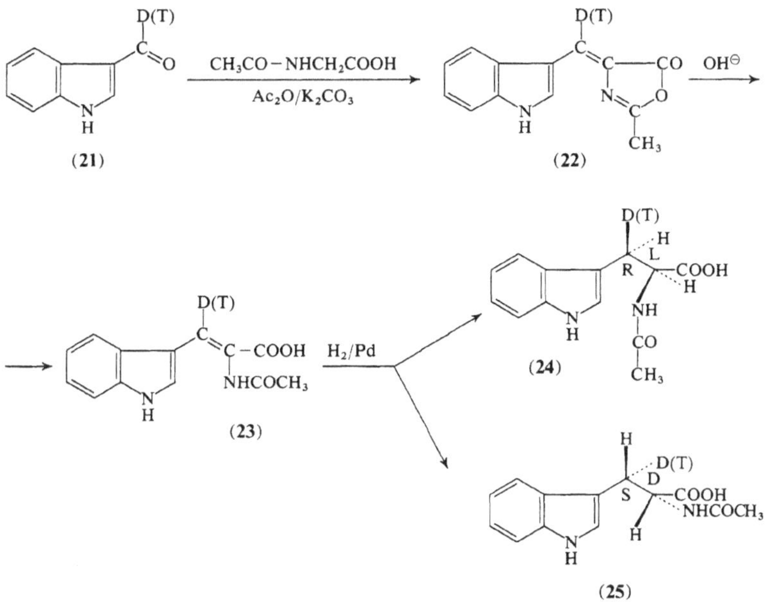

(**24**) and (**25**) was found to take place *cis* with high stereospecificity. The racemate was resolved by treatment with (–)-1-phenylethylamine (*292*) and after deacylation the corresponding diastereoisomeric labelled tryptophan was obtained.

All chemical procedures to synthesize tryptophan lead to the D,L-amino acid, so that optical resolution is needed to isolate the more important L-isomer. Although the D-isomer is partly converted to the L-form by inversion *in vivo* (*181, 331*), its nutritional value is much lower than that of the L-form.

Although a number of methods for optical resolution of D,L-tryptophan have been reported, most of them employed chemical or enzymatic procedures and only a few reports on preferential crystallization procedures have appeared [see ref. (*156*) for a review]. Recently YAMADA *et al.* (*448*) made use of the benzenesulfonate of D,L-tryptophan for optical resolution by preferential crystallization. Seeding a supersaturated solution of the sulfonate salt with the crystals of L-tryptophan brought about preferential crystallization of the L-isomer, while the D-isomer remained in the mother liquor.

3. Biosynthesis

Tryptophan can be obtained by microbiological fermentation using simple carbon sources such as sugar and some nitrogen source. Extensive research has been carried out on the microbial production of L-tryptophan with the aim of developing useful procedures alternative to the synthetic ones. The fermentation procedure has also the advantage that only L-tryptophan is produced and separation of the two isomers is not needed.

The biosynthetic pathway of tryptophan in several microorganisms has been extensively studied. A detailed review of the microbial production of tryptophan by TERUI (*406*) has recently appeared.

There are two types of production of tryptophan by fermentation; the first uses anthranilic acid as a precursor and the second uses indole or its derivatives. It has been shown that anthranilic acid can be converted into indole microbiologically.

The biosynthesis of tryptophan occurs by condensation of L-serine with indole, this reaction is catalyzed by tryptophan synthetase. The enzyme is a pyridoxal-phosphate containing enzyme which catalyzes nucleophilic β-substitution reactions of amino acids. The β-hydroxyl group of serine is substituted by indole by the action of the enzyme. The reaction is thought to proceed *via* a ketimine intermediate (**27**) which undergoes elimination to give an aminoacrylate-pyridoxal phosphate Schiff base (**28**). Addition at the β-carbon of indole followed by reversal of the process constitutes the enzymatic synthesis of L-tryptophan.

The structural and catalytic characteristics of tryptophan synthetase are under active investigation. The enzyme has a molecular weight of 135,000 and is composed of four subunits held together by noncovalent bonds. FLOSS and coworkers (*361*) determined the steric course of the tryptophan synthetase catalyzed reaction between serine and indole, using the enzyme isolated from *Neurospora crassa*. The reaction was

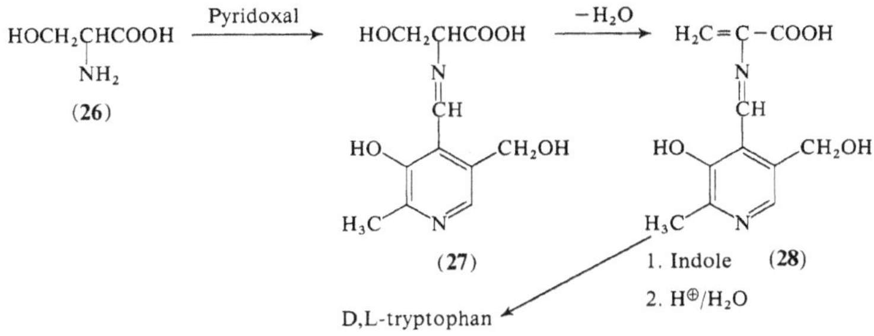

shown to proceed with *retention* of configuration at C-3 of (2S,3R)-serine-3-T (**29**), yielding (2S,3S)-tryptophan-3-T (**30**). The configuration of the enzymatically prepared tritiated tryptophan was checked by chemical degradation as well as by direct comparison with a chemically synthesized tritium labelled tryptophan (*225*).

The *non-enzymatic* reaction of indole, L-serine and pyridoxal leads to D,L-tryptophan. This steric course points to the pyridoxylidene aminoacrylic acid as an intermediate (*228*).

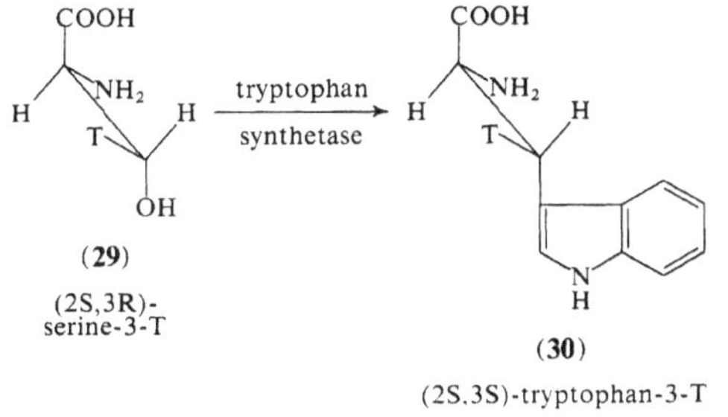

Recently, YAMADA *et al.* (*449*) reported a new method for the microbial production of L-tryptophan employing cells of *Proteus rettgeri* with a high level of tryptophanase activity. The method is simple and certainly is one of the most economical processes to-date for large scale production of L-tryptophan.

Tryptophanase is an enzyme which catalyzes the stoichiometric interconversion of L-tryptophan and pyruvate, ammonia and indole; it requires pyridoxal phosphate as a cofactor. The enzyme from *Proteus rettgeri* has been isolated in a crystalline state and its catalytic properties were investigated. The enzyme catalyzes a series of α,β-elimination, β-replacement and the reversal of α,β-elimination reactions.

The yields of L-tryptophan produced by this method are impressive: from 6 g of indole and 8 g of pyruvate in 100 ml of culture broth, 10 g of amino acid have been obtained. It was found important to add to the reaction mixture inosine, which form an insoluble complex with the synthesized L-tryptophan. When indole was replaced by 5-hydroxy-, 5-methyl- or 5-amino-indole, the corresponding 5-substituted L-tryptophan was synthesized (*449*).

4. Aspects of Metabolism

The metabolism of tryptophan has been of interest to investigators in many different fields of biochemistry and medicine. The biological transformations of this indole derivative are very complex and involve a series of enzymes, which cause i) changes of the side chain (transaminases, decarboxylases, etc.); ii) splitting of the indole nucleus from the side chain (tryptophanase); iii) changes in the pyrrole nucleus (tryptophan pyrrolase, kynureninase, etc.); iiii) changes in the benzene ring (hydroxylases, etc.).

It is beyond the scope of this review article to discuss details of tryptophan metabolism, since detailed reviews are available (*8, 36, 59, 445*), while summaries dealing with biochemical transformations of the indole nucleus of tryptophan can be found also in books of general biochemistry*.

Tryptophan is metabolized by several different pathways (Fig. 1) each yielding biologically important substances such as tryptamine and in particular serotonin (5-hydroxytryptamine), which seems to be involved in certain mental disorders. Indole-3-acetic acid is a plant growth hormone; its precursor is tryptamine or indole-3-pyruvic acid. In humans, the microorganisms of the large intestine can further degrade indole-3-acetic acid to yield indole, skatole (3-methyl-indole) and other substances.

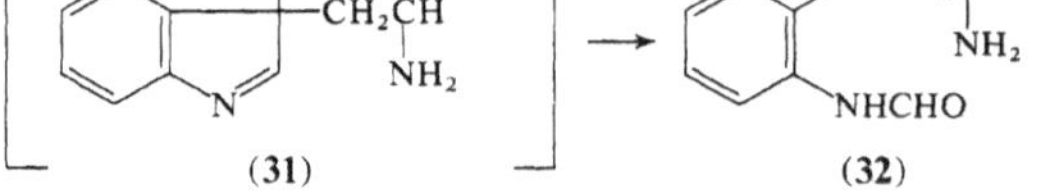

(31) (32)

* Recent studies concerning this subject have been reported at the First International Meeting on Tryptophan Metabolism held at Padua University in May 1974; the papers presented have been published in *Acta Vitam. Enzymologica* (Milano), 1975.

The major and most important metabolic pathway of tryptophan is, however, the kynurenine pathway (Fig. 2). Tryptophan undergoes oxidative cleavage between C-2 and C-3, giving formylkynurenine (**32**) by the action of tryptophan pyrrolase (*164*). 3H-Indolylhydroperoxides (**31**) have been proposed (*109, 442*) as intermediates in the formation of formylkynurenine and reviews discussing this oxidative cleavage are available (*8, 165, 402*).

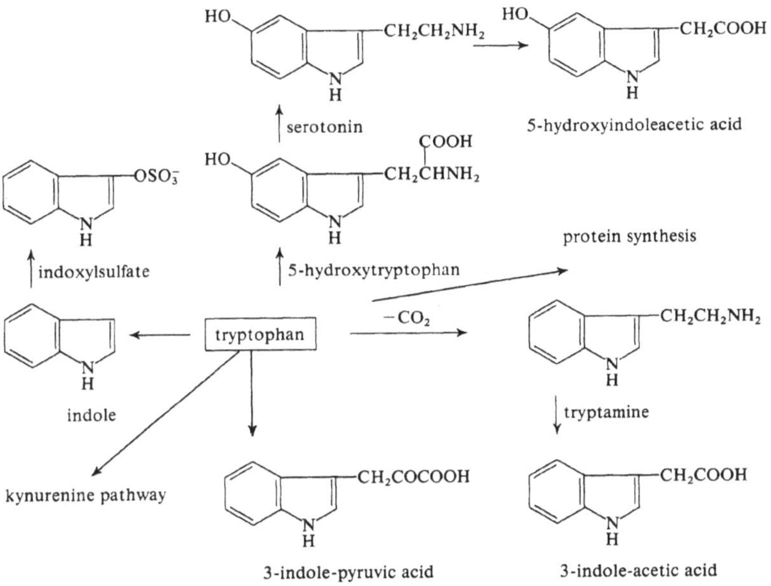

Fig. 1. Some metabolic transformations of tryptophan

Kynurenine, which is formed from the formyl derivative by the action of formamidase, may be metabolized to a quinoline derivative, kynurenic acid. Alanine may be removed from kynurenine by kynureninase to form anthranilic acid.

Most of the kynurenine formed is, however, hydroxylated to xanthurenic acid and 8-hydroxyquinaldic acid. The enzyme kynureninase yields from 3-hydroxykynurenine 3-hydroxyanthranilic acid, which is then degraded *via* the glutarate pathway.

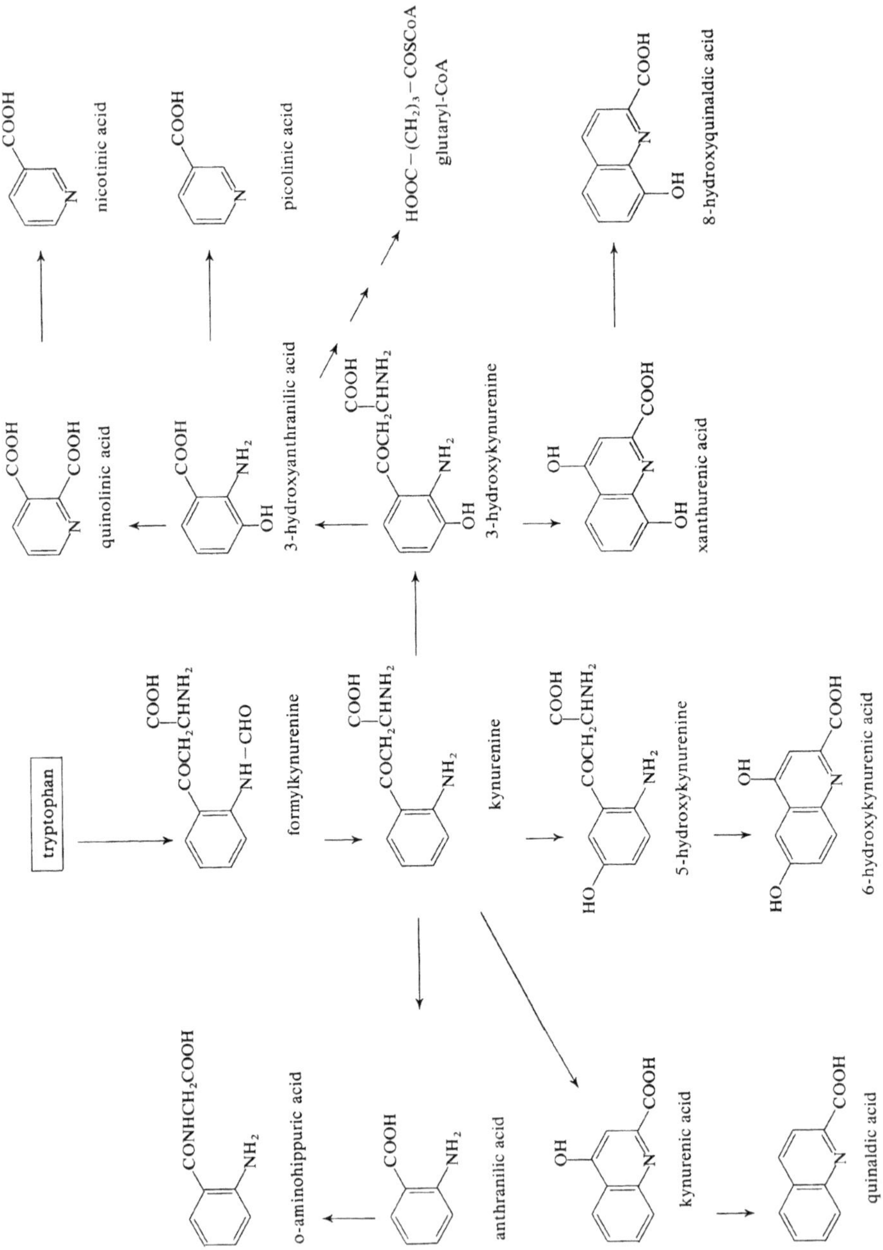

Fig. 2. Outline of kynurenine pathway of tryptophan metabolism

The biological hydroxylation of the indole nucleus of tryptophan at the 5-position has been extensively studied (*402*). Arene oxides (**33**) seem to be involved as intermediates in the biological hydroxylation (*51, 95, 209*).

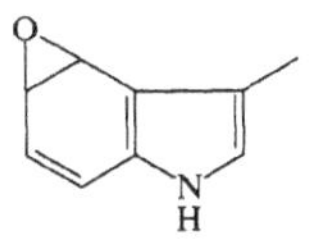

(**33**)

III. Chemistry of the Modification and Cleavage

In one approach to the elucidation of structure-function relationships in an enzyme the effects on biological activity of specific covalent alterations in the native structure of the protein in question are noted. The field of protein modification has been the subject of extensive review in recent years. Authoritative discussions and compilations are available, in which the vast information on the effects of different reagents on a great number of proteins and enzymes is summarized (*81, 169, 230, 413*). For a critical discussion on the ways in which chemical modification can be used to acquire information regarding structure and function of enzymes or proteins in general, the reader is referred to the excellent article of COHEN (*82*).

Chemical modification of amino acid side chain functionalities will also serve to cleave specific peptide bonds selectively. Chemical cleavage of a polypeptide chain exploits the unique reactivity of chemically modified side chains of particular amino acids in the labilization of adjacent peptide bonds by neighbouring group participation (*68*). The residues investigated so far for this purpose have been methionine, cysteine and the aromatic amino acids including tryptophan (*438–440, 443*).

Many aspects of the degradation of polypeptides are discussed in the reviews of FONTANA and SCOFFONE (*125*) and of HEINRIKSON and KRAMER (*169*), while specific experimental procedures are given in the volume of Methods in Enzymology, edited by HIRS (*183*). A comprehensive presentation of selective chemical cleavage of polypeptide chains can be found in the excellent review of SPANDE *et al.* (*395*).

Great impetus to the research on the chemical modification and fragmentation of proteins was given by the basic studies of WITKOP who in 1961 surveyed the subject and delineated chemical principles and criteria for modification and cleavage of peptide chains (*438*).

The great reactivity of the indole nucleus of tryptophan has evoked much interest in finding techniques for its selective chemical modification. However, in spite of the variety of reagents introduced over the years for this purpose, only few meet the requirements of an ideal modifying reagent which should be specific, that is, it should react stoichiometrically with the target function to the exclusion of all other functional groups and yield a single well-defined product.

In the following, the most frequently used reagents for tryptophan modification are discussed, while the effects of modifications on biological properties of enzymes are presented in Section VII.

1. Chemical Oxidation

From the historical point of view, interest in the chemical oxidation of tryptophan and related indole compounds was stimulated by attempts to elucidate the fate of tryptophan in animal metabolism. Biological hydroxylations and oxidative degradations of the indole nucleus represent main steps in the metabolic pathway of tryptophan (see Section II.4).

Among the techniques used to modify the tryptophan residue in proteins, most oxidative methods show poor selectivity since several other functional groups are also susceptible to oxidation. Under particular conditions, however, some reagents have been found to attack the indole nucleus preferentially; these techniques are examined in this Section. Halogenating agents like N-bromosuccinimide, hypochlorite and others, have also an oxidative action on tryptophan, but these reagents are discussed separately in Section III.3.

1.1. Peracids

Oxidation with performic acid, prepared by mixing formic acid with H_2O_2, is used in protein chemistry to carry out oxidative cleavage of disulfide bonds to cysteic acid residues. The reagent is a powerful oxidizing agent and other amino acids are also attacked, including methionine, tyrosine and histidine (*182*). Tryptophan is also oxidized, but the nature of the reaction products has not been clarified in detail. Studies have been performed on the products arising from the performic acid oxidation of free tryptophan, whereas little is known on the oxidation products of tryptophan linked in a peptide moiety. The oxidation behaviour of the indole nucleus could be different in free or bound tryptophan, because of the presence of free amino and carboxyl groups in the former.

In 1947 WITKOP (*437*) oxidized tryptophan with peracetic acid and isolated in low yield 2-hydroxytryptophan (**34**). This was later shown to be (**35**) by spectroscopic techniques and called oxindolylalanine* (*85*, *338*).

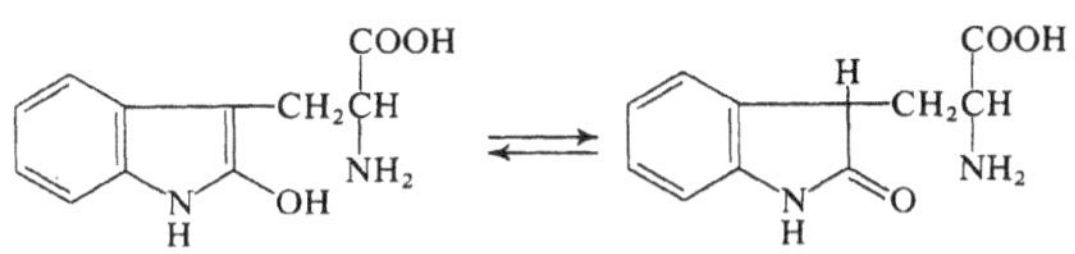

(**34**) (**35**)

Cleavage of the pyrrole ring on the other hand has been demonstrated in subsequent studies. SAKAMURA and OBATA (*332*) isolated kynurenine as a product of both performic and peracetic acid oxidations of tryptophan. BENASSI and coworkers (*37*, *38*) showed that performic acid oxidation of tryptophan gives a complex mixture of products, indicating that pyrrole ring fission is accompanied by hydroxylation of the benzene ring together with secondary reactions. They suggest that oxidation may proceed by various alternative pathways as shown below (*36* to *44*).

FINLAYSON (*117*) has reported that performic acid oxidation of lysozyme gives a higher conversion of tryptophan to kynurenine than that given by oxidation of free tryptophan. The tryptophan residues can also be converted into an oxidized form which does not give kynurenine on hydrolysis.

Various attempts to oxidize oxindolylalanine to formylkynurenine by chemical or biochemical methods all proved unsuccessful. In alkaline solution (**35**) is readily oxidizable by iodine (*218*) or air (*340*) to diastereomeric dioxindolylalanines (**45**). Diastereoisomerism is possible because of the presence of a second asymmetric center at C-3 of the

* In this article the term "oxindole" has been used instead of "2-hydroxy-indole", in accordance with DALGLIESH and coworkers (*85*), who reported that the absorption data suggest that the structure of the hydroxy derivative of tryptophan is best represented by a lactam structure; the term "hydroxytryptophan" is therefore misleading and should be replaced by "oxindolylalanine".

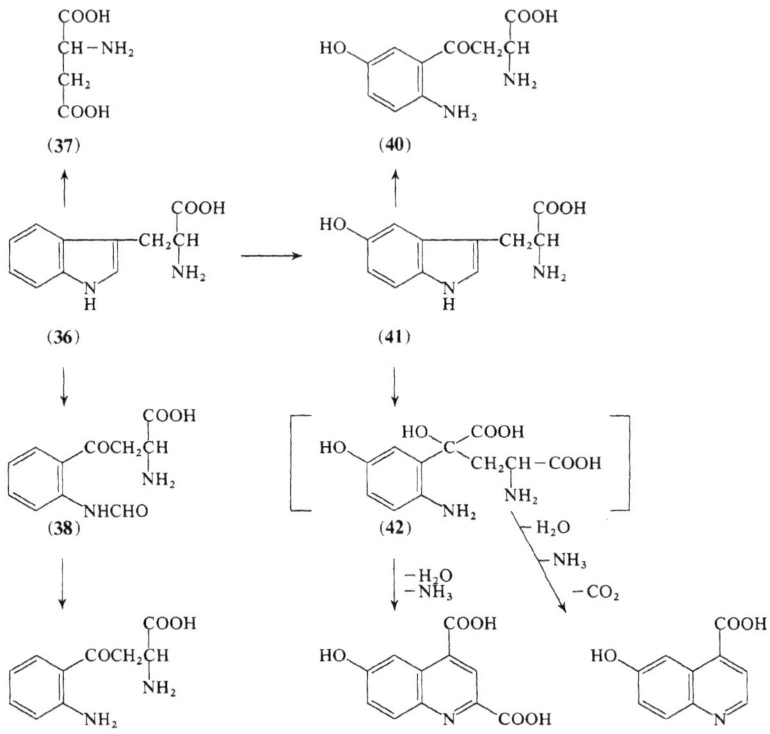

(37) (40) (36) (41) (38) (42) (39) (43) (44)

indole. Presumably diastereoisomerism is precluded in the case of oxindolylalanine because of keto-enol tautomerism. It is also worthy of mention that all evidence suggests that neither (**35**) nor (**45**) is an intermediate in the tryptophan-kynurenine metabolic pathway (*164*).

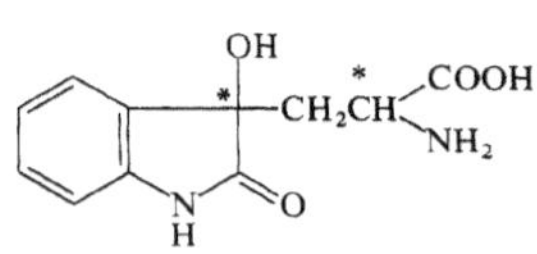

(**45**)

On the basis of the results cited above pyrrole ring oxidation of tryptophan can seemingly occur by either of two pathways, depending on the nature of the oxidants and reaction conditions. One path leads to oxindolylalanine or dioxindolylalanine, while the other gives kynurenine and derivatives.

Recent studies carried out by Savige (*340*) seem to clarify the alternative pathways of oxidation, since an intermediate common to both pathways seems to have been isolated. Peracetic acid oxidation of tryptophan has been carried out in water using one mole of preformed oxidant at 0° C. Under these reaction conditions, the hexa-

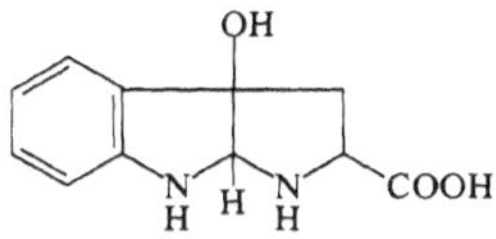

(**46**)

hydropyrroloindole compound (**46**) was isolated in 40% yield. Oxindolylalanine could not be detected as an oxidation product. However, (**46**) was found to undergo rearrangement to oxindolylalanine on heating in dilute mineral acid; this conversion clarifies the earlier isolation of this compound by Witkop (*437*). The pathway leading to pyrrole ring cleavage has been also clarified by Savige (*340*), since the hexahydropyrroloindole (**46**) undergoes further oxidation with peracetic acid to give a mixture of formylkynurenine and a compound of probable benzoxazinone structure (**47**), which has been preferred to the alternative hydroperoxy structure (**48**). This compound undergoes decarboxylation

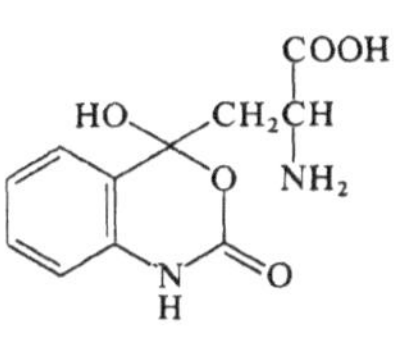

(**47**)

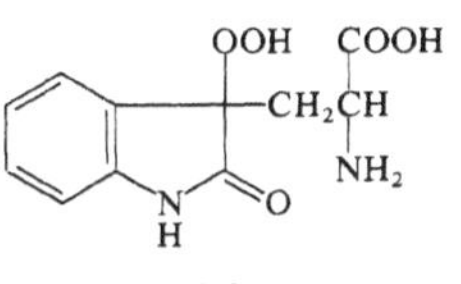

(**48**)

to kynurenine on heating in aqueous solution and undergoes autoxidative decomposition to unidentified products in cold neutral or alkaline solution.

Later, Savige (*339*) reported that the hexahydropyrroloindole (**46**) reacted readily with thiols in aqueous acetic acid to give the correspond-

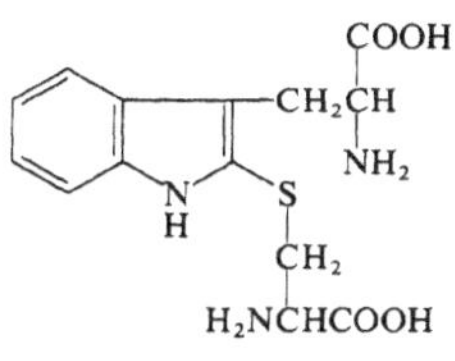

(**49**)

ing 2-thioether-indole compounds. Reaction of (**46**) with cysteine is a convenient new route for synthesis of tryptathionine (**49**) (see also Section III.5).

1.2. Ozone

In 1944 WITKOP and GRASER (*441*) found that ozonolysis of skatole (**50**) resulted in cleavage of the pyrrole double bond to give *o*-formamido-acetophenone in good yield. The reaction was further investigated by SCOFFONE and PREVIERO (*307, 313*) with tryptophan peptides. In

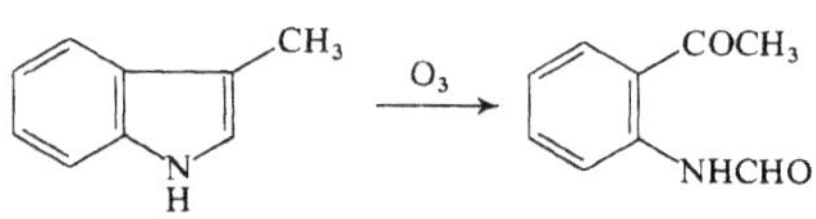

(**50**) (**51**)

anhydrous formic acid (*307*) or preferably in ethyl acetate (*419*), simple tryptophan derivatives and peptides were converted to the corresponding formylkynurenine compounds. The formyl group was subsequently removed by mild acid treatment.

Generally, peptide bond cleavage has not been observed during ozonolysis and only recently a preliminary report (*333*) has appeared in which high yields of cleavage have been described when the ozonolysis was carried out at low temperatures (−70° C).

Procedures have been devised whereby the kynurenine or formylkynurenine peptide bond can be selectively cleaved. When formylkynurenine peptides are heated in bicarbonate buffer for periods up to 4 hours, the peptide bond is selectively hydrolyzed in yields of 49–59% (*311*). Intramolecular catalysis involving the neighbouring γ-keto group of formylkynurenine peptides was thought to be responsible for the labilization of the peptide bond.

Reduction of kynurenine peptides can be achieved with $NaBH_4$ (*308*) or electrolytic reduction at controlled potential (*415*) to γ-(*o*-aminophenyl)-homoserine derivatives (**53**). Both N- and C-peptide bonds were cleaved

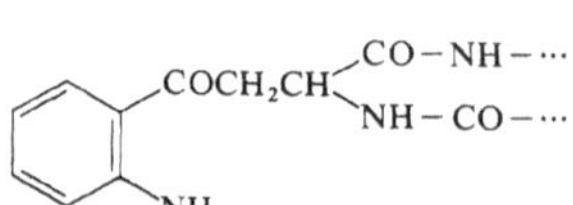

(**52**)

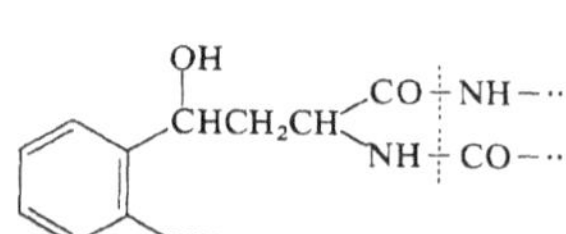

(**53**)

in this instance in 20–40% yields. VERONESE *et al.* (*415*) made an extensive study of the mechanism of this cleavage reaction.

MORISHITA *et al.* (*265, 266*) obtained cleavage of formylkynurenyl peptides by heating with hydrazine at pH 3.6. The hydrazones (**54**) undergo ring-closure to tetrahydropyridazones (**55**) with concomitant

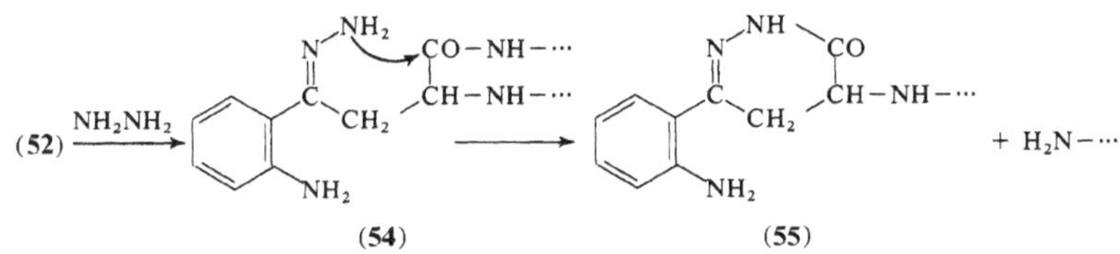

cleavage. The yields of cleavage products were 35–68%, but some non-specific peptide bond cleavage by the action of hydrazine was observed.

Ozonolysis in formic acid in the presence of resorcinol (*306, 307*) has been used to modify tryptophan residues in proteins. The reaction seems quite specific; however the use of formic acid severely limits the utility of the technique for structure-function studies in enzymes.

1.3. Miscellaneous Oxidations

The oxidation of tryptophan with periodate under acid, neutral and alkaline conditions has been studied (*18, 221, 326*). Yields of isolated products were low due to occurrence of deamination and tar formation (*78*). Products detected (8% yield) include formylkynurenine under neutral conditions and dioxindolylalanine under acid conditions. Traces of serine and aspartic acid were also detected. Oxidation in the presence of ammonia gave the quinazoline (**56**) previously identified (*338*) as a photooxidation product of tryptophan. By contrast, indole-3-

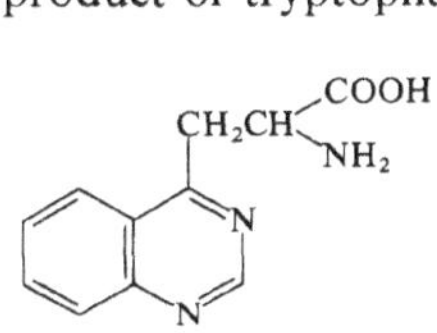

(**56**)

propionic acid, the de-α-amino analogue of tryptophan, was readily oxidized by sodium periodate in aqueous methanolic solution to give the acid (**57**) in 68% yield (*78*). When indole-3-propionyl-glycine was

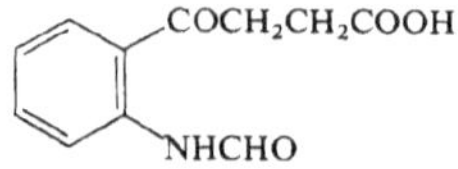

(57)

oxidized under similar conditions, partial cleavage of the peptide bond occurred and free glycine was detected (*18*). The same cleavage was also reported by WITKOP (*438*). ATASSI (*13*) reported that oxidation of sperm whale myoglobin with periodate at pH 5 and at 0° results in specific cleavage of the tryptophan peptide bond.

Various organic peroxides have been tested as oxidizing reagents for tryptophan residues in proteins (*160, 161, 279, 455*). A suitable system is hydrogen peroxide in bicarbonate buffer containing 10% dioxan. The rate of oxidation is dependent on pH, dioxan concentration and type of buffer ion. Indications have been obtained that dioxan and bicarbonate are involved in the oxidation, probably through formation of an organic peroxide with greater oxidizing power than hydrogen peroxide (*160, 161*).

Oxydation of tryptophan with alkaline persulfate was reported (*52*) to give kynurenine, formic acid and the sulfate esters of 3- and 5-hydroxykynurenine, showing that pyrrole ring fission is accompanied by hydroxylation of the benzene ring.

Studies have been made on the effect on tryptophan of a system (ferrous sulfate, ascorbic acid, EDTA, pH 6.7, 37° C) which supposedly effects hydroxylation of aromatic compounds in a manner closely analogous to *in vivo* hydroxylations. DALGLIESH (*94*) detected 5- and possibly 7-hydroxy-tryptophan using this oxidizing system, while kynurenine gave 3- and 5-hydroxy-kynurenine as principal products. However, PFAENDER *et al.* (*301*), using a similar system, obtained principally oxindolylalanine (13%) together with low yields of 5- and 6-hydroxytryptophan and 6-hydroxy-oxindolylalanine.

Recent studies (*385, 386, 387*) have established that tryptophan is susceptible to free radical autoxidation in both acid and alkaline solution. In acidic solution, the autoxidation at 100° C is preceded by protonation of the indole nucleus to produce a more easily degraded species and so the rate is proportional to the pH, while in basic solution impurities present in the base increase the reaction rate. In the pH range 2–7 the maximum rate of decomposition occurs at pH 5.3. Two protonations seem to be involved in the autoxidation and these take place in the indole and carboxyl function of the molecule. That the carboxyl group is involved in the oxidation process has been proposed (*381*).

References, pp. 426—449

2. Photooxidation

Irradiation with light is a potentially useful tool for effecting modification of amino acid side chains in polypeptides and proteins, since light, apart from ultraviolet light of short wave-length, should be of insufficiently high energy to bring about fission of peptide bonds. For example, irradiation of a protein with ultraviolet light of wave-length 280 nm might be expected to bring about specific modification of tyrosine and tryptophan residues only, since they are principally responsible for absorption of light in this region in proteins. Other potential sites for sensitized photooxidation include histidine, methionine, cysteine and cystine residues (for reviews see *131, 132, 224, 255, 319, 426*).

The literature on the effect of photooxidation of proteins in somewhat confusing, since it is apparent that insufficient attention has been paid to the factors that control photooxidation of amino acid side chains. A wide variety of light sources has been used by different authors. Clearly the pathways of photoreactions will be related to the wave-length of light used and it is particularly difficult to interpret results of irradiation of proteins when mixed ultraviolet/visible light sources such as sunlight have been used.

The fate of the tryptophan residue when proteins are irradiated with ionizing radiation is discussed subsequently.

2.1. Irradiation by Ultraviolet Light

Irradiation of both soluble and fibrous proteins with ultraviolet light leads to considerable destruction of cystine and aromatic acid residues, in particular tryptophan. However, apart from a small degree of conversion to formylkynurenine and/or kynurenine residues, all attempts to ascertain the nature of the tryptophan photoproducts have so far been unsuccessful (*12, 63, 303, 304*). PIRIE and DILLEY (*304*) have proposed that photo-oxidation of tryptophan to formylkynurenine is followed by splitting out of N-formylanthranilic acid (**60**). RIVETT's

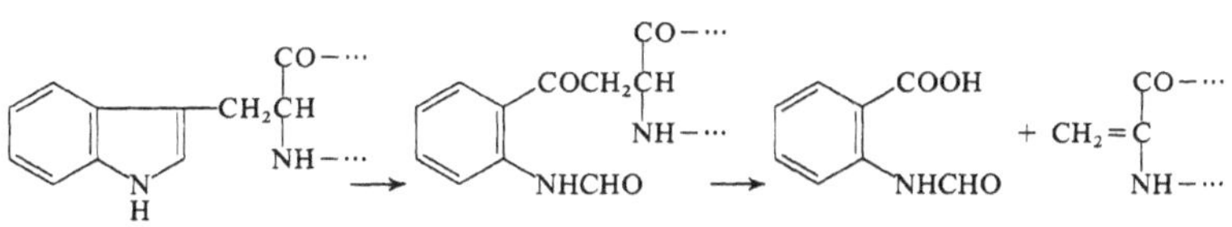

(**58**) (**59**) (**60**) (**61**)

(*327*) view that tryptophan residues may be converted largely to amino-acrylic acid residues is consistent with this concept.

The findings with respect to ultraviolet irradiation of tryptophan itself do not greatly assist in clarifying our understanding of the fate of tryptophan residues in proteins. Irradiation of free tryptophan leads to considerable deamination and/or decarboxylation (*12, 98*), which is not possible for tryptophan residues in proteins. Furthermore, photopolymerization can occur for tryptophan and/or a small model compound, but is much less possible for tryptophan residues in proteins because of conformational factors.

ASQUITH and RIVETT (*12*) irradiated tryptophan in oxygenated aqueous solution with a medium pressure mercury lamp under neutral, acidic and basic conditions and identified several degradation products including formylkynurenine, kynurenine, aspartic acid, serine, glycine, alanine and β-alanine. They suggest that the main degradative pathway of tryptophan is conversion to kynurenine *via* the indolenine hydroperoxide (see Section III.1.1). The formation of amino acids could arise from further degradation of kynurenine as shown below.

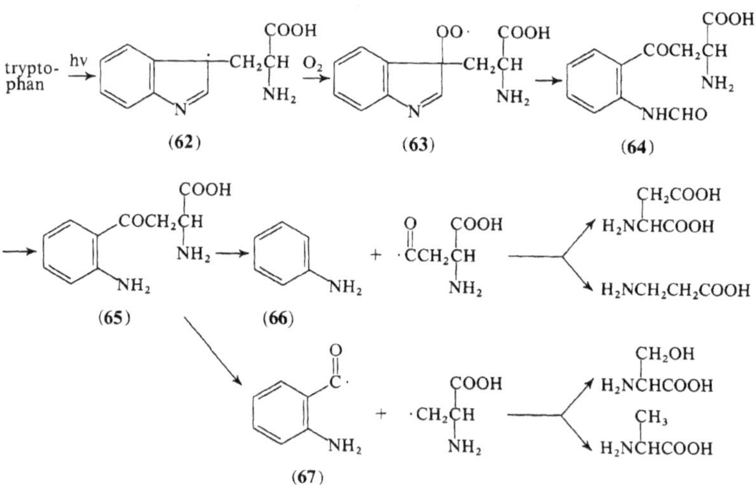

An important observation in the course of this work (*12*) was that neither oxindolylalanine nor dioxindolylalanine were detected in irradiation products. This fact is in accord with the results discussed previously (Section III.1.1.) in that these compounds are unlikely intermediates in the oxidation of tryptophan to kynurenine.

References, pp. 426—449

2.2. Dye-Sensitized Photooxidation

Dye-sensitized photooxidation with visible light to identify residues critically involved in the biological activity of enzymes has been introduced by WEIL *et al.* (*425*). The procedure is mild and moderately selective for some amino acid side chains; in addition to tryptophan, tyrosine, histidine, methionine, cysteine and cystine are also photooxidizable. It has been found possible, by an appropriate choice of dye, to bring about specific modification of a particular residue. There is, however, considerable variation in the effects of different dyes and of particular experimental conditions [see ref. *112, 255, 319, 426*) for reviews].

Unfortunately, the photooxidative technique has been often used under arbitrarily chosen conditions and the results obtained in some cases are unreliable due to uncritical interpretations of the data and incomplete chemical analysis of the photooxidized protein samples. Interpretation of the results is hopeless without complete and accurate analysis of all oxidizable amino acids. Furthermore, the utility of the work so far reported is limited by ignorance of the chemical nature of the products derived from amino acids susceptible to photooxidation.

Despite the extensive use to which photochemical oxidations of proteins have been put very little attention has been given to identifying the end products of the photoreactions. Apart from methionine which is converted to methionine sulfoxide and more slowly to its sulfone, the fate of the other amino acid residues of irradiated proteins remains largely unknown. On the basis of limited chemical studies, it is clear, however, that photoreactions usually lead to a mixture of products. The multiplicity of products formed by photooxidation of enzymes will undoubtedly limit the utility of the technique. There are in fact numerous examples from work dealing with the chemical modification of enzymes which indicate that modification of a particular residue with different chemical reagents leads to enzyme derivatives with different biological and physico-chemical properties.

Usually the destruction of a particular residue has been followed by checking the disappearance on the chromatogram on the automatic amino acid analyzer after acid hydrolysis (*369*). This procedure could be also misleading, since the possibility exists that the photo-induced modification can be reversed during hydrolysis, as in the case of methionine sulfoxide which is reduced back to methionine. This observation suggests that some other *unsuspected* modifications might occur with other amino acids (*82*).

It is also noteworthy that the initial photo products from histidine, tyrosine and tryptophan seem to be carbonyl compounds which can subsequently condense with amino groups to give inter- and intramolecular crosslinks within the protein (*426*).

Despite the above limitations, conditions have been found which result in specific oxidation of particular residues. These newer approaches include the use of chromophoric inhibitors or coenzymes to achieve effective sensitized photooxidation. In this case, since only the amino acid residues which are in the vicinity of the sensitizer are modified, the possibility exists that the residues which are, or are near the active site of an enzyme (*426*) can be identified.

The fate of tryptophan residues on sensitized photoirradiation of a protein is largely unknown. It seems certain that tryptophan is converted to a number of different reaction products, including formylkynurenine, depending upon experimental conditions used, wavelength of light, type and concentration of sensitizer, buffer and other parameters (*426*).

Apart from the results reported by JORI *et al.* (*211–214*) who claimed *quantitative* conversion of tryptophan to formylkynurenine residues when proteins are photooxidized in aqueous buffered solution with proflavine as sensitizer, no other author has reported similar quantitative conversion. On the other hand it has been found that loss of tryptophan clearly does not match the formation of formylkynurenine (*63, 304*). GOMYO and FUJIMAKI (*148*) detected kynurenine and 3-hydroxykynurenine as products of the lumiflavine sensitized photooxidation of lysozyme in buffered aqueous solution. KRAVCHENKO and LAPUK (*229, 232*), careful purified a photooxidized sample of lysozyme and isolated a group of monooxidized enzymes in which a single tryptophan entity was oxidized. Even this relatively homogeneous product could be separated into different species of enzyme with different degrees of residual activity and different ultraviolet spectra (*229*).

Actually, the figures quoted by JORI *et al.* for formylkynurenine formation upon photooxidation of proteins were determined in the intact protein on the basis of an absorbancy at 318 nm of $\varepsilon = 1200$ (*211, 212, 214*). This value is considerably less than half that reported by others. For example, DALGLIESH (*93*) found $\varepsilon = 3750$ (λ 321 nm), while more recently WALRANT and SANTUS (*423, 424*) found $\varepsilon = 3180$ (λ 318 nm).

JORI *et al.* indicate that the estimation of formylkynurenine can be made by acid hydrolysis to kynurenine followed by analysis with an amino acid analyzer, but it is not clear from the paper (*212*) which of the two methods were used to estimate the actual figures reported. In addition, JORI *et al.* have as yet given no evidence in support of their claim that formylkynurenine is the single end-product of tryptophan photooxidation, but merely quote (*212*) in a footnote unpublished experiments of RHODES and GARDNER that photooxidation of tryptophan leads to formylkynurenine. Curiously, it was reported earlier that proflavine-sensitized photooxidation of peptide bound tryptophan in 98–100% formic acid (and not aqueous formic acid as stated in the above mentioned footnote) leads to kynurenine, which absorbs near 360 nm, and not to formylkynurenine (*212*).

In the light of these considerations, the results of JORI *et al.* are questionable and should be considered with utmost caution. All available literature clearly shows that formylkynurenine is not the only product formed upon photooxidation of tryptophan, although seemingly it represents the major reaction pathway. Initial attack by oxygen at indole C-3 is implicated in all current interpretation of indole oxidations, and decomposition of the resulting hydroperoxide usually leads to opening of the indole ring (see also previous Section III.1.1.). One reason for the lower yields of products obtained from free tryptophan than from a derivative is the occurrence of deamination in the case of the former.

SAVIGE reported that photooxidation of tryptophan in water, using methylene blue, eosin, riboflavin or Rose Bengal as sensitizer, gave formylkynurenine in low yield, together with other products (*338*). GURNANI *et al.* (*159*) reported the possible formation of dioxindolylalanine as a product of methylene blue sensitized photooxidation of tryptophan, but the properties of the product, as SAVIGE (*338*) has pointed out, do not correspond to those of true dioxindolylalanine (*218*). Photooxidation of N-acetyltryptophan in bicarbonate buffer, using Rose Bengal as sensitizer is reported to give principally N-acetyl-N′-formylkynurenine in 60% yield (*338*).

Relatively numerous chemical studies have been reported on the photooxidation of free tryptophan, rather than of tryptophan-containing model peptides, the chief reason being that from a practical aspect it is usually easier to detect and identify the corresponding photoproducts. However, as previously stated, in relation to the chemical oxidation of tryptophan (Section III.1.1), the oxidation path of free tryptophan should not necessarily parallel that of tryptophan in proteins.

Results of work on the photochemistry of the free amino acid have however led to the discovery of three noteworthy photoreactions, which are described below.

2.2.1. Conversion of Tryptophan to Hexahydropyrroloindole

The isolation of the hexahydropyrroloindole derivative (**46**) on photooxidation of tryptophan irradiated in water in the presence of methylene blue is of interest (*339*). This product was proved to be an intermediate in the oxidation of tryptophan to formylkynurenine by peracetic acid (*340*). However, although the hexahydropyrroloindole was found together with formylkynurenine in the photoproducts of tryptophan, the compound could not be further photooxidized to formylkynurenine. These results would seem to exclude the compound as an intermediate in the photoconversion of tryptophan to formylkynurenine. It is noteworthy that (**46**) can be made to undergo isomerization to oxindolylalanine by heating or ultraviolet irradiation in acidified solution (*339*).

Recent findings by NAKAGAWA *et al.* (*274, 275*) have thrown light on the role of hexahydropyrroloindoles in the oxidation of tryptophan analogues. Photooxidation of tryptophan methyl ester (**68a**) in benzene solution in the presence of Rose Bengal gave the corresponding hexahydropyrroloindole in low yield. It was proposed that the tryptophan derivative (**68a**) reacts with singlet oxygen to give the hydroperoxyindolenine (**68b**) which instead of undergoing the rearrangement to formylkynurenine ester (**70**) (path A), is converted to (**71**) *via* path B.

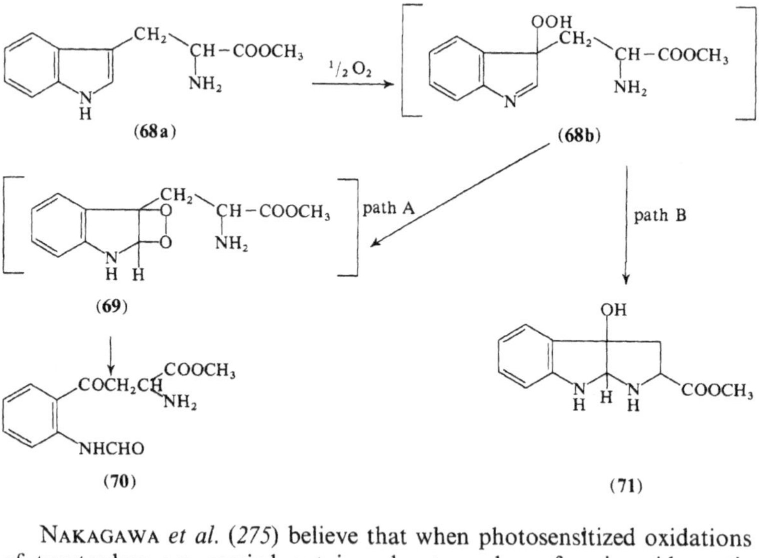

NAKAGAWA *et al.* (*275*) believe that when photosensitized oxidations of tryptophan are carried out in solvents such as formic acid, acetic acid and water, participation of the ethylamino side chain is unfavorable and that the reaction consequently proceeds *via* the alternative path A of pyrrole ring cleavage. However, SAVIGE (*339*) recently found that photooxidation of tryptophan in water, in the presence of methylene blue gives the hexahydropyrroloindole in reasonable yield along with formylkynurenine.

Photooxidation of the tryptamine derivative (**73**) in the presence of pyridine N-oxide in methylene chloride solution using 2537Å light affords the analogous hexahydropyrroloindole (**75**), the epoxide (**74**) being the proposed intermediate (*274, 275*).

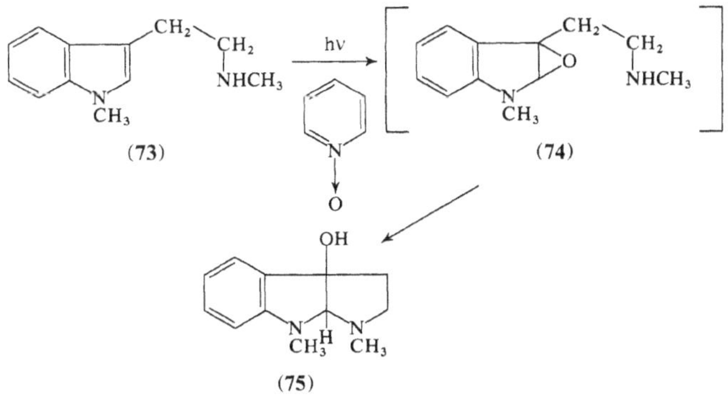

2.2.2. Conversion of Tryptophan to Quinazolines

The isolation by Savige (*338*) of the quinazoline derivative (**56**) on photooxidation of tryptophan in very dilute aqueous ammonia deserves comment. The quinazoline (**56**) was subsequently identified as an irradiation product of tryptophan in aqueous ammonia with a ultraviolet light source (*12*) and also as a product of periodate oxidation of tryptophan in aqueous ammonia (*326*).

It has recently been reported by Maeda *et al.* (*242*) that photooxidation of 2,3-disubstituted indoles in alcohol solution containing ammonium acetate in the presence of methylene blue gives analogous quinazolines. If one accepts the concept that dye sensitized oxidations of an indole to the corresponding aminoacylketone proceed by reaction with singlet oxygen, giving the hydroperoxyindolenine and dioxetan (*112, 242, 256*), then the photoconversion of indole to quinazoline likely occurs by reaction of either the hydroperoxyindolenine (**77**) or dioxetan (**78**) with ammonia.

Although Savige (*338*) earlier found no reaction between formylkynurenine and ammonia in the very dilute aqueous ammonia used in the photoconversion of tryptophan to (**80**), he has since found that formylkynurenine on treatment with 2N aqueous ammonia in

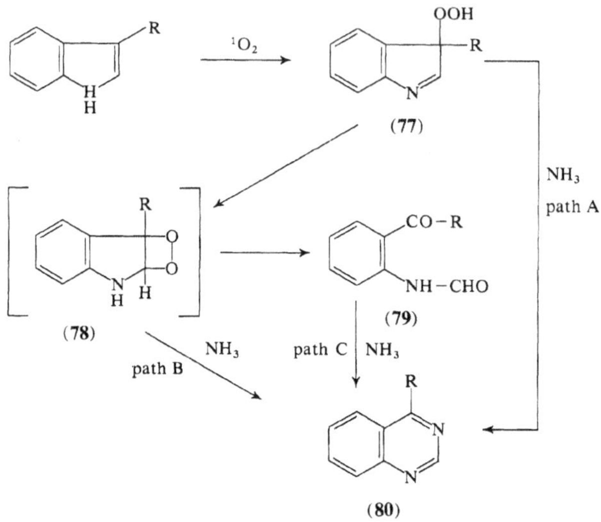

where R = $-CH_2CH(NH_2)COOH$

the dark at room temperature is slowly converted to the quinazoline (**80**) (*339*).

The possibility that conversion of tryptophan into the quinazoline occurs also in biological systems containing ammonia should be considered. It is noteworthy that a quinazoline pathway of tryptophan metabolism has been proposed (*245*).

2.2.3. Conversion of Tryptophan to Carbolines

Irradiation of tryptophan with visible light in acetic or formic acid solution in the presence of methylene blue gives good conversion to a β-carboline (*215*). The reaction probably proceeds *via* ˙OH abstraction from the solvent molecule by photoexcited methylene blue, with formation of an acyl $R\dot{C}=O$ radical, which attacks the 2-position of the indole substrate. The 2-acyl derivative cyclizes to the dihydro-β-carboline (**82**) which is aromatized by methylene blue photosensitized dehydrogenation to (**83**). β-Carbolines variously substituted in the 1- or 3-position have been obtained by an appropriate choice of the solvent and indole side chain.

References, pp. 426—449

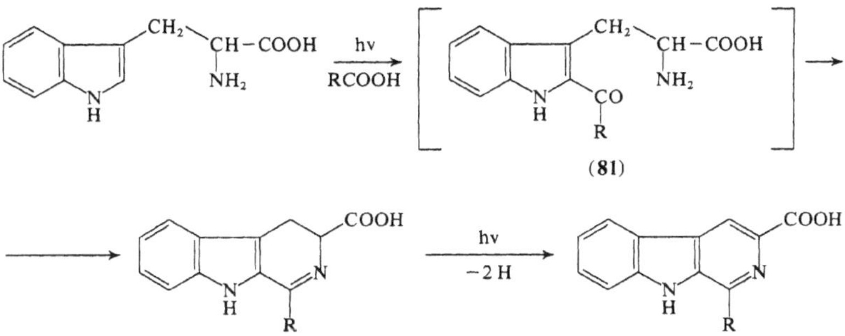

(81)

(82) (83)

Evidence supporting the above reported mechanism has been adduced, since photooxidation of skatole under conditions used for tryptophan was found to give 2-acetyl-3-methyl-indole (**84**) in 70% yield (*215*).

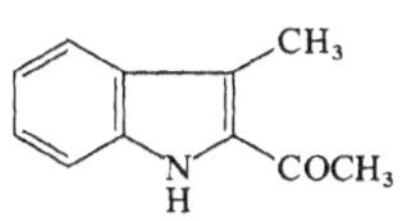

(84)

The high yields of (**84**) would indicate that a similar acylation reaction in the 2-position of the indole nucleus of tryptophan containing proteins has to be expected when irradiation is carried out in acetic acid, but without conversion to the β-carboline since the amino group is aminoacylated.

2.3. Ionizing Radiation

The effect of ionizing radiation (x or γ rays) on tryptophan or on tryptophan-containing proteins in aerated aqueous solution is seemingly to bring about partial conversion of tryptophan to formylkynurenine (*5, 206, 300*). The oxidizing species are presumed to be hydroxy and hydroperoxy radicals. Oxindolylalanine does not seem to be an intermediate in the reaction. Suggested mechanisms (*206, 300*) propose 2-hydroxy-3-hydroperoxyindoline (**86**) as an intermediate. However, in the light of recent theories on tryptophan oxidation, which were discussed in other sections a modification of the above discussed mechanism can be also proposed which involves hydroperoxyindolenine as an intermediate.

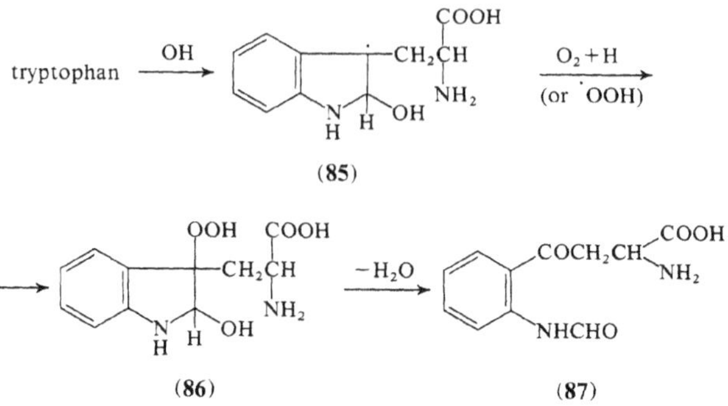

ALDRICH and CUNDALL (*5*) studied the radiation-induced inactivation of lysozyme with ^{60}CO γ-rays. Variation of pH, the presence or absence of oxygen and nitrous oxide and other additives has made it possible to investigate the separate role of hydroxyl radicals, solvated electrons, and hydrogen atoms as inactivating species. The evidence suggests that reaction of the hydroxyl radical with tryptophan residues is largely responsible for the radiation-induced inactivation. These authors (*5*) believe that γ-ray irradiation brings about conversion of tryptophan to oxindolylalanine, but the evidence is not conclusive.

3. Brominating Agents

3.1. N-Bromosuccinimide

Halogenation of indoles has been performed with a great variety of halogenating agents, which include, in addition to halogens, dioxan dibromide, pyridinium bromide perbromide, N-bromosuccinimide, N-iodosuccinimide, ICl, chloramine T-KI, hypochlorite and *t*-butyl hypochlorite (*394*).

The use of brominating agents like N-bromosuccinimide (NBS), N-bromoacetamide, N-bromophthalimide and bromine has been introduced by WITKOP and coworkers (*248, 295, 438*) for the chemical modification of proteins. These agents will cause oxidative halogenation of reactive side chain functions of amino acids. The oxidation is not limited to the indole nucleus of tryptophan, but other amino acids including histidine, tyrosine, methionine, cysteine, cystine and lysine can be modified by halogenating agents and in particular by NBS. Side reactions involving other side chains have been generally assumed to be minimal by comparison with the oxidation of tryptophan if a low molar ratio of NBS to protein is maintained. However, the results of several investigators, in particular those of KRONMAN and ROBBINS (*230, 231*), clearly indicate that selectivity of reaction toward tryptophan is hardly achieved. In addition, the action of NBS accomplishes not only modification of amino acid side chains, but also peptide bond cleavage involving tyrosine, histidine as well as of tryptophan residues (*316, 372, 374, 377, 438*).

The chemistry of the NBS reaction with indoles has been extensively investigated. It seems relevant to discuss this here together with some aspects related to the bromination of tryptophan and related 3-substituted indoles (*175, 218, 295, 443*).

NBS reacts with the indole nucleus following the normal pattern of electrophilic substitution; attack at the 3-position is favored. When this position is blocked substitution occurs at the 2-position. Indole and 3-alkylated indoles exposed to one mole of NBS under anhydrous conditions yield 3-bromoindole and 2-bromo-3-alkylindoles respectively (*175, 394*). By contrast, halogenation in hydroxylic solvents often leads directly to oxindoles and both 2- and 3-halo-indoles readily form 2-oxindole (*394*).

HINMAN and BAUMAN (*175*) have investigated the reaction of 3-substituted indoles with NBS in *t*-butyl alcohol. When 1 : 1 molar ratios of indole and brominating agent are used and this is followed by treatment with water, yields up to 50% of the corresponding oxindole can be obtained. Use of a second mole of NBS results in the formation of 3-bromooxindoles. On the other hand, when the NBS reaction was performed in aqueous media, 5-bromooxindole was formed. Subsequent bromination in the 5-position is apparently greatly retarded in *t*-butyl alcohol relative to water.

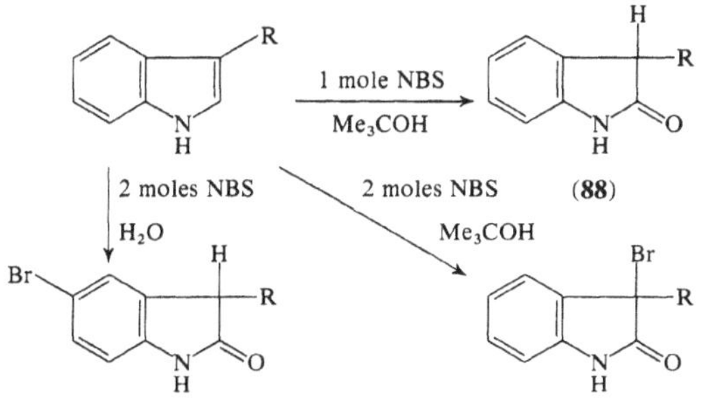

(**88**)

(**89**)

(**90**) where R = $-CH_2COOH$

The reaction of 3-methylindoles with NBS in pyridine at room temperature results in a quantitative yield of the pyridinium salt (**91**), which by acid hydrolysis is converted to 3-methyloxindole (**92**), (*77, 226*).

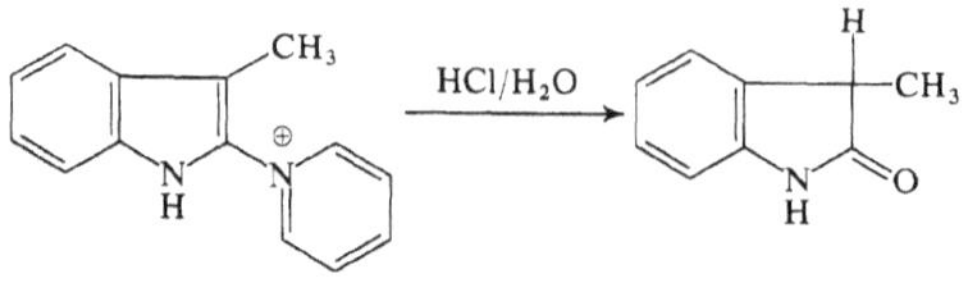

(91) (92)

$Br^{\ominus}$

The mechanism proposed (*177*) for the reaction involves bromination at the most nucleophilic 3-position of the indole nucleus, nucleophilic addition of pyridine at the carbon atom of the reactive indolenine and subsequent elimination of HBr.

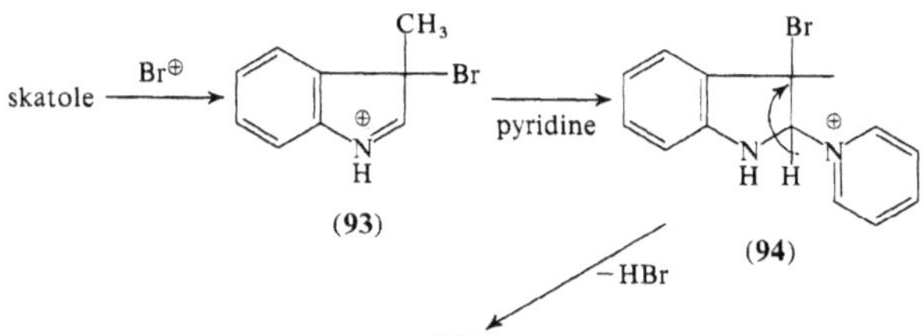

(93) (94) (91)

Treatment of indole-3-propionic acid with 3 equiv. of NBS in aqueous media leads to the cyclic product 5-bromolactone (**96**) (*295*). Similar spirolactone formation has been found also with indole 3-propionyl

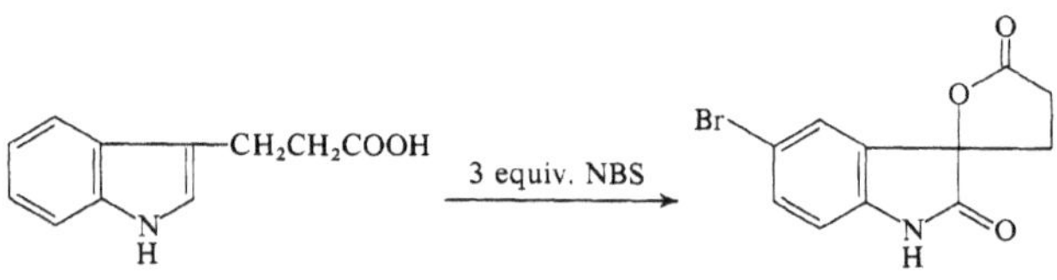

(95) (96)

peptides or tryptophan peptides, *e.g.*, with N-carbobenzoxy-tryptophanylglycine. In these cases, lactonization is accompanied by peptide bond cleavage (*295, 438, 439, 440, 443*).

The most plausible mechanism put forward to account for the formation of brominated spirolactones of oxindoles is shown below (**97**→**104**) (*175, 295*). According to this mechanism, up to 3 equiv. of oxidative bromine participate in the modification-cleavage reaction (*295*,

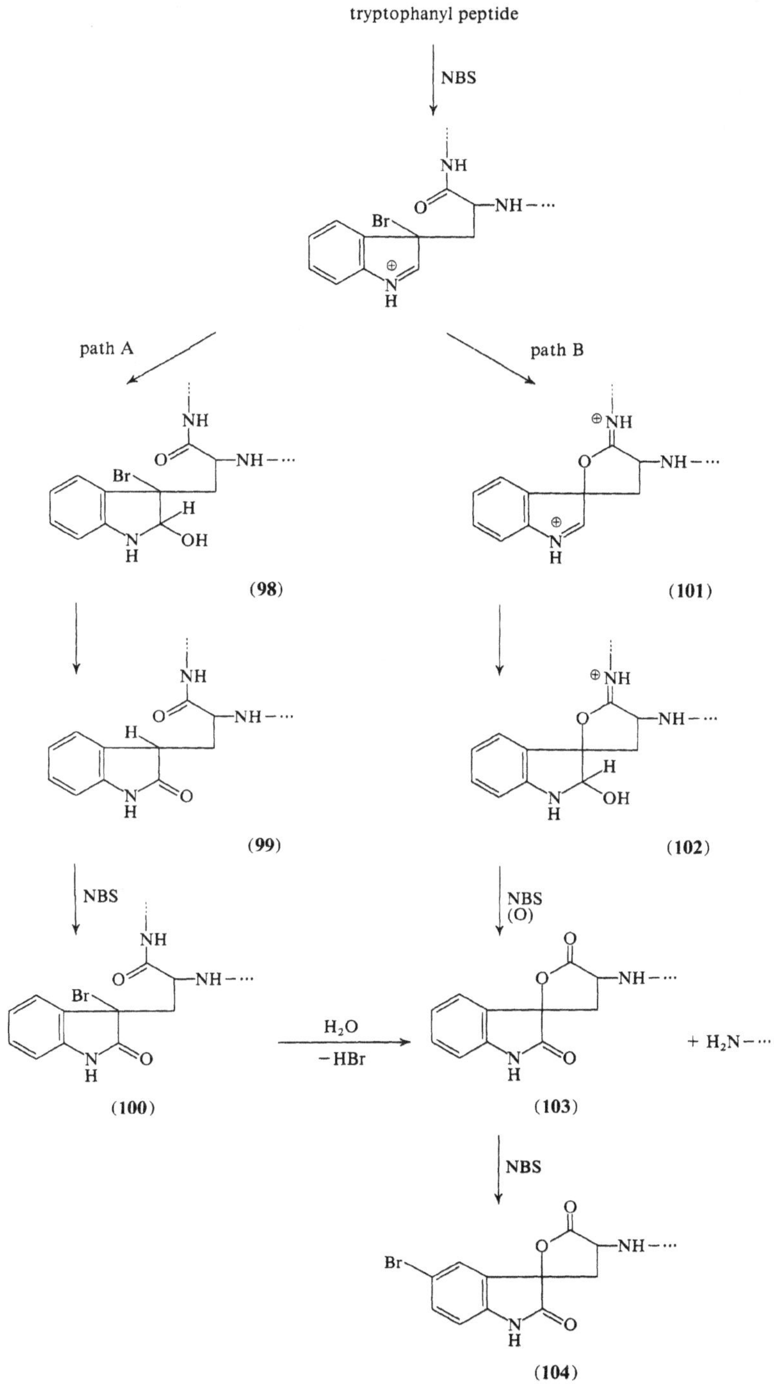
tryptophanyl peptide
NBS
path A
path B
(98)
(101)
(99)
(102)
NBS
NBS (O)
H_2O
$-HBr$
$+ H_2N-\cdots$
(100)
(103)
NBS
(104)

316). On the basis of the results discussed above for the reaction of NBS with simple 3-substituted indoles, a 3-bromoindolenine appears to be the first initial intermediate, which, following path A or B, undergoes spontaneous debromination through a series of oxidation and hydrolysis reactions. These reactions lead to the formation of an oxindole derivative, which promotes the cleavage reaction. The third equiv. of NBS forms a stable 5-bromooxindole derivative.

Although the yields of cleavage with model tryptophan peptides ranged between 50 to 80% (*440, 443*), the NBS cleavage reaction has not been successfully applied to the preparation of peptide fragments from proteins to be used for sequence work (*125, 169*).

N-Terminal tryptophanyl peptide bonds are not cleaved by the action of NBS. The reason for the failure of the cleavage reaction may be related to the possibility that the free α-amino group of tryptophan derivatives and not the carboxamido function interact preferentially with the bromoindolenine. Similar amino group participation has been found with N-terminal tyrosyl peptides, which are not cleaved by NBS (*434*).

Tryptophan cleavage with NBS can be blocked by reaction of the tryptophan peptide or protein with 2-hydroxy-5-nitrobenzyl bromide the Koshland reagent (see Section III.4.) or by ozonolysis of tryptophan residues to formylkynurenine (*313, 419*). These modification reactions of the indole nucleus restrict NBS cleavage to tyrosyl and histidyl peptide bonds.

More recently, WITKOP and coworkers (*284*) have further studied the reaction of halogenating agents with tryptophan compounds in alkaline media. N-Acetylated derivatives of tryptophan and tryptamine

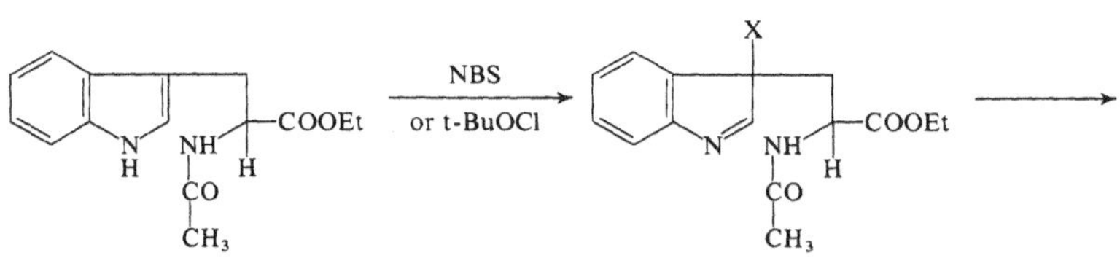

(105) (106)

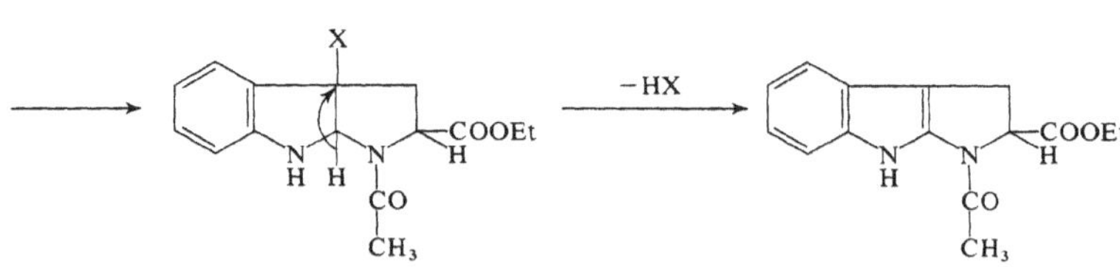

(107) (108)

in phosphate buffer at pH 9 with NBS or in methylene chloride containing triethylamine with *t*-butyl hypochlorite give acid-labile tricyclic tetrahydropyrroloindoles* characterized by λ_{max} 308 nm. These cyclizations presumably occur *via* β-haloindolenines, which ring close with dehydrohalogenation.

The formation of the tetrahydropyrroloindole derivatives requires the presence of a 3-(2-acetamidoethyl) side chain or, in the case of tryptophan derivatives, a carboxyl-blocked, N_{α}-acetylated derivative.

Tetrahydropyrroloindoles have been found by Ohno *et al.* (*285*) to be suitable intermediates for the synthesis of oxindolylalanine and its N-acyl esters. The method consists in the acidic hydrolysis of (**108**) in sealed tubes at 110° C in 6N HCl, giving a 75% yield of oxindolylalanine. Hydrolysis of (**108**) with 20% acetic acid at room temperature provided the N-acetyl ethyl ester (**110**). Treatment with a 50% excess of

(**108**)

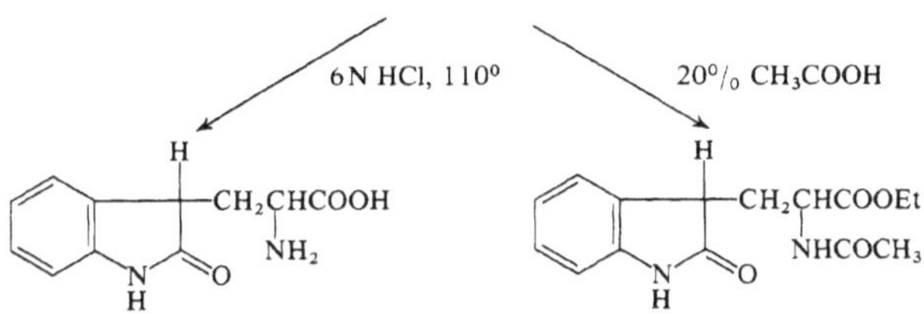

(**109**) (**110**)

t-butyl hypochlorite in buffered methylene chloride and subsequent refluxing in aqueous ethanol containing an excess of sodium acetate, aromatized (**108**) with deacetylation to the acid stable ethyldihydropyrrolo-[2,3-b]-indole-2-carboxylate (**111**) in 40% yields. This 14-π-electron

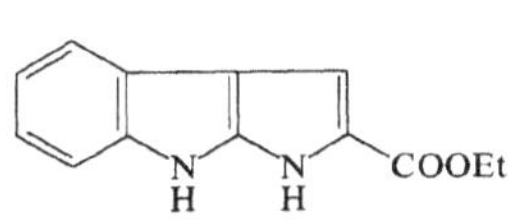

(**111**)

system constitutes a new aromatic system, which is present in anhydrodethiosporidesmin, a degradation product of sporidesmin (*284*).

* The authors (*284*) named these compounds dihydropyrrolo [2,3-b] indoles, but in this article the nomenclature according to Chemical Abstracts has been used.

Iodinating agents, like N-iodosuccinimide, ICl and chloramine T-KI similarly oxidize and cleave simple tryptophan peptides as well as tyrosine peptides. Methionine is oxidized to its sulfoxide and cysteine oxidation proceeds ultimately to cysteic acid (*6, 7*).

Oxidative cleavage can be obtained by "active iodine" that is generated with H_2O_2, iodide and a peroxidase (*7*). All these three constituents are required for oxidation and fission to proceed. The oxidation and the oxidative cleavage probably occur by the mechanism discussed above for brominating agents. Oxidation concomitant with tryptophan peptide bond cleavage is observed at acidic pH and up to pH 5.0, with yields ranging from 30 to 40%, whereas at pH 7.0 no significant cleavage occurs, because the formation of the iminolactone is hindered.

Whether the iodinating agents can be employed as selective reagents for peptide bond cleavage in proteins remains to be established.

3.2. BNPS-skatole

A very mild brominating agent has been introduced recently for modification of the tryptophan residue in proteins (*119, 289*). On treatment of 2-(2-nitrophenylsulfenyl)-3-methyl-indole (NPS-skatole) (**112**) with one equiv. of NBS in 80% acetic acid, the positive bromide-containing compound 2-(2-nitrophenylsulfenyl)-3-methyl-3-bromoindolenine (BNPS-skatole) (**113**) was obtained.

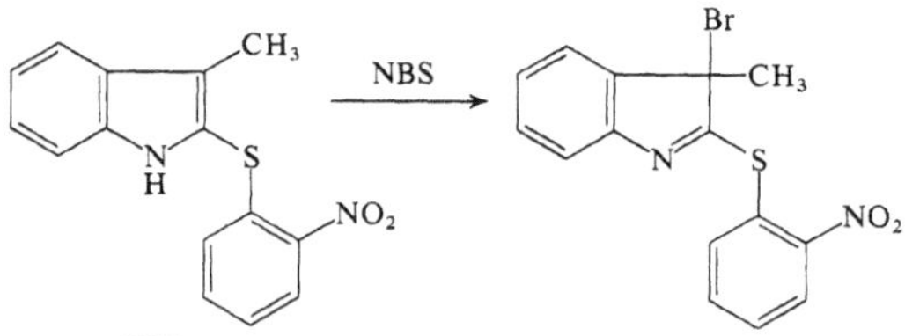

(**112**) (**113**)

BNPS-skatole is a much more selective agent than NBS. After exposure of an amino acid mixture to 10 equiv. of BNPS-skatole in 50% acetic acid for 30 min at room temperature, tryptophan was completely absent, methionine was converted to methionine sulfoxide, and the other amino acids were recovered quantitatively. By contrast, if the amino acid mixture is oxidized with NBS, not only tryptophan, but also methionine, tyrosine, histidine, and cystine were completely absent. Cysteine, as expected, is easily oxidized by BNPS-skatole to cystine and to some extent, by excess reagent, to cysteic acid. If present, sulfhydryl groups must be reversibly protected, as in selective modification with many other classes of reagents (*375*).

As anticipated from the action of NBS on tryptophan peptides (*295*), BNPS-skatole was proved useful also for selective cleavage of tryptophanyl peptide bonds (*289*). With a variety of tryptophan containing peptides, 10-fold excess of reagent in 20 hours gave cleavage yields (30–59%) comparable to those obtained with NBS. However, BNPS-skatole is a selective reagent and does not cleave at sites other than tryptophanyl peptide bonds (*119, 289*).

Since cleavage is time-dependent and since tryptophan reacts rapidly with the reagent, modification without any significant cleavage was checked when the reaction was carried out with a low excess of reagent and for shorter times. This was clearly proved with staphylococcal nuclease (*289*). Reaction with BNPS-skatole and subsequent reduction of the 4 partially oxidized methionine residues produced a derivative of nuclease with modification only of residue 140, the single tryptophan unit.

When nuclease was incubated with 100 equiv. of BNPS-skatole in 50% acetic acid for 28 hours, cleavage at the tryptophanyl peptide bond occurred in addition to modification. The released COOH-terminal nonapeptide was identified as a single spot on paper electrophoresis and from the amino acid recovery on the analyzer. The yield of cleavage product was estimated as approximately 15%. No additional peptides were released by cleavage at any other site in the sequence on the basis of analysis by electrophoresis.

The important feature of BNPS-skatole as modifying reagent of tryptophan residues lies in its higher selectivity in comparison to NBS; BNPS-skatole (2–3 equiv., 30 min in 50% acetic acid) rapidly achieves quantitative oxidation of the indole ring of tryptophan and the sulfur atom of methionine. Since methionine sulfoxide can be reduced back to methionine, it is possible to obtain a protein derivative selectively modified at tryptophan. Modification can be carried out under conditions in which there is little risk of cleavage of peptide bonds. Much longer reaction time and excess reagent are needed for cleavage, as clearly indicated by the studies with nuclease.

The use of BNPS-skatole for the *selective cleavage* of the tryptophanyl peptide bond has been proved useful in sequence studies. The cleavage yields so far obtained (about 50–70%) with peptides (*289, 371*) and proteins (*119*) indicate that the reagent is the best so far available for the cleavage of the tryptophanyl peptide bond and approaches in utility the cyanogen bromide cleavage of methionine peptide bonds (*375*). The reagent has been successfully applied to staphylococcal nuclease

(*289*), horse heart cytochrome c (*128*), the encephalitogenic protein A_1 (*41, 65*), chicken glutamate dehydrogenase (*263*), trypsin (*171*) and bovine factor X_1 (*110*).

3.3. *Tribromocresol (TBC)*

BURSTEIN and PATCHORNIK (*66*) reported the use of 2,4,6-tribromo-4-methylcyclohexadienone (tribromocresol, TBC) (**114**) for the selective

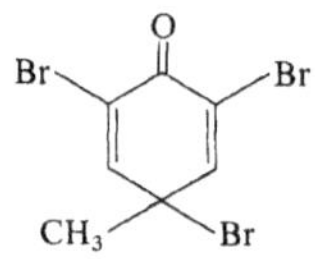

(**114**)

modification and cleavage of tryptophanyl peptide bonds in peptides and proteins.

Tyrosyl and hystidyl peptide bonds, which are usually cleaved by other brominating agents such as NBS, are sensitive to oxidation also but are not cleaved by TBC. Tyrosine was converted by TBC to 3,5-dibromotyrosine, cysteine to cysteic acid, methionine to methionine sulfoxide and tryptophan most probably converted to a bromooxindole derivative. Optimal conditions for the cleavage of the tryptophanyl peptide bond were found to be 3 equiv. of TBC at pH 3 for 5–15 min at room temperature. The model peptide N-carbobenzoxy-tryptophanyl-glycine was cleaved in about 50% yield; NBS also cleaved this peptide to the same extent. In lysozyme, selective cleavage of the expected tryptophanyl peptide bonds (yields 5–60%) was obtained.

TBC is a much more selective agent than NBS and other brominating agents. In addition, it does not react with free amino groups, and excess of TBC can be used in the cleavage reaction. The mechanism of the cleavage reaction of tryptophanyl peptide bonds was thought to be similar to the one proposed with NBS (*295*).

Analogously to TBC, N-bromo-benzotriazole (**115**) was also found an effective reagent for the modification and cleavage of tryptophanyl peptide bond (*129*). The reagent was found less reactive than NBS

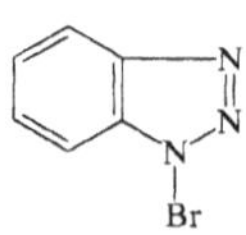

(**115**)

and to cause effects on tryptophan and other amino acids similar to those of TBC.

In comparing NBS, TBC, N-bromo-benzotriazole and BNPS-skatole, it appears that the latter may offer certain advantages. BNPS-skatole is much less destructive toward tyrosine, histidine and cystine than the others, and this may be an important factor in choosing between them for a particular purpose. In addition, by exposing a tryptophan-containing protein to low amounts of BNPS-skatole and for a short time, it is possible to obtain modification and not cleavage of the tryptophanyl peptide bond.

4. 2-Hydroxy-5-nitrobenzyl Bromide (Koshland's Reagent)

4.1. Chemistry of the Reaction

Koshland and coworkers (*25, 193, 194*) introduced 2-hydroxy-5-nitrobenzyl bromide (HNB-Br) (**116**) for the modification of tryptophan residues in proteins. The reagent has been found to possess striking selectivity in modifying tryptophan residues at acid pH, although the

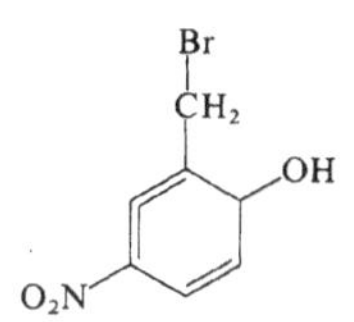

(**116**)

sulfhydryl group of cysteine also reacts at a much slower rate. At alkaline pH, cysteine and tyrosine react at about the same rate as tryptophan.

The extensive use of HNB-Br in protein modification studies has evoked much interest in its chemistry, and several reports have appeared from different laboratories (*236, 253, 378*). The utility of the reagent clearly depends on a thorough understanding of the mechanism of its reaction with tryptophan residues in proteins.

The extremely high reactivity of HNB-Br with the indole nucleus seems to be related to a quinone methide type structure (**117**), which attacks the indole moiety involving a charge transfer complex (*236*). Indoles are known to be donors in various complexes that exhibit the characteristics expected of charge-transfer interaction (*364*). Theoretical

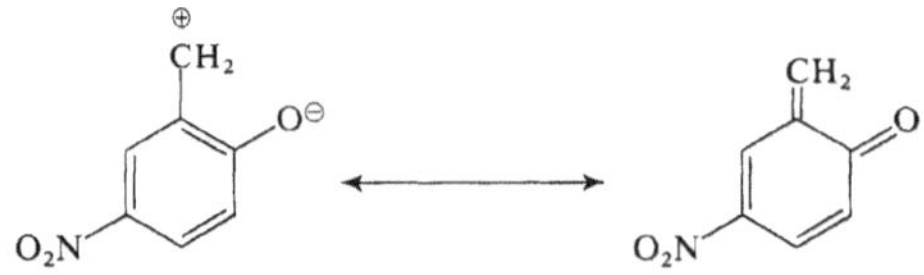

(**117a**) (**117b**)

considerations indicate that the indole position 3 and the methylene group of the quinone methide could interact in a charge-transfer complex. These are precisely the loci which react to form an HNB-modified indole (*236*).

SPANDE, WILCHEK and WITKOP (*378*) allowed an equimolar mixture of the HNB-Br reagent and 3-methylindole to react in aqueous acetone. The carbinolamine (**119**) was obtained in 70% yield as a result of alkylation at the 3-position. The isomeric indolenine structure (**118**) was ruled out on the basis of nmr data, but was believed to be an intermediate in the reaction. The imine function of an indolenine

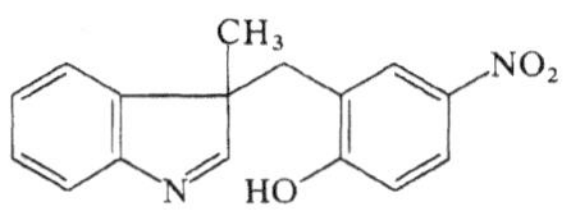

(**118**)

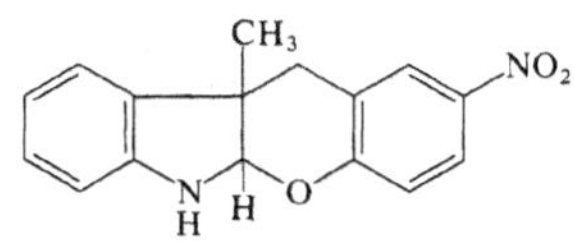

(**119**)

easily undergoes nucleophilic attack and the nucleophile in this case, namely the phenolic oxygen, is readily available in the same molecule, leading to cyclization.

MCFARLAND *et al.* (*253*) studied the reaction of free tryptophan with HNB-Br. The major product was a monosubstituted compound resulting from initial attack of the reagent at the 3-position of the indole nucleus followed by cyclization. In this case the cyclizing nucleophile in the initial indolenine adduct is the free amino group. Assuming that the product is formed almost exclusively with the *cis* ring juncture, there should be two possible isomers, in which the carboxyl- and HNB-groups bear either a *cis* or *trans* relationship, as indicated by structures (**120**) and (**121**).

LOUDON *et al.* (*236*) reported a detailed study of the chemistry of the *monosubstitution* products formed by the reaction between HNB-Br and tryptophan ethyl ester. These authors confirmed the formation of two diastereomers analogous to (**120**) and (**121**) as a result of the cyclization by the amino group and succeeded in separating them. When these compounds were subjected to acidic conditions, a re-

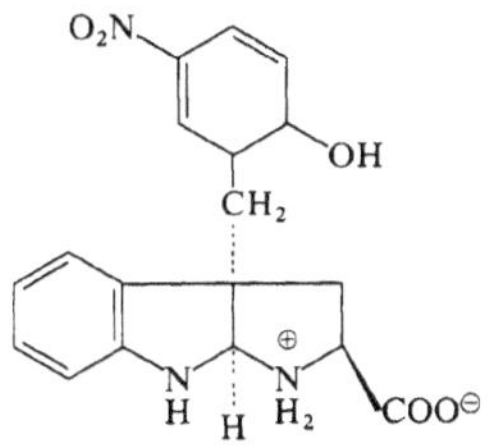

(120)

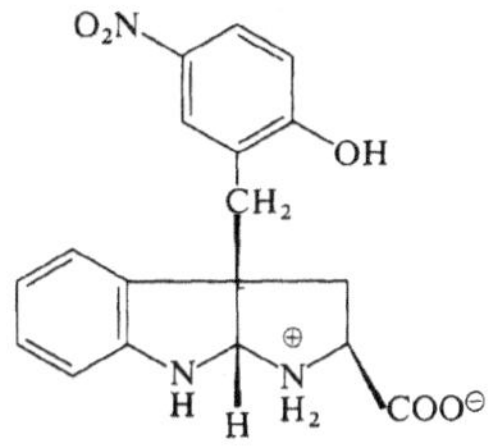

(121)

arrangement occurred with migration of one of the two substituents at position 3 to position 2 to yield the 2,3-disubstituted indole (**123**). Such a reaction destroys the second asymmetric center and yields a single compound. The hydroxynitrobenzyl group is a very effective migrating group, since it can revert to a quinone methide type

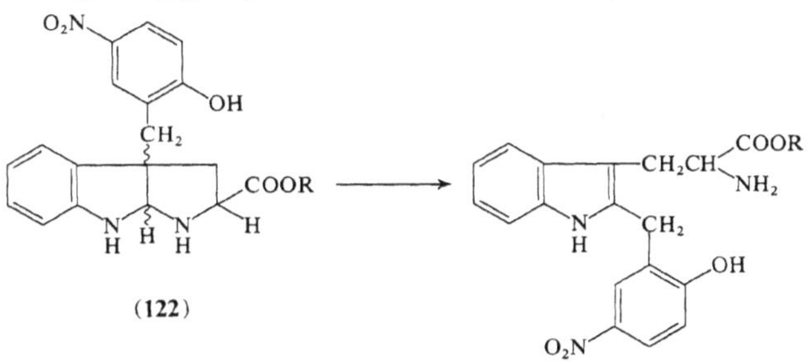

(122) (123)

structure merely by the loss of a proton, and would easily support a positive charge.

On the basis of the results reported by SPANDE *et al.* (*378*) the amino group leading to cyclization does not need to be free, since N-acylated

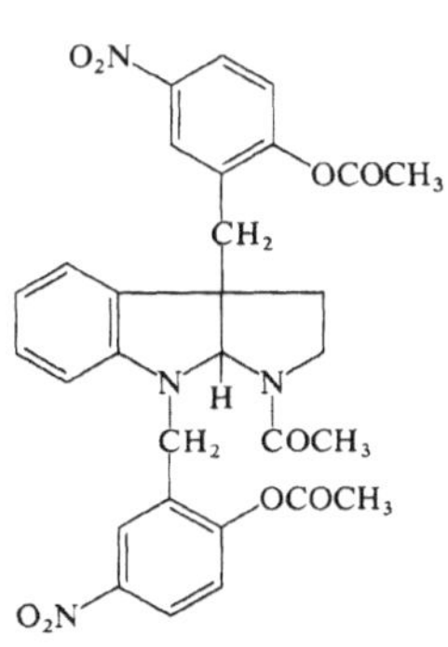

(124)

tryptamine and N-acetyl-tryptophan methyl ester also give hexahydropyrroloindoles similar to (**122**). N-Acetyl-tryptamine was found to combine with two equivalents of HNB-Br giving in good yields a product which was isolated and analyzed as its *o,o'*-diacetyl derivative, found to be (**124**).

CASNATI and coworkers (*69, 70*) found that γ, γ-dimethylallyl bromide reacts with N-acetyl-tryptamine similarly to the Koshland's reagent. Two compounds were isolated and their structure (**125**) and (**126**) were determined (*70*).

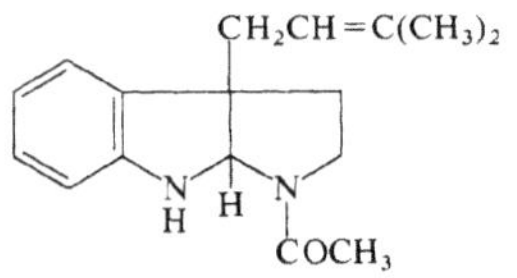

(**125**)

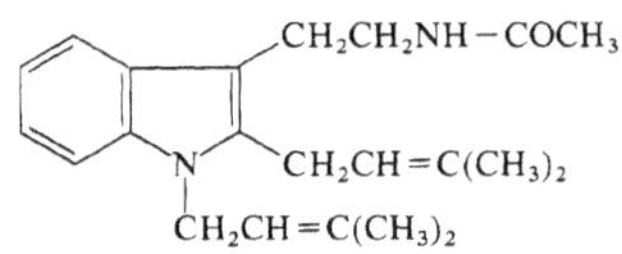

(**126**)

More recently (*216, 320*) it has been found that the HNB-group linked at the indole nucleus of tryptophan in a protein can be reversibly removed. REISS and LUKTON (*320*) studied the reaction of HNB-Br with lysozyme and found that the formation of HNB-lysozyme resulted in enzyme inactivation by modification of one tryptophan residue. The residue which carried the HNB group was found to be tryptophan-62, which with other reagents such as 2-nitrophenylsulfenyl chloride (*341*) and N-bromosuccinimide (*162*) was found to be very reactive. However, it was found that when HNB-lysozyme was kept at pH 2.8 for several hours, the modified enzyme regained its activity, although not completely, and formation of HNB-OH was noticed. This fact would imply that regeneration of lysozyme activity is related to reversible modification of the 62-tryptophan residue (*320*).

A mechanism for the elimination of the HNB-label from the indole nucleus of a tryptophan residue was proposed, which involves protonation of the indolenine-HNB complex followed by cleavage of the protonated complex to yield the unmodified tryptophan residue and a quinone methide, which by solvation yields HNB-OH. The indolenine structure responsible for the elimination reaction was also confirmed by reduction with $NaBT_4$. One mole of tritium was incorporated in the HNB-protein, and after reduction of the $C=N-$ bond by the hydride neither release of HNB-OH nor concomitant reactivation of the enzyme could be effected.

Despite the complex nature of the reaction of HNB-Br with the indole nucleus of tryptophan, the reagent has been extensively used in protein modification studies to assess the role of tryptophan residues

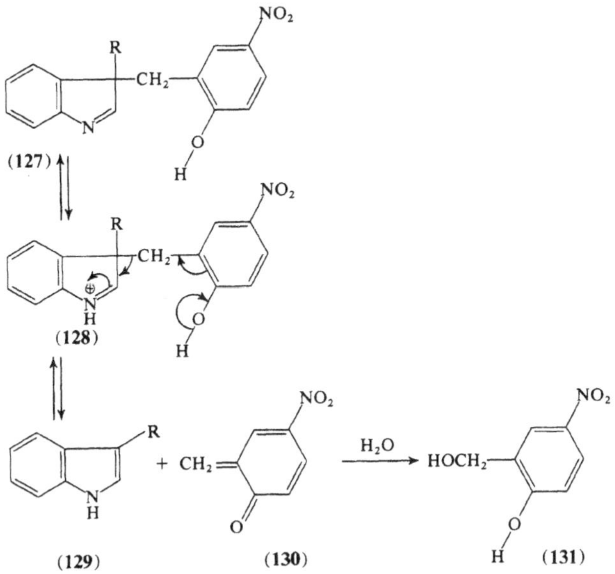

in enzymes (see Section VII.) and also for the quantitative determination of tryptophan in proteins (see Section IV.4.3.).

4.2. Other Nitrobenzyl Reagents

(2-Hydroxy-5-nitrobenzyl)-Sulfonium Bromide (*132*). Satisfactory labelling of tryptophan residues in proteins can be achieved by using HNB-Br as a solution in a dry, water-miscible organic solvent such as acetone, dioxan, methanol or dimethoxy-ethane, to reach a final concentration of organic solvent of 5–10%. However, organic solvents can cause denaturation of enzymes, so that if denaturation of the enzyme under study occurs in the reaction conditions employed, the reagent cannot be used in assessing the role of tryptophan moieties in the biological activity of enzymes.

To circumvent these difficulties, HORTON and TUCKER (*196*) synthesized a water-soluble reagent, dimethyl (2-hydroxy-5-nitrobenzyl)-sulfonium bromide (**132**) which, like HNB-Br, can react selectively with tryptophan residues of proteins in aqueous solution, but in the absence of organic solvent.

Although the sulfonium reagent is rapidly hydrolyzed in neutral and alkaline solutions, it is relatively stable in acidic aqueous solutions.

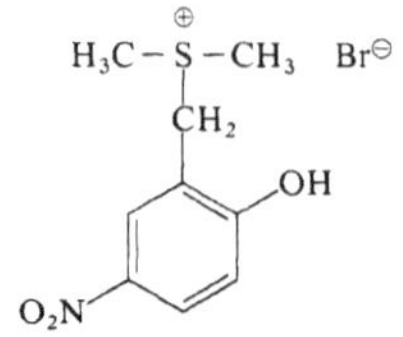

(132)

Consequently, it has been found convenient to dissolve an appropriate excess of the reagent in water prior to its addition to a buffered solution of the protein. An alternative is simply to add the reagent as a solid.

However, even more complex alkylations of the indole ring by the water soluble sulfonium reagent had to be expected, since methylation reactions have to be taken into account.

HEINRICH *et al.* (*167*) studied the reaction of (**132**) with the model compound N-acetyl-tryptophan amide at a constant pH of 4.7. At least *ten* different reaction products could be detected; some of them were expected on the basis of the results obtained by previous workers. However, in the case of N-acetyl-tryptophan amide, the most stable and abundant product was found to be a compound (**134**) bearing a 2-piperidone ring system, which could arise from a cyclization of the presumed intermediate (**133**) in the reaction.

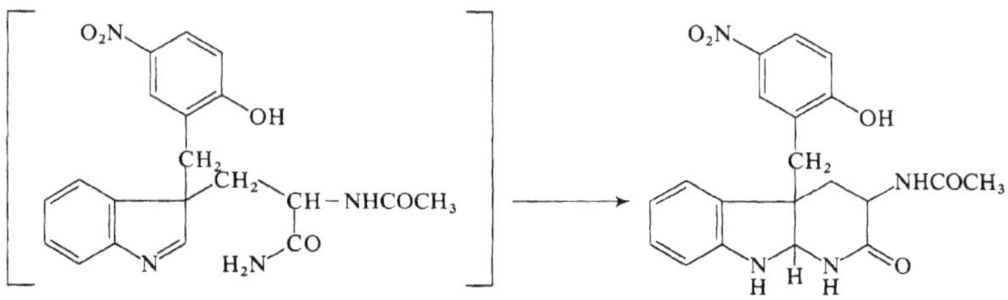

(133) (134)

2-Methoxy-5-nitrobenzyl Bromide (**135**). This reagent has been used (*193*) to modify tryptophan residues in a manner similarly to that

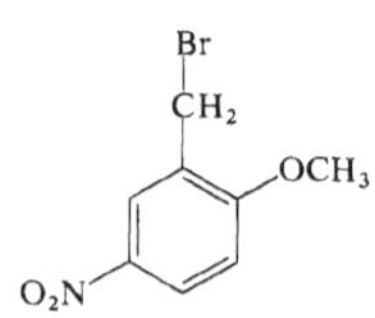

(135)

effected by the 2-hydroxy-compound (HNB-Br). It reacts in a fashion similar to a classical benzyl halide so that, in addition to tryptophan, the thioether group of methionine is also attacked. It has the advantage that its absorption is not sensitive to changes in pH but is sensitive to changes in the polarity of the environment. In fact it has been used as an environmentally sensitive "reporter group" which provides spectral data useful in interpreting interactions with residues in its vicinity.

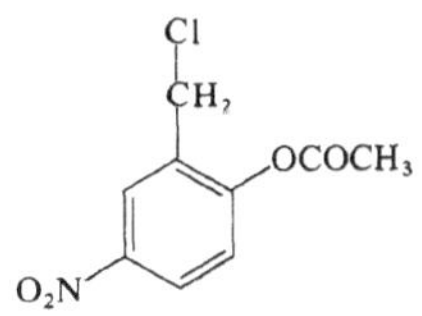

(136)

2-Acetoxy-5-nitrobenzyl Chloride (**136**). The reagent has been utilized to incorporate the *p*-nitrophenolic reporter group into enzymes which possess nitrophenyl ester hydrolyzing activities (*195*). The 2-acetoxy-reagent is hydrolyzed at an active site of the enzyme and the 2-hydroxy-reagent is generated *in situ*, providing an affinity labelling of tryptophan residues which are near the active site of the enzyme.

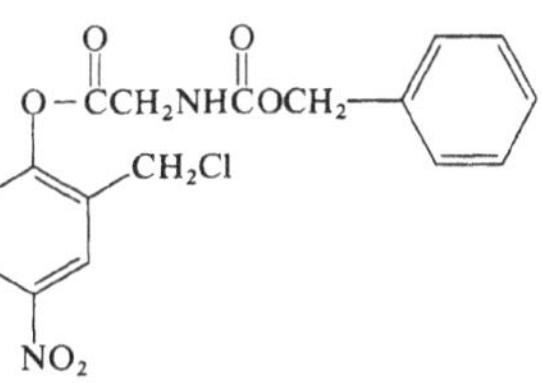

(137)

2-Chloromethyl-4-nitrophenyl (N-carbobenzoxy)glycinate (**137**). This reagent resembles the *p*-nitrophenyl derivative of carbobenzoxyglycine, which has been shown to be an excellent substrate for the proteolytic enzyme papain (*436*). Enzymic hydrolysis of (**137**) by papain releases the highly reactive HNB-Cl *in situ*, to provide selective alkylation of one or more tryptophyl residues present at the active site of papain (*260*).

Reaction of papain, at pH 5.0 with a 200-fold molar excess of (**137**) resulted in the incorporation of one HNB-group per molecule of enzyme. Spectral properties of the hydroxynitrobenzylated enzyme provided information concerning the nature of the environment of the reporter group (*260*).

5. Sulfenyl Halides

5.1. General

Sulfenyl halides have been found to be specific, mild reagents for modification of tryptophan containing peptides and proteins (*123, 124*). By reaction with a sulfenyl halide like 2-nitrophenylsulfenyl chloride

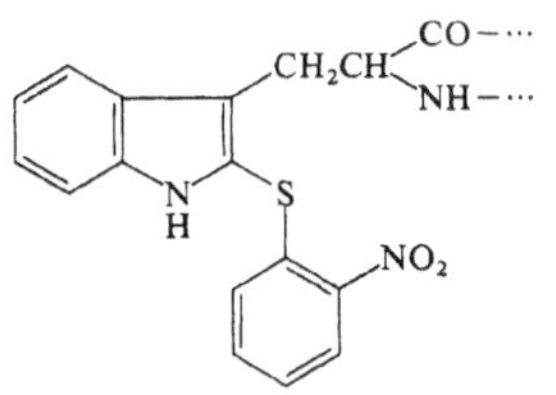

(138)

tryptophan is converted into a derivative containing a thioether function in the 2-position of the indole nucleus (**138**). WIELAND *et al.* (*428, 430*) used this reaction to synthesize tryptathionine peptides related to phalloidin (see Section VI.2.4.). The only other amino acid affected is cysteine, which is converted to a mixed disulfide (**139**) (*126*).

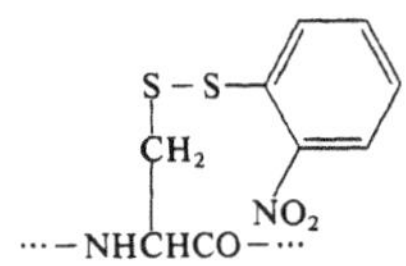

(139)

In principle, any member of the large class of sulfenyl halides could be used for labelling of tryptophan in proteins. However, these reagents are highly unstable in aqueous solvent with the notable exception of 2-nitrophenyl compounds (*123, 354*). In the case of the easily hydrolyzable sulfenyl halides, an aqueous medium can still be employed, provided that an excess of reagent dissolved in an organic solvent is used and is added dropwise under stirring to an aqueous solution of the protein.

At the present time the most widely used sulfenyl halide is 2-nitrophenylsulfenyl chloride (*124*). Other compounds tested are 4-nitrophenyl- (*354*), 2,4-dinitrophenyl- (*354*), 2-nitro-4-carboxyphenyl- (*416*) and 2,4-dinitrophenyl-1,5-disulfenyl chloride (*417*). The bifunctional reagent has been used to introduce selectively intramolecular cross-links between tryptophan units (**140**).

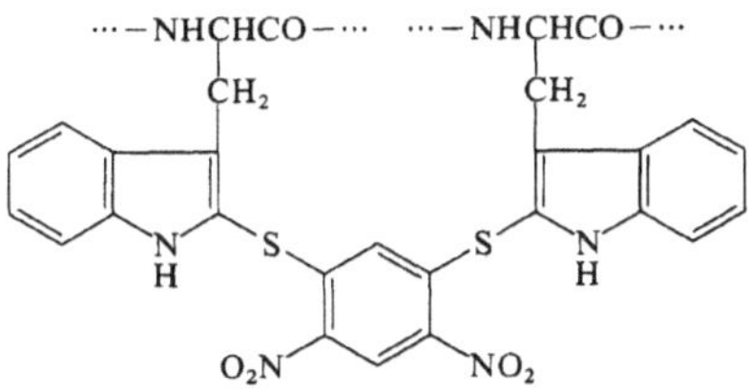

(**140**)

Quantitative modification of tryptophan in proteins can be achieved readily by using a low molar ratio of reagent to protein. The tryptophan residues of lysozyme, trypsin, and chymotrypsin were fully modified by reacting the protein with 20 equivalents of a freshly prepared solution of 2-nitrophenylsulfenyl chloride in 50% acetic acid (*354*). Acetate buffer pH 4–5 has been used in the modification of human growth hormone (*60*), lysozyme (*341*) and α-lactalabumin (*357*).

Sulfenyl halides possess several particularly valuable features for protein modification studies. Their high selectively for tryptophan in proteins lacking SH-groups is especially remarkable. Furthermore, even if cysteine residues are present, selectivity for tryptophan can still be achieved since the disulphides formed by reaction of the halide with cysteine SH-groups can be easily reconverted to cysteine by subsequent reduction with mercaptoethanol. Other techniques for tryptophan modification are less specific and in several cases the reaction products are undefined.

The sulfenylation reaction increases the number of procedures available for modification of tryptophan in the study of biological function of tryptophan-containing enzymes. One of the main advantages of this technique is the possibility of using a variety of sulfenyl halides, thus leading to a change in size, polarity, or other physico-chemical properties of the labelled enzymes.

Apart from the valuable data about the involvement of tryptophan in enzymic processes (Section VII.), which can be obtained by use of sulfenyl halides, the analytical utility of this reaction is another major application. By reaction with nitrophenylsulfenyl halides a chromophore is generated in a protein which absorbs in the visible part of the spectrum. This permits easy quantitation of the reaction, and thereby allows determination of the tryptophan content of a protein (see Section IV.4.2. for the analytical application of the reaction).

Furthermore, nitroaryl chromophoric groups, if included in optically active compounds, may exhibit characteristic extrinsic Cotton effects as additional probes of protein conformation. A conformational study on poly-L-tryptophan modified with 2-nitrophenylsulfenyl chloride has been reported (*298*).

Finally, the yellow color of the tryptophan derivatives is a convenient chromatographic marker for peptides (*278*). In particular, 2-nitro-4-carboxyphenylsulfenyl chloride has been used for identifying tryptophan peptides in the diagonal electrophoresis technique (*283*).

An elegant application of the sulfenylation reaction was devised by WILCHEK and MIRON (*432*) in the one-step isolation of tryptophan peptides by affinity chromatography. The method consists of using an affinity column packed with anti-dinitrophenyl antibodies, through which a mixture of peptides, previously treated with 2,4-dinitrophenylsulfenyl chloride is passed. Only the tryptophan peptides bearing the dinitrophenyl-moiety covalently linked at the 2-position of the indole nucleus were specifically sequestered by the column. The method was checked successfully with human serum albumin. In this case the protein was reacted with DNPS-Cl, digested with trypsin and passed through the affinity column. A single yellow peptide [Ala-Trp(DNPS)-Ala-Val-Ala-Arg] was retained on the column and subsequently eluted with 10% formic acid.

Sulfenyl halides have so far been used in studies on the role of tryptophan in biological function of enzymes and also for the analytical determination of this amino acid. Looking beyond its use on problems connected with enzymology and protein chemistry, since the sulfenylation reaction is generally valid the ubiquity of the indole nucleus will prompt its application to other naturally occurring indole substances.

5.2. Chemistry of the 2-Thioether-indole Function

The chemistry of the 2-thioether indole moiety has been studied in connection with the use of sulfenyl halides as modifying reagents for tryptophan peptides and proteins. This functions also occurs naturally, being present in the highly poisonous cyclic peptides from *Amanita phalloides*, such as phalloidin, in which the –SH group of cysteine is linked to a tryptophan residue at the 2-position through a thioether bridge (see also Section VI.2.4.). During the structural and synthetic work on these peptides, WIELAND and coworkers (*427, 430, 431*) made significant contributions to this field.

Oxindole-derivatives. Hydrolysis of an indole 2-thioether in 20% acetic acid at 95–120° C in a sealed tube gives the corresponding oxindole. The reaction was studied with several indole 2-thioethers, the results of this study (*420*) can be summarized as follows. Various 2-sulfenyl derivatives of tryptophan and tryptamine were each quantitatively converted to the oxindole in 30 minutes at 120° C, whereas the corresponding N-acetylated substrates were completely *resistant* to hydrolysis, under the same conditions. Peptides containing tryptophan 2-thioether residues were also unaffected by heating in aqueous acetic acid. Analogously, 2-thioether-3-methylindole was not hydrolyzed after prolonged heating. Finally, the reaction has been found to be acid catalyzed, the optimum pH being abount 1.5 (*420*).

A mechanism such as that shown below rationalizes the data. Protonation at the 3 position facilitates nucleophilic attack by the free amino group of the side chain on C-2 with concomitant displacement of the thioaryl-function and formation of a tetrahydropyrroloindole derivative (**144**), which is further hydrolyzed to an oxindole derivative.

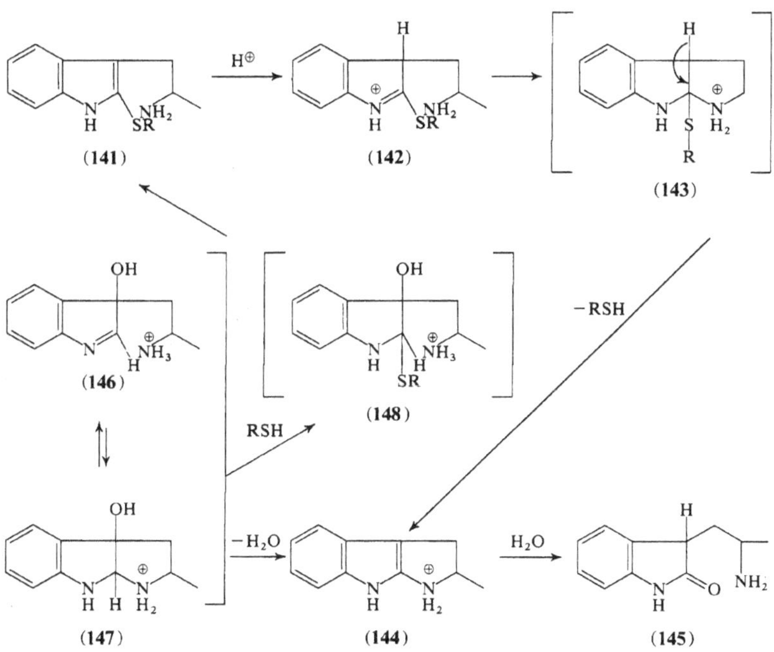

Conversion of (**144**) to (**145**) has been previously shown to occur on treatment of the hexahydropyrroloindole (**147**) obtained by peracetic acid oxidation of tryptophan (see Section III.1.1.). (**147**) is converted by 2N HCl to the oxindole (**145**); in dilute acetic acid (**147**) can also add thiols giving 2-thioether indole compounds (**141**) (*339*).

The above reaction offers a convenient synthetic pathway to oxindoles from parent indoles bearing an aminoethyl side chain at the 3-position. The yields are almost quantitative, as checked with tryptophan, 5-fluoro-tryptophan, tryptamine and serotonin (*127*).

*Thioxindole-derivatives**. Upon reaction of tryptophan or another indole with the bifunctional sulfenyl halide, sulfur dichloride, in glacial acetic acid, dimers are obtained with a disulfide bridge at the 2-position (**149**). By reduction with excess mercaptan (mercaptoethanol,

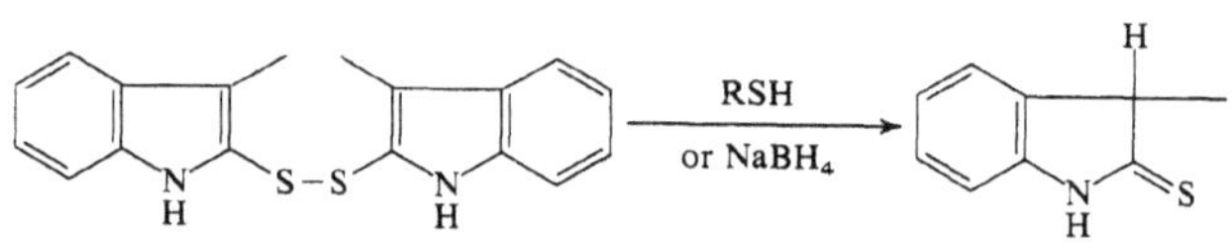

(**149**) (**150**)

thioglycolic acid) or sodium borohydride, the corresponding thioxindole derivatives (**150**) are formed.

The reaction has been used to introduce a thiol group in the 2-position of a number of indole-containing compounds, in particular tryptamine, serotonin, skatole and tryptophan, the yields ranging from 65 to 95 per cent (*127, 431*). The single tryptophan residue of the pancreatic peptide hormone glucagon (29 amino acids) has also been modified by this method to the thioxindolylalanine derivative (*127*).

WILCHEK and MIRON (*433*) described a method for quantitative removal of 2,4-dinitrophenyl groups from 2-(2,4-dinitrophenylsulfenyl)-tryptophan peptides and proteins to yield thioxindolylalanine derivatives. Cleavage of the nitroaryl group is carried out using mercaptoethanol as a thiolysis agent. The pentapeptide Gly-Ile-Thr-Trp(DNPS)-Lys was

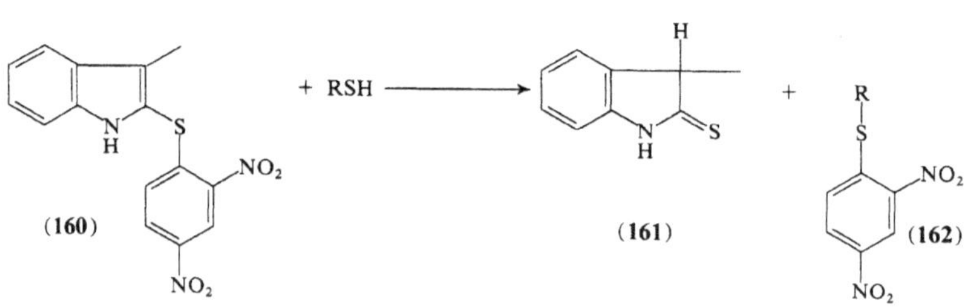

(**160**) (**161**) (**162**)

* The term thioxindole has been used in analogy to the oxindole terminology. Previously (*431, 433*), these compounds have been termed 2-thiol- or 2-mercapto derivatives. The right structure of the compounds is an indole-2-thione (*273*).

reacted with 50-fold molar excess of thiol for several hours at room temperature and the pentapeptide containing thioxindolylalanine was isolated. The thiolysis was also carried out successfully on two proteins, *viz.*, human serum albumin and subtilisin (*453*).

In view of the mild conditions under which thiolysis proceeds, the method of Wilchek and Miron (*433*) may find application in the study of tryptophan-containing enzymes and proteins. However, in comparing the two methods above discussed for preparing thioxindolylalanine compounds, the use of sulfur dichloride and subsequent reduction by thiols seems the most useful and appropriate.

Reaction with N-Bromosuccinimide (NBS). The behaviour of 2-sulfenylated-tryptophan peptides towards NBS is of particular interest, since NBS can be used to modify this amino acid residue in proteins (see Section III.2.1.). In the following scheme, the reactions occurring upon treatment of a 2-sulfenylated tryptophan peptide with NBS are shown. Bromination at the 3-position of the indole moiety is followed by displacement of the halogen atom by the carbonyl function of the side chain amide, with concomitant cyclization to an imido-lactone (**165**).

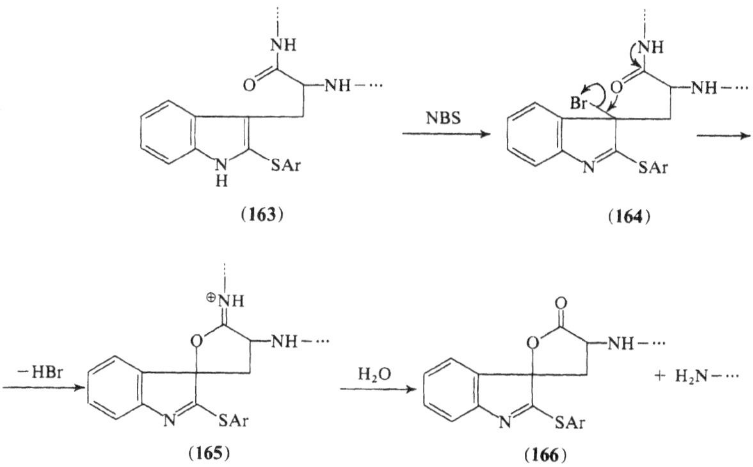

(**163**) (**164**) (**165**) (**166**)

Hydrolysis of (**165**) leads to cleavage of the tryptophanyl peptide bond in yields ranging from 50 to 70 per cent (*123, 371*).

The proposed mechanism of peptide bond cleavage was checked by isolating the 3-bromo-indolenine (**113**) (Section III.3.2.) from the model compound 2-(2-nitrophenylsulfenyl)-3-methylindole and the spirolactone (**167**) from 2-(2-nitrophenylsulfenyl)-indolyl-3-propionic acid, when these compounds were each treated with NBS (*371*).

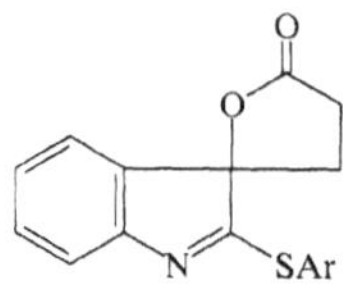

(167)

When the spirolactone (**167**) was reduced with $NaBH_4$ in anhydrous monoglyme, regeneration of the indole moiety was observed, with concomitant release of thiol (*371*). These results indicate that sulfenylation of the indole nucleus can be reversed by two consecutive reactions using NBS and $NaBH_4$.

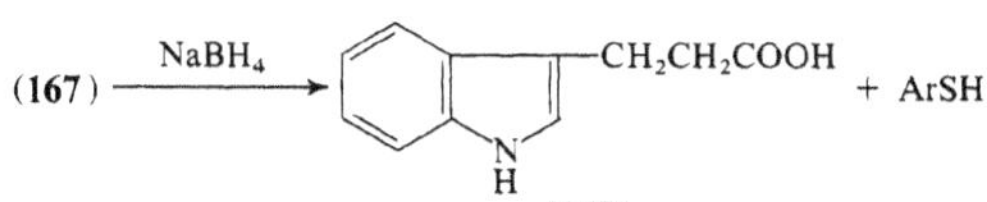

(168)

Such a *reversible* blocking of the 2-position by sulfenylation was of particular interest in the synthesis of bromo-derivatives of indolyl-3-propionic acid. Specific bromination or nitration of the benzene ring of an indole is normally difficult to achieve, since the pyrrole ring also will react with electrophilic reagents. As an example, 5-bromoindole-3-propionic acid has been prepared using this reversible protection, that is by bromination of the 2-sulfenylated compound and removal of the arylthio group by the two consecutive reactions with NBS and $NaBH_4$ (*127*).

Oxidation. Oxidation of an indole 2-thioether is of interest since this would increase the number of derivatives of tryptophan obtainable from the sulfenylation reaction, thus providing new permutations in the study of the structure-function relationships of tryptophan-containing peptides and proteins. In addition, the indole 2-sulfoxide function occurs naturally in the poisonous cyclic peptides amatoxins (*114, 429*).

Oxidation of the model compound 2-ethylthio-3-methyl-indole has been studied by HINO *et al.* (*178, 179, 273*). This compound is easily oxidized by molecular oxygen, oxidation taking place spontaneously when a solution of (**169**) in benzene is stirred in an open vessel at room temperature for several hours. Indications have been given that oxidation proceeds by autoxidation at the 3-position giving the intermediate 3-hydroperoxyindolenine (**170**) to afford quantitatively the oxidized derivatives (**171**) and (**172**) in 1 : 1 ratio (*273, 276*). When hydrogen peroxide was used as the oxidizing agent in the presence

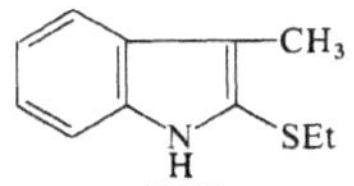

(169)

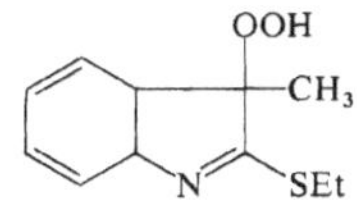

(170)

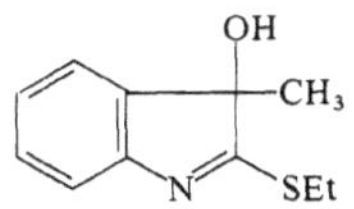

(171)

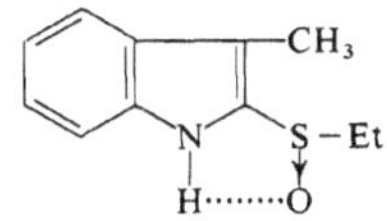

(172)

of acetic acid the sulphones (**173**) and (**174**) were also obtained (*180*).

WIELAND and coworkers (*64, 114, 429*) independently reported oxidation of the indole-2-thioether function with H_2O_2 in acetic acid and applied the oxidative reaction to phalloidin. The isolation of the

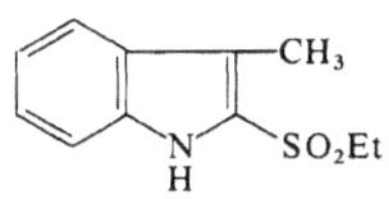

(173)

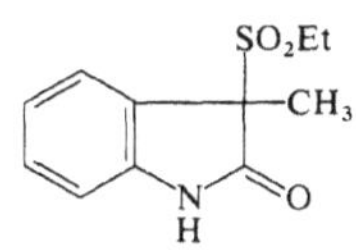

(174)

two diastereoisomeric sulfoxides from 2-thioethyltryptophan and X-ray structural analysis allowed determination of the absolute configurations of the toxic and nontoxic phalloidin sulfoxides (see also Section VI.2.2.). The indole system of 2-ethylsulfinyltryptophan is not exactly planar, due to the effects of the substituents at C-1 and C-2 (*429*).

6. Isotopic Labelling

Labelling of tryptophan, either free or incorporated in a peptide, will be of value in determining, in precise chemical and stereochemical terms, the fate of this amino acid along biosynthetic or degradative pathways. In addition, biologically important molecules specifically labelled in the tryptophan units could find important applications in physiological and endocrinological studies (*31*).

The substitution of one or several hydrogen atoms by deuterium or tritium can lead to a molecule having the same biological properties as that of the original material. Deuterium labelling has its own particular advantages, and the label can be followed by mass spectrometry and NMR techniques. However, when heavy dilution with unlabelled material occurs in the biochemical system, tritium by its nuclear characteristics appears to be the label of choice.

Tritiated tryptophan has been prepared by catalytic dehalogenation of the 5-bromo-derivative by BIRKOFER and HEMPEL (*44*). Labelling with tritium at the α- (*62*) and β-carbon (*318*) of tryptophan has been reported. KIRBY and VARLEY (*225*) prepared tryptophan stereoselectively labelled with tritium and deuterium at the β-methylene group by using labelled 3-formyl indole as a starting compound (see also Section II.2.).

Labelling of tryptophan bound in a peptide moiety can be achieved by the boron trifluoride complex of tritiated acetic acid (*151*), but the labelling is not specific for tryptophan, since tyrosine, phenylalanine and serine are labelled. Tritiated trifluoroacetic acid has been found by BAK *et al.* (*22, 24*) and HOLT *et al.* (*191* to be a selective labelling agent for tryptophan in peptides and proteins. The selectivity for the indole nucleus has been checked by showing that tritium label was present only in the tryptophan peak of the ion-exchange chromatogram of the enzyme hydrolyzate of the treated proteins (myoglobin, lysozyme and wool keratin). Optimal conditions for labelling were found to be a 15 minute treatment in tritiated trifluoroacetic acid in the presence of mercaptoethanol to minimize tryptophan destruction (*158, 248*). Although all five hydrogen atoms in the aromatic system of tryptophan exchange in tritiated trifluoracetic acid, only two or three of the labelled sites are stable to back-exchange in aqueous solution. Tritiation of tryptophan, followed by addition of water and lyophilization, led to labelling of only 2.4 hydrogen atoms per molecule.

A new method for tritiation of tryptophan and tryptophan peptides has been introduced by FROMAGEOT and coworkers (*246*). The procedure consists of reacting the tryptophanyl compound with 2-nitrophenylsulfenyl chloride (see Section III.5.) and then removing the sulfenyl label at the 2-position of the indole nucleus by catalytic hydrogenation in the presence of palladium and tritium gas. The method permits the selective replacement of the 2-thioether function by a tritium atom. By this procedure, tryptophan, N_α-acetyl-tryptophanamide and a tryptophan-containing pentapeptide have been tritiated. It is worth noting that under the experimental conditions employed the methionine thioether was not split. The method, because of its high specificity for the amino acid residue and position, appears to be of great value, being particularly useful in the labelling of biologically important tryptophan derivatives including peptide hormones and enzymes.

N-Acetyl-tryptophan and indole-3-propionic acid have been labelled with deuterium and tritium by SPANDE and FONTANA (*371*), who employed the corresponding 2-sulfenyl derivatives. The spirolactones (**176**) obtained from (**175**) by treatment with N-bromosuccinimide were reduced with $NaBD_4$ or $NaBT_4$ in good yields back to the corresponding indole compounds specifically labelled with deuterium or tritium at the 2-position. The method seems to offer a new general procedure which permits introduction of label specifically at the 2-position of indoles bearing a propionic acid side chain at the 3-position.

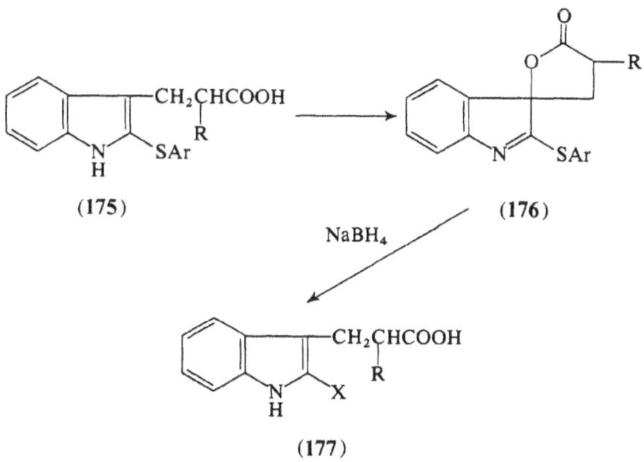

where R = $-H$; $-NHCOCH_3$

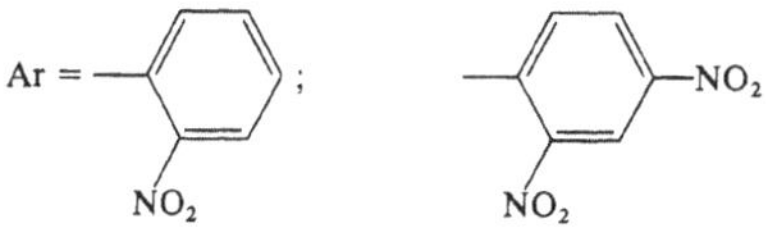

X = D, T

7. C-Acylation

PREVIERO *et al.* (*312*) reported that carboxylic acid chlorides or anhydrides are able to C-acylate the tryptophan indole ring under relatively mild conditions provided that the reaction is carried out in anhydrous trifluoroacetic acid, which is also a good solvent for peptides and proteins. Acylation of the indole ring occurs mainly at the 2-position; the resulting indolyl-ketones (**180**) can be easily converted in quantitative yields to 1-substituted 3,4-dihydro-β-carbolines (**182**).

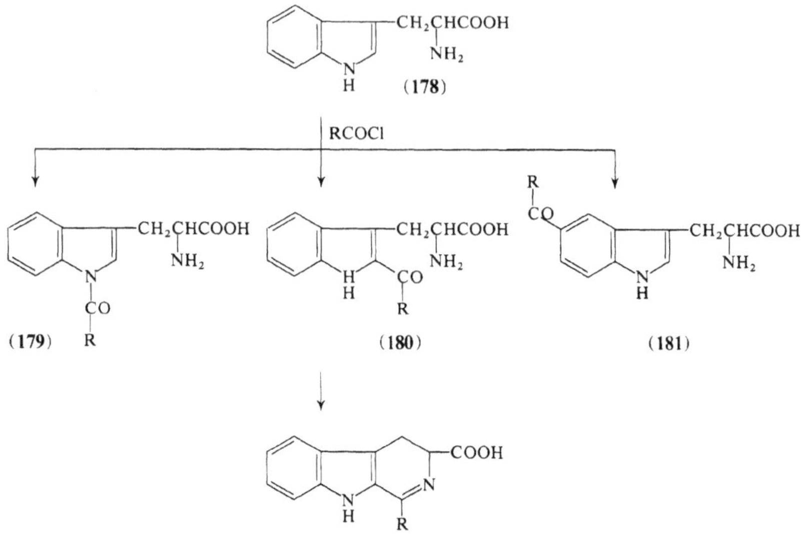

(182)

When a solution of DL-tryptophan was treated with acetyl chloride at room temperature, a yellow substance was isolated as main product and identified as 1-methyl-3,4-dihydro-β-carboline-3-carboxylic acid. Under similar conditions D- or L-tryptophan reacted analogously yielding the corresponding optically active β-carbolines. Minor products, corresponding to (**179**) and (**181**) were also isolated. The formation of the three compounds (**179**), (**180**) and (**181**) can be viewed as an acylation of a reactive ring and the general order of positional reactivity of the indole moiety can be deduced as $2 < 1 < 5$.

When tryptophan peptides were treated with acetyl chloride in trifluoroacetic acid the cyclization reaction to a β-carboline did not occur and the 2-acylated compound (**183**) was detected (*312*).

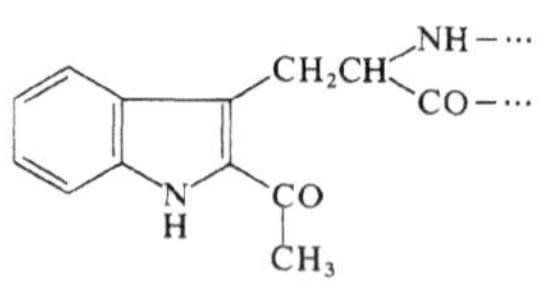

(183)

Lysozyme, a protein containing six tryptophan residues, was analogously treated with CH_3COCl in trifluoroacetic acid. Acid hydrolysis gave in 70% yield the acid stable β-carboline (**185**), the formation of which can be rationalized by the reaction path (**184**) to (**185**).

References, pp. 426—449

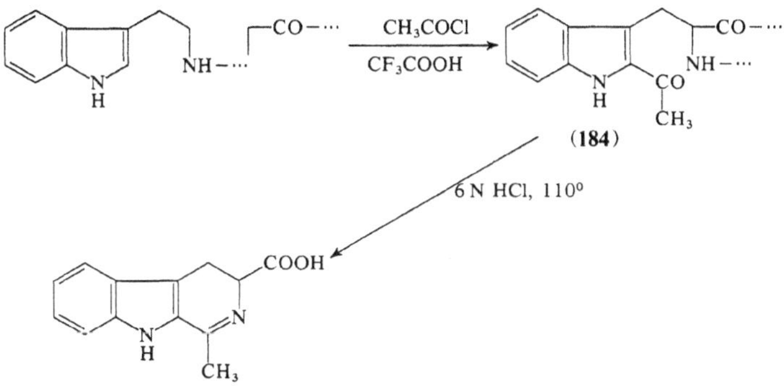

(184)

(185)

Previero *et al.* (*312*) suggested that advantage might be taken of the stability of the β-carboline system toward acid hydrolysis and of its spectral properties (λ_{max} 352 nm) for the quantitative analysis of tryptophan in proteins, assuming that the 70% yield could be a chemical constant of the reaction. This constancy was found in the case of glucagon, lysozyme, trypsinogen and chymotrypsinogen. The analytical application of the C-acylation reaction seems quite interesting, although it remains to be verified with a larger set of proteins.

8. N-Formylation

Tryptophan and tryptophan-containing peptides and proteins specifically react in formic acid containing dry HCl to give the corresponding 1-formyl-tryptophan-derivative (**186**) (*310*). This modification can

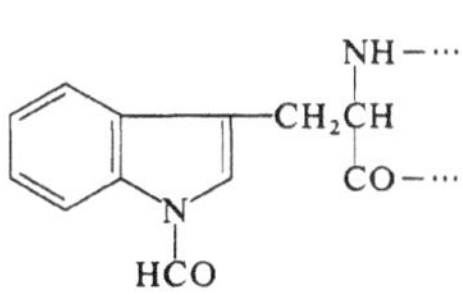

(186)

by easily reversed at pH values greater than 8.5–9. (**186**) is rapidly converted back to tryptophan, as determined by spectrophotometric and chromatographic analysis.

The ultraviolet spectrum of 1-formyl tryptophan shows a maximum of absorption near 300 nm. The high yield of specific formylation of tryptophan in proteins can be utilized for the spectrophotometric determination of tryptophan, since tyrosine chromophores do not interfere at acid pH with the chromophore of 1-formyltryptophan.

9. Cyclodehydration

Polyphosphate ester (PPE) is a good dehydrating agent in condensation reactions and is also a mild agent in some acid-catalyzed rearrangements (*220, 454*). N-Formyl- and N-acetyl-tryptophan and their esters were cyclized in the presence of PPE under mild conditions to to give the corresponding 3,4-dihydro-β-carboline-3-carboxylic acid derivatives (**188**) (*220*).

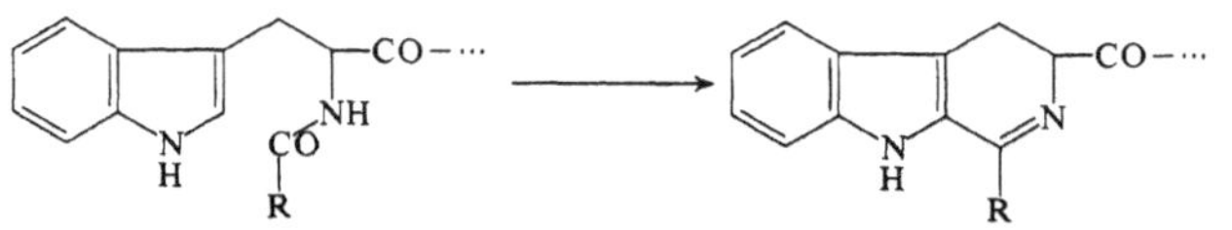

(**187**) (**188**)

Tryptophan peptides have also been found to be readily cyclized by PEE, for example, N-formyl-tryptophanyl-glycine methyl ester. When N-(N-acetylglycyl)-tryptamine (**189**) was treated with PPE, the inter-

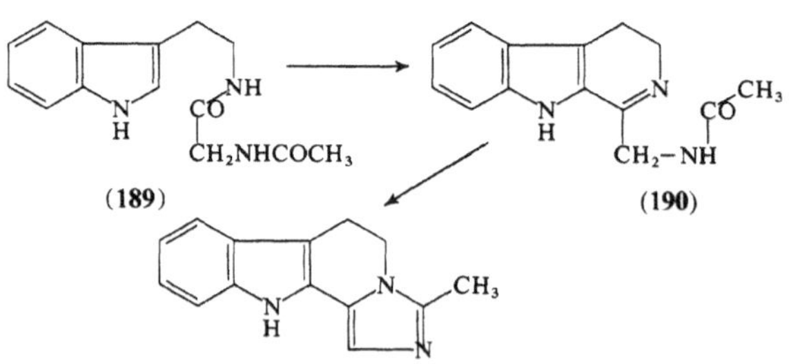

(**189**) (**190**) (**191**)

mediate β-carboline compound (**190**) was isolated. This is further converted to 5H- imidazo-(1'.5':1.2)-pyrido-[3.4-b]-indole (**191**). Similarly acetyl-glycyl-tryptophan methyl ester upon heating with PPE gave the tetracyclic compound (**192**).

References, pp. 426—449

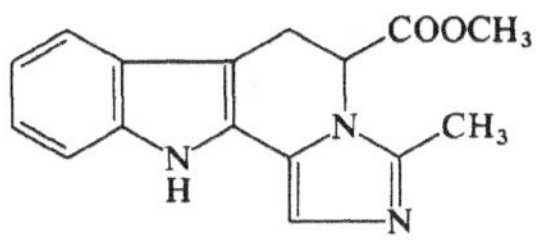

(192)

10. Reaction with Nitrite

Although the reaction of nitrous acid with peptides and proteins has been extensively used for analytical purposes (van Slyke estimation), the nature of the organic products has often been overlooked. Recently, BONNETT and HOLLEYHEAD (*49*) reported that treatment of N-acyl-tryptophan derivatives with sodium nitrite under mildly acidic conditions furnishes nitrosamines with the nitroso group located at the indolic nitrogen.

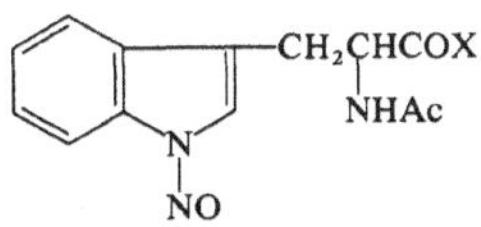

(193)

X = $-OCH_3$; $-OH$; $-NHCH_2COOC_2H_5$

The reaction takes place smoothly in about 50% yield in aqueous acetic acid at room temperature with simple tryptophan derivatives and also with tryptophan peptides. The N-nitrosamine bond in the tryptophan derivatives is cleaved by mineral acids, such as HBr in acetic acid. The derivatives are also photolabile, so that the electronic spectra cannot be measured accurately.

Nitrosation of the indole nucleus was found to be a drawback in the synthesis of tryptophan peptides when the azide coupling procedure was used (see Section VI.1. for more details). The nitrosation reaction has been also found useful in devising an analytical method for determination of tryptophan and related indole compounds (see Section IV.6.).

11. Reaction with Tetranitromethane

Tetranitromethane (TNM) (**194**) has been used in protein chemistry for the selective chemical modification of tyrosine residues (*367*). It was shown that the reagent can be highly specific for nitration of tyrosine when it is employed under mild conditions. However, if it is used under more vigorous conditions, for example, in high molar excess, other residues such as histidine, methionine and tryptophan will react.

CUATRECASAS *et al.* (*92*) observed that TNM can cause nitration of the single tryptophan residue in staphylococcal nuclease, but the nature of the nitro derivative was not determined. The inactivation of papain by TNM has also been thought to be due to tryptophan modification (*264*). Nitration has been observed also with lactoferritin (*407*).

SOKOLOVSKY *et al.* (*368*) studied the reaction of TNM with amino acids other than tyrosine and found that cysteine, methionine, histidine and tryptophan could be modified by the reagent. By using tryptophan model compounds, it was demonstrated that a number of nitrated derivatives of tryptophan of unknown structure could be obtained.

SPANDE, FONTANA and WITKOP (*370*) studied the reaction of TNM with the model compound skatole. When this compound was allowed to react with TNM in ethyl ether, a yellow compound was isolated whose structure was shown to be the 2-dinitromethylene-3-nitro derivative (**197**). A rationale for the formation of (**197**) is illustrated by the sequence (**195**)→(**197**). Reduction of (**197**) with excess $Na_2S_2O_4$ led to 2-cyano-3-methylindole (**198**).

It must be emphasized that the dinitromethylene compound was obtained when the reaction was carried out in an organic solvent. The usual nitration conditions with TNM in aqueous buffers could lead to an alternate reaction pathway.

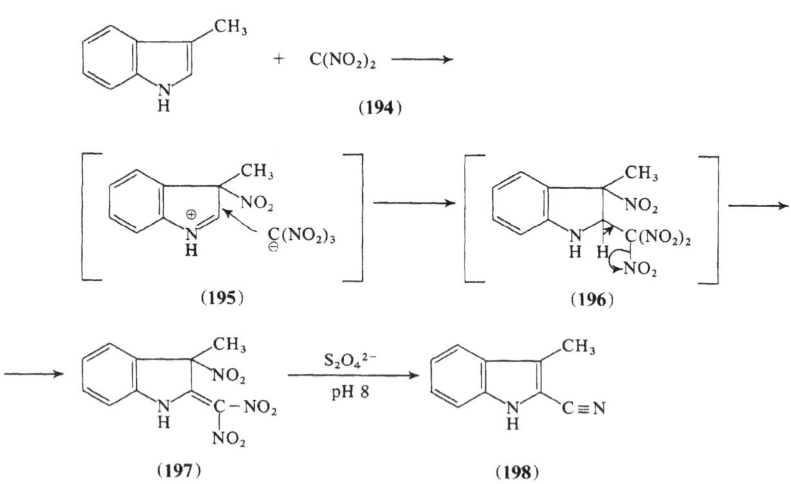

12. Reaction with Diazonium Salts

Recently SPANDE and GLENNER (*376*) reinvestigated the reaction of 3-substituted indole derivatives with diazonium salts. In earlier work, it was proposed that the reaction of diazotized sulfanilic acid with tryptophan takes place at the 3-position. On the other hand, reaction of skatole with diazotized arsanilic acid was assumed to give a 2-substituted derivative, solely by analogy to the pyrroles (*259, 395*).

When skatole was treated with the diazonium salt derived from *p*-nitro-*o*-anisidine (Fast Red B) in 50% aqueous dioxan at pH 3.0 a crystalline 1:1 azo compound (**199**) was isolated in 66% yield. An

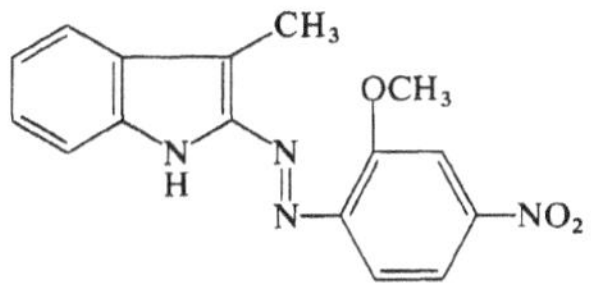

(**199**)

unusual dihydrotriazene (**200**) was also isolated as a minor product, which is reversibly transformed by alkali to the triazene anion (**201**).

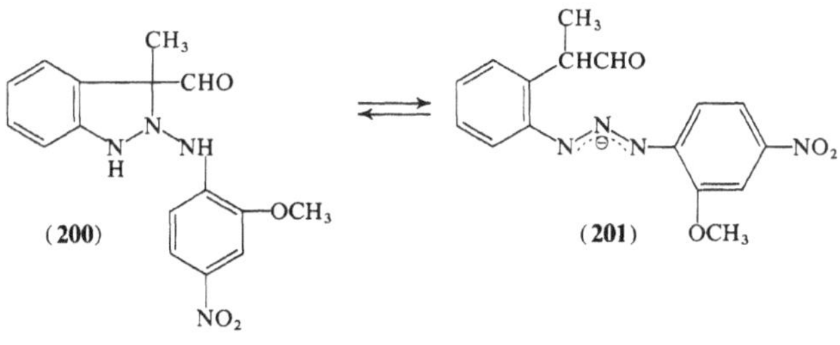

(**200**) (**201**)

This reaction constitutes the first example of indole ring cleavage by the action of a diazonium ion.

13. Reaction with 3-Acetoxyxanthine

Two products obtained by reaction of L-tryptophan with the activated oncogen 3-acetoxyxanthine (**202**) at neutral pH and at 60° C have been characterized as the diastereoisomeric [3-(2-amino-2-carboxy-

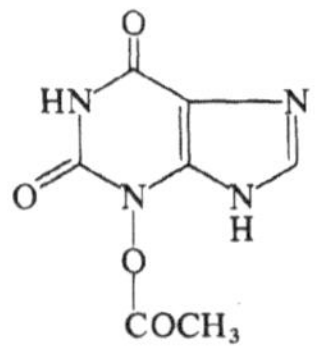

(202)

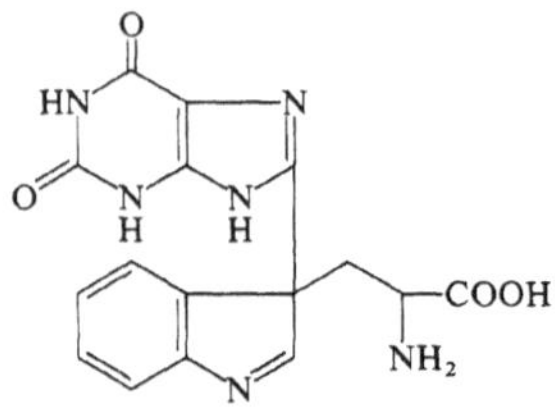

(203)

ethyl)-3-(8-xanthinyl)] indolenines (**203**). The two isomers are isolated in a combined yield of 25%. Simultaneously, a portion of tryptophan is oxidized to unidentified colored products and a portion (30%) is recovered (*388*).

This reaction with tryptophan could have specific effects on mechanism of cellular control, since aromatic amino acids play a role in the specific binding of proteins to DNA.

14. Reaction with Trichloroisocyanuric Acid

The reaction of trichloroisocyanuric acid (**204**) with tryptophan led to the formation of an oxindole derivative (*14*). This oxidation which is complete at pH 7.0 in proteins may be useful for cleavage at

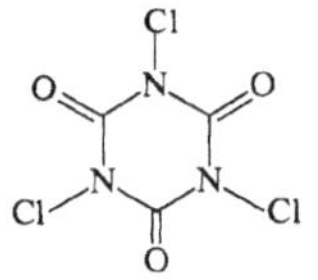

(204)

tryptophan peptide bonds if oxidation is followed by mild acid treatment. However, the cleavage reaction remains to be verified. The reagent should be of limited utility in protein modification studies, because of its broad reactivity. Lysozyme was modified by the action of the reagent not only at tryptophan, but also at the methionine, tyrosine and cystine units.

15. Miscellaneous

The reagents that are used most widely for the chemical modification of tryptophan in proteins are limited to simple, readily available compounds. Perhaps more sophisticated and more useful reagents remain to be uncovered and could be applied to solve the variety of problems encountered in peptide and protein chemistry.

A critical reading of the book by SUNDBERG (*394*) on the chemistry of indoles could suggest to the reader useful and in principle new techniques for the modification of tryptophan containing peptides and

Table 1. *Miscellaneous Reactions of the Indole Nucleus of Tryptophan*

Reagent	Results	Reference
n-Hexanal	Upon heating with tryptophan for 2 hours at 50° yields 1-pentyl-2,3,4,9-tetrahydro-1H-pyrido-[3,4-b]indole-3-carboxylic acid	(*11*)
Xylose	Reaction with tryptophan at 160° in aqueous solution affords β-carboline derivatives	(*58*)
Bisulfite	In the presence of bisulfite and oxygen, tryptophan gives three unidentified products. The reaction shows a free radical mechanism	(*201*)
Diethylpyrocarbonate	The reagent affords ethoxycarbonylation of the indole nucleus	(*330*)
Dimethylsulfoxide	Photoaddition of the compound to the indole nucleus occurs, when poly-tryptophan is subjected to photolysis with a high pressure mercury lamp	(*111*)
Croconic Acid	In the presence of H_2SO_4 reacts with tryptophan to form red-blue colored complexes. The reagent has found analytical application	(*421*)
Trichlorotriazine	Gives an unidentified colored product upon reaction with N-acylated-tryptophan	(*403*)
4-Chloro-3,5-dinitrophenacyl bromide	The compound has been used as a polyfunctional protein reagent and reacts with various functional groups. With tryptophan it forms a charge-transfer complex and reaction occurs with substitution of Cl or Br	(*110*)
Tetracyanoethylene	The reagent reacts with tryptophan giving not simply a charge-transfer complex, but CN transfer to the amino acid occurs	(*40*)
Metal nitrilotriacetates	React with tryptophan leading to mixed complexes containing the metal, nitrilotriacetate and tryptophan in the ratio 1 : 1 : 1	(*2*)

proteins. Many reagents have been used and thoroughly studied with simple 3-substituted indoles, the simple models of the tryptophan moiety.

In Table 1 are listed some reagents used to modify the tryptophan residue, which have not been referred to in the previous Sections.

IV. Analytical Determination

Much of the remarkable progress in protein chemistry over the past years has been due to the development of accurate and convenient methods for the qualitative and quantitative analysis of amino acids. The most common procedure for amino acid analysis of a protein involves acid hydrolysis in the absence of oxygen, resolution of the amino acids by ion-exchange chromatography, and colorimetric estimation with ninhydrin (*369*).

Although the amino acid composition of many proteins has been reported, evaluation of the number of tryptophan residues often has been lacking. This situation reflects the difficulty associated with quantitative determination of tryptophan residues in a protein, since acid hydrolysis usually results in the extensive destruction of tryptophan. Basic hydrolysis is less destructive, but suffers also from limitations (*197,*, *281*).

Enzymatic hydrolysis of proteins is an attractive alternative to the hydrolytic procedures and has been used to avoid destruction of tryptophan (*174*). A mixture of proteolytic enzymes, such as chymotrypsin, thermolysin, papain, leucine aminopeptidase and pronase has been used (*135, 192, 353*). Upon completion of the enzymatic hydrolysis of the protein, the hydrolyzate is rendered free of proteolytic enzymes by precipitation and centrifugation and the extract directly analyzed with the amino acid analyzer. The method however may not be generally valid, because of the possibility of artefacts due to self-digestion of the enzymes used.

Satisfactory alternatives to the hydrolytic procedures have been sought for many years. Methods have been tried to produce colored derivatives by exploiting the reactivity of the indole nucleus in the intact protein (*135, 353*). Sodium hypochlorite, ferric chloride, cupric sulfate, bromine, and sodium nitrite, usually in acid solution, have been found to produce color with tryptophan-containing proteins. Later workers have formed other colored derivatives by reaction with tryptophan-specific reagents, while still others have used the characteristic absorption at 280 nm to estimate tryptophan spectrophotometrically or spectrofluorometrically.

Recently magnetic circular dichroism (MCD) has been used to determine the tryptophan content of proteins (*190*). The intense positive band near 293 nm, uniquely characteristic of tryptophan, allows direct and quantitative measurement of this amino acid with high sensitivity and accuracy in the intact protein (see Section V.2.1., for a more detailed discussion of the method). However, although the technique seems to be very promising, it has the disadvantage of requiring highly specialized equipment.

Because of uncertainties in the present methods for tryptophan analysis, it is obvious that a satisfactory general procedure remains to be found (*135, 353*). Acid hydrolysis with *p*-toluenesulfonic (*233, 234*) or mercaptoethanesulfonic acid (*299*) seems to be the most reliable known method, even though the analytical figures for tryptophan have to be checked by different methods.

In this section, methods for the quantitative determination of tryptophan, free or bound in a polypeptide chain, are discussed in more detail.

1. Acid Hydrolysis

The lability of tryptophan under the conditions usually employed for acid hydrolysis of proteins (6 N HCl, 24 hrs, in sealed tubes under vacuum) is not due to instability of the indole nucleus under such conditions, but to side reactions involving non-proteinaceus material, like carbohydrates, or to the presence of particular amino acids in the acidic hydrolysis mixture. In fact, gramicidin A, a pentadecapeptide containing, apart from tryptophan, only "purely" aliphatic amino acids like glycine, alanine, leucine and valine, gives a quantitative recovery of tryptophan after acid hydrolysis in 6 N HCl (*120*).

Mondino and Bongiovanni (*262*) observed that cystine, tryptophan and to a lesser extent threonine and serine were degraded when a mixture of all the common amino acids found in proteins was hydrolyzed in 6 N HCl by an open-flask technique and analyzed on an automatic amino acid analyzer. More recently, Gruen (*157*) carefully studied the separate effect of each of the other commonly occuring amino acids on the recovery of tryptophan, using the standard procedure for amino acid analysis, whereby tryptophan is eluted from the short column along with the basic amino acids as a well resolved peak appearing before lysine (*135, 369*).

GRUEN (*157*) found that the presence of most of the amino acids tested during hydrolysis does not affect the tryptophan recovery. Of the hydroxyamino acids, threonine had no effect at all, while tyrosine has a small effect. On the other hand serine was found to reduce the recovery of tryptophan by about 10% due to a deamination reaction of serine. This produces pyruvic acid which, as a ketoacid, interacts with the indole nucleus of tryptophan (*287*). Whereas cysteine had no effect, cystine caused a substantial loss of tryptophan (about 40%), and was itself recovered on the analyzer partly in the reduced cysteine form. These results would indicate that a major factor in the loss of tryptophan during acid hydrolysis of a protein is degradative oxidation by cystine.

Tryptophan recovery from a protein could be greatly enhanced when the acid hydrolysis was performed under reducing conditions or in the presence of an antioxidant. MATSUBARA and SASAKI (*249*) discovered that the presence of 2–4% thioglycolic acid in the 6N HCl has a protective effect on tryptophan. Recovery of tryptophan for some proteins reported by these authors was 80–90%. However, when glucose was present in the protein sample, the protective effect of thioglycolic acid was lost and no tryptophan could be recovered from the hydrolyzate. Analogous results have been obtained using mercaptoethanol as protective agent (*158*).

LIU and CHANG (*234*) found that hydrolysis of a protein with *p*-toluenesulfonic acid gave rather less destruction than HCl. Hydrolysis in the 3N acid with 0.2% 3-(2-aminoethyl)-indole as a scavenger (*vide infra*) gave an 88–98% recovery of tryptophan. This procedure provides a newer method for the determination of tryptophan in proteins, giving results comparable with those obtained by other non-hydrolytic methods. An advantage of the method lies also in the fact that the hydrolyzate can be placed on the column for analysis without prior removal of the solvent, as is required when 6N HCl is used. In addition, the recovery of all other amino acids is comparable to that obtained from a 6N HCl hydrolyzate.

Recently, the utilization of mercaptoethanesulfonic acid (*299*) as catalyst for hydrolysis of proteins was found to give very high tryptophan recoveries; the results indicate improvement on the methods of MATSUBARA and SASAKI (*249*) and LIU and CHANG (*234*). The acid has been used at a concentration of 3N and hydrolysis time of 22 hrs at a temperature of 110° C are sufficient to give complete hydrolysis of protein molecules. Under these conditions, total cleavage of the highly resistant valine and isoleucine peptide bonds has also been obtained. However, if the protein sample contains carbohydrates, the decomposition of tryptophan is increased, even using mercaptoethanesulfonic acid.

GRUEN and NICHOLLS (*158*) examined in detail the relative effectiveness of various additives in the 6N HCl hydrolysis of proteins. Acid hydrolysis was carried out in the presence of a reducing agent or a substituted indole compound such as tryptamine or 3-indole propionic acid, which lack the carboxyl or amino function, respectively, of tryptophan. The rationale was that these molecules would be degraded in preference to tryptophan.

Indole-3-propionic acid was found the most effective scavenger, giving a more improved tryptophan recovery than tryptamine while exerting no deleterious effects on the other amino acids. It was also found that hydrolysis of proteins under reflux with nitrogen bubbling through boiling 6N HCl gave higher tryptophan recovery, than the usual procedure of carrying out the hydrolysis in carefully sealed and evacuated tubes. Tryptamine could also be employed successfully as a scavenger with HCl. Although ninhydrin positive, it is not eluted from the column during a normal short column analysis and thus causes no interference. Tryptamine does not affect the recovery of nearly all the other amino acids, only cystine being partially reduced.

Among the reducing agents, thioglycolic acid, mercaptoethanol and H_2S have been tested in the hydrolysis with 6N HCl. Thioglycolic acid was found to have some advantages over the other reducing agents, although this reagent invalidates proline and cystine analyses on the amino acid analyzer (*158*).

Generally, only ion-exchange techniques have been employed for the determination of tryptophan in proteins following either basic or acidic hydrolysis. Recently GEHRKE and TAKEDA (*142*) proposed the use of gas-liquid chromatography (GLC) for the analysis of tryptophan in proteins using an acidic hydrolysis method with 6N HCl containing 5% thioglycolic acid at 110° C for 20 hours.

To avoid chromatographic interferences it has been found necessary to remove the thioacid and associated products from the hydrolyzate prior to GLC analysis. Since the procedure of GLC analysis of amino acids involves an esterification (*t*-butylation) and an acylation (N-trifluoroacetylation) step (*305*) removal of these substances was found to be mandatory for achieving completeness of these reactions. Analyses of four tryptophan containing tripeptides showed that destruction of tryptophan was less than 5%. The method was also successfully applied to lysozyme and trypsinogen (*142*).

2. Basic Hydrolysis

Basic hydrolysis of proteins by either $Ba(OH)_2$ or NaOH has been considered preferable to acid hydrolysis for tryptophan analysis; unfortunately, the method does not always give satisfactory results. SPIES and CHAMBERS (*380*) studied factors affecting the stability of tryptophan in 5 N NaOH. They found that tryptophan is more labile when the amino acid is peptide-linked than in the free form and that losses depend on protein composition. OELSHLEGEL *et al.* (*281*) hydrolyzed proteins with NaOH in the presence of thioglycolic acid, the recoveries of tryptophan being 89–96%.

More recently, HUGLI and MOORE (*197*) made a careful study of the variables in technique for the alkaline hydrolysis of proteins and for the chromatographic analysis of the protein hydrolyzate. Quantitative recoveries of tryptophan were obtained when hydrolysis of the protein was carried out at 110° C or 135° C in 4.2 N NaOH in the presence of starch. NaOH has been preferred to $Ba(OH)_2$, since Ba^{++} will interfere in the chromatographic analysis of the hydrolytic mixture and has to be precipitated by sulfate or carbonate. During this manipulation, adsorption of tryptophan on to the precipitate could occur. In order to avoid silicate formation during alkaline hydrolysis, the hydrolysis is performed in polypropylene liners sealed inside evacuated glass tubes. The protective effect of starch could be related to its oxidative degradation in alkaline solution in the presence of oxygen. The degradative reaction apparently removes the residual traces of oxygen in the hydrolysis mixture by incorporating the oxygen into products that do not react with tryptophan. An analogous protective effect has been obtained when thiodiglycol was used as antioxidant; this compound combines with oxygen to give the sulfoxide.

The recovery of tryptophan after basic hydrolysis could be improved by storing the sample at pH 4.25 instead of at pH 2.2, since it has been found that tryptophan undergoes a slow destruction in citrate buffer at acid pH values (*197*).

Carbohydrate does not interfere in the tryptophan analysis using basic hydrolysis. This allows use of the method in the analysis of samples with appreciable carbohydrate content, including the analysis of tryptophan in foods.

3. Spectrophotometry

In the absence of prosthetic groups or nucleic acids, the absorption of proteins in the ultraviolet region between 250 and 310 nm is due to the contribution of the three aromatic residues, namely, tryptophan, tyrosine and phenylalanine (*104*). Since phenylalanine has a low molar extinction and the absorption maxima of the other two aromatic residues occur at higher wavelengths, the absorption of proteins at 280 to 295 nm is due only to tryptophan and tyrosine. Therefore, if the small contribution of phenylalanine is neglected, the absorption of the solution of a protein can be considered as arising from a mixture of the two amino acids. This fact was used by HOLLIDAY in 1936 (*185*) and later by GOODWIN and MORTON (*150*) as the basis of a spectrophotometric determination of both these amino acids in proteins.

The procedure is to dissolve the protein in 0.1 N NaOH and make up to a convenient volume in order to obtain an absorption of 0.2 to 0.5 near 290 nm. The total molar concentration (a) of tryptophan and tyrosine is obtained [Eq. (1)] from the absorption at 295 nm, one of the wavelengths at which the absorption curves of the two amino acids intersect and therefore have the same extinction (ε).

$$A_{295} = a\,\varepsilon \tag{1}$$

From the extinction at another wavelength, 280 nm, the concentrations of the two components can be calculated. If at 280 nm the molar extinction of tryptophan (ε_1) and of tyrosine (ε_2) is known, the observed absorption A_{280} is given by Eq. (2)

$$A_{280} = b\,\varepsilon_1 + (a-b)\,\varepsilon_2 \tag{2}$$

(where b is the concentration of tryptophan and $a-b$ the concentration of tyrosine) and b by Eq. (3).

$$b = \frac{A_{280} - a\,\varepsilon_2}{\varepsilon_1 - \varepsilon_2} \tag{3}$$

After determination of the protein concentration by standard methods, the tryptophan residues/protein (mole/mole) is obtained from the ratio of the two concentrations.

Analysis of the tryptophan content by spectrophotometry is simple and does not require specialized equipment. However, caution should be exercized in applying this method. Tyrosine residues in proteins sometimes behave abnormally, and incubation at pH 13 is necessary for maximal ionization of the phenolic group (*230*). Therefore, the method must be tested after exposure of the protein to pH 13 for varying lengths of time. On the other hand it was found by SPIES and CHAMBERS (*380*) that slight destruction of tryptophan could occur in 0.1 N alkaline solution.

The original application of the method to proteins was based on the assumption that the absorption of amino acids is unaffected when they are linked in a polypeptide chain. Later work has suggested that in preparing the standard solutions the use of amino acids in peptide combination is more satisfactory, *e.g.*, glycyl-tryptophan (*106*), since a red shift (1 to 3 nm) in the absorption maximum of a free amino acid occurs when it is incorporated in proteins. BENCZE and SCHMIDT (*39*) modified the procedure of GOODWIN and MORTON (*150*) in order to overcome this problem.

The spectrophotometric method has been used with reasonable success, but great difficulties can be encountered in the determination of tryptophan by the spectrophotometric method when this amino acid is present in the protein at a low level or when the tyrosine content is high.

The older methods used 0.1 N NaOH to "normalize" the environment of the tryptophan and tyrosine residues. More recently EDELHOCH (104) further improved the method by performing the absorbance measurements at 288 and 280 nm on the protein, dissolved in 6 M guanidine hydrochloride, in order to destroy its secondary and tertiary structure. Blocked tryptophanyl and tyrosyl compounds (N-acetyl-tryptophan amide, glycyl-tyrosyl-glycine) were selected as model substances to represent the absorbance of the two chromophores in proteins. However, although the method was shown to give values of tryptophan content in very good agreement with those known from sequence studies (trypsinogen, chymotrypsinogen, lysozyme), an imperfect fit of the spectra of the solutions containing the model compounds to those of the proteins was observed. The latter are shifted to the red in all three cases examined.

EDELHOCH (*104*) stressed that contaminants, cofactors and turbidity contribute to absorbance at 280–290 nm. A good method for checking for the presence of absorbing impurities is to examine the absorbance of the protein in neutral 6 M guanidine hydrochloride at 315 nm, where tyrosine and tryptophan do not absorb.

Of the methods proposed to correct absorption values at 280 to 290 nm, the one most commonly used is linear extrapolation of the absorption curve between 320 and 360 nm back to the 280 nm region where measurements are made. However, this correction can be inaccurate because the absorption curve above 320 nm is usually slightly curvilinear, making tangent selection difficult.

BREDDERMAN (*56*) recently proposed an alternative mode of making such a correction, by measuring absorption not only at 280 and 288 nm, as suggested by EDELHOCH (*104*), but also at 275 nm. This

References, pp. 426—449

third wavelength is on a shoulder of the tryptophan curve and near the maximum of tyrosine.

The fluorometric method of DENKLA and DEWEY (*97*) for assay of tryptophan is the most suitable and the one most commonly used for the determination of tryptophan concentrations in tissue samples because of its sensitivity, specificity and simplicity. Tryptophan is converted to a fluorophore, norharman, by heating it in acid solution with formaldehyde and ferric chloride. A revised procedure of the original method has been also reported (*46*).

4. Colorimetry

4.1. p-Dimethylaminobenzaldehyde

The long known reaction of the indole nucleus of tryptophan with aldehydes under strongly acidic conditions to give colored products (*135, 287, 397*) was employed by SPIES and CHAMBERS (*379*) in a colorimetric procedure for the determination of tryptophan. In this method, the tryptophan-containing compounds (peptides or proteins) are dissolved in concentrated sulfuric acid containing *p*-dimethylaminobenzaldehyde (DAB) and a blue color is obtained after a prescribed time by adding sodium nitrite solution.

The nature of the chromophore obtained upon reaction of the tryptophan residue and DAB is so far not completely elucidated. Current evidence does not permit a choice between structures (**205**) (derived from 1 mole of tryptophan and 1 mole of DAB) and (**206**), but (**206**) seems more likely since its formation requires oxidation (*135*).

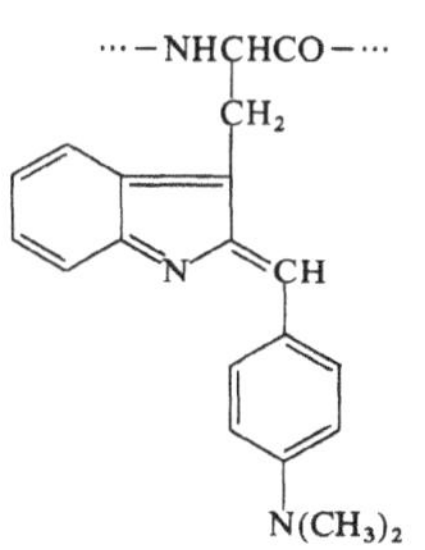

(**205**)

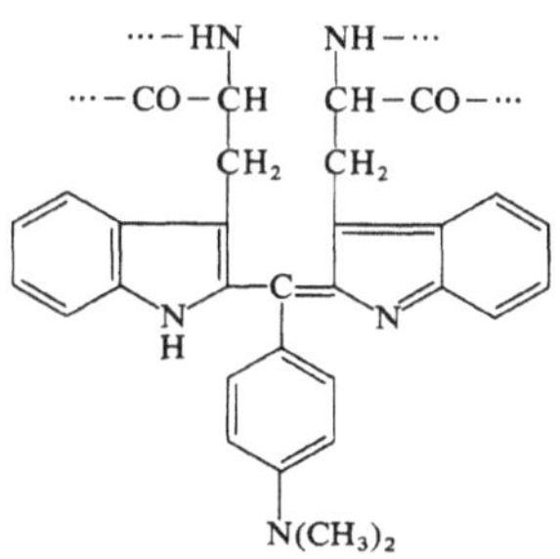

(**206**)

Although the method of SPIES and CHAMBERS has been extensively used and normally gives reliable results, the method has some limitations. Variations in the absorption maximum (from a normal value of 590–

600 nm to 545–460 nm) have been verified with different protein samples (*379*). Intense light, oxidizing and reducing agents, including H_2S, bisulfite and chloride ions interfere with the method (*135, 379*). Interference has been eliminated to some extent by appropriate addition of silver sulfate to the assay mixture (*379*). Low tryptophan values are obtained due to reaction of sulfur amino acids, serine and threonine with the indole nucleus of tryptophan. The nature of these reactions remains to be established.

FRIEDMAN *et al.* (*136*) eliminate interference due to cysteine and cystine residues by reducing the protein disulfide bonds and alkylating the native and generated-SH groups with vinyl derivatives. JUNEJA *et al.* (*219*) evaluated in detail the interference by sulfur amino acids; they estimated the tryptophan content of wool colorimetrically after hydrolysis with 6N $Ba(OH)_2$ for 5 hours at 125° C followed by removal of liberated H_2S and barium with silver sulfate and sulfuric acid respectively.

Reaction of the indole nucleus of tryptophan with xanthydrol (*99*) or glyoxylic acid (*57, 356*) has also been employed for analytical purpose, but the procedures have received only limited application.

GAITONDE (*139, 140*) has found that treatment of either free or protein-bound tryptophan with ninhydrin in a mixture of formic acid and hydrochloric acid for 10 min at 100° C results in a chromophore with λ_{max} at 390 nm. FRIEDMAN and FINLEY (*135*) propose structure (**207**)

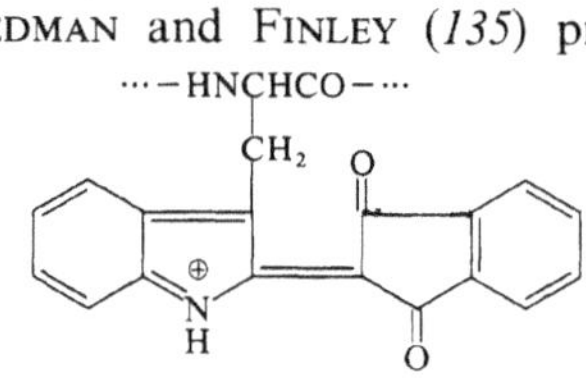

(**207**)

for the chromophore. The reaction, however, seems not to be unique, since the extinction coefficient of the chromophore is dependent upon the nature of the model compound and reaction conditions.

ZAHNLEY and DAVIS (*456*) found that colorimetric determination of tryptophan in intact proteins by the ninhydrin reaction gives high apparent tryptophan contents for proteins having high tyrosine/tryptophan ratios. An equation was proposed to correct for this interference by tyrosine. To apply the correction, the tyrosine content must be known, but it can be determined readily by other methods. Application of such a correction to chicken ovoinhibitor, which contains 17 tyrosine residues and one tryptophan residue per molecule, gave results that agree with the tryptophan content determined by other methods such as fluorescence, oxidation with NBS and magnetic circular dichroism.

References, pp. 426—449

4.2. Sulfenyl Halides

Sulfenyl halides have been found (*47, 123, 124*) to be specific, mild reagents for modification of the tryptophan and cysteine residues of polypeptides and proteins in acidic media. Tryptophan is attacked at the 2-position of the indole nucleus to give a thioether, whereas cysteine is converted to a mixed disulfide (see Section III.5.).

The unsymmetrical disulfide at the cysteine residue can be easily cleaved by means of reducing agents, such as mercaptoethanol, thioglycolic acid or sodium borohydride, or by simply exposing it to alkaline treatment, thereby allowing specific labelling only of the tryptophan residue.

By using nitrophenylsulfenyl halides a chromophore is generated in the protein which absorbs in the visible part of the spectrum. This permits easy quantitation of the reaction by spectrophotometry. At the present time the most widely used sulfenyl halide is 2-nitrophenylsulfenyl chloride (NPS-Cl). The 2-thio-(2-nitrophenyl)-tryptophan residue, obtained by reaction of NPS-Cl with tryptophan-containing proteins, absorbs at 365 nm with a molar absorptivity $\varepsilon = 4\,000\ M^{-1} \cdot cm^{-1}$ in aqueous acetic acid (*122, 354*).

It was shown (*447*) that the sulfenyl halide reacts with the indole nucleus of tryptophan much faster when the latter is part of a protein than with the free amino acid or with a small peptide. This enhanced reactivity enables one to perform the sulfenylation reaction in aqueous acetic acid in spite of the fact that the reagent decomposes rapidly (2 to 3 min), provided that the halide is first dissolved in glacial acetic acid. Quantitative modification of tryptophan residues in proteins is achieved readily by using a low molar ratio of reagent to protein (*47, 354*).

The labelled protein is separated by precipitation or by gel filtration and then the 2-NPS-tryptophan content determined spectrophotometrically at 365 nm. Protein concentration is determined by conventional methods, the most accurate being acid hydrolysis of an aliquot of the solution and subsequent quantitative amino acid analysis.

More recently, SANTARIUS and BELITZ (*335*) proposed a simplified procedure of tryptophan assay using NPS-Cl without isolation of the NPS-labelled protein and at the same time improving the sensitivity of the method.

4.3. Koshland's Reagent

KOSHLAND and coworkers (*25, 193, 194*) have used 2-hydroxy-5-nitrobenzyl bromide (HNB-Br) as a specific reagent for the analytical determination of tryptophan. Upon reaction with tryptophan-containing proteins, a chromophore is generated in the protein which absorbs in the visible part of the spectrum. The reaction of Koshland's reagent with the indole nucleus of tryptophan is complex and has been presented above in detail (see Section III.5.).

Assuming that the chromophore when covalently bound to the tryptophan residue has the same molar extinction as the model compound 2-hydroxy-5-nitrobenzyl alcohol ($\varepsilon = 18{,}000$ at 410 nm in alkali solution), the reaction can be quantitated by spectrophotometry.

However KOSHLAND's analytical method appears to have some disadvantages. The possibility of disubstitution products has been confirmed by SPANDE *et al.* (*378*) and others (*236, 253*). This finding is consistent with the known polyalkylation reaction of the indole ring by alkylating agents (*396*) and explains the excessive incorporation of chromophore into the protein when a high concentration of reagent is employed (*25, 102, 329, 444*). The possibility of forming polysubstitution products suggests that the visible spectral characteristics of the HNB group should not be used as the sole criterion for group-counting in native proteins. It would seem mandatory therefore that the estimation of tryptophan content by HNB-Br should be confirmed by another method.

The reaction of HNB-Br takes place also with cysteine and the alkylation is irreversible. Therefore a procedure for blocking this residue by carboxymethylation is needed prior to treatment of the SH-containing protein with the reagent. Alternatively, specific analyses must be performed to determine the extent to which cysteine residues may have reacted with HNB-Br. Finally, quantitative reaction is achieved only under conditions of unfolding of the protein by dissolving it in 10 M urea or, best, by cleavage of the S–S bridges. The use of 5 M guanidine hydrochloride as an unfolding reagent causes incorporation of the yellow label not only in tryptophan containing proteins, but also in ribonuclease, a tryptophan- and cysteine-free protein (*25*).

5. Other Methods

5.1. N-Bromosuccinimide

The use of N-bromosuccinimide (NBS) as a specific reagent for modification of tryptophan in proteins has led to a rapid and convenient spectrophotometric method for determining the tryptophan content of a protein (*373, 377*). The procedure is based on the large decrease in absorbance at 280 nm which accompanies the oxidation of tryptophan to an oxindole chromophore.

The optical density decrease at 280 nm can be related to the optical density of tryptophan in the sample using the empirical factor 1.31, which corrects for the absorption at 280 nm of the oxidation product of tryptophan. The amount of tryptophan in the solution is calculated using an extinction coefficient for tryptophan at 280 nm of 5,500 (*373*).

Oxidation is conducted in an acidic medium, usually an acetate or acetate-formate buffer, by using a controlled amount of NBS. Side reactions usually encountered with NBS and proteins, such as oxidation of methionine, cysteine, cystine, histidine and tyrosine, have been generally assumed to be minimal and not to interfere with the tryptophan determination. Experiments with model compounds confirm that the reactivity toward NBS of tryptophan is much higher than that of other amino acid residues. Oxidation of residues other than tryptophan is revealed by increased consumption of NBS by the protein. Usually, 2–4 equivalents of reagent are needed for complete oxidation of tryptophan, but with some proteins, an excess of up to 20 equivalents may be needed (*230, 372, 374*). Some proteins possess tryptophan residues which react very slowly or not at all under the usual conditions of assay; the reaction should then be carried out in 8 M urea.

The modest material requirements and the ease with which the determination can be made make the NBS method attractive and it has been used as an alternative to the other available methods. However, the method suffers from definite limitations.

The studies of KRONMAN *et al.* (*231*) on the NBS oxidation of lysozyme demonstrate clearly that the reagent is not specific for tryptophan, since amino acid analyses of oxidized samples of lysozyme revealed losses of tyrosine and histidine even at relatively moderate concentrations of NBS. In addition, spectral analyses for tryptophan oxidation following the usual procedure (*373*) were lower than those obtained by amino acid analysis. The authors (*231*) suggested that this difference is the result of a systematic error in the spectral procedure due to oxidation of tyrosyl groups.

It is evident from the results obtained with lysozyme that it is impossible to generalize on the behaviour of NBS with proteins. Analyses of tryptophan by the NBS procedure require checking by an alternative analytical technique.

5.2. *Miscellaneous*

SLUMP and SCHREUDER (*365*) described a chromatographic method for tryptophan determination, using an automatic amino acid analyzer, which is especially useful for tryptophan analysis in foods. Tryptophan was separated from the other amino acids by means of gel filtration on Sephadex G-25 columns. This separation is ascribed to the property of the dextran gels to adsorb aromatic compounds to a larger extent than aliphatic compounds.

The method seems promising and useful also for protein analysis, as checked with α-chymotrypsin. Hydrolysis was carried out with 2.7 N $Ba(OH)_2$ at 120° C, which had been carefully boiled to remove dissolved air. An aliquot of the hydrolyzate was applied to the Sephadex column, which was connected to the automatic analyzer. The tryptophan peak was completely resolved from those of other amino acids.

Conversion of tryptophan or other indole derivatives into a nitrosamine by treatment with nitrous acid followed by diazotization with N-1-naphthylethylene diamine dihydrochloride has been recently studied by GOSWAMI (*152*). The method seems to be sensitive and to follow Beer's law over of wide range of concentrations of tryptophan. Selective nitrosation of tryptophan is achieved at room temperature in 1 M HCl and diazotization at 10° C or below. Sodium chloride was found to interfere with the assay. The temperature must be controlled with care, since the diazotization reaction is exothermic and diazonium compounds are stable only at low temperatures. This fact would explain the optimum temperature of 10 or 6° C in the diazotization reaction.

A sensitive spectrophotometric method for the determination of free or bound tryptophan has been proposed by MESSINEO and MUSARRA (*258*). The method utilizes the reaction of fructose with cysteine in sulfuric acid to give a green chromophore (λ_{max} 415 nm), which upon the addition of tryptophan or other indole derivative forms a pink chromophore (λ_{max} 518 nm). Once the pink chromophore is formed, it is stable for more than 24 hours. The exact nature of the reactions involved is not known. The method offer advantages, since it does not require previous hydrolysis of the protein, and is simple and quite sensitive.

A tentative procedure for determining the tryptophan content in a protein was proposed by PREVIERO *et al.* (*310*). The method involves formylation of the indole nitrogen of tryptophan by dissolving the protein in formic acid-HCl solution (see Section III.7.). The high yield (over 90%) of specific formylation of tryptophan, concomitant with the shift of absorption to longer wavelengths (298 nm), allows the spectrophotometric determination of tryptophan in proteins. The method was shown to be fairly accurate with several test proteins (*310*).

The formation of acid stable β-carboline derivatives from tryptophan peptides and proteins was also suggested as a useful method for the analytical determination of tryptophan (see Section III.6.) (*312*).

V. Structure Determination

This section attempts to introduce the reader to some present and future potentialities of spectroscopic techniques that can be applied to polypeptide molecules containing tryptophan residues. Information concerning the environment and location of tryptophan side chains and the changes in environment of these residues due to structural transformations can be gained by ultraviolet difference spectroscopy and fluorescence measurements. However, since these two techniques are well known to biochemists and have been reviewed extensively (*55, 101, 116, 172, 410, 411*), it was decided to limit the following presentation to other modern spectroscopic tools. Among those particularly worthy of mention appear to be circular dichroism, magnetic circular dichroism, circular polarization of fluorescence, nuclear magnetic resonance (^{1}H, ^{13}C, ^{14}N) and X-ray diffraction.

1. Ultraviolet Absorption

The absorption spectra of tryptophan derivatives in the region 310 to 250 nm are composed of two different types of $\pi \rightarrow \pi^*$ electronic transitions. One is the $^{1}A-{}^{1}L_b$, while the other is the $^{1}A-{}^{1}L_a$ type. Unfortunately, the intensities and shapes of the bands have been expecially difficult to determine owing to their extensive overlap.

Using, *inter alia*, N-stearyl-L-tryptophan *n*-hexyl ester as a model compound, Strickland and colleagues (*390, 391*) recently showed that in apolar solvents the distribution of oscillatory strengths is 17.4% for $^{1}L_b$ and 82.6% for $^{1}L_a$. The $^{1}L_b$ transition has well-resolved vibronic bands at 289.5, 282.5, 278.5, and 270.5 nm, whereas the $^{1}L_a$ transition has relatively vibronic bands at 289.5, 283.5, 279, 273, and 267 nm. According to Yamamoto and Tanaka (*450*) the transition moment direction of the $^{1}L_b$ makes an angle of 54° with the long axis of the tryptophan-containing molecule, while that of $^{1}L_a$ lies at an angle of −38° to the same axis.

The effect of hydrogen bonding upon the near-UV spectrum of tryptophan derivatives has been investigated (*390, 391*). Hydrogen bonding of the indolyl NH the 0–0 ^1L_a band induces red shifts of 3–9 nm depending on the nature of the hydrogen acceptor. In contrast the red shift of the 0–0 ^1L_b band is only about 1 nm. The results obtained indicate that in proteins the ^1L_a band may be notably red shifted due to hydrogen bonding of the indolyl NH to other protein groups. Hydrogen bonding of the indolyl NH to solvent molecules does not alter the total ($^1L_a + {}^1L_b$) oscillator strength (*390*). However, changes in bulk solvent properties do have measurable effects upon the indolyl oscillator strength.

The influence of polarizable organic molecules, *e.g.* organic acids and ionic detergents, on the light absorption of tryptophan and its derivatives has been studied (*10, 34, 35, 108*). The nature of the spectral changes depends upon the character of the interacting solvent and solute molecules: a net positive electrostatic field induces a red shift, whereas a negative one causes a blue shift. The absorption changes increase with the size of the non-polar part of the polarizable molecule. The spectral changes can be assigned to the electrostatic attraction or repulsion of electrons of the indolyl nitrogen according to the electrostatic character of the polarizable molecule interacting with the chromophore. The calculated interaction energy suggests that in aqueous solution van der Waals interactions are operative between the chromophore and the compounds containing polarizable or charged groups.

The far-UV absorption bands of tryptophan derivatives have also been investigated (*16*). Two bands are evident, which are centered at about 220 nm (^1B_b) and 195 nm (^1B_a), respectively. The ^1B_b has somewhat greater intensity and more discernible vibrational fine structure or more resolvable components than the ^1B_a band. The effect of changing the state of charge of tryptophan derivatives on their far-UV absorption spectra was assessed. It was found that the ^1B_a band is shifted between 0.5–2.5 nm to the blue, and the ^1B_b band between 1.0–2.8 nm to the red, when the charge on the molecule is made less negative or more positive. In addition, as the refractive index of the solvent increases, the two spectral bands suffer concurrent red shifts. The experimental polarizations of the ^1B_b and ^1B_a transitions are still unknown. However, the orientation angles between the transition moment directions of the two bands and the long axes of the tryptophanyl derivatives have been calculated and found to be 18° and −61° for ^1B_b and ^1B_a, respectively (*450*).

2. Circular Dichroism

Many biologically active polypeptides have been shown to possess CD bands in the wavelength region above 270 nm. In peptide hormones and antibiotics and in non-conjugate proteins these bands are due to tryptophan and tyrosine residues, the only chromophores which absorb in this spectral region. Because of the weakness and spectral overlap of these Cotton effects, their structure is usually imperfectly resolved. However, in recent years some progress in analyzing the fine structure of the CD spectra of tryptophan and derivatives has been reported (*149*).

By using a computer read-out program with multiple scans of the spectrum MYER and McDONALD (*271*) were able to show that the CD pattern of L-tryptophan above 250 nm in water (pH 7.0) consists of at least three positive bands centered at 291, 281, and 272 nm. The positions of these bands are approximately the same as those of the absorption peaks of the indole chromophore, thus suggesting their origin.

The near-UV CD fine structure of tryptophan and seven of its derivatives were examined by employing high-resolution spectra recorded at 77° K (*389*). The spectra may be grouped into four classes: i) ^1L_b bands intense, ii) ^1L_a bands intense, iii) both ^1L_a and ^1L_b bands intense, iv) fine structure whose origin was not readily identified. Both the 0–0 and $0 + 850\ cm^{-1}$ ^1L_b transitions occur together (near 290 and 283 nm, respectively) and have the same CD sign. A number of ^1L_a transitions were identified, but their relative intensities varied greatly.

The ellipticity bands in the near-UV of the diketopiperazines of tryptophan are several times stronger than those of the corresponding linear dipeptides (*105, 389*). It was therefore possible to observe the vibronic structure of the CD spectra of these compounds. The relative strengths of the two major ^1L_b bands are similar in the CD spectra of diketopiperazines containing a single tryptophan residue. In the cyclo-L-Trp-L-Trp spectra, however, a different ratio of ^1L_b intensities was observed, suggesting that this molecule may also have a small exciton CD contribution. In the tryptophanyl diketopiperazines having only one aromatic side chain the ^1L_b CD bands are positive and relatively weak. By contrast, the ^1L_b CD bands are strongly negative in cyclo-L-Trp-L-Trp. These findings suggests that coupling between electric dipole transitions may be a major source of CD intensity for tryptophanyl moieties in biologically active polypeptides.

In the CD spectrum of L-tryptophan below 250 nm, of the four ellipticity bands, only two seem to reflect their counterparts in the absorption spectrum (*271*). The shoulder at about 218 nm corresponds to the exceedingly strong indole absorption peak at 218 nm, and the negative dichroic band centered at about 200 nm possibly reflects, at least in part, the $\pi \rightarrow \pi^*$ carboxyl transition. There seems to be no directly perceivable absorption band corresponding to the relatively large positive ellipticity maximum at about 224 nm. Subsequently, the far-UV CD bands of tryptophan residues were investigated by AUER (*16*). Four types of spectra were found. All have intense positive ellipticity near 223 nm, but have differing patterns of zero or negative ellipticity, in one or more bands, below about 215 nm. Resolution of CD spectra into Gaussian components yielded a consistent set of four bands in the ^1B_b region, while only one broad band is needed in the ^1B_a region. Solvent effects and vicinal charge effects were also characterized. The possibility that parts of the experimental CD spectra arise from coupling with the $n \rightarrow \pi^*$ and $\pi \rightarrow \pi^*$ transitions of carboxyl, amide or ester chromophores was discounted.

The CD spectrum of poly-L-tryptophan was examined by FASMAN and colleagues in films cast from dimethylformamide (*384*) and by PEGGION *et al.* in glycol monomethyl ether solution (*297*). Below 250 nm the CD pattern obtained in the solid state is not consistent with the more recent data obtained in solution. In fact a negative intense band at 232 nm has been reported for the solid film (*384*), while, in solution a positive Cotton effect at 230 nm has been found, accompanied by a negative one at 219 nm (*297*). Quite clearly the observed CD pattern arises from strong overlapping of indole and peptide contributions. Furthermore, in the near UV region, the CD spectrum appears to be strongly solvent dependent (*297*).

Investigations of chiroptical properties of tryptophan-containing copolymers in solution led to the conclusion that in trifluoroethanol and in glycol monomethyl ether, poly-L-tryptophan assumes the right-handed α-helical conformation (*86*).

Several peptide antibiotics and hormones, and proteins exhibit CD spectra which are dominated by the contribution of the indolyl chromophore of tryptophan residues; among others, the antibiotic gramicidin A (*204, 408, 414*), the growth hormones from different species (*186*), the luteinizing- and follicle-stimulating-hormone releasing factor (*247*), the proteins avidin and streptavidin (*155*) and homologous lysozymes (*270*), give CD spectra which have been investigated in detail in recent years.

2.1. Magnetic Circular Dichroism

Although the phenomenon of magnetically induced optical activity was discovered by FARADAY in 1845, it has only been during the last five years that magnetic circular dichroism (MCD) has been applied to peptide molecules.

Natural and magnetic optical activity have separate physical origins and, consequently, provide fundamentally different information. In natural optical activity the differential absorption of left and right circularly polarized light which gives rise to the observed Cotton effects is caused by the intrinsically dissymmetric perturbing molecular field of the chiral substance, whereas in magnetic optical activity the dissymmetric perturbing field is external to the molecule. The consequence of this separate physical origin of the two phenomena is that CD is sensitive to subtle aspects of molecular stereochemistry while MCD provides information about the electronic structure of the component chromophores.

The MCD properties of tryptophan and tryptophan containing model peptides have been extensively investigated (*29, 30, 32, 138, 190*). The MCD spectra show two oppositely signed bands which are the

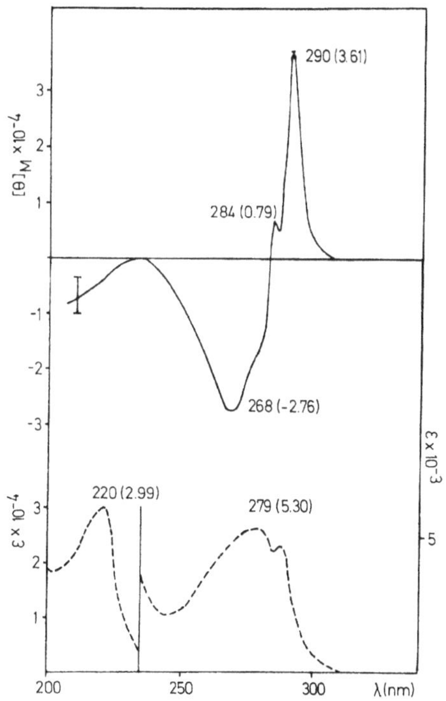

Fig. 3. MCD (———) and UV (------) spectra of L-leucyl-L-tryptophanamide acetate in 0.01 M phosphate buffer (pH 6.9) (*30*)

result of the mixing of both ^1L_a and ^1L_b electronic transitions by the magnetic field. The longer wavelength ^1L_b transition exhibits two partially resolved positive MCD bands at 290 nm and 284 nm associated with the 0–0 and $0+730\ cm^{-1}$ vibrational components, respectively, whereas the ^1L_a band gives a broad negative band with some vibrational fine structure (Fig. 3).

Tryptophan is the only naturally occurring amino acid that gives positive MCD bands. Furthermore the 290 nm band is intense and almost completely free of overlapping contributions by other bands associated with chromophoric side chains of other amino acid residues. This fortuitous situation suggested the possibility of using the 290 nm MCD band as an analytical and non-destructive technique to determine the tryptophan content in proteins (see Section IV.). MCD can also monitor the consequences of certain chemical modifications of tryptophan which in turn can aid in studying the participation of this residue in catalysis and/or binding of substrates, inhibitors and coenzymes.

The intensity of the 290 nm MCD band is a linear function both of concentration and magnetic field strength. Compared to its MCD, the CD of L-tryptophan is negligibly small. The D-isomer of tryptophan, tryptophan containing peptides, and other amino and carboxyl substituted derivatives of tryptophan display MCD characteristics closely similar to those of L-tryptophan itself, both in regard to sign, wavelength maxima and band intensities.

The 290 nm band intensity is unaffected by the conformation of the peptide molecule. Whereas no change was observed in the MCD spectrum of tryptophan between pH 1 and 12, the L-tyrosine MCD spectrum is highly sensitive to pH changes. At pH values higher than 8, a shoulder appears on the long-wavelength side of the main band which increases in intensity and leads to the negative band at 292 nm at pH 12. These changes are caused by the ionization of the phenolic chromophore. In order to keep the overlapping contribution of L-tyrosine to the 290 nm L-tryptophan band at a minimum, it is obviously necessary to carry out the measurements at a pH below 8. The error due to overlapping contributions from the negative L-tyrosine MCD at neutral pH values is comparatively small, but it cannot be neglected in those cases for which the tyrosine/tryptophan ratio is high.

Solvent perturbation and denaturation studies indicate that the environment of the indole chromophore minimally affects the intensity of the 290 nm MCD band but significantly shifts the wavelength maximum necessitating its identification in the quantitation process.

In order to test the applicability of the method tryptophan determinations were carried out on a number of proteins whose molecular weight and amino acid composition were known (*29, 30, 73, 254*). The resulting values are in close agreement with those given in the literature. The molar magnetic ellipticity per tryptophan residue of these proteins and model compounds has recently served to quantitate the tryptophan content of xanthine oxidase (*32*).

MCD has also provided an effective means for following the course of the reaction of free tryptophan with either N-bromosuccinimide (NBS) or with 2-nitrophenylsulfenyl chloride (NPS-Cl) (*190*). Sequential addition of NBS to free tryptophan rapidly abolishes the 290 nm band. This phenomenon seems characteristic of the oxidative process. Although the complexity of the reaction precludes detailed characterization of all reaction products by MCD, it allows identification of the general nature of this reaction as an oxidation process and application to other chemical modifications involving the oxidation of tryptophan such as photoxidation, ozonization, and iodination (*103*).

By contrast, reaction of NPS-Cl with free tryptophan and lysozyme, which results in the formation of the 2-substituted derivative (see Section III.3.), eliminates the 290 nm band but generates a new band at 303 nm with a magnetic ellipticity somewhat greater than that of the parent tryptophan band. In this instance, both the disappearance of the substrate (tryptophan residue), and the appearance of product can be quantitated by MCD (*190*).

Quantitative determination of the tryptophan content of glutamate dehydrogenase by MCD yields four tryptophan residues per polypeptide chain (*444*) contrary to the three residues previously suggested by structural analysis. It should be mentioned that the modification of this enzyme with Koshland's reagent gave consistently higher figures, suggesting the occurrence of double modification of some tryptophan residues.

In short, MCD is an exceptionally sensitive, accurate, and nondestructive analytical method for tryptophan quantitation and modification. Its extension promises to be of great value to many aspects of composition, structure, and function of protein molecules.

2.2 Circular Polarization of Luminescence

In addition to ORD and CD, the optical activity of a dissymmetric molecule is manifested in an additional spectroscopic property if the molecule is luminescent. Circular polarization of luminescence (CPL) is the emission analog of CD and is related to the molecular conformation in the electronically excited state in the same way that CD is related to the molecular conformation in the ground state. The spectral characteristics of CPL make this method very useful for the exploration of the conformational properties of peptides, polypeptides, proteins, and protein-ligand complexes (*381*).

While there is an analogy between CD and CPL, it should be pointed out that unique information can be obtained from CPL. Since only luminescent molecules contribute to the CPL spectrum, the information obtained is more specific than that yielded by CD. For proteins this means that the conformation of the molecule is probed by CPL only in the region of those tryptophan and tyrosine residues which are both fluorescent and are situated in a dissymmetric environment when they are in their electronically excited states. The spectral resolution between tyrosine and tryptophan, and among different residues of each type, is significantly greater in emission than in absorption. In condensed media, CPL is usually associated with a single electronic transition ($S_1 \rightarrow S_0$) whereas, for example, the tryptophan chromophore is associated with two transitions (from the ground state to the excited states ^1L_a and ^1L_b) in its longest wavelength absorption band. It is thus clear that the CPL of polypeptide molecules may provide specific information which may be difficult to deduce from their CD spectra.

The extent of CPL, $\Delta f/f$, for a single allowed electronic transition in condensed phases should be approximately constant across the whole of its corresponding emission band. Marked deviation from such behaviour may therefore serve as a strong indication of "heterogeneity" of emitting chromophores in the system studied, the different chromophores having different emission spectra and different optical activity, thus causing $\Delta f/f$ to assume different values across the observed composite emission spectrum. The heterogeneity in the emitting chromophores may stem from the presence of different kinds of emitting species or differences in the environment of chromophores of the same kind. This aspect of CPL renders this technique a rather powerful tool for the discrimination among various fluorescing groups in peptide molecules.

The CPL of various cyclic dipeptides containing tryptophan residues has been studied (*346*). The optical rotatory power deduced from these measurements vanished upon electronic excitation in fluid media but is essentially the same in the ground and electronically excited states in highly viscous media. A change in conformation thus occurs upon excitation in fluid solution essentially before light emission takes place. The change responsible for the disappearance of the optical activity is faster than the rate of rotational relaxation as monitored by the linear polarization of the fluorescence. These results may be of value in the theoretical calculation of the conformation of cyclic dipeptides in the excited state.

In a preliminary communication TURNER *et al.* (*409*) reported that the signal-to-noise ratios in lysozyme CPL spectrum is several times larger than for L-tryptophan spectrum. This was not unexpected since one can find an analogous behaviour in the corresponding CD spectra.

The CPL of tyrosine and tryptophan residues in derivatives of subtilisin types Carlsberg and Novo were resolved (*347*). The tyrosine residues in the Carlsberg protein, and both tyrosine and tryptophan residues in the Novo protein, were found to be heterogeneous with respect to their optical activity and emission spectra. Changes in the environment of the emitting tyrosine residues in both proteins and in the tryptophan residues in the Novo protein were found on changing the pH from 5.0 to 8.3. The pH dependence of the enzymatic activity of these two proteins may be due, at least in part, to local conformational changes in the molecules. CPL also revealed that covalently bound inhibitors at the active site of subtilisin Novo effect the environment of the emitting aromatic side chains, presumably *via* changes in conformation.

A CPL study indicated that the conformation of the indole side chain of the single tryptophan residue of staphyloccocal nuclease and its interaction with the environment is very similar in the ground state and in its first singlet excited state (*381*). Also the environment of the tryptophan residue is not affected noticeably by the binding of calcium ions that are known to be necessary for the activity of this enzyme.

In marked contrast to the behaviour of staphylococcal nuclease and azurin, which also has its emission anisotropy factor nearly constant across the emission band, the CPL properties of human serum albumin reflect a heterogeneity in the population of the molecules of this protein, different molecules having different emission spectra and different dissymmetric environments for their single tryptophan residue (*381*).

In chicken pepsinogen, a protein that contains six tryptophan residues, the various indole side chains have different environments and thus showed somewhat different emission spectra and different optical activity (*381*).

In summary, CPL has characteristic potentialities and limitations as a spectroscopic tool. It is applicable only to luminescent molecules, and then only to a limited number of electronic transitions (generally, to a single transition). In complex systems, however, these limitations may turn out to be advantageous, supplying more specific information about the emitting species.

3. Nuclear Magnetic Resonance

3.1. ^{1}H-NMR

Tryptophan has the most complicated ^{1}H-NMR spectrum of the aromatic amino acids, although resonances can be assigned fairly readily (*23, 50, 83, 244, 277, 358*). At 220 MHz in deuterated dimethyl sulfoxide all aromatic resonances are resolved (Fig. 4); the C_2 proton gives essentially a singlet, the C_7 and C_4 protons give doublets, and the C_5

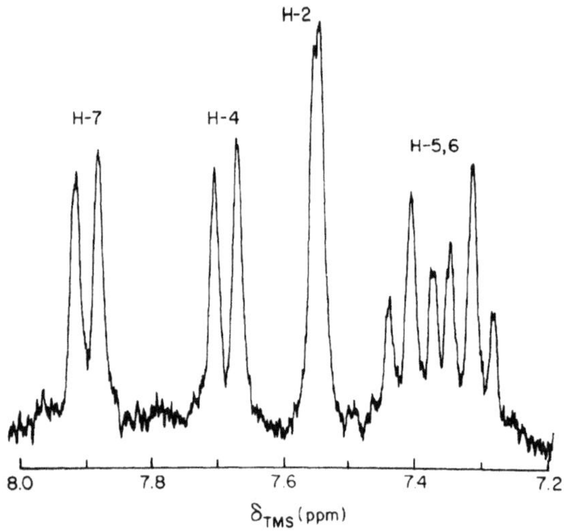

Fig. 4. 220 Mhz ^{1}H NMR spectrum of the aromatic resonances of tryptophan in [$^{2}H_6$] dimethyl-sulfoxide (*83*)

and C_6 proton signals are considered to be doublets of doublets. The problem which arises in dealing with tryptophan lies in selective changes in this multi-peak pattern due to local charge and environmental effects, as evidenced by Trp-Trp. The C_2 proton is the proton most affected by substitution and pH effects, and this must be attributed to the adjacent tryptophan imino group. In particular, the C_2 proton resonance is observed to behave differently in Trp-Ala and Ala-Trp. The chemical shift change appears to reflect the titration of the free amino group in the latter case. While these effects are not large, they considerably complicate any attempt of using simulated spectra of the free tryptophan for curve-fitting of spectra of denaturated proteins. The chemical shift changes of tryptophan resonances as a function of the nature of solvent also indicate their sensitivity to environmental effects.

Chemical shifts and coupling constants for the aliphatic protons of tryptophan in basic aqueous solutions have been obtained as a function of temperature and concentration (*73*). At low concentrations the vicinal coupling constants diverge with increasing temperature while at high concentration they converge. These results were interpreted in terms of variations with temperature and concentration of the relative energies of the three classical staggered rotamers, the two less favorable becoming less stable with increasing temperature. The rotamer energy variations appear to reflect both solute-solvent and solute-solute interactions.

The resonances of the indole NH protons for five of the six tryptophan residues of hen egg-white lysozyme were resolved by observing the ^1H NMR spectrum in water (*147*). Resonances were assigned to specific tryptophan residues by differential deuterium exchange rates, inhibitor perturbation and chemical modification. Well resolved indole NH resonances were obtained for all five tryptophan residues of human lysozyme.

In oxymyoglobin and myoglobin, Patel and colleagues (*296*) were able to observe resonances of the NH indole protons of the two tryptophans moieties in the range 10 to 12 ppm downfield from 3-(trimethylsilyl)-propanesulfonic acid (sodium salt). These signals were identified by chemical modifications coupled with considerations of crystallographic structure and the dependence of the signals on the species from which the myoglobin was obtained and on spin, pH, and temperature.

A high-resolution ^1H NMR study of the reversible complex formed between N-formyl-L-tryptophan and α-chymotrypsin in aqueous solution indicated that the complex has the same structure as that determined for the crystalline state by X-ray methods (*145*). The resonances from the aromatic protons of the inhibitor were shifted upfield considerably in the enzyme-inhibitor complex, but the alkyl and formyl resonances were only slightly affected.

The ^1H NMR spectrum of gramicidin A was recorded in several solvents (*414*). Signals of the formyl proton, peptide NH's, and indole NH's were assigned. Only the formyl proton was useful in assigning the polypeptide backbone conformation.

Resonances of the single tryptophan residue in the ^1H NMR spectrum of horse heart cytochrome c have been assigned (*252*). In particular the C_2 signal is somewhat at higher field than expected. However, the position of this resonance is strongly influenced by the nature of the hydrogen bonding of the indole NH proton. This hydrogen bond appears to be due to the carboxyl group of the propionic acid substituent of porphyrin ring 4 in cytochrome c.

^1H NMR spectra of diastereomeric cyclo-dipeptides containing tryptophan and histidine residues were examined in solutions of D_2O and dimethyl sulfoxide as function of pH and temperature (*15*). An upfield shift of the imidazole protons of the LL compound was observed, in contrast to the LD compound. This result was explained by the existence of a preferred conformation of the LL isomer in which the indole ring of the tryptophan residue faces the diketopiperazine ring, while the imidazole moiety lies away from it. Calculated values of the indole ring current explain the measured upfield shifts.

3.2. ^{13}C-NMR

Recently, the first ^{13}C-NMR studies of amino acids, peptides and proteins have been reported. The spectrum of the aromatic region of L-tryptophan (*53, 422*) is very similar to that of 3-methyl-indole in dimethyl sulphoxide, the main difference being a downfield shift of the C_2 resonance in the amino acid, which is probably caused by the proximity of the α-amino group. In aqueous solution the effects of peptide bond formation on the tryptophan aromatic resonances are small, as found for the other amino acids, C_3 being shifted most upon formation of Gly-L-Trp, followed by C_2. The C_2 shift propably results from removal of the positively charged α-amino group from the vicinity of the 2-position. Replacement of the free α-carboxyl of L-tryptophan with a peptide bond has less effect on the aromatic resonances than replacement of the α-amino group. In L-Trp-L-Trp separate peaks, corresponding to the two indole rings, are observed for four of the aromatic carbons, while the remaining four aromatic resonances are broader, due to a slight chemical shift difference between equivalent carbons of the two rings. Clearly, shifts caused by the interaction of the indole rings in L-Trp-L-Trp are small, in agreement with recent findings which show that ring current shifts in ^{13}C NMR are of minor importance.

Single-carbon resonances in the aromatic region of the natural abundance ^{13}C NMR spectra of native proteins have been assigned. Some resonances, most notably those of C_3 of tryptophan residues, are readily identified because they do not normally overlap with any other type of carbon peaks. On the other hand, resonances of C_8 and C_9 are not easily identified because they are found in the same range as some resonances of tyrosine, phenylalanine and histidine residues. However, OLDFIELD and ALLERHAND (*288*), using partially relaxed

Fourier transform spectra, were able to identify the resonances of C_8 and C_9 of tryptophan residues in horse heart cytochrome c and hen egg white lysozyme.

FOSSEL *et al.* (*134*) reported the ^{13}C NMR spectrum of gramicidin A in dimethyl sulfoxide, a solvent in which the molecule is probably a monomer having little or no secondary structure, and in methanol, a solvent which supports the dimer structure. Of the four L-tryptophans in gramicidin A, two have D-leucine both preceeding and following them in the sequence, one has D-leucine preceding and ethanolamine following, and one has D-valine preceding and D-leucine following. In dimethyl sulfoxide, while all of the rest of the indole carbon resonances appear as single lines and are within 0.3 ppm of their model compound value, the indole C_3 carbon appears as three nonequivalent resonances in gramicidin A. As noted above, there are three types of L-tryptophan as distinguished by the chemical nature of adjacent groups. This finding exemplifies that differences in chemical shift could result from differences in the nature of adjacent groups if the conformations are random. In methanol the indole resonances appear as single but broadened lines. The indole C_3 carbon resonance no longer appears as a multiple resonance. This phenomenon may be due to overlapping of broadened lines. Four of the five protonated carbons of the indole ring which appear as single resonances in dimethyl sulfoxide break into double resonances in methanol. The fifth remains as a single line. Thus, ^{13}C chemical shift studies reveal much about the structure of gramicidin A in solution.

3.3 ^{14}N-NMR

Broad lines, low sensitivity, and low resonance frequency have for a long time prevented ^{14}N NMR spectroscopy from being applied to the study of amino acids and peptides. However, with the advent of high field superconducting solenoids and accumulation techniques, these difficulties are less formidable.

In a recent study (*321*) the shifts and linewidths of aqueous solutions of amino acids, peptides, and derivatives have been correlated. It was seen that the groups directly bonded to the α-nitrogen atom determine the region in which the shift occurs, but other groups in the molecule, or intermolecular interactions, may cause small perturbations within that region. The extreme of the observed shifts in the α-amino acids are 332 and 346 ppm (with trimethyl-phenyl-ammonium chloride chosen as external secondary standard). In the case of tryptophan at pH 7.15 the shift was observed at 343 ppm.

The secondary amino group of the indole ring in tryptophan has a low field shift (293 ppm), probably because of rapid tautomerism with the C = N indolenine form, in which the nitrogen has a π-bonded structure. However, the equilibrium favours the indole form, the deshielding contribution from the π-electrons in the indolenine form is, therefore, small, moving the chemical shift of the indole nitrogen atom in tryptophan to much higher field than in imidazole, but to lower field than in the saturated counterpart, proline.

In summary, ^{14}N resonances of amino acids and peptides show large chemical shifts and can often be distinguished for chemically different nuclei in molecules of biochemical interest at sufficiently high applied magnetic field strengths. However, most nitrogen resonances appear as broad lines which often overlap and are sometimes too weak to be observed in dilute solution. Resonances from quaternary nitrogen atoms give very narrow lines and may be expecially useful in the study of chemical exchange between a small molecule and its complex with a larger one.

4. X-Ray and Conformational Energy Calculation Studies

4.1. X-Ray Structure Determination

The first structure of a tryptophan-containing substance, namely glycyl-L-tryptophan dihydrate, was determined early in 1956 (*294*). The atoms of the indole ring system with the methylene carbon attached to it were found to be coplanar. The bonds within the benzene ring, excluding C_5–C_6 are 1.39–1.41 Å long; it is doubtful if they are significantly longer than in the isolated benzene ring. The bond angles do not deviate significantly from 120°. In the pyrrole ring the bond angles and bond lengths are nearly the same as those reported for pyrrole except for C_8–N, which was found to be unexpectedly short (1.31 Å). The plane of the indole ring forms a dihedral angle of 73° with the plane of the peptide group and an angle of 30° with that of the carboxyl group. The centres of the indole rings in the crystal packing, which are related by screw axes, form approximately a face-centered array. The weakly basic nitrogen atom of the pyrrole ring does not take part in hydrogen bond formation, and the layer is held together by van der Waals forces only.

Crystals of L-tryptophan hydrochloride and hydrobromide are isomorphous with each other and the conformation of the amino acid cations is essentially the same in both crystals (*405*). The plane of the

References, pp. 426—449

indole moiety forms a dihedral angle of 70.6° with that of the carboxyl group. The average of six C–C bond lengths in the benzene ring is 1.396, while that of C–N bond lengths in the pyrrole ring is 1.384 Å, respectively. The bond lengths in the indole ring of L-tryptophan hydrohalides agree well with the theoretical values. There are no significantly short C–N bonds, as reported for Gly-L-Trp dihydrate (*294*). This suggest that in this dipeptide the contribution from the resonance of *ortho*- and *para*-benzoquinone types, having C_8–N double bond, increases. Bond angles in the indole ring of L-tryptophan hydrohalides are normal. In these molecules the C_γ (C_3) lies at gauche position with both -COOH and $-^+NH_3$ groups. In contrast, in Gly-L-Trp dihydrate the C_γ (C_3) lies at gauche position with C_α-hydrogen and peptide groups. The crystal packing of the L-tryptophan hydrohalides is analogous to that of Gly-L-Trp dihydrate.

The crystal structure of DL-tryptophan formate was reported (*67*). There are no significant differences from the situation above described for L-tryptophan hydrohalides.

In cyclo-Gly-Trp the diketopiperazine ring is planar and hydrogen bonded in ribbons (*269*). In contrast to the results obtained in solution, which indicate that the tryptophyl side chain lies folded over the diketopiperazine ring (the C_α–C_β angle is near 60°), in the solid state the aromatic side chain is well separated from the diketopiperazine ring (the C_α–C_β angle is −77°). This difference from the solution structure may be a result of the extensive hydrogen-bonding network in the solid state or may just indicate a variety of low energy conformations.

Recently, the structure of the first N- and C-protected tryptophan derivative, *i.e.*, N-acetyl-L-tryptophan methyl ester, was solved by Cotrait and Barrans (*91*). There is one hydrogen bond NH ... O = C of 2.86 Å between the NH indole group and the acetyl C = O of different molecules. Molecular packing in the crystal is established by van der Waals forces between the two hydrophobic methyl groups and the aromatic ring. The conformation of the tryptophanyl side chain is very close to that of the L form of DL-tryptophan formate. Fig. 5 shows bond lengths and angles of N-acetyl-L-tryptophan methyl ester.

An X-ray analysis carried out on oriented films of poly-L-tryptophan cast from solution of the polymer in dimethylformamide indicated conclusively that the solid-state conformation of the polymer is that of an α-helix (*297*). Such a conformation was shown to be present also in solution, given the appropriate solvents and temperature conditions.

A detailed X-ray investigation at 2.5 Å of the binding of N-formyl-L-tryptophan to the active site of the proteolytic enzyme α-chymotrypsin

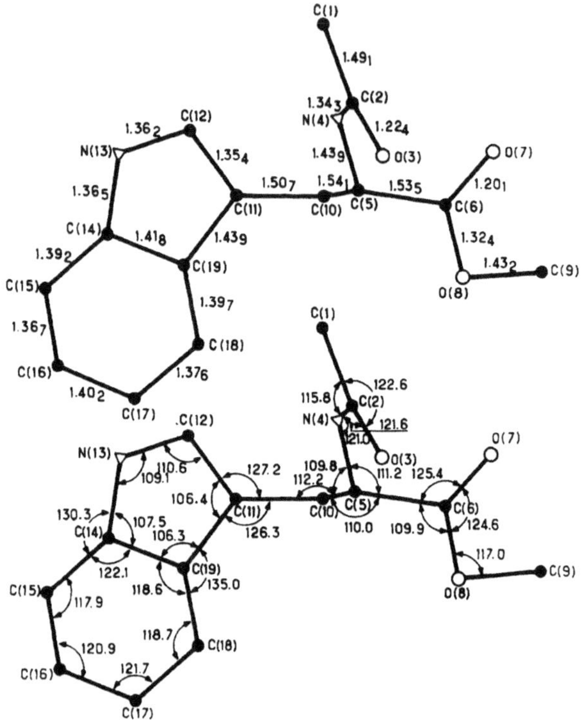

Fig. 5. Bond lengths and angles of N-acetyl-L-tryptophan methyl ester (*91*)

suggested that the indolyl side chain lies in a hydrophobic pocket near Ser-195 (*382*). Its most prominent interactions are with peptide bonds 190–191 and 191–192 above the ring and 215–216 below it. The binding of N-formyl-L-tryptophan does not alter the conformation of the enzyme's active site at all.

4.2. Conformational Energy Calculations

Quantum-mechanical calculations on the conformational energy map of tryptophanyl residues by an all-valence-electrons method (PCILO) were carried out by PULLMAN and coworkers (*243*). The conformational map for the ground state of N-acetyl-N′-methyl-tryptophanamide revealed as the most stable form the C_7^{ax}, which corresponds to a seven-membered hydrogen-bonded ring, the hydrogen bond taking place between the NH of the methylamide group and the carbonyl of the acetyl group. However, there are a number of other energetically significant conformations. No experimental results are available for the model "dipeptide". The number of experimental conformations of tryptophanyl residues from globular proteins is low, but they fall satisfactorily within the allowed conformational space.

By using classical energy expressions backbone-side group conformations of tryptophanyl residues in a "dipeptide" unit were calculated (*314*, *377*). The planar aromatic side group prefers a nearly perpendicular conformation with the plane through the atoms C_α, C_β, and C_γ (C_3). The preferred orientations of the two planes, *viz.*, the plane through the atoms C_α, C_β and C_γ (C_3) and that through the atoms C_β, C_γ (C_3), and C_2, respectively, is decided by the steric interaction between the C_2–H atoms and the carbonyl carbon and the nitrogen at the C_α, and is characteristic of the position of the C_γ (C_3) atom. The conformational features of the side chain discussed for the "dipeptide" unit were found to be nearly the same for the free amino acid and terminal tryptophanyl residues.

5. Charge Transfer Interactions

It is well established that indole derivatives act as donors in forming charge transfer complexes. Various modes of donation have been suggested: π-donation, n-donation, and charge complimentarity or localized donation from the C_3 position. Probably all these do occur under the appropriate conditions and the different explanations are correct in their context. Tryptophan which contains both the indole ring and an amino group can in the right conditions act also as an n-electron donor from the nitrogen of the α-amino group (*363*, *364*).

The first studies on the complexing ability of tryptophan were those with flavins. The apparent association constants are of the order 20–60 M^{-1}. Freezing these complexes produces a very marked intensification of color which disappears on rewarming. Similar colors can be produced by preparing solid complexes. The complexes of flavins with tryptophanyl residues have been utilized to distinguish between buried and exposed groups in proteins (*118*, *130*).

Tryptophan shows spectral change rather different from that exhibited by other amino acids in water containing oxygen dissolved under high pressure. The new absorption with oxygen lies to longer wavelength than with the aliphatic amino acids; in addition there is a superimposed band on the new absorption believed to be the singlet-triplet transition of tryptophan at 360 nm. It may well be that tryptophan in this case functions as a π-donor instead of, or in addition to, being an *n*-donor. Perturbation of the π-electron system would be required to make the singlet-triplet transition allowed, but whether the proximity of the paramagnetic oxygen is sufficient to promote this transition without π-electron being involved in charge transfer to the oxygen is a debatable point.

A series of studies has been carried out on the interaction of tryptophan with chloranil (tetrachlorobenzoquinone). This molecule was chosen as a model for the biological quinones. An absorption band at *ca.* 350 nm is apparent in the complex; also a shift of the carbonyl band from 1690 cm^{-1} to 1633 cm^{-1} is evident. The latter is unequivocal evidence that tryptophan forms 1:1 charge-transfer complex with chloranil. Moreover tryptophan normally regarded as a good π-electron donor also behaves as a *n*-donor in complexing with chloranil.

Coloured compounds are formed when phthalyl-tryptophan is incorporated into the same molecules as phenylalanine, the colour being ascribed to an intramolecular charge transfer complex. Optically active molecules have CD bands associated with the charge transfer bands. Using the area under the bands as a measure of complexing, association constants have been evaluated which agree with those obtained from absorption changes.

The interaction of tryptophan with various pteridines has been examined. Solutions at pH 7 show slight colour due to absorption in the region near 420 nm. Freezing such solutions produces intensification of the colour, a good test of weak complexing. A tryptophan-containing protein, bovine serum albumin, complexes in a similar manner.

The crystal structure of the red picric acid salt of D,L-tryptophan-methanol was determined by X-ray diffraction methods (*141*). The indole and picrate planes are stacked, with interplanar spacing of 3.3–3.5 Å. The stacked pairs are relatively isolated and without π–π interactions between adjacent pairs. The stacking interactions appears to be of the donor-acceptor (charge-transfer) type. The vibrational spectrum of tryptophan picrate contains a strong band at 1740 cm^{-1}, which is not observed in the spectra of either of the components, and is attributed to the C=O stretching vibration (*250*).

Tryptophan residues in lysozyme form charge transfer complexes with imidazole, the residues being the charge donors. Studies of intramolecular indole-imidazolium complexes have been carried out using natural and synthetic peptides containing tryptophanyl and hystidyl residues.

The interaction between tryptophan (and tryptophan peptides) and various nucleic acids has been extensively studied by HELÈNE and coworkers (*170*). Combination of various spectroscopic techniques (fluorescence, absorption, CD, NMR) was very fruitful. The formation of stacked charge-transfer complexes between aromatic amino acids and nucleic acid bases in either frozen aqueous solutions or concentrated fluid solutions was demonstrated. Tryptophan could play a special role in the formation of protein-nucleic acid complexes, due to direct interaction between the aromatic ring and nucleic acid bases. Intercalation and /or hydrogen bonding could help to maintain the nucleic acid in a definite position with respect to the protein molecule. Theoretical studies support the conclusion that the purines and pyrimidines may act as charge acceptors with the pyrimidines being better acceptors.

The addition of tryptophan to NAD gives a 1 : 1 complex presumably of the charge transfer type. A marked increase of absorption near 400 nm was observed. The site of electron acceptance of the coenzyme is thought to be primarily on the nicotinamide moiety of NAD. The formation of complexes between tryptophan and pyridinium ions was used for detecting exposed tryptophan residues in proteins. Some insight into the solution topography of lysozyme (*54, 176, 198*), α-lactalbumin (*54, 328*), and acetylcholinesterase (*360*) was gained by analyzing this type of charge-transfer interaction.

VI. Synthesis of Tryptophan Peptides

1. Synthesis in Solution and in the Solid Phase

Synthesis of tryptophan-containing peptides has been complicated by the presence of the highly reactive indole side chain. Oxidative destruction of tryptophan during the synthetic manipulations has generally been thought to occur during the acid treatment used for removal of most of the classical protecting groups (*248*). Cleavage of the benzyl ester bond connecting the peptide chain to the styrene-divinylbenzene copolymer support in the solid-phase peptide synthesis to give the free carboxyl group is normally accomplished by HBr in trifluoroacetic acid, or by anhydrous HF. With both methods tryptophan degradation has been detected (*248, 257*). N-*t*-Butylation of the indole moiety during removal of the *t*-butyloxycarbonyl and *t*-butyl blocking groups constitutes a serious danger (*4*).

The addition of antioxidants (thiols) and cation scavengers (indole, anisole, water) has been found to minimize side reactions in the removal of acid-labile N- and C-protecting groups and in the cleavage of peptides from the resin in the solid-phase synthesis by anhydrous HF (*248*). The HF procedure, *inter alia,* was found generally more suitable for the solid-phase synthesis of complicated polypeptides regardless of the presence of tryptophan. The recoveries of intact tryptophan are higher with HF cleavage than with HBr in trifluoroacetic acid (*248*).

Under mild conditions of temperature and pH sodium nitrite reacts with tryptophan peptides to give N^i-nitroso derivatives. Such substitution on tryptophan occurs when peptide synthesis is carried out *via* the acid azide coupling method (*3, 49*).

Treatment of tryptophan peptides with HCl in formic acid for the cleavage of some acid-labile protecting groups results in formylation of the indole nitrogen. The degree of transformation of tryptophan into N^i-formyl tryptophan depends on the concentration of HCl and the time of reaction (*286, 451*). It has been shown that N^i-formyl-tryptophan is stable in solution under the conditions used for peptide synthesis including treatment with HF. In particular, the results demonstrated that pre-introduction of the formyl group into the indole nucleus is very effective as regards the protection of tryptophan from oxidation and N^i-

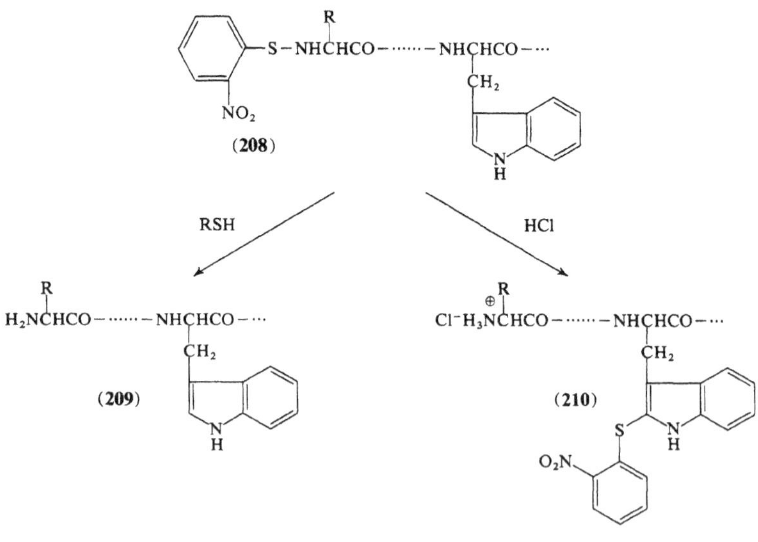

substitution. The formyl group can be quantitatively removed by mild alkali treatment or by various nucleophiles. Also, the uncharged free amino groups eventually present in a peptide can serve as nucleophiles for the removal of the formyl group. However, conditions were found which reduce the extent of this side reaction to a minimum (*286, 451*).

Sulfenyl halides have recently aroused great interest in peptide chemistry (*123*). Reversible protection of the amino function through sulfenamide formation has achieved such a popularity that in particular the 2-nitrophenylsulfenyl (NPS) protecting group is now widely employed in peptide synthesis. However, acid deblocking of the NPS group from tryptophan peptides (**208**) gives an unwanted rearrangement resulting in isomeric peptides in which the NPS-group is covalently bound at the 2-position of the indole nucleus of tryptophan (**210**) (*121, 122*).

Various nucleophilic reagents, especially mercaptans, have been suggested for the selective removal of the NPS moiety without side reactions involving tryptophan, thus rendering this N-protecting group suitable also for the synthesis of tryptophan peptides (*121, 123*). As an alternative useful procedure, the NPS group was cleaved with mineral acids in the presence of an excess of indole compounds as scavengers (*447*) (see below the special case of the synthesis of glucagon).

Successful syntheses of several long chain peptides containing tryptophan, including ACTH (*21, 143, 350, 351*), β-MSH (*352*), tyrocalcitonin (*324*), gastrin (*33, 96, 267*), and of peptide fragments covering the entire sequence of ribonuclease T (*453*) and baker's yeast iso-1-cytochrome C (*268*), proteins containing a single tryptophan residue, have been reported. Some other examples are discussed below.

2. Special Cases

2.1. Glucagon

Glucagon is an oligopeptide pancreatic hormone of 29 amino acids with a single tryptophan residue in position 25 of the sequence. This peptide constituted a special synthetic problem, because of the

His-Ser-Gln-Gly-Thr-Phe-Thr-Ser-Asp-Tyr-Ser-Lys-Tyr-Leu-Asp-
1 5 10 15

-Ser-Arg-Arg-Ala-Gln-Asp-Phe-Val-Gln-Trp-Leu-Met-Asn-Thr
20 25 29

Glucagon

simultaneous presence, in addition to tryptophan, of all other trifunctional amino acid residues, *i.e.*, methionine, histidine, arginine, lysine, carboxy- and hydroxy-amino acids, which all contributed to particular synthetic difficulties. The successful synthesis was achieved by WÜNSCH and coworkers in 1967 by the conventional solution technique (*446*) (Fig. 6).

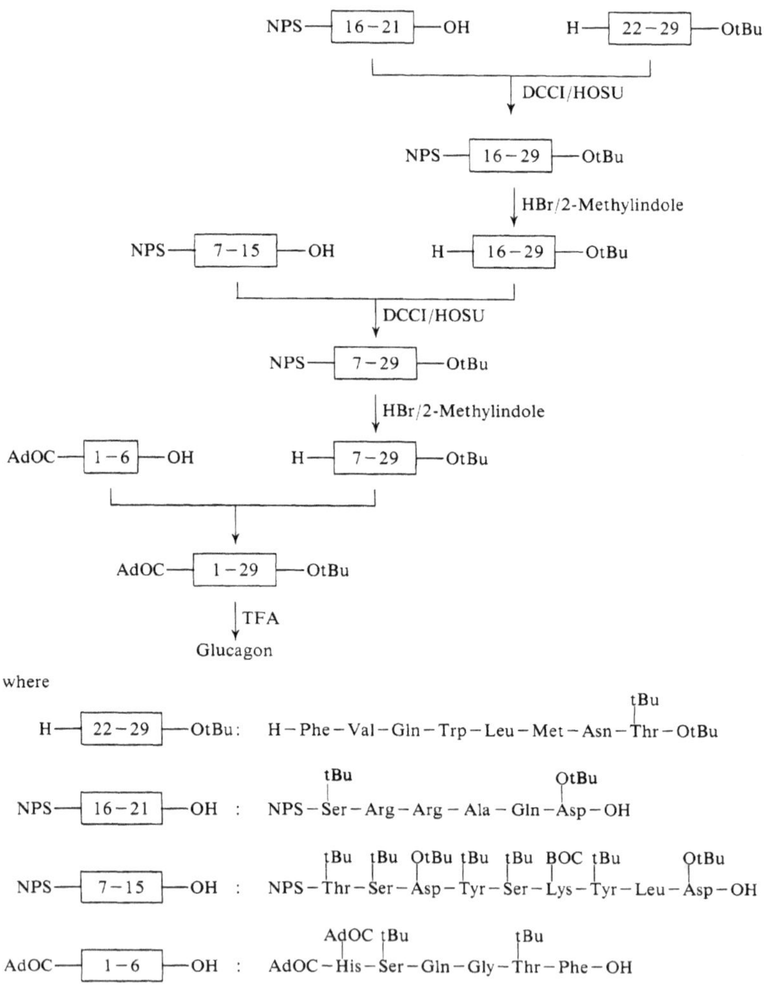

Fig. 6. Outline of the synthesis of glucagon by the solution procedure. DCCI, dicyclohexylcarbodiimide; HOSU, N-hydroxysuccinimide; NPS, 2-nitrophenylsulfenyl; tBu, *t*-butyl; AdOC, adamantyloxycarbonyl; TFA, trifluoroacetic acid

A synthetic scheme was employed which employed an appropriate selection of protecting groups. Protection of the side chain carboxyl, hydroxyl and amino functions was achieved with acid labile protecting groups (*t*-butyl esters, *t*-butyl ethers and *t*-butyloxycarbonyl group), while the α-amino function was protected with the 2-nitrophenylsulfenyl group (*123*). The sulfenamide function of the N_α-protecting group was cleaved selectively and without side reactions involving the single tryptophan-25 residue by using mercaptans as cleaving agents (*121*) or by HBr in organic solvents with an excess of an indole derivative (1- or 2-methyl-indole) as a scavenger (*447*).

2.2. *Valine-gramicidin A*

The pentadecapeptide antibiotic valine-gramicidin A shows the

HCO-L-Val-Gly-L-Ala-D-Leu-L-Ala-D-Val-L-Val-D-Val-L-Trp-D-
1 5

-Leu-L-Trp-D-Leu-L-Trp-D-Leu-L-Trp-$NHCH_2CH_2OH$
10 15

Valine-gramicidin A

unusual feature of an alternating structure of L and D amino acid residues and a tryptophan content as high as four residues. The synthesis has been accomplished through the conventional solution method by SARGES and WITKOP (*336*) and through the solid phase method by FONTANA and GROSS (*120*).

The outline of the solid-phase synthesis is depicted in Fig. 7.

t-Butyloxycarbonyl-amino acids were used throughout the synthesis, the coupling step being carried out in methylene chloride with N,N′-dicyclohexylcarbodiimide and the N-protecting moiety removed by acid treatment with 50% trifluoroacetic acid in methylene chloride containing 1% β-mercaptoethanol. The N-formylated pentadecapeptide ethanolamide was obtained by cleavage of the N-free pentadecapeptide-resin with ethanolamine and subsequent N-formylation with *p*-nitrophenylformate. After purification by countercurrent distribution, the synthetic antibiotic behaved exactly like the natural compound in its physical chemical properties and possessed *full antibiotic activity*.

The successful synthesis of this pentadecapeptide by the solid phase technique is particularly remarkable in view of the high tryptophan content. All four residues were completely resistant to repetitive acid treatment with trifluoroacetic acid used to remove the N-protecting *t*-butyloxy-

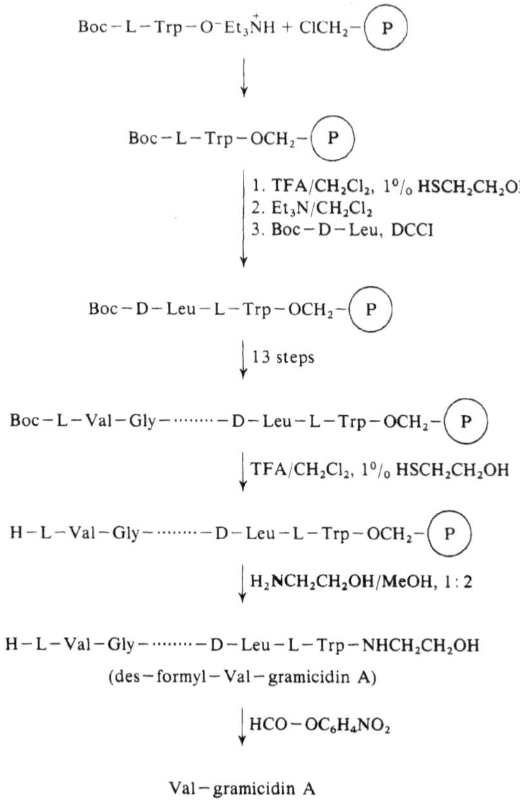

Fig. 7. Outline of the synthesis of Val-gramicidin A by the solid-phase procedure; P, polystyrene-2% divinylbenzene copolymer; Boc, *t*-butyloxycarbonyl; TFA, trifluoroacetic acid; DCCI, N,N′-dicyclohexylcarbodiimide

carbonyl group, provided that β-mercaptoethanol was added to the acid medium as an antioxidant. However, it is worth noting that the high efficiency of the countercurrent distribution technique in separating extremely similar compounds made possible the purification of the crude synthetic valine-gramicidin A, yielding a synthetic fully active peptide antibiotic.

2.3. Luteinizing Hormone-Releasing Hormone

Extensive research has been carried out on establishing structure-activity relationships for the luteinizing hormone-releasing hormone

References, pp. 426—449

(LH-RH) in order to discover related compounds which might be

pGlu-His-Trp-Ser-Tyr-Gly-Leu-Arg-Pro-Gly-NH_2
1 5 10

Luteinizing Hormone-Releasing Hormone (LH-RH)

useful for developing new methods of birth control. This decapeptide, containing a single tryptophan residue and a pyroglutamyl-residue at the N-terminal position, has been synthesized in various laboratories by both the classic solution (*144, 362, 452*) and solid phase techniques (*89, 251, 261, 325*). For a recent review on the subject the reader is referred to the article of SCHALLY and coworkers (*89*).

An outline of the solid phase synthesis of LH-RH is presented in Fig. 8. As in the case of valine-gramicidin A, after the *t*-butyloxycarbonyl-tryptophan residue was coupled to the peptide-resin, cleavage of the N-protecting group was achieved by trifluoroacetic acid in the presence

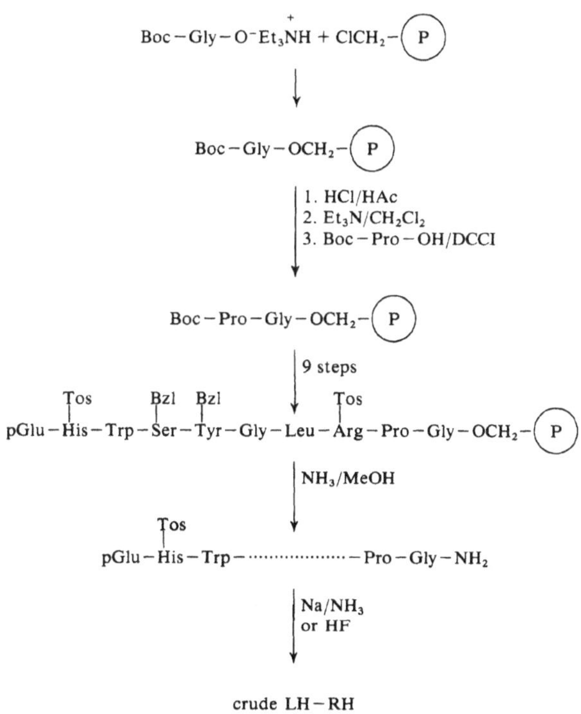

Fig. 8. Outline of the solid phase synthesis of LH-RH; P, polystyrene-2% divinylbenzene copolymer; Boc, *t*-butyloxycarbonyl; DCCI, N,N′-dicyclohexylcarbodiimide

of β-mercaptoethanol in order to protect the tryptophan residue from side reactions. A drawback was found in the use of the sodium reduction deprotection method, which causes cleavage of the single prolyl-peptide bond to some extent. The pure hormone was obtained by purification of the crude compound by ion-exchange chromatography. Several analogs of LH-RH have been synthesized and tested for biological activity.

Synthesis and determination of biological activities of analogs of LH-RH has revealed that tryptophan in position 3 is one of the most important residues for biological activity (*19*). Recently COY *et al.* (*90*) have synthesized analogs in which the tryptophan residue has been replaced successively by pentamethylphenylalanine (**211**), 5-fluorotryptophan, *p*-aminophenylalanine, *p*-nitrophenylalanine and histidine

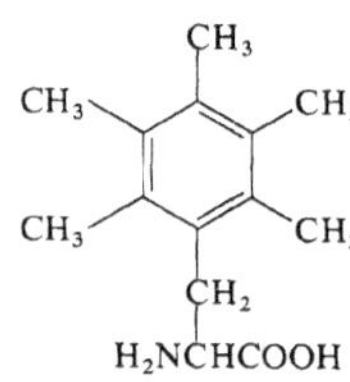

(**211**)

in an attempt to ascertain the factors responsible for the contribution of the tryptophan residue to biological activity.

The pentamethylphenylalanine analog of LH-RH was found to have follicle stimulating hormone and luteinizing hormone releasing potencies of 34–70% when compared with natural hormone. Other LH-RH analogs had little biological activity. Since pentamethylphenylalanine is very dissimilar both structurally and chemically to the indole ring system of tryptophan, it is remarkable that this replacement result in retention of high levels of biological activity. Since the only similarity in the properties of the two amino acids, apart from aromaticity, is their ability to form π–π electron complexes with certain aromatic molecules, it was suggested that similar electronic interactions occur between tryptophan or pentamethylphenylalanine and an aromatic moiety on the pituitary receptor (*90*).

It is worth noting here that HOFFMAN *et al.* (*184*) recently found that a phenylalanine analog of ACTH was devoid of enzymatic activity, but exhibited affinity for the ACTH receptor. For these reasons, COY *et al.* (*90*) suggested that the essential feature of tryptophan in maintaining hormonal action was due to its electron-donating properties.

Phallotoxins

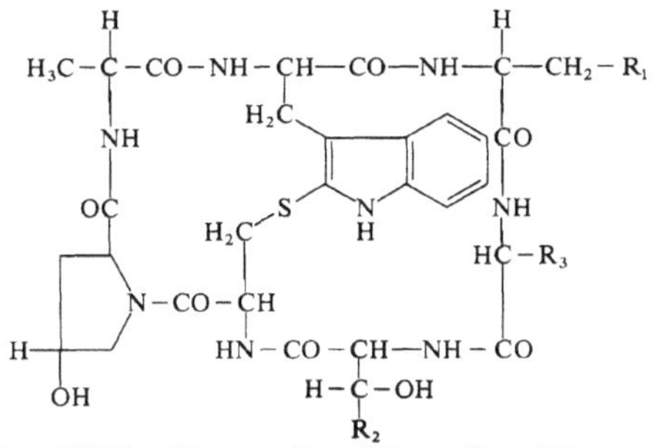

Phalloin	$R_1 = C(OH)-CH_3$ CH_3	$R_2 = CH_3$	$R_3 = CH_3$
Phalloidin	$R_1 = C(OH)-CH_2OH$ CH_3	$R_2 = CH_3$	$R_3 = CH_3$
Phallisin	$R_1 = C(OH)-CH_2OH$ CH_2OH	$R_2 = CH_3$	$R_3 = CH_3$
Phallacidin	$R_1 = C(OH)-CH_2OH$ CH_3	$R_2 = CO_2H$	$R_3 = CH(CH_3)_2$
Ketophalloidin	$R_1 = CO-CH_3$	$R_2 = CH_3$	$R_3 = CH_3$

Amatoxins

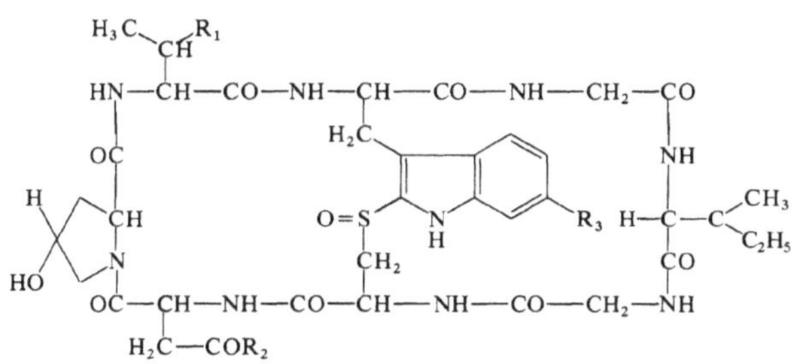

α-Amanitin	$R_1 = CH-CH_2OH$ OH	$R_2 = NH_2$	$R_3 = OH$
β-Amanitin	$R_1 = CH-CH_2OH$ OH	$R_2 = OH$	$R_3 = OH$
γ-Amanitin	$R_1 = CH-CH_3$ OH	$R_2 = NH_2$	$R_3 = OH$
Amanullin	$R_1 = CH_2-CH_3$	$R_2 = NH_2$	$R_3 = OH$
O-Methylaldoamanitin	$R_1 = CHO$	$R_2 = NH_2$	$R_3 = OCH_3$
O-Methyl-desmethyl-γ-amanitin	$R_1 = CH_2OH$	$R_2 = NH_2$	$R_3 = OCH_3$

Fig. 9. Chemical structures of Phallotoxins and Amatoxins

2.4. Toxins of Amanita Phalloides

Among the homodetic, polycyclic, biologically-active tryptophan peptides, the toxins of the *Amanita phalloides* represent a special case [see reference (*427*) and (*349*) for a review]. In fact, the poisonous components of this deathcap, the phallotoxins and amatoxins (Fig. 9), both contain a cycloheptapeptide structure with an additional intramolecular covalent bond connecting the β-sulfur atom of a L-cysteine residue with the carbon atom in position 2 of the indole ring of a L-tryptophan residue. In the amatoxins the thioether bridge is oxidized to the sulfoxide (Fig. 9).

Total synthesis of these toxins presented peptide chemistry with several problems, in particular formation of the sulfur bridge. The latter was explored with model compounds (*430, 431*). It was found (*428*) that the coupling reaction between the sulfur of N-acetyl-L-cysteine and the 2-position of the indole nucleus of N-acyl-L-tryptophan was reproducible and profitable if glacial acetic acid was used as a solvent. The formation of the intermediate S-chloride from the cysteine derivative and N-chlorosuccinimide at room temperature was followed colorimetrically, and the tryptophan component added at the moment of maximum extinction.

The main representative of the phallotoxins, phalloidin, also contains, like all members of this group, a γ-hydroxy-leucine residue. It has been found extremely difficult to introduce this amino acid residue as a component of a peptide chain, because of its tendency to form a γ-lactone ring. When the norvaline analog (norphalloin; Fig. 10) was proved to be even more toxic than the natural toxins (*113*), the first synthesis of a phallotoxin had been achieved. Later on, a number of synthetic analogs of norphalloin were synthesized in order to obtain information on structure-activity relationships (*115, 166, 427*).

Phalloidin has been oxidized to the diastereoisomeric sulfoxides by H_2O_2 in acetic acid (*64, 427*). After separation of the two isomers, it has been found that toxicity is strictly dependent on chirality of the trisubstituted sulfur atom, since only one isomeric component was shown to be poisonous. WIELAND and coworkers (*429*) succeeded in oxidizing 2-ethylthio-L-tryptophan by H_2O_2 in acetic acid and separating the resultant diastereomeric sulfoxides from each other. One of these sulfoxide components was subjected to X-ray analysis and found to have the R-configuration at the chiral SO center. By analogy with the ORD curve of this diastereoisomeric sulfoxide, the same configuration was assigned to the SO of the toxic phalloidin sulfoxide.

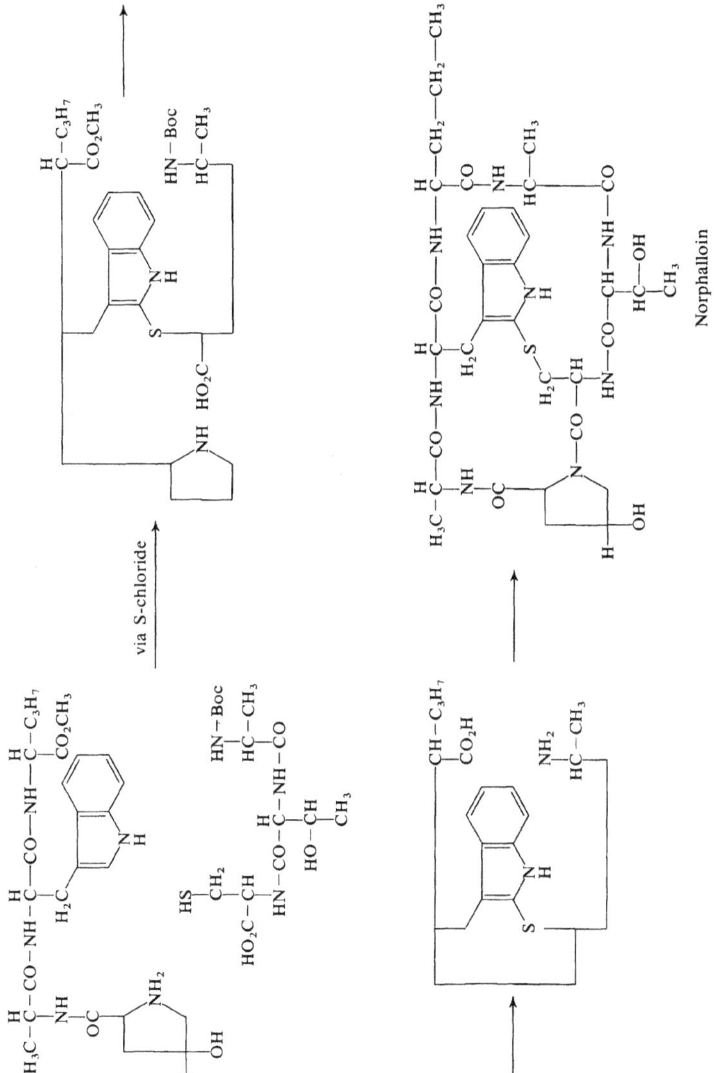

Fig. 10. Outline of the synthesis of norphalloin

VII. The Role of Tryptophan in the Biological Function of Enzymes

Chemical modification of amino acid side chains in enzymes has been carried out to induce changes in the biological activity, in the conformation and the physical properties of an enzyme or to introduce special-purpose groups into a protein (*81, 82, 146, 169, 183, 230, 375, 413*).

Chemical modification has generated the concept of "essential" residues, residues whose modification leads to loss of the biological function of an enzyme. Essential is a neutral term, involving no assumptions about how the biological activity of an enzyme depends on the "essential" residues.

Modification of an amino acid side chain resulting in changes or losses of activity could be related to i) modification of a residue involved in the active site of the enzyme, ii) changes in the binding affinity for the substrate or iii) conformational changes of the gross structure of the enzyme. Meaningful results of modification studies will be achieved only if the effect of the modifier can be assigned to a catalytic site, binding site or conformation-controlling site (*81, 82*).

Chemical reactions can also be used to obtain information on the tertiary structure of proteins, since by chemical methods is possible to label an exposed group in the amino acid sequence, thereby providing determination of its *exposure* and ultimately assessing the nature of its environment (for a detailed discussion on this subject the reader is referred to the review of KRONMAN and ROBBINS (*230*).

Before complete protein structures were available from X-ray crystallography, considerable effort was expended in determining whether particular groups in proteins were on the surface or the interior of the molecule. Although considerable success has been achieved in crystallographic analysis of proteins, relationships between protein conformation in the crystal and in solution are still uncertain (*82, 230*). Therefore covalent structural analysis and side chain modifications will continue to be a major task of the protein chemist.

Several studies of proteins have been made on the assumption that specific reactions occur between the reagents used and the tryptophan residues in proteins. This ideal condition is rarely fulfilled. It is worth recalling that, although extensively used for tryptophan modification studies, NBS is by no means selective for the indole nucleus of tryptophan and also the final modification products cannot be defined unambiguously (*438*). A similar situation is also presented by the Koshland reagent, which was shown to give, upon reaction with model tryptophanyl compounds, up to *ten* different products (*168, 236, 253, 378*).

In conclusion, investigations on the chemical reactivity of tryptophan residues in enzymes in which the possible effect of the reagent used on other side chains in the protein was neglected should be regarded as purely qualitative.

The literature dealing with the chemical modification of tryptophan residues in proteins and enzymes is extensive (*169*). It is beyond the scope of this review article to summarize in detail the results obtained. A compilation of those reagents which are applied must frequently to different enzymes is presented in Table 2. The great number of investigations on the role of tryptophan in enzymes is certainly related to the great reactivity of the indole nucleus and to the numerous available modification techniques. In addition, tryptophan contributes significantly to the optical, stereooptical and fluorescence properties of proteins and these properties are very useful indeed in following the structural changes associated with modification (see Section V.).

Table 2. *Modification of Tryptophan in Enzymes and Proteins*

Protein	Results	Reference
	2-Hydroxy-5-nitrobenzyl bromide	
Malate dehydrogenase	Complex alkylation; 1-Trp-residue per subunit is modified in 8 M urea with inactivation of the enzyme	(*342*)
Cobratoxin	The modification of the single Trp-residue leads to a derivative with lowered protein toxicity	(*75, 76*)
Ribonuclease T_1	The modification of the single Trp-residue abolishes enzyme activity	(*404, 202*)
Erabutoxin a	Alkylation of Trp-29 residue; loss of toxicity	(*355*)
Myelin encephalitogenic basic protein	Single Trp-residue alkylated; loss of neurological activity	(*77*)
Bovine pancreatic deoxyribonuclease	2 of the 3 Trp-residues reacted in the native enzyme, with 25% loss of activity	(*315*)
Aspergillus saitoi Ribonuclease	Complete modification of Trp-residues achieved in 6 M guanidine hydrochloride	(*203*)
Transferrin	Upon modification of the Trp-residue the metal binding efficiency is decreased	(*133*)
α-Lactalbumin	Alkylation with enzyme inactivation occurs at Trp-26, Trp-104 and Trp-118	(*26, 27*)

Table 2 (continued)

Protein	Results	Reference
Neuramidase	Alkylation of Trp-residues determines inactivation; substrate protection has been checked	(*20*)
Lysozyme	Modification of Trp-62 causes loss of activity; reversible inactivation has been observed	(*26–28, 42, 320*)
Porcine pituitary hormone	The single Trp-residue is modified, with complete loss of activity	(*19*)
Heavy meromyosin	Two moles of reagent bound per mole of protein	(*173*)
Glutamate dehydrogenase	After modification of 2 Trp-residues, the association between the subunits is strongly reduced, while activity is only slightly reduced; the modification is followed also by MCD; polyalkylation of the indole nucleus occurs	(*444*)
Carbonic anhydrase	The degree of exposure of Trp-residues is determined using the reagent; only 1 of the 6 Trp-residues is exposed at the surface of the protein	(*235*)
	Dimethyl (2-hydroxy-5-nitrobenzyl) sulfonium salt	
Bovine chymotrypsinogen A	Modification of Trp-215 leads to zymogen inactivation	(*222*)
α-Lactalbumin	Reaction occurs with Trp-26, Trp-104 and Trp-108; 1 residue does not react	(*27–28*)
Yeast transketolase	3 of the 18 Trp-residues have been modified, leading to a decrease in Km	(*167*)
Transferrin	Alkylation of the single Trp-residue has no effect on metal binding	(*133*)
	2-Nitrophenylsulfenyl Chloride	
Staphylococcal nuclease	Modification of the single Trp-residue results in 50% loss of activity; various nitrophenylsulfenyl chlorides are used	(*290*)
Transferrin	The single Trp-residue modification decreases metal binding efficiency	(*133*)
Lysozyme	Selective modification of Trp-62 causes complete loss of activity	(*341*)

Table 2 (continued)

Protein	Results	Reference
α-Lactalbumin	Reaction in 0.1 N acetic acid results in modification of 1 Trp-residue; loss of biological activity in the lactose synthesis reaction	*(357)*
Cobratoxin	Modification of Trp decrease the lethal toxicity of the protein	*(76)*
Myoglobin	The modified Trp-residues are used as convenient markers in chromatography	*(270)*
α-Chymotrypsin	Complete modification of the 8 Trp-residues is achieved in aqueous acetic acid	*(354)*
Bacterial proteinase	All 4 Trp-residues modified in aqueous acetic acid	*(329)*
Cytochrome	Modification of the single Trp-residue together with cleavage of the heme-group	*(432, 128)*
Myelin encephalitogenic basic protein	The single Trp-residue modified; loss of neurological activity	*(77)*
Trypsin	All 4 Trp-residues modified in aqueous acetic acid	*(354)*
Apomyoglobin	After modification of the 2 Trp-residues, enzymic hydrolysis with a mixture of proteolytic enzymes leads to a decreased cleavage of the modified tryptophanyl peptide bond	*(192)*
Human chorionic somatotropin	Modification of Trp-residue brings to a complete loss of lactogenic activity	*(43)*
Human growth hormone	The single Trp-residue is modified in acetate buffer pH 4.0; full growth promoting activity is retained	*(43, 60)*
Luteinizing hormone	Hormonally active derivatives are prepared with modified Trp-residue	*(223)*
Pituitary lactogenic hormone	Selective modification of Trp-149 at pH 3.5–4.0; in aqueous acid Trp-90 and Trp-149 are modified; both derivatives show reduced lactogenic potency	*(223)*
Neocarzinostatin	One of the 2 Trp-residues can be modified with retention of activity; the other residue is buried inside the protein and its modification causes loss of biological activity	*(240, 241)*
Lysozyme	Modification of Trp-108 can be obtained in aqueous acetic acid; almost complete loss of activity	*(48, 418)*

Table 2 (continued)

Protein	Results	Reference
	N-Bromosuccinimide	
Papain	At pH 4–4.5, 3 Trp-residues are modified, while in 8 M urea all 5 residues are oxidized	(*237*)
Cytochrome c	Modification of the single Trp-residue is not selective, but also methionine and tyrosine are attacked by the reagent; spectral studies are performed on the oxidized protein	(*272, 282*)
Lysozyme	Trp-residues are being modified together with other functional groups, including tyrosine	(*231*)
Thioredoxin from *E. coli*	Modification of the 2 Trp-residues causes loss of activity	(*187–189*)
Erabutoxin	The single Trp-residue is modified with loss of lethal activity	(*355*)
	H_2O_2	
α-Chymotrypsin	5 of the 8 Trp-residues are modified	(*161*)
Lysozyme	The degree of exposure of the Trp-residues is determined; different concentrations of H_2O_2 in dioxane are used	(*160*)
Cytochrome c	The single Trp-residue is oxidized; the spectral properties of the modified protein is investigated	(*412*)
Trypsinogen	2 of the 4 Trp-residues are modified	(*160*)
Peroxidase	Oxidation occurs at tryptophan and tyrosine residues, with concomitant loss of activity	(*87*)
Thioredoxin from *E. coli*	The Trp-residue is modified and the spectral properties of the modified protein investigated	(*188–189*)
	Ozone	
α-Chymotrypsin	1 Trp-residue modified without effect on enzymic activity	(*307*)
Trypsin	2 Trp-residues modified with retention of activity	(*307*)

References, pp. 426—449

Table 2 (continued)

Protein	Results	Reference
Lysozyme	Modification of Trp-106 and Trp-108 does not affect the biological activity of the enzyme	(*309*)
Apomyoglobin	After ozonization the tryptophan peptide bond is partially resistant to the enzymic hydrolysis with a mixture of proteolytic enzymes	(*192*)
	Formic acid-HCl	
Cobratoxin	Upon modification of the single Trp-residue the lethal toxicity is reduced	(*75*)
Myelin encephalitogenic basic protein	Modification of the single Trp-residue causes loss of neurological activity	(*77*)
Thioredoxin	Modification of the 2 Trp-residues causes complete inactivation	(*187*)
Cytochrome c	The single Trp-residue is modified, with inactivation	(*17, 282*)
Trypsin	Modification of Trp-residues leads to the loss of proteolytic activity and retention of esterasic activity	(*84*)
	Tetranitromethane	
Staphylococcal nuclease	With excess reagent, the single Trp-residue is nitrated as well as tyrosine; the exact nature of the product(s) is unknown	(*92*)
Lactoferritin	In addition to tyrosine, tryptophan is modified also	(*407*)
Lysozyme	Higher ratios of the reagent to protein lead to the modification of tryptophan and also polymerization of the protein	(*323*)
Papain	Inactivation has been related to the nitration of Trp-residues	(*264*)

Tryptophan contains the largest aromatic side chain found in proteins; its role is presumably limited to structural interactions such as hydrophobic bonds. However, in a few cases this residue was found to be involved in enzyme catalysis and substrate binding. The role of tryptophan residues in the biological activity of lysozyme and dehydrogenases is discussed below in more detail.

1. Lysozyme

The molecule of hen egg-white lysozyme contains six tryptophan residues, three of which (tryptophan 62, 63, and 108) have been located by X-ray crystallography in the region of the binding site of saccharides to the enzyme (*302*). The mechanism put forward for lysozyme catalysis based upon the crystal structure, assumes a catalytic role for glutamic acid 35 (*45*). The tryptophan residue 108 interacts with this glutamic acid residue, the distance between the hetero nitrogen of the indole ring of tryptophan 108 and one of the carboxyl oxygen atoms of glutamic acid 35 has been calculated as 4.9 Å (*45*).

Iodination of lysozyme brings about a complete loss of enzymatic activity (*163*). Recently RUPLEY *et al.* (*163, 199, 200*) have carried out a detailed study of the reaction and identified tryptophan-108 as the single residue modified and therefore responsible for the loss of activity. Reaction of lysozyme with iodine at pH 5.5 gives an enzymatically inactive single oxidized derivative. This intermediate can be converted to an oxindole-lysozyme, its proposed structure being an ester derivative having the carboxyl of glutamic acid 35 covalently bonded to the indole C-2 of tryptophan 108 (*200*).

Tryptophan-62 in lysozyme was found to be very sensitive toward chemical modification. This residue could be selectively modified by N-bromosuccinimide (*162*), iodine (*162*), 2-hydroxy-5-nitrobenzyl bromide (*26*) and 2-nitrophenylsulfenyl chloride (*341*). All these modifications cause inactivation of the enzyme, thus indicating the importance of this residue to its activity. X-Ray studies of crystals of lysozyme and lysozyme-inhibitor complexes indicated that tryptophan-62 takes part in the binding of the substrate saccharide (*45*). Therefore, modification of tryptophan-62 would prevent binding of the substrate, thus inactivating the enzyme.

2. Dehydrogenases

Studies with tritium labelled substrates have shown that labelling occurs at the methylene group of the tryptophan residue in pyridine nucleotide-linked enzymes such as alcohol, lactate and malate dehydrogenase (*9, 343, 344, 345*). The tritium labelling was considered compatible with tryptophan involvement in the enzymatic reaction through dehydrogenation to an indolenine (**213**) cation in the presence of substrate and coenzyme. The mechanistic significance of this tritium labelling would seem to indicate that tryptophan residues in the above mentioned enzymes are implicated in enzymatic catalysis (*344*).

References, pp. 426—449

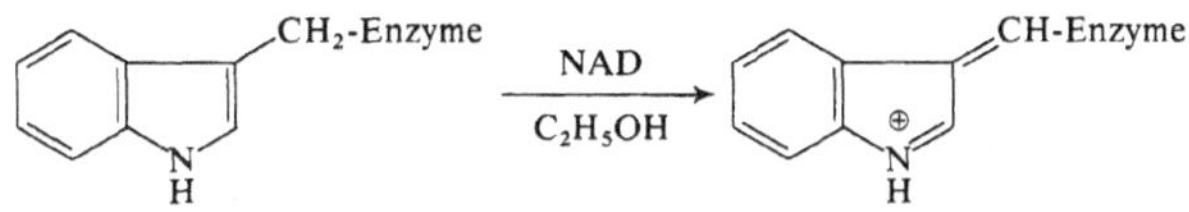

(212) (213)

Hino and Nakagawa (*179*) demonstrated the mechanism suggested above using model reactions. The indolenine (**215**), prepared by

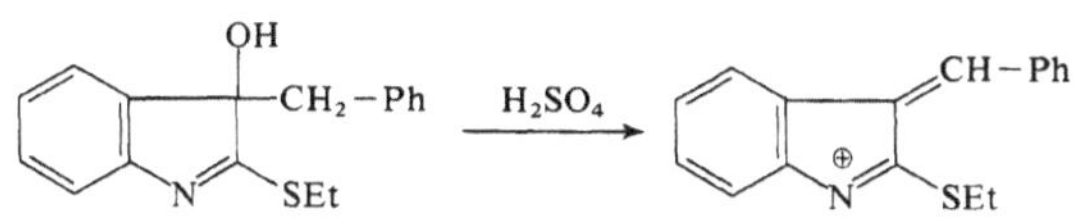

(214) (215)

dehydration of (**214**), was cleanly reduced by the Hantzsch ester (**216**), a model NAD compound, in acetonitrile at room temperature to the

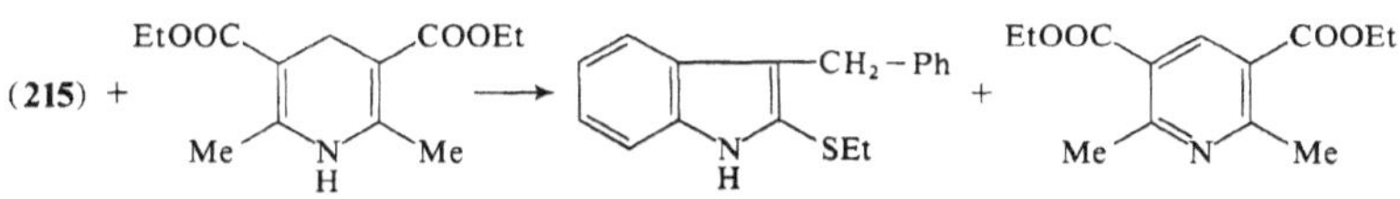

(216) (217) (218)

2-ethylthio-3-benzyl-indole (**217**) and the pyridine derivative (**218**). Oxidation of ethanol to acetaldehyde with the indolenine met some difficulties, but the oxidative power of an indolenine was proved by using the 2-methyl-indolenine (**219**) which was back reduced to the indole by aluminium alkoxide. The formation of acetaldehyde was

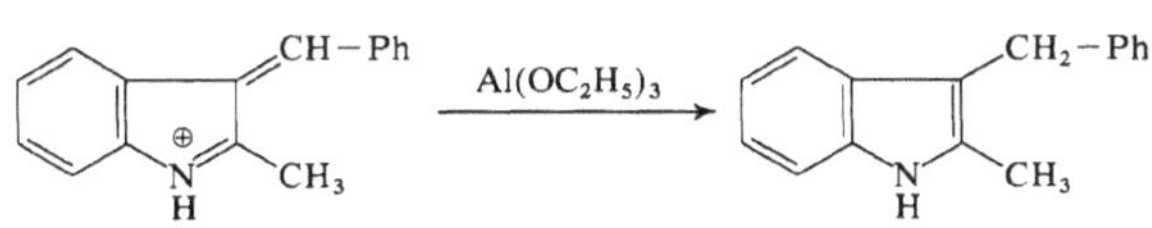

(219) (220)

proved by isolation of its 2,4-dinitrophenylhydrazone. Further evidence for hydrogen transfer from ethanol was demonstrated by reduction of (**219**) with deuterated alkoxide to a deuterium labeled indole compound.

VIII. Conclusion

In this chapter the chemistry, biochemistry and structural characteristics of tryptophan in peptides and proteins have been reviewed with the interests of both biochemists and organic chemists in mind.

The chapter is intended to provide a compact and fairly complete record of the literature on the topics presented. The techniques involved in studying the chemistry of tryptophan in proteins have been discussed with emphasis on the relative advantages and disadvantages of the various available methods and the factors which enable them to be successfully applied.

The chemical techniques used to modify and cleave the tryptophanyl peptide bond are clearly open to development or improvement. The application of indole chemistry (*394*) to complex protein molecules must take account of the presence of a multitude of other reactive functional groups, recognising that chemical events at the indole nucleus might be accompanied by a host of satellite reactions.

It is generally recognised that the scientific frontier of organic chemistry has moved from small molecules to the large macromolecular entities. This chapter will possibly stimulate interest among a wider audience of organic chemists in the problems of contemporary protein chemistry.

Note Added in Proof

The use of BNPS-skatole (**113**) for the selective chemical cleavage of tryptophanyl peptide bonds has been further reported (*459–462, 467*). The selectivity of the reagent and the good yields of peptide bond fission clearly indicate that BNPS-skatole is of general utility for fragmentation of polypeptides and proteins, especially for sequence analysis.

A new general method for oxidation of 3-substituted indoles to the corresponding 2-oxindoles has been recently found (*464*). The method, which is considerably superior to those in current use, consists in treating the indole compound with a mixture of dimethylsulphoxide and concentrated HCl. The yields are high both for simple indoles, such as indolyl-3-propionic acid and tryptophan, and for tryptophan peptides and proteins. The oxidizing species is believed to be the dichlorosulfide $(CH_3)_2SCl_2$ or the derived sulfonium ion $(CH_3)_2S^+Cl$ formed by the interaction of dimethylsulfoxide and HCl (*463, 465*). This should provide a positive halogen source effecting chlorination to an unstable chloroindole which is easily hydrolyzed to the oxindole compound (*395*). The peptide Phe-Val-Gln-Trp-Leu and valine-gramicidin A, a pentadecapeptide containing 4 tryptophan residues per molecule, have been *quantitatively* oxidized to the corresponding oxindolylalanine derivatives. The 6 tryptophan residues of lysozyme have been also successfully modified and the enzymatic and conformational properties of the lysozyme derivative have been investigated.

A mixture of dimethylsulfoxide and HBr caused, on the other hand, oxidation of the indole nucleus to the dioxindole level, the reaction being accompanied by cleavage of the tryptophanyl peptide bond (*464*). Horse heart cytochrome C, containing a single tryptophan residue in position 59 of the polypeptide chain, was treated with this mixture to give modified and uncleaved cytochrome C, fragment 1–59 and fragment 60–104. The yields of cleavage were 60–70%. This new procedure appears particularly promising, in view of the highly selective and effective cleavage.

The reaction of hexahydropyrroloindole (HPI) (**46**) with thiols to give the corresponding 2-thioether-tryptophan compounds has been further investigated (*464*). Reaction of cysteine with HPI (1.2 equiv) in 25% trifluoroacetic acid produces quantitatively tryptathionine, an amino acid contained in the toxic peptides of *Amanita phalloides* (see Section VI.2.4.). Reduced ribonuclease, a protein containing 8 cysteine residues per molecule, was treated with HPI, and the modified protein purified by gel filtration. The completeness of the reaction was confirmed by hydrolysis with *p*-toluenesulfonic acid (*233*) and analysis of the hydrolyzate. A value of 7.6 (theory 8) residues per mole of protein of oxindolylalanine, the product of hydrolysis of the tryptathionine residues (*431*) (see Section III.4.2.), was obtained. This new reaction of cysteine residues should be of value in peptide synthesis, providing a simple method for linking tryptophan and cysteine as a basic step in the chemical synthesis of the peptides of *Amanita phalloides*.

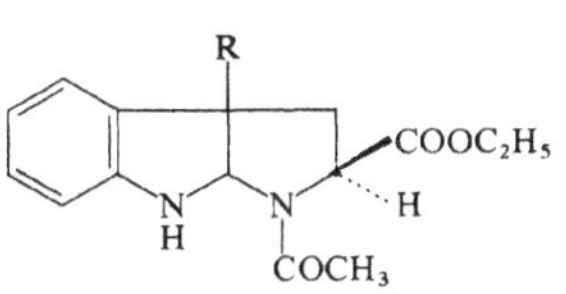

(**221**)

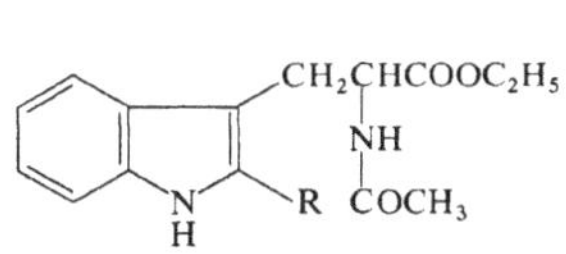

(**222**)

The reaction of the diazonium salt "Fast Red B" (*376*) (see also Section III.12.) with N-acetyl-L-tryptophan ethyl ester has been investigated (*466*). The tricyclic compound (**221**) was isolated, together minor amounts of (**222**) and (**223**). In contrast to the reactions with

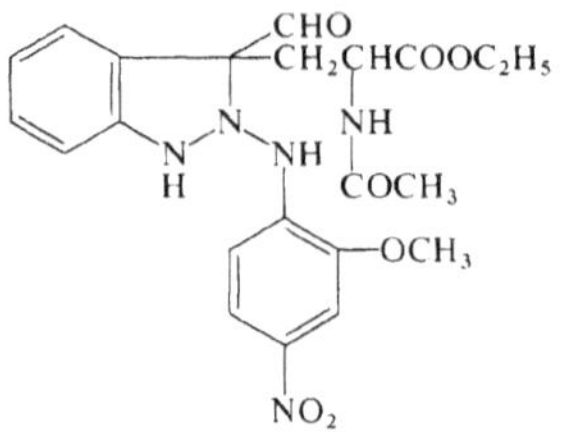

(**223**)

where R = 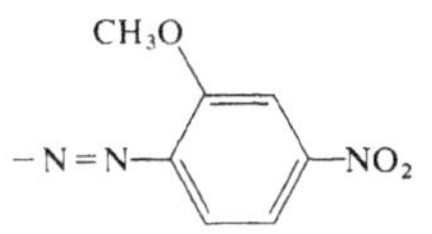

skatole at pH 3 (*376*), in this case the reaction with "Fast Red B" leads also to 5-substituted derivatives as well as 3α,5-disubstituted intermediates.

References

1. ACHESON, R. M.: An Introduction to the Chemistry of Heterocyclic Compounds. Interscience Publ., J. Wiley (1973).

2. AFGHAN, B. K., and J. ISRAELI: Reaction of Metal Nitrilotriacetates with (−)-Tryptophan. Bull. Soc. Chim. Fr. **4**, 1393–1394 (1969).

3. AGARWAL, K. L., G. W. KENNER, and R. C. SHEPPARD: Peptides. XXVIII. Synthesis of Ovine-Bovine Gastrin I. J. Chem. Soc. C **6**, 954–958 (1969).

4. ALAKHOV, YU. B., A. A. KIRYUSHKIN, V. M. LIPKIN, and G. W. A. MILNE: Butylation of the Tryptophan Indole Ring: a Side-Reaction during the Removal of *t*-Butyloxycarbonyl and *t*-Butyl Protecting Group in Peptide Synthesis. J. C. S. Chem. Commun. 406–407 (1970).

5. ALDRICH, J. E., and R. B. CUNDALL: The Radiation-induced Inactivation of Lysozyme. Int. J. Rad. Biol. **16**, 343–358 (1969).

6. ALEXANDER, N. M.: Oxidation and Oxidative Cleavage of Tryptophanyl Peptide Bonds during Jodination. Biochem. Biophys. Res. Commun. **54**, 614–619 (1973).

7. — Oxidative Cleavage of Tryptophanyl Peptide Bonds during Chemical- and Peroxidase-catalyzed Iodinations. J. Biol. Chem. *249*, 1946–1952 (1974).

8. ALLEGRI, G., and A. DE ANTONI: Transamination of Kynurenine. Acta Vitamin. Enzymol. (Milan) **28**, 223–242 (1974).

9. ALLEN, L. M., and W. R. GROVER: Tryptophan Tritiation in Supernatant Malate Dehydrogenase from Pig Heart. Biochem. Biophys. Res. Commun. **41**, 1518–1522 (1970).

10. ANDREWS, L. J., and L. S. FORSTER: Protein Difference Spectra. Effect of Solvent and Charge of Tryptophan. Biochemistry **11**, 1875–1879 (1972).

11. ARAI, S., M. ABE, M. YAMASHITA, H. KATO, and M. FUJIMAKI: Applying Proteolytic Enzymes on Soybean. VIII. Formation of an Indole Derivative by Condensation between Tryptophan and n-Hexanal. Agr. Biol. Chem. **35**, 552–559 (1971).

12. ASQUITH, R. S., and D. E. RIVETT: Studies on the Photooxidation of Tryptophan. Biochim. Biophys. Acta **252,** 111–116 (1971).
13. ATASSI, M. Z.: Specific Cleavage of Tryptophyl Peptide Bonds with Periodate in Sperm Whale Myoglobin. Arch. Biochem. Biophys. **120,** 56–59 (1967).
14. ATASSI, M. Z.: Reaction of Amino Acids and Proteins with Trichloroisocyanuric Acid. Tetrahedron Letters **49,** 4893–4896 (1973).
15. AUDERN, M., and M. SHEINBLATT: NMR Studies of Cyclic Dipeptides containing Histidine and Tryptophan Residues, in: Peptides, Polypeptides and Proteins (E. R. BLOUT, F. A. BOVEY, M. GOODMAN, and N. LOTAN, Eds.), p. 293–299. New York: J. Wiley (1974).
16. AUER, H. E.: Far-Ultraviolet Absorption and Circular Dichroism Spectra of L-Tryptophan and Some Derivatives. J. Amer. Chem. Soc. **95,** 3003–3011 (1973).
17. AVIRAM, I., and A. SCHEJTER: Modification of the Tryptophanyl Residue of Horse Heart Cytochrome C. Biochim. Biophys. Acta **229,** 113–118 (1971).
18. AVRUTSKAYA, I. A., K. K. BABIEVSKII, M. M. BELIKOV, E. V. ZAPOROZHETS, and M. YA. FIOSHCHIN: Electrochemical Reduction of Indolylnitroacrylate in Alcohol Solution. Elekrokhimiya **9,** 163–1368 (1973).
19. BABA, Y., A. ARIMURA, and A. V. SCHALLY: Tryptophan Residue in Porcine LH and FSH-releasing Hormone. Biochem. Biophys. Res. Commun. **45,** 483–487 (1971).
20. BACHMAYER, H.: Effect of Tryptophan Modification on the Activity of Bacterial and Viral Neuramidase. FEBS Lett. **23,** 217–219 (1972).
21. BAJUSZ, S., K. MEDZIHRADSZKY, Z. PALEY, and ZS. LANG: Total Synthesis of Human Corticotropin (α-ACTH). Acta Chim. Acad. Sci. Hung. **52,** 335–341 (1967).
22. BAK, B., C. DAMBMANN, and F. NICOLAISEN: Hydrogen-deuterium Exchange in Tryptophan. Acta Chem. Scand. **21,** 1674–1675 (1967).
23. BAK, B., C. DAMBMANN, F. NICOLAISEN, and E. J. PEDERSEN: Proton Magnetic Resonance Spectra at 220 MH_z of Amino Acids, Porcine and Bovine Insulin and the A and B Chains of Bovine Insulin. J. Mol. Spectr. **26,** 78–97 (1968).
24. BAK, B., J. LED, and E. J. PEDERSEN: Isotopic Labeling of Tryptophan and Tryptophan Residues in Polypeptides. Acta Chem. Scand. **23,** 3051–3054 (1969).
25. BARMAN, T. E., and D. E. KOSHLAND, Jr.: A Colorimetric Procedure for the Quantitative Determination of Tryptophan Residues in Proteins. J. Biol. Chem. **242,** 5771–5776 (1967).
26. BARMAN, T. E.: Reactivities of the Tryptophan Residues of α-Lactalbumin and Lysozyme to 2-Hydroxy-5-Nitrobenzyl Bromide. J. Mol. Biol. **52,** 391–394 (1970).
27. BARMAN, T. E., and W. BAGSHAW: Modification of the Tryptophan Residues of Bovine α-Lactalbumin with 2-Hydroxy-5-nitrobenzyl Bromide and with Dimethyl (2-hydroxy-5-nitrobenzyl) Sulfonium Bromide. II. Effect on the Specific Protein Activity. Biochim. Biophys. Acta. **278,** 491–500 (1972).
28. BARMAN, T. E.: Modification of the Tryptophan Residues of Bovine α-Lactalbumin with 2-Hydroxy-5-nitrobenzyl Bromide and with Dimethyl (2-hydroxy-5-nitrobenzyl) Sulfonium Bromide. I. Characterization of the Modified Protein. Biochim. Biophys. Acta **257** (2), 293–313 (1972).
29. BARTH, G., E. BUNNENBERG, and C. DJERASSI: Magnetic Circular Dichroism Studies. XIX. Determination of the Tyrosine: Tryptophan Ratio in Proteins. Anal. Biochem. **48,** 471–479 (1972).
30. BARTH, G., W. VOELTER, E. BUNNENBERG, and C. DJERASSI: Magnetic Circular Dichroism Studies. XVII. Magnetic Circular Dichroism Spectra of Proteins. A New Method for the Quantitative Determination of Tryptophan. J. Amer. Chem. Soc. **94,** 1293–1298 (1972).
31. BATTERSBY, A. R.: Some Applications of Tritium Labeling for the Exploration of Biochemical Mechanisms. Accounts Chem. Res. **5,** 148–154 (1972).

32. BAYER, E., BACHER, P. KRAUSS, W. VOELTER, G. BARTH, E. BUNNENBERG, and C. DJERASSI: Investigation of Xanthine Oxidase. Magnetic Circular Dichroism Studies. Eur. J. Biochem. **22**, 580–584 (1971).
33. BEACHAM, J., P. H. BENTLEY, R. A. GREGORY, G. W. KENNER, J. K. MACLEOD, and R. C. SHEPPARD: Human Gastrin: Isolation, Structure and Synthesis. Synthesis of Human Gastrin I. Nature **209**, 585–586 (1966).
34. BELLO, J.: Thermal Perturbation Difference Spectra of Proteins Containing Tryptophyl Residues. Biochemistry **9**, 3562–3568 (1970).
35. BELLO, J., and H. R. BELLO: Spectral Studies of Interactions of Detergents with Tryptophyl Compounds. Eur. J. Biochem. **34**, 535–538 (1973).
36. BENASSI, C. A.: Aspects of Tryptophan Metabolism in Man. Res. Progress in Organic-Biological and Medicinal Chemistry, Soc. Editoriale Farm. (Milan) **1**, 45–81 (1964).
37. BENASSI, C. A., E. SCOFFONE, and F. M. VERONESE: Products Arising in the Performic Oxidation of Tryptophan. Tetrahedron Letters **49**, 4389–4393 (1965).
38. BENASSI, C. A., F. M. VERONESE, A. DE ANTONI, and P. PAJETTA: Products Arising from the Performic Acid Oxidation of Tryptophan. Gazz. Chim. Ital. **97**, 1–17 (1967).
39. BENCZE, W. L., and K. SCHMID: Determination of Tyrosine and Tryptophan in Proteins. Anal. Chem. **25**, 1193–1196 (1957).
40. BEN LADKAR, T., M. H. BRIERI, and J. BOLARD: Induced Optical Activity in Donor-Acceptor Complexes. C. R. Acad. Sci., Sez. B **271**, 1201–1203 (1970).
41. BERGSTRAND, H.: Isolation and Partial Characterization of Some Proteolytically and Chemically Derived Fragments of Bovine Encephalitogenic Protein. Eur. J. Biochem. **21**, 116–124 (1971).
42. BEWLEY, T. A., and C. H. LI: Reactivity of the Tryptophan Residues in Lysozyme. Nature **206**, 624 (1965).
43. BEWLEY, T. A., H. KAWAUCHI, and C. H. LI: Human Pituitary Growth Hormone. XXXIII. Comparative Studies of the Single Tryptophan Residue in Human Chorionic Somatotropin and Human Pituitary Growth Hormone. Biochemistry **11**, 4179–4187 (1972).
44. BIRKOFER, L., and K. HEMPEL: Synthesis of Tritium-Labeled Amino Acids of High Specific Activity. Chem. Ber. **96**, 1373–1381 (1963).
45. BLAKE, C. C. F.: A Crystallographic Study of the Oxidation of Lysozyme by Iodine. Proc. Roy. Soc. B **167**, 435–438 (1967).
46. BLOXAM, L. D., and W. H. WARREN: Error in the Determination of Tryptophan by the Method of Denkla and Dewey. A Revised Procedure. Anal. Biochem. **60**, 621–625 (1974).
47. BOCCÙ, E., F. M. VERONESE, A. FONTANA, and C. A. BENASSI: Sulfenyl Halide as Modifying Reagents for Polypeptides and Proteins. VII. Quantitative Evaluation of Tryptophan and Cysteine Residues in Proteins. Eur. J. Biochem. **13**, 188–192 (1970).
48. — — — — Studies on the Function of Tryptophan-108 on Lysozyme. Acta Vitam. Enzymologica, (Milan), **28**, 75–78 (1975).
49. BONNETT, R., and R. HOLLEYHEAD: Reaction of Tryptophan Derivatives with Nitrite. J. Chem. Soc., Perkin I, 962–964 (1974).
50. BOVEY, F. A., and G. V. D. TIERS: Proton N.M.R. Spectroscopy. V. Studies of Amino Acids and Peptides in Trifluoroacetic Acid. J. Amer. Chem. Soc. **81**, 2870–2878 (1959).
51. BOYD, D. R., J. W. DALY, and D. M. JERINA: Rearrangement of [1-^{2}H]- and [2-^{2}H]-Naphthalene 1,2-Oxides to 1-Naphthol. Mechanism of the NIH Shift. Biochemistry **11**, 1961–1966 (1972) and reference therein.

52. BOYLAND, E., P. SIMS, and D. C. WILLIAMS: The Oxidation of Tryptophan and Some Related Compounds with Persulphate. Biochem. J. **62,** 546–550 (1956).
53. BRADBURY, J. H., and R. S. NORTON: Carbon-13 NMR Spectra of Tryptophan, Tryptophan Peptides and of Native and Denaturated Proteins. Biochim. Biophys. Acta **328,** 10–19 (1973).
54. BRADSHAW, R. A., and D. A. DERANLEAU: Use of N-methylnicotinamide Chloride as a Conformational Probe in Proteins. Identification of the Binding Sites in Chicken Egg-white Lysozyme and a Comparison with Bovine α-Lactalbumin. Biochemistry **9,** 3310–3315 (1970).
55. BRAND, L., and B. WITHOLT: Fluorescence Measurements, in: Methods in Enzymology – Enzyme Structure (C. H. W. HIRS, Ed.). **11,** 776–856. London-New York: Academic Press. 1973.
56. BREDDERMAN, P. J.: Tryptophan Analysis of Proteins in 6M Guanidine Hydrochloride: Modification for More General Application. Anal. Biochem. **61,** 298–301 (1974).
57. BRIESKORN, C. H., and K. DANZIGER: Color Reaction of Tryptophan and Glyoxilic Acid Sulfuric Acid. Z. Lebensm.-Unters. Forsch. **137,** 362–370 (1968).
58. BROENTIGAM, K. H., and T. SEVERIN: Maillard Reaction. IX. Reaction of Tryptophan with Xylose. Z. Lebensm.-Unters. Forsch. **154,** 80–83 (1974).
59. BROQUIST, P. H., and J. S. TRUPIN: Amino Acid Metabolism, Ann. Review Biochem. **35,** 231–274 (1966).
60. BROVETTO-CRUZ, J., and C. H. LI: Human Pituitary Growth Hormone. Studies of the Tryptophan Residues. Biochemistry **8,** 4695–4700 (1969).
61. BROWN, R. D., and B. A. W. COLLER: A Theoretical Study of the Chemistry of Furan, Pyrrole, Benzofuran, Indole, Dibenzofuran, and Carbazole. Australian J. Chem. **12,** 152–165 (1959).
62. BRUNDISH, E., D. F. ELLIOTT, and R. WADE: Synthesis of L-Amino Acids and their Derivatives Labelled with Tritium. J. Labelled Compd. **7,** 473–493 (1971).
63. BUCKINGHAM, R. H., and A. PIRIE: Effect of Light on Lens Proteins in Vitro. Exp. Eye Res. **14,** 297–299 (1972).
64. BUKU, A., R. ALTMANN, and TH. WIELAND: Components of the Green Deathcap *Amanita phalloides*. XLVI. The Nontoxic Sulfoxide Diastereoisomeric of O-Methyl-α-amanitin. Liebigs Ann. Chem. **1974,** 1580–1586.
65. BURNETT, P. R., and E. H. EYLAR: Allergic Encephalomyelitis. Oxidation and Cleavage of the Single Tryptophan Residue of the A_1 Protein from Bovine and Human Myelin. J. Biol. Chem. **246,** 3425–3430 (1971).
66. BURSTEIN, Y., and A. PATCHORNIK: Selective Chemical Cleavage of Tryptophanyl Peptide Bonds in Peptides and Proteins. Biochemistry **11,** 4641–4650 (1972).
67. BYE, E., and C. ROEMMING: Crystal Structure of DL-Tryptophan Formate. Acta Chem. Scand. **27,** 471–484 (1973).
68. CAPON, B.: Neighbouring Group Participation. Quarterly Reviews **18,** 45–111 (1964).
69. CASNATI, G., A. DOSSENA, and A. POCHINI: Electrophilic Substitution in Indoles: Direct Attack at the 2-Position of 3-Alkylindoles. Tetrahedron Letters 5277–5280 (1972).
70. CASNATI, G., M. FRANCIONI, A. GUARESCHI, and A. POCHINI: Insertion of Isoprene Units into Indole Systems. Tetrahedron Letters 2485–2487 (1969).
71. CASNATI, G., R. MARCHELLI, and A. POCHINI: Rearrangement of 3-alkyl-1-alkyl-Indoles: A Model Reaction for the Biogenesis of Echinulin-type Compounds. J. Chem. Soc. Perkin I, 754–757 (1974).
72. CAUZZO, G., and G. JORI: Methylene Blue-photosensitized Conversion of 3-substituted Indoles to β-Carboline Derivatives. J. Org. Chem. **37,** 1429–1433 (1972).
73. CAVANAUGH, J. R.: The Rotational Isomerism of the Aromatic Amino Acids by Nuclear Magnetic Resonance. J. Amer. Chem. Soc. **92,** 1488–1493 (1970).

74. CHAN, T. L., and K. A. SCHELLENBERG: Studies on the Presence and Role of Tryptophan in Pig Heart Mitochondrial Malate Dehydrogenase. J. Biol. Chem. **243,** 6284–6290 (1968).
75. CHANG, C., and K. HAYASHI: Chemical Modification of the Tryptophan Residue in Cobratoxin. Biochem. Biophys. Res. Commun. **37,** 841–846 (1969).
76. CHANG, C. C., and C. C. YANG: Immunochemical Studies on Tryptophan Modified Cobratoxin. T'ai-Wan I Hsueh Hui Tsa Chih **71,** 381–382 (1972). Chem. Abstr. **78,** 14382y (1973).
77. CHAO, L., and E. R. EINSTEIN: Localization of the Active Site Through Chemical Modification of the Encephalitogenic Protein. J. Biol. Chem. **245** (23), 6397–6403 (1970).
78. CHATTERJEE, A. K., G. J. DURANT, H. HENDRICKSON, Y. C. LEE, and R. MONTGOMERY: Periodate Oxidation of Biopolymers. Biochem. Biophys. Res. Commun. **4,** 425–430 (1961).
79. CHIBATA, I., and S. YAMADA: Amino Acids. III. Synthesis of DL-tryptophan. Bull. Agr. Chem. Soc. Japan. **21,** 58–67 (1957).
80. CHIBATA, I., and S. YAMADA: Amino Acids. IV. Enzymic Resolution of DL-tryptophan. Bull. Agr. Chem. Soc. Japan. **21,** 62–66 (1967).
81. COHEN, L.: Group-specific Reagents in Protein Chemistry. Ann. Rev. Biochem. **37,** 695–726 (1968).
82. COHEN, L.: Chemical Modifications as a Probe of Structure and Function, in: The Enzymes (P. D. BOYER, Ed.), Vol. I, 148–212. London-New York: Academic Press. 1970.
83. COHEN, J. S.: Proton Magnetic Resonance Studies of Oligopeptides Containing Aromatic Residues. Biochim. Biophys. Acta. **229,** 603–609 (1971).
84. COLETTI-PREVIERO, M. A., A. PREVIERO, and E. ZUCKERKANDL: Separation of the Proteolytic and Esterasic Activities of Trypsin by Reversible Structural Modifications. J. Mol. Biol. **39,** 493–501 (1969).
85. CORNFORTH, J. W., R. H. CORNFORTH, C. E. DALGLIESH, and A. NEUBERGER: D,L-β-3-oxindolylalanine (DL-Hydroxytryptophan). I. Synthesis. Biochem. J. **48,** 591–597 (1951).
86. COSANI, A., E. PEGGION, A. S. VERDINI, and M. TERBOJEVICH: Far-Ultraviolet Optical Rotatory Properties of Poly-L-Tryptophan. Part. I. Biopolymers **6,** 963–971 (1968).
87. COULSON, A., and T. YONETANI: Oxidation of Cytochrome c Peroxidase with Hydrogen Peroxide. Identification of the Endogenous Donor. Biochem. Biophys. Res. Commun. **49,** 391–398 (1972).
88. COY, D. H., E. J. COY, and A. V. SCHALLY: Effect of Simple Amino Acid Replacements on the Biological Function of Luteinising Hormone-Releasing Hormone. J. Med. Chem. **16,** 1140–1143 (1973).
89. COY, D. H., E. J. COY, and A. V. SCHALLY: Solid Phase Synthesis of Luteinizing Hormone-Releasing Hormone and its Analogues, in: Methods in Enzymology (B. W. O'MALLEY and J. G. HARDMAN, Eds.), **37,** part B, 416–423. London-New York: Academic Press. 1975.
90. COY, D. H., E. J. COY, Y. HIROTSU, J. A. VILCHEZ-MARTINEZ, A. V. SCHALLY, J. W. VAN NISPEN, and G. I. TESSER: Investigation of the Role of Tryptophan in the Luteinizing Hormone-Releasing Hormone. Biochemistry **13,** 3550–3553 (1974).
91. COTRAIT, M., and Y. BARRANS: Structure Cristalline de l'Ester Methylique du N-Acetyl-L-tryptophane. Acta Cryst. **B 30,** 510–513 (1974).
92. CUATRECASAS, P., S. FUCHS, and C. B. ANFINSEN: Tyrosyl Residues at the Active Site of Staphylococcal Nuclease. Modifications by Tetranitromethane. J. Biol. Chem. **243,** 4787–4798 (1968).

93. DALGLIESH, C. E.: The Synthesis of N′-Formyl-DL-Kynurenine and Related Compounds, and Observations on the Synthesis of Kynurenine. J. Chem. Soc. **1952,** 137–141.

94. — Nonspecific Formation of Hydroxylated Metabolites of the Aromatic Amino Acids. Arch. Biochem. Biophys. **58,** 214–226 (1955).

95. DALY, J. W., D. M. JERINA, and B. WITKOP: Migration of Deuterim During Hydroxylation of Aromatic Substrates by Liver Microsomes. I. Influence of Ring Substituents. Arch. Biochem. Biophys. **128,** 517–527 (1968).

96. DAVEY, J. M., A. H. LAIRD, and J. S. MORLEY: Polypeptides. Part III. The Synthesis of the C-Terminal Tetrapeptide Sequence of Gastrin, its Optical Isomers, and Acylated Derivatives. J. Chem. Soc. (**C**), 555–566 (1966).

97. DENCKLA, W. D., and H. K. DEWEY: The Determination of Tryptophan in Plasma, Liver and Urine. J. Lab. Chim. Med. **69,** 160–169 (1967).

98. DESCHREIDER, A. R., and M. RENARD: The Influence of Ultraviolet Rays on the Absorption Spectrum of L-Tryptophan. Bull. Inst. Agron. et Stas. Rech. Gembloux **23,** 151–165 (1955), Chem. Abstr. **49,** 15482e (1955).

99. DICKMAN, S. R., and A. L. CROCKETT: Reactions of Xanthydrol. IV. Determination of Tryptophan in Blood Plasma and Proteins. J. Biol. Chem. **220,** 957–965 (1956).

100. DIOPOH, J., and M. OLOMUCKI: New Protein Reagents. II. 4 Chloro-3,5-dinitrophenacyl Bromide. Biochim. Biophys. Acta **263,** 220–225 (1972).

101. DONOVAN, J. W.: Ultraviolet Difference Spectroscopy. New Techniques and Applications, in: Methods in Enzymology (C. H. W. HIRS and S. N. TIMASHEFF, Eds.). **27,** 497–525. London-New York: Academic Press. 1973.

102. DOPHEIDE, T. A. A., and W. M. JONES: Studies on the Tryptophan Residues in Porcine Pepsin. J. Biol. Chem. **243,** 3906–3911 (1968).

103. ECKSTEIN, H., G. BARTH, R. E. LINDER, E. BUNNENBERG, and C. DJERASSI: Untersuchungen des Magnetischen Zirkulardichroism. XXIV. Kinetische Untersuchungen der Tryptophanoxidation in Proteinen. Liebigs Ann. Chem. **1974,** 990–995.

104. EDELHOCH, H.: Spectroscopic Determination of Tryptophan and Tyrosine in Proteins. Biochemistry **6,** 1948–1954 (1967).

105. EDELHOCH, H., R. E. LIPPOLDT, and M. WILCHEK: The Circular Dichroism of Tyrosyl and Tryptophanyl Diketopiperazines. J. Biol. Chem. **243,** 4799–4805 (1968).

106. EDWARDS, B. G.: Ultraviolet Spectra of Some Indole Derivatives, Including Tryptophan and Gramicidin. Arch. Biochem. Biophys. **21,** 103–108 (1949).

107. ELLINGER, A., and C. FLAMAND: Constitution of the Indole Group in Egg Albumin. IV. Preliminary Communication. Synthesis of Racemic Tryptophan. Ber. **40,** 3029–3033 (1907).

108. ELÖDI, P., and S. LAKATOS: Anomalous Absorption of Tryptophan. Eur. J. Biochem. **36,** 45–52 (1973).

109. EK, A., H. KISSMAN, J. B. PATRICK, and B. WITKOP: Chemical Contributions to the Mechanism of the Biological Oxidation of Tryptophan. Experientia **8,** 36–40 (1952).

110. ENFIELD, D. L., L. H. ERICSSON, K. A. WALSH, H. NEURATH, and K. TITANI: Bovine Factor X_1 (Stuart Factor). Primary Structure of the Light Chain. Proc. Natl. Acad. Sci. U.S. **72,** 16–19 (1975).

111. ESKINS, K., and M. FRIEDMAN: Photoaddition of Dimethyl Sulfoxide to Polypeptides. Photochem. Photobiol. **12,** 245–247 (1970).

112. EVANS, N. A.: Reaction of 3-methylindole with Singlet Oxygen. Aust. J. Chem. **24,** 1971–1973 (1971).

113. FAHRENHOLZ, F., H. FAULSTICH, and TH. WIELAND: Components of the Green Deathcap Toadstool, XVII; Peptide Syntheses, XLVIII. Synthesis of Norphalloin and of a Monocyclic Compound with 18-Membered Ring. Liebigs Ann. Chem. **743,** 83–94 (1971).

114. FAULSTICH, H., TH. WIELAND, and C. JOCHUM: Über die Inhaltsstoffe des Grünen Knollenblätterpilzes. XXXIV. Amanin und die Amanitine sind Sulfoxide. Liebigs Ann. Chem. **713,** 186–195 (1968).

115. FAULSTICH, H., E. NEBELIN, and TH. WIELAND: Peptide Syntheses, LIV. Syntheses of Analogs of Norphalloin. Liebigs Ann. Chem. **1973,** 50–58.

116. FASMAN, G. D., and B. BABLOUZIAN: Differential Spectrofluorometry, in: Methods in Enzymology – Enzyme Structure (C. H. W. HIRS and S. N. TIMASHEFF, Eds.). **27,** Part D, 811–822. London-New York: Academic Press. 1973.

117. FINLAYSON, A. J.: Performic Acid Oxidation of Egg-white Lysozyme. Can. J. Biochem. **47,** 31–37 (1969).

118. FOERY, W., R. E. MAC KENZIE, F. Y. WU, and D. B. MC CORMIERE: Flavinyl Peptides. III. Studies of Intramolecular Interactions in Flavinyl Aromatic Amino Acids by Proton Magnetic Resonance. Biochemistry **9,** 515–525 (1970).

119. FONTANA, A.: Modification of Tryptophan with BNPS-Skatole (2-(2-Nitrophenyl-sulfenyl)-3-Methyl-3-Bromoindolenine), in: Enzyme Structure. Methods in Enzymology (C. H. W. HIRS and S. N. TIMASHEFF, Eds.). **25,** 419–423. New York: Academic Press. 1972.

120. FONTANA, A., and E. GROSS: Solid Phase Synthesis of the Pentadecapeptide Val-Gramicidin A. Peptides 1972 (H. HANSON and H. D. JAKUBKE, Eds.). North Holland Publ. Co., 229–234 (1972).

121. FONTANA, A., F. MARCHIORI, L. MORODER, and E. SCOFFONE: New Removal Conditions of Sulfenyl Group in Peptide Synthesis. Tetrahedron Letters **26,** 2985–2987 (1966).

122. FONTANA, A., F. MARCHIORI, R. ROCCHI, and P. PAJETTA: On the Protection of the Amino Group by Sulfenyl Residues. I. Reaction of Tryptophan with Sulfenyl Halides. Gazz. Chim. Ital. **96,** 1301–1312 (1966).

123. FONTANA, A., and E. SCOFFONE: Sulfenyl Halides as Reagents in Peptide and Protein Chemistry. Mechanisms of Reactions of Sulfur Compounds (N. KHARASCH, Ed.), Intra-Science Res. Foundation, Santa Monica, Calif. **4,** 15–24 (1969).

124. — — Sulfenyl Halides as Modifying Reagents for Polypeptides and Proteins, in: Enzyme Structure. Methods in Enzymology (C. H. W. HIRS and S. N. TIMASHEFF, Eds.). **25,** 482–494. New York: Academic Press. 1972.

125. — — Degradation of Polypeptides and Proteins. Elucidation of Organic Structures by Physical and Chemical Methods (K. W. BENTLEY and G. W. KIRBY, Eds.). **4 (II),** 451–548. New York: J. Wiley. 1973.

126. FONTANA, A., E. SCOFFONE, and C. A. BENASSI: Sulfenyl Halides as Modifying Reagents for Polypeptides and Proteins. II. Modification of Cysteinyl Residue. Biochemistry **7,** 980–986 (1968).

127. FONTANA, A., F. M. VERONESE, and E. BOCCÙ: Labelling of the Indole Nucleus of Tryptophan at the 2-Position. Acta Vitam. Enzymologica (Milan) **28,** 79–83 (1975).

128. FONTANA, A., C. VITA, and C. TONIOLO: Selective Cleavage of the Single Tryptophanyl Peptide Bond in Horse Heart Cytochrome C. FEBS Letters **32,** 139–142 (1973).

129. FONTANA, A.: unpublished results (1975).

130. FÖNY, W., R. E. MC KENZIE, F. Y. WU, and D. B. MC CORMICK: Flavinyl Peptides. III. Studies of the Intramolecular Interactions in Flavinyl Aromatic Amino Acids by Proton Magnetic Resonance. Biochemistry **9,** 515–525 (1970).

131. FOOTE, C. S.: Mechanisms of Photosensitized Oxidation. Science **162,** 963–970 (1968).

132. — Photosensitized Oxygenation and the Role of Singlet Oxygen. Acc. Chem. Res. **1,** 104–110 (1968).

133. FORD-HUTCHINSON, A. W., and D. J. PERKINS: Chemical Modification of the Tryptophan Groups of Transferrin. Eur. J. Biochem. **25,** 415–419 (1972).

134. FOSSEL, E. T., W. R. VEATCH, YU. A. OVCHINNIKOV, and E. R. BLOUT: A C^{13} Nuclear Magnetic Resonance Study of Gramicidin A in Monomer and Dimer Films. Biochemistry **13**, 5264–5275 (1974).

135. FRIEDMAN, M., and J. W. FINLEY: Methods of Tryptophan Analysis. Agr. Food Chem. **19**, 626–631 (1971).

136. FRIEDMAN, M., L. H. KRULL, and J. F. CAVINS: Reactions of Amino Acids, Peptides and Proteins with α,β-unsaturated Compounds. XIV. Chromatographic Determination of Cystine and Cysteine Residues in Proteins as S-β-(4-pyridylethyl)-cysteine. J. Biol. Chem. **245**, 3868–3871 (1970).

137. FUKUI, K., T. YONEZAWA, C. NAGATA, and H. SHINGU: Molecular Orbital Theory of Orientation in Aromatic, Heteroaromatic and Other Conjugated Molecules. J. Chem. Phys. **22**, 1433–1442 (1954).

138. GABRIEL, M., D. LARCHIER, H. RINNERT, and C. THIRION: Magnetic Circular Dichroism of Tyrosine and Tryptophan. Variation as a Function of pH. C. R. Acad. Sci. Ser. B **276**, 39–41 (1973).

139. GAITONDE, M. K., and T. DOVEY: A Rapid and Direct Method for the Quantitative Determination of Tryptophan in the Intact Protein. Biochem. J. **117**, 907–911 (1970).

140. GAITONDE, M. K.: A Fluorimetric Method for the Determination of Tryptophan in Animal Tissues. Biochem. J. **139**, 625–631 (1974).

141. GARTLAND, G. L., G. R. FREEMAN, and C. E. BUGG: Crystal Structures of Tryptamine Picrate and D,L-Tryptophan Picrate-Methanol, Two Indole Donor-Acceptor Complexes. Acta Cryst. B **30**, 1841–1849 (1974).

142. GEHRKE, C. W., and H. TAKEDA: Gas-Liquid Chromatographic Analysis of Tryptophan in Proteins. J. Chromat. **76**, 77–89 (1973).

143. GEIGER, R., K. STURM, and W. SIEDEL: Synthesis of a Biologically Active Tricosapeptide Amide with the Amino Acid Sequence 1–23 of Corticotropin (ACTH). Ber. **97**, 1207–1213 (1964).

144. GEIGER, R., W. KÖNIG, H. WISSMANN, K. GEISEN, and F. ENZMANN: Synthesis and Characterisation of a Decapeptide Having LH-RH/FSH-RH Activity. Biochem. Biophys. Res. Commun. **45**, 767–773 (1971).

145. GERIG, J. T. C., and R. A. RIMERMAN: N-Formyl-L-Tryptophan – Chymotrypsin Complex in Solution. Biochem. Biophys. Res. Commun. **40**, 1149–1154 (1970).

146. GLAZER, A. N.: Specific Chemical Modification of Proteins. Ann. Rev. Biochem. **39**, 101–130 (1970).

147. GLICKSON, J. D., C. C. MC DONALD, and W. D. PHILLIPS: Assignment of Tryptophan Indole NH Proton Resonance of Lysozyme. Biochem. Biophys. Res. Commun. **35**, 492–498 (1969).

148. GOMYO, T., and M. FUJIMAKI: Changes of Protein by Dye-sensitized Photooxidation. III. Photodecomposition Products of Lysozyme. Agric. Biol. Chem. Tokyo **34**, 302–309 (1970).

149. GOODMAN, M., and C. TONIOLO: Conformational Studies of Proteins with Aromatic Side Chain Effects. Biopolymers **6**, 1673–1689 (1968).

150. GOODWIN, T. W., and R. A. MORTON: Spectrophotometric Determinations of Tyrosine and Tryptophan in Proteins. Biochem. J. **40**, 628–632 (1946).

151. GOSZTONYI, T., J. MARTON, and A. KOVACS: The Labeling of β-lipoprotein with Tritium. Nature (London) **208**, 381–382 (1965).

152. GOSWAMI, A. K.: Determination of Tryptophan and Indole Substances by a Colorimetric Diazotisation Method. Analyst **99**, 657–660 (1974).

153. GREEN, N. M., and B. WITKOP: Oxidation Studies of Indoles and the Tertiary Structure of Proteins. Trans. N. Y. Acad. Sci. **26**, 659–669 (1964).

154. GREEN, J. P., and J. P. MABRIEN: Quantum Chemical Studies of Charge Transfer Complexes of Indoles. Proc. Natl. Acad. Sci. U.S.A. **54**, 659–664 (1965).

155. GREEN, N. M., and M. D. MELAMED: Optical Rotatory Dispersion, Circular Dichroism and Far-Ultraviolet Spectra of Avidin and Streptavidin. J. Biochem. **100,** 614–621 (1966).
156. GREENSTEIN, J. P., and M. WINITZ: Chemistry of the Amino Acids **3,** 2316–2347. New York: J. Wiley. 1961.
157. GRUEN, L. C.: Effect of other Amino Acids on Recovery of Tryptophan Following Acid Hydrolysis. Aust. J. Biol. Sci. **26,** 287–290 (1973).
158. GRUEN, L. C., and P. W. NICHOLLS: Improved Recovery of Tryptophan Following Acid Hydrolysis of Proteins. Anal. Biochem. **47,** 348–355 (1972).
159. GURNANI, S., M. ARIFUDDIN, and K. T. AUGUSTI: Effect of Visible Light on Amino Acids. I. Tryptophan. Photochem. Photobiol. **5,** 495–505 (1966).
160. HACHIMORI, Y., H. HORINISHI, K. KURIHARA, and K. SHIBATA: States of Amino Acid Residues in Proteins. V. Different Reactivities with H_2O_2 of Tryptophan Residues in Lysozyme, Proteinases, and Zymogens. Biochim. Biophys. Acta **93,** 346–360 (1964).
161. HACHIMORI, Y., K. KURIHARA, H. HORINISHI, A. MATSUSHIMA, and K. SHIBATA: States of Amino Acid Residues in Proteins. VIII. Tyrosine, Histidine, and Tryptophan Residues in Chymotrypsin in the Presence of Substrate and in Diisopropylphosphorylchymotrypsin. Biochim. Biophys. Acta **105,** 167–177 (1965).
162. HARTDEGEN, F. J., and J. A. RUPLEY: Inactivation of Lysozyme by Iodine. Oxidation of a Single Tryptophan. Biochim. Biophys. Acta **92,** 625–627 (1964).
163. HARTDEGEN, F. J., and J. A. RUPLEY: Oxidation of Lysozyme by Iodine: Identification of Oxindolylalanine 108. J. Mol. Biol. **80,** 649–656 (1973).
164. HAYAISHI, O., S. ROTHBERG, A. H. METLER, and Y. SAITO: Studies on Oxygenases. Enzymatic Formation of Kynurenine from Tryptophan. J. Biol. Chem. **229,** 889–896 (1957).
165. HAYAISHI, O.: Enzyme Hydroxylation. Ann. Rev. Biochem. **38,** 21–44 (1969).
166. HEBER, H., H. FAULSTICH, and TH. WIELAND: Syntheses of Further Analogues of Norphalloin. Gly^1-, L-Val^1- and D-Abu^2-Norphalloin and (β-Trideutero)-Ala^5-Norphalloin. Intern. J. Peptide Protein Res. **6,** 381–389 (1974).
167. HEINRICH, C. P., K. NOACK, and O. WISS: Chemical Modification of Tryptophan at the Binding Site of Thiamine Pyrophosphate in Transketolase from Baker's Yeast. Biochem. Biophys. Res. Commun. **49,** 1427–1432 (1972).
168. HEINRICH, C. P., S. ADAM, and W. ARNOLD: The Reaction of Dimethyl (2-Hydroxy-5-Nitrobenzyl) Sulfonium Bromide with N-Acetyl-L-Tryptophan Amide. FEBS Lett. **33,** 181–183 (1973).
169. HEINRIKSON, R. L., and K. J. KRAMER: Recent Advances in the Chemical Modification and Covalent Structural Analysis of Proteins, in: Progress in Biorganic Chemistry (E. T. KAISER and F. J. KÉZDY, Eds.), p. 141–250. New York: J. Wiley. 1974.
170. HÉLÈNE, C., J. L. DIMICOLI, H. N. BORAZAN, M. DURAND, J. C. MAURIZOT, and J. J. TAULME: in: Conformation of Biological Molecules and Polymers. Interactions of Aromatic Amino Acids with Nucleic Acids (E. D. BERGMANN and B. PULLMANN, Eds.), 361–380. Jerusalem: The Israel Academy of Sciences and Humanities. 1973.
171. HERMODSON, M. A., L. H. ERICSSON, H. NEURATH, and K. A. WALSH: Determination of the Amino Acid Sequence of Porcine Trypsin by Sequenator Analysis. Biochemistry **12,** 3146–3153 (1973).
172. HERSKOVITS, T. T.: Difference Spectroscopy in Methods in Enzymology, Enzyme Structure (C. H. W. HIRS, Ed.) **11,** 748–775. London-New York: Academic Press. 1967.
173. HIDENARI, Y., F. MORITA, and K. YAGI: Interaction of Heavy Meromyosin from Substrate. VI. Difference Absorption Spectra Induced by ATP Analogs and

Chemical Modification of Tryptophanyl Residue with 2-hydroxy-5-nitrobenzyl Bromide. J. Biochem. (Tokyo) **72,** 1227–1236 (1972).

174. HILL, R. L., and W. R. SCHMIDT: Complete Enzymic Hydrolysis of Proteins. J. Biol. Chem. **237,** 389–396 (1962).

175. HINNAN, R. L., and C. P. BAUMANN: Reactions of N-Bromosuccinimide and Indoles. A Simple Synthesis of 3-Bromo-oxindoles. J. Org. Chem. **29,** 1206–1215 (1964).

176. HINMAN, L. M., C. R. COAN, and D. A. DERANLEAU: Solution Topography of Proteins by Charge Transfer. Model Complexes, Ribonuclease, and Lysozyme. J. Amer. Chem. Soc. **96,** 7067–7073 (1974).

177. HINO, T., and M. NAKAGAWA: 1-(3-Methyl-2-Indolyl) Pyridinium Bromide. Synthesis and Autoxidation of its Catalytic Hydrogenation Product. Tetrahedron **23,** 1441–1450 (1967).

178. HINO, T., M. NAKAGAWA, and S. AKABOSHI: 2-Ethoxyindoles and 2-Ethylthioindoles: Their Autoxidation and Reactions with Piperidine. J. C. S. Chem. Commun. 656–658 (1967).

179. HINO, T., and M. NAKAGAWA: Oxidation-Reduction of 2-Substituted 3-Benzylindoles and 3-Benzylidene-3 H-indoles. Model Reactions for Alcohol Dehydrogenase. J. Amer. Chem. Soc. **91,** 4598–4599 (1969).

180. HINO, T., H. YAMAGUCHI, and M. NAKAGAWA: Oxidation of 2-Ethylthioindoles with Hydrogen Peroxide. Oxidative Migration of the Ethylsulphinyl Group. J. C. S. Chem. Commun. 473–474 (1972).

181. HIROSHI, O.: Microbial Formation of L-Tryptophan from D-Tryptophan. Agr. Biol. Chem. **32,** 254–256 (1968).

182. HIRS, C. H. W.: Performic Acid Oxidation, in: Methods in Enzymology (C. H. W. HIRS, Ed.) **11,** 197–199. London-New York: Academic Press. 1967.

183. — Methods in Enzymology **11.** London-New York: Academic Press. 1967.

184. HOFMANN, K., J. A. MONTIBELLER, and F. M. FINN: ACTH Antagonists. Proc. Natl. Acad. Sci. U.S. **71,** 80–83 (1974).

185. HOLIDAY, E. R.: Spectrophotometry of Proteins. I. Absorption Spectra of Tyrosine, Tryptophan and Their Mixtures. Biochem. J. **30,** 1795–1799 (1936).

186. HOLLADAY, L. A., R. G. HAMMONDS, JR., and D. PUETT: Growth Hormone Conformation and Conformational Equilibria. Biochemistry **13,** 1653–1661 (1974).

187. HOLMGREN, A.: Reversible Chemical Modification of the Tryptophan Residues of Thioredoxin from *Escherichia coli B*. Eur. J. Biochem. **26,** 528–534 (1972).

188. HOLMGREN, A.: Tryptophan Fluorescence Study of Conformational Transition of the Oxidized and Reduced Form of Thioredoxin. J. Biol. Chem. **247,** 1992–1998 (1972).

189. — Effects of Oxidation of Tryptophan Residues in Thioredoxin from *Escherichia coli* by N-Bromosuccinimide. J. Biol. Chem. **248,** 4106–4111 (1973).

190. HOLMQUIST, B., and B. L. VALLEE: Tryptophan Quantitation by Magnetic Circular Dichroism in Native and Modified Proteins. Biochemistry **12,** 4409–4417 (1973).

191. HOLT, L. A., B. MILLIGAN, and D. F. RIVETT: Tritiation of Tryptophyl Residues in Proteins. Biochemistry **10,** 3559–3564 (1971).

192. HOLT, L. A., and B. MILLIGAN: Application of Enzymic Hydrolysis and Tritium Labelling to a Study of the Modification of Tryptophyl Residues in Proteins. Aust. J. Biol. Sci. **26,** 871–876 (1973).

193. HORTON, H. R., and D. E. KOSHLAND, JR.: Environmentally Sensitive Groups Attached to Proteins, in: Methods in Enzymology (C. H. W. HIRS, Ed.) **11,** 856–870. London-New York: Academic Press. 1967.

194. — — Modification of Proteins with Active Benzyl Halides, in: Methods in Enzymology (C. H. W. HIRS and S. N. TIMASHEFF, Eds.) **11,** 468–482. London-New York: Academic Press. 1972.

195. HORTON, H. R., and G. YOUNG: 2-Acetoxy-5-nitrobenzyl Chloride. A Reagent Designed to Introduce a Reporter Group Near the Active Site of Chymotrypsin. Biochim. Biophys. Acta **194,** 272–278 (1969).

196. HORTON, H. R., and W. P. TUCKER: Dimethyl (2-hydroxy-5-nitrobenzyl) sulfonium Salts. Water Soluble Environmentally Sensitive Protein Reagents. J. Biol. Chem. **245,** 3397–3401 (1970).

197. HUGLI, T. E., and S. MOORE: Determination of the Tryptophan Content of Proteins by Ion Exchange Chromatography of Alkaline Hydrolysates. J. Biol. Chem. **247,** 2828–2834 (1972).

198. IKEDA, K., and K. HAMAGUCHI: Binding of N^1-Methylnicotinamide Chloride to Trp-62 Linked with the Ionization of Glu-35 of Hen Egg-white Lysozyme. J. Biochem. (Tokyo) **74,** 221–230 (1973).

199. IMOTO, T., F. J. HARTDEGEN, and J. A. RUPLEY: Oxidation of Lysozyme by Iodine: Isolation of an Inactive Product and its Conversion to an Oxindolealanine-Lysozyme. J. Mol. Biol. **80,** 637–648 (1973).

200. IMOTO, T., and J. A. RUPLEY: Oxidation of Lysozyme by Iodine: Identification and Properties of an Oxindolyl Ester Intermediate: Evidence for Participation of Glutamic Acid 35 in Catalysis. J. Mol. Biol. **80,** 657–667 (1973).

201. INOUE, M., and H. HAYATSU: Interactions between Bisulfite and Amino Acids. Formation of Methionine Sulfoxide from Methionine in the Presence of Oxygen. Chem. Pharm. Bull. **19,** 1286–1289 (1971).

202. IRIE, M.: State of Tryptophan Residue in Ribonuclease T_1 and Carboxymethyl Ribonuclease T_1. J. Biochem. **68,** 31–37 (1970).

203. IRIE, M., M. HARADA, and F. SAWADA: State of Tryptophan Residues in Ribonuclease from *Aspergillus Saitoi.* J. Biochem. **72,** 1351–1359 (1972).

204. ISBELL, B. E., C. RICE-EVANS, and G. H. BEAVEN: The Conformation of Gramicidin A in Solution. FEBS Letters **25,** 192–196 (1972).

205. JACOBSEN, C.: Chemical Modification of the High-Affinity Bilirubin Binding Site of Human Serum Albumin. Eur. J. Biochem. **27,** 513–519 (1972).

206. JAYSON, G. G., G. SCHOLES, and J. WEISS: Formation of Formylkynurenine by the Action of X-Rays on Tryptophan in Aqueous Solution. Biochem. J. **57,** 386–390 (1954).

207. JACKSON, A. H., B. NAIDOR, and P. SMITH: Electrophilic Substitution in Indoles. IV. The Cyclization of Indolyl-butanole to Tetrahydrocarbazole. Tetrahedron **24,** 6119–6129 (1968).

208. JACKSON, A. H., and A. E. SMITH: Electrophilic Substitution in Indoles. I. Model Experiments Related to the Synthesis of Echinulin. Tetrahedron **21,** 989–1000 (1965).

209. JERINA, D. M., J. W. DALY, B. WITKOP, P. ZALTZMAN-NIREMBERG, and S. UDENFRIEND: The Role of Arene Oxide-Oxepin System in the Metabolism of Aromatic Substrates. Formation of 1,2-Naphthalene Oxide from Naphthalene by Liver Microsomes. J. Amer. Chem. Soc. **90,** 6525–6527 (1968).

210. JORKASKY, D. K., S. E. PEARSON, and C. L. BORDERS, Jr.: Reversible Inactivation of Avian Lysozymes by Dimethyl (2-Hydroxy-5-Nitrobenzyl)-Sulfonium Bromide. Biochem. Biophys. Res. Commun. **52,** 987–991 (1973).

211. JORI, G., M. FOLIN, G. GENNARI, G. GALIAZZO, and O. BUSO: Photooxidation of Lanthanide Ion-Lysozyme Complexes. A New Approach to the Evaluation of Intramolecular Distances in Proteins. Photochem. Photobiol. **19,** 419–434 (1974).

212. JORI, G., and G. GALIAZZO: Proflavine-Sensitized Selective Photooxidation of the Tryptophyl Residues in Papain. Photochem. Photobiol. **14,** 607–619 (1971) and references therein.

213. — — The Contribution of the Single Tryptophyl Residues to the Fluorescence Emission of Porcine Elastase. Z. Naturforsch. **29b,** 261–265 (1974).

214. Jori, G., G. Galiazzo, and O. Buso: Photosensitized Oxidation of Elastase. Selective Modification of the Tryptophyl Residues. Arch. Biochem. Biophys. **158,** 116–125 (1973).

215. Jori, G., G. Galiazzo, and G. Gennari: Photosensitized Conversion of Tryptophan to β-carboline Derivatives. Photochem. Photobiol. **9,** 179–181 (1969).

216. Jorkasky, D. K., S. E. Pearson, and C. L. Borders: Reversible Inactivation of Avian Lysozymes by Dimethyl(2-hydroxy-5-nitrobenzyl)sulfonium Bromide. Biochem. Biophys. Res. Commun. **52,** 987–991 (1973).

217. Julian, P. L., E. W. Meyer, and H. C. Printy: The Chemistry of Indoles, in Heterocyclic Compounds (R. C. Elderfield) **3,** 1–271. New York: J. Wiley. 1952.

218. Julian, P. L., E. E. Dailey, H. C. Printy, H. L. Cohen, and S. Hamashige: Studies in the Indole. Series XVI. Oxindole-3-alanine and Dioxindole-3-alanine. J. Amer. Chem. Soc. **78,** 3503–3508 (1956).

219. Juneja, K. K., A. D. Sule, and V. B. Chipalkatti: Estimation of Tryptophan Content of Wool by the Barium Hydroxide Hydrolysis Method. Text. Res. J. **38,** 461–466 (1968).

220. Kanaoka, Y., E. Sato, and O. Yonemitsu: Amino Acids and Peptides. II. Cyclodehydration of Some Tryptophan-Dipeptides and their Derivatives with Polyphosphate Ester. Tetrahedron **24,** 2591–2594 (1968).

221. Kantouch, A. A., and A. Bendak: Action of Periodic Acid and its Salts on Wool. II. Amino Acids in Periodate-oxidized Wool. Text. Res. J. **39,** 851–858 (1969).

222. Katrukha, G. S., N. Zylber, and V. Keil-Dlouha: Influence of the Substitution of Tryptophan-215 in Bovine Chymotrypsinogen A on its Potential Enzymic Activity. FEBS Lett. **29,** 25–30 (1973).

223. Kawauchi, H., T. A. Bewley, and C. H. Li: Selective Modification of Tryptophan-149 in Ovine Pituitary Lactogenic Hormone. Biochemistry **12,** 2124–2130 (1973).

224. Kearns, D. R.: Physical Chemical Properties of Singlet Molecular Oxygen. Chem. Reviews **71,** 395–427 (1971).

225. Kirby, G. W., and M. J. Varley: Synthesis of Tryptophan Stereoselectively Labelled with Tritium and Deuterium in the β-Methylene Group; the Steric Course of Hydroxylation in Sporidesmin Biosynthesis. J. C. S. Chem. Comm. 833–834 (1974).

226. Kobayashi, T., and N. Inokuchi: A New Method for the Preparation of Oxindoles from Parent Indoles. Tetrahedron **20,** 2055–2058 (1964).

227. Komachiya, Y., S. Suzuki, T. Yamada, H. Miyayashiki, and S. Sakurai: DL-Tryptophan. Nippon Kagaki Zasshi **86,** 856–860 (1965).

228. Korte, K., and U. Schmidt: Syntheses in the Adermine Series IV. Mechanism of the Cysteine and Tryptophan Biosynthesis. Monatsh. Chem. **102,** 207–213 (1971).

229. Kravchenko, N. A., and V. K. Lapuk: Isolation and Characteristics of the Products of Selective Photooxidation of Lysozyme Through Tryptophan. Biokhimiya **34,** 832–838 (1969).

230. Kronman, M. J., and F. M. Robbins: Buried and Exposed Groups in Proteins, in: Fine Structure of Proteins and Nucleic Acids (G. D. Fasman and S. N. Timasheff, Eds.), Marcel Dekker **1970,** 271–408.

231. Kronman, M. J., F. M. Robbins, and R. E. Andreotti: Reaction of N-Bromosuccinimide with Lysozyme. Biochim. Biophys. Acta **147,** 462–472 (1967).

232. Lapuk, V. K., L. A. Chistyakora, and N. A. Kravchenko: Tryptophan Determination in Lysozyme and its Photooxidation Products. Anal. Biochem. **24,** 80–89 (1968).

233. Liu, T.-Y.: Determination of Tryptophan, in: Methods in Enzymology (C. H. W. Hirs and S. N. Timasheff, Eds.) **25,** 44–55, London-New York: Academic Press. 1972.

234. Liu, T. Y., and Y. H. Chang: Hydrolysis of Proteins with p-Toluenesulfonic Acid. Determination of Tryptophan. J. Biol. Chem. **246,** 2842–2848 (1971).

235. LIUDSKOG, S., and H. NILSSON: Location of Tryptophanyl Groups in Human and Bovine Carbonic Anhydrases. Ultraviolet Difference Spectra and Chemical Modification. Biochim. Biophys. Acta **295,** 117–130 (1973).
236. LOUDON, G. M., and D. E. KOHSLAND, JR.: Chemistry of a Reporter Group: 2-Hydroxy-5-nitrobenzyl Bromide. J. Biol. Chem. **245,** 2247–2254 (1970).
237. LOWE, G., and A. S. WHITWORTH: A Kinetic and Fluorimetric Investigation of Papain Modified at Tryptophan-69 and -177 by N-Bromosuccinimide. Biochem. J. **141,** 503–515 (1974).
238. MAEDA, I., and R. YOSHIDA: The Reaction Products of the Hydroformylation of Acrolein Acetals and Acetates. Bull. Chem. Soc. Japan **41,** 2969–2974 (1968).
239. — — Synthesis of DL-Tryptophan from Acrolein. Bull. Chem. Soc. Japan **41,** 2975–2978 (1968).
240. MAEDA, H., and J. MEIENHOFER: On the Tryptophan Content of Neocarzinostatin. Int. J. Protein Res. **2,** 135–136 (1970).
241. MAEDA, H.: Chemical Reactivity and State of Tryptophan Residues in Neocarzinostatin under Physiological Conditions. J. Antibiot. **26,** 776–777 (1973).
242. MAEDA, K., T. MISHIMA, and T. HAGACHI: The Formation of Substituted Quinazolines from Substituted Indoles by the Sensitized Photo-oxygenation in the Presence of Ammonium Acetate. Bull. Chem. Soc. Japan **47,** 334–338 (1974).
243. MAIGRET, B., B. PULLMAN, and D. PERAHIA: Molecular Orbital Calculations of the Conformation of Polypeptides and Proteins. II. Conformational Energy Maps and Stereochemical Rotational States of Aromatic Residues. Biopolymers **10,** 107–128 (1971).
244. MANDEL, M.: Proton Magnetic Resonance Spectra of Some Proteins. I. Ribonuclease, Oxidized Ribonuclease, Lysozyme and Cytochrome C. J. Biol. Chem. **240,** 1586–1592 (1965).
245. MANN, S.: Quinazoline Derivatives in *Pseudomonas Aeruginosa.* Arch. Mikrobiol. **56,** 324–329 (1967).
246. MARCHE, P., J.-P. GIRMA, J.-L. MORGAT, and P. FROMAGEOT: Specific Tritiation of Indole Derivatives by Catalytic Desulfenylation. Application to the Labelling of Tryptophan Containing Peptides. Eur. J. Biochem. **50,** 375–382 (1975).
247. MARCHE, P., J. L. MORGAT, and P. FROMAGEOT: Solvent Effects on Luteinizing- and Follicle-Stimulating-Hormone Releasing Factor. Polymorphism Studied by Circular Dichroism. Eur. J. Biochem. **40,** 513–518 (1973).
248. MARSHALL, G. R.: Studies with Tryptophan in Solid-Phase Synthesis. In: Adv. Exp. Med. Biol. (N. BACK, L. MARTINI, and R. PAOLETTI, Eds.), **II,** Pharmacology of Hormonal Polypeptides and Proteins, 48–58. New York: Plenum Press. 1968.
249. MATSUBARA, H., and R. M. SASAKI: High Recovery of Tryptophan from Acid Hydrolyzates of Proteins. Biochem. Biophys. Res. Commun. **35,** 175–181 (1969).
250. MATSUNAGA, Y.: Charge-Transfer and Proton-Transfer in the Formation of Molecular Complexes. V. Tryptophan Picrate. Bull. Chem. Soc. Japan. **46,** 998–999 (1973).
251. MATSUO, H., A. ARIMURA, R. M. G. NAIR, and A. V. SCHALLY: Synthesis of the Porcine LH- and FSH-releasing Hormone by the Solid-phase Method. Biochem. Biophys. Res. Commun. **45,** 822–827 (1971).
252. MC DONALD, C. C., and W. D. PHILLIPS: Proton Magnetic Resonance Studies of Horse Cytochrome c. Biochemistry **12,** 3170–3186 (1973).
253. MC FARLAND, G., B. Y. INOUE, and K. NAKANISHI: The Reaction of Koshland's Protein Reagent with Tryptophan. Tetrahedron Letters 857–860 (1969).
254. MC FARLAND, T. M., and J. E. COLEMAN: Magnetic Circular Dichroism of Tryptophanyl Residues in Proteins. Azurin and Carbonic Anhydrase. Eur. J. Biochem. **29,** 521–527 (1972).

255. McLaren, A. D.: Photochemistry of Enzymes, Proteins and Viruses. Adv. Enzymology **9**, 75–170 (1969).
256. McLean, S., and G. I. Dimitrienko: Autooxidation of 2,3-Dialkylindoles. Can. J. Chem. **49**, 3642–3647 (1971).
257. Merrifield, R. B.: Solid-phase Peptide Synthesis. Adv. in Enzymology **32**, 221–296 (1969).
258. Messineo, L., and E. Musarra: A Sensitive Spectrophotometric Method for the Determination of Free or Bound Tryptophan. Intern. J. Biochem. **3**, 700–704 (1972).
259. Mingoia, Q.: On Arsenical Derivatives of Pyrrole and Indole. Gazz. Chim. Ital. **60**, 134–140 (1930).
260. Mole, J. E., and H. R. Horton: 2-Chloromethyl-4-nitrophenyl (N-Carbobenzoxy) glycinate. A New Reagent Designed to Introduce an Environmentally Sensitive Conformational Probe near the Active Site of Papain. Biochemistry **12**, 5278–5285 (1973).
261. Monahan, M. W., and J. Rivier: Luteinizing Hormone Releasing Factor. The Total Solid-phase Synthesis on a Benzhydrylamine Polymer. Biochem. Biophys. Res. Commun. **48**, 1100–1105 (1972).
262. Mondino, A., and G. Bongiovanni: Experimental Study of Amino Acid Degradation Under Open Flask Hydrolytic Conditions. J. Chromat. **52**, 405–413 (1970).
263. Moon, K., and E. L. Smith: Sequence of Bovine Liver Glutamate Dehydrogenase. III. Peptides Produced by Specific Chemical Cleavages; the Complete Sequence of the Protein. J. Biol. Chem. **238**, 3082–3088 (1973).
264. Morihara, K., and K. Nagami: Tryptophan Residue in the Active Site of Papain. J. Biochem. (Tokyo) **65**, 321–323 (1969).
265. Morishita, M., and F. Sakiyama: The non-enzymatic Cleavage of Peptide Bonds. II. The Chemical Cleavage of the Tryptophyl Bond in Several Synthetic Peptides. Bull. Chem. Soc. Japan **43**, 524–530 (1970).
266. Morishita, M., F. Sakiyama, and K. Narita: The Chemical Cleavage of the Tryptophyl Bond. Bull. Chem. Soc. Japan **40**, 433–435 (1967).
267. Morley, J. S.: Polypeptides. Part IV. Synthesis of Human Gastrin (HI). J. Chem. Soc. **c 22**, 2410–2421 (1967).
268. Moroder, L., G. Borin, F. Marchiori, and E. Scoffone: Synthetic Peptides Related to the Entire Sequence of Yeast Iso-1-cytochrome c. Biopolymers **11**, 2191–2194 (1972).
269. Morris, A. J., A. J. Geddes, and B. Sheldrick: Cyclo-glycyl-tryptophyl, $C_{13}H_{13}N_3O_2$. Cryst. Struct. Commun. **3**, 345–349 (1974).
270. Mulvey, R. S., and S. Beychok: Further Studies on the Circular Dichroism of Human Lysozyme. Implications for Structure and Comparison of Predicted Secondary Structures in Homologous Lysozymes. Biochemistry **13**, 2980–2985 (1974).
271. Myer, Y. P., and L. H. MacDonald: The Circular Dichroism of L-Tryptophan by an Improved Dichrograph. J. Amer. Chem. Soc. **89**, 7142–7144 (1967).
272. Myer, Y. P., and P. K. Pal: Conformation of Cytochromes. VIII. Spectroscopic Properties of N-Bromosuccinimide-modified Horse Heart Cytochrome C. Biochemistry **11**, 4205–4216 (1972).
273. Nakagawa, M., and T. Hino: 2-Ethoxy and 2-Ethylthioindoles. Autoxidation and Nucleophilic Substitutions. Tetrahedron **26**, 4491–4503 (1970).
274. Nakagawa, M., T. Kaneko, H. Yamagughi, T. Kawashima, and T. Hino: Photoinduced Oxygenation of Tryptamines by Aromatic Amine N-Oxides. Tetrahedron **30**, 2591–2600 (1974).
275. Nakagawa, M., T. Kaneko, K. Yoshikawa, and T. Hino: Photosensitized Oxygenation of Tryptophan Methyl Ester and N_b-Methyltryptamine. Isolation and Identi-

fication of 3_a-Hydroxypyrroloindole and 4_a-Hydroxy-1,2-oxazinoindole. J. Amer. Chem. Soc. **96,** 624–625 (1974).

276. NAKAGAWA, M., H. YAMAGUCHI, and T. HINO: Autoxidation of Indoles Having a Hetero-Atom at 2-Position. Decomposition of the Intermediate 3-Hydroperoxide. Tetrahedron Letters **47,** 4035–4038 (1970).

277. NAKAMURA, A., and O. JARDETZKY: Systematic Analysis of Chemical Shifts in the Nuclear Magnetic Resonance Spectra of Peptide Chains. I. Glycine-Containing Peptides. Proc. Nat. Acad. Sci. U.S. **58,** 2212–2219 (1967).

278. NEDKOV, P., and B. MELOUN: Modification of Tryptophan and Tyrosine Residues of Dolphin Myoglobin. Collect. Czech. Chem. Commun. **34,** 2021–2029 (1969).

279. NEUMANN, N. P.: Oxidation with Hydrogen Peroxide, in: Methods in Enzymology (C. H. W. HIRS and S. N. TIMASHEFF, Eds.) **25,** 393–401. New York: Academic Press. 1972.

280. NORMAN, R. O. C., and R. TAYLOR: Electrophilic Substitution in Benzenoid Compounds, p. 282. Amsterdam: Elsevier. 1965.

281. OELSHLEGEL, F. J., Jr., J. R. SCHROEDER, and M. A. STAHMANN: Simple Procedure for Basic Hydrolysis of Proteins and Rapid Determination of Tryptophan Using a Starch Column. Anal. Biochem. **34,** 331–337 (1970).

282. O'HERN, D. J., P. K. PAL, and Y. P. MYER: Conformational and Functional Studies of Chemically Modified Cytochromes: N-Bromosuccinimide- and Formyl-Cytochromes C. Biochemistry **14,** 382–391 (1975).

283. OHNO, M.: personal communication, reported in ref. [124].

284. OHNO, M., T. F. SPANDE, and B. WITKOP: Cyclization of Tryptophan and Tryptamine Derivatives to 2,3-Dihydropyrrolo [2,3-b] indoles. J. Amer. Chem. Soc. **92,** 343–348 (1970).

285. — — — A New Practical Synthesis of L-2-Hydroxytryptophan and Its Derivatives. J. Org. Chem. **39,** 2635–2637 (1974).

286. OHNO, M., S. TSUKAMOTO, S. SATO, and N. IZUMIYA: Improved Solid-Phase Synthesis of Tryptophan-Containing Peptides. II. Use of N^{α}-*t*-Butyloxycarbonyl-N^{i}-formyltryptophan. Bull. Chem. Soc. Japan. **46,** 3280–3285 (1973).

287. OLCOTT, H. S., and H. FRAENKEL-CONRAT: Formation and Loss of Cysteine During Acid Hydrolysis of Proteins. Role of Tryptophan. J. Biol. Chem. **171,** 583–594 (1947).

288. OLDFIELD, E., and A. ALLERHAND: Identification of Tryptophan Resonance in Natural Abundance Carbon-13 Nuclear Magnetic Resonance Spectra of Proteins. Application of Partially Relaxed Fourier Trasform Spectroscopy. J. Amer. Chem. Soc. **97,** 221–224 (1975).

289. OMENN, G. S., A. FONTANA, and C. B. ANFINSEN: Modification of the Single Tryptophan Residue in Staphylococcal Nuclease by a New Mild Oxidizing Agent. J. Biol. Chem. **245,** 1895–1902 (1970).

290. OMENN, G. S., and I. PARIKH: Modification of Staphylococcal Nuclease with Nitrophenylsulfenyl Halides. Biochemistry **10,** 1173–1177 (1971).

291. OTTENHEYM, J. H.: Status and Prospects for Amino Acid Production by Chemical Synthesis, in: Amino Acid Fortification Protein Foods, Rep. Intern. Conf. **2,** 449–466 (1969).

292. OVERBY, L. R.: Reciprocal Resolution of DL-tryptophan and DL-α-phenylethylamine. J. Org. Chem. **23,** 1393–1394 (1958).

293. PAILTHOPE, M. T., and C. H. NICHOLLS: Indole Nitrogen-Hydrogen Bond Fission During the Photolysis of Tryptophan. Photochem. Photobiol. **14,** 135–145 (1971).

294. PASTERNAK, R. A.: The Crystal Structure of Glycyl-L-Tryptophan Dihydrate. Acta Cryst. **9,** 341–349 (1956).

295. PATCHORNIK, A., W. B. LAWSON, E. GROSS, and B. WITKOP: The Use of N-

Bromosuccinimide and N-Bromoacetamide for the Selective Cleavage of C-Tryptophyl Peptide Bonds in Model Peptides and Glucagon. J. Amer. Chem. Soc. **82**, 5923–5927 (1960).

296. Patel, D. J., L. Kampa, R. G. Shulman, T. Yamane, and B. J. Wyluda: Proton Nuclear Magnetic Resonance Studies of Myoglobin in Water. Proc. Nat. Acad. Sci. U.S. **67**, 1109–1115 (1970).

297. Peggion, E., A. Cosani, A. S. Verdini, A. Del Pra, and M. Mammi: Conformational Studies on Poly-L-Tryptophan: Circular Dichroism and X-Ray Diffraction Studies. Biopolymers **6**, 1477–1486 (1968).

298. Peggion, E., A. Fontana, and A. Cosani: Conformational Studies on a Modified Poly-L-Tryptophan: Circular Dichroism and Optical Rotatory Dispersion of Poly-2-(2-nitrophenylsulfenyl)-L-tryptophan and of Random Copolymers of L-Tryptophan and 2-(2-Nitrophenylsulfenyl)-L-tryptophan. Biopolymers **7**, 517–526 (1969).

299. Penke, B., R. Ferenczi, and K. Kovacs: A New Acid Hydrolysis Method for Determining Tryptophan in Peptides and Proteins. Anal. Biochem. **60**, 45–50 (1974).

300. Peter, G., and B. Rajewsky: The Indirect Action of X-Rays on Amino Acids. II. Irradiation of Tryptophan. Z. Naturforsch. **18b**, 110–114 (1969).

301. Pfaender, P., G. Feige, and Th. Wieland: Indole zum Vergleich mit Amanita-Giften. VI. Hydroxylierung von Tryptophan und Tryptophyl-peptiden. Z. Naturforsch. **23b**, 926–930 (1968).

302. Phillips, D. C.: The Hen Egg-white Lysozyme Molecule. Proc. Nat. Acad. Sci. U.S. **57**, 484–495 (1967).

303. Pirie, A.: Formation of N′-Formylkynuremine in Proteins from Lens and Other Sources by Exposure to Sunlight. Biochem. J. **125**, 203–208 (1971).

304. Pirie, A., and K. J. Dilley: Photo-Oxidation of N′-Formylkynurenine and Tryptophan Peptides by Sunlight or Simulated Sunlight. Photochem. Photobiol. **19**, 115–118 (1974).

305. Pisano, J. J.: Gas-Liquid Chromatography (GLC) of Amino Acid Derivatives, in: Methods in Enzymology (C. H. W. Hirs and S. N. Timasheff, Eds.) **25**, 27–44. London-New York: Academic Press. 1972.

306. Previero, A., and E. Bordignon: Controlled Modification of Tryptophan, Methionine, Cystine and Tyrosine in Peptides and Proteins. Gazz. Chim. Ital. **94**, 630–638 (1964).

307. Previero, A., M. A. Coletti, and L. Galzigna: Modification of Tryptophan Residues in Trypsin, α-Chymotrypsin and Pepsinogen. Biochem. Biophys. Res. Commun. **16**, 195–198 (1964).

308. Previero, A., M. A. Coletti-Previero, and P. Jollès: Nonenzymic Cleavage of Tryptophyl Peptide Bonds in Peptides and Proteins. II. Release of a Chemically Modified Tryptophan Residue from Model Peptides by a Mild Basic Treatment. Biochim. Biophys. Acta **124**, 400–402 (1966).

309. — — — Localization of Non-Essential Tryptophan Residues for the Biological Activity of Lysozyme. J. Mol. Biol. **24**, 261–268 (1967).

310. Previero, A., M. A. Coletti-Previero, and J. C. Cavadore: A Reversible Chemical Modification of the Tryptophan Residue. Biochim. Biophys. Acta **147**, 453–461 (1967).

311. Previero, A., M. A. Coletti-Previero, and P. Jollès: Nonenzymatic Cleavage of Tryptophyl Peptide Bonds in Peptides and Proteins. I. Cleavage of C-tryptophyl Peptide Bonds in Model Peptides through their Conversion into N′-Formylkynurenine Derivatives. Biochem. Biophys. Res. Commun. **22**, 17–21 (1966).

312. Previero, A., G. Prota, and M. A. Coletti-Previero: C-Acylation of the Tryptophan Indole Ring and its Usefulness in Protein Chemistry. Biochim. Biophys. Acta **285**, 269–278 (1972).

313. PREVIERO, A., E. SCOFFONE, C. A. BENASSI, and P. PAJETTA: Modification of Tryptophan Residues in Peptides. Gazz. Chim. Ital. **93**, 849–858 (1963).
314. PONNUSWAMY, P. K., and V. SASISEKHARAN: Conformation of Amino Acids. V. Conformation of Amino Acids with δ-Atoms. Intern. J. Protein Res. **3**, 9–18 (1971).
315. PAULOS, T. L., and P. A. PRICE: Identification of a Tryptophan Residue Essential to the Catalytic Activity of Bovine Pancreatic Deoxyribonuclease. J. Biol. Chem. **246**, 4041–4045 (1971).
316. RAMACHANDRAN, L. K., and B. WITKOP: N-Bromosuccinimide Cleavage of Peptides, in: Methods in Enzymology (C. H. W. HIRS, Ed.) **11**, 283–299. London-New York: Academic Press. 1967.
317. RAMACHANDRAN, J., and V. LEE: Preparation and Properties of the o-Nitrophenyl Sulfenyl Derivative of ACTH: An Inhibitor of the Lipolytic Action of the Hormone. Biochem. Biophys. Res. Commun. **38**, 507–512 (1970).
318. RAMAMURTHY, T. V., and L. PICHAT: Preparation of High Specific Activity DL-Tryptophan-2,3-T and Its Resolution to L-Tryptophan-2,3-T. J. Label. Compounds **9**, 325–330 (1973).
319. RAY, W. J., JR.: Photochemical Oxidation, in: Methods in Enzymology (C. H. W. HIRS, Ed.) **11**, 490–497. London-New York: Academic Press. 1967.
320. REISS, A., and A. LUKTON: Reactivation of Lysozyme from Inactive HNB-Lysozyme. Biorganic Chem. **4**, 1–21 (1975).
321. RICHARDS, R. E., and N. A. THOMAS: A Nitrogen-14 Nuclear Magnetic Resonance Study of Amino Acids, Peptides and other Biologically Interesting Molecules. I. Chem. Soc., Perkin II, 368–374 (1974).
322. RIDD, J. H.: Heteroaromatic Reactivity. Phys. Methods Heterocyclic Chem. **1**, 109–160 (1963).
323. RIGGLE, W. L., J. A. LANG, and C. L. BORDERS, JR.: Reaction of Turkey Egg-white Lysozyme with TNM. Modification of Tyrosin and Tryptophan. Can. J. Biochem. **51**, 1433–1439 (1973).
324. RINIKER, B., M. BRUGGER, B. KAMBER, P. SIEBER, and W. RITTEL: Thyrocalcitonin. IV. Die Totalsynthese des α-Thyrocalcitonins. Helv. Chim. Acta **52**, 1058–1073 (1969).
325. RIVAILLE, R., A. ROBINSON, M. KAMEN, and G. MILHAND: Synthèse en Phase Solide de l'Hormone de Liberation de l'Hormone Lutéotrophique (LM-RH). Helv. Chim. Acta **54**, 2772–2775 (1971).
326. RIVETT, D. E., and J. F. K. WILSHIRE: The Periodate Oxidation of Tryptophan and Some Related Compounds. Aust. J. Chem. **24**, 2717–2722 (1971).
327. RIVETT, D. E.: personal communication.
328. ROBBINS, F. M., and L. G. HOLMES: Binding of N-Methylnicotinamide Chloride by Tryptophan Residues of α-Lactalbumin. J. Biol. Chem. **247**, 3062–3065 (1972).
329. ROBINSON, G. W.: Reaction of a Specific Tryptophan Residue in Streptococcal Proteinase with 2-Hydroxy-5-nitrobenzyl Bromide. J. Biol. Chem. **245**, 4832–4841 (1970).
330. ROSEN, C. G., T. GEJUALL, and L. O. ANDERSSON: Reaction of Diethyl Pyrocarbonate with Indole Derivatives with Special Reference to the Reaction with Tryptophan Residues in a Protein. Biochim. Biophys. Acta. **221**, 207–213 (1970).
331. RUBAN, E. L., N. S. MAMULINA, and N. I. YUKHANOVA: Transformation of D-Tryptophan into L-Tryptophan by *Flavobacterium Butyri 331*. Izv. Akad. Nauk SSSR, Ser. Biol. 150–153 (1972).
332. SAKAMURA, S., and Y. OBATA: A New Chemical Method for the Preparation of Kynurenine by Oxidation with Peracetic Acid. Nippon Nogei-Kagaku Kaishi (Chem. Abstr. **52**, 20296a, 1955) **29**, 817–819 (1955).
333. SAKIYAMA, F., and N. MASUDA: Chemical Cleavage of the Peptide Bond at the Tryptophan Residue. Improved Method. Chem. Lett. **9**, 949–952 (1973).

334. SAMY, T. S. A., M. ATREY, H. MAEDA, and J. MEIENHOFER: Selective Tryptophan Oxidation in the Antitumor Protein Neocarzinostatin and Effects on Conformational and Biological Activity. Biochemistry **13**, 1007–1014 (1974).

335. SANTARIUS, K., und H. D. BELITZ: Eine einfache Methode zur Tryptophan-Bestimmung in Proteinen. Z. Lebensmittel-Unters. und Forsch. **150**, 280–282 (1973).

336. SARGES, R., and B. WITKOP: Gramicidin A. VI. The Synthesis of Valine- and Isoleucine-gramicidin A. J. Amer. Chem. Soc. **87**, 2020–2027 (1965).

337. SASISEKHARAN, V., and P. K. PONNYSWAMY: Studies on the Conformation of Amino Acids. X. Conformations of Norvalyl, Leucyl and Aromatic Side Group in a Dipeptide Unit. Biopolymers. **10**, 583–592 (1971).

338. SAVIGE, W. E.: Isolation and Identification of Some Photo-Oxidation. Products of Tryptophan. Aust. J. Chem. **24**, 1285–1293 (1971).

339. — Paper presented at the 5th Int. Wool Text. Res. Conference, Aachen (1975), Schriftenreihe, in press.

340. — New Oxidation Products of Tryptophan. Aust. J. Chem. **28**, 2275–2283 (1975).

341. SCHECHTER, Y., Y. BURSTEIN, and A. PATCHORNIK: Sulfenylation of Tryptophan-62 in Hen Egg White Lysozyme. Biochemistry **11**, 653–660 (1972).

342. SCHELLENBERG, K. A., and G. W. MC LEAN: Titration of Malate Dehydrogenase with 2-Hydroxy-5-nitrobenzyl Bromide. Biochim. Biophys. Acta **191**, 727–728 (1969).

343. SCHELLENBERG, K. A.: Evidence for the Participation of Tryptophan as Intermediate in Transfer of Hydrogen between Diphosphopyridine Nucleotide and Substrate in Yeast Alcohol Dehydrogenase. J. Biol. Chem. **240**, 1165–1169 (1965).

344. — Participation of Tryptophan Residues in Dehydrogenase Action. II. Position of Tritium-labelled Hydrogen in Yeast Alcohol Dehydrogenase after various Means of Inactivation and Hydrolysis. J. Biol. Chem. **241**, 2446–2452 (1966).

345. — Tryptophan as Intermediate in Dehydrogenase Action. III. Evidence for Complete Cycle of Hydrogen Transfer between Substrate and Tryptophanyl Residues in Rabbit Muscle Lactate Dehydrogenase. J. Biol. Chem. **242**, 1815–1820 (1967).

346. SCHLESSINGER, J., A. GRAFUI, and I. Z. STEINBERG: Optical Rotatory Power in the Ground State and Electronically Excited State of Diketopiperazines Containing Aromatic Side Chains. J. Amer. Chem. Soc. **96**, 7396–7400 (1974).

347. SCHLESSINGER, J., R. S. ROCHE, and I. Z. STEINBERG: A Study of Subtilisin Type Novo and Carlsberg by Circular Polarization of Fluorescence. Biochemistry **14**, 255–262 (1975).

348. SCHMIR, G., L. COHEN, and B. WITKOP: The oxidative Cleavage of Tyrosyl-Peptide Bonds. I. Cleavage of Dipeptides and Some Properties of the Resulting Spirodienone-Lactones. J. Amer. Chem. Soc. **81**, 2228–2233 (1959).

349. SCHRÖDER, E., and K. LÜBKE: Synthese von Peptiden und Peptidwirkstoffen, in: Progress in the Chemistry of Organic Natural Compounds (L. ZECHMEISTER, Ed.) **26**, 48–119. Wien-New York: Springer. 1968.

350. SCHWYZER, R., and P. SIEBER: Total Synthesis of Adrenocorticotropic Hormone. Nature **199**, 172–174.

351. SCHWYZER, R., and H. KAPPELER: Synthesis of a Tetracosapeptide with high corticotropic Activity: β^{1-24}Corticotropin. Helv. Chim. Acta **46**, 1550–1572 (1963).

352. SCHWYZER, R., B. ISELIN, H. KAPPELER, B. RINIKER, W. RITTEL, and H. ZUBER: Synthesis of β-melanotropin (β-MSH) with Amino Acid Sequence of the Bovine Hormone. Helv. Chim. Acta **46**, 1975–1996 (1963).

353. SCOFFONE, E., and A. FONTANA: Identification of Specific Amino Acid Residues. Protein Sequence Determination. Methods and Techniques (S. B. NEEDLEMAN, Ed.), p. 185–210. New York: Springer. 1970.

354. SCOFFONE, E., A. FONTANA, and R. ROCCHI: Sulfenyl Halides as Modifying Reagents

for Polypeptides and Proteins. I. Modification of Tryptophan Residue. Biochemistry **7**, 971–979 (1968).

355. SETO, A., S. SATO, and N. TAMIYA: Properties and Modification of Tryptophan in a Sea Snake Toxin Erabutoxin a. Biochim. Biophys. Acta **214**, 483–489 (1970).

356. SHAW, J. L. D., and W. D. MCFARLANE: The Determination of Tryptophan by Modified Glyoxylic Acid Method Employing Photoelectric Colorimetry. Can. J. Res. **16 B**, 361–368 (1938).

357. SHECHTER, Y., A. PATCHORNIK, and Y. BURSTEIN: Selective Sulfenylation of Tryptophan Residues in α-Lactalbumin of Bovine Milk. J. Biol. Chem. **249**, 413–419 (1974).

358. SHEINBLATT, M.: Determination of Amino Acid Sequence in Di- and Tripeptides by Nuclear Magnetic Resonance Techniques. J. Amer. Chem. Soc. **88**, 2845–2848 (1966).

359. SHELTON, K. R., and K. S. ROGERS: Tryptophanyl Fluorescence of Sodium Dodecyl Sulfate Treated and 2-Mercaptoethanol Reduced Proteins: A simple Assay for Tryptophan. Anal. Biochem. **44**, 134–142 (1971).

360. SHINITZKY, M., Y. DUDAI, and I. SILURAN: Spectral Evidence for the Presence of Tryptophan in the Binding Site of Acetylcholinesterase. FEBS Lett. **30**, 125–128 (1973).

361. SKYE, G. F., R. POTTS, and H. G. FLOSS: Stereochemistry of the Tryptophan Synthetase Reaction. J. Amer. Chem. Soc. **96**, 1593–1595 (1974).

362. SIEVERTSSON, H., J. K. CHANG, C. BOGENTOFT, B. L. CURRIE, K. FOLKERS, and C. Y. BOWERS: Synthesis of the Luteinizing Relasing Hormone of the Hypothalamus and its Hormonal Activity. Biochem. Biophys. Res. Commun. **44**, 1566–1571 (1971).

363. SLIFKIN, M. A.: in: "Charge Transfer Interactions of Biomolecules" (M. A. SLIFKIN, Ed.), Chapter 3, Amino Acids and Proteins, p. 52–75. London-New York: Academic Press. 1971.

364. — in: "Charge Transfer Interactions of Biomolecules" (M. A. SLIFKIN, Ed.), Chapter 5, Indoles and Imidazoles, p. 97–117. London-New York: Academic Press. 1971.

365. SLUMP, P., and H. A. W. SCHRENDER: Determination of Tryptophan in Foods. Anal. Biochem.**0**, 182–186 (1968).

366. SOKOLOVSKY, M., D. HARELL, and J. F. RIORDAN: Reactions of Tetranitromethane with Sulfhydryl Groups in Proteins. Biochemistry **8**, 4740–4745 (1969).

367. SOKOLOVSKY, M., J. F. RIORDAN, and B. L. VALLEE: Tetranitromethane. Reagent for the Nitration of Tyrosyl Residues in Proteins. Biochemistry **5**, 3582–3589 (1966).

368. SOKOLOVSKY, M., M. FUCHS, and J. F. RIORDAN: Reaction of Tetranitromethane with Tryptophan and Related Compounds. FEBS Lett. **7**, 167–170 (1970).

369. SPACKMAN, D. H., W. H. STEIN, and S. MOORE: Automatic Recording Apparatus for Use in the Chromatography of Amino Acids. Anal. Chem. **30**, 1190–1205 (1958).

370. SPANDE, T. F., A. FONTANA, and B. WITKOP: An Unusual Reaction of Skatole with Tetranitromethane. J. Amer. Chem. Soc. **91**, 6199–6200 (1969).

371. SPANDE, T. F., and A. FONTANA: unpublished results, reported in ref. *123* and *375*.

372. SPANDE, T. F., and B. WITKOP: Reactivity toward N-Bromosuccinimide as a Criterion for Buried and Exposed Tryptophan Residues in Proteins, in: Methods in Enzymology (C. H. W. HIRS, Ed.). **11**, p. 522–528. London-New York: Academic Press. 1967.

373. — — Determination of the Tryptophan Content of Proteins by N-Bromosuccinimide, in: Methods in Enzymology (C. H. W. HIRS, Ed.). **11**, p. 498–506. London-New York: Academic Press. 1967.

374. — — Tryptophan Involvement in Binding Sites of Proteins and in Enzyme-Inhibitor

Complexes as Determined by Oxidation with N-Bromosuccinimide, in: Methods in Enzymology (C. H. W. HIRS, Ed.). **11**, p. 506–522. London-New York: Academic Press. 1967.

375. SPANDE, T. F., B. WITKOP, Y. DEGANI, and A. PATCHORNIK: Selective Cleavage and Modification of Peptides and Proteins. II. Advan. Protein Chem. **24**, 97–260 (1970).

376. SPANDE, T. F., and G. G. GLENNER: The Reaction of Indoles with a Diazonium Salt (Fast Red B). J. Amer. Chem. Soc. **95**, 3400–3401 (1973).

377. SPANDE, T. F., N. M. GREEN, and B. WITKOP: The Reactivity toward N-bromosuccinimide of Tryptophan in Enzymes, Zymogens, and Inhibited Enzymes. Biochemistry **5**, 1926–1933 (1966).

378. SPANDE, T. F., M. WILCHEK, and B. WITKOP: The Reaction of Derivatives of Tryptophan, Tryptamine, and Other Indoles with 2-Hydroxy-5-nitrobenzyl Bromide (Koshland's Reagent). J. Amer. Chem. Soc. **90**, 3256–3258 (1968).

379. SPIES, J. R., and D. C. CHAMBERS: Determination of Tryptophan with p-Dimethylaminobenzaldehyde Using Photochemical Development of Color. Anal. Chem. **22**, 1209–1210 (1950).

380. — — Chemical Determination of Tryptophan in Proteins. Anal. Chem. **21**, 1249–1266 (1949).

381. STEINBERG, I. Z., J. SCHLESSINGER, and A. GAFIN: Application of Circular Polarization of Luminescence to the Study of Peptides, Polypeptides and Proteins, in: Peptides, Polypeptides, and Proteins (E. R. BLOUT, F. A. BOVEY, M. GOODMAN, and N. LOTAN, Eds.), p. 351–369. New York-London: J. Wiley. 1974.

382. STEITZ, T. A., R. HENDERSON, and D. M. BLOW: Structure of Crystalline α-Chymotrypsin. III. Crystallographic Studies of Substrates and Inhibitors Bound to the Active Site of α-Chymotrypsin. J. Mol. Biol. **46**, 337–348 (1969).

383. STELLWAGEN, E., and S. VAN ROOYAN: The Structural Environement of the Tryptophanyl Residue of Horse Heart Ferricytochrome C. J. Biol. Chem. **242**, 4801–4805 (1967).

384. STEVENS, L., R. TOWNEND, S. N. TIMASHEFF, G. D. FASMAN, and J. POTTER: The Circular Dichroism of Polypeptide Films. Biochemistry **7**, 3717–3720 (1968).

385. STEWART, M., and C. H. NICHOLLS: Kinetic Study of the Acid Autoxidation of Tryptophan. Aust. J. Chem. **25**, 2139–2144 (1972).

386. STEWART, M., and C. H. NICHOLLS: A Kinetic Study of the Alkali Degradation of Tryptophan. Aust. J. Chem. **25**, 1595–1599 (1972).

387. STEWART, M., and C. H. NICHOLLS: A kinetic Study of the Autoxidation of Tryptophan at 100° C in the pH Range 2–7. Aust. J. Chem. **27**, 205–208 (1974).

388. STÖHRER, G., G. SALEMNICK, and G. B. BROWN: A Chemical Adduct of Tryptophan and the Oncogen 3-Acetoxyxanthine. Biochemistry **12**, 5084–5086 (1973).

389. STRICKLAND, E. H., M. WILCHEK, J. HORWITZ, and C. BILLUPS: Low Temperature Circular Dichroism of Tyrosyl and Tryptophanyl Diketopiperazines. J. Biol. Chem. **245**, 4168–4177 (1970).

390. STRICKLAND, E. H., C. BILLUPS, and E. KAY: Effects of Hydrogen Bonding and Solvents upon the Tryptophanyl ^1L_a Absorption Band-Studies Using 2,3-Dimethylindole. Biochemistry **11**, 3657–3662 (1972).

391. STRICKLAND, E. H., and C. BILLUPS: Oscillatory Strengths of the ^1L_a and ^1L_b Absorption Bands of Tryptophan and Several Other Indoles. Biopolymers **12**, 1989–1995 (1973).

392. STRICKLAND, E. H., J. HORWITZ, and C. BILLUPS: Fine Structure in the Near-Ultraviolet Circular Dichroism and Absorption Spectra of Tryptophan Derivatives and Chymotrypsinogen A at 77° K. Biochemistry **8**, 3205–3213 (1969).

393. SUMPTER, W. C., and M. MILLER: The Chemistry of Heterocyclic Compounds (A. WEISSBERGER, Ed.), Interscience **8**. New York: J. Wiley. 1954.

394. SUNDBERG, R. J.: The Chemistry of Indoles. London-New York: Academic Press. 1970.

395. — The Chemistry of Indoles, p. 17–18. London-New York: Academic Press. 1970.

396. — The Chemistry of Indoles, p. 19–31. London-New York: Academic Press. 1970.

397. — The Chemistry of Indoles, p. 39–56. London-New York: Academic Press. 1970.

398. — The Chemistry of Indoles, p. 78–83. London-New York: Academic Press. 1970.

399. — The Chemistry of Indoles, p. 83–85. London-New York: Academic Press. 1970.

400. — The Chemistry of Indoles, p. 142–163. London-New York: Academic Press. 1970.

401. — The Chemistry of Indoles, p. 230–235. London-New York: Academic Press. 1970.

402. — The Chemistry of Indoles, p. 282–312. London-New York: Academic Press. 1970.

403. SUZUKI, S., Y. HACHIMORI, and U. YAOEDA: Spectrophotometric Determination of Glycine with 2,4,6-trichloro-5-triazine. Anal. Chem. **42,** 101–103 (1970).

404. TAKAHASHI, K.: Structure and Function of Ribonuclease T_1. VIII. Reaction of 2-hydroxy-5-nitrobenzyl Bromide with the Single Tryptophan Residue in Ribonuclease T_1. J. Biochem. (Tokyo) **67,** 541–547 (1970).

405. TAKIGAWA, T., T. ASHIDA, Y. SASADA, and M. KAKUDO: The Chrystal Structures of L-Tryptophan Hydrochloride and Hydrobromide. Bull. Chem. Soc. Japan **39,** 2369–2378 (1966).

406. TERUI, G.: Tryptophan, in: The Microbial Production of Amino Acids (YAMADA, K., S. KINOSHITA, T. TSUNODA, and K. AIDA, Eds.), p. 515–531. New York: J. Wiley. 1974.

407. TEUWISSEN, B., P. L. MASSON, P. OSINSKI, and J. F. HERMANS: Metal-combining Properties of Human Lactoferritin. Effect of Nitration of Lactoferritin with Tetranitromethane. Eur. J. Biochem. **35,** 366–371 (1973).

408. TONIOLO, C.: Relationship among Primary Structures, Conformations, and Biological Activities of Polypeptide Antibiotics. Il Farmaco, Ed. Sci. **26,** 741–769 (1971).

409. TURNER, D. H., I. TINOCO, JR., and M. MAESTRE: Fluorescence Detected Circular Dichroism. J. Amer. Chem. Soc. **96,** 4340–4342 (1974).

410. UDENFRIEND, S.: Proteins and Peptides, in: Fluorescence Assay in Biology and Medicine, **2,** Chapter 6, p. 248–283. London-New York: Academic Press. 1969.

411. — Amino Acids, Amines, and their Metabolites, in: Fluorescence Assay in Biology and Medicine. **2,** Chapter 5, p. 195–247. London-New York: Academic Press. 1969.

412. ULMER, D. D.: Tyrosine and Tryptophan in Cytochrome C. Biochemistry **5,** 1886–1892 (1966).

413. VALLEE, B. L., and J. F. RIORDAN: Chemical Approachs to the Properties of Active Sites of Enzymes. Ann. Rev. Biochem. **38,** 733–794 (1969).

414. VEATCH, W. R., E. T. FOSSEL, and E. R. BLOUT: The Conformation of Gramicidin A. Biochemistry **13,** 5249–5256 (1974).

415. VERONESE, F. M., E. BOCCÙ, C. A. BENASSI, and E. SCOFFONE: Preferential Hydrolysis of Kynurenine Peptides. Z. Naturforsch. **24,** 294–300 (1969).

416. VERONESE, F. M., E. BOCCU, and A. FONTANA: Sulfenyl Halides as Modifying Reagents for Polypeptides and Proteins. V. The Use of 2-Nitro-4-Carboxyphenylsulfenyl Chloride. Ann. Chim. (Rome) **58,** 1309–1319 (1968).

417. — — — Sulfenyl Halides as Modifying Reagents for Polypeptides and Proteins. VI. 2,4-Dinitrophenyl-1,5-disulfenyl Chloride, a Cross-linking Reagent for Tryptophan Residues. Int. J. Protein Res. **2,** 67–74 (1970).

418. — — — Modification of Tryptophan-108 of Lysozyme with 2-Nitro-4-carboxyphenylsulfenyl Chloride. FEBS Lett. **21,** 277–280 (1972).

419. VERONESE, F. M., A. FONTANA, E. BOCCÙ, and C. A. BENASSI: Synthesis of Kynurenine Peptides. Gazz. Chim. Ital. **97,** 321–331 (1967).

420. — — — — Sulfenyl Halides as Modifying Reagents for Polypeptides and Proteins.

IV. On the Conversion of 2-Thioaryl-Tryptophan to 2-Hydroxy-tryptophan. Z. Naturforsch. **23b,** 1219–1325 (1968).

421. Voelkl, A., and G. Quadbeck: New Color Reaction to Identify Indoles. Ärztl. Lab. **19,** 45–50 (1973).

422. Voetter, W., G. Jung, E. Breitmaier, and E. Bayer: Carbon-13 NMR Chemical Shifts of Amino Acids and Peptides. Z. Naturforsch. **26b,** 213–222 (1971).

423. Walrant, P., and R. Santus: N-Formylkynurenine, a Tryptophan Photooxidation Product, as a Photodinamic Sensitizer. Photochem. Photobiol. **19,** 411–417 (1974).

424. Walrant, P., and R. Santus: Ultraviolet and N-Formyl-kynurenine Sensitized Photoinactivation of Bovine Carbonic Anhydrase: an Internal Photodinamic Effect. Photochem. Photobiol. **20,** 455–460 (1974).

425. Weil, L., W. G. Gordon, and A. R. Buchert: Photooxidation of Amino Acids in the Presence of Methylene Blue. Arch. Biochem. Biophys. **33,** 90–109 (1951).

426. Westhead, E. K.: Dye-Sensitized Photooxidation, in: Methods in Enzymology (C. H. W. Hirs and S. N. Timasheff, Eds.), **25,** p. 401–409. London-New York: Academic Press. 1972.

427. Wieland, T. H.: Aspects on Syntheses of Biologically Active Peptides, in: Peptides, Proc. 12th Eur. Symp. Reinhardsbrunn Castle 1972 (H. Hanson and H.-D. Jakubke, Eds.), p. 38–49. Amsterdam: North Holland Publ. Co. 1973.

428. Wieland, Th., C. Jochum, and H. Faulstich: Optimization of the Synthesis of 2-Indole Thioethers from Derivatives of Tryptophan and Cysteine. Liebigs Ann. Chem. **727,** 138–142 (1969).

429. Wieland, Th., M. P. Jordan de Urries, H. Indest, H. Faulstich, A. Gieren, M. Sturm, and W. Hoppe: Components of the Green Deathcap Toadstoal, *Amanita phalloides*, XLV. The Absolute Configurations of the Toxic and Nontoxic Phalloidin Sulfoxide and of the Amatoxins. Liebigs Ann. Chem. **1974,** 1570–1579.

430. Wieland, Th., and B. Sarges: Synthesis of Tryptathionine Peptides. Liebigs Ann. Chem. **658,** 181–193 (1962).

431. Wieland, Th., O. Weiberg, W. Dilger, and E. Fischer: Synthesis of 2-hydroxy-tryptophan and 2-mercaptotryptophan. Liebigs Ann. Chem. **592,** 69–80 (1955).

432. Wilchek, M., and T. Miron: The Isolation of Tryptophan-containing Peptides by Affinity Chromatography. Biochim. Biophys. Acta **278,** 1–7 (1972).

433. — Conversion of Tryptophan to 2-Thioltryptophan in Peptides and Proteins. Biochem. Biophys. Res. Commun. **47,** 1015–1020 (1972).

434. Wilchek, M., T. Spande, G. Milne, and B. Witkop: The Nonenzymatic Conversion of Tyrosine into Mono- and Dihydroxyindoles. Biochemistry **7,** 1777–1786 (1968).

435. Wilchek, M., and B. Witkop: Protection of Tryptophan During Cleavage of Tyrosine Peptide Bonds by N-Bromosuccinimide. Biochem. Biophys. Res. Commun. **26,** 296–300.

436. Williams, D. C., and J. R. Whitaker: Kinetics of Papain-catalyzed Hydrolyses of Neutral Substrates. Biochemistry **6,** 3711–3717 (1967).

437. Witkop, B.: Directed Oxidations in the Indole Series. II. A New Method of Reaction of Perbenzoic Acid. Liebigs Ann. Chem. **558,** 91–98 (1947).

438. — Selective Modification and Cleavage of Proteins. Advan. Protein Chem. **16,** 221–321 (1961).

439. — Miscellaneous Nonenzymatic Cleavages of the Peptide Bond, in: Methods in Enzymology (C. H. W. Hirs, Ed.), **11,** 308–315. London-New York: Academic Press. 1967.

440. — Chemical Cleavage of Proteins. Selective Fragmentations Reveal Structure. Science **162,** 318–322 (1968).

441. Witkop, B., and G. Graser: Decomposition of Indole Derivatives by Means of Ozone. Liebigs Ann. Chem. **556,** 103–114 (1944).

442. WITKOP, B., and J. B. PATRICK: Mechanism of Oxidation. I. New Type of Peroxide Rearrangement Catalyzed by Acid. J. Amer. Chem. Soc. **73**, 2196–2200 (1951).

443. WITKOP, B., and L. K. RAMACHANDRAN: Progress in Nonenzymic Selective Modification and Cleavage of Proteins. Metabolism **13**, 1016–1025 (1964).

444. WITZEMANN, V., R. KOBERSTEIN, H. SUND, I. RASCHED, H. JORNVALL, and K. NOACK: Studies on Glutamate Dehydrogenase: Chemical Modification and Quantitative Determination of Tryptophan Residues. Eur. J. Biochem. **43**, 319–325 (1974).

445. WOLF, H.: Studies on Tryptophan Metabolism in Man, Virum Costers Bogstrykkeri (1974).

446. WÜNSCH, E.: Die Totalsynthese des Pankreas-Hormons Glucagon. Z. Naturforsch. **226**, 1269–1276 (1967).

447. WÜNSCH, E., A. FONTANA und F. DREES: Zur Entacyclierung von Nα-(2-Nitrophenylsulfenyl)-Peptiden bei Anwesenheit von Tryptophan in der Peptidsequenz. Z. Naturforsch. **22b**, 607–609 (1967).

448. YAMADA, S., M. YAMAMOTO, and I. CHIBATA: Optical Resolution of DL-Amino Acids by Preferential Crystallization Procedure. J. Org. Chem. **38**, 4408–4412 (1973).

449. YAMADA, H., H. YOSHIDA, H. NAKAZAWA, and H. KUMAGAI: Microbiological Synthesis of Tryptophan and its Related Amino Acids. 1st Inter. Symp. on Tryptophan Metabolism, Acta Vitam. et Enzymologica (Milan) (1975), in press.

450. YAMAMOTO, Y., and J. TANAKA: Polarized Absorption Spectra of Crystals of Indole and its Related Compounds. Bull. Chem. Soc. Japan. **45**, 1362–1366 (1972).

451. YAMASHIRO, D., and C. H. LI: Protection of Tryptophan with the Formyl Group in Peptide Synthesis. J. Org. Chem. **38**, 2594–2597 (1973).

452. YANAIHARA, N., T. HASHIMOTO, C. YANAIHARA, K. TSUJI, Y. KENMOCHI, F. ASHIZAWA, T. KANEKO, H. OKA, S. SAITO, A. ARIMURA, and A. V. SCHALLY: Syntheses and Biological Evaluation of Analogs of Luteinizing Hormone-Relasing Hormone (LH-RH) modified in Position 2, 3, 4 or 5. Biochem. Biophys. Res. Commun. **52**, 64–73 (1973).

453. YANAIHARA, N., C. YANAIHARA, G. DUPUIS, J. BEACHAM, R. CAMBLE, and K. HOFMANN: Studies on Polypeptides. XLII. Synthesis and Characterization of Seven Fragments Spanning the Entire Sequence of Ribonuclease T_1. J. Amer. Chem. Soc. **91**, 2184–2185 (1969).

454. YONEMITSU, O., K. MIYASHITA, Y. BAN, and Y. KANAOKA: Polyphosphate Ester as a Synthetic Agent. XI. A Novel Synthesis of Ethylindolenines by Ethylation of 2,3-Disubstituted Indoles with Polyphosphate Ester. Tetrahedron **25**, 95–100 (1969).

455. YUTAKA, H., H. HIROO, K. KENZO, and S. KAZOO: States of Amino Acid Residues in Proteins. V. Different Reactivities with H_2O_2 of Tryptophan Residues in Lysozyme, Proteinases, and Zymogens. Biochim. Biophys. Acta **93**, 346–360 (1964).

456. ZAHNLEY, J. C., and J. G. DAVIS: Effect of High Tyrosine Content on the Determination of Tryptophan in Protein by the Acidic Ninhydrin Method. Application to Chicken Ovoinhibitor. Biochem. J. **135**, 59–61 (1973).

457. ZAPOROZHETS, E. V., I. A. AVRUTSKAYA, K. K. BABIEVSKII, V. M. BELIKOV, and M. FIOSHIN: Electrochemical Reduction of the Methyl Ester of α-nitro-β(3-indolyl)-acrylic acid. Elektrokhimiya **8**, 1243–1245 (1972).

458. ZAPOROZHETS, E. V., I. A. AVRUTSKAYA, M. YA. FIOSHIN, K. K. BABIEVSKII, and V. M. BELIKOV: Effect of Electrode Material on the Electroreduction of Indolylnitroacrylate. Elektrokhimiya **9**, 72–76 (1973).

459. ENFIELD, D. L., L. H. ERICSSON, K. A. WALSH, H. NEURATH, and K. TITANI: Bovine Factor X_1 (Stuart Factor). Primary Structure of the Light Chain. Proc. Nat. Acad. Sci. USA **72**, 16—19 (1975).

460. JESTY, J., A. K. SPENCER, Y. NAKASHIMA, Y. NEMERSON, and W. KONIGSBERG: The Activation of Coagulation Factor X. Identity of Cleavage Sites in the Alternative

Activation Pathways and Characterization of the COOH– Terminal Peptide. J. Biol. Chem. **250,** 4497—4504 (1975).

461. KARKHANIS, Y. D., D. J. CARLO, S. W. BROSTOFF, and E. H. EYLAR: Allergic Encephalomyelitis. Isolation of an Encephalitogenic Peptide Active in the Monkey. J. Biol. Chem. **250,** 6141—6150 (1975).

462. LEVY, P. L., M. K. PANGBURN, Y. BURSTEIN, L. H. ERICSSON, H. NEURATH, and K. A. WALSH: Evidence of Homologous Relationship between Thermolysin and neutral Protease A of *Bacillus subtilis*. Proc. Nat. Acad. Sci. USA **72,** 4341—4345 (1975).

463. RYNBRANDT, R. M.: The Reaction of Sulfoxides with Hydrogen Chloride. Tetrahedron Letters **381,** 3553—3555 (1971).

464. SAVIGE, W. E., and A. FONTANA: paper in preparation (1975).

465. SCORRANO, G.: Equilibria and Reactions of Organic Sulfoxides in Moderately Concentrated Acids. Accounts Chem. Res. **6,** 132—138 (1973).

466. SPANDE, T. F., and G. G. GLENNER: The Reaction of a Tryptophan Derivative with the Diazonium Salt, "Fast Red B", unpublished work (1975).

467. TITANI, K., L. H. ERICSSON, K. A. WALSH, and H. NEURATH: Amino Acid Sequence of Bovine Carboxypeptidase B. Proc. Natl. Acad. Sci. USA **72,** 1666—1670 (1975).

(Received June 1, 1975)

The Present Status of Flavin and Flavocoenzyme Chemistry

By P. HEMMERICH, Fachbereich Biologie der Universität Konstanz, Federal Republic of Germany

With 3 Figures

Contents

I. Introduction: History and Outlook

The term "flavin" stands for the yellow redox-active subgroup of the first coenzyme ever to be elucidated in terms of molecular structure. The "heroic" period of redox-enzymology of the early thirties ended in HUGO THEORELL'S (*172—175*) description of the first enzyme ever to be split reversibly to yield coenzyme and apoprotein. This was the NADPH-oxidizing "Old Yellow Enzyme" from yeast, whose biological function is still unknown even nowadays. The structure of the coenzyme has been shown to be riboflavin-5′-phosphate, viz. a phosphate derivative of vitamin B_2, which had been synthesized somewhat earlier by KARRER's group (*83*), who based this work upon even more important chemical precursor studies of KUHN and coworkers (*102*, *100*) and on the biochemical work of WARBURG and CHRISTIAN (*188*).

This initial period of flavin research which has been reviewed most thoroughly by BEINERT in 1960 (*4*), had an afterglow in the late thirties, when flavosemiquinones and, thereby, the first stable radicals of potential relevance in biology, were detected in halfreduced flavin systems by MIACHAELIS and SCHWARZENBACH (*129*) and, somewhat later, isolated in the crystalline state by KUHN and STRÖBELE (*101*). After this, further interest in flavins remained dormant for nearly three decades owing not only to the second World War and to the relatively slow development of flavin enzymology as compared to flavin chemistry, but also due to the failure of "flavin pharmacology" for the reasons discussed below.

The present review is written in a series devoted to "Progress in the Chemistry of Natural Products". Historically, it has become a habit (perhaps not a good one) to consider as "natural products" only or preferably those stable molecules of organic origin which exhibit *low molecular weight* and either accumulate as plant or animal metabolites or as end products of biopolymer degradation. German chemical terminology, in particular, carries along the burden of a *defacto* nonexisting antagonism between "Naturstoff" and "Kunststoff", where the latter term stands exclusively for "synthetic polymer". Since the biological relevance of a "natural product", let alone its "mode of action", is not defined by its origin or method of isolation, "natural product chemistry" has been largely a matter of crystallisation, structural elucidation and synthesis of such products which are, perhaps only incidentally, crystallisable.

After these three goals had been achieved during the "heroic period" of flavin research, flavin had to await its emancipation as natural product and its establishment as an ubiquitous redox-cofactor which plays a key role, frequently even the rate limiting role, in all biological redox chains. Detrimental to further development of flavin chemistry were, however, not only the lack of biochemical knowledge, but also the fact that one did not know what to do with a pigment which showed no structural variations throughout the plant and animal kingdom, while it was supposed that "natural products" should exhibit source specificity and should occur as groups of derivatives based on a given prototype. Last and not least, in spite of its being a vitamin or coenzyme, flavin failed completely in the hands of the — at least financially — most potent group of promoters, the pharmacologists. This failure was not ameliorated by the circumstance that the vitamin became cheap and was widely used

References, pp. 517—527

as a food additive of totally innocuous character (*153*) — a rare quality in food additives which removed all doubts concerning its usefulness.

If, in the present review, flavin is rediscovered as a "natural product", it is for two reasons:

a) The progress of biochemistry has in the interval resulted in the discovery of several families of "new flavocoenzymes" which exhibit characteristic variations of the flavin chromophore.

b) Redefinition of the scope and limitations of "natural product chemistry" is necessary in order to include those biocatalytic species which, because of their high chemical activity, do not comply with the irrational criteria outlined above, but have at the same time a more direct biological relevance.

An excellent demonstration of the validity of this point is at hand in the flavin family. In flavocoenzymes, viz. the so-called flavin nucleotides FMN and FAD, the polyhydroxylic chain is not on the ribose level, the flavin linkage is not a nucleosidic one and cannot be split reversibly by hydrolysis. Flavin side chain degradation occurs photolytically by a process whose mechanism needs further attention. The products of this photolysis, lumiflavin and lumichrome, are, therefore, not natural products in the strict meaning of the term and were found to be biologically inactive. From their being inactive, however, it does not follow that they are biologically irrelevant, for the following simple reasons: Upon detachment from the protein, the flavocoenzymes develop "new", i.e. largely "unbiological" or at least nonspecific qualities, namely

- strongly hydrophilic character,
- light sensitivity,
- radical instability (towards disproportionation),
- a *general* tendency to form π-charge-transfer complexes with aromatic donor-acceptor molecules, in particular selfstacking complexes of oxidized with reduced flavin ("flavoquinhydrones").

Furthermore, removal of the anionic (phosphate) groups from the free coenzymes does not greatly reduce their hydrophilic character, although the newly formed neutral riboflavin or vitamin B_2 develops a general insolubility, owing to "pairing" of strong intermolecular hydrogen bridges in the "pyrimidone" part of the heteroaromatic nucleus.

These phenomena can be classified as "unnatural" properties of "natural products", and this explains why commercially available "natural" flavocoenzymes are largely unsuitable as chemical "models" for the flavo-holoprotein active sites.

In order to obtain better "chemical models" for protein-bound flavin, one has to either acetylate or cut the ribityl side chain of riboflavin down to a methyl group (but not further) *and* to remove the "pairing" N(3)-hydrogen by alkylation. These derivatives, i.e. 3,10-dialkyl-isoalloxazines (or 3-alkyl "lumiflavins", if $R(10) = CH_3$) exhibit the same light stability and combination of moderately lipophilic *and* hydrophilic character as the protein-bound flavin chromophore. They are still devoid of radical stability, of course, and still exhibit "stacking" tendencies that might give rise to unsound biological conclusions.

To summarize, "good" flavin "models" and protein-bound flavin must differ from free flavocoenzymes and vitamins by their lack of self-stacking, self-pairing, (photolytic) "self-deleting" and pronounced hydrophilic properties.

Consequently, good flavin models are a failure as regards pharmacology while, on the other hand, hardly any biochemical disclosures of value can be expected from chemical studies of pharmacologically active synthetic flavin antimetabolites (*194*). This discrepancy has been one of the main reasons, why there was practically no progress in the "molecular biology" of flavoprotein action for the twenty years between 1938 and 1956, until BEINERT started to tackle the most important chemical problem inherent in flavins, viz. the structure, spectral properties, kinetic and thermodynamic stability of flavin radicals (*3*). These studies by a famous biochemist marked the beginning of a new era of flavin research which was characterized by a lag of chemical knowledge due largely to the special mentality of the classical "natural product chemist", for whom all problems inherent in a given molecule were solved after structure elucidation and confirmation by synthesis of one and only one "Kekulé" structure. Since nowadays this job has been most efficiently taken over by the crystallographer, even the most classical natural product chemist can recognize that flavin remained the great unknown among the coenzymes in spite of the achievements of the heroic thirties and in spite of its cheap synthesis by TISHLER in 1947 (*176*).

The second era of flavin research carried out under the leadership of biochemists such as V. MASSEY, T. P. SINGER, and E. C. SLATER, culminated in the "First International Flavin Symposium" at Amsterdam in 1965 (*167*). This symposium created a tradition; since then a continuous account of "progress in flavinology" has been given by the Proceedings of Flavin Symposia in Amsterdam 1965 (*167*), Nagoya 1967 (*196*), Durham 1968 (*82*), Konstanz 1971 (*201*), and San Francisco 1975 (*163*). The chemical and physico-chemical contributions to these symposia have increased from meeting to meeting although up to now they have not outweighed the biochemical and biophysical ones. In the light of these developments, flavin has gained new interest as a "natural product"; meanwhile flavin chemistry, and chemical physics reveals that Nature itself is more sophisticated than traditional "natural

References, pp. 517—527

product chemistry", but still widely conceivable in terms of "Kekulé" structures.

Two remarks may serve to clarify this statement:

1. It is now widely recognized that biological dehydrogenation ("CH-activation") starts with a two-electron transfer, i.e. between alcohols and nicotinamide. Hence there is no doubt that a "hydride" equivalent is transformed to yield two electron equivalents and a proton and that this process as well as its reversal is uniquely flavin-dependent all over the range of biochemistry. The transformation of the σ-pair of the CH-bond to be dehydrogenated into single π-electrons is mediated by $(4n+2)\pi$-electron (i.e. aromatic) systems such as $NAD(P)^+$ $(n=1)$, which yields a "homoaromatic" reduced form, and finally flavoquinone $(n=4)$ which yields upon reduction a $4n\pi$-system $(n=5)$, i.e. flavohydroquinone, which if planar and extensively π-delocalized would turn out to be "antiaromatic". Such a species would be a most potent donor of radical electrons, but any deeper understanding of the process requires a thorough comprehension of conformation (planarity), energy (electron delocalisation) and multiplicity (spin conservation). Such problems have been much studied in connection with "simple" aromatic hydrocarbons, but the demonstration of biological relevance is still awaited.

2. Most flavoproteins appear to be strongly membrane-oriented enzymes. Hence, the flow of flavin substrates does not seem to be entirely controlled by plasma diffusion and is liable to steric restrictions which induce additional asymmetry into the „active site" of the reduced flavin molecule. This asymmetry may or may not be conserved upon removal of the flavoprotein from the membrane, but must be lost upon removal of the flavocoenzyme from the protein, a situation which is different in the case of nicotinamides. This raises the question of how many "active sites" there are in the flavin nucleus; if there are several sites, as it turns out, we are left with the question of chemical site specificity. Hence, the solution chemistry of flavin, which is already very different for water and aprotic solvents, will separate, under suitable conditions, into several types of "anisotropic flavin chemistry", e.g. if flavin is fixed in a membrane interphase in a way which may differ from enzyme to enzyme and from organism to organism.

It must be emphasized that the present review is not intended to be just one more piece of secondary literature about flavoproteins. Its aims are, quite on the contrary, the following ones:

– Interpretation of the flavoprotein manifold and its biological relevance in the light of recent progress on the structural chemistry of flavins.

– Closure of the gap between chemistry and enzymology, which is becoming more and more apparent in recent years all over the range of biochemistry, and which has indeed contributed to the rise of a new "separate" discipline "bio(in)organic chemistry".

– Critical outline of "chemical models" for the protein-bound flavin.

– Revival of the "natural product" aspect of flavocoenzymes in the light of newly discovered flavocoenzymes.

A similar effort towards a reconciliation of chemistry and enzymology of flavins is presently being undertaken by BRUICE (*12*). In the opinion of the present reviewer, however, his article suffers from two deficiencies: First, no effort is made to scrutinize flavin chemistry for its biological relevance. In fact, chemistry is not biologically relevant, "because it is" – this is a very common mistake among chemists predigesting their own experimental data for the use of biologists. Second, the dynamic aspects of flavin chemistry are overemphasized. This is again a very common mistake among chemists trying to understand (bio)catalysis. Intermediates in biocatalysis are by definition shortlived. This does not necessarily mean that they could only be mimicked by chemical models which are similarly shortlived. Biochemistry is, owing to the common scarcity of material and lack of stability, largely confined to kinetic studies. It is, however, the aim of "model chemistry" to provide chemical samples of comparable physical properties in higher quantity and stability in order to create a *structural* basis for the interpretation of biodynamic data.

Earlier reviews of purely chemical character, but meant to be more comprehensive, are scarce: After the aforementioned "classical" review of BEINERT (*4*) there has been a synopsis of chemical flavin synthesis by LAMBOOY (*104*). Flavin reactivity has been non-critically reviewed by BEREZOVSKII *et al.* (*6*). Three earlier attempts have been made to establish partial surveys of "bioorganic flavin chemistry" (*74, 79, 146*). The "nutritional aspects" of the flavin vitamin as well as the unsatisfactory state of flavin pharmacology have been documented in a recent monography by RIVLIN (*153*).

II. Nomenclature

Flavin nomenclature (see Scheme 1) abounds in atrocities, which have become so commonly used that it seems an idle task to discuss them, but nevertheless this problem must be tackled.

The worst example is given by the "flavin nucleotides" FMN and FAD, the first of which is no nucleotide at all and the second no dinucleotide. The term FMN should simply be changed into FMP, viz. (ribo)flavin-monophosphate, while the abbreviation FAD might even be preserved as representing "flavin-adenine-diphosphate".

References, pp. 517–527

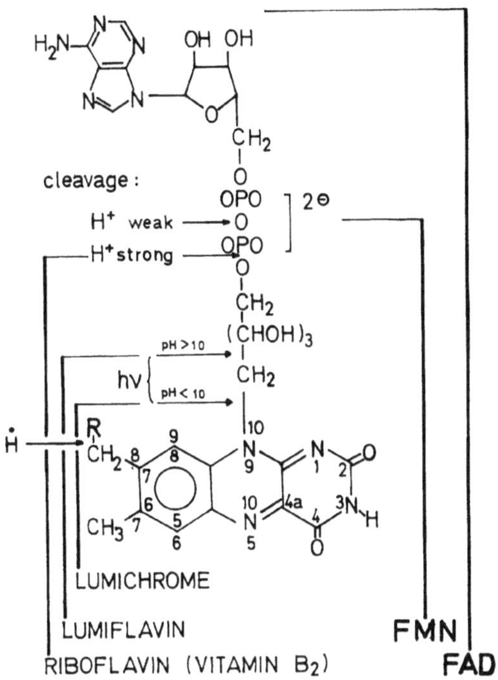

Scheme 1. Flavin and Flavocoenzyme nomenclature and modes of breakdown by mild and strong protolysis, photolysis and hydrogenolysis (R = H: Loosely bound; R = Protein [His, Cys]: Covalently bound flavin)

We may continue with the question what is meant at all by the term flavin: Is it synonymous with isoalloxazine (*101*) or with 7,8-dimethyl-isoalloxazine (*139*)? If the term flavin is meant to comprise the 7,8-methyl groups as seems most reasonable, any flavin modified in such a position must be termed 8-X-8-*nor*-flavin, in contrast to additions to the same position, which must lead to 8-X-1,8-dihydroflavin.

Furthermore, the term flavin becomes disputable as a coenzyme, since the "covalently fixed" flavin of several flavoproteins has turned out to be a prosthetic group (see below). In the opinion of this author, however, a prosthetic group which undergoes chemical modifications during catalysis might well be termed coenzyme, irrespective of covalent fixation.

Another unfortunate creation is the term "isoalloxazine", a result of the early flavin synthesis from alloxane (*103*). Since flavin biosynthesis runs through pteridine precursors of the lumazine type (*150*), the term isoalloxazine should be dropped entirely (as was done by Chemical Abstracts who use the systematic name 7,8-dimethyl-benzo-[g]-pteridine-2,4(3H, 10H)-dione) in favour of benzolumazine.

A further discrepancy which concerns the numbering system continues to befog the literature. The old German system (*101*) derived from anthracene is still encountered even in parts of the United States, maybe as a means of biochemists' disregard for "Chemical Abstracts", which use the more reasonable IUPAC (*104*) numbering. The latter permits a correlation of flavin with its precursor pteridine by applying the same number 5 to the "active" nitrogen site and number 4a to the active bridge carbon (*cf.* Scheme 1).

Furthermore, we have specified the flavin redox states by the terms flavoquinone, flavosemiquinone and flavohydroquinone in view of the polarographic reversibility, i.e. the quinonoid character, of the oxido-reduction. These terms are definitely superior to "oxidized flavin, flavin radical, fully reduced flavin" which are most widely used in biochemistry. In this context, we have tried to outline chemical evidence for the existence and potential biological relevance of several dihydroflavin isomers. The terms "fully reduced" and certainly "leucoflavin" should be dropped definitively in favour of "dihydroflavin". Leucoflavin is in fact not leuco (i.e. colourless) but even more deeply (though, of course, less intensely) coloured (*79*) than oxidized flavin. Hence, this term should be abandoned for "1,5-dihydroflavin", which is identical with the product of the reversible reduction of flavoquinone, namely flavohydroquinone.

The capital letter F (*12*) should remain quite definitely a domain of fluorine chemists, even though it can be argued that the abbreviation "Fl" for flavin may be confused, if written undiligently, with Fe (for iron). Hence, we have been trying in recent years to adhere to the abbreviation "Fl_{ox}" for flavoquinone and "$Fl_{red}H_2$" for dihydroflavin. Caution must be employed so that these abbreviations are used consistently: The term "Fl_{ox}" interferes with the definition of the flavoquinone anion. A more precise shortening would, therefore, be $Fl_{ox}H$ (*158*) for flavoquinone which affords, however, some trouble with the various H-atoms in the reduced species which then would read $Fl_{red}H_3$. Hence, we adhere to the simple Fl_{ox} for common use.

Finally, the useful subscripts "ox" and "red" must be clearly defined as to their correlation either with true redox states or with flavin chromophores. For example, the addition of water to an alkyl-flavoquinonium ion 5-RFl^{+}_{ox} (*181*) leads to a dihydroflavin chromophore, though the redox state did not change: The true redox state of 5-RFl-4a-OH formed in this fashion is indicated by the oxygen atom, though it stems from hydroxyl ion. The picture becomes indeterminate in the case of redox-active substituents with a potential approximating flavin potential, *cf.* the sequence of X in Scheme 2.

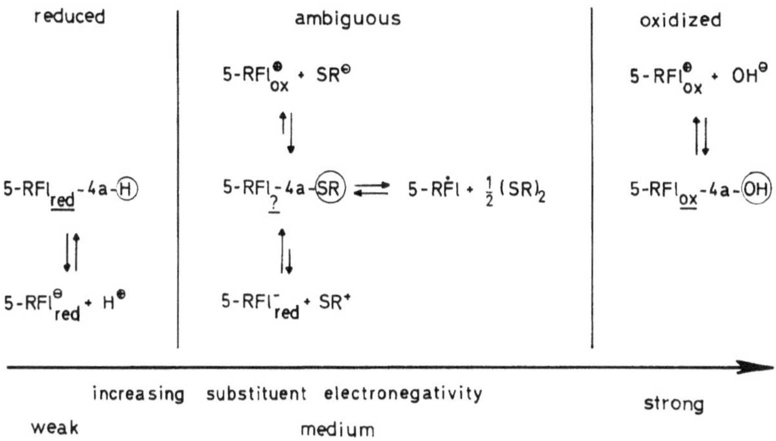

Scheme 2. (Dihydro)flavin substituent electronegativity and "redox state". Note that the RFlX-chromophore is that of dihydroflavin, irrespective, of the eletronegativity of X = H, SR, OH $\lambda_{max} \sim 370$ nm

In this sequence, the subscripts "ox", "red" determine the mode of preferable 4a-bond dissociation, which is ambiguous in the case of SR (*64, 200*) viz. a matter of the environment. The chromophore, however, remains the same which means that the change of redox-state implied in this sequence cannot easily be followed by spectrophotometry. In chemical formulae this appears trivially true, but not so for a biochemist dealing with structurally unknown and perhaps shortlived spectral flavin intermediates. Analogous problems arise with flavin complexed by a redox active metal such as $Fe^{II}/^{III}$, where the redox shuttle can either occur between flavoquinone and -semiquinone, or between the latter and the flavohydroquinone. In such a case the use of dots as a formal marker for the radical state is questionable, since it might be mistaken for the spin state which may or may not be known.

A similarly wide-spread abuse is the confusion of "half reduced" and "flavin radical". The term "half reduced" does not imply any information about spin concentration in a flavin system, as the electrons might be anything between 100% paired and 100% unpaired in a half reduced flavin system as demonstrated by the equilibrium (1):

$$\mathrm{Fl_{ox} + Fl_{red}H_2 \rightleftarrows 2\,\dot{F}lH}. \tag{1}$$

The term "radical" should only be allotted to an observed flavin species, if doublet character has been demonstrated *quantitatively*, as hard as this might be in a given situation.

III. Mechanism of Flavin-Dependent Oxido-Reduction

It has already been stated that flavin fulfils the requirements of a quinonoid system. It exhibits thermodynamic reversibility of one-electron as well as two-electron transfer. *Basing his terms on electrochemical data,* MICHAELIS (*128*) was right in postulating that a transfer of redox equivalents to and from flavin occurs stepwise, one electron at a time. It must, therefore, be emphasized that flavin *can* react this way, but need not do so, and this defines the uniqueness of flavin among redox-biocatalysts. This uniqueness becomes more convincing, if we consider electron carriers located next to flavin in biological redox chains, such as nicotinamides and ubiquinones (*78*):

– *Nicotinamides* have been accepted as obligatory $2e^-$-transfer agents. Radicals can be formed only by processes such as pulse radiolysis but they are neither chemical nor biological intermediates (*105*). Caution must be employed to avoid consideration of "false models". Typical cases of "false models" in relation to nicotinamide are all pyridinium systems substituted in positions 2 or 4 instead of in position 3 (*99*). Such a substituent induces considerable radical stability, since it allows spin delocalisation towards the substituent which is forbidden for the 3-substituted analogs.

– *Ubiquinones* are derivatives of *p*-benzoquinone, which is known to be a relatively poor acceptor of $2e^-$-equivalents, be it hydride (even borohydride), carbanions, arsenite etc., This relative lack of electrophilicity will be further decreased by the methoxyl substituents in ubiquinone. Biologically, the electron donor-acceptor activity of ubiquinones is strictly radical, since they serve as carriers between two iron systems, viz. the ferredoxin-like iron of the respiratory iron-flavoproteins as input site and cytochrome as output site.

References, pp. 517—527

Hence, flavin systems are unique in acting in $1e^-$- as well as $2e^-$-transfer. They can be separated into the subclasses of 1/1-, 1/2-, 2/1- and 2/2-flavoproteins (*80*) according to the number of redox equivalents (not absolutely electrons!) transferred in their input and output reactions, respectively (Scheme 3 and Table 1).

Nowadays, a characteristically modified flavin, 5-deazaflavin (*33, 142*), has gained much interest in flavin enzymology because it can replace natural flavin at many enzyme sites. This derivative, however, is no quinoid system and its radical has no thermodynamic stability. It is, therefore, a mutilated flavin retaining the $2e^-$-transfer (i.e. nicotinamide-like) properties only (*cf.* Table 2). The thia-analog, however, is an O_2-stable dihydro derivative, which exhibits a stable radical, but no reversibly reducible oxidized form. Hence, it is also a mutilated flavin, but one which retains the e^--transferase activity only.

As to the mechanism of these oxido-reductions, we have to ask the following questions:

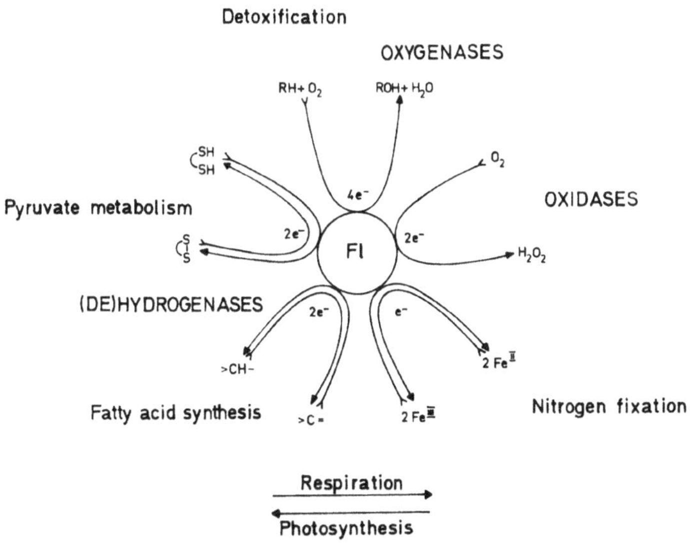

Scheme 3. Flavoprotein functions and redox-stoichiometry

Table 1. *Flavoprotein Subclasses*

Flavoprotein data largely taken from MASSEY *et al.* (*121*), MASSEY and HEMMERICH (*120*), ENTSCH *et al.* (*29*), HASTINGS (*57*) and coworkers, and their references. The system might have to be completed by adding a class of transhydrogenases = $2e^-$-input/$2e^-$-output, "non-radical" flavoproteins. Metalloflavoproteins are regarded as dehydrogenase/e^--transferases with internal $1e^-$-acceptor. Flavodoxins must be considered as 1 : 1-input-output or "pure" e^--transferases.

The bold lines separate the main subclasses of *e^--transferases* (including flavodoxins as "*pure* e^--transferases), *oxidases* (including oxygenases in the absence of oxene-acceptor) and, finally, *oxygenases* (including luciferase) in the presence of oxene acceptor)

Block position	Redox-equiv. input	Redox-equiv. output	Class name	Radical type	Radical yield	O_2-Reduction rate	O_2-Reduction product	Fl-O_2-"Complex" structure	Fl-O_2-"Complex" absorption (nm)
5	2	1	e^--transferase	blue	high	slow	$\dot{O}_2^-$	5-HFl-4a-OOH	~370
1 or 2α	2	2	oxidase	red	high	fast	HOO^-	⊖Fl-9a-OOH	~420
1 or 2α 5	2	2	oxygenase without substrate	blue	low	slow	H_2O_2	5-HFl-4a-OOH	~370
	2	4	oxygenase with substrate	red	low	fast	$[O] + H_2O$	2α-HFl-9a-OOH	~420
1 or 2α 5 1 ?	2	2	luciferase resting (cold)	none		slow	H_2O_2	5-HFl-4a-OOH	~370
	2	4	luciferase active (20° C)	none		fast	$[O] + H_2O$	1-Fl-10a-OOAc ?	~610 ?
1 or 2α	2	2	free flavin	red	low	slow	polar H_2O_2 unpol. $O + H_2O$	1,2α-R-Fl-9a-OOH	420
	preferred								
5	1	1	free flavin	blue	high	slow	$pH > 8$: $\dot{O}_2^-$ $pH < 8$: H_2O_2	5-R-Fl-4a-OOH	365
unblocked	1 or	2	free flavin	$pH < 8.5$: blue $pH > 8.5$: red	low	fast	$\dot{O}_2^-$ H_2O_2	isomer equil. pH and polarity dependent	

Table 2. *Flavin Analogs Modified in Position 5 and Their Redox-Characteristics*
Note that the sulfonium form of oxidized thiaflavin is stable only in strong acid, upon dilution it adds OH^- yielding the sulfoxide, irreversibly. In the "mutilated analogs" only one of the three "redox-shuttles" encountered in flavins is working

	Flavins (X = N)	Deazaflavins (X = CH)	Thiaflavins ($X = \overset{\backslash //}{S_+}$)
radical			
free	revers. disprop.	irrevers. disprop.	stable
protein bound	stable	unstable	?
oxidized	quinoid	quinoid (flavin-analog.)	non-quinoid (sulfoxide)
reduced	hydroquinoid	non-hydroquinoid (NADH-analog.)	hydroquinoid (flavin-analog.)
redox-shuttles			
$2e^-$	+(dehydrogenase)	+	–
upper $1e^-$	+(artificial?)	–	–
lower $1e^-$	+(e^--transferase)	–	+

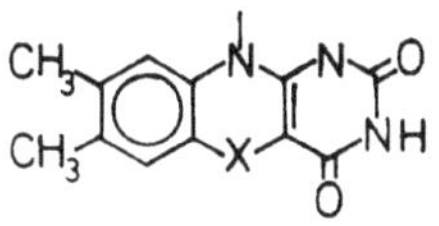

X = NH: Dihydroflavin
$X = CH_2$: Dihydrodeazaflavin
X = S: Thiaflavin

– If $1e^-$-transfer is involved it must be specified which one of the two one-electron "shuttles" is concerned? Present knowledge points unequivocally to the fact that only the "lower shuttle" (between hydroquinone and semiquinone) is biocatalytically relevant (*124*), in the electron output of flavins.

– If $2e^-$-transfer is involved what is the "carrier" of the electron pair, H^- (hydride) or R^- (carbanion) in (de)hydrogenation, OOH^- in O_2-activation? And in the same way,

– what is the geometry of the bond in the "Michaelis" complexes $[Fl_{ox}, RH] \leftrightarrow RFl_{red}H$, $[Fl_{red}H_2, O_2] \leftrightarrow HFlOOH$. Is it a "covalent" (i.e. σ-type) interaction or a "charge-transfer" (π-type) complex? Are potential σ-bonds fixed within the flavin nucleus or are they fluctuating, or is it an ion pair held together by "environmental (hydrophobic) forces" rather than by Coulomb attraction or charge transfer?

In Schemes 4 and 5 we are comparing flavin with the known bond geometry of a less complex, but somewhat related residue, cyclopentadienyl. The latter might, just like flavin, form every possible type of bond, (fixed σ-, fluctuating σ-, π-, and ionic) with the only restriction that it cannot form cations, i.e. it cannot be "fully oxidized" while flavin (Scheme 5), of course, can: Fl_{ox} is aromatic (*cf.* above), while cyclopentadienyl cation is antiaromatic — as is reduced flavin (*cf.* below), if ever it assumes coplanarity.

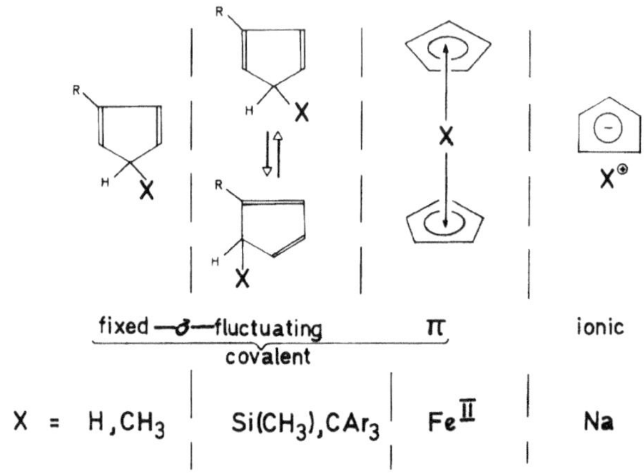

Scheme 4. Nature of bonding in cyclopentadienyl-X species. For fluctuating X see (*95, 106*)

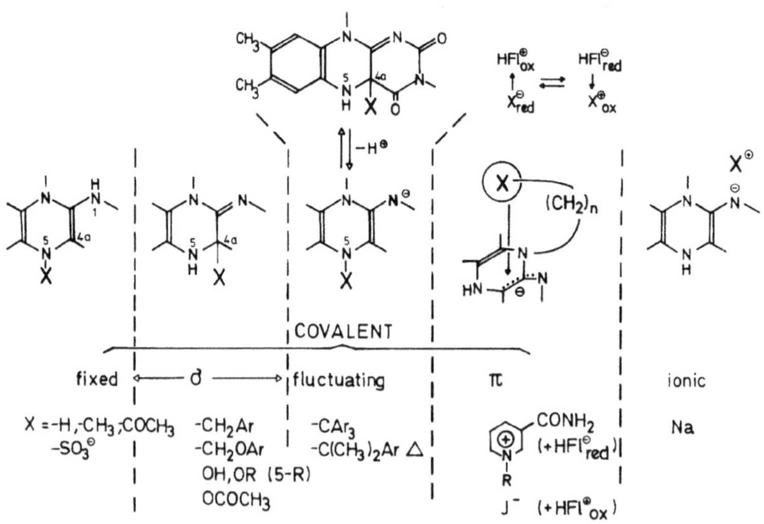

Scheme 5. Nature of bonding in dihydroflavinyl-X species = fluctuating at elevated temperature (*68*). (5R) means 4-OH, OR only stable if N-5-alkylated (*22*)

– Finally, one has to ask: Is flavin-dependent oxido-reduction a ping-pong mechanism (eq. 2)

$$\begin{aligned} &\text{a) Donator} + Fl_{ox} \rightarrow \text{oxid. Donator} + Fl_{red} \\ &\text{b) } Fl_{red} + \text{Acceptor} \rightarrow \text{red. Acceptor} + Fl_{ox}, \end{aligned} \quad (2)$$

which would involve only *binary* complexes donator-Fl and acceptor-Fl, or is there an occurrence of ternary complexes, which would require combination of the right hand side of 2a with the left hand side of 2b.

It will be shown that it is possible to develop chemical model ideas for the binary interactions, but obviously the ternary ones are biologically most important and very hard to simulate in terms of structural geometry.

After some remarks on flavoquinone the structural discussion of possible reaction intermediates will start with the radical (= "complex" of $Fl_{ox} + e^-$) and continue with dihydroflavin (= "complex" of $Fl_{ox} + H^-$), flavin-substrate complexes ($Fl_{ox} + R^-$) and flavin-oxygen complexes ($Fl_{ox} + OOH^-$). Scheme 6 illustrates the "σ-input and -output" positions for formation and decay of these intermediates.

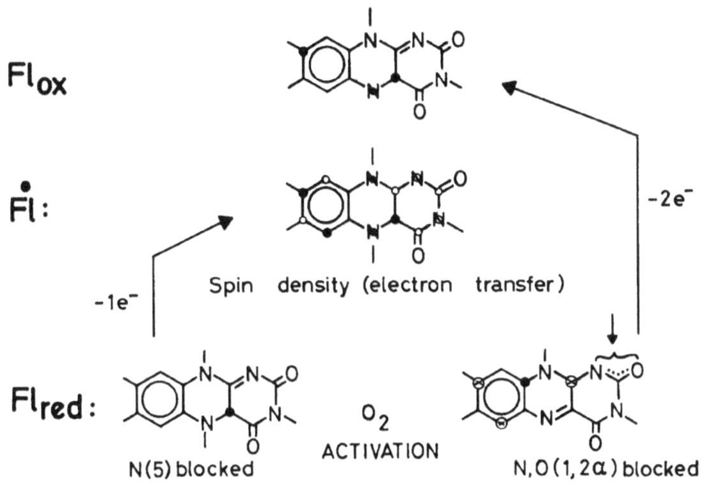

Scheme 6. "Active sites" within the flavoquinone (dehydrogenation),-semiquinone (e^--transfer) and -hydroquinone (O_2-activation) nucleus. Full circles denote sites of high activity. Low $1e^-$-transfer activity is predicted at sites of low spin density (open circles) (*26*). Note that the activity of the reduced system depends strictly on non-bonding nitrogen pairs blocked by H-bridges (apoproteins) or alkyl groups (chemical models). Crossed circles denote sites of possible, but hitherto not verified activity (*80*)

IV. Flavoquinone

Oxidized flavin, i.e. flavoquinone, is the best known redox state of flavins, owing to its characteristic light absorption, its strong fluorescence and its stability to air. Binding to protein, however, might "shift" the absorption (within +25 nm for the first 450 nm) transition, more drastically for the second one (*56*), an effect which has been traced by HARBURY *et al.* (*56*) and WEBER (*193*) to a single vibrational mode of Fl_{ox} being favoured in the partially immobilized enzyme bound state. Similar "shifts" are found in the absorption spectra of crystalline flavoquinone powders (*41*). Furthermore, protein may entirely or partially quench the Fl_{ox}-fluorescence by excitone transfer. This is mimicked by the well-known excitone transfer within the flavin adenine hairpin in the free coenzyme (*193*). This latter hairpin, however, turns out to be a chemical artefact, and all hitherto available information on FAD conformation in FAD-flavoproteins points to the conclusion that, within the holoprotein, the adenine is "folded away" from the flavin. The observed fluorescence quenching must, in these cases, be due to interaction with protein prosthetic groups such as tyrosine and tryptophane.

Because of these properties flavoquinone lends itself to all kinds of physico-chemical studies, quite irrespective of its limited biological relevance as compared with the other flavin redox-states. The situation is not as bad as with, e.g., folic acid, where the air-stable form is an artefact without biocatalytic power. Nevertheless, if we summarize the multiple "flavin activities" roughly as (de)hydrogenation, O_2-activation and electron transfer, it appears that flavoquinone plays an active role only in dehydrogenation, while it is only the final product of flavin-dependent O_2-activation and is artificial with respect to electron transfer which takes place between semiquinone and hydroquinone.

It is important to note, however, that flavoquinone is protonated at N(1) (and not at N(5)) with a pK as low as zero (*21*). This results in the coincidence of the 450 and 370 nm transitions of Fl_{ox} to form a single band centered at 395 nm which has an extinction of >20.000 M^{-1} cm^{-1}. It is a main characteristic of "covalently (protein) bound" flavocoenzymes (see chapter below), that this band remains split ($\lambda_{max} \approx 375$ nm, shoulder ≈ 410) as a consequence of 8α-substitution. The cations 1-RFl_{ox} are fluorescent with an emission around 490 nm, provided R is kinetically stable over the lifetime of the excited state. This is trivially true for R = alkyl, but it can be reached also for R = H in rigid media (*28*). In flavin-dependent bioluminescence of photobacteria, 1-RFl_{ox}^* (and certainly not Fl_{ox}^*) is most probably the emitting species, while it remains to be proved whether R in the „luciferase" case is a proton immobilized by the apoprotein or a substrate (acyl) residue (see below).

On the contrary, the excited flavoquinone triplet ($^3Fl^*_{ox}$) is protonated at N(5), as predicted by SONG (*169*) and verified by SCHREINER *et al.* (*160*), and has the surprisingly high pK of ≈ 5, as will be discussed in the Photochemistry section below.

V. Flavin Radicals

As mentioned above, BEINERT (*3*) was first to "see" flavin radicals in neutral solution, while MICHAELIS and SCHWARZENBACH (*129*) could only infer their existence in the largely disproportionated state of the half reduced neutral flavin system. There are several means for shifting equilibrium between oxidized and reduced flavin towards the radical side:

1. *Protonation:* The neutral radical, $\dot{F}lH$, will accept a proton with a pK of about 3.

$$\begin{array}{ccc} Fl_{ox} + Fl_{red}H_2 & \overset{H^+}{\rightleftharpoons} & HFl^+_{ox} + Fl_{red}H_3^+ \\ \uparrow\downarrow & & \downarrow\uparrow \\ 2\ \dot{F}lH & & 2\ \dot{F}lH_2^+ \end{array} \tag{3}$$

This brings about comproportionation as shown in eq. 3, since the respective pK-values of Fl_{ox} and $Fl_{red}H_2$ are smaller by more than two orders of magnitude. The red cations $\dot{F}lH_2^+$ formed in this manner can in many cases be crystallized, under strongly acidic conditions, as was done first by KUHN and STROEBELE ("rhodoflavin") (*101*). $\dot{F}lH_2^+$ can serve as a very sensitive measure for the presence of flavins even in complex mixtures, if measured by EPR; by analysis of the hyperfine structure of the EPR-signal, modifications in the high spin density centers of the flavin nucleus can thus be recognised. This was done for the first time by HEMMERICH *et al.* (*67*) in the elucidation of the "covalently protein-bound flavin of succinic dehydrogenase" (*cf.* below). "Rhodoflavin" has, however, no biological relevance by itself, owing to its high acidity.

2. *Alkylation:* Introduction of alkyl groups into the flavin nucleus is a complex matter, which will be extensively reviewed in the "dihydroflavin" section below.

In the half reduced system, 5-*alkylation* brings about comproportionation to the radical. Optical as well as EPR comparison of this stabilized radical, 5-R$\dot{F}$l (*21, 136*), with BEINERT's unstable H$\dot{F}$l (*3*) leads to the conclusion that the neutral flavosemiquinones are blue and have the tautomeric structure 5-HḞl. The pK of this ṄH-group has been found to be around 8.5 (*19, 26*) depending somewhat on the nature of flavin side chains.

The second possible tautomer type of neutral flavosemiquinone, 1-HḞl, is, therefore, anticipated to be rather acidic at N(1)H and rather basic at N(5). Its formation by direct neutralization of a radical cation would, consequently, require blocking of N(1) or O(2α) by alkylation. This, however, results (*69*) in destabilization of the radical state to an even greater extent than encountered in the free flavin system (Scheme 7).

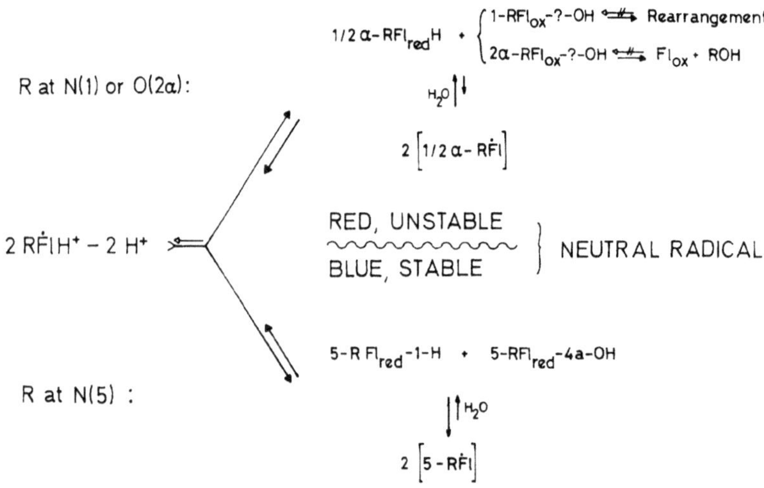

Scheme 7. Thermodynamic stability (disproportionation equilibria) of flavosemiquinone tautomers. Upon deprotonation of the stable, red (but because of high acidity biologically irrelevant) cations red unstable or blue stable neutral radicals are formed according to which site is blocked

Recomproportionation to yield 1-RFl could only be brought about by further removal of the proton from N(5) in 1-RFl$_{red}$H, which requires very strong base, since it is a typical pyrrole NH. Such a strong base first interferes, however, with the equally present 1- or 2α-RFl$^+_{ox}$ by formation of "adducts" RFlX (*cf.* below) which decompose *irreversibly:*

$$\text{1-RFl-9a-X} \rightleftharpoons \text{1-RFl-10a-X} \xrightarrow{\text{X=OH}} \text{ring concentration } (\textit{23}) \quad (4)$$

$$2\alpha\text{-RFl-9a-X} \rightleftharpoons 2\alpha\text{-RFl-2-X} \xrightarrow{\text{X=OH}} \text{Fl}_{\text{ox}} + \text{ROH } (\textit{22}) \quad (5)$$

Hence, blocking of the lone pair at N(5) *by alkyl stabilizes the radical and suppresses the* $2e^-$*-transfer system, while blocking of the* N(1) *or* O(2α) *lone pairs destabilizes the radical and suppresses* $1e^-$*-transfer.* $1e^-$*- and* $2e^-$*-transfer can thus be switched on and off in flavoproteins by formation of strongly directed hydrogen bridges between apoprotein and* N(5)*- or* N/O(1/2α)*-sites of the flavin, respectively.*

3. *Deprotonation:* Since blue $H\dot{F}l$ ($\equiv$5-$H\dot{F}l$, pK≈8.5, *cf.* above) is more acidic than Fl_{ox} (pK≈10) (for chemical studies this dissociation should in any case be blocked by N(3)-prealkylation), alkalinization of the half-reduced free flavin system will again enhance radical formation (*70*) according to:

$$Fl_{ox} + HFl^-_{red} \underset{pH>8.5}{\overleftarrow{\longrightarrow}} 2\,\dot{F}l^- + H^+ . \qquad (6)$$

But stoichiometric amounts of the red anion radical $\dot{F}l^-$ — as well as detectable amounts of neutral 1/2α $R\dot{F}l$ (*cf.* Scheme 7) — can only be obtained under strictly aprotic, strongly basic conditions and in the absence of 8α-CH_3 groups in Fl_{ox}, which are liable to slow deprotonation by base upon prolonged standing (*76*).

Considering the full scale of pH-dependence of flavin radical stability, one must conclude that for the free flavin system there is a gap of radical stability in the neutral range (*26*), which is further deepened upon blocking of N(1) or O(2α) and eliminated upon blocking of N(5). The stable 5-$H\dot{F}l$ is blue (λ_{max}≈580 nm), while the unstable 1-$H\dot{F}l$ and the stable $\dot{F}l^-$ (pH>9) are red (λ_{max}≈470 nm).

4. 5-Blocking can also be achieved by *metal 4α,5-chelation* (*70*). The biological relevance of direct flavin-metal interaction is, however, still questionable. The chelates $Me\dot{F}l$ are dark red with an additional absorption band of still unknown character in the near infrared (*70*).

5. *Protein fixation*: Massey *et al.* (*121*) were the first to investigate systematically bioradical formation by "comparative enzymology of flavoproteins". It turned out that flavoproteins can be divided into two classes according to whether their radicals are blue or red. Moreover and quite surprisingly, it became obvious that all "red radical flavoproteins" were 2/2-flavoproteins (Table 1, *cf.* above), while the "blue radical flavoproteins" happened to be either 1/1-, 2/1- or 1/2-flavoproteins. In other words: All flavoproteins involved essentially with $1e^-$-transfer (either by input or by output or by both) have blue radicals. Massey and Palmer referred to the "blue radical flavoproteins" as "dehydrogenases" and to

the "red radical flavoproteins" as "oxidases". In the opinion of this author, however, this terminology seems to be liable to misunderstanding, for two reasons:

(1) The 1/1-flavoproteins, i.e. the well-known flavodoxins, are not dehydrogenases — and still they are "blue radical flavoproteins".

(2) All flavoprotein (output) oxidases are at the same time (input) dehydrogenases.

Hence, the term "electron transferases" as contrasted with "oxidases" appears preferable over "dehydrogenases" for the "blue radical flavoproteins" (*80*).

The full spectrum of characteristic differences between the subclasses of flavoproteins is outlined in Table 1. To summarize, there are three classes of flavoproteins, which permits one to establish a correlation between catalytic function as well as radical properties, namely flavoproteins which give blue and essential radicals, flavoproteins which give red and non-essential radicals and, third, flavoproteins which up to now exhibit only low yields of radical, if any, in the native state (*121*). This might be due to cooperative effects between flavin and a second redox-active center in the molecule, which will render the uptake of the second electron much faster than that of the first one. Such a situation can be imagined for example for the interaction of flavin and "active disulfide" in lipoamide dehydrogenase, but it could also happen through flavin-flavin interaction in a "2 Fl per mole"-flavoprotein (*74*). Such an "interflavin" contact need not have the same geometry as in aqueous solution, where Fl_{ox} and $Fl_{red}H_2$ form the flavoquinhydrone π-complex practically at diffusion-controlled rates with the appearance of a typical charge transfer absorption. Intermolecular electron exchange is for example rather fast in flavodoxins (*116*), though no intermediate charge transfer spectrum can be observed; the x-ray structure shows that the intermolecular contact is confined to the "benzenoid" site of the flavin (position 6—9) (*109*). Chemically, a similar contact was found to occur and to be stabilized during the autophotolysis of lumiflavin in concentrated phosphate buffer (*157*) (*cf.* Scheme 8). This stabilization is, of course, due to the fact that two flavoquinone molecules are interacting, and not flavoquinone plus flavohydroquinone. But it shows that redox interaction can occur between two flavins units perpendicular to each other.

Most recently, a kinetically labile "biflavin" intermediate in chemical "interflavin electron transfer" has been found by FAVAUDON (*32*) which is different from the flavoquinhydrone π-complex.

This does not yet solve the mystery of how the "buried" flavin in flavodoxins transfers its electron, but it shows a way out of the dilemma one encounters with the assumption of electrons "tunnelling" through the major part of the apoprotein.

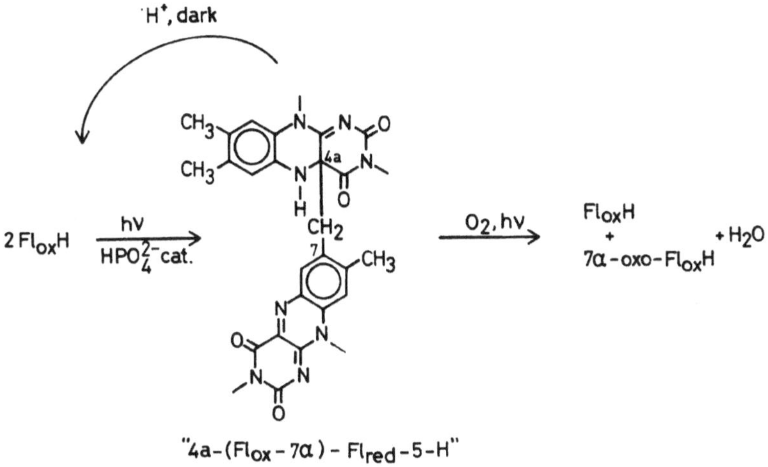

Scheme 8. Phosphate dependent intermolecular photolysis of lumiflavin and decay of covalent bi-flavin intermediate

VI. Dihydroflavins

The study of "fully reduced flavin" was largely neglected up to the late sixties (*53*) for reasons which belong to the realm of "science psychology" rather than elsewhere. Reduced flavin, unlike Fl_{ox}, appears to be colourless (but only in dilute solution) and nonfluorescent (but only in the protein free state and in liquid solution). Unlike $\dot{F}l$, it appears to be diamagnetic (but maybe only in the vibrational ground state), and is above all, extremely oxygen-sensitive (but only in the free planar or "planarizable" state and in the oxidase and oxygenase type flavoproteins). Hence, much expensive spectrometric apparatus cannot be used for dihydroflavin studies, painstaking precautions against air must be taken at all times.

However, like in other coenzyme systems, e.g. in (hydro)folate, it turns out that Fl_{ox} is not involved in all catalytic functions of flavoproteins, e.g. not in electron transfer, which is shuttling between $\dot{F}l$ and Fl_{red}, and in the same way $\dot{F}l$ is not involved in dehydrogenation, but at the time of this writing there is no well-known biological function of flavin which does not involve Fl_{red}. The term "well-known" has been used in this context, because there is indeed a number of flavin systems in nature which at first sight appear to be redox-inactive, starting with such enzymes such as carboligase (*54*) which catalyse the formation of new CC-bonds,

$$2\,\mathrm{OHCCOOH} \longrightarrow \mathrm{OHCCH(OH)COOH} + \mathrm{CO_2} \tag{7}$$

and oxynitrilase (*40*), which catalyses reaction 8

$$\mathrm{C_6H_5CH(OH)CN} \rightleftharpoons \mathrm{C_6H_5CHO} + \mathrm{HCN}, \tag{8}$$

and ending up with "photobiologically active flavins" which are assumed to shuttle between excited and ground state of Fl_{ox}. However, time may change this picture, as it did in the case of bacterial luciferase, which is now known to cycle between Fl_{ox}, Fl_{red} and RFl^{*+}_{ox} (see below). The main problems connected with dihydroflavin structure and activity are outlined briefly as follows:

1. Alkylation of Dihydroflavins

Alkyl groups can be introduced into the flavin nucleus, in principle, at positions N(1), O(2α), N(3), O(4α), C(4a), N(5), and C(8), depending on the flavin redox state, the (electrophilic or nucleophilic) nature of the alkyl residue, the solvent polarity and potential steric overcrowding. Except for N(3), which is redox-inactive, the alkylation sites exert a steering effect on oxidoreduction, since the redox potential is strictly dependent on all pK-values of any flavin species involved (*21*):

- C(8)- and C(4a)-alkylation block oxidoreduction entirely.
- N(1), O(2α or 4α)-alkylation suppress the radical level.
- N(5)-alkylation favours the radical level. In detail, the following rules apply:

1) N(1) is the site of strongest basicity, but not of strongest nucleophilicity in the dihydroflavin anion HFl^{-}_{red}. This is due to steric hindrance by the peri-substituent at N(10). This hindrance can only be overcome in the case of *intra*molecular N(1)-alkylation, e.g. (*66*):

$$\mathrm{HOCH_2CH_2\text{-}10\text{-}Fl_{ox}} + \mathrm{SOCl_2} \rightarrow \left.\begin{matrix}\mathrm{CH_2\text{-}10\text{-}} \\ | \\ \mathrm{CH_2\text{-}1\text{-}}\end{matrix}\right\} \mathrm{Fl^{+}_{ox}Cl^{-}} + \mathrm{SO_2} + \mathrm{HCl}. \tag{9}$$

In all other cases,

2) O(2α) and, to minor degree, O(4α) will be attacked after N(3) in the reduced state, if N(5) is protected by an acyl group (*77*).

3) The "redox-inactive" N(3)-position reacts selectively in Fl_{ox} under mild aprotic base catalysis (*77*). Stronger protic base would hydrolyze the pyrimidine subnucleus of flavin irreversibly (*12*), while stronger aprotic base would deprotonate 8α-CH with concomitant irreversible flavin dimerization (*76*). In order to avoid complex mixtures in reductive

alkylations of flavins, one is well advised to start with N(3) pre-alkylated in the oxidized state.

4) N(5) *is,* surprisingly, *the first position to react,* if unprotected, *in* $Fl_{red}H^-$ (*43*) though the charge in dihydroflavin anion resides entirely in the pyrimidine subnucleus. Consequently, one must assume that either charge control is entirely lost in this case in favour of frontier orbital control (*96*) or that the very rapid reaction runs through slight, but rapidly reacting admixtures of radical anion $\dot{F}l^-$, where N(5) is high in both charge and spin density. This electrophilic N(5)-alkylation is paralleled by the nucleophilic N(5)-photoalkylation of Fl_{ox} with "active" carboxylates R_3C-COO^-, with elimination of CO_2 (*71*).

5) Additionally, another important site of alkyl attack is the *bridge carbon* C(4a), which in most cases, electrophilic (*43*) and nucleophilic (*71*) ones, competes with N(5). The question as to which site is preferred by a given alkyl residue is still badly understood. Some hints may be taken from the observation that cyclopentadienyl, the nucleophile prototype, attacks always at C(4a) (*66*), while acyl residues as electrophile prototypes can only be fixed at N(5) (*98*). Less pronouncedly "-philic" substituents can occupy either position and, above a given temperature, migrate or even fluctuate between them (*53*) (*cf.* below).

6) C(8), finally, is attacked in $^3Fl^*_{ox}$ under alkaline conditions by a nucleophilic benzyl residue, as outlined in the Photochemistry section. Upon acidification, 8-$BzFl^-_{red}$ rearranges by migration of Bz towards N(5) and C(4a).

2. Structural Isomers (Tautomers) of Dihydroflavin

Many structural isomers (tautomers) of dihydroflavin can exist theoretically. Again, we may differentiate between two "subclasses" according to whether blocking groups (H-bonds from the apoprotein in the enzymatic, alkyl groups in the chemical system) are retained at N, O(1, 2a) or N(5). If N(1) or O(2α) is blocked, the second hydrogen or alkyl substituent may be situated at either of the positions 5,6,8,9a or 10a. If N(5) is blocked, the corresponding positions may be 4a or 1. Common to both subclasses is the only isomer, which is a N,N-dihydroflavin and, therefore, subject to thermodynamic reversibility of oxidoreduction. This 1,5-dihydroflavin is termed, consequently, "flavohydroquinone" and has been assumed, since the early thirties (*101*) and up to the late sixties, to be the only dihydroflavin isomer capable of existence in biological as well as chemical systems. Its properties will be discussed in the following paragraph.

Derivatives of other 1,n-dihydroflavins will be discussed in the chapter on "flavin oxygen adducts" (see below), where position n means a carbon atom substituted by a strongly electronegative group such as OH, OR, OOH, etc. The dilemma encountered in the determination of the redox state (chromophore = dihydroflavin, redox balance = flavoquinone) of such compounds, e.g. "hydroxydihydroflavins", has been discussed in the Nomenclature chapter (*cf.* above).

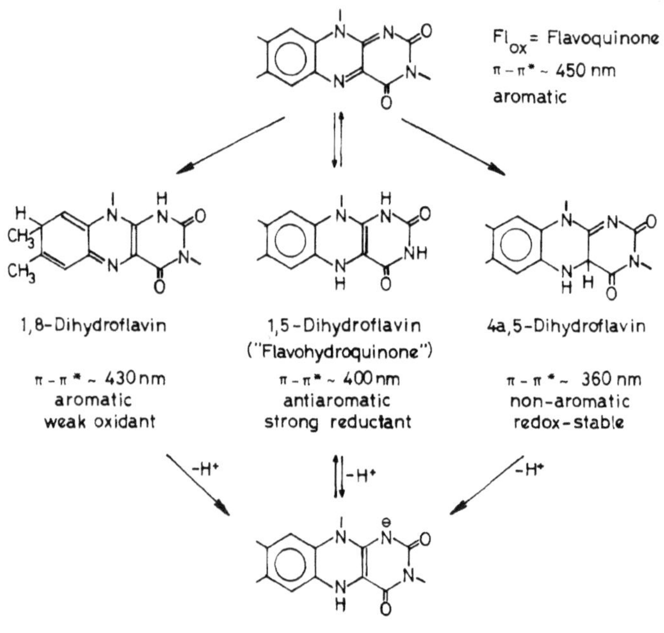

Scheme 9. Oxidized and reduced flavin chromophores. Note that the term "antiaromatic" applies only to the entirely flat (vibrationally excited) state of the bent flavohydroquinone

As "true reduced flavins", however, and apart from flavohydroquinone, one example of each subclass needs discussion, viz. 4a,5-dihydroflavin("5-blocked") (*181, 182*) and 1,8-dihydroflavin("1/2α-blocked") (*13*). Both isomers exhibit chromophores which are essentially different from flavohydroquinone, as outlined in Scheme 9. The three dihydro isomers exhibit, furthermore, totally different chemical reactivity: 1,8-dihydroflavin is aromatic with respect to electron distribution in the center (pyrazine) subnucleus, 1,5-dihydroflavin is, contrarily, antiaromatic (unless the pyrazine subnucleus is deplanarized, see below). It contains

2 π-electrons more than the other two isomers. Consequently, 1,8-dihydroflavin is still an oxidant, though a weak one, which may lead to "overreduction" of flavoquinone to yield tetrahydro- and hexahydroflavins (*60*), while 1,5-dihydroflavin is a strong reductant which cannot be reduced further by any means whatsoever. Its formation is, however, reversible and (irreversible) overreduction and can, therefore, proceed via reoxidation and isomerization. 4a,5-Dihydroflavin, on the other hand, is non-aromatic and, therefore, electrochemically stable. Its potential activity as hydride donor is an unresolved problem.

The 1,8- and 4a,5-isomers are hitherto only known as 8- (*181, 182*) or 4a-alkyl-derivatives (*13*). The alkyl groups migrate intramolecularly between these positions and position 5, depending on pH, (*53 8*). "Active"residues such as benzyl and benzyl homologs $C_6H_5CR_2$- (R = H or CH_3) reach isomerization equilibrium within seconds at room temperature. This demonstrates that flavin is an active surface which is capable of digesting σ-pairs as well as single or paired π-electrons, at specific sites (*cf.* Scheme 5).

Critical evaluation of the nature of alkyl residues stabilized at C(4a) and/or N(5) allows one to conclude that nucleophilic residues prefer C(4a), while electrophilic ones prefer N(5) (*98*). Hence, it has been proposed to consider the following reaction sequence for flavin-dependent (de)hydrogenation:

$$Fl_{ox} + RH \overset{A}{\rightleftarrows} 4a\text{-}R\text{-}Fl_{red}\text{-}5\text{-}H \overset{B}{\rightleftarrows} 5\text{-}R\text{-}Fl^{-}_{red} + H^{+} \overset{C}{\rightleftarrows} 5\text{-}H\text{-}Fl^{-}_{red} + R^{+}. \qquad (10)$$

A = nucleophilic group transfer
B = isomerization and dissociation
C = electrofugic protolysis
A + B + C = "hydride" transfer

This is schematically outlined for R = NAD in Scheme 10. While for "simple" residues R such as benzyl, step B constitutes a true equilibrium of sigmatropic group shift plus subsequent deprotonation, it must be assumed in the case of R = NAD that steps B and C are synchronous, because of the pronounced leaving group aptitude inherent in the stable cation NAD^+ as compared to simple carbonium ions. "Charge transfer" complexes (*7*) may or may not be formed preceding step A and following step C, depending on the extent of possible π-overlap between Fl_{ox} and RH on the one hand and HFl_{red} and R^+ on the other hand. The second type of CT-complex will, of course, be more stable, since it is made up of ions of opposite charge. In any case it is not the "charge transfer energy" which is responsible for π-complex stability, but hydrophobic stacking forces in the first case and Coulomb attraction in the second one.

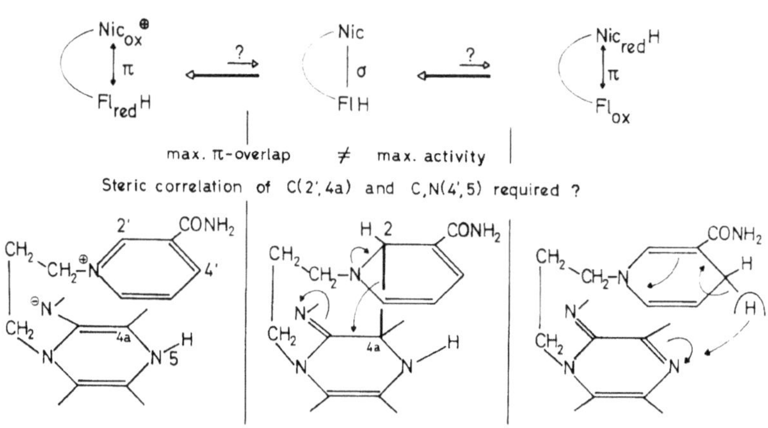

Scheme 10. Nicotinamide-Flavin e^-(sigma)-transfer through π-complex. The transfer rate is meant to be not proportional to total π-overlap, but rather to active site contact yielding synchronous $2e^-(2'\text{-}4a)$ and $H^+(4'\text{-}5)$ transfer. For the covalently linked flavin-nicotinamide biscoenzymes *cf.* (*7*)

The sequence A→B→C implies loss of H-label, provided hydrogen at position 5 is mobile. 5-Deazaflavins (*33*) are, therefore, not suitable models to study this effect. If the 5-center is nitrogen, the mobility of 5-H is still a question which needs to be resolved thoroughly. Hydrogen exchange at nitrogen centers usually operates by way of either protonation or deprotonation via hydrogen bridges. In the present case, however, the pyrrole type N(5) is neither basic (pK_a of $Fl_{red}H_3^+ < 0$) nor acidic. If, therefore, reversible reoxidation towards the N(5)H-acid $\dot{F}lH$ (*cf.* above) is efficiently excluded (which is difficult in chemical, but easy in biological systems), N(5)H exchange may be slowed down considerably and even more so in nonpolar environment. Thus, steps B, C might occur without exchange of hydrogen at N(5).

3. Planarity of Dihydroflavin

Next to isomerism, the planarity of a dihydroflavin is the most important structural problem (*93*). For 1,8- and 4a,5-isomers this question is relatively easy to answer in that these chromophores are planar, apart from the one tetrahedral carbon center where the hydride equivalent is localized. In the 1,5-isomer, or flavohydroquinone, however, the two electron equivalents of reduction are delocalised to an extent which depends on the "pyramidality" of the N-centers 5 and 10. And this pyramidality is dependent on the nature of the substituent at N(5), being minimal for hydrogen and high for an alkyl group R in 5-R-Fl_{red}H. Furthermore, it depends on the state of ionisation, the anion RFl^{-}_{red} formed at a pK of 6-7 being more bent than the conjugate acid. The degree of bending is reflected by the long wave transition (*21*).

Obviously, it is not primarily the size of the substituent which governs the hypsochromic absorption shift, since there is such a drastic difference between R=H and R=CH_3, although CH_3 can be easily accommodated in the flavin plane as evidenced by the stability of the semiquinone 5-CH_3-$\dot{F}l$ as compared to the unstable 5-$C(CH_3)_3$-$\dot{F}l$. Correspondingly, 5-acyl-dihydroflavins exhibit no amide character in the C-5-N bond (*171*). On the contrary, the N(5) in these compounds is highly pyramidal, which in turn renders the acyl group active towards solvolysis (Fig. 1).

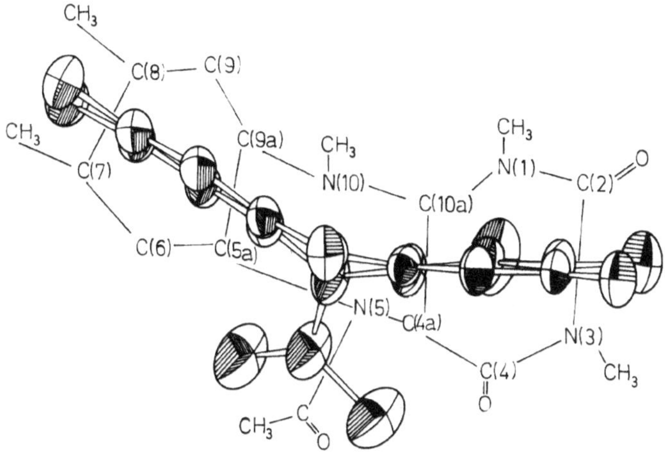

Fig. 1. Stereoformula of 1,3,10-trimethyl-5-acetyl-1,5-dihydroflavin (*93*) showing the N5-N10 folding and the noncoplanarity of the 5-Acetyl-group (with kind permission of Dr. Kierkegaard)

Recent x-ray studies on reduced flavodoxins (*109*), on the other hand, reveal a dihydroflavin moiety which appears largely planar at 2.5 Å resolution. But there is one main fact one should not forget in the context of flavodoxins: These proteins do serve as ferredoxin substitutes and are unique in being involved exclusively with one-electron transfer, i.e. they are "1/1-flavoproteins" (see above). Furthermore, delocalisation of the N(5)-nonbonding electrons in flavohydroquinone can be inhibited in a largely planar flavin nucleus, if only the N(5)-H and N(10)-R bonds are kept out of the plane. Of course, apoproteins may either planarize or deplanarize flavohydroquinone to a greater extent than is found in the free molecule. The more planar the molecule, the more negative the redox-potential; this is in keeping with flavodoxin function as an electron sink in biological redox chains.

Free flavohydroquinones have been shown to undergo rapid inversion of the N(5) and N(10) atom even in the presence of bulky substituents, which demonstrates that the "butterfly wing" conformations can interconvert without ever going through the totally planar state (*171*). Hence, the question of flavohydroquinone planarity requires careful examination in each single flavoprotein. This can be done most reliably by detailed analysis of the absorption spectra and the recently discovered dihydroflavin fluorescence (*46*). The free molecule is nonfluorescent in liquid solution, probably owing just to the "butterfly wing motion". Freezing of this motion results in the appearance of this fluorescence in protein-free solid solutions as well as in certain flavoproteins. In these enzymes, dihydroflavin motion seems to be arrested, which creates at least one new center of chirality in the reduced enzyme. Using this idea, the retention of label (*20, 108*) observed in a variety of flavoprotein-dependent hydrogen transfers, may be explained in a way which is more specific and satisfying than the assumption of "reduced water accessibility" at the active site.

The bent state of flavohydroquinone is more prone to substitution than the flat state, which in turn is more susceptible to oxidation. This is borne out by the following observations:

1. The bent 5-R-Fl_{red}^- accepts a second alkyl equivalent (to yield the zwitterion 5,5-$^+R_2Fl_{red}^-$) more readily than 5-H-Fl_{red}^- the first one (*43*).

2. Alkyl substituents at N(5) as well as N(1) strongly reduce the O_2-affinity of dihydroflavin (see below), while they delete at the same time the long wave end absorption of the Fl_{red}-chromophore (*43*).

VII. Flavin-Substrate Complexes

Unsubstituted dihydroflavin-anions might, as has been stated previously, be considered as "Flavin-hydride complexes", though one might infer that the term "complex" is abused if used in this way. In fact there are two ways to define the term "complex". Historically, this term means that one does not know how to explain the structure of a given species in terms of chemical bonds. Nowadays, however, pure chemistry attempts to comprehend practically everything, including Amor and Psyche, in terms of chemical bonds, and the term "complex" is confined, in pure chemistry, to a species in which the coordination number of the central atom exceeds its oxidation number. In biochemistry, however, the smell of mystery still adheres to the term "complex", and if there is any definition at hand, except for the "ignoramus"-confession, it means a dissociable "species of high thermodynamic and low kinetic stability". And since σ-bonds are felt (rather than thought) to exhibit high kinetic stability under all conditions, a biological "complex" means two halves kept together by anything but σ-bonds. This is, of course, more a feeling than an idea, and it should be discarded, wherever it turns up, since there exist, of course, many types of kinetically unstable σ-bonds, even such ones which are unstable towards *homo*lysis.

A σ-bond should be defined as a pair of electrons delocalised (to a first approximation) over two atomic centers in a distribution of axial symmetry, having maximal charge density at some point of the axis. There are, of course, gradual transitions between σ- and π- or ionic bonds, as evidenced by the phenomena of hyperconjugation, homoaromaticity, valence tautomerisation, proton delocalisation, three-center or fluctuating bonds (*95, 106*) etc. Apart from these, biology offers two bonding phenomena as solutions for all pending problems of "complex" structure: Hydrophobic bonding and charge transfer (*118*), the first one usually accounting for the energy, the second one for spectral behavior.

Hydrophobic bonding is largely a matter of water structure. In nonpolar systems hydrophobic bonding is negligible. Since it has been established that flavoprotein active sites are frequently nonpolar, hydrophobic bonding is reduced to a purely entropic "key-in-the-hole"-phenomenon. If a key is put into a hole, charge transfer will conceivably occur between key and hole, unless both are made from insulating material. Nobody, however, will claim that the key fits into the hole because charge transfer occurs subsequently. In the present context, the term "charge transfer" shall be confined strictly to π-overlapping systems of "reasonable" donor-acceptor strength. The question stands, whether charge transfer is indispensable: Charge transfer precedes electron transfer, which transforms the donor-acceptor pair into a spin-paired radical ion pair, which, if mobile protons are available as in flavins, can rapidly tautomerize to yield a spin-paired diradical π-complex (Sequence A):

$$0)\ A_{ox} + B_{red}H_2 \rightleftarrows (A_{ox} \overset{\pi}{—} B_{red}H_2) \quad \pi\text{-complex formation}$$

$$1)\ (A_{ox} \overset{\pi}{—} B_{red}H_2) \rightleftarrows (\dot{A}^- \overset{\pi}{—} \dot{B}H_2^+) \quad e^-\text{-transfer}$$

$$2)\ (\dot{A}^- \overset{\pi}{—} \dot{B}H_2^+) \rightleftarrows (H\dot{A} \overset{\pi}{—} \dot{B}H) \quad \text{tautomerisation}$$

$$3)\ (H\dot{A} \overset{\pi}{—} \dot{B}H) \begin{cases} \longrightarrow H\dot{A} + \dot{B}H & \text{radical dissociation forbidden} \\ \rightleftarrows (HA^{-}_{red} \overset{\pi}{—} B^{+}_{ox}H) & e^{-}\text{-transfer} \end{cases}$$

$$4)\ (HA^{-}_{red} \overset{\pi}{—} B^{+}_{ox}H) \rightleftarrows (H_2A_{red} \overset{\pi}{—} B_{ox}) \quad \text{tautomerisation}$$

$$5)\ (H_2A_{red} \overset{\pi}{—} B_{ox} \rightleftarrows H_2A_{red} + B_{ox} \quad \pi\text{-complex dissociation}$$

Sequence A. Double $1e^{-}(\pi)$-transfer

Two criticisms apply to this concept:

1) Such "invisible" (because of diamagnetism) diradicals, even short-lived ones, are unknown for the lighter elements. Spin correlation in the absence of charge correlation has not been satisfactorily established up to now.

2) Such electron transfer is easily established, but it is hard to imagine, how nature could avoid at least occasional spin decoupling and subsequent homolysis of the diradical (*cf.* A3) strictly in favour of the rapid transfer of a second radical electron from the donor in the same way (*cf.* A4). Hence, $1e^{-}$-transfer catalyzed by π-complexation is easy to conceive, while catalysis of "synchronous" $2e^{-}$-transfer by π-complexes is not.

Two-electron transfer seems more easily conceivable by reaction Sequence B. When applied to enzymes, this scheme should, furthermore, include formation of "hydrophobic" pre-complexes, since the Sequence B 1—5 is meant to happen within the protein in the same way as a chemical reaction is happening in a solvent cage — the protein "cage" being, of course, much more stereospecific than any solvent cage.

$$1)\ B_{red}H_2 \rightleftarrows B_{red}H^{-} + H^{+} \quad \text{deprotonation}$$

$$2)\ B_{red}H^{-} + A_{ox} \rightleftarrows (HB \overset{\sigma}{—} A)^{-} \quad \sigma\text{-complex formation}$$

$$3)\ (HB \overset{\sigma}{—} A)^{-} \rightleftarrows (B \overset{\sigma}{—} AH)^{-} \quad \text{tautomerisation}$$

$$4)\ (B \overset{\sigma}{—} AH)^{-} \rightleftarrows B_{ox} + A_{red}H^{-} \quad \text{dissociation}$$

$$5)\ A_{red}H^{-} + H^{+} \longrightarrow A_{red}H_2 \quad \text{reprotonation}$$

Sequence B. $2e^{-}(\sigma)$-transfer

Sequence B 1—5 remains easily applicable, when the starting complex has charge-transfer character. On the contrary, the alternative picture of essential charge-transfer and double $1e^-$-transfer complexes (Sequence A) remains useless in all those cases, where the substrates do not lend themselves to the formation of charge transfer complexes — and such is the majority of flavoprotein substrates.

Step B 1, i.e. deprotonation, has been verified recently for CH-substrates of flavoproteins by WALSH and coworkers (*187*), thus giving rise to the term "carbanion mechanism", though this mode of reaction has been predicted as early as 1963 by the present author (*62*). σ-Complex formation is to be understood as nucleophilic attack of a substrate carbanion at the flavin site. Depending on the nature of the substrate residue, this σ-contact might be very short-lived. Suitable "substrate models", however, yield much more stable "σ-complexes", namely *alkyldihydroflavins,* as discussed in the Dihydroflavin section above.

Another problem encountered in flavin-substrate complexes is the determination of the flavin redox state, as outlined above. It is, of course, the spectral behavior, which decides the assignment. If the complex is of σ-type, its spectral behavior will simulate a dihydroflavin. Determination of protonation state is then the problem to be resolved. If on the other hand it is a π-complex, two possibilities are open according to Sequence A. In the case of nicotinamide substrates, both types of π-complexes, the red Fl_{ox}-$Nic_{red}H$ and the green Nic^+_{ox}-Fl^-_{red}, can be verified in chemical as well as enzymic systems (*7*). But it remains to be seen whether they are *indispensable* prerequisites for exchange of redox equivalents between nicotinamides and flavins. Maximal π-overlap is apparently favourable for oxidoreduction, but it is an open question whether any other steric conditions might do as well. Scheme 10 accounts for what may be called "σ-transfer" of $2e^-$-equivalents within a π-complex.

VIII. Flavin-Acceptor-Complexes

Mechanisms of flavohydroquinone oxidation are the least understood feature of dihydroflavins. The principal biological electron acceptors from reduced flavin are the following ones:

a) *Ferric iron* in cytochromes and ferredoxins connected with flavoprotein e^--transferases. Fe^{III} is trivially one-electron accepting, but leaves the question as to the site of electron transfer in the flavin as well as the acceptor molecule entirely unresolved.

b) *Flavin itself* e.g. in ETF, Sulfite Reductase (*126, 162*). In this case it is probable that only one electron equivalent is transferred at a time at the output site, the semiquinone thus being essential. This has at least been ascertained for the free flavin system in an aprotic solvent (DMF) by recent studies of FAVAUDON and LHOSTE (*32*). These authors demonstrated that flavoquinhydrone, i.e. the charge transfer complex $Fl_{ox} \overset{\pi}{\leftarrow} Fl_{red}H_2$, is not on the pathway of this reaction. Hence, "interflavin" oxidoreduction must occur by models of interaction more refined than those involving a mere π-overlap. Such contact apparently constitutes physically well defined, though short-lived "flavin dimers", linked by a σ-bond, $HFl \overset{\sigma}{—} FlH$, which tends to homolyze towards $2\,H\dot{F}l$. It has been suggested earlier, that such an interaction might occur through a hydrazine-like N(5)-N(5′) bond (*74*), but this has been disproved by FAVAUDON and LHOSTE (*32*).

c) *Molecular oxygen* in all flavoprotein oxidases and oxygenases (*120*). Since even those flavoproteins specifically linked with metal protein acceptors, i.e. the "e^--transferases" (see above), can react — if only slowly — with O_2, it is obvious that O_2 can, in connection with flavin, react in a threefold mode:

– as *1e⁻-acceptor*, producing superoxide, in the "false" reoxidation of e^--transferase flavoproteins,

– as *2e⁻-acceptor*, producing hydrogen peroxide, in the reoxidation of flavoprotein oxidases, and

– as *4e⁻-acceptor*, producing water and one "oxene" equivalent for incorporation into substrate, in the action of flavoprotein oxygenases.

A clue to the understanding of these biological oxidations was found by GIBSON and HASTINGS (*50*), who observed in the first rapid kinetics study of flavohydroquinone autoxidation that this reaction was *autocatalytic*.

This peculiar autocatalysis was ascribed to the formation of flavin radicals, which were thought to accelerate the oxygen consumption by comproportionation

$$HFl_{red}^- + Fl_{ox} \underset{+H^+}{\overset{-H^+}{\rightleftarrows}} 2\,\dot{F}l^- \xrightarrow{2\,O_2} 2\,Fl_{ox} + 2\,\dot{O}_2^- . \tag{11}$$

This is indeed true for the free, aqueous flavin system under alkaline conditions, as was shown by MASSEY *et al.* (*122*), who provided direct evidence for superoxide formation.

The aqueous, alkaline flavin system, however, turns out to be a "non-model" for oxidase and oxygenase action in the light of the following data:

1. TOLLIN *et al.* (*179*) showed by flash photolysis studies, as was most convincingly confirmed by pulse radiolysis (*31*), that in contrast to $\dot{F}l^-$, neutral $F\dot{H}l$ is inert towards oxygen. This is consistent with the high

kinetic stability of many "blue flavoprotein radicals" towards oxygen. Hence, the autoxidative decay of $H\dot{F}l$ can only occur via deprotonation or disproportionation.

2. A fast reaction sequence of 100% one-electron autoxidation could, consequently, only apply to flavohydroquinone in aqueous *alkaline* solution:

$$HFl_{red}^{-} + 2\,O_2 \longrightarrow Fl_{ox} + 2\,\dot{O}_2^{-} + H^{+}, \tag{12}$$

which is not consistent with the comparatively low yield of superoxide found by rapid freeze experiments (*122*). This low yield decreases practically to zero with decreasing pH, an observation which was explained by the fact that superoxide selfdismutation is accelerated at decreasing pH. Since, at the same time, superoxide was assumed to arise from homolytic cleavage of a "flavin-oxygen complex" HFlOOH, it seems more obvious to assume that this homolysis is suppressed with decreasing pH (*66*).

$$HFl\text{-}OOH \underset{pH<8}{\overset{pH>8}{\rightleftarrows}} \dot{F}l^{-} + \dot{O}_2^{-} + 2\,H^{+}. \tag{13}$$

For there is no reason, why this radical formation should be irreversible, provided one of the radical reactants is efficiently protonated. Hence, we assume that the "one-electron" reaction is, even under the most suitable aqueous alkaline conditions, only partial and that there is efficient competition by the "two-electron" reaction

$$HFl\text{-}OOH \longrightarrow Fl_{ox} + H_2O_2. \tag{14}$$

This reaction should be pH-independent for HFlOOH, unless HFlOOH has an acidic NH, which would favour formation of an anion $^{-}$FlOOH at neutral pH (see below).

3. The autocatalysis of flavohydroquinone autoxidation is still found, though to somewhat smaller extent, in the presence of dismutase (*122*) (which removes superoxide) and in systems, which totally disfavour radical formation (*66*) (see above, 1/2α-blocked systems). Hence, one is left with the choice of either accepting Fl_{ox} as the proper autocatalyst, or of postulating an autocatalytic reaction of the intermediate HFlOOH with the starting H_2Fl_{red}. Such a reaction has indeed been found as early as 1965 by Mager *et al.* (*110—115*), who provided evidence for the stoichiometry

$$2\,H_2Fl_{red} + O_2 \longrightarrow 2\,Fl_{ox} + 2\,H_2O, \tag{15}$$

excluding any intermediate formation of free H_2O_2, which would practically be inert towards H_2Fl_{red} at the observed concentrations. This stoichiometry applies to the reaction in *nonpolar media* and it implies the third mode of cleavage (see above) of the intermediate HFlOOH. While aqueous alkaline conditions favour one-electron reduction of O_2, neutral to acid aqueous conditions favour two-electron reduction and nonpolar conditions, finally, lead to four-electron reduction:

$$HFl\text{-}OOH \longrightarrow Fl_{ox} + [O] + H_2O. \quad (16)$$

By this transfer of an "oxene" equivalent is meant to occur within a ternary complex made up from HFlOOH and a substrate to be oxygenated. In the free system, this happens to be a second molecule of starting H_2Fl_{red}, unless a ternary component, such as e.g. phenylalanine (*115*), is added.

In Scheme 11 we have tried to assign to each mode of preferred HFlOOH-cleavage a specific HFlOOH-isomer.

– HOO- at C(4a) is assumed to lend itself preferably to homolytic fission, because of the high spin density found at C(4a) in $H\dot{F}l$ (*26*) thus resulting in $1e^-$-reduction of O_2. The chromophore is known from 5-R-Fl-4a-OH (*181*) to have $\lambda_{max} \approx 370$ nm.

– HOO at C(9a) (*80*) is thought to prefer splitting of OOH^-, since it is the only HFlOOH-isomer which can be expected to form the anion $^-$FlOOH readily at neutral pH. Its pK compares well with that of a 7,8-dihydrolumazine derivative (*27*).

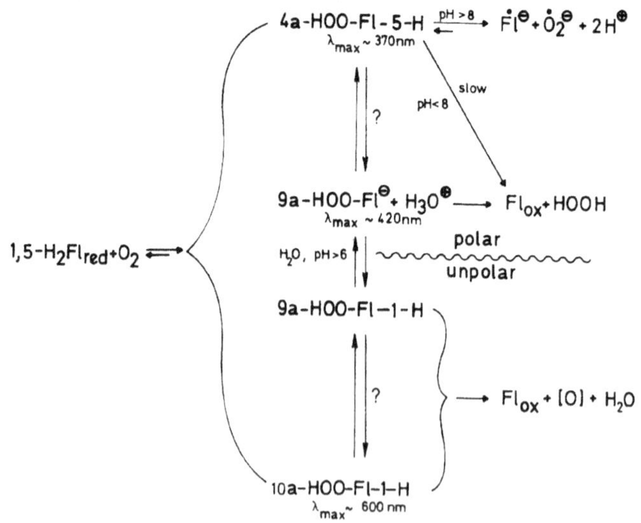

Scheme 11. Formation and cleavage of HFlOOH-isomers

Furthermore, the 1-H-Fl-9a-OOH chromophore is, in our opinion, known from 1-R-Fl-9a-OR′, which is the structure we assign to the "alcohol adducts" obtained first by F. MÜLLER (*133*). This author preferred originally and rather intuitively, like many other colleagues (*112, 121*), the isomeric 1-R-Fl-10a-OR′ structure. The following arguments oppose this assignment:

1. Transfer of the blocking substituent from N(1) towards O(2α) shifts the absorption by 25 nm bathochromically (*80, 134*). Hence, the iminoether group N(1)=C(2)-OR must be conjugated with the main chromophore, which applies to the 9a-, but not to the 10a-isomer (*cf.* Scheme 12).

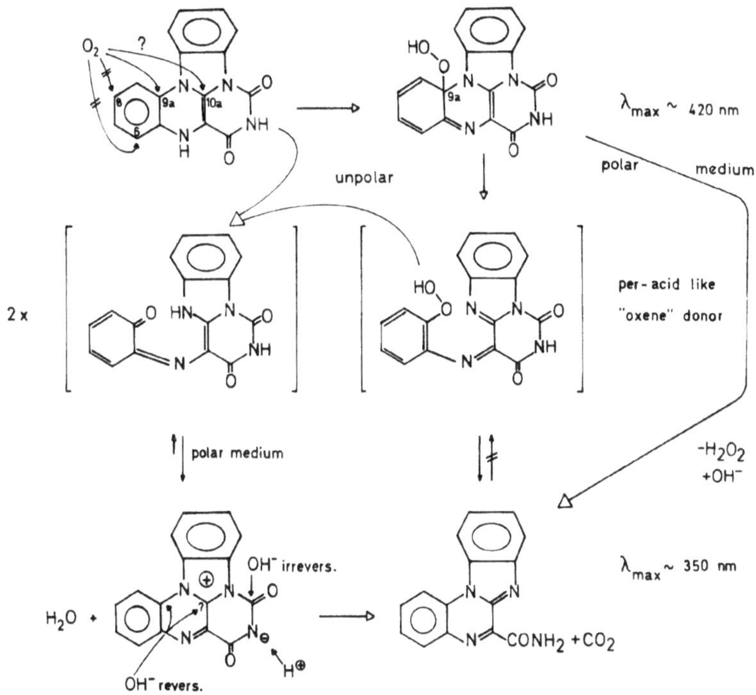

Scheme 12. C(9a)-linked "oxene" ($4e^-$) activation of O_2. Note that oxygenation is taking place only in unpolar media and that "self-oxygenation" of starting dihydroflavin is competing with oxene-transfer upon extraneous substrates

2. NMR-spectroscopy does not discriminate between 9a- and 10a-isomers, though it eliminates potential 6- and 8-analogs, contrary to prior expectations (*80, 134*).

3. Basing the structure of HFlOOH on the structure of irreversible degradation (*110*) products is not permissible, since all possible HFlOOH-isomers must be assumed to equilibrate easily, and the question is as to the energetically most favourable isomer irrespective of side reactions which start from the other ones. Hence, the irreversible decay of 1-R-Fl-10a-OH to yield "spirohydantoin" speaks, if anything, against the 10a-isomer being the biologically relevant one.

4. 10a-additions require deletion of an amidine resonance, while 9a-addition concerns a quinone-imine carbon of much higher electrophilicity.

5. Satisfactory "models" are at hand for the 9a-chromophore, e.g. 7,8-dihydropterin-6-carboxylates, (*27*), which absorb around 400 nm. The 10a-chromophore, however, can at best be mimicked by a model which appears at first sight far-fetched, i.e. alloxane-dimethyl-aminoanil (*72*) (*cf.* Scheme 13). But in fact 1-R-Fl-10a-OOH is nothing but an *ortho*-dialkylamino-anil of a 3,4-dihydro-alloxane. Such alloxane derivatives are well-known to have deep blue colours, λ_{max} 600 nm (*94, 149*). In fact we know such blue flavin derivatives, as will be outlined below.

6. The 1,10-benzo-bridged dihydroflavin derivative of KNAPPE (*97*) yields a 420-nm-intermediate upon autoxidation, and even a relatively stable one, though addition at 10a must clearly be disfavoured in this

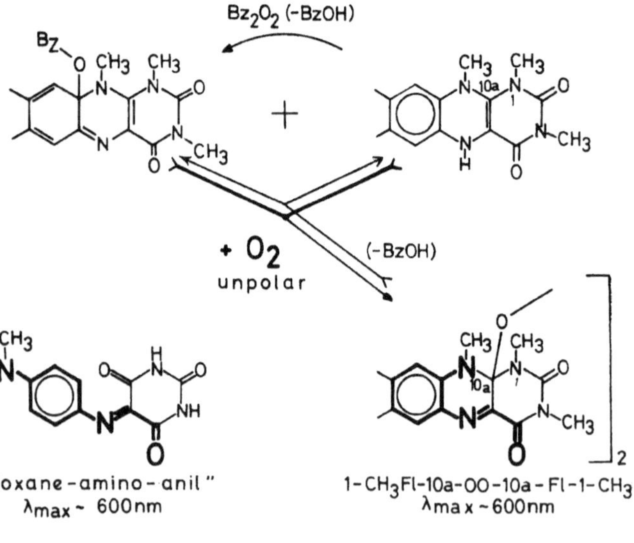

Scheme 13. Formation of "Blue Flavin-Oxygen Complex": Bz = benzoyl. The overall reaction is $2\,CH_3Fl_{red}H + Bz_2O_2 + O_2 \rightarrow (CH_3Fl\text{-}O\text{-})_2 + 2\,BzOH$. The known blue alloxane-amino-anil chromophore is given for comparison

compound owing to additional resonance in the bridging benzimidazole subnucleus.

In view of all these arguments we tend to believe that not position 10a, but only position 9a in 1/2α-blocked flavins can be the site of nucleophilic attack in the "Müller-adducts" 1/2α RFl-?-OR′ (*133*). As for electrophilic attack by O_2 in flavohydroquinone, however, position 10a might be worth some consideration, if the following conditions can be met:

1) The hydrogen activity of the environment should be low for 1-R-Fl-10a-OOH, in order to avoid irreversible rearrangement (*23*).

2) 10a-adducts may occur, if at all, with N(1)-blocked flavins, *cf.* the only known (to our knowledge) heterocyclic o-amidine bridge structure in the so-called "uric acid glycole" (*72*), but not with (2α)-blocked flavins.

3) The chromophore should be blue, not red.

These three conditions are fulfilled by a blue derivative observed as early as 1960 (*77*) upon autoxidation of 1,3,10-trimethyl-1,5-dihydroisoalloxazines: This blue intermediate was mistaken at the time for the neutral radical 1-$CH_3\dot{F}l$, which in fact is red, not blue, as was found later on, and extremely unstable (*cf.* above). In the mid sixties we found in collaboration with Dr. EHRENBERG, that this blue intermediate was EPR-silent and diamagnetic on the magnetic microbalance. This unresolved problem was reinvestigated in our laboratory in 1973 in the context of MAGER's findings about flavin-dependent $4e^-$-reduction of O_2. To our great surprise, we established structure 1-CH_3Fl-10a-O-O-10a′-FlCH_3-1′, for the blue intermediate as outlined elsewhere (*80*).

As to the mechanism of its formation, it becomes obvious that a first intermediate, reactive CH_3FlOOH is trapped by $CH_3Fl^+_{ox}$, while otherwise it would attack a second $CH_3Fl_{red}H$ to yield the colourless decay product CH_3FlOH.

It is now an open question, whether the blue 10a-structure occurs at all in nature, since up to now it has not be seen in flavoproteins. It seems that the red 9a-intermediate is used in both oxidases and oxygenases, depending on additional steering by the weaker H-bridge at O(2α), which leaves a partial negative charge for splitting of OOH^-, or the stronger H-bridge at N(1), which leads to 9a-10 ring opening (see below).

Two questions remain to be settled:

1) How can nature circumvent the spin conservation law which opposes rapid formation of the singlet HFlOOH from $HFl_{red}H$ and triplet O_2?

2) How can a peroxide of the type HFlOOH, bearing the OOH function at a saturated, tetrahedral carbon center, be an "oxene donator",

a capacity which is, chemically, strictly reserved for per-acids and their vinylogs, i. e. OOH at unsaturated, trigonal carbon centers?

Question 1 is considered more or less problematic by molecular physicists, since relaxation of multi-spin systems is a matter which is still not fully explained by theory. We have advanced an hypothesis (*78*), however, which is based on the fact, that O_2 reacts with *flat* reduced flavin, not the bent one. The slower N-inversion is in flavohydroquinones, the more slowly autoxidation proceeds. Flat flavohydroquinone is, however, an energy-rich conformation, at least in the diamagnetic singlet ground state, as we found (*171*). It might be that there is a lowlying triplet state which could, in contact with O_2, be populated and rearrange immediately to yield one of three possible HFlOOH isomers.

The second question has been tackled first by HAMILTON (*55*) and then by MCCAPRA (*127*), who made it clear that any "saturated" HFl-OOH had to undergo rearrangement first to yield a per-acid vinylog before "oxene-transfer" could occur. MCCAPRA proposed ring opening at 1-10a in the pyrimidine subnucleus in order to account for this requirement, neglecting the fact that opening of this bond is irreversible because of "spiro rearrangement". HAMILTON proposed a 4a,5-ring opening which, however, must lead to rapid internal oxidoreduction between the strongly reducing o-phenylene diamine and the strongly oxidizing alloxane-oxide halves.

We prefer, therefore, such an intermediate cleavage to occur at 10-10a or 10-9a in the center pyrazine subnucleus, which is well known to be rapidly reversible from the old chemical synthesis of the flavin nucleus.

Furthermore, KNAPPE's model (Scheme 12) shows that oxygenation is still taking place even if 10,10a-ring opening becomes most improbable.

IX. New Flavocoenzymes

1. Historical Background

Since THEORELL's discovery of FMN as the prosthetic group of "Old Yellow enzyme" (*172—175*) and WARBURG's recognition of FAD as the coenzyme of D-amino acid oxidase (*189—191*), flavin has been recognized as the cofactor of a rapidly growing number of redox enzymes. In most flavoenzymes investigated so far, the coenzyme is tightly bound to the (apo)protein ($K_d < 10^{-6}$ M), but can be released by most protein-denaturing procedures (e. g. heat, methanol, or acid treatment). In many cases reversible dissociation can be achieved by dialysis at high ionic strength (e.g. 2M KBr (*117, 123*)) or acid ammonium sulfate precipitation of the apoprotein (*191*). These characteristics became the basis for analytical methods to estimate the flavin content in biological tissues (*145*).

It took twenty more years until Boukine (*9*) and Green (*52*) independently observed that a substantial part of the total tissue flavin could in fact not be extracted by the conventional methods mentioned above and required proteolytic digestion for its solubilization. It thus became apparent that two fundamentally different types of flavocoenzymes existed, the first species being solubilizable, the second "covalently bound" to the protein. At the same time it was recognized that a main fraction of "insoluble flavin" was associated with the succinate dehydrogenase of inner mitochondrial membranes (*87, 165*). Subsequent efforts concentrated on the isolation and characterization of small flavin peptides from succinate dehydrogenase, with the aim of elucidating the chemical nature of the link between the flavin moiety and the protein backbone and, at the same time, established the point of substitution on the flavin nucleus. Kearney, in 1960, was the first to isolate a homogeneous peptide upon trypsin/chymotrypsin digestion of beef heart succinate dehydrogenase (*84*) which contains a covalently bound FAD moiety. Alkaline photolysis removed the nucleotide part of the molecule and yielded a "lumiflavin peptide" by means of which the direct link between the protein and the heteroaromatic part of the flavocoenzyme was established. The diphosphate bridge of the FAD peptide could be removed enzymatically or by mild acid hydrolysis (*84*), but the isolated "riboflavin peptide" turned out to be even more light sensitive than free riboflavin. This increase in photosensitivity is due to removal of the protecting adenine "hairpin" configuration, which accelerates the dehydrogenation of its own ribityl side chain by the excited flavin triplet state. As a consequence of attack on the peptide or ribityl side chain, this photolysis leads under all conditions to a mixture of products; accordingly the yield during Kearney's "lumiflavin degradation" of the FAD peptide was very small. In a similar way, tyrosine residues in the neighbourhood of the flavin-peptide link contribute to photostability. Hence, further enzymatic or acid degradation of a flavin oligopeptide enhances photosensitivity to such an extent that all work done with preparations at $pH > 2$ has to be carried out under absolute exclusion of light.

Mainly for these reasons, progress came to a halt, except for a report by Chi *et al.* (*18*) in 1965 who carried out alkaline hydrolysis of a "Kearney type" peptide and found unmodified urea as the only low molecular weight product from cleavage of the pyrimidine subnucleus of the peptide linked flavin. Hence, the peptide residue could not be linked to the flavin N(3) position.

Since position 10 was excluded by KEARNEY, position 3 by CHI *et al.* and links to positions 1, 2, 4 and 5 would have yielded chromophores with spectral properties drastically different from those of succinate dehydrogenase flavin (SD-flavin), (*21, 43, 137*), we concluded in 1969 (*67*) that the peptide had to be connected to the benzene subnucleus of the flavin, i.e. to positions 6—9. Position 8α was the most probable candidate as it had been shown to possess the highest chemical activity in the benzene subnucleus (see below) and on biosynthesis of SD-flavin was expected to involve unmodified riboflavin as precursor even at this early time. Furthermore, modifications of the 8-CH_3 group were found to have the strongest influence of the ESR hyperfine pattern of the flavin radical (*135*).

The first goal was, therefore, a "SD-flavin" preparation, which would yield direct evidence for 8α-substitution. This was first achieved by drastic acid hydrolysis of a flavin peptide preparation and subsequent direct inspection of the radical cation obtained by reduction of the hydrolysate with $TiCl_3$ in 6 N HCl. At pH < 1 flavin photolysis is negligible because of dynamic proton quenching of the photoexcited flavoquinone cation (*107*) and the semiquinoid state is highly stable. This allows hyperfine analysis of "SD-flavin" even in the presence of large amounts of non-flavin peptides and of mixtures of flavin peptides if the attachment of the peptides to the flavin nucleus is the same. The ESR spectra obtained pointed conclusively to 8α as the position of substitution (*67*).

Next, the peculiar pH dependence of the fluorescence which shows a pK ≈ 4.7, not found in "normal" flavins (Fig. 2), and was first observed by KEARNEY (*84*), had to be explained. Studies with 8α-substituted models led to the conclusion that a basic function of suitable pK had to be connected directly to C(8α) and that among all commonly encountered basic protein residues only imidazole could reasonably fit the experimental observations (*67, 154*).

The biochemical aspects of 8α-modified flavins have been reviewed recently (*166*), whereas the approaches used to identify the site of attachment of the flavin have been discussed in detail by SINGER and EDMONDSON (*164*). Hence, the present chapter concentrates on the chemical and spectral properties of biologically relevant modified flavocoenzymes which have been discovered recently.

The salient characteristic of these "new flavocoenzymes" is a nucleus modified at positions 8α, 8 or 6 (*cf.* below). These modifications have profound effects on the spectral and chemical properties of the flavin and therefore the chemistry of this part of the molecule will be considered first. This short discussion is also intended to help answer some of the questions related to a) the possible role of these modifications in the

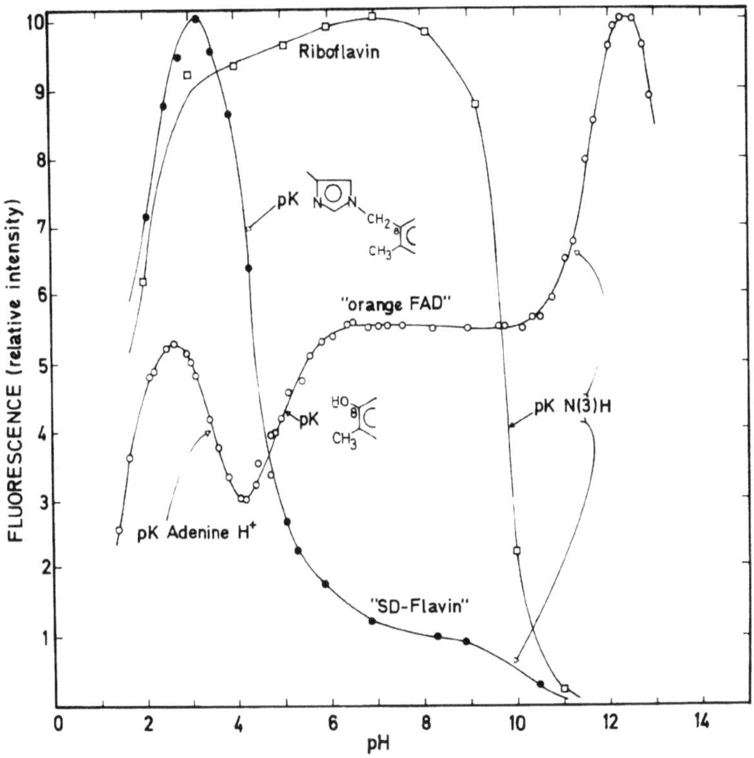

Fig. 2. pH Dependence of the fluorescence emission intensity of FAD (□), 8α-histidyl-riboflavin (●) and 8-hydroxy-FAD (○).
The flavin concentrations were approx. 10^{-6} M in 0.05 *M* NaCl at the pH values given. For further details *cf.* (*154*) and (*67*)

mechanism of flavin catalysis, b) the reactivity (stability) of 8α-substituted flavins, c) the peculiarities of the absorption and fluorescence spectra of 8- and 8α-modified flavins, and d) the biosynthesis of the new flavins.

In the preceding chapters the flavin has been presented as a system composed of an electron deficient (polar) pyrimidine subnucleus and a benzene moiety which are connected through the redox-active pyrazine ring. Following a comprehensive series of chemical and physicochemical investigations by our group and others (*6, 79, 82, 104*) the electron distribution in the molecule was the subject of extensive theoretical calculations mainly by Song *et al.* (*170*). These calculations accord in general with the experimental findings; they confirm in particular that position 8 is a center of low π-electron density and thus lead to the inference that the C(8)-methyl group is strongly hyperconjugated and relatively acidic:

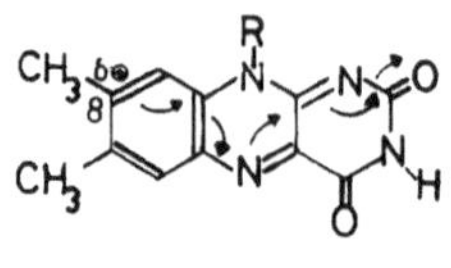

In a previous report (*79*) we have compared the activation of the (8)-CH_3 group in flavins to the reactivity of the methyl function in p-nitrotoluene. But the electron deficiency of position 8 is perhaps best demonstrated by the acidity of the hydroxyl function in 8-OH-(iso)alloxazines (*48, 49*). The pK of this group in the isoalloxazine (**1**) ($R_1 = CH_3$, no R_2, $R_3 = H$) is 4.6 and thus close to the pK of 2,4-dinitrophenol (pK = 3.96), and in the alloxazine (**1**) (no R_1, $R_2 = R_3 = H$) (pK = 6.60) it compares with the pK of p-nitrophenol (pK = 7.15).

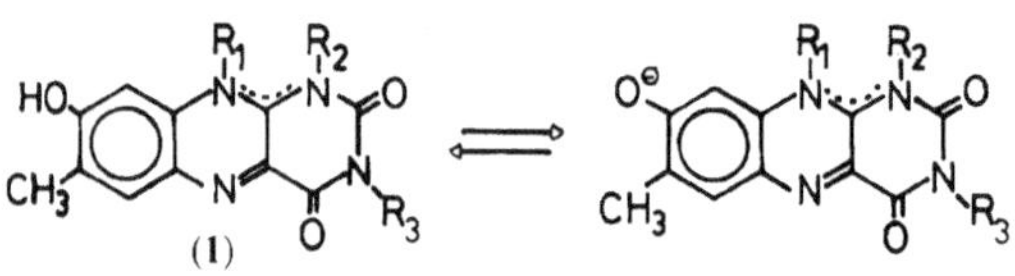
(**1**)

If these pK shifts are viewed as resulting from the electron deficiency of the system, the "activating" effect of the pyrimidine nucleus becomes evident and is comparable to that of two nitro groups. The electrophilicity of position 8 is further reflected in a series of facile reactions of which only the most typical will be reported here.

BULLOCK and JARDETZKY (*14*) demonstrated that the 8-CH_3 group selectively exchanges its protons in D_2O of a rate which is maximal at pH 6.8 in phosphate buffer:

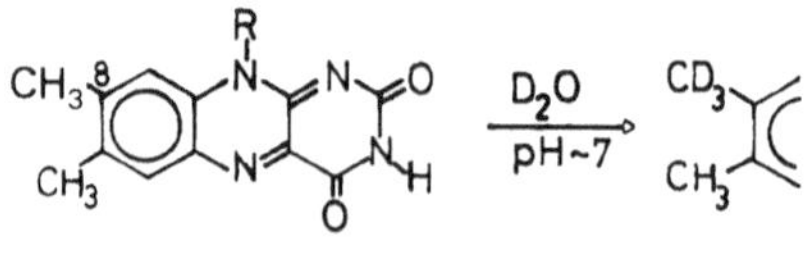

Oxidation with the anion of hydrogen peroxide ($HO-O^{\ominus}$) converts isoalloxazine-8-sulfonic acids (**2**) into 8-hydroxy-alloxazines (**3**). This reaction, which has been described recently by SMITH *et al.* (*168*), is typical for electron deficient sulfonic acids (*178*) and apparently does not occur at the less activated position 6.

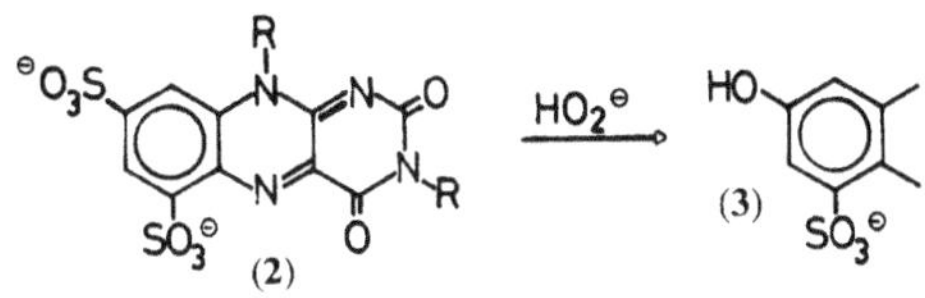

Flavins condense with activated aldehydes in the presence of sulfuric acid and dehydrating agents to yield deeply coloured condensation products (*61*). Similarly the dimerization of some substituted flavins to yield "biflavins" has been described (*76*). The loss of an 8α-proton has been proposed as the initial key step for both condensation reactions (*76*).

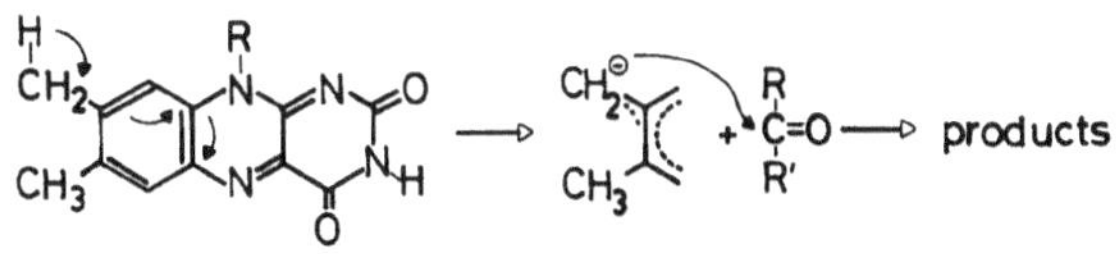

The methyl group at C(8) can be selectively oxidized with NO_2^- in acetic acid to the corresponding carboxylic acid (**4**) (*76*). The acid (**4**) is stable only at neutral pH and undergoes facile decarboxylation under acid conditions to yield the isoalloxazine (**5**):

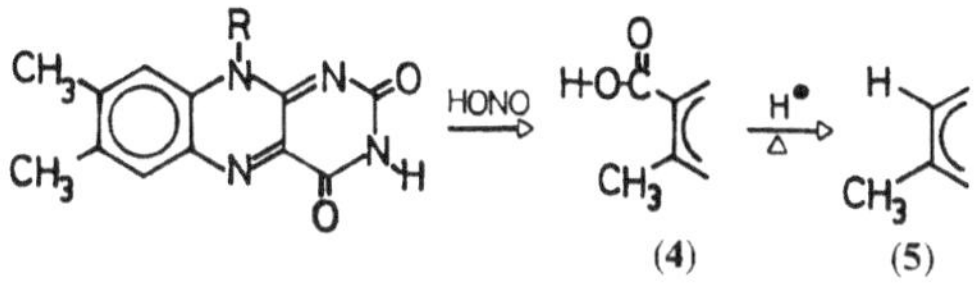

A further typical feature is the high reactivity of 8α-halogenated flavins, which can be compared to that of p-nitrobenzyl-halogenides. The facile replacement of bromine at position 8α by nucleophile functions of amino acids was the key step in the synthesis of the modified coenzymes which will be detailed below. Similarly halogen or $-N_2^+$ at position 8 can be replaced by a variety of nucleophiles for example by amines, alcohols or mercaptide (*76, 152, 177*).

2. 8α-Monosubstituted Flavocoenzymes

As in the case of many other types of natural products, the structural elucidation of the first member of this new class, the coenzyme of succinate dehydrogenase (SD-flavin), proved to be the most arduous and time-consuming (see introduction). Following Kearney's early work with SD-flavin peptides, Singer's group obtained on digestion with trypsin a flavin peptide containing 23 amino acids which were sequenced (*92*). Degradation with trypsin plus chymotrypsin yielded as main product a pentapeptide (*91*), whereas with aminopeptidase M a flavin derivative was obtained in which the flavin is covalently linked to a single amino acid. This very light-sensitive "SD-flavin" was identical in most respects with the product of anaerobic hydrolysis of different flavin peptides in 6 N HCl at 95° for 16 hrs (see also Introduction). Under more drastic conditions (6 N HCl, 125°, 16 hrs) histidine and riboflavin were liberated in rather poor yield (*154*). This is probably the result of decay reactions which produce reducing equivalents and partially reduced the flavin, the latter acting as an intermolecular reducing agent in a reductive cleavage of the flavinyl-peptide bond. Similarly, hydrolysis in the presence of external reducing agents (see below) led to formation of one molecule of histidine and of riboflavin (*186*). Thus the structure of the pentapeptide was established as Ser-His(FAD)-Thr-Val-Ala. N(3) of histidine was then determined as the site of the covalent link between the two moieties by comparison with synthetic hystidyl-flavins in the following way (Scheme 14).

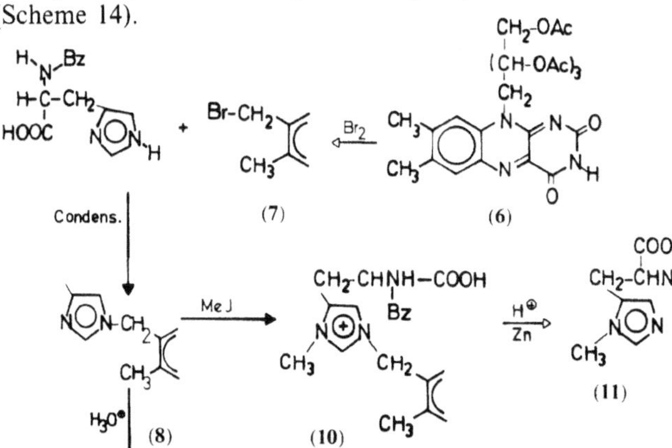

SD-flavin-peptides

Scheme 14. Pathway for the synthesis and degradation of 8α-histidylflavins

At first riboflavin was protected by acetylation of the ribityl side chain to yield the tetraacetyl derivative (**6**) and subsequently monobrominated at position 8α with bromine in dioxane and radical catalysis to form (**7**) in 80% yield. The monobromo derivative was allowed to react with N-benzoylhistidine to yield (**8**) in 50% yield. Of the two possible N-histidyl isomers, practically only the less hindered N(3) substituted one is present at this stage. Acid hydrolysis of the protecting groups (6 N HCl, 100°) first yields a single isomer 9a, however under these conditions a slow isomerisation occurs yielding finally a product mixture consisting mainly of the second isomer (**9b**). Interestingly, the reverse isomerisation (**9b**→**9a**) seems to occur much more slowly or not at all (*42, 186*) (*cf.* Note Added in Proof on page 517).

For the identification of the isomer, which was formed first, the protected flavin derivative (**8**) was methylated at the histidine function to yield the quaternary salt (**10**). Upon reductive cleavage of the histidine-flavin link with zinc dust in trifluoroacetic acid-acetic acid acid hydrolysis of the intermediary products yielded N(1)-methyl-histidine (**11**) and riboflavin (*cf.* Scheme 14). The mixture of the N(1)- and N(3)-histidylflavins 9a and 9b separated into the components by thin-layer chromatography on paper or on cellulose, or by column chromatography or, alternatively, by paper electrophoresis in pyridine-acetate buffer (*186*). The product obtained by acid hydrolysis of succinate dehydrogenase peptides was found to consist of the same isomer mixture obtained during the hydrolysis of the synthetic intermediate (**8**). On the other hand, enzymatic digestion of SD-flavin peptides with aminopeptidase M produced the pure N(3)-hystidyl-flavin isomer (**9a**), which underwent partial rearrangement upon heating with acid.

Since this work was done, 8α-N(3)-histidyl-FAD (*cf.* **9a**) has been identified in D-6-hydroxynicotine oxidase from *Arthrobacter oxidans* (*10, 11, 131*). It is probably also present in sarcosine dehydrogenase from *pseudomonas* (*143*) and from rat liver (*144*), and is possibly also the coenzyme of dimethylglycine dehydrogenase from rat liver (*37*).

The absorption spectra of 8α-histidyl-flavins (**9**) are, in their general features, very similar to those of unmodified flavins, e. g. FAD, FMN or lumiflavin (*186*). In the neutral state (pH 7), the lowest energy band at 445 nm ($\varepsilon \sim 12.000$) and the UV band at 268 nm are unaltered, whereas the second band exhibits a hypsochromic shift to 355 nm as compared to riboflavin (372 nm). Upon protonation of the histidine moiety ($pK \sim 4.7$) the same band is shifted further to 345 nm. In the SD-flavin cation (protonation at N(1)) the band centered at ~400 nm in normal flavins, is now partially resolved and has a maximum at 370 nm and a pronounced shoulder at 412 nm.

The most characteristic feature of 8α-histidyl flavins however, is the pH dependence of fluorescence, *viz.* the difference in emission intensity between pH 3.5 and 7. The emission of riboflavin is practically constant in the range pH 3 to pH 9 (Fig. 2). At low pH values it is decreased by dynamic proton quenching (*107*), whereas the pK at ~ 10 is attributed to the dissociation of N(3)-H (*180*). With SD-flavin (**9**) an additional pK is observed at ~4.7, which corresponds to deprotonation of the histidine moiety and goes along with an

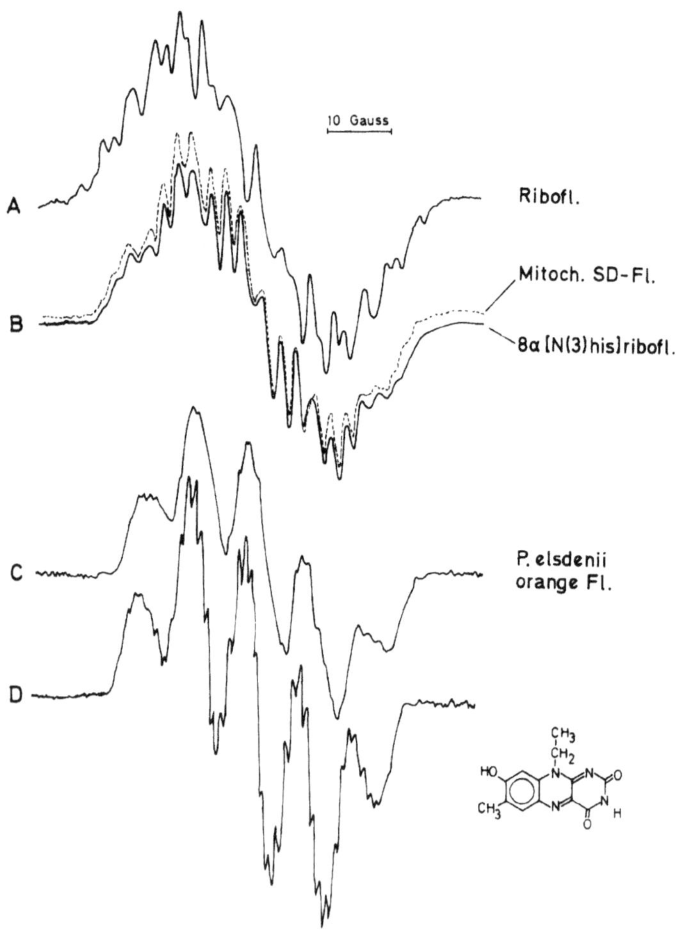

Fig. 3. Comparison of the ESR spectra of the radical cations of riboflavin, 8α-substituted flavins and 8-OH-flavins. Curves A and B taken from (*86*) and C and D were provided by S. Ghisla and S. G. Mayhew (*49*). For experimental details *cf.* these references

emission decrease of ~90%. This quenching is due to interaction of the excited flavin chromophore with 8α-substituents carrying π- or non-bonding electrons of relatively high energy and is even more pronounced with e.g. 8α-flavinyl-thioethers (*cf.* coenzyme of monoamino oxidase below).

The ESR signal of SD-flavin (Fig. 3) clearly differs in its structure from that of riboflavin; moreover, its total width is considerably smaller (46 G) as compared to riboflavin (52 G) (*154, 186*). This indicates that two ESR-active protons, which otherwise contribute to the signal of riboflavin, are absent, one proton being replaced by a substituent and the interaction of the radical spin with the remaining proton (at position 8α) strongly reduced as a consequence of inhibition of free rotation around the flavin-C(8α)histidyl-N(3)-bond (*67*). While the ESR spectra fail to differentiate between the two SD-flavin isomers (**9a**) and (**9b**), the NMR spectra (*186*) show basically the same chemical shifts for the histidine signals, but different values for the signals of the C(6) and C(9) protons and also a different shape for the signals of the ribityl side chain. This probably reflects a somewhat different orientation of the ribityl side chain towards the flavin nucleus, possibly as the result of its interaction with the histidine moiety. An extensive ENDOR investigation of SD-flavin peptides and a comparison with a series of models has been carried out (*154*); this basically confirms the deductions and structural assignments discussed above.

The second modified flavin of natural origin to be discovered was 8α-S-cysteinyl-FAD, the coenzyme of monoamino oxidase from liver and kidney outer mitochondrial membranes. Taking their departure from investigations of Yasunobu (*81*) and Hellerman (*30*), which indicated the presence of covalently bound flavin in preparations of this enzyme, Singer and his group (*85, 185*) isolated the flavinyl peptide by degradation of MAO with trypsin-chymotrypsin and identified cysteine as the amino acid residue bound next to the flavin moiety (*184*). The absorption spectrum of the flavin peptide from monoamino oxidase is readily differentiated from that of riboflavin by a hypsochromic shift of the second absorption band (360 nm, compare with 372 for riboflavin), in the neutral oxidized state (*44, 184*). It is similar to that of 8α-histidyl-riboflavin in the cationic state in that the band centered around 400 nm (abs. max. 375 nm, shoulder at 410 nm) is partially resolved. The fluorescence emission (*4, 30*) is only 10% of that of riboflavin, but oxidation with peracids raises it to 90% of riboflavin emission.

The ESR spectra of the radical cation of MAO-flavin deserve some comment. The general hyperfine pattern is well resolved, thus contrasting with the patterns of SD-flavin or riboflavin: This might be ascribed to the higher mobility of the ribityl side chain in MAO-flavin peptides as compared to SD-flavin. In its central part the spectrum is practically identical with the ESR spectra of 8-nor-8-carboxy-riboflavin (*164*), or of 8α-oxo-riboflavin (*45*): The signal width measured between the apparent outermost lines is 37 G. From a comparison with riboflavin (52 G) or SD-flavin (46 G), this would appear to correspond to removal of more than one proton from position 8α. The spectrum of MAO-flavin is, as should be expected, very similar to the spectra of 8α-OH- or 8α-S-(n-octyl)-riboflavin (*45*). In these spectra, two signals of weak intensity can be observed at the outermost part of the ESR spectrum and result in a total width of 46-47 G. Strong magnification of the "shoulders" of the ESR spectrum of MAO-flavin reveals similar, weak signals which were found to vary in intensity depending on the preparation of the coenzyme. A similar pattern is observed in the ESR spectra of 8α-modified lumiflavin in the series 8-CH_3, 8-CH_2-OH, 8-CH_2-SR, 8-$CH=O$, 8-$CH(OR)_2$, 8-COOH.

From the results mentioned above, it could be concluded that in MAO-flavin cysteine sulfur is connected to position 8α of the flavin. Final structural proof was provided by comparison with synthetic material. This was obtained by condensation of 8α-Br-tetraacetylriboflavin (**7**) with cysteine, and subsequent mild acid hydrolysis of the acetyl protecting groups (*44*).

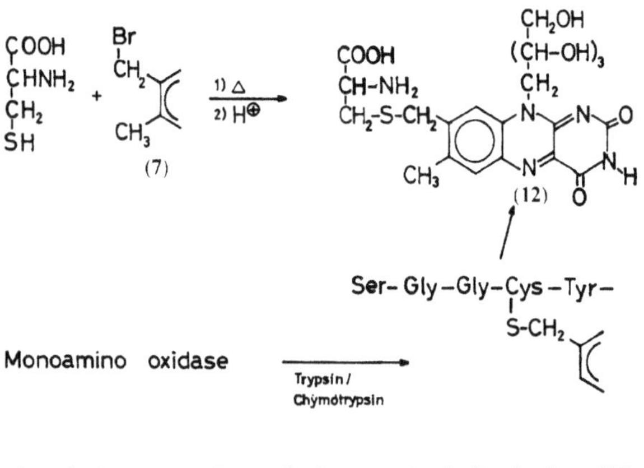

The chemical properties of 8α-cysteinyl-riboflavin (**12**) closely resemble those of activated thioethers and have been investigated mainly with model compounds carrying n-alkyl derivatives in position 8β (*44*). Reaction of (**13**), for example, with peracids leads to the sulfones (**14**), which in turn can be cleaved by reduction with dithionite to yield unsubstituted flavin and the corresponding sulfinate (*25*). Whereas the flavin C(8α)/histidyl-N(3) bond in (**9**) can be reduced only in rather poor yield with e.g. zinc dust in glacial acetic acid (*85, 186*) the same reaction with the cysteinyl thioethers produces cysteine and riboflavin almost quantitatively (*44*).

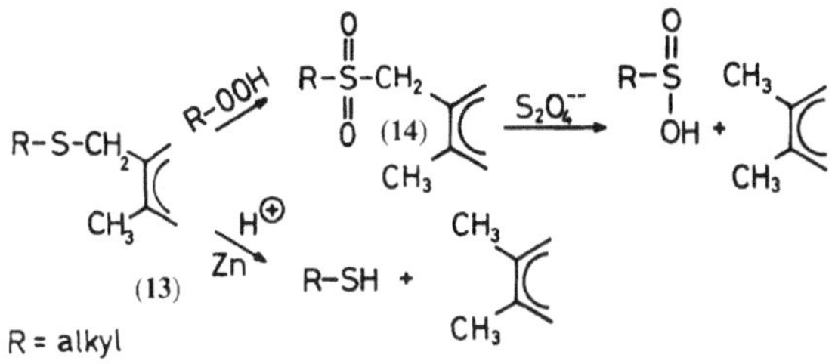

The redox properties of 8α-monosubstituted flavins have been the object of an extensive and comparative investigation by EDMONDSON and SINGER (*25*). A series of such derivatives of potential biological relevance has been synthetized starting from 8α-bromo-tetraacetylriboflavin (**7**) (*90, 154*).

3. 8α-Disubstituted Flavocoenzymes, Hemiacetals of 8α-Oxoflavins

In the last few years a further type of modified flavocoenzyme has beeen discovered which is structurally related to the coenzymes of succinate dehydrogenase and monoamine oxidase, but differs considerably in its chemical properties. Early literature reports indicated the presence of a flavin in cytochrome c_{552} from Chromatium which could not be "extracted" with trichloroacetic acid or acidic ammonium sulfate (*1*), but could be released, for example by trypsin digestion or by incubation with saturated urea solutions (*2*). Absorption, fluorescence and ESR behaviour were closely similar to those of 8α-cysteinyl-riboflavin (**12**) and indicated the presence of a covalent link to the protein, through position 8α (*185*). Strong acid hydrolysis of these peptides liberated the flavin as mixture of riboflavin derivatives; oxidation with performic acid and acid dephosphorylation yielded a homogeneous riboflavin derivative, which was identical with 8-nor-8-carboxy-riboflavin (*185*).

Hence the difference between 8α-cysteinyl-riboflavin (**12**) and the flavin from cyt c_{552} must be in the oxidation state of the 8α-carbon; it should be higher than 8α-CH_2-SR and lower than 8α-COOH, namely 8-CH=O. A flavin mixture at the 8α-oxo- level was first obtained in the lumiflavin series upon acid hydrolysis of 8α-dibromo-lumiflavins (*67, 154*). This strongly electron deficient aldehyde exists in aqueous solution preferably in the hydrated form. The equilibration is slow and both forms can be separated on thin layer chromatography (*45*). Urea may obviously add to such an aldehyde function and this is probably the reason, in our opinion, why cyt c_{552} is cleaved by urea to yield a free flavin of "ill defined chemical nature" (*185*). The Cys-sulfur is, according to the chloroplatinate test, in the sulfide state and since no second sulfur atom is present, a reasonable structural proposal is a thio-hemiacetal

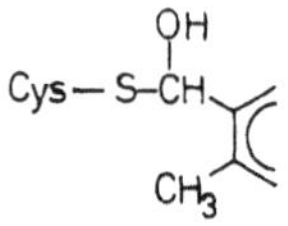

The fact however, that this group cannot be cleaved by nucleophiles such as RS^- or NH_2OH remains very puzzling. Edmondson (*24*) has shown that 8α-oxoriboflavin, in analogy to 8α-oxo-lumiflavin mentioned above, is in slow equilibrium with its 5′-hemiacetal. This in turn is identical with the "mixture of riboflavin derivatives" obtained from acid hydrolysis of cyt c_{552} flavin.

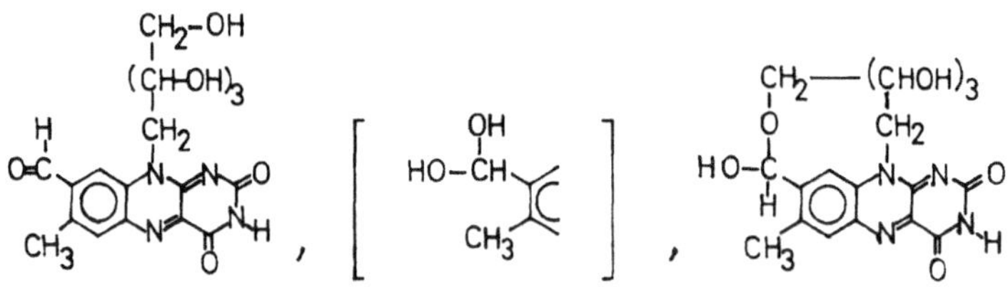

In nucleotides however, position 5′ is not free for acetal formation and the additional hydroxyl functions of the side chain can not form a macrocyclic ring for steric reasons (*24*). The anomalous thiohemiacetal stability towards hydrolysis cannot be simulated at the lumiflavin or riboflavin level, while experiments with 8α-oxo-FMN or -FAD are lacking.

This stability is confined to the FAD-peptide Tyr-Thr-Cys(FAD)-Tyr and it has been suggested (*89*), that either adenylate or tyrosine residues, or both, bring about stabilization through complexation in a way which has not yet been defined in chemical terms. In the opinion of the present author, the thermodynamic stabilization due to formation of "molecular complexes" should be minimal, while kinetic stabilization seems not to apply. Thus, formation of a macrocyclic thioacetal, possibly with the phenolic hydroxyl function of tyrosine (or with the threonine hydroxyl group) would better explain the unusual stability.

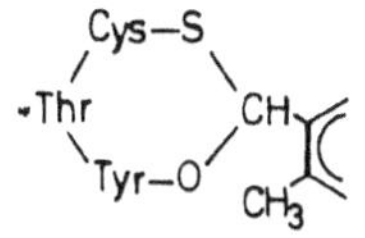

Formation of a thioacetal with the hydroxyl functions of the adenylate moiety, analogous to the thiohemiacetal described by EDMONDSON with 8α-oxo-riboflavin (*24*), is also possible, but seems improbable in view of the rather weak CD signal of the flavin peptide (*89*). In order to explain the unusually low fluorescence emission intensity "parallel" stacking of tyrosine and flavin is not a requirement. Strong quenching has been observed upon substitution of position 8α with imidazole or morpholine (*154*); these molecules clearly are not coplanar with the flavin moiety. Furthermore flavin derivatives containing two substituents at position 8α, as for example imidazole or morpholine, have been isolated in pure form and show practically no fluorescence emission, as compared with riboflavin (*45*).

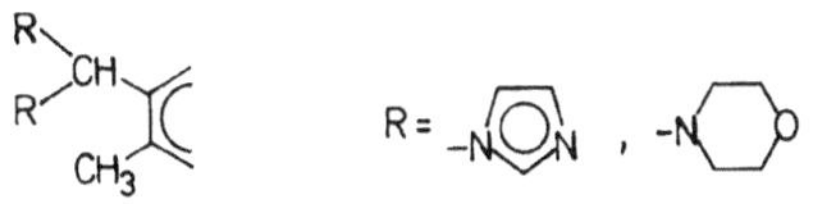

A "thiohemiacetal" function is possibly contained in a related enzyme, flavocytochrome c_{553} from *Chlorobium* thiosulfatophilum (*2*), where in fact a flavocoenzyme can be released from the protein under condition similar to those employed for the Chromatium cytochrome c_{552}, and has properties very similar to those of the flavin from Chromatium cytochromes c_{552}.

A similar type of linkage between coenzyme and protein in which position 8α of the flavin is at the oxidation level of carbonyl was found in thiamine dehydrogenase (from soil bacterial (*138*) and in β-cyclopiazonate oxidocyclase from *Penicillium* cyclopium (*88, 155*). In these cases, however, it is probable that N(1) of a histidine residue constitutes the bridge to the protein backbone. The oxidation level of position 8α was deduced from decay data similar to those mentioned above for the 8α-thiohemiacetals. Of particular interest is the fluorescence profile of the flavin peptide obtained by trypsin-chymotrypsin hydrolysis of β-cyclopiazonate oxidocyclase. Fluorescence quenching similar to that observed with SD-flavin (**9**) is observed, but in this case the pK attributed to protonation of the histidine imidazole is shifted from 4.7 to 5.4, and the maximal fluorescence obtained is only 20% of that of FMN (*164*). Upon performic acid oxidation of the peptide the emission intensity is increased to 60% of that of FMN and the pK is further shifted to 5.8, thus indicating that an amino acid residue other than the 8α-substituent might be partially responsible for the quenching. This is supported by the fact acid hydrolysis of the peptide (N HCl, 95° C) releases a histidyl-flavin which resembles FMN in its maximal fluorescence emission intensity, has a fluorescence pK shifted to 4.80 and is no longer affected by oxidation with peracids. Recent results (D. EDMONDSON and T. P. SINGER, private communication) indicate that this flavin is indistinguishable from a byproduct obtained by hydrolysis of 8α-(N-benzoylhistidyl)-tetraacetylriboflavin (**8**). Thus these results appear to exclude disubstitution at position 8α since the two protons of the 8-CH_2-R group are revealed by the NMR spectra, and since the ESR spectra are practically indistinguishable from those of SD-flavin (**9**). The clarification of the exact chemical nature of this coenzyme will have to await further analytical work. (See Note Added in Proof on page 517.)

4. Flavocoenzymes Bearing Functional Modifications in the Chromophore

Recently, a further type of flavocoenzyme has been found in preparations of biological origin showing a type of structure modification basically different from 8α-substituted, covalently bound flavins. MAYHEW and MASSEY (*125*) first presented evidence for the occurrence of a prosthetic group characterized by an unusual orange colour, which was associated with a NADH dehydrogenase from *Peptostreptococcus elsdenii*. This enzyme was subsequently isolated in pure form and its function was found to be that of a electron transfer flavoprotein (ETF) similar to the mammalian ETF of BEINERT (*4*). It is a most unusual enzyme since it consists of two different protein subunits, one containing "normal FAD" and the second varying amounts of an orange coenzyme, a green coenzyme and FAD. The three coenzymes could be removed from the protein by conventional methods (e.g. heat or methanol denaturation, or trichloroacetic acid treatment) and thus were not covalently bound to the protein. The mixture could be separated on DEAE cellulose into the components, the coenzyme fractions were hydrolyzed with phosphodiesterase, and AMP was identified as one product of the reactions. The coloured components of the hydrolysate were purified through binding to apoflavodoxin, a flavoprotein specific for flavin-5′-monophosphates (*48*, *126*). This suggested a FMN-related structure for both new chromophores.

The *orange coenzyme* was strongly fluorescent and showed a typical pH-dependent emission which is reproduced in Fig. 2 (8) for the compound at the FAD level. Among the "pK's" observed, the lowest one (pK ~ 2) can be attributed to dynamic proton quenching as with normal flavins (*107*) and the second at ~3.4 to deprotonation of the adenine moiety, since it disappeared after treatment with phosphodiesterase (which hydrolyzes pyrophosphate bonds selectively). These two extra pK's are not present in normal flavin and must be ascribed to the presence of one or more new functions. These pK values are also reflected in the absorption spectra, which are characterized by a band of high extinction centered at 472 nm at pH 7 ($\varepsilon \sim 41.000\ M\ cm^{-1}$, compare to ~12400 for riboflavin at 445 nm) (*195*). After acid dephosphorylation, this orange chromophore behaved as an anion at pH 7 and as a neutral molecule at pH 3. The NMR spectra (*49*) showed that one CH_3 group was lacking as compared to riboflavin. The ESR spectra (*49*) of the radical cation showed poor hyperfine resolution and a total line width of 44 G, which compares with 46 G for 8α-histidyl-riboflavin and 52 G for riboflavin (*186*); the total width is further reduced to 38 G in deuterated solvents. From this it was concluded that:

a) one methyl group in the flavin chromophore had been replaced by a strongly acidic function,

b) the missing CH_3 was the one which contributes strongly to the ESR hyperfine structure in riboflavin, i.e. C(8)-CH_3.

c) the new 8-substituent was presumably small (binding to apoflavodoxin),

d) the new substituent must be in direct conjugation with the heteroaromatic flavin nucleus.

The orange chromophore exhibited light sensitivity similar to that of FMN and was degraded to a pale yellow lumichrome analog, which turned orange-red upon deprotonation (pK 6.6). This compound was compared with synthetic 7-methyl-8-hydroxy-alloxazine (**17**), which in turn was obtained by straightforward condensation of violuric acid with 4-aminocresol in the presence of borate and indeed found to be identical.

Subsequently, 8-nor-8-hydroxy-flavin compounds were synthetized by an analogous procedure and were shown to simulate the spectral parameters of the natural product at the riboflavin level as, for example, in the ESR (Fig. 3), the pH dependence of fluorescence emission and absorption spectra. Thus the structure of the orange chromophore was established as the 8-nor-8-hydroxy-derivative of FAD (**18**, R = Rib-Ⓟ-AMP). Alternatively, a hydroxy function can be introduced into position 8 of the flavin ring by hydrolysis of 8-nor-8-diazonium-flavin compounds (*152*).

The chromophore of the orange flavin can be represented by two tautomeric forms (**18a** and **18b**), which should possess different chemical properties and electronic spectra. Synthesis of the appropriate model compounds (*49*), which were alternatively substituted at the positions

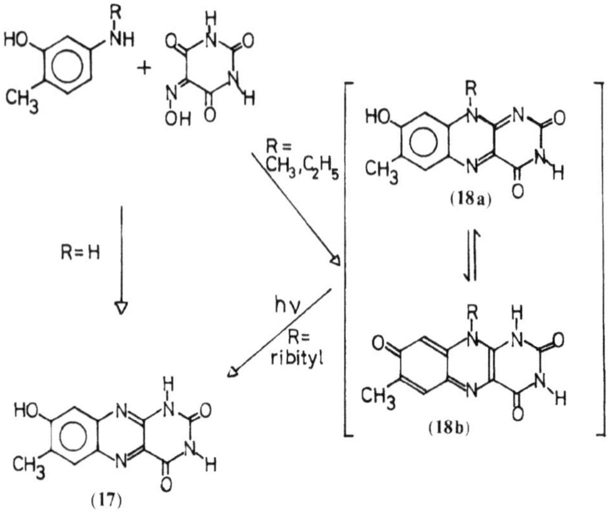

(17)

N(1), N(3), and C(8), indicate that the neutral species are best represented by the canonical formula (**18a**), whereas the anion formed upon deprotonation of C(8)-OH (pK = 4.7) probably has an electronic structure close to (**18b**). The extra pK's could also be explained and reflect protonation of (**18**) at position N(1) (pK = 0.7) and deprotonation at position N(3)-H (pK = 11.0). Due to the complexity of the ETF and to the fact that modified flavins were found in varying amounts, the question as to the origin of 8-OH-nor-flavins has not yet been answered; the possibility exists that they are formed in a side reaction of the normal enzymatic catalytic cycles or upon prolonged exposure of the cell to unnatural conditions, e. g. high oxygen concentration (*49*).

A closely related flavin has been found in cultures of *Streptomyces davawensis* by Japanese researchers. The structure of this compound, which was called "roseoflavin" because of its deep red colour (abs. max. 496 nm), was ascertained by oxidative degradation of the ribityl side chain and subsequent X-ray crystallography of the 2′-acetyl derivative (*130*). It represents the 8-nor-8-dimethylamino derivative of riboflavin (**19**) and is thus an analog of 8α-aminoflavin types synthetized earlier by HEMMERICH *et al.* at the lumiflavin (*75*) and BEREZOWSKII at the riboflavin level (*5*).

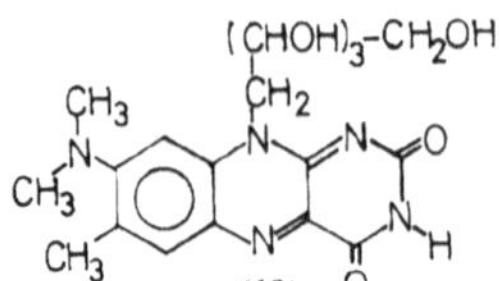

(19)

References, pp. 517—527

The *green chromophore* occurring together with the orange one was nonfluorescent over the whole pH range and not light sensitive. Because of lack of material no ESR or NMR spectra could be recorded, but this would have turned out to be useless, since the model compounds (see below) are very similar in these respects to riboflavin (*158*). Hence, absorption characteristics being the only guide for structural elucidation, the problem would have been very hard to resolve, if, by coincidence, a synthetic flavin chromophore showing practically identical absorption patterns had not been obtained on the lumiflavin level at the same time (*158*). Illumination of suitably substituted lumiflavin (but not riboflavin) cations leads in high yield to addition of water yielding hydroxydihydroflavins. Subsequent dehydrogenation leads to a mixture of 6- and 9-hydroxy-flavins; the former turned out to have the same chromophore as the green coenzyme from *P. elsdenii* (*126*):

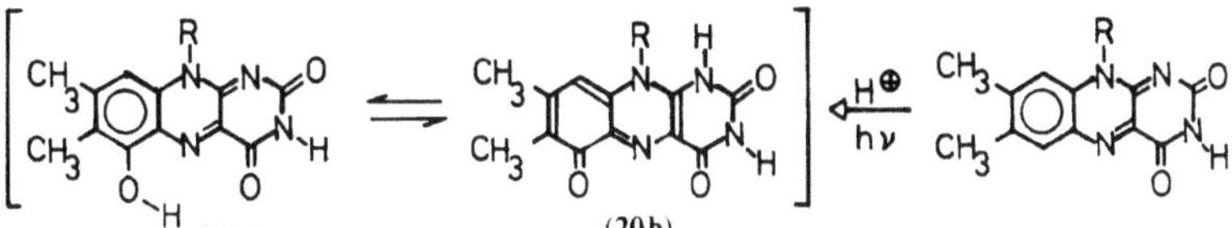

(20a) (20b)

R = ribityl-Ⓟ-AMP in ETF

R = ribityl-Ⓟ in glycolate oxidase

R = CH_3, (photochemical synthesis)

As in the case of the "orange flavin", the 6-hydroxy-flavin-chromophore can exist in two tautomeric forms (**20a** and **20b**); the anion formed upon deprotonation of the C(6)-OH function can similarly be represented by the mesomeric forms derived from (**20a** and **20b**). Synthesis of N(1), resp. C(6)-O blocked model compounds indicated (*158*) that the green colour (band at 600 nm) must in fact be attributed to the quinoid structure (**20b**) and not to (**20a**).

The same type of coenzyme was isolated from pig liver glycolate oxidase, but as a derivative of FMN (*126*). Also in this case the modified flavocoenzyme appears to be present in the enzyme in varying quantities as a companion of normal FMN. Hence, even more than in the case of the "orange flavin" (**18**) the origin of the modification is difficult to explain or to verify experimentally. The enzymes, from which the green chromophores has been isolated, originate from completely different sources, namely aerobic mammalian tissue (pig liver) and a strict anaerobe (*Peptostreptococcus elsdenii*), and catalyze basically different reactions.

A *photochemical* "hydroxylation" of enzyme bound FAD appears unlikely, since it could not be reproduced by illumination of pure preparations of ETF (*126*). Furthermore, acid photolysis of riboflavin leads exclusively to hydroxylation at position 9 (*161*). It is more probable that during aerobic enzyme purification, "false turnovers" of the reduced enzyme with O_2 may lead to "oxygenase type" reactions, which in the absence of extraneous substrate lead to "self-oxygenation" of the flavin (*cf.* above).

Finally it should be mentioned that the green and orange flavocoenzymes can serve as sensitive indicators when bound to the active site of (apo)flavoenzymes. Interaction with the protein can stabilize either of the tautomeric forms (**18a, 18b**) or (**20a, 20b**), and cause shifts of the pK's which can be readily monitored by changes in the absorption spectra. This method has been used successfully in studies of the active sites of the apoenzymes of flavodoxin, D-amino acid oxidase, lactate oxidase and "Old Yellow Enzyme" (*47, 49*).

X. Flavin Photochemistry

The chemistry of the electronically excited state of flavin has up to now concentrated on flavoquinone. Investigation of photochemistry inherent in the radical and reduced flavin is very difficult, since flavoquinone would have to be expected as a product or by-product and, hence, it would be very difficult to differentiate between e.g., flavohydroquinone photochemistry and flavoquinone-sensitized ground state chemistry of flavohydroquinone. But this should not discourage further work, which is badly needed in this area.

From the "natural product" point of view, flavoquinone photochemistry is relevant for three reasons:

1) The isolation and characterisation of flavocoenzymes (see above) is hampered by their photosensitivity. This applies particularly to the monophosphate coenzymes (FMN's) and the flavin vitamin. The photosensitivity is a property inherent in the free solvated form, not the protein bound form for reasons which can be imagined easily (see "photolysis", below).

2) Surprisingly, the ground state chemistry of the protein-bound flavoquinone and the excited state chemistry of the free, solvated flavoquinone do have much in common (see below). This is not because flavin excited states occur in normal flavoprotein catalysis, notwithstanding most recent speculations concerning "photo(bio)chemistry without light"

(*195a*); it is rather due to the change in the acidity pattern of the flavin nucleus upon excitation which, in $^3Fl^*_{ox}$, renders N(5) a basic site of pyridine strength (see below).

3) Flavin photochemistry is an obligatory forerunner of flavin photobiology which, in the recent past has developed into a field of its own and will apparently become more and more important (see below).

Flavin photo*chemistry* must be divided into four categories: photoaddition, photodehydrogenation, photodealkylation and photoelectron transfer.

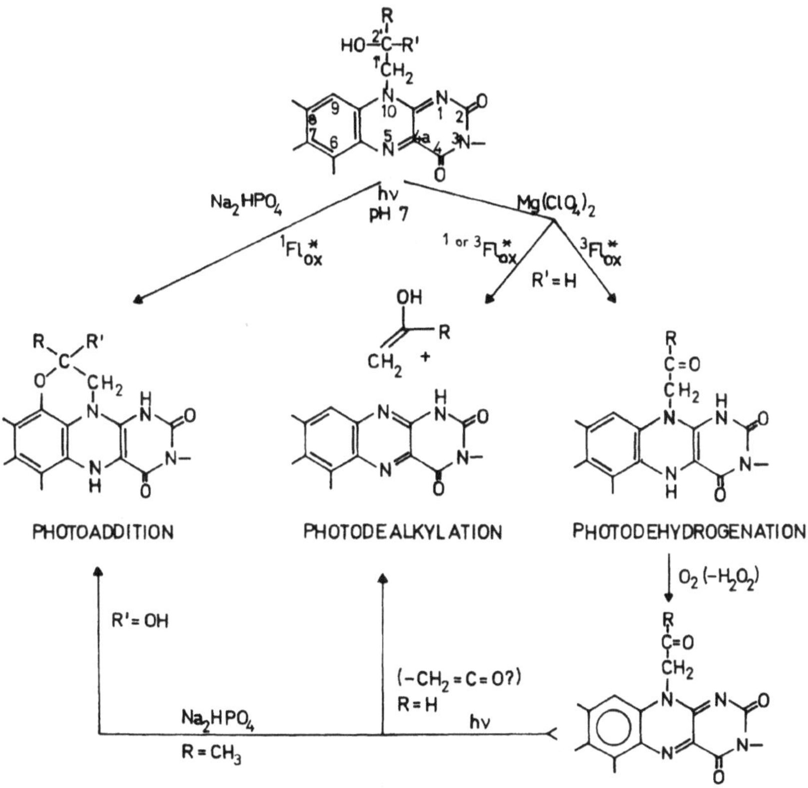

Scheme 15. The three modes of *intra*molecular flavin photochemistry. The *inter*molecular course is possible with solvent ROH in the "addition" mode, with a wider variety of active CH- and C-COO$^-$-bonds in the "dehydrogenation" mode. Obligatorily intermolecular for natural flavins is the fourth mode of flavin photochemistry: photoelectron transfer

1) *Photoaddition* follows equation 17 (*cf.* Scheme 15)

$$\mathrm{Fl_{ox} + ROH \xrightarrow{h\nu} RO-Fl_{red}H} \qquad (17)$$

and yields hydroxy- or alkoxydihydroflavins as first intermediates. Flavin reduction is observed spectrally, although the overall reaction involves no net reduction. R can either be a residue of solvent ROH (R = H or alkyl, intermolecular addition), or it may be contained within the flavin molecule as a side chain hydroxyl in a sterically suitable position (intramolecular addition). In any case, the local "concentration" of ROH must be high, since photoaddition occurs through the excited singlets $^1Fl^*_{ox}$ (*161*) and 1/2 α-HFl^{+*}_{ox} (*158*).

2) The second and strictly intramolecular type of flavin photochemistry, *N(10)-photodealkylation*, is both, singlet or triplet, dependent and has been studied systematically only very recently (*51*). It is much less general, as it does not involve a reduction at all, not even a spectrally apparent one under anaerobic conditions, and it requires synchronous breakage (fragmentation) of a N(10)-C(1′) and a C(2′)-H bond (eq. 18) in *cis*-periplanar conformation with direct proton transfer yielding lumichrome and an olefin (*cf.* Scheme 15):

$$\mathrm{>N(10)-CH_2-CR_2 \xrightarrow{h\nu} >NH + CH_2{=}CR_2} \qquad (18)$$

3) *Photodehydrogenation* (or, entirely equivalent to it, photodecarboxylation) follows equation (19), the most common course of flavin photochemistry. It involves a net reduction, yielding free dihydroflavins or alkyl derivatives thereof

$$\mathrm{^3Fl^*_{ox}} + \overbrace{\left\{\begin{array}{l} \mathrm{RH} \xrightarrow[-\mathrm{H^+}]{} \\ \text{"group transfer"} \\ \mathrm{RCOO^-} \xrightarrow[-\mathrm{CO_2}]{} \end{array}\right\} \mathrm{R\text{-}Fl^-_{red}}}^{\text{"hydride transfer"}} \xrightarrow{\mathrm{H_2O}} \mathrm{HFl^-_{red} + ROH} \qquad (19)$$

Structure, stability and essentiality of the intermediates R-$Fl_{red}H$, depend very much on the nature of the residue R (see below). Since H_2Fl_{red} undergoes fast autoxidation (*122*), photodehydrogenations are in general reversible with respect to the flavin chromophore, while photoaddition and photodealkylation are not. Furthermore, photoadditions are necessarily nucleophilic, whereas the character of photodehydrogenations is still much disputed though the radical course postulated in the earlier literature (see below) is now widely abandoned in favour of the "carbanion mechanism", which also seems to apply to reductions of flavoproteins by CH-substrates (*187*). Photodehydrogenation occurs only through the long-lived triplet $^3\mathrm{Fl}^*_{\mathrm{ox}}$ (*148*); the intermolecular path can, therefore, compete efficiently.

The intramolecular path of photodehydrogenation (and concomitant photodealkylation) which involves the N(10)-side chain of 10-hydroxyalkylflavins (for review *cf. 15*) has previously been designated as *photolysis* (because of its irreversibility), although mechanistically there appears to be no difference in the primary steps of inter- or intramolecular dehydrogenations and photolysis.

Intramolecular addition and dealkylation involve peri-attack at the flavin nucleus, thus favouring reactions at C(2′) of the N(10)-side chain and six-membered cyclic transition states. On the other hand (as demonstrated in Scheme 15 for the case of riboflavin) 3′- and 4′-CHOH appear to be preferred in side chain (intramolecular) dehydrogenation. This follows from the fact that the methine group next to N(10), i.e. HC(2′)-OH, does not dehydrogenate faster than HC(n′)-OH for $n > 2$ (*199*). Hence, dehydrogenation appears to involve attack of the back-bent side chain upon the azomethine subgroup C(4a)=N(5) of the flavin nucleus, rather than with peri-attack at N(1), whenever sterically feasible. Obviously, N(1) is not a good "intake" position for redox equivalents, in agreement with spin distribution in flavin radicals (*26*).

However, flavin *photolysis*, i.e. photo-induced selfdestruction of any type, need not be confined to intramolecular modes of reaction, although such reactions have attracted most attention owing to their high velocity and relevance to vitamin-B_2 photo-instability. But quite generally, any aliphatic CH-center attached to a flavin nucleus is susceptible to intermolecular photodehydrogenation by a second flavin in the excited triplet state. SCHÖLLNHAMMER *et al.* (*156*) have recently studied the photolysis of lumiflavin and find that C(7)-CH_3 can be the preferred center of attack (see above).

4) The fourth mode of $\mathrm{Fl}^*_{\mathrm{ox}}$-chemistry is photoelectron transfer, e.g.

$$\mathrm{ArOH} + \mathrm{Fl}_{\mathrm{ox}} \xrightarrow{h\nu} \mathrm{Fl}^*_{\mathrm{ox}} + \mathrm{ArOH} \longrightarrow \dot{\mathrm{Fl}}^- + \mathrm{Ar}\dot{\mathrm{O}}\mathrm{H}^+ \longrightarrow \mathrm{H}\dot{\mathrm{Fl}} + \mathrm{Ar}\dot{\mathrm{O}}. \tag{20}$$

(slow back-reaction: $\mathrm{H}\dot{\mathrm{Fl}} + \mathrm{Ar}\dot{\mathrm{O}} \xrightarrow{\text{slow}} \mathrm{ArOH} + \mathrm{Fl}_{\mathrm{ox}}$)

If the oxidized substrate does not decay by its own paths of disproportionation or dimerisation, the reaction becomes "reversible in the dark", as for example, in the case of the phenoxyl radical (*179*). Photoelectron transfer is confined strictly to π-electrons of suitably low potential, such as occurring in phenol or indol derivatives (tryptophan).

Under most conditions several modes of photolysis will compete in a given flavin system (*161*). If, however, addition and dealkylation do not compete, the most common form of irreversible selfdestruction of flavin, viz. intramolecular dehydrogenation, can be easily suppressed by extraneous photosubstrates, which either induce photo-electron transfer, as for example phenol, which is reversible by itself, or substrate photodehydrogenation, as for example EDTA, which is reversible with oxygen, with respect to Fl_{ox} (*36*). In aprotic solvents 1,4-cyclohexadiene performs the same service as EDTA in aqueous environment (*97*).

Two questions which deal with the much disputed and biologically most relevant mechanism of photodehydrogenation require careful discussion:

1. What is actually transferred: Hydrogen or residue of the substrate RH?

2. Can the overall $2e^-$-process involve a rapid sequence of two radical processes?

As to question 1, no analysis of flavin-dependent photodehydrogenation can, unfortunately, provide a clearcut differentiation between "hydride transfer" and "group transfer followed by hydrolysis" (*cf.* eq. 19). This complication arises because depending on the nature of R, the alkyldihydroflavin $RFl_{red}H$ will easily undergo dark protolysis to yield $HFl_{red}^- + R^+$, e.g. for $R = CH_3 - CH(OH)-$. In aprotic solution, many photodehydrogenation reactions yield "adducts" $RFl_{red}H$ (absorbing around 360 nm), where aqueous systems yield $HFl_{red}H$ (absorbing 330 nm) and 400 nm (sh)) (*98*). Elsewhere (*98*) we have listed these products of such group-transferring photoreactions which have so far been isolated. The selection of the addition site by the residues R has been discussed in the "Dihydroflavin" chapter (see above). In no single instance up to date has it been possible to demonstrate a hydride transfer unequivocally, i.e. under exclusion of "group transfer plus subsequent protolysis". The suspicion, therefore, remains that flavin-dependent (photo)-dehydrogenation quite generally involves group transfer and protolysis, in contrast to nicotinamide-dependent dehydrogenation, which means true hydride transfer.

While this question remains open, many data are available which disfavour the *radical* nature of flavin-dependent photodehydrogenation:

1. Flavoquinone cations $1\text{-}RFl_{ox}^+$ initiate photodehydrogenation even more easily than neutral Fl_{ox} (*158*), thus arguing for a polar mechanism.

2. Photodehydrogenation is entirely independent of the stability of the potentially intermediate radicals. 5-Deazaflavins, which are known to resist $1e^-$-uptake, undergo analogous photoreductions as smoothly as flavins (*33*).

3. The stability of the flavin radicals does not speak for, but rather against a radical mechanism of photoreduction, since only very small amounts of radicals (if at all) are observed in these reactions *in spite* of the stability of these well-known species. This is one of the remarkable common features of flavin photo- and biochemistry. Hence, if in a reaction $^3Fl^*_{ox} + RH \rightarrow R\dot{F}l + \dot{H}$ practically no free radical $R\dot{F}l$ can be observed in spite of its stability, recombination of the solvated radical pair must be faster than diffusion, and faster than its formation. This turns the question of a radical path into semantics.

4. In keeping with point 3 no "self decay" products of $\dot{R}$, e.g. R_2 or chain derivatives, have ever been found in flavin dependent dehydrogenations.

5. Transfer of bulky groups such as t-butyl or even α,α-dimethylbenzyl to Fl_{ox} to yield stable non-planar $RFl_{red}H$ occurs smoothly, although the corresponding radical $R\dot{F}l$ undergoes immediate elimination to yield $H\dot{F}l$ + olefin, since the bulky substituent R cannot be accommodated in the radical plane (*68*).

6. Since the singlet-induced photoaddition of water, for example, to Fl_{ox} cannot be a radical reaction for energetic reasons, it is hardly conceivable that the reaction of the triplet with RH, i.e. photodehydrogenation, should inevitably follow a radical mechanism. Quite generally, polar reactions of excited states are in the present author's opinion somewhat underrated in photochemistry. Thus the distinction of $2 \times 1e^-$- and $1 \times 2e^-$-processes serves to no purpose other than to ascertain spin conservation in the overall reaction

$$^3Fl^*_{ox} + RH \rightarrow R^1Fl_{red}H. \qquad (21)$$

This dilemma might, however, be overcome by assumption of a low lying dihydroflavin triplet, as outlined above for the analogous case of the reaction between dihydroflavin and $^3O_2$.

Finally, two very special flavin-dependent photoreactions need to be mentioned:

1. *Photofragmentation* is induced by flavin in sulfur-containing systems according to the following equations

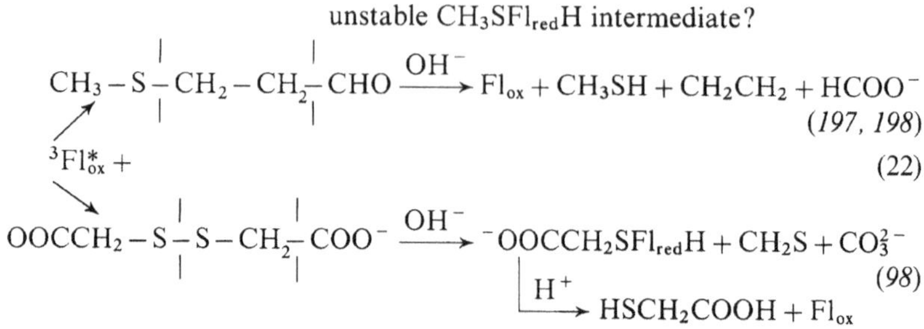

(197, 198)

(22)

(98)

The stoichiometry or redox-balance of these rather analogous reactions is difficult to work out, since CH_3SH and, of course, CH_2S are unstable under the reaction conditions. In any case, these badly understood, but highly interesting reactions require further attention, especially the question why the first reaction remits free flavin, while the second one yields a stable adduct.

2. Instead of attack at a given carbon-center (eq. 23),

$$^3Fl_{ox}^* + R-\left\{\begin{matrix} COO^- & \longrightarrow & CO_2 \\ H & & H^+ \end{matrix}\right\} + RFl_{red}^- \qquad (23)$$

flavin-dependent *photodehydrogenation* can also proceed *via 1,3-allyl shifts* in the case of suitable substrates.

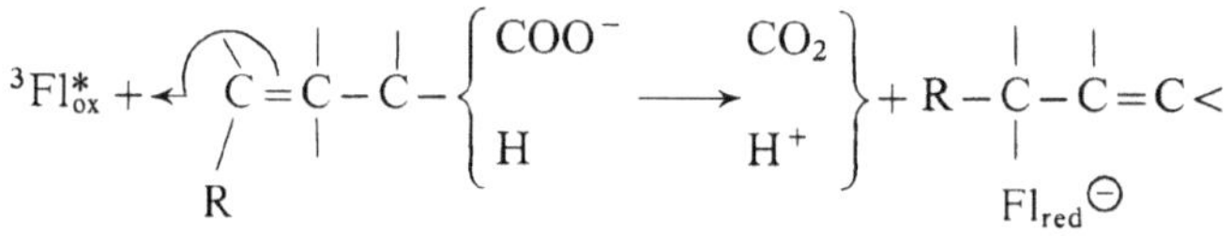

(24)

With propargylamines, for example, both types of attack are taking place competitively, leading to different stable products (Scheme 16) (*39, 202*).

The "monoamine oxidase (MAO) model" path (*119*) of the reaction implies formation of a characteristic new chromophore, a dihydroflavin-trimethincyanine, which can be split by base with back formation of Fl_{ox} and the debris of malonic dialdehyde. The overall reaction, therefore is, without consumption of flavin, as follows:

$$HC{\equiv}C-CH_2NR_2 \xrightarrow[+2OH^-]{-2e^-} O{=}CH-CH{=}CH-OH + HNR_2 \qquad (25)$$

The "lactate oxidase (LO) model" path (*159*) involving secondary addition at C(4a), is irreversible because of subsequent 4,4a-ring opening which is enforced by aromatization of the substrate residue.

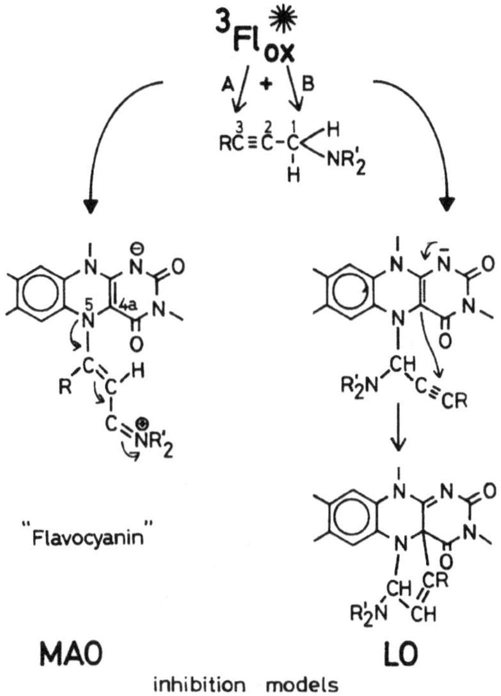

Scheme 16. Flavin photochemistry with propargylamine substrate. Competing substrate carbon atoms 1 and 3 lead to new adduct chromophores, which are identical with the chromophores of enzyme (Monoamine Oxidase = MAO and Lactate Oxidase = LO) inhibitor adducts

As is borne out by these data, uncertainty about the site of substrate or inhibitor addition to the flavin in photochemistry and in enzymology is very similar. Moreover the adduct chromophores are absolutely identical in the photochemical and in the enzymic cases (*159, 119*). We may be far away from explaining these phenomena as functions of the molecular structures and physical properties of flavin, substrate and environment, but the "model" relevance of flavin photochemistry to a molecular understanding of flavin enzymology cannot be denied.

A final statement is required about the *pH-dependence of flavin photochemistry*. It was discovered early by various authors (*16, 65, 147, 192*) and studied systematically by HAAS and HEMMERICH (*53*) that triplet mediated photodehydrogenations of most, but not all substrates exhibit a characteristic rate decrease with an "action-pK" of around 5. This pK is the same regardless of whether the actual reaction involves decarboxylation or dehydrogenation or whether the product is $RFl_{red}H$ or $HFl_{red}H$.

It may be overlapped by pK's of the substrate ground state, but it is also obtained with entirely non-acidic and non-basic substrates such as benzyl alcohol. Neither starting Fl_{ox} ground state nor any (alkyl)dihydroflavin products have pK's in this range. Hence, this action pK could reflect the pK of an excited flavoquinone, but since the pK of 1-HFl^{+}_{ox} is around zero and the pK of its excited singlet even lower, the corresponding triplet pK cannot be as high as 5 unless there is a violation of FOERSTER's rules (*35*).

SCHREINER *et al.* (*160*) however, demonstrated by flash photolysis experiments that, nevertheless, the pK of $^3(HFl_{ox})^*$ was as high as 5. From this they concluded, in agreement with calculations done by SONG (*169*) that the flavoquinone triplet was, unlike the ground state, protonated at N(5). This is a very important finding, since it explains why flavin-dependent photodehydrogenation is fast at acidic pH, because the active species is a cation. This again argues in favour of a polar mechanism.

XI. Flavin-Dependent Photobiology

1. Although the involvement of FMN in bacterial luciferase action has been known for a long time, the recent past has experienced the beginning of an understanding of this bioluminescence in terms of chemical bonding. The following hypotheses have been established:

– The emitter chromophore is not neutral Fl^{*}_{ox}, but cationic flavoquinone, 1-RFl^{*+}_{ox} (*28*). The N(1)-blocking residue has hitherto been assumed to be an apoprotein-linked rigid hydrogen bridge, in agreement with the fact that 1-HFl^{+}_{ox} becomes fluorescent as soon as the rate of proton exchange is slowed down sufficiently by cooling or increase of solvent viscosity. In view of the pK$\approx$0 of this N(1)H-group, however, it is still surprising that the apoprotein should preserve the excited cation to the extent that emission occurs prior to dissociation. Hence, in Scheme 17 R is assigned, hypothetically, to the product, i.e. acyl, residue, which would give a better explanation of the relatively long lifetime of RFl^{*+}_{ox}.

– The overall reaction has been definitely established (*59*) as

$$HFl_{red}H\text{-enzyme} + RCHO + O_2 \rightarrow Fl_{ox} + RCOOH + H_2O \quad (26)$$

– At low temperature and in the absence of substrate, $FMNH_2$-luciferase reacts with oxygen to yield a HFlOOH-enzyme which absorbs at around 370 nm (*58*). Upon warming up in the presence of substrate, luminescence along with substrate "oxygenation" (RCHO → RCOOH) is seen, while in the absence of substrate, O_2 is converted into H_2O_2. By the present author this oxygen adduct is assumed to be a 5-H-Fl-4a-OOH "storage isomer", which undergoes shift of OOH from the inactive C(4a) to the active C(10a) upon addition of substrate.

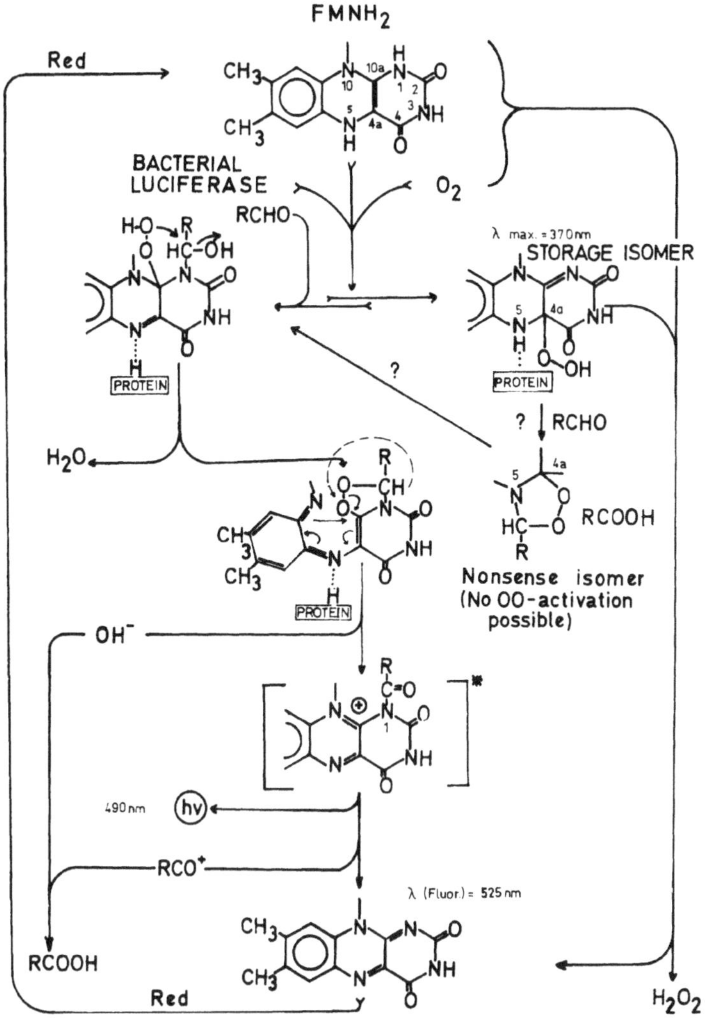

Scheme 17. Hypothetical course of bacterial luciferase action. The distinction between the two covalent "quarternary $FMNH_2$-O_2-Substrate-Protein Complexes", the 4a-5 isomer being inactive and the 1-10a isomer being active, is based upon the possible activation of the latter by 10-10a ring opening, which creates an unsaturated peroxide, while the "nonsense isomer" remains saturated. If this assumption is wrong, and O-O activation of the nonsense isomer is brought about by the protein in some way or another, the emitting species must be considered to be 5-HFl-4a-OH ($\lambda_{abs} \approx 370$ nm) which would be an alternative to the above Scheme

The need for such a storage isomer is obvious, since in the case of active 9a- or 10a-isomer being stored, the danger of flavin-self-oxygenation would be imminent. Such oxygenated flavins (*48*) have recently been found in enzyme preparations, presumably as a consequence of too many oxidase turnovers during prolonged purification (*cf.* Scheme 11 and section on "New Flavocoenzymes").

It is obvious from what has been said on HFlOOH-isomers (*cf.* above) that "oxene guns" can be derived from opening of either 9a-10 or 10-10a bonds, as long as no (irreversible) 10a-1 bond opening happens. In the case of the HFl-10a-OOH intermediate outlined in Scheme 17, this "false" cleavage is avoided by substrate addition at N(1) and subsequent cyclization of the peroxide function. It is assumed, that the distortion of the dihedral angle from 97° in H_2O_2 to about 180° in the 5-membered ring adds further to activation.

2. Investigations of FRÖHLICH and DIEHN (*38*) as well as CHECCUCCI and coworkers (*17*) revealed flavin as the photoactive receptor chromophore in Euglena algae. Since this flavin is rigid with respect to the geometry of the whole organism, direct fixation to a lipid membrane seems probable. Correspondingly, the action spectrum reveals the shape of flavoquinone in nonpolar medium, the vibration modes of the first transition being well resolved.

3. Similarly there are strong hints in favour of flavin-dependent phototropy in *Phycomyces* fungi (*151*) as well as in corn seedlings: All this points to the fact that within the near future, flavin-dependent energy and signal transfer may develop into an important problem of "molecular sensory physiology".

XII. Concluding Remarks

It may have become apparent from the content of this review that flavin is a "somewhat omnipotent" molecule. After 20 years of active involvement in flavin chemistry, the present reviewer still feels that he stands at the very beginning, though twenty years ago flavin was held to be a "solved problem" in contemporary "natural product chemistry".

Thanks are due to numerous members of the world-wide "flavin community" for communication of recent data and critical evaluation, sometimes denial of ideas brought forth in this article. In spite of this, the reviewer did not hesitate to articulate what he thinks are "chemically sound speculations on the molecular biology of flavin-dependent oxidoreduction". The future will decide.

Special mention must be made of Dr. SANDRO GHISLA's outstanding contributions to chapter IX and of the assistance in the assembling of data given by ALBERT WESSIAK.

References, pp. 517—527

Special thanks are due to Prof. Dr. PEDER KIERKEGAARD, University of Stockholm, for communication of x-ray data and drawings of stereoformulae.

If someone complains that undue preference has been given by the author to his own work, his attention is directed to the parallel review by T. C. BRUICE (*12*) for comparison. It was not the aim of the present article to produce a comprehensive and neutral survey of earlier literature which would have resulted merely in listing of many arguable opinions. Instead, the author set himself the task of trying to illuminate an area of more or less outspoken disagreements in order to outline pathways towards an understanding of flavin biocatalysis in straight terms of chemical structure.

Note Added in Proof

Most recent investigations by EDMONDSON and KENNEY (Biochem. Biophys. Res. Comm. **68**, 242 [1976]) have shown, that an assignment made on p. 494 is incorrect: Compound **9b**, obtained upon prolonged heating of SD-flavin (natural or synthetic) under acid conditions (**9a**), is not the N(1)-histidyl-isomer of **9a**, but a derivative of **9a** modified in the ribityl side chain. The true N(1)-histidyl-isomer is obtained synthetically as a byproduct of the synthesis of **9a** and has been identified with the covalently bound flavoprotein from β-cyclopiazonate oxidocyclase.

References

1. BARTSCH, R. G.: Properties of c-Typ Cytochromes of Chromatium. Feder. Proc. **20**, 43 (1961).

2. BARTSCH, R. G., T. E. MEYER, and A. B. ROBINSON: Complex c-Typ Cytochromes with Bound Flavin. In: "Structure and Function of Cytochromes" (K. OKUNI, M. C. D. KAMEN, I. SEZUKU, eds.), p. 443. Tokyo: University of Tokyo Press. 1968.

3. BEINERT, H.: Spectral Characteristics of Flavins at the Semiquinoid Oxidation Level. J. Amer. Chem. Soc. **78**, 5323 (1956).

4. — Flavin Coenzymes. In: "The Enzymes" (P. D. BOYER, H. LARDY, K. MYRBÄCK, eds.), Vol. 2, 2nd ed., p. 339. New York: Academic Press. 1960.

5. BEREZOVSKII, V. M., L. S. TUL'CHINSKAYA, and N. A. POLYAKOVA: Alloxazine and Isoalloxazine Derivatives. XIII. Synthesis of 7-Aminoalloxazine, 7-Aminodimethylriboflavin, and their Derivatives. Zh. Obsh. Khim. **35**, 673 (1965).

6. — — — Reactivities of Alloxazines and Isoalloxazines. Russ. Chem. Rev. **41**, 574 (1972).

7. BLANKENHORN, G.: Flavin-Nicotinamide Biscoenzymes: Models for the Interaction between NADH (NADPH) and Flavin in Flavoenzymes. Eur. J. Biochem. **50**, 351 (1975).

8. BLANKENHORN, G., and P. HEMMERICH: Unpublished Results.

9. BOUKINE, V. N.: Compounds of Soluble Vitamins with Proteins in Fats. In: Resumés de Communications, 3ème Congrès International de Biochimie, Bruxelles, p. 61 (1955).

10. BRÜHMÜLLER, M., and K. DECKER: Covalently Bound Flavin in D-6-Hydroxynicotine Oxidase from Arthrobacter oxidans. Amino-Acid Sequence of the FAD-Peptide. Eur. J. Biochem. **37**, 256 (1973).

11. BRÜHMÜLLER, M., H. MÖHLER, and K. DECKER: Covalently Bound Flavin in D-6-Hydroxynicotine Oxidase from Arthrobacter oxidans. Purification and Properties of D-6-Hydroxynicotine Oxidase. Eur. J. Biochem. **29**, 143 (1972).
12. BRUICE, T. C.: Models and Flavin Catalysis. Submitted to "Progress in Bioorganic Chemistry" (E. T. KAISER, F. J. KEZDY, eds.). New York: Wiley.
13. BRÜSTLEIN, M., W.-R. KNAPPE und P. HEMMERICH: Neue Photoalkylierungsreaktionen am Flavinkern. Angew. Chem. **83**, 854 (1971).
14. BULLOCK, F. I., and O. JARDETZKY: An Experimental Demonstration of the Nuclear Magnetic Resonance Assignments in the 6,7-Dimethylalloxazine Nucleus. J. Org. Chem. **30**, 2056 (1965).
15. CAIRNS, W. L., and D. E. METZLER: Photochemical Degradation of Flavins. VI. A New Photoproduct and Its Use in Studying the Photolytic Mechanism. J. Amer. Chem. Soc. **93**, 2772 (1971).
16. CARR, D. O., and D. E. METZLER: The Oxidation of Ethyl-1,2-Dihydro-2-naphtoate by Flavins and its Stimulation by Light. Biochem. Biophys. Acta **205**, 63 (1970).
17. CHECCUCCI, A., G. COLOMBETTI, G. DEL CARRATORE, R. FERRARA, and F. LENCI: Red Light-Induced Accumulation of Euglena gracilis. Photochem. Photobiol. **19**, 223 (1974).
18. CHI, T. F., Y. L. WANG, C. L. TSOU, Y. C. WANG, and C. H. YU: Scientia Sinica **14**, 1193 (1965).
19. DRAPER, R. D., and L. L. INGRAHAM: A Potentiometric Study of the Flavin Semiquinone Equilibrium. Arch. Biochem. Biophys. **125**, 802 (1968).
20. DRYSDALE, G. R., M. J. SPIEGEL, and P. J. STRITTMATTER: Flavoprotein-catalyzed direct Hydrogen Transfer between Pyridine Nucleotides. J. Biol. Chem. **236**, 2323 (1961).
21. DUDLEY, K. H., A. EHRENBERG, P. HEMMERICH und F. MÜLLER: Spektren und Strukturen der am Flavin-Redoxsystem beteiligten Partikeln. Studien in der Flavinreihe IX. Helv. Chim. Acta **47**, 1354 (1964).
22. DUDLEY, K. H. und P. HEMMERICH: Stabile Dihydroflavine und quartäre Flaviniumsalze. Studien in der Flavinreihe, 12. Mitt. Helv. Chim. Acta **50**, 355 (1967).
23. — — Flavins XIII. Rearrangement Reactions of 1,3,10-Trialkylflavinium Salts. J. Org. Chem. **32**, 3049 (1967).
24. EDMONDSON, D. E.: Intramolecular Hemiacetal Formation in 8-Formylriboflavine. Biochemistry **13**, 2817 (1974).
25. EDMONDSON, D. E., and T. P. SINGER: Oxidation-Reduction Properties of the 8α-Substituted Flavins. J. Biol. Chem. **248**, 8144 (1973).
26. EHRENBERG, A., F. MÜLLER, and P. HEMMERICH: Basicity, Visible Spectra and Electron Spin Resonance of Flavosemiquinone Anions. Eur. J. Biochem. **2**, 286 (1967).
27. EISELE, R.: Ph. D. Thesis, University of Konstanz (1974).
28. ELEY, M., J. LEE, J. M. LHOSTE, C. Y. LEE, M. J. CORMIER, and P. HEMMERICH: Bacterial Bioluminescence. Comparisons of Bioluminescence Emission Spectra, the Fluorescence of Luciferase Reaction Mixtures, and the Fluorescence of Flavin Cations. Biochemistry **9**, 2902 (1970).
29. ENTSCH, B., D. P. BALLOU, and V. MASSEY: The Role of Oxygenated Flavins in the Catalytic Reaction of p-Hydroxy-Benzoate Hydroxylase. In: "Flavins and Flavoproteins" (T. P. SINGER, ed.). Amsterdam: Elsevier, in press.
30. ERWIN, V. G., and L. HELLERMAN: Mitochondrial Monamino Oxidase. I. Purification and Characterisation of the Bovine Kidney Enzyme. J. Biol. Chem. **242**, 4230 (1967).
31. FARAGGI, M., P. HEMMERICH, and I. PECHT: O_2-Affinity of Flavin Radical Species as Studied by Pulse Radiolysis. FEBS-Lett. **51**, 47 (1975).

32. Favaudon, V., and J.-M. Lhoste: The Kinetics of Flavine Oxidation-Reduction I. Biochemistry **14**, 4734 (1975).
33. Fenner, H., H. H. Roessler, and H. J. Duchstein: Structure and Reactivity of 5-Deazaflavins. In: "Flavins and Flavoproteins" (T. P. Singer, ed.). In Press.
34. Forrest, H. S., C. van Baalen, M. Viscontini und M. Piraux: Reaktion von CN^- mit hydriertem 2-Amino-6-hydroxy-pteridin. Helv. Chim. Acta **43**, 1005 (1960).
35. Förster, T.: Elektrolytische Dissoziation angeregter Moleküle. Z. Elektrochem. **54**, 42 (1950).
36. Frisell, W. R., C. W. Chung, and C. G. Mackenzie: Catalysis of Oxidation of Nitrogen Compounds by Flavin Coenzymes in the Presence of Light. J. Biol. Chem. **234**, 1297 (1959).
37. Frisell, W. R., and C. G. Mackenzie: Separation and Purification of Sarcosine Dehydrogenase and Dimethylglycine Dehydrogenase. J. Biol. Chem. **237**, 94 (1962).
38. Fröhlich, O., and B. Diehn: Photoeffects in a Flavin-containing Lipid Bilayer Membrane and Implications for Algal Phototaxis. Nature **248**, 802 (1974).
39. Gärtner, B., und P. Hemmerich: Zur Propargylaminhemmung der Monoaminooxidase: Struktur des Inhibitor-Komplexes. Angew. Chem. **87**, 137 (1975).
40. Gerstner, E. und E. Pfeil: Zur Kenntnis des Flavinenzyms Hydroxynitril-Lyase (D-Oxynitrilase). Hoppe-Seyler's Z. Physiol. Chem. **353**, 271 (1972).
41. Ghisla, S.: Personal Communication.
42. Ghisla, S., U. Hartmann und P. Hemmerich: Die Synthese des Succinat-Dehydrogenase-Riboflavins. Angew. Chem. **82**, 669 (1970).
43. Ghisla, S., U. Hartmann, P. Hemmerich und F. Müller: Die reduktive Alkylierung des Flavinkerns, Struktur und Reaktivität von Dihydroflavinen. XVIII Mitt. Liebigs Ann. Chem. **1973**, 1388.
44. Ghisla, S., and P. Hemmerich: Synthesis of the Flavocoenzyme of Monoamine Oxidase. FEBS-Lett. **16**, 229 (1971).
45. — Unpublished observations.
46. Ghisla, S., V. Massey, J.-M. Lhoste, and S. G. Mayhew: Fluorescence and Optical Characteristics of Reduced Flavins and Flavoproteins. Biochemistry **13**, 589 (1974).
47. Ghisla, S., V. Massey, and S. G. Mayhew: Studies on the Active Centers of Flavoproteins; Binding of 8-Hydroxy-FAD and 8-Hydroxy-FMN to Apoproteins. In: "Flavins and Flavoproteins" (T. P. Singer, ed.). Amsterdam: Elsevier, in press.
48. Ghisla, S., and S. G. Mayhew: Identification and Structure of a Novel Flavin Prosthetic Group Associated with Reduced Nicotinamide Adenine Dinucleotide from Peptostreptococcus elsdenii. J. Biol. Chem. **248**, 6568 (1973).
49. — — Eur. J. Biochem., in press.
50. Gibson, Q. H., and J. W. Hastings: The Oxidation of Reduced Flavin Mononucleotide by Molecular Oxygen. Biochem. J. **83**, 368 (1962).
51. Gladys, M. und W.-R. Knappe: Photochemie des (Iso)Alloxazins III. Intramolekulare Photodealkylierung von 10-Alkylisoalloxazinen, eine Modellreaktion für den Riboflavinphotoabbau. Chem. Ber. **107**, 3658 (1974).
52. Green, O. E., S. Mii, and P. M. Kohout: Studies on the Terminal Electron Transport System. I. Succinic Dehydrogenase. J. Biol. Chem. **217**, 551 (1955).
53. Haas, W., and P. Hemmerich: pH-Dependence, Isotope Effects and Products of Flavin-sensitized Photodecarboxylation and Photodehydrogenation. Z. Naturforsch. **27b**, 1035 (1972).
54. Hall, R. L., B. Vennesland, and F. J. Kézdy: Glyoxylate Carboligase of Escherichia coli. J. Biol. Chem. **244**, 3991 (1969).
55. Hamilton, G. A.: The Proton in Biological Redox Reactions. In: "Progress in Bioorganic Chemistry" (E. T. Kaiser, F. J. Kézdy, eds.), Vol. 1, p. 83. New York: Wiley. 1971.

56. HARBURY, H. A., K. F. LA NOUE, P. A. LOACH, and R. M. AMICK: Molecular Interaction of Isoalloxazine Derivatives II. Proc. Nat. Acad. Sci. USA **45**, 1708 (1959).
57. HASTINGS, J. W., *et al.*: Several Papers. In: "Flavins and Flavoproteins" (T. P. SINGER, ed.). Amsterdam: Elsevier, in press.
58. HASTINGS, J. W., C. BALNY, C. LEPEUCH, and P. DOUZOU: Spectral Properties of an Oxygenated Luciferase – Flavin Intermediate Isolated by Low-Temperature Chromatography. Proc. Natl. Acad. Sci. U.S.A. **70**, 3468 (1973).
59. HASTINGS, J. W., and Q. H. GIBSON: Intermediates in the Bioluminescent Oxidation of Reduced Flavin Mononucleotide. J. Biol. Chem. **238**, 2537 (1963).
60. HEIZMANN, C., P. HEMMERICH, R. MENGEL, and W. PFLEIDERER: Pteridine Synthesis from Riboflavin and other Isoalloxazines. In: "Chemistry and Biology of Pteridines" (K. IWAO, M. AKINO, M. GOTO, Y. JWANAMI, eds.), p. 105. Tokyo: Int. Acad. Printing Co. Ltd. 1970.
61. HEMMERICH, P.: Studien in der Lumiflavin-Reihe VIII. Die Kondensation von 8-Methyl-isoalloxazinen mit Aldehyden. Helv. Chim. Acta **43**, 1942 (1960).
62. — Die Koordinationschemie der Flavokoenzyme und die Bedeutung der Nicht-Häm-Metallionen in der Atmungskette. In: „Mechanismen enzymatischer Reaktionen" (14. Colloquium der Gesellschaft für Physiologische Chemie, Mosbach/Baden 1963). Berlin-Göttingen-Heidelberg-New York: Springer. 1964.
63. — Flavosemichinon-Metallchelate: Modelle zur Erklärung der „active site" in mitochondrialen Flavoenzymen. Zum Verhalten des Riboflavins gegen Metallionen III. Helv. Chim. Acta **47**, 464 (1964).
64. — Free Radicals of Biological Interest as Studied by ESR. Proc. Roy. Soc. **A302**, 335 (1968).
65. — Discussion remark. In: "Flavins and Flavoproteins" (H. KAMIN, ed.), p. 52. Baltimore: University Park Press. 1971.
66. HEMMERICH, P., A. P. BHADURI, G. BLANKENHORN, M. BRÜSTLEIN, W. HAAS, and W.-R. KNAPPE: Model Studies towards Demonstration of Covalent $2e^-$-Transfer Intermediates and Their Structure in Flavin Dependent CH- and O_2-Activation. In: "Oxidases and Related Redox Systems" (T. E. KING, H. S. MASON, M. MORRISON, eds.), p. 3. Baltimore: University Park Press. 1973.
67. HEMMERICH, P., A. EHRENBERG, W. H. WALKER, L. E. G. ERIKSSON, J. SALACH, P. BADER, and T. P. SINGER: On the Structure of Succinate Dehydrogenase Flavocoenzyme. FEBS-Lett. **3**, 37 (1969).
68. HEMMERICH, P., and W. HAAS: Recent Developments in the Study of "Fully Reduced Flavin". In: "Reactivity of Flavins" (K. YAGI, ed.). Tokyo: University Press, 1975.
69. HEMMERICH, P., and W.-R. KNAPPE: Flavin-Dependent Substrate Dehydrogenation; Model Studies and Mechanisms. In: "Structure and Function of Oxidation Reduction Enzymes" (A. AKESON, A. EHRENBERG, eds.), p. 367. Oxford: Pergamon Press. 1971.
70. HEMMERICH, P., and J. LAUTERWEIN: The Structure and Reactivity of Flavin-Metal Complexes. In: "Inorganic Biochemistry" (G. EICHHORN, ed.), Vol. 2, p. 1168, Amsterdam: Elsevier. 1973.
71. HEMMERICH, P., V. MASSEY, and G. WEBER: Photo-Induced Benzyl Substitution of Flavins by Phenylacetate: A Possible Model for Flavoprotein Catalysis. Nature **213**, 728 (1967).
72. HEMMERICH, P., and F. MÜLLER: Flavin-O_2-Interaction Mechanism and the Function of Flavin in Hydroxylation Reactions. Ann. N. Y. Acad. Sci. **212**, 13 (1973).
73. HEMMERICH, P., F. MÜLLER, and A. EHRENBERG: The Chemistry of Flavin-Metal Interaction. In: "Oxidases and Related Redox Systems" (T. E. KING, H. S. MASON, M. MORRISON, eds.), p. 157. New York: Wiley. 1965.
74. HEMMERICH, P., G. NAGELSCHNEIDER, and C. VEEGER: Chemistry and Molecular Biology of Flavins and Flavoproteins. FEBS-Lett. **8**, 69 (1970).

75. HEMMERICH, P., B. PRIJS, and H. ERLENMEYER: Synthesen in der Lumiflavinreihe IV. Helv. Chim. Acta **42,** 1604 (1959).
76. — — — Studien in der Lumiflavinreihe V. Spezifische Reaktivität 8-ständiger Substituenten am Isoalloxazinkern; Flavin-Dimere. Helv. Chim. Acta **42,** 2164 (1959).
77. — — — Studien in der Lumiflavinreihe VI. Alkylierungs- und Desalkylierungsreaktionen an (Iso)alloxazinen; 1,3,10-Trimethylflavosemichinone. Helv. Chim. Acta **43,** 372 (1960).
78. HEMMERICH, P., and M. SCHUMAN-JORNS: Mechanisms of Hydrogen Transfer in Redox Enzymes. In: "Enzymes: Structure and Function" (J. DRENTH, R. A. OOSTERBAAN, C. VEEGER, eds.), p. 95. Amsterdam: North Holland Pub. Corp. 1972.
79. HEMMERICH, P., C. VEEGER und H. C. S. WOOD: Fortschritte in der Chemie und Molekularbiologie der Flavine und Flavocoenzyme. Angew. Chem. **77,** 699 (1965).
80. HEMMERICH, P., and A. WESSIAK: The Structural Chemistry of Flavin Dependent O_2-Activation. In: "Flavins and Flavoproteins" (T. P. SINGER, ed.). Amsterdam: Elsevier, in press.
81. IGAUE, I., B. GOMES, and K. T. YASUNOBU: Beef Mitochondrial Monoamine Oxidase, a Flavin Dinucleotide Enzyme. Biochem. Biophys. Res. Commun. **29,** 562 (1967).
82. KAMIN, H., ed.: Flavins and Flavoproteins. Baltimore: University Park Press. 1971.
83. KARRER, P., K. SCHÖPP und F. BENZ: Synthesen von Flavinen IV. Helv. Chim. Acta **18,** 426 (1935).
84. KEARNEY, E. B.: Studies on Succinic Dehydrogenase XII. Flavin Component of the Mammalian Enzyme. J. Biol. Chem. **235,** 865 (1960).
85. KEARNEY, E. B., J. I. SALACH, W. H. WALKER, R. L. SENG, W. KENNEY, E. ZESZOTEK, and T. P. SINGER: The Covalently-Bound Flavin of Hepatic Monoamino Oxidase. 1. Isolation and Sequence of a Flavin Peptide and Evidence for Binding at the 8α-Position. Eur. J. Biochem. **24,** 321 (1971).
86. KEARNEY, E. B., J. I. SALACH, W. H. WALKER, R. SENG, and T. P. SINGER: Structure of the Covalently Bound Flavin of Monoamine Oxidase. Biochem. Biophys. Res. Commun. **42,** 490 (1971).
87. KEARNEY, E. B., and T. P. SINGER: On the Prosthetic Group of Succinic Dehydrogenase. Biochem. Biophys. Acta **17,** 596 (1955).
88. KENNEY, W. C., D. E. EDMONDSON, and T. P. SINGER: A Novel Form of Covalently Bound Flavin from Thiamine Dehydrogenase. Biochem. Biophys. Res. Commun. **57,** 106 (1974).
89. — — — The Covalently Bound Flavin of Chromatium Cytochrome c_{552}. 2. Sequence of Flavin Peptides and Flavin-Tyrosine Interaction. Eur. J. Biochem. **48,** 449 (1974).
90. KENNEY, W. C., and W. H. WALKER: Synthesis and Properties of 8α-Substituted Riboflavins of Biological Importance. FEBS-Lett. **20,** 297 (1972).
91. KENNEY, W. C., W. H. WALKER, E. B. KEARNEY, E. ZESZOTEK, and T. P. SINGER: Amino Acid Sequence at the Active Center of Succinate Dehydrogenase. Biochem. Biophys. Res. Commun. **41,** 488 (1970).
92. KENNEY, W. C., W. H. WALKER, and T. P. SINGER: Studies on Succinate Dehydrogenase XX. Amino Acid Sequence around the Flavin Site. J. Biol. Chem. **247,** 4510 (1972).
93. KIERKEGAARD, P., R. NORRESTAM, P. WERNER, I. CSÖREGH, M. GLEHN, R. KARLSSON, M. LEIJONMARCK, O. RÖNNQUIST, B. STENNESLAND, O. TILLBERG, and L. TORBJÖRNSSON: X-Ray Structure Investigations of Flavin Derivatives. In: "Flavines and Flavoproteins" (H. KAMIN, ed.), p. 1. Baltimore: University Park Press. 1971.
94. KING, T. E., and J. W. CLARK-LEWIS: The Structures of Some Supposed Azetid-2:4-

diones. Part III. The "Alloxan-5-o-dimethylaminoanil" of Rudy and Cramer, and its Alkali Hydrolysis Product. J. Chem. Soc. **1951,** 3080.
95. KLÄUI, W.: Untersuchungen über den Einfluß koordinativ gebundener Übergangsmetalle auf Reaktionen von ungesättigten Carbocyclen in Übergangsmetall-Komplexen. Ph. D. Thesis, University of Zürich: 1975.
96. KLOPMAN, G.: Chemical Reactivity and the Concept of Charge and Frontier-Controlled Reactions. J. Amer. Chem. Soc. **90,** 223 (1968).
97. KNAPPE, W.-R.: Photochemie des 10-Phenylisoalloxazins. Intramolekulare Singulett- und intermolekulare Triplett-Reaktionen. Chem. Ber. **107,** 1614 (1974).
98. KNAPPE, W.-R., and P. HEMMERICH: Covalent Intermediates in Flavin-sensitized Photodehydrogenation and Photodecarboxylation. Z. Naturforsch. **27b,** 1032 (1972).
99. KOSOWER, E. M.: Molecular Biochemistry, p. 166. New York: McGraw-Hill. 1962.
100. KUHN, R., K. REINEMUND, and F. WEYGAND: Synthese des Lumi-lactoflavins. Ber. **67,** 1460 (1934).
101. KUHN, R. und R. STRÖBELE: Über Verdo-, Chloro- und Rhodoflavine. Ber. **70,** 753 (1937).
102. KUHN, R., and T. WAGNER-JAUREGG: Über die aus Eiklar und Milch isolierten Flavine. Ber. **66,** 1577 (1933).
103. KUHN, R., and F. WEYGAND: Synthese des 9-Methyl-isoalloxazins. Berg. **67,** 1409 (1934).
104. LAMBOOY, J. P.: The Alloxazines and Isoalloxazines. In: "Heterocyclic Compounds" (R. C. ELDERFIELD, ed.), Vol. 9, p. 118. N. Y. 1967.
105. LAND, E. J., and A. J. SWALLOW: One-Electron Reactions in Biochemical Systems as Studied by Pulse Radiolysis I. Nicotinamide — Adenine Dinucleotide and Related Compounds. Biochim. Biophys. Acta **162,** 327 (1968).
106. LARRABEE, R. B.: Fluxional Main Group IV Organometallic Compounds. The Implications for Orbital Symmetry Rules. J. Organomet. Chem. **74,** 313 (1974).
107. LASSER, N., and I. FEITELSON: Excited State pK-Values from Fluorescence Measurements. J. Phys. Chem. **77,** 1011 (1973).
108. LOUIE, D. D., and N. KAPLAN: Stereospecifity of Hydrogen Transfer Reaction of Pseudomonas aeruginosa Pyridine Nucleotide Transhydrogenase. J. Biol. Chem. **245,** 5691 (1970).
109. LUDWIG, M., R. M. BURNETT, G. D. DARLING, S. R. JORDAN, D. S. KENDALL, and W. W. SMITH: The Structure of Clostridium MP. Flavodoxin as a Function of Oxidation State: Some Comparisons. In: "Flavins and Flavoproteins" (T. P. SINGER, ed.). Amsterdam: Elsevier, in press.
110. MAGER, H. I. X.: Nonenzymic Activation and Transfer of Oxygen by Reduced Alloxazine. In: "Flavins and Flavoproteins" (T. P. SINGER, ed.). Amsterdam: Elsevier, in press.
111. MAGER, H. I. X., and W. BERENDS: Hydroperoxides of Partially Reduced Quinoxalines, Pteridines and (Iso)alloxazines: Intermediates in Oxidation Processes. Rec. Trav. Chim. **84,** 1329 (1965).
112. — — The Structure of Transient Hydroperoxides in the Autoxidation of Reduced Flavins. Biochem. Biophys. Acta **118,** 440 (1966).
113. — — Activation and Transfer of Oxygen. V. Spontaneous Oxidation of 1,3,10-trimethyl-5,10-dihydroalloxazine in Acidic Media in Connection with its Hydroxylating Ability. Rec. Trav. Chim. **91,** 611 (1972).
114. — — Activation and Transfer of Oxygen. VI. The Influence of Hydrochloric Acid on the Autoxidation of Dihydroalloxazines. Rec. Trav. Chim. **91,** 630 (1972).
115. — — Activation and Transfer of Oxygen. IX. Nonenzymic Hydroxylation of Phenylalanine by Model Systems of Dihydroalloxazine/O_2, Dihydroalloxazine/H_2O_2 and Alloxazinium Cation/H_2O_2. Tetrahedron **30,** 917 (1974).

116. MASSEY, V.: Unpublished.

117. MASSEY, V., and B. CURTI: A New Method of Preparation of D-Amino Acid Oxidase Apoprotein and a Conformational Change after its Combination with Flavin Adenin Dinucleotide. J. Biol. Chem. **241,** 3417 (1966).

118. MASSEY, V., and S. GHISLA: Role of Charge-Transfer Interactions in Flavoprotein Catalysis. Ann. N. Y. Acad. Sci. **227,** 446 (1974).

119. MASSEY, V., S. GHISLA, D. BALLOU, C. T. WALSH, Y. F. CHENG, and R. H. ABELES: Rapid Reaction Studies on Dehydrogenation and Elimination Reactions of D-Amino Acid Oxidase and Lactate Oxidase. In: "Flavins and Flavoproteins" (T. P. SINGER, ed.). Amsterdam: Elsevier, in press.

120. MASSEY, V., and P. HEMMERICH: Flavine and Pteridine Monooxygenases. In: "The Enzymes" (P. D. BOYER, ed.), 3rd ed., Volume XII B, 191. N. Y.: Academic Press. 1976.

121. MASSEY, V., F. MÜLLER, R. FELDBERG, M. SCHUMAN, P. A. SULLIVAN, L. G. HOWELL, S. G. MAYHEW, and R. H. MATTHEWS: The Reactivity of Flavoproteins with Sulfite. Possible Relevance to the Problem of Oxygen Reactivity. J. Biol. Chem. **244,** 3999 (1969).

122. MASSEY, V., G. PALMER, and D. BALLOU: On the Reaction of Reduced Flavins with Molecular Oxygen. In: "Oxidases and Related Redox Systems" (T. E. KING, H. S. MASON, and M. MORRISON, eds.), Vol. 1, p. 25. Baltimore: University Park Press. 1973.

123. MAYHEW, S. G.: Studies of Flavin Binding in Flavodoxins. Biochem. Biophys. Acta **235,** 289 (1971).

124. MAYHEW, S. G., G. P. FOUST, and V. MASSEY: Oxidation-Reduction Properties of Flavodoxin from Peptostreptococcus elsdenii. J. Biol. Chem. **244,** 803 (1969).

125. MAYHEW, S. G., and V. MASSEY: Evidence for a Novel Flavin Prosthetic Group Associated with NADH Dehydrogenase from Peptostreptococcus elsdenii. Biochim. Biophys. Acta **235,** 303 (1971).

126. MAYHEW, S. G., C. D. WHITFIELD, S. GHISLA, and M. SCHUMAN-JORNS: Identification and Properties of New Flavins in Electron Transferring Flavoprotein from Peptostreptococcus elsdenii and Pig-Liver Glycolate Oxidase. Eur. J. Biochem. **44,** 579 (1974).

127. MCCAPRA, F., and D. W. HYSERT: Bacterial Bioluminescence. Identification of Fatty acid as Product, its Quantum Yield, and a Suggested Mechanism. Biochem. Biophys. Res. Comm. **52,** 298 (1973).

128. MICHAELIS, L., M. P. SCHUBERT, and C. V. SMYTHE: Potentiometric Study of the Flavins. J. Biol. Chem. **116,** 587 (1936).

129. MICHAELIS, L., and G. SCHWARZENBACH: The Intermediate Forms of Oxidation-Reduction of the Flavins. J. Biol. Chem. **123,** 527 (1938).

130. MIURA, R., K. MATSUI, K. HIROTSU, A. SHIMADA, M. TAKATSU, and S. OTANI: X-Ray Crystallographic Determination of a Derivative of a New Flavin Compound, Roseoflavin. J. Chem. Soc. Chem. Commun. **1973,** 703.

131. MÖHLER, H., M. BRÜHMÜLLER, and K. DECKER: Covalently Bound Flavin in D-6-Hydroxynicotine Oxidase from Arthrobacter oxidans. Identification of the 8α-(N-3-Histidyl)-riboflavin-Linkage between FAD and Apoenzyme. Eur. J. Biochem. **29,** 152 (1972).

132. MORRIS, J. R., H. L. CRESPI, and J. J. KATZ: Characterization of a new Photo-ESR Signal Assotiated with Photosynthesis. Biochem. Biophys. Res. Commun. **49,** 139 (1972).

133. MÜLLER, F.: On the Reaction of Flavins with Alcohols. In: "Flavins and Flavoproteins" (H. KAMIN, ed.), p. 363. Baltimore: University Park Press. 1971.

134. — In: "Flavins and Flavoproteins" (T. P. SINGER, ed.). Amsterdam: Elsevier, in press.

135. MÜLLER, F., P. HEMMERICH, and A. EHRENBERG: ESR-Spectra of Flavin Radical Cations. In Preparation.

136. MÜLLER, F., P. HEMMERICH, A. EHRENBERG, G. PALMER, and V. MASSEY: The Chemical and Electronic Structure of the Neutral Flavin Radical as Revealed by Electron Spin Resonance Spectroscopy of Chemically and Isotopically Substituted Derivatives. Eur. J. Biochem. **14,** 185 (1970).

137. MÜLLER, F., V. MASSEY, G. HEIZMANN, P. HEMMERICH, J. M. LHOSTE, and D. C. GOULD: The Reduction of Flavins by Borohydride: 3,4-Dihydroflavin Structure, Absorption and Luminescence. Eur. J. Biochem. **9,** 392 (1969).

138. NEAL, R. A.: Bacterial Metabolism of Thiamine III. Metabolism of Thiamine to 3-(2′-Methyl-4′-Amino-5′-Pyrimidylmethyl)-4-Methyl-thiazole-5-acetic Acid (Thiamine Acetic Acid) by a Flavoprotein isolated from a soil Microorganism. J. Biol. Chem. **245,** 2599 (1970).

139. NEUMÜLLER, O.-A., ed.: Römpps Chemie-Lexikon. Stuttgart: Franck'sche Verlagshandlung. 1973.

140. NICOLI, M. Z., and J. W. HASTINGS: Bacterial Luciferase. The Hydrophobic Environment of the Reactive Sulfhydryl. J. Biol. Chem. **249,** 2393 (1974).

141. NICOLI, M. Z., E. M. MEIGHEN, and J. W. HASTINGS: Bacterial Luciferase. Chemistry of the Reactive Sulfhydryl. J. Biol. Chem. **249,** 2385 (1974).

142. O'BRIEN, D. E., L. T. WEINSTOCK, and C. C. CHENG: Synthesis of 10-Deazariboflavin and Related 2,4-Dioxopyrimido (4,5-b) quinolines. J. Heterocycl. Chem. **7,** 99 (1970).

143. PATEK, D. R., C. R. DAHL, and W. R. FRISELL: Isolation of Acid-Nonextractable Flavins from a Bacterial Sarcosine Oxidase. Biochem. Biophys. Res. Commun. **46,** 885 (1972).

144. PATEK, D. R., and W. R. FRISELL: Purification and Characterisation of the Flavin Prosthetic Group of Sarcosine Dehydrogenase. Arch. Biochem. Biophys. **150,** 347 (1972).

145. PEARSON, W. N.: Riboflavin. In: "The Vitamins" (P. GYÖRGY, W. N. PEARSON, eds.), Vol. VII, p. 99. N. Y.: Academic Press. 1967.

146. PENZER, G. R., and G. K. RADDA: The Chemistry and Biological Function of Isoalloxazines (Flavins). Quart Rev. **21,** 43 (1967).

147. PENZER, G. R., and G. K. RADDA: The Chemistry of Flavins and Flavoproteins. Photoreduction of Flavins by Amino Acids. Biochem. J. **109,** 259 (1968).

148. — — Photochemistry of Flavins. In: "Vitamins and Coenzymes" (D. B. MCCORMICK, L. D. WRIGHT, eds.), Vol. XVIII of "Methods in Enzymology", p. 479. N. Y.: Academic Press. 1971.

149. PILOTY, O., and K. FINCKH: Über die Harnsäuregruppe. Liebigs Ann. Chem. **333,** 37 (1904).

150. PLAUT, G. W. E.: Water-Soluble Vitamins, Part II. Riboflavin and Folic Acid. Ann. Rev. Biochem. **30,** 409 (1961).

151. POFF, K. L., and W. L. BUTLER: Absorbance Changes induced by Blue Light in Phycomyces blakesleeanus and Dietyostelium discoideum. Nature **248,** 799 (1974).

152. POLYAKOVA, N. A., L. S. TUL'CHINSKAYA, L. G. ZAPESOCHNAYA, and V. M. BEREZOVSKII: Alloxazines and Isoalloxazines XXXI. Synthesis of Hydroxy Analogs of Natural Flavines. Zhur. Obsh. Khim. **42,** 465 (1972).

153. RIVLIN, R. S., ed.: Riboflavin. N. Y.: Plenum Press. 1975.

154. SALACH, J., W. H. WALKER, T. P. SINGER, A. EHRENBERG, P. HEMMERICH, S. GHISLA, and U. HARTMANN: Studies on Succinate Dehydrogenase. Site of Attachment of the Covalently-Bound Flavin to the Peptide Chain. Eur. J. Biochem. **26,** 267 (1972).

155. SCHABORT, I. C., and D. J. J. POTGIETER: β-Cyclopiazonate Oxidocyclase from Pennicillum Cyclopium. II. Studies on Electron Acceptors, Inhibitors, Enzyme Kinetics,

Amino Acid Composition, Flavin Prosthetic Group and other Properties. Biochim. Biophys. Acta **250**, 329 (1971).

156. SCHÖLLNHAMMER, G., W. HAAS, and P. HEMMERICH: Unpublished.

157. SCHÖLLNHAMMER, G., and P. HEMMERICH: Unpublished.

158. — — Nucleophilic Addition at the Photoexcited Flavin Cation: Synthesis and Properties of 6- and 9-Hydroxy-Flavocoenzyme Chromophores. Eur. J. Biochem. **44**, 561 (1974).

159. SCHONBRUNN, A., C. WALSH, S. GHISLA, H. OGATA, V. MASSEY, and R. ABELES: The Structure of the Flavin Inhibitor Adduct from Lactate Oxidase. In: "Flavins and Flavoproteins" (T. P. SINGER, ed.). Amsterdam: Elsevier, in press.

160. SCHREINER, S., U. STEINER, and H. E. A. KRAMER: Determination of the pK Values of the Lumiflavin Triplet State by Flash Photolysis. Photochem. Photobiol. **21**, 81 (1975).

161. SCHUMAN-JORNS, M., G. SCHÖLLNHAMMER, and P. HEMMERICH: Intramolecular Addition of the Riboflavin Side Chain: Anion Catalyzed Neutral Photochemistry. Eur. J. Biochem. **57**, 35 (1975).

162. SIEGEL, L. M., H. KAMIN, D. C. RUEGER, R. P. PRESSWOOD, and O. H. GIBSON: An Iron-Free Sulfite Reductase Flavoprotein from Mutants of Salmonella tryphimurium. In: "Flavins and Flavoproteins" (H. KAMIN, ed.), p. 523. Baltimore: University Park Press. 1971.

163. SINGER, T. P., ed.: Flavins and Flavoproteins. Amsterdam: Elsevier, in press.

164. SINGER, T. P., and D. E. EDMONDSON: 8α-Substituted Flavins of Biological Importance. FEBS-Lett. **42**, 1 (1974).

165. SINGER, T. P., E. B. KEARNEY, and V. MASSEY: Observations on the Flavin Moiety of Succinate Dehydrogenase. Arch. Biochem. Biophys. **60**, 255 (1956).

166. SINGER, T. P., and W. C. KENNEY: To be published.

167. SLATER, E. C., ed.: Flavins and Flavoproteins. Amsterdam: Elsevier. 1966.

168. SMITH, S. B., M. BRÜSTLEIN, and T. C. BRUICE: Electrophilicity of the 8-Position of the Isoalloxazine (Flavine) Ring System. Comment on the Mechanism of Oxidation of Dihydroalloxazine. J. Amer. Chem. Soc. **96**, 3696 (1974).

169. SONG, P.-S.: On the Basicity of the Excited State of Flavins. Photochem. Photobiol. **7**, 311 (1968).

170. SUN, M., T. A. MOORE, and P.-S. SONG: Molecular Luminescence Studies of Flavins. I. The Excited States of Flavins. J. Amer. Chem. Soc. **94**, 1730 (1972).

171. TAUSCHER, L., S. GHISLA, and P. HEMMERICH: NMR-Study of Nitrogen-Inversion and Conformation of 1,5-Dihydro-Isoalloxazines (Reduced Flavins). Helv. Chim. Acta **56**, 630 (1973).

172. THEORELL, H.: Reindarstellung (Kristallisation) des gelben Atmungsferments und die reversible Spaltung desselben. Biochem. Z. **272**, 155 (1934).

173. — Über die Wirkungsgruppe des gelben Ferments. Biochem. Z. **275**, 37 (1935).

174. — Reindarstellung der Wirkungsgruppe des gelben Ferments. Biochem. Z. **275**, 344 (1935).

175. — Das gelbe Oxydationsferment. Biochem. Z. **278**, 263 (1935).

176. TISHLER, M., K. PFISTER, R. D. BABSON, K. LADENBURG, and A. J. FLAMING: The Reaction between o-Aminoazo Compounds and Barbituric Acid. A New Synthesis of Riboflavin. J. Amer. Chem. Soc. **69**, 1487 (1947).

177. TUL'CHINSKAYA, L. S., N. A. POLYAKOVA, and V. M. BEREZOVSKII: Alloxazine and Isoalloxazine Compounds XXVI. Azo Coupling of 7-Amino and 7-Hydroxyalloxazines. Zh. Obsh. Khim. **40**, 1859 (1970).

178. VAINSHTEIN, F. M., E. I. KUKHTENKO, E. I. TOMILENKO, and E. A. SHILOV: Nucleophilic Replacement of the Sulfonato Group in Aromatic Compounds with the Participation of Oxidizing Agents. Zh. Org. Khim. **3**, 1654 (1967).

179. VAISH, S. P., and G. TOLLIN: Flash Photolysis of Flavins. V. Oxidation and Disproportionation of Flavin Radicals. Bioenergetics **2**, 61 (1971).

180. WALAAS, E., and O. WALAAS: Kinetics and Equilibria in Flavoprotein Systems. V. The Effects of pH, Anions and Partial Structural Analogues of the Coenzyme on the Activity of D-Amino Acid Oxidase. Acta Chem. Scand. **10**, 122 (1956).

181. WALKER, W. H., P. HEMMERICH, and V. MASSEY: Reductive Photoalkylierung des Flavinkerns und flavin-katalysierte Photodecarboxylierung von Phenylacetat. Studien in der Flavinreihe XV. Helv. Chim. Acta **50**, 2269 (1967).

182. — — — Light-Induced Alkylation and Dealkylation of the Flavin Nucleus. Stable Dihydroflavins: Spectral Course and Mechanism of Formation. Eur. J. Biochem. **13**, 258 (1970).

183. WALKER, W. H., E. B. KEARNEY, R. L. SENG, and T. P. SINGER: Sequence and Structure of a Cysteinyl Flavin Peptide from Monamine Oxidase. Biochem. Biophys. Res. Commun. **44**, 287 (1971).

184. — — — — The Covalently-Bound Flavin of Hepatic Monamine Oxidase. 2. Identification and Properties of Cysteinyl Riboflavin. Eur. J. Biochem. **24**, 328 (1971).

185. WALKER, W. H., W. C. KENNEY, D. E. EDMONDSON, T. P. SINGER, J. R. CRONIN, and R. HENDRIKS: The Covalently Bound Flavin of Chromatium Cytochrome c_{552}. I. Evidence for Cysteine Thiohemiacetal at the 8α-Position. Eur. J. Biochem. **48**, 439 (1974).

186. WALKER, W. H., T. P. SINGER, S. GHISLA, and P. HEMMERICH: Studies on Succinate Dehydrogenase. 8α-Histidyl-FAD as the Active Center of Succinate Dehydrogenase. Eur. J. Biochem. **26**, 279 (1972).

187. WALSH, T., A. SCHONBRUNN, and R. H. ABELES: Studies on the Mechanism of Action of D-Amino Acid Oxidase. Evidence for Removal of Substrate α-Hydrogen as a Proton. J. Biol. Chem. **246**, 6855 (1971).

188. WARBURG, O., and W. CHRISTIAN: Über das gelbe Ferment und seine Wirkungen. Biochem. Z. **266**, 377 (1933).

189. — — Das Coferment der d-Alanin Oxidase. Biochem. Z. **296**, 294 (1938).

190. — — Isolierung der prosthetischen Gruppe der d-Aminosäureoxidase. Biochem. Z. **298**, 150 (1938).

191. — — Bemerkung über gelbe Fermente. Biochem. Z. **298**, 368 (1938).

192. WEATHERBY, G. D., and D. O. CARR: Riboflavin-Catalyzed Photooxidative Decarboxylation of Dihydrophtalates. Biochemistry **9**, 344 (1970).

193. WEBER, G.: Fluorescence of Riboflavin and Flavin-Adenin-Dinucleotide. Biochem. J. **47**, 114 (1950).

194. WEYGAND, F., R. LÖWENFELD, and E. F. MÖLLER: Über die Spezifität von 6.7-Dichlor-9-d-ribo-flavin als Antagonist des Lactoflavins. Chem. Ber. **84**, 101 (1951).

195. WHITBY, L. G.: A New Method for Preparing Flavin Adenine Dinucleotide. Biochem. J. **54**, 437 (1953).

195a. WHITE, E. H., J. D. MIANO, C. J. WATKINS, and E. J. BREAUX: Chemisch erzeugte angeregte Zustände. Angew. Chem. **86**, 292 (1974).

196. YAGI, K., ed.: Flavins and Flavoproteins. Tokyo: University of Tokyo Press. 1968.

197. YANG, C. S.: Photosensitized Conversion of Ethionine to Ethylene by Flavin Mononucleotide. Photochem. Photobiol. **12**, 419 (1970).

198. YANG, C. S., and P. HEMMERICH: Unpublished.

199. YANG, C. S., and D. MCCORMICK: The Photochemical Degradation of Flavins as Influenced by the Length and Extent of Hydroxylation of the Side Chain. J. Amer. Chem. Soc. **87**, 5763 (1965).

200. YOKOE, I., and T. C. BRUICE: Oxidation of Thiophenyl and Nitroalkanes by an Electron Deficient Isoalloxazine. J. Amer. Chem. Soc. **97**, 450 (1975).

201. Zeitschrift für Naturforschung: **27b**, Heft 9 (1972).

202. Zeller, E. A., B. Gärtner, and P. Hemmerich: 4a,5-Cycloaddition Reactions of Acetylenic Compounds at the Flavoquinone Nucleus as Mechanisms of Flavoprotein inhibitors. Z. Naturforsch. **27b,** 1050 (1972).

(Received May 26, 1975)

Namenverzeichnis. Author Index

Kursiv gedruckte Seitenzahlen beziehen sich auf Literaturverzeichnisse

Page numbers printed in *italics* refer to References

Sachverzeichnis. Subject Index

Von · By

A. Siegel, Wien

Fortschritte der Chemie organischer Naturstoffe
Progress in the Chemistry of Organic Natural Products

Published up to now:

Volume 1: 41 figures. VI, 371 pages. 1938. Cloth DM 85,—

Volume 2: 24 figures. VII, 366 pages. 1939. Cloth DM 85,—

Volume 3: 10 figures. VI, 252 pages. 1939. Cloth DM 64,—

Volume 4: 47 figures. VIII, 499 pages. 1945. Cloth DM 115,—

Volume 5: 34 figures. VIII, 417 pages. 1948. Cloth DM 96,—

Volume 6: 32 figures. VIII, 392 pages. 1950. Cloth DM 90,—

Volume 7: 12 figures. VII, 330 pages. 1950. Cloth DM 76,—

Volume 8: 47 figures. XI, 400 pages. 1951. Cloth DM 92,—

Volume 9: 20 figures. XI, 535 pages. 1952. Cloth DM 123,—

Volume 10: 19 figures. IX, 529 pages. 1953. Cloth DM 122,—

Volume 11: 67 figures. VIII, 457 pages. 1954. Cloth DM 105,—

Volume 12: 15 figures. X, 550 pages. 1955. Cloth DM 127,—

Volume 13: 48 figures. XII, 624 pages. 1956. Cloth DM 144,—

Volume 14: 38 figures. VIII, 377 pages. 1957. Cloth DM 87,—

Volume 15: 81 figures. VI, 244 pages. 1958. Cloth DM 56,—

Volume 16: 27 figures. VI, 226 pages. 1958. Cloth DM 52,—

Volume 17: 57 figures. X, 515 pages. 1959. Cloth DM 118,—

Volume 18: 65 figures. X, 600 pages. 1960. Cloth DM 138,—

Volume 19: 16 figures. VIII, 420 pages. 1961. Cloth DM 97,—

Further volumes see next page

Volume 20: 33 figures. XIII, 509 pages. 1962.
Cloth DM 117,—

Cumulative Index / Generalregister 1–20. 1938–1962. XVI, 369 pages. 1964.
Cloth DM 85,—

Volume 21: 14 figures. VII, 362 pages. 1963.
Cloth DM 88,—

Volume 22: 8 figures. VII, 370 pages. 1964.
Cloth DM 103,—

Volume 23: 58 figures. VIII, 397 pages. 1965.
Cloth DM 109,—

Volume 24: 25 figures. VIII, 475 pages. 1966.
Cloth DM 130,—

Volume 25: 25 figures. VII, 348 pages. 1967.
Cloth DM 92,—

Volume 26: 97 figures. IX, 456 pages. 1968.
Cloth DM 144,—

Volume 27: 47 figures. VIII, 412 pages. 1969.
Cloth DM 130,—

Volume 28: 14 figures. XII, 503 pages. 1970.
Cloth DM 155,—

Volume 29: 18 figures. VIII, 554 pages. 1971.
Cloth DM 181,—

Volume 30: 28 figures. VIII, 666 pages. 1973.
Cloth DM 225,—

Volume 31: 60 figures. IX, 693 pages. 1974.
Cloth DM 231,—

Volume 32: 31 figures. VIII, 560 pages. 1975.
Cloth DM 220,—

Contents: W. K. SEIFERT, Carboxylic Acids in Petroleum and Sediments – P. G. SAMMES, Naturally Occurring 2,5-Dioxopiperazines and Related Compounds – R. J. HIGHET and E. A. SOKOLOSKI, Structural Investigations of Natural Products by Newer Methods of NMR Spectroscopy – P. M. SCOPES, Applications of the Chiroptical Techniques to the Study of Natural Products – H. C. VAN HUMMEL, Chemistry and Biosynthesis of Plant Galactolipids – H. KÖSSEL and H. SELIGER, Recent Advances in Polynucleotide Synthesis – Author Index – Subject Index.

Further information will be sent on request

Price reduction for subscribers / Preisermäßigung für Subskribenten: 10%.

Special price reduction (20% of the list price) for the Vols. 1—30 plus Cumulative Index. / Vorzugspreis (20% Nachlaß) bei Bezug der Bände 1—30 inklusive Generalregister.

GPSR Compliance
The European Union's (EU) General Product Safety Regulation (GPSR) is a set of rules that requires consumer products to be safe and our obligations to ensure this.

If you have any concerns about our products, you can contact us on

ProductSafety@springernature.com

In case Publisher is established outside the EU, the EU authorized representative is:

Springer Nature Customer Service Center GmbH
Europaplatz 3
69115 Heidelberg, Germany

www.ingramcontent.com/pod-product-compliance
Lightning Source LLC
Chambersburg PA
CBHW071406230926
51846CB00001B/23
9783709132647